AF670378

# MODERN AGRICULTURE AND THE ENVIRONMENT

Developments in Plant and Soil Sciences
Volume 71

This volume is dedicated to the late David Rosen, Professor of Entomology at the Faculty of Agricultural, Food and Environmental Quality Sciences of the Hebrew University of Jerusalem. Prof. Rosen, who passed away on January 8, 1997 after a prolonged illness, was an internationally renowned authority on biological and integrated control of insect pests.

David was born in Tel Aviv on April 20, 1936. After completing his academic training at the Faculty of Agriculture of the Hebrew University of Jerusalem, he joined the faculty as a lecturer in 1967, rapidly rising through the academic ranks to be appointed full professor in 1978. His years on the faculty were marked by tireless activity for the promotion of a biological approach to insect pest management. His many and varied research projects encompassed both his own area of expertise, the systematics of parasitic hymenoptera, and numerous related fields in collaboration with experts throughout the world.

As an outstanding teacher, David shared his vast knowledge and his personal views with countless students, and many of those who completed graduate studies under his guidance are today at the forefront of plant protection in Israel. He was also a prolific writer, authoring or co-authoring some 200 research papers and book chapters, and authoring or editing 9 books during his scientific career.

David's cheerful optimism, his sense of humor, and his personal and professional integrity impressed all those who knew him. With his untimely death, his many colleagues, students and friends will miss his inspiration, his leadership, and his warm personality.

David was deeply involved in the organization of the First International Conference on Modern Agriculture and the Environment. In spite of his illness, he invested a tremendous amount of time and energy to make this book a high quality contribution to our understanding and advancement of environmental quality. This volume is dedicated to his memory.

**Yona Chen, Elish Tel-Or and Yithak Hadar**

# Modern Agriculture and the Environment

Proceedings of an International Conference, held in Rehovot, Israel, 2–6 October 1994, under the auspices of the Faculty of Agriculture, the Hebrew University of Jerusalem


Edited by

DAVID ROSEN
Professor of Entomology

ELISHA TEL-OR
Professor of Botany

YITZHAK HADAR
Professor of Microbiology

YONA CHEN
Professor of Soil Science and Dean
The Hebrew University of Jerusalem,
Faculty of Agriculture, Rehovot 76100, Israel

SPRINGER-SCIENCE+BUSINESS MEDIA, B.V.

A C.I.P. Catalogue record for this book is available from the Library of Congress

ISBN 978-94-010-6279-4 ISBN 978-94-011-5418-5 (eBook)
DOI 10.1007/978-94-011-5418-5

---

*Printed on acid-free paper*



Hartnolls Limited, Bodmin, Cornwall

# Contents

## SECTION IV: MODELLING OF AGRICULTURAL POLLUTION



## SECTION V: WASTEWATER TREATMENT AND IRRIGATION



## SECTION VI: RECYCLING OF MUNICIPAL AND AGRICULTURAL WASTES

## SECTION VIII: BIOREMEDIATION



## SECTION IX: ECONOMIC AND POLICY ASPECTS OF NATURAL RESOURCES

# Preface

This volume comprises the proceedings of the First International Rehovot Conference on Modern Agriculture and the Environment, held at the Rehovot Campus of the Faculty of Agriculture, the Hebrew University of Jerusalem, Israel, 2–6 October 1994. The conference, first in a series intended to be convened in Rehovot at 4–5 year intervals to address various aspects of the interaction of agriculture and the environment, was initiated, organised and carried out under the auspices of the Faculty of Agriculture, the leading academic institution in agricultural and environmental studies in Israel. It featured four keynote addresses, 39 invited lectures, 40 submitted papers, and 62 posters. Of these, 51 articles, written by 122 contributing authors from 14 countries, were selected by the editors to be presented in this book.

All through the twentieth century, and especially ever since the advent of the Green Revolution, modern agriculture has been striving to feed and clothe the ever-increasing multitudes of the human species through improved technology, relying heavily on tremendous inputs of fertilisers, pesticides, and various other agrochemicals. Undoubtedly, this has been a great blessing to mankind, and enormous strides have indeed been made in the never-ending struggle against starvation, but these have been achieved at a very steep price of increased environmental deterioration. In fact, modern agriculture has become one of the major factors contributing to the degradation of the world's fragile biosphere. This has been manifested in mounting rates of soil, water and air pollution, increased emission of greenhouse gases, depletion of the ozone layer in the stratosphere, accumulation of manure and other solid wastes, excessive exploitation of forests and open lands, elimination of the natural habitats of many plants and animals, an almost unprecedented mass extinction of living species, and an alarming destruction of biodiversity. Obviously, this should not be allowed to go on. Urgent steps have to be taken, on a national and regional as well as on a global level, to stop, avoid and mitigate these detrimental processes.

This is what this book is all about. It is divided into nine sections, addressing various problems of, and suggested solutions to, the complex interaction between modern agriculture and the environment. **Section I** is devoted to chemical pesticides, means to reduce their utilisation, and some of the main alternatives to their overuse and misuse in pest management; **Section II** deals with chemical fertilisers and how to minimise pollution caused by them; **Section III** addresses some of the environmental aspects of aquaculture; **Section IV** presents approaches to the modelling of agricultural pollution; **Section V** covers the treatment of wastewater and its utilisation in irrigation; **Section VI** deals with the recycling of municipal and agricultural wastes; **Section VII** presents pollution problems caused by heavy metals, and some means for their mitigation; **Section VIII** is devoted to bioremediation of agricultural pollutants; and, finally, **Section IX** presents discussions of economic and policy aspects of natural resources. We do hope that this broad coverage will help focus the interest of agriculturists, policy makers and environmentalists alike on the all-important environmental issues facing modern agriculture.

The editors are grateful to numerous colleagues all over the world, who graciously served as peer reviewers for the various chapters of this book. Special thanks are due to Ms Ruth Ann Yonah, Faculty of Agriculture, Rehovot, who provided invaluable editorial services.

*David Rosen*
*Elisha Tel-Or*
*Yitzhak Hadar*
*Yona Chen*

# The First Rehovot Conference on Modern Agriculture and the Environment

**Organizing Committee**

Prof. E. Tel-Or, Chairman
Prof. B. Rubin, Deputy Chairman
Prof. Y. Chen
Prof. Y. Hadar
Prof. E. Hochman
Prof. U. Mingelgrin
Prof. D. Rosen
Dr J. van Rijn
Mr D. Lavie, Coordinator
Ms G. Lavon, Administrative Secretary
Ms R.A. Yonah, Assistant to the Editorial Board

**International Advisory Board**

Prof. Y. Chen, Chairman, The Hebrew University of Jerusalem, Rehovot, Israel
Prof. A. Banin, The Hebrew University of Jerusalem, Rehovot, Israel
Dr Y. Bar-Or, Ministry of the Environment, Jerusalem, Israel
Prof. J.E. Casida, University of California, Berkeley, USA
Prof. O. Franzle, University of Kiel, Germany
Prof. Y. Folman, Ministry of Agriculture, Bet Dagan, Israel
Prof. F.A.M. de Haan, Wageningen Agricultural University, The Netherlands
Prof. H.A.J. Hoitink, Ohio State University, Wooster, USA
Prof. J.C. van Lenteren, Wageningen Agricultural University, The Netherlands
Prof. W. Verstraete, University of Ghent, Belgium
Prof. R.J. Wagenet, Cornell University, Ithaca, USA
Dr M. Waldman, Ministry of Science and the Arts, Jerusalem, Israel
Prof. Q.G. Zhao, Institute of Soil Science, Nanjing, China

**Sponsors**

Israel Academy of Sciences and Humanities
The Hebrew University of Jerusalem
Israel National Commission for UNESCO
BARD – The United States – Israel Binational Research and Development Fund
Israel Ministry of Education, Information Center
Israel Ministry of the Environment

Israel Ministry of Foreign Affairs, Division for International Cooperation
Israel Ministry of Science and the Arts
Israel Ministry of Agriculture, The Chief Scientist
Israel Ministry of Tourism
KKL – Jewish National Fund
Embassy of the United States, Tel Aviv
British Embassy, Tel Aviv
Milchan Bros Ltd
Machteshim Chemical Works Ltd
Dead Sea Works Ltd
Bromine Compounds Ltd
Haifa Chemicals Ltd

# List of contributors

**M. Abraham** Department of Cell and Animal Biology, Institute of Life Sciences, The Hebrew University of Jerusalem 91904, Israel

**A. Acher** Institute of Soils and Water, Water Treatment Laboratory, The Volcani Center, ARO, PO Box 6, Bet Dagan 50250, Israel

**C. Alabouvette** Station de Recherches sur la Flore Pathogene dans le Sol, INRA, 17 Rue Sully, 21024 Dijon Cedex, France

**M. Altstein** Department of Entomology, The Volcani Center, ARO, PO Box 6, Bet Dagan 50250 Israel

**D. Amir-Shapira** Israel Ministry of the Environment, Division of Solid Waste Management, PO Box 34033, Jerusalem 95464, Israel

**V. Andreoni** Department of Food Science and Microbiology, Faculty of Agriculture, State University, Via Celoria, 2, 20133 Milan, Italy

**K.R.S. Ascher** Institute of Plant Protection, The Volcani Center, ARO, PO Box 6, Bet Dagan 50250, Israel

**Y. Avnimelech** Laboratory for the Management of Environmental Systems, Faculty of Agricultural Engineering, The Technion, Haifa 32000, Israel

**M. Bahadir** Institute für ökologische Chemie und Abfallanalytik, Technische Universität Braunschweig, Hagenring 30, 38106 Braunschweig, Germany

**J. Barg** Cardiovascular and Hypertension Research Laboratory, Wolfson Medical Center Tel Aviv University Medical School, Holon, Israel

**F. Beese** Institut für Bodenkunde und Waldernährung, Georg-August Universität Göttingen Büsgenweg 2, D-37077 Göttingen, Germany

**O. Ben Aziz** Department of Entomology, The Volcani Center, ARO, PO Box 6, Bet Dagan 50250, Israel

**S. Boussiba** Microalgal Biotechnology Laboratory, Jacob Blaustein Institute for Desert Research, Ben-Gurion University at Sede-Boker, 84990 Israel

**A. Bouzaher** World Bank, 600 19th St, Room H8021, Washington, DC 20433, USA

**E. Brill** Department of Agricultural Economics, Faculty of Agriculture, The Hebrew University of Jerusalem, PO Box 6, Rehovot 76100, Israel

**L. Cadot** Unite de Science du Sol, INRA, 78850 Thiverval-Grignon, France

**R. Carzaniga** Institute of Plant Pathology, Faculty of Agriculture, State University, Via Celoria 2, 20133 Milan, Italy

**J.E. Casida** Environmental Chemistry and Toxicology Laboratory, Department of Environmental Science, Policy and Management, College of Natural Resources, University of California, 115 Wellman Hall, Berkeley, CA 94720-3112, USA

**B. Chefetz** Department of Soil and Water Science, Faculty of Agriculture, The Hebrew University of Jerusalem, PO Box 12, Rehovot 76100, Israel

**N. Chejanovsky** Department of Entomology, Institute for Plant Protection, Volcani Center, ARO, PO Box 6, Bet-Dagan 50250, Israel

**Y. Chen** Department of Soil and Water Science, Faculty of Agriculture, The Hebrew University of Jerusalem, PO Box 12, Rehovot 76100, Israel

**A. Chlopecka** Department of Crop and Soil Sciences, University of Georgia, Athens, Georgia 30606, USA

**M. Colombo** Department of Food Science and Microbiology, Faculty of Agriculture, State University, Via Celoria, 2, 20133 Milan, Italy

**C. Crecchio** Istituto di Chimica Agraria, Universià di Bari, Via Amendola 165/a, 70126 Bari, Italy

**M. Curci** Istituto di Chimica Agraria, Università di Bari, Via Amendola 165/a, 70126 Bari, Italy

**F.S. Davies** Horticultural Sciences Department, PO Box 110690, University of Florida, Gainesville, Florida 32611, USA

**M. DeNobili** Dipartimento di Produzione Vegetale e Technologie Agrarie, Via delle Scienze, 208 University of Udine, 33100 Udine, Italy

**D. Di Simone** Department of Food Science and Microbiology, Faculty of Agriculture, State University, Via Celoria 2, 20133 Milan, Italy

**R.S. Dias** Laboratorio Quimico Agricola Rebelo da Silva, Tapada da Ajuda Apartado 3228, P-1306 Lisboa, Portugal

**E. Díaz-Barrientos** Instituto de Recursos Naturales y Agrobiologia, Apartado 1052, 41080 Sevilla, Spain

**J. Drozd** Institute of Soil Science and Agricultural Environment Protection, Agricultural University of Wroclaw, Grunwaldzka 53, Wroclaw 50357, Poland

**N.J. Dudley** Centre for Water Policy Research, The University of New England, Armidale, NSW 2351, Australia

**J. Dumanski** Centre of Land and Biological Resources Research, Agriculture and Agri-Food Canada, Central Experimental Farm, Bldg. 74, Ottawa, Ontario K1A 0C6, Canada

**E. Dunkelblum** Department of Entomology, The Volcani Center, ARO, PO Box 6, Bet Dagan 50250, Israel

**J.J. Ferguson** Horticultural Sciences Department, PO Box 110690, University of Florida, Gainesville, Florida 32611, USA

**C. Finoli** Department of Food Science and Microbiology, Faculty of Agriculture, State University Via Celoria 2, 20133 Milan, Italy

**E. Fischer** Institute of Soils and Water, Water Treatment Laboratory, The Volcani Center, ARO, PO Box 6, Bet Dagan 50250, Israel

**O. Fränzle** Geographisches Insitut der Christian-Albrechts-Universität, D-24098 Kiel, Germany

**O. Froy** Department of Botany, Life Sciences Institute, Tel Aviv University, Ramat Aviv, Tel Aviv 69978, Israel

**T. Gabay** Department of Entomology, The Volcani Center, ARO, PO Box 6, Bet Dagan 50250, Israel

**P.W. Gassman** Center for Agricultural and Rural Development, Iowa State University, 578 Heady Hall, Ames, IA 50011-1070, USA

**Y. Gazit** Department of Entomology, The Volcani Center, ARO, PO Box 6, Bet Dagan 50250, Israel

**V. Gnanalingam** Unite de Science du Sol, INRA, 78850 Thiverval-Grignon, France

**E.R. Graber** Institute of Soils and Water, The Volcani Center, ARO, PO Box 6, Bet Dagan 50250, Israel

**M.E. Grebus** Department of Plant Pathology, Ohio Agricultural Research and Development Center, Ohio State University, 1680 Madison Avenue, Wooster, OH 44691, USA

**M. Gurevitz** Department of Botany, Life Sciences Institute, Tel Aviv University, Ramat Aviv, Tel Aviv 69978, Israel

**Y. Hadar** Department of Plant Pathology and Microbiology, Faculty of Agriculture, The Hebrew University of Jerusalem, PO Box 12, Rehovot 76100, Israel

**B.D. Hammock** Department of Entomology and Toxicology, University of California, Davis, California 95616-8584, USA

**R.E. Hantschel** GSF – Forschungszentrum für Umwelt und Gesundheit, Institut für Bodenökologie, Neuherberg, Ingolstädter Landstr. 1, Postfach 1129, D-85758 Oberschlei*β*üheim, Germany

**Y. Henis** Department of Plant Pathology and Microbiology, Faculty of Agriculture, The Hebrew University of Jerusalem, PO Box 12, Rehovot 76100, Israel

**R. Herrmann** Department of Entomology and Toxicology, University of California, Davis, California 95616-8584, USA

**E. Hochman** Department of Agricultural Economics, Faculty of Agriculture, The Hebrew University of Jerusalem, PO Box 12, Rehovot 76100, Israel

**H.A.J. Hoitink** Department of Plant Pathology, Ohio Agricultural Research and Development Center, Ohio State University, 1680 Madison Avenue, Wooster, OH 44691, USA

**J.K. Horowitz** Department of Agricultural and Resource Economics, University of Maryland, College Park, MD 20742, USA

**S. Houot** Unite de Science du Sol, INRA, 78850 Thiverval-Grignon, France

**J.L. Hutson** Department of Soil, Crop and Atmospheric Sciences, Cornell University, Bradfield and Emerson Halls, Ithaca, NY 14853, USA

**Y. Iger** Faculty of Agricultural Engineering, The Technion, Haifa 32000, Israel

**Y. Inbar** Division of Solid Waste Management, Israel Ministry of the Environment, PO Box 34033, Jerusalem 95464, Israel

**R. Indiati** Ministry of Agriculture, Food and Forestry Resources, Experimental Institute for Plant Nutrition, Via della Navicella 2, 00184 Rome, Italy

**R.C. Izaurralde** Department of Renewable Resources, University of Alberta, 4-42 Earth Sciences Building, Edmonton, Alberta T6G 2E3, Canada

**A. Jezierski** Institute of Chemistry, University of Wroclaw, Poland

**R.E. Just** Department of Agricultural and Resource Economics, University of Maryland, College Park, MD 20742, USA

**U. Kafkafi** Department of Field and Vegetable Crops, Faculty of Agriculture, The Hebrew University of Jerusalem, PO Box 12, Rehovot 76100, Israel

**J. Katan** Department of Plant Pathology and Microbiology, Faculty of Agriculture, The Hebrew University of Jerusalem, PO Box 12, Rehovot 76100, Israel

**L. Kautzki** Soil and Water Institute, The Volcani Center, ARO, PO Box 6, Bet Dagan 50250, Israel

**E. Kimmel** Faculty of Agricultural Engineering, The Technion, Haifa 32000, Israel

**J.R. Kiniry** Grasslands, Soil and Water Research Laboratory, USDA-ARS, 808 E. Blackland Road Temple, TX 76503-6112, USA

**M. Kochva** Laboratory for the Management of Environmental Systems, Faculty of Agricultural Engineering, The Technion, Haifa 32000, Israel

**F. Korte** Institut für Chemie Weihenstephan, Technische Universität Munchen, Lehrstuhl für ökologische Chemie, Schulstrabe 10, D-85356 Freising, Germany

**P.G. Laksminarayan** Center for Agricultural and Rural Development, Iowa State University, 578 Heady Hall, Ames, IA 50011-1070, USA

**R. Laster** Laster and Gouldman Law Offices, 10 Hanassi St, Jerusalem 92188, Israel

**L. Leita** Dipartimento di Produzione Vegetale e Technologie Agrarie, Via delle Scienze, 208 University of Udine, 33100 Udine, Italy

**D. Levanon** Institute of Soils and Water, The Volcani Institute, ARO, PO Box 6, Bet Dagan 50250, Israel

**M. Lješevic** Faculty of Agriculture, University of Belgrade, Nemanjina 6, 11081 Belgrade, Yugoslavia

**E. Loffredo** Instituto di Chimica Agraria, Università di Bari, Via Amendola 165/A, 70126 Bari, Italy

**L. Madrid** Instituto de Recursos Naturales y Agrobiologia, Apartado 1052, 41080 Sevilla, Spain

**R.T. Mandelbaum** Soil and Water Institute, The Volcani Center, ARO, PO Box 6, Bet Dagan 50250, Israel

**Y. Manor** Central Virology Laboratory, Sheba Medical Center, Ramat Gan, Israel

**L. Marchiol** Dipartimento di Produzione Vegetale e Technologie Agrarie, Via delle Scienze, 208 University of Udine, 33100 Udine, Italy

**R. Martens** Institut für Bodenbiologie, Bundesforschungsanstalt fur Landwirtschaft, Bundesallee 50, 38116 Braunschweig, Germany

**J. Milivojević** Faculty of Agriculture, University of Belgrade, Nemanjina 6, 11081 Belgrade, Yugoslavia

**C. Mondini** Istituto Sperimentale per la Nutrizione della Piante, Rome, Italy

**H. Moskowitz** Departments of Entomology and Toxicology, University of California, Davis, CA 95616-8584, USA

**G. Muhlbachova** Research Institute of Crop Production, Prague, Czech Republic

**J.F. Muir** Institute of Aquaculture, University of Stirling, Stirling FK9 4LA, UK

**L. Muszkat** Department of Pesticide Chemistry, The Volcani Center, ARO, PO Box 6, Bet Dagan 50250, Israel

**M. Nedić** Faculty of Agriculture, University of Belgrade, Nemanjina 6, 11081 Belgrade, Yugoslavia

**S. Netanyahu** Department of Agricultural and Resource Economics, University of Maryland, College Park, MD 20742, USA

**I. Nissim** Division of Solid Waste Management, Israel Ministry of the Environment, PO Box 34033, Jerusalem 95464, Israel

**G.A. O'Connor** Soil and Water Science Department, University of Florida, Gaineseville, FL 32611, USA

**G. Origgi** Department of Food Science and Microbiology, Faculty of Agriculture, State University, Via Celoria, 2, 20133 Milan, Italy

**M.D.R. Pizzigallo** Istituto di Chimica Agraria, Università di Bari, Via Amendola 165/a, 70126 Bari, Italy

**J.H. Prueger** National Soil Tilth Laboratory, 2150 Pammel Drive, Ames, Iowa 50011, USA

**S. Ravid** Department of Agricultural Botany, Faculty of Agriculture, The Hebrew University of Jerusalem, PO Box 12, Rehovot 76100, Israel

**D. Rosen** Department of Entomology, Faculty of Agriculture, The Hebrew University of Jerusalem, PO Box 12, Rehovot 76100, Israel

**F. Rouxel** Station de Pathologie Vegetale, INRA, BP. 29, 35650 Le Rheu, France

**P. Ruggiero** Istituto di Chimica Agraria, Università di Bari, Via Amendola 165/a, 70126 Bari, Italy

**M. Sela** Department of Agricultural Botany, Faculty of Agriculture, The Hebrew University of Jerusalem, PO Box 12, Rehovot 76100, Israel

**N. Senesi** Istituto di Chimica Agraria, Università di Bari, Via Amendola 165/A, 70126 Bari, Italy

**P. Sequi** Experimental Institute for Plant Nutrition, Ministry of Agriculture, Food and Forestry Resources, Via della Navicella, 2, 00184 Rome, Italy

**C. Serra-Wittling** Unite de Science du Sol, INRA, 78850 Thiverval-Grignon, France

**N. Shafir** Soil and Water Institute, The Volcani Center, ARO, PO Box 6, Bet Dagan 50250, Israel

**J.C. Soveral-Dias** Laboratorio Quimico Agricola Rebelo da Silva, Tapada da Ajuda Apartado 3228, P-1306 Lisboa, Portugal

**A.G. Stone** Department of Plant Pathology, Ohio Agricultural Research and Development Center, Ohio State University, 1680 Madison Avenue, Wooster, OH 44691, USA

**J. Tajek** Alberta Land Resource Unit, Agriculture and Agri-Food Canada, Suite 1295, Royal LePage Building, 10130 103 Street, Edmonton, Alberta T5J 3N9, Canada

**E. Tel-Or** Department of Agricultural Botany, Faculty of Agriculture, The Hebrew University of Jerusalem, PO Box 12, Rehovot 76100, Israel

**P.M. Tillotson** Environmental Studies Section, Dupont Agricultural Products, Experiment Station, Wilmington, DE 19880, USA

**D. Urbach** Department of Botany, Life Sciences Institute, Tel Aviv University, Ramat Aviv, Tel Aviv 69978, Israel

**J.C. van Lenteren** Laboratory of Entomology, Wageningen Agricultural University, PO Box 8031, 6700 EH Wageningen, The Netherlands

**J. van Rijn** Department of Animal Science, Faculty of Agriculture, The Hebrew University of Jerusalem, PO Box 12, Rehovot 76100, Israel

**A. Vecchio** Department of Food Science and Microbiology, Faculty of Agriculture, State University, Via Celoria, 2, 20133 Milan, Italy

**Z. Vogel** Department of Neurobiology, The Weizmann Institute of Science, Rehovot 76100, Israel

**R.J. Wagenet** Department of Soil, Crop and Atmospheric Sciences, Cornell University, Bradfield and Emerson Halls, Ithaca, New York 14853, USA

**S.E. Wendelaar** Bonga Department of Animal Physiology, Faculty of Science, University of Nijmegen, 6525 ED Nijmegen, The Netherlands

**M. Wolter** Institute für Bodenbiologie, Bundesforschungsanstalt für Landwirtschaft, Bundesallee 50, 38116 Braunschweig, Germany

**Wu Xiaoqiang** Microalgal Biotechnology Laboratory, Jacob Blaustein Institute for Desert Research, Ben-Gurion University at Sede-Boker, 84990 Israel

**F. Zadrazil** Institute für Bodenbiologie, Bundesforschungsanstalt für Landwirtschaft, Bundesallee 50, 38116 Braunschweig, Germany

**A. Zaritsky** Department of Life Sciences, Ben-Gurion University of the Negev, PO Box 653, Be'er Sheva 83105, Israel

**M. Zhang** Department of Renewable Resources, 4-42 Earth Science Building, University of Alberta, Edmonton, Alberta T6G 2E3, Canada

**N. Zilberberg** Department of Botany, Life Sciences Institute, Tel Aviv University, Ramat Aviv, Tel Aviv 69978, Israel

**M. Zilberbrand** Institute of Earth Sciences, The Hebrew University of Jerusalem, Jerusalem 91904, Israel

**D. Zilberman** Department of Agricultural and Resource Economics, University of California, Berkeley, CA 74720, USA

**E. Zlotkin** Department of Cell and Animal Biology, Institute of Life Sciences, The Hebrew University of Jerusalem, Jerusalem 91904, Israel

# SECTION I

# Towards Environmentally Acceptable Pest Management

# 1
# Safer and more effective insecticides for the future

JOHN E. CASIDA AND GARY B. QUISTAD

## 1. Introduction

Synthetic organic insecticides have served for the past half century as the principal means to control insect-borne diseases and minimize losses in food and fibre production from pest insect attack. Pests destroy about one-third of the world food supply during growth, harvesting and storage, even with the use of pesticides and other control methods (Ware, 1994). Without pesticides, especially in the tropics, there would be dramatic losses of crops, greatly exacerbating existing food shortages in developing countries (Forget, 1991). The world population is expected to grow by more than one billion people or about 20% in the next decade, which in turn also requires expanded use of agrochemicals for efficient agriculture. This can only be achieved by the most effective use of the current insecticides and the development of better compounds for the years ahead.

Pesticides represent a very large business in terms of dollars, pounds and number of compounds. The world market for pesticides was $25 billion in 1992 (McDougall and Phillips, 1993) of which insecticides made up about $8 billion (Perrior, 1993). In amounts, this is about one billion pounds of pesticides per year in the US and several billion pounds worldwide (Ware, 1994). In numbers, there are about 900 active ingredients in current use.

Four problems are created by the extensive use of pesticides. The first is adverse health effects in humans. They have a greater likelihood of occurring with insecticides than other pesticides because insect control agents are often inherently more toxic, sharing common sites of action in insects and mammals. Pesticides account for more than 20 000 fatalities (including suicides) yearly with most occurring in developing countries and, while 85% of the world agrochemical production is used in industrialized nations, the incidence of human poisoning may be 13 times higher in Third World countries than in the US (Forget, 1991). The second problem is the potential for environmental pollution in light of the amounts used, e.g. four pounds per person annually in the US. The third problem is selection of resistant strains, particularly of insects and mites, which have less sensitive target sites or enhanced detoxification capabilities thereby requiring higher doses or even resulting in control failures and necessitating replacement compounds (Georghiou, 1990). Finally, there is the perception – presented most vividly by Rachel Carson (1962) and sporadically perpetuated in the press – that 'applied entomology ... has armed itself with the most modern and terrible

*D. Rosen* et al. *(eds): Modern Agriculture and the Environment, 3–15.*

weapons, and that in turning them against the insects it has also turned them against the earth'.

This review considers the quest for safer and more effective pesticides. Enhanced potency for recent pesticides has led to their use at rates only 1 to 10% of those common only a few years ago. Research is being focused on new biochemical targets to circumvent resistance. Insecticides are used for illustration but there are almost comparable developments for herbicides and fungicides.

## 2. Four nerve targets for the current major insecticides

The nervous system is the target for every one of the 20 most important insecticides based on worldwide sales (Voss *et al.*, 1995) and for at least 88% of the insecticide sales volume (Leicht, 1993). There are four principal nerve targets. Most insecticides (62%) inhibit acetylcholinesterase (AChE) since this is the molecular target for the organophosphorus compounds (OPs) and methylcarbamates (MCs) with sales in a ratio of 2:1, respectively (Perrior, 1993). The voltage-dependent sodium channel and the $\gamma$-aminobutyric acid (GABA)-gated chloride channel are the sites of action for 24% of the current insecticides (by sales volume) with a 3:1 preference for the former target. Although the nicotinic acetylcholine receptor (nAChR) now accounts for only 2% of sales, its role in pest control is projected to grow rapidly in the near future. The remaining 12% of present insecticide sales are for compounds with widely diverse modes of action. Potent natural products also act at the major targets of current insecticides and occasionally have served as synthetic prototypes for commercial products.

*AChE* inhibitors provide the greatest versatility in insect control programmes (Naumann, 1989; Voss *et al.*, 1995). The discoveries of OPs as insecticides and their mode of action, inhibition of AChE, occurred in the late 1930s. Natural product OPs with similar activities were found only recently (e.g. from *Streptomyces* in 1987) and were not the origin of this chemical class of insecticides (reviewed by Naumann, 1989). In contrast, the MC insecticides were based on the Calabar bean toxicant physostigmine which served as a principal probe early in this century to understand the biochemical, physiological and toxicological role of AChE. Although dimethylcarbamates are less important as insecticides, they were the first effective MCs discovered serendipitously in a programme designed to find insect repellents (Kuhr and Dorough, 1976). AChE is now one of the best understood enzymes in structure, function and kinetics of inhibition (Fukuto, 1990), in part due to the importance of this knowledge in the design and development of OP and MC insecticides. AChE is critically important in both mammals and insects, i.e., they have the same target site, so selectivity is often provided by differences in metabolism, e.g. phosphorothionate to phosphate oxidation as a bioactivation prior to phosphorylation of AChE. This enzyme typically recovers quickly from inhibition by MCs but slowly with the OPs, i.e. minutes versus hours or days. Ageing of the phosphoryl moiety of inhibited AChE can also occur and involves dealkylation. This process further slows reactivation of the enzyme and, with certain OP insecticides, may impart additional undesirable properties in action at another nerve target, neuropathy target esterase, leading to delayed neurotoxicity in vertebrates.

Half of the 20 most important insecticides in worldwide sales are OPs, i.e. acephate, chlorpyrifos, diazinon, dimethoate, fenitrothion, methamidophos, methidathion, monocrotophos, profenofos and terbufos (Voss *et al.*, 1995). Additional important OPs in market volume are fonofos, malathion, methyl parathion and parathion. Four MCs (aldicarb, carbaryl, carbofuran and methomyl) are also among the most important insecticides, so AChE inhibitors account for 70% of the top 20 compounds based on sales. Their importance is attributable to the ease of synthesis and the large number of compounds examined in optimizing for potency, selective toxicity, systemic action, biodegradability, and chemical and biological properties for each of a myriad of uses. MCs have some advantages in reduced chronic toxicity, but, as with the OPs, high acute mammalian and bird toxicity for some compounds, along with cross-resistance, can be significant disadvantages.

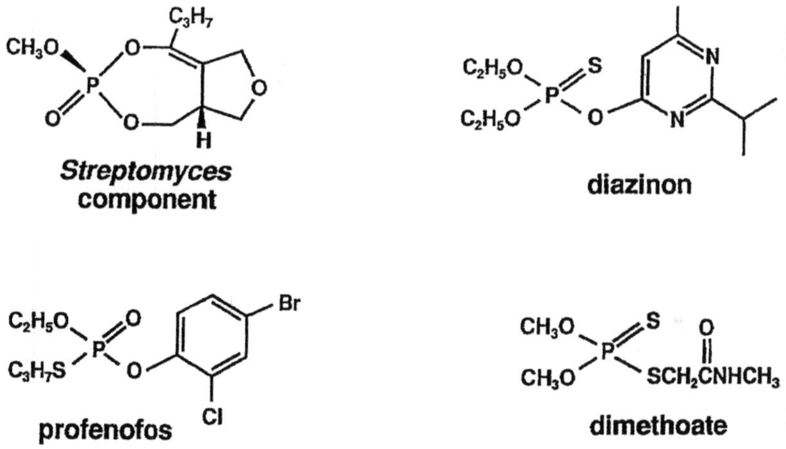

*Figure 1.* Organophosphorus insecticide target: acetylcholinesterase.

The *voltage-dependent sodium channel* is the target for DDT and the pyrethroids. DDT, the most important of all insecticides in historical impact for pest control, pollution and persistence, is still used extensively in developing countries (Forget, 1991) and remains a major environmental concern. Pyrethrum is the most important botanical insecticide with about 500 000 pounds of active esters (the pyrethrins) used annually in primarily non-agricultural applications. Amazingly, DDT and the pyrethrins work at the same target, one of six distinct sites in the sodium channel, and cross-resistance was reported in 1953. The chemical structure of the natural pyrethrins also served as the basis for extensive synthetic modifications that resulted in the economically-prominent pyrethroids (Elliott, 1995). Virtually every region of the pyrethrin structure has been manipulated to produce a vast array of synthetic analogues which, although they have only vague similarity to the original pyrethrins (e.g. silafluofen), remarkably seem to act in the same way so cross-resistance from

physostigmine

carbaryl

aldicarb

alanycarb

*Figure 2.* Methylcarbamate target: acetylcholinesterase.

an insensitive target site is a potential problem. Major commercial pyrethroids at this (the second most important) site of insecticide action include cyhalothrin, cypermethrin, deltamethrin and esfenvalerate, i.e. one-fifth of the 20 most important insecticides in sales (Voss *et al.*, 1995). Their high potency importantly allows much lower use rates, e.g. 1 to 10% of that for DDT which they essentially replaced in many pest control situations. Environmental and metabolic lability of the pyrethroids also provide substantial selective toxicity to insects relative to mammals although toxicity to fish is often a problem.

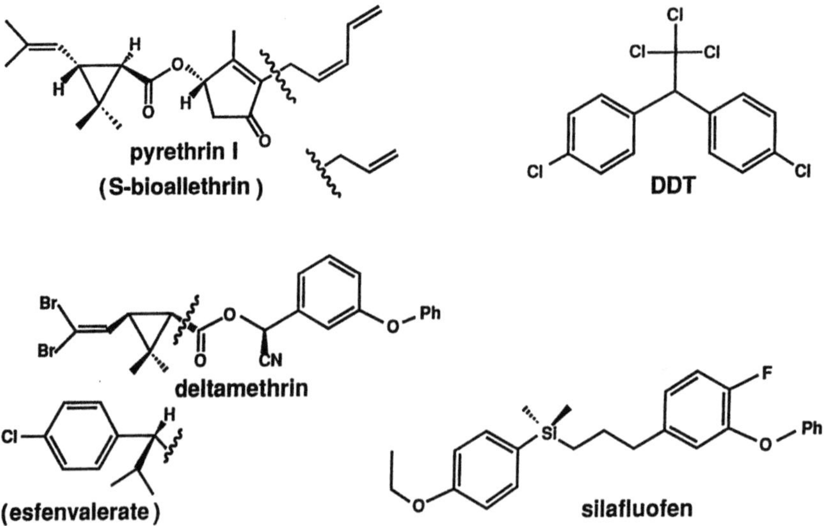


*Figure 3.* Voltage-dependent sodium channel target.

The *GABA-gated chloride channel* is one of the most important targets for insecticide action. This channel was recognized as the target first for picrotoxinin, the insecticidal and piscicidal constituent of the fishberry plant, and then for three major classes of chlorinated hydrocarbon insecticides (chlorinated cyclodienes, chlorobornanes in toxaphene, and hexachlorocyclohexane) which were used to the extent of three billion pounds before being restricted for reasons of adverse toxicology and persistence (Casida, 1993). Today, there is some use of lindane (the gamma isomer of hexachlorocyclohexane) and the major remaining chlorinated cyclodiene is endosulfan, used extensively to control cotton pests and representing most of the 6% share of world sales for compounds acting at this channel. Target site resistance is a potential problem because of the earlier extensive use of cyclodienes, toxaphene and hexachlorocyclohexane but is alleviated somewhat by the relatively small selection pressure from current compounds. Rapid recurrence of resistance may compromise the utility of new insecticides such as the arylpyrazole fipronil, a channel blocker, but this should not affect avermectin (and emamectin), chloride channel activators, which bind to a different site. Insecticides acting at the GABA-gated chloride channel have distinct advantages of high potency often coupled with a broad spectrum of insect control. The channel blockers also have potential disadvantages of persistence, the stigma of being chlorinated hydrocarbons, and action at a channel that is also essential in mammals so adequate selectivity is difficult to achieve.

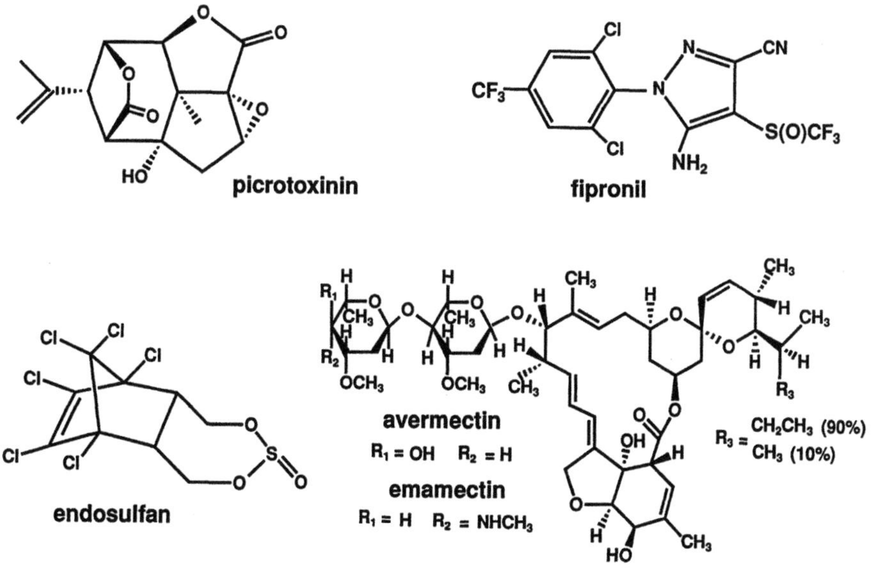

*Figure 4.* GABA-gated chloride channel target.

The *nAChR* is an old target for pest control since nicotine has been used as an insecticide for over 300 years. Utilization of this target is rather meagre (only 2% of world sales) with cartap as the principal commercial product (Leicht, 1993). This is changing rapidly with the recent introduction of imidacloprid and the likelihood that acetamiprid and related insecticides will also become important. These new agonists of the nAChR have significant advantages such as low mammalian toxicity and systemic activity against sucking insects (by virtue of increased polarity). However, they are generally less active or ineffective on contact and ingestion by chewing insects.

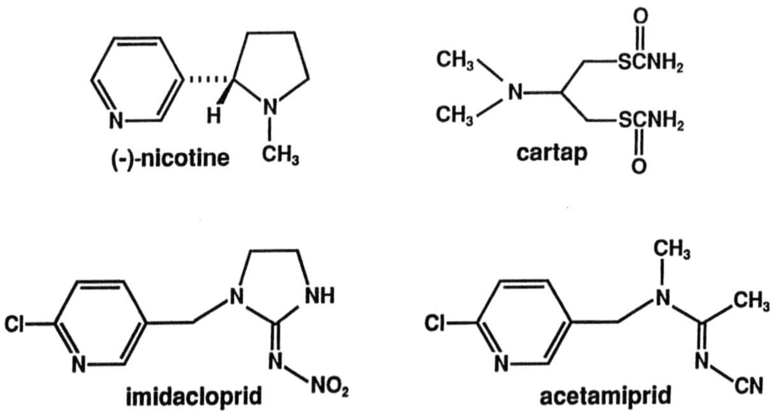

*Figure 5.* Acetylcholine receptor target.

## 3. Historical status of current types of major neuroactive insecticides

When viewed as the number of new active ingredients on an annual basis from 1930–1995, insecticides are a mature commercial field. The median time for introduction of new chlorinated hydrocarbons was 1950 while that of AChE inhibitors was about 1960 (1957 for the OPs and 1962 for the MCs) (graph based on 1930–1985 data for developmental and market compounds by priority year; Naumann, 1989). Two to four OPs are currently under development and the last MC shown (1984) is the proinsecticide alanycarb. Pyrethroids are the only large class of current neuroactive agents that continues to expand with a median time for introduction of 1981 for the period up to 1990 (data for priority years 1953–1990 from Elliott, 1989, 1995). Several new pyrethroids have been introduced since 1991 or are still under development.

Some of the emerging insecticides illustrate the utilization of an old nerve target by a new chemical class. Thus, in the past MCs followed OPs as inhibitors of AChE, and the pyrethroids reused the DDT site. More recently the GABA-gated chloride channel has been identified as the site of action for fipronil and the nAChR is reutilized by imidacloprid (Moffat, 1993).

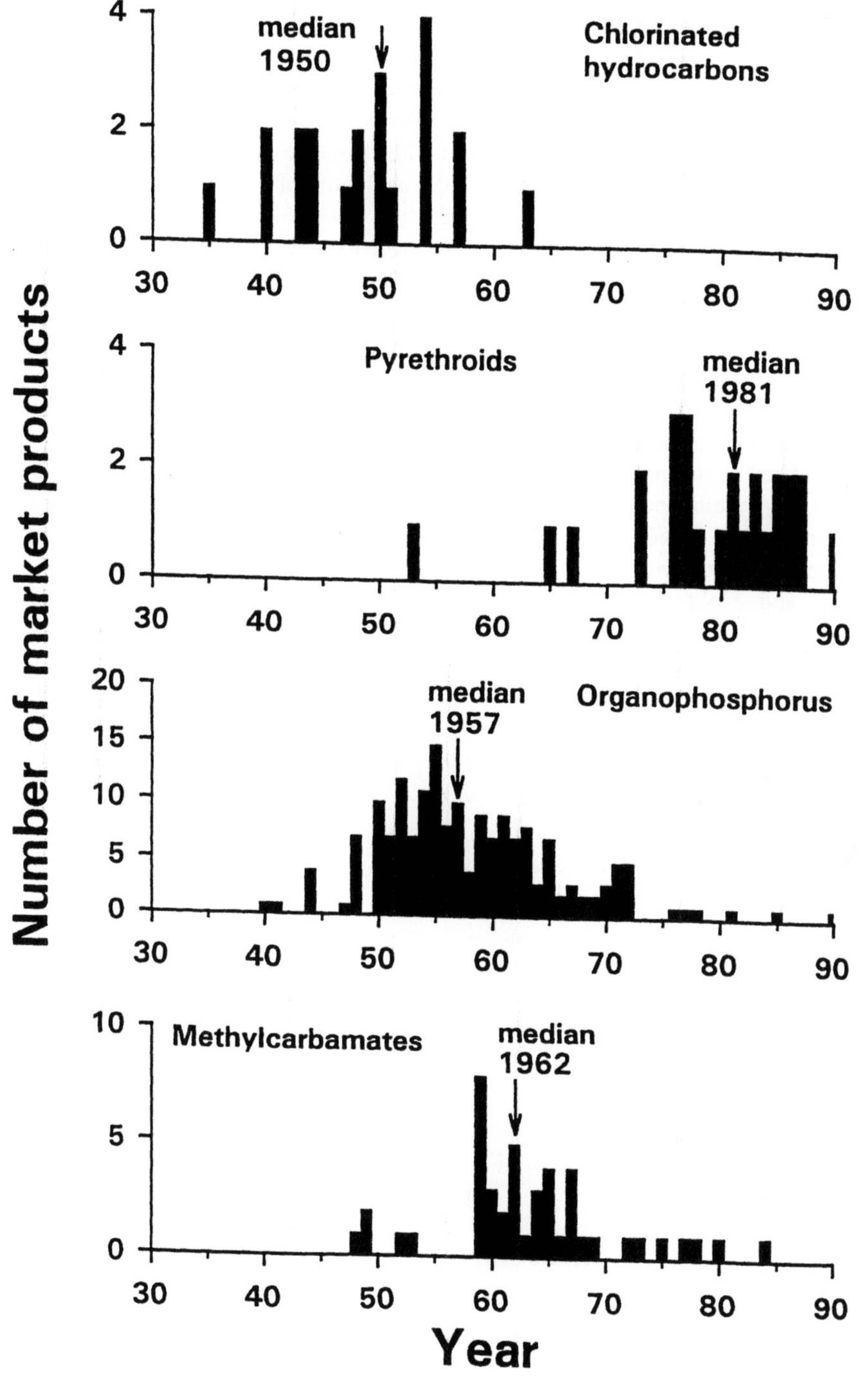


*Figure 6.* Priority years of new insecticides.

## 4. New or underutilized nerve targets

Several insecticidal natural products (i.e. veratridine and lipid amides related to pipercide) and synthetic chemicals (e.g. dihydropyrazoles) disrupt nerve function at the sodium channel by binding to sites unrelated to pyrethroid action. Such products should have significant advantages in avoiding cross-resistance from earlier use of the pyrethroids. The potential for insecticidal analogues of these compounds has not been realized as yet on a commercial basis.

The *glutamate receptor* is an attractive target for new insecticides because this amino acid is the major transmitter ligand at the insect neuromuscular junction but not at that of mammals which utilizes acetylcholine. Certain spider toxins and philanthotoxin (and synthetic analogues) act at this site and are lethal to insects when injected. Analogues active by topical application or by ingestion have not been reported nor have any compounds with potency sufficient for development. *Insect neuropeptides* are products of the nervous system which control a wide range of essential functions such as diuresis, juvenile hormone titre and muscle movement. These polypeptides have been proposed as insect control agents, if they could be delivered to the insect target by a vector such as a baculovirus, but commercial products have not evolved yet. *Octopamine* is the main aromatic biogenic amine in insects and evokes complex hormonal, transmitter and modulator actions. The practical utility of pesticidal compounds with octopaminergic properties was demonstrated by chlordimeform prior to its removal from the US market in 1989 for reasons of adverse toxicology. Amitraz is the only pesticide currently utilizing this mode of action, but the potential of the site for insect control is reflected by the disappointment of growers when chlordimeform was removed from widespread usage.

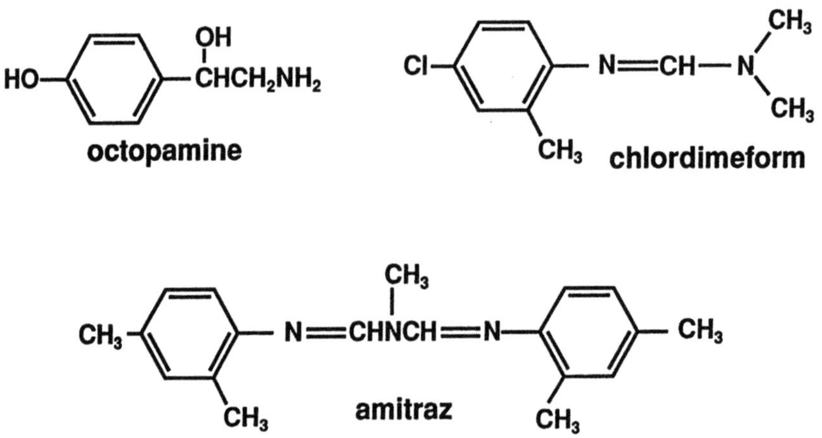

*Figure 7.* Underutilized target: octopamine receptor.

## 5. Growth regulators

Structure elucidation of insect juvenile hormone in 1967 stimulated great interest in synthetic analogues (juvenoids) as insect growth regulators (IGRs) which prolong larval development, prevent maturation, and affect fertility. Mammalian safety was a paramount feature of the first commercial IGR, methoprene, and subsequent juvenoids such as fenoxycarb and pyriproxyfen by virtue of targets in insects which are absent in mammals. Unfortunately juvenoids act very slowly, which farmers are reluctant to accept, and can produce larger, more destructive caterpillars. The early juvenoids were limited to niche markets (mosquitoes, fleas and cockroaches), but fenoxycarb and pyriproxyfen are becoming increasingly important for agricultural applications.

The counterpart to insect juvenile hormone is ecdysone which in excess amount prematurely accelerates metamorphosis with toxic consequences. Although ecdysone by itself is too complicated for practical utility, simpler ecdysone agonists (e.g. tebufenozide) offer potential as do juvenile hormone antagonists which present a theoretically appealing opportunity for insect control that has been probed for 25 years but not exploited successfully. Disruption of the action of insect hormones remains a desirable goal in agricultural pest control.

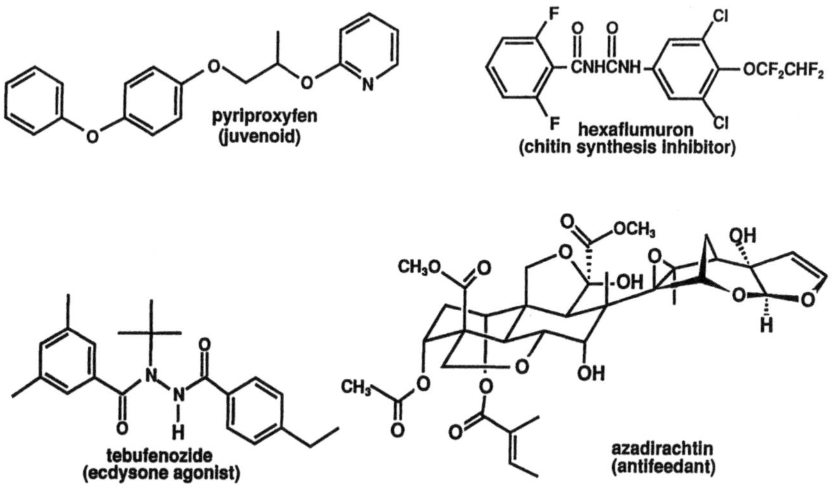

*Figure 8.* Growth regulators.

The chitin deposition inhibitor diflubenzuron appeared in the early 1970s and while the first analogues were difficult to formulate, they were active on insects but non-toxic to mammals. Later compounds such as teflubenzuron and hexaflumuron (for termite control) were of greatly improved potency. The benzoylphenylureas play an important role in some integrated pest management programmes, e.g. for cotton pests, but their general use is limited by slow action. Nikkomycin, an antibiotic from *Streptomyces*, uses chitin synthetase as a target and is very effective but not marketed for mite and insect control. Neem extract is the latest IGR registered (1993) for food use in the US, acting more as a feeding deterrent than a latent toxicant. The active ingredient azadirachtin is non-toxic to mammals and with minimal environmental impact.

## 6. Energy and muscle targets

Oxidative phosphorylation and respiration are the targets for six new insecticides announced since 1990 (Perrior, 1993). Many of the newer compounds are also potent acaricides (Dekeyser and Downer, 1994).

The pyrrole Pirate is a proinsecticide activated by metabolic *N*-dealkylation to a potent insecticide acting as an uncoupler of oxidative phosphorylation (Hunt, 1994; Treacy *et al.*, 1994). Safety is conferred by selective bioactivation by the pest. Historically this proinsecticide approach for uncouplers (such as nitrophenols and 2-trifluoromethylbenzimidazoles) has been prone to difficulty since at least some beneficial animals are usually sensitive.

Fenazaquin (a quinazoline derivative) and three other heterocyclic compounds (fenpyroximate, pyridaben and tebufenpyrad) are inhibitors of NADH/ubiquinone oxidoreductase, the key enzyme in the mitochondrial electron transport chain in Complex I which generates energy in the form of ATP (Hackler *et al.*, 1994). These new insecticides/acaricides presumably bind to the same site as rotenone, a botanical insecticide of historical importance in elucidating the mechanisms of mitochondrial respiration but with limited commercial utility because of structural complexity and environmental lability.

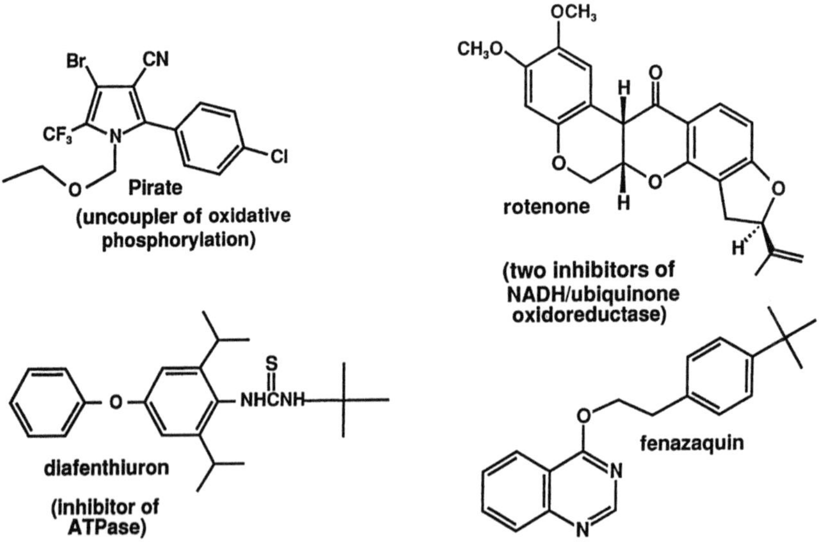

*Figure 9.* Energy metabolism.

The thiourea diafenthiuron, also a proinsecticide, inhibits mitochondrial ATPase both *in vitro* and *in vivo* via its carbodiimide degradation product formed metabolically or abiotically (Ruder and Kayser, 1994). This carbodiimide is a reactive intermediate (presumably somewhat stabilized by steric constraints) that covalently derivatizes a mitochondrial protein proposed to be the $F_o$ proteolipid portion of ATPase.

Ryania, as the powered stemwood of *Ryania speciosa*, has been used in small amounts for about 50 years to control lepidopterous larvae and other chewing pests. The insecticidal ingredient, ryanodine, is a complex alkaloid that acts at nanomolar levels to open the $Ca^{2+}$-release channel of skeletal muscle. Major advances have been made on the chemistry and action of ryanodine, but practical synthetic insecticides from this important biochemical probe are unlikely in the near future.

## 7. Discovery of totally new targets

The number of potential insecticide targets seems limitless if one merely considers all of the essential insect enzymes, channels and receptors. However, there may be a finite number of practical targets, i.e. that are suitable for agrochemicals of acceptable potency and selectivity (Casida, 1990). Three general routes to the discovery of new targets or optimization of pest control chemicals include new biology, new chemistry and serendipity. New biology involves genetic engineering of biological agents and crops to control insect pests without employing insecticides. While biopesticides are expected to increase in importance, they currently represent less than 1% of the agrochemical market (Rodgers, 1993) with *B. thuringiensis* contributing 90% of total sales. Recombinant DNA technology has improved the potency of *Bacillus thuringiensis* endotoxins. It has also allowed the recent introduction of cotton, corn and potatoes expressing levels of endotoxin sufficient to significantly reduce pest-induced crop damage; hopefully these crops will maintain their effectiveness without rapid selection of resistant pest strains. Attempts to improve natural strains of insect-specific baculoviruses, such as by speeding their action through inclusion of insecticidal proteins (e.g. juvenile hormone esterase and toxins from scorpions, spiders and insects), have been largely basic research endeavours with the fruits of labour still uncertain and complicated by potential regulatory problems. New targets typically are discovered by elucidating the mode of action of novel chemical prototypes which may evoke unusual symptoms of toxicity in target pests. Pymetrozine (a pyridine azomethine) is an example of a new insecticide discovered through the 'chemically directed concept' which emphasizes novel chemistry rather than a biorational approach (Kristinsson, 1994). It acts at an unknown target, possibly in the nervous system, that regulates feeding of sucking pests. Natural products have often served as probes for target sites by virtue of their chemical diversity relative to bioactivity. Combinatorial chemistry is a new technique which involves the use of simple molecular building blocks that can be linked like polymers to make combinations to form libraries of compounds used in conjunction with mass screenings. This potentially huge library will contain a wide range of structurally-diverse compounds for bioassay at known targets (e.g. receptor binding or enzyme assay) or in new toxicological systems. Serendipity continues to be a major source of new targets resulting from either keen scientific perception or pure luck with a success rate somewhat related to the number of compounds examined (Casida, 1987).

## 8. Newer, safer and more effective insecticides

Established commercial insecticides (about 300 at present) are generally effective and often cheaper than the new compounds. The expense of reregistering some of the older pesticides in the US (e.g. mesocosm studies) is forcing the discontinuation of compounds whose niche market will not sustain them economically. Other insecticides will be lost for lack of proprietary status, questionable toxicology, or as possible victims of resistance.

The evolution of a new pesticide is a tedious and expensive process, requiring 8–10 years and $50–100 million for discovery, optimization, registration and commercialization. Since 1990 only about five new insecticides/acaricides have been announced annually and, while representing the successful sieving of 20 000 to 50 000 synthetic chemicals in each case, these candidate control agents still face the rigours of the marketplace. Although insecticides today represent about one-third of all newly-released pesticides, this proportion is likely to decrease as the research emphasis continues to shift towards herbicides and will shrink further because market-driven changes in the agrochemical industry have resulted in numerous consolidations in recent years to fewer companies and scientists.

The challenges to discover safer and more effective insecticides are substantial but probably will be met as long as there are economic or human welfare benefits. The continuing pressure on the AChE target of OPs and MCs and the sodium channel target of the pyrethroids is likely to exacerbate problems of resistance which will necessitate further quests for new chemical types and biochemical targets. The public's current negative perception of pesticides will mandate that new insecticides must be compatible with integrated pest management and acceptable in their human and environmental toxicology. One method to reduce environmental impact is to increase potency another 10-fold since smaller quantities are then actually applied. Additional challenges include the need to achieve target-site selectivity in pest *vs.* non-target species and to balance the persistence needed for pest control with the biodegradability required for safety. Application amounts of insecticides have dropped as inherent toxicity and specificity to the pest have increased. These trends must continue to ensure the availability of acceptable insecticides to produce and protect the food and fibre for an expanding world population.

## Acknowledgement

Preparation of this review was supported by Grant Number PO1 ES00049 from the National Institute of Environmental Health Sciences, NIH. We thank Ephraim Cohen for helpful suggestions.

## References

Carson R 1962 Silent Spring. The Riverside Press, Cambridge, MA.

Casida J E 1987 Serendipity in pesticide mode of action and metabolism. In A New Turn in Pesticide Sciences. Ed. M Eto. pp 115–144. Soft Science Publications, Tokyo, Japan.

Casida J E 1990 Pesticide mode of action: Evidence for and implications of a finite number of biochemical targets. In Pesticides and Alternatives: Innovative Chemical and Biological Approaches to Pest Control. Ed. J E Casida. pp 11–22. Elsevier Science, New York.

Casida J E 1993 Insecticide action at the GABA-gated chloride channel: Recognition, progress and prospects. Arch. Insect Biochem. Physiol. 22, 13–23.

Dekeyser M A and Downer R G H 1994 Biochemical and physiological targets for miticides. Pestic. Sci. 40, 85–101.

Elliott M 1989 The pyrethroids: Early discovery, recent advances and the future. Pestic. Sci. 27, 337–351.

Elliott M 1995 Chemicals in insect control. In Pyrethrum Flowers: Production, Chemistry, Toxicology, and Uses. Eds J E Casida and G B Quistad. pp 3–31. Oxford University Press, New York.

Forget G 1991 Pesticides and the third world. J. Toxicol. Environ. Health 32, 11–31.

Fukuto T R 1990 Mechanism of action of organophosphorus and carbamate insecticides. Environ. Health Perspect. 87, 245–254.

Georghiou G P 1990 Overview of insecticide resistance. In Managing Resistance to Agrochemicals. Eds M B Green, H M LeBaron and W K Moberg. pp 18–41. ACS Symp. Ser. 421. Amer. Chem. Soc., Washington, DC.

Hackler R E, Suhr R G, Sheets J J *et al.* 1994 Chemistry and miticidal activity of fused pyrimidine derivatives of fenazaquin. In Advances in the Chemistry of Insect Control III. Ed. G G Briggs. pp 70–84. The Royal Society of Chemistry, Cambridge, UK.

Hunt D A 1994 2-Arylpyrroles: Novel uncouplers of oxidative phosphorylation. In Advances in the Chemistry of Insect Control III. Ed. G G Briggs. pp 127–140. The Royal Society of Chemistry, Cambridge, UK.

Kristinsson H 1994 Pymetrozine – A new insecticide. In Advances in the Chemistry of Insect Control III. Ed. G G Briggs. pp 85–102. The Royal Society of Chemistry, Cambridge, UK.

Kuhr R G and Dorough H W 1976 Carbamate Insecticides: Chemistry, Biochemistry and Toxicology. CRC Press, Cleveland, OH.

Leicht W 1993 Imidacloprid – A chloronicotinyl insecticide. Pestic. Outlook 4, 17–21.

McDougall J and Phillips M 1993 The world agrochemical market. Chem. Ind. 888–891.

Moffat A S 1993 New chemicals seek to outwit insect pests. Science 261, 550–551.

Naumann K 1989 Acetylcholinesterase inhibitors. In Progress and Prospects in Insect Control. Ed. R McFarlane. pp 21–41. Monograph No. 43, British Crop Protection Council, Farnham, UK.

Perrior T R 1993 Chemical insecticides for the 21st century. Chem. Ind. 883–887.

Rodgers P B 1993 Potential of biopesticides in agriculture. Pestic. Sci. 39, 117–129.

Ruder F and Kayser H 1994 The thiourea insecticide diafenthiuron inhibits mitochondrial ATPase *in vitro* and *in vivo* by its carbodiimide product. Biochem. Soc. Trans. 22, 241–244.

Treacy M, Miller T, Black B, Gard I, Hunt D and Hollingworth R M 1994 Uncoupling activity and pesticidal properties of pyrroles. Biochem. Soc. Trans. 22, 244–247.

Voss G, Neumann R and Kobel W 1995 Economy-society-environment: How fit are organophosphorus insecticides? In Progress and Prospects of Organophosphorus Agrochemicals. Eds M Eto and J E Casida. pp. 31–41. Kyushu University Press, Fukuoka, Japan.

Ware G W 1994 The Pesticide Book. 4th ed. Thomson, Fresno, CA.

## 2
# From *Homo economicus* to *Homo ecologicus:* towards environmentally safe pest control

J.C. VAN LENTEREN

## 1. Introduction

The word insect is often considered synonymous with the term noxious organism. Insects are regularly mentioned as being harmful, cruel and dangerous creatures. Upon asking people to list some insect names, they will start with such examples as mosquitoes, flies, wasps, locusts, and ants. Beneficial, beautiful, or benign insects do not usually occur among the first five species mentioned.

The number of animal species is estimated to be between 20 and 50 million, out of which, to date only 1.7 million have been identified and named. Eighty per cent of all animal species are insects. Among the insects, plant eaters are our main competitors and they comprise about half of the insect world, i.e. some millions of species. The other half consists of insects which live off other insects (predators and parasites), insects which live off other animals, and insects feeding on dead organic material. Although the number of plant-eating insects is very large, the number of species regarded as pests is small. Estimates of the number of pest species vary between 500 and 10 000 species worldwide (van Lenteren, 1993). The last figure includes harmful insects which attack man or cattle, which damage man's dwellings or are generally a nuisance.

Thus, a balanced view should picture most insects as playing a very positive role in nature, while only a minimal fraction of insect species – less than 0.01% – may cause problems for humans. Due to its activities in controlling this small group of pest insects – a development which started only 50 years ago – the entomological profession has come under severe criticism.

In this chapter I will first discuss the role played by insects and the problematic situation with regard to insect control. Then I will make suggestions on how to improve this situation. Finally, I will turn to a more general question: I will contemplate on how mankind views nature and how this influences our future.

## 2. The vast majority of insects are useful

Even the most conservative estimate of 10 000 species makes it clear that very few insects are problematic. Why is this number so small? Ecologists have ample evidence relevant to answer this question. First of all, not all plant-eating insects

*D. Rosen* et al. *(eds): Modern Agriculture and the Environment, 17–31.*

occur on the plants we produce for human use. Ninety per cent of our food consists of only 300 plant species. Many insects feed on the other 300 000 plant species and are thus not directly detrimental to us. Further, we rarely see these non-commercial plants being completely defoliated by insects. In nature, only 1% of the plant biomass is eaten by insects and other herbivores, which is in strong contrast to the potentially 20–100% consumption of food crops by insects. What has gone wrong? Comparison of population dynamics of potential pest species in natural ecosystems and agro-ecosystems points towards several general explanations with regard to the difference in degree of consumption.

- Wild plants possess resistance against a variety of insects, which results in no or slow population development.
- In natural ecosystems, host plants specific for certain insect species are not easily discovered, leading to high insect mortality during dispersal. Only a few insects are able to locate a host plant and reproduce.
- In natural ecosystems, a myriad of natural enemy species maintain plant-eating insects at low population densities and, even in agro-ecosystems, a number of potential pests are held at non-damaging levels by natural enemies which occur naturally.

These combined factors result generally in low population densities of plant-eating insects in nature. Agriculture brought the creation of monocultures of genetically identical crop plants which are maintained in an excellent condition and are often grown in the same geographical area over many years. This has led to the occurrence of maximal survival and reproduction rates for pest insects, whereas many species of their natural enemies have problems of surviving in monocultures. Thus, man has created first-rate favourable environments for plant-eating insects through the provision of more accessible and/or more nutritious food or lush growth, as well as by the decimation of their natural enemies (Huffaker *et al.*, 1976). Extensive studies have been made to find out how the agro-ecosystems influenced pest population dynamics and how these situations could be changed to profit in a better way from the pest control mechanisms which nature provides freely (review: van Lenteren, 1987).

Millions of species of beneficial insects occur in nature in the form of natural enemies of pest insects. Besides beneficial insects as predators and parasites, we know other large groups of useful insects in the role of essential elements to keep ecosystems functioning, such as pollinators and recyclers (insects eating dead organic material), and smaller groups of beneficial insects producing nectar, silk, lac and wax (review: Smith *et al.*, 1973). Entomology has also played an eminent role in culture and recreation (e.g. Hogue, 1987; Southwood, 1977). Further, insects are used as food, although this was more general in the past than today. Large orthopterans are still an important item of the diet, especially for children, in certain parts of Africa (review of entomophagy: DeFoliart, 1989).

*Table 1.* Estimated money spent on crop protection (in billions of US$).

| Control agent | US$ billion |
|---|---|
| Conventional synthetic pesticides | 26.00 |
| Plant material resistant to pests and diseases | 6.00 |
| Biological control | 0.050 |
| Bacterial and fungal-derived toxins | 0.110 |
| Botanical pesticides | 0.090 |
| Behaviour-modifying chemicals | 0.050 |

## 3. The role of chemical control is strongly overrated

A generally accepted idea is that most pests are kept under control with synthetic chemical pesticides. I will make clear that this is an incorrect viewpoint. The world market for chemical pesticides is estimated at US$26 billion (Klassen, 1993; Figure 1), but some biologically based technologies already make up a significant part of the crop protection market, particularly host-plant resistance to pests and diseases (Table 1). The world market for seeds and propagules, most of which have genetic resistance to at least several harmful organisms, is about US$6 billion. Money spent on other methods such as biological control, behaviour-modifying chemicals (e.g. pheromones) and botanicals is still relatively modest in the developed world and amounted to US$255 million in 1991.

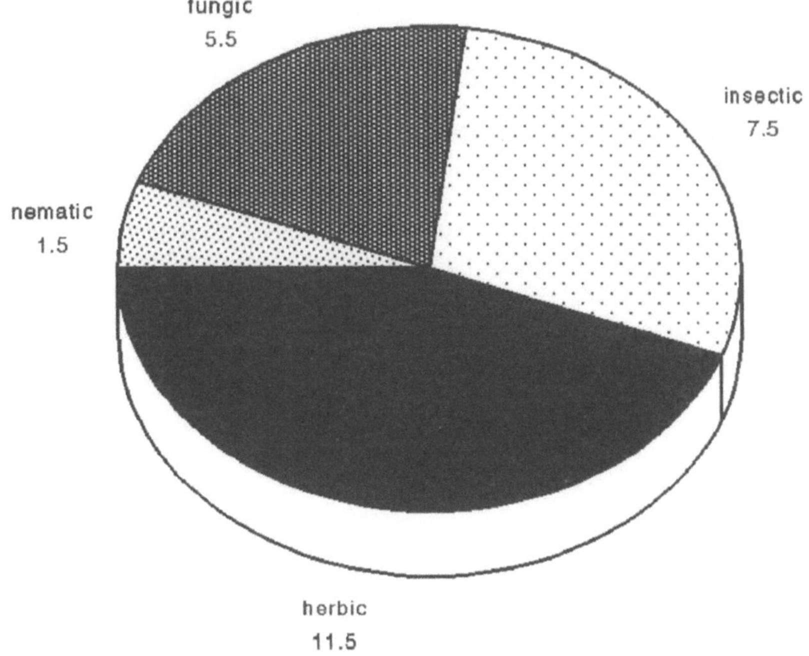


*Figure 1.* Estimated world market for chemical pesticides in 1991 (in billions of US$).

Biologically-based pest control methods seem to play a minor role in the total crop protection picture. But such reasoning is based on actual amounts of money spent on control agents, and contains some inherent mistakes. First of all, most of the potential pests are kept at densities below the treatment thresholds by host-plant resistance and natural biological control (DeBach and Rosen, 1991). For example, the continuous profits obtained by these forms of control are never accounted for. Second, when permanent control is obtained with inoculative biological control, where a natural enemy is introduced once to control a pest forever, the annual benefits resulting year after year from this approach are not considered in cost/benefit calculations. Therefore, the importance of biologically-based pest control methods is very seriously underscored in comparisons of chemical and other controls. It is, of course, short-sighted and wrong to use the investments actually spent on a certain crop protection technique to indicate its importance. A correct means of evaluation would be to calculate cost-benefit ratios based on the costs to develop a method and the economic benefits obtained after application. Conservative calculations of cost-benefits indicate a more than ten times better ratio for biological control than for chemical control.

## 4. Research in pest control needs to be redirected to biologically-based, long-term, environmentally safe solutions

To combat insect pests, some 200 different chemical ingredients are used in an array of formulations. Insecticides form the most hazardous category of the pesticides because, unlike for example fungicides, they are aimed at killing animal life. The majority of insecticides can be characterized as having a broad-spectrum activity, with well-known risks for producers, appliers, consumers and the environment. These risks are of general concern. However, the main problem for the chemical industry, at present, is the development of insect-resistance against pesticides. The exponential increase of resistance leads to a dramatic rise in human disease problems (e.g. malaria, due to insect-vector resistance) and a decrease in the yields of some crops (Metcalf, 1980). Furthermore, the development of new pesticides has become increasingly difficult. As many more potential chemicals need to be screened, the overall production costs are rocketing and more research is necessary before new pesticides are legislated. The rate at which insects are developing resistance to new and complex pesticides is, however, not decreasing. Chemical pest control has resulted in more than 500 insect species becoming resistant to one or more pesticides. All attempts to eradicate pest insects have failed. Harmful insects survived all chemical tactics invented in order to destroy them.

The above factors will lead to ever-increasing costs for chemical control. As a result, a dramatic decrease in the number of newly marketed insecticides appearing per year has already been experienced over the last two decades: 20 new active ingredients were registered yearly in the 1960s, which is in strong contrast to the less than one being registered per year at present. Therefore, the role of agricultural entomologists in pest control will have to change from studying opportunistic

*Table 2.* Methods to prevent or reduce development of pests.

*Prevention*
Prevent introduction of new pests (inspection and quarantine)
Start with clean seed and plant material (thermal disinfection)
Start with pest free soil (steam sterilization and solarization)
Prevent introduction from neighbouring crops

*Reduction*
Apply cultural control (crop rotation)
Use plants which are (partly) resistant to pests
Apply one of the following control methods:

- Mechanical control (mechanical destruction of pest organisms)
- Physical control (heating)
- Control with attractants, repellents and antifeedants
- Control with pheromones
- Control with hormones
- Genetic control
- Biological control
- (Selective) chemical control

Control based on sampling and spray thresholds: guided or supervised control
Control based on the integration of methods which cause the least disruption of ecosystems: integrated control

methods of (chemical) pest control – whereby problems are awaited, solutions are hoped for and the aforementioned indicators of resistance development and the low number of new pesticides available are ignored – to studying biologically-based methods of pest control. Many alternative methods are already available (Table 2) and we can perceive an important change in research approach. The development of conventional chemical pesticides, which is generally based on empiricism, is in the process of being replaced by research for new methods based on in-depth understanding of biological mechanisms and relationships (Klassen, 1993). For an extensive review of developments in the area of biologically-based technologies, I refer to Lumsden and Vaughn (1993). These technologies represent an enormous and still largely untapped biological and genetic resource and create excellent possibilities for pest control methods to be used in integrated pest management approaches. Here I will shortly discuss only two biologically-based methods: host-plant resistance and biological control.

### 4.1 *Host-plant resistance*

Although host-plant resistance has quite often been exploited for the control of some diseases, it can be used to a much larger extent. Klassen (1993) provides examples of the use of resistant varieties for suppression of diseases and insect pests of some major field crops in the USA, where host-plant resistance is employed on vast areas to control both pests and diseases. It seems that in North America the development of insect-resistant cultivars is more common than in Europe. The combination of partial host-plant resistance and biological control remains to be exploited yet. Also, breeding of plants which augment natural enemy behaviour or which attract natural enemies is a just-emerging approach (review: Dicke, 1996).

### *4.2 Biological control*

Pest species are generally attacked by an array of natural enemies, from viruses to predatory mammals. Quantitatively, insects form the largest group of natural enemies. There are no accurate estimates of the numbers of natural enemies of insects and mites, but their numbers will be in the order of millions of species. Most pest insects carry many species of natural enemies. Some pest species have well over 100 recorded, rather specific enemies. Besides these specific enemies, many generalist predators occur.

During the past 100 years of modern biological insect control some 5000 introductions of natural enemies have been made to new areas worldwide. In 1500 cases this has led to substantial, long-term reduction of the pest insect and to complete control in 450 instances (Waage, 1994). Hence, biological insect control has an excellent record of success. In the age of modern biological control, empirical evidence and a large number of case studies (e.g. Bellows, 1997; Huffaker and Messenger, 1976) have shown that biological control practised by experts is a very cost-effective method of pest control. Benefit/cost analyses for 30 inoculative biological control programmes show that benefits are between 18 and 1500 to one, better than for any other pest control method. Even though no concerted efforts have been made to measure the global economic benefits of natural and biological control, the total benefits must be reckoned in billions of dollars (Tisdell, 1990). Besides its good pest reduction capacity, biological control has highly positive social, ecological and economic benefits. Additionally, once a good natural enemy has been identified, it can be used indefinitely as development of resistance to natural enemies comparable to pesticide resistance does not occur. Therefore, it should often be the first pest control tactic to be explored. As biological control compares very favourably with chemical control (Table 3), there are certainly no reasons to stifle the investments in research based on economics, success ratio or number of agents available for evaluation. Funding of this work is still very limited, especially if one

*Table 3.* Comparison of aspects related to the development and application of chemical and biological control in 1992.

| | Chemical control | Biological control |
|---|---|---|
| Number of 'ingredients' tested | > 1 million | 5500 |
| Success ratio | 1:50 000 | 1:20 |
| Developmental costs | US$100 million | US$2 million |
| Developmental time | 10 years | 10 years |
| Benefit per unit of money invested | 2.5–5 | 30 |
| Risk of resistance | Great | Nil/small |
| Specificity | Small | Large |
| Harmful side-effects | Many | Nil/few |

realises the complications of this type of research. In my view, however, the slow increase in the use of biological control is more a matter of politics and policy than of any inherent weakness of the method itself. This also applies to several other methods mentioned in Table 2, which have been developed but receive too little support to be implemented.

## 5. Chemical control is irresponsibly cheap and, as a result, frustrates implementation of alternatives

Costs of pest control are very low and amount to only 3% of the investment for producing a crop. The costs of pesticides are in part so low because the chemical industries are often not being held responsible for any indirect costs inflicted upon the environment or man (e.g. for cleaning up chemical wastes or for reduction in diversity of flora and fauna). The availability of very cheap pesticides has resulted in a decrease of treatment thresholds. This makes the application of other than chemical control harder than before these chemicals were on the market and more damage had to be tolerated.

Appreciation of the indirect costs (social and environmental costs) and inclusion of these costs in their sales price results in more realistic costs of pesticides and will automatically lead to an increased probability that alternatives will be used. However, the extreme complexity of estimating indirect costs related to pesticide use and the scarcity of data have delayed an assessment until recently. Pimentel and his co-workers have been pioneers in this area (see e.g. Pimentel *et al.*, 1980). They estimated the annual indirect costs of pesticides to be US$840 million. Adding this to the direct costs of US$2800 million results in a return of US$3 per dollar invested in pesticides, rather than the US$4 without indirect costs taken into account. Pimentel *et al.* (1980) warn against the obvious oversimplification of this estimate. How can one place a monetary value on human lives lost or disabled due to pesticide poisoning? What do wildlife losses cost? Some environmentalists state that the estimate of US$840 million is only a small portion of the environmental and social costs. A more complete cost/benefit analysis would, therefore, further reduce the profitability of pesticides and lead to a better appreciation of the usefulness of other control methods.

To be able to at least partially regain such secondary or social costs, prices of pesticides should be doubled or tripled through a system of levies, whereby levies should be highest for the most environmentally-damaging pesticides. A re-evaluation of more realistic cost/benefit ratios for crop protection methods is also essential to obtain a change in crop protection policy and a more honest appraisal of the usefulness of alternatives.

## 6. Implementation of sustainable pest control is not stimulated

### *6.1 Funding of research on biologically-based pest control methods*

The results obtained in non-chemical pest control are, of course, in the first instance dependent on the amount of research and development work. During the past five decades, little, and usually only public money was invested for the development of non-chemical control methods. The money was well spent, because considerable results were obtained (Klassen, 1993). Implementation was more hindered by constraints mentioned below, than purely by financial limitations.

### 6.2 *Farmers' attitudes*

Until very recently, only few farmers asked for development of non-chemical control methods. Initiatives for development of integrated pest management (IPM) programmes were previously made by and must still mainly come from researchers. But being unable to control a pest with chemicals becomes a strong drive for farmers to change their ideas on IPM. Currently, the attitude of several groups of farmers is changing. European fruit growers and producers of glasshouse vegetables, for example, have experienced the positive aspects of IPM and seriously worry about the increasing public concern on pesticide usage. Therefore they now generally prefer to use IPM methods (van Lenteren *et al.*, 1992).

### 6.3 *The viewpoint of the chemical industries*

In general, we can state that any complication in a simple chemical pest control programme is appreciated as a negative development by the large industries. Alternatives such as biological control not only complicate chemical control programmes, but they seem to be unattractive commercially as well because of a combination of the impossibility of patenting natural enemies, the complicated mass production, their short shelf-life, their specificity, and the more complicated guidance for growers. Chemical industries will not start the production of other than broad-spectrum pesticides on their own initiative, unless the use of those pesticides is prohibited or when pest organisms substantially develop resistance – but time is on our side! We cannot blame the chemical industry for this attitude because they provide pesticides, the use of which is permitted by a government's registration policy.

### 6.4 *Role of the governments*

Therefore, it is only governmental bodies who are able to change the pest control picture through measures that make some kinds of chemical control less attractive or impossible (by measures concerning registration, taxation, side-effect labelling etc.), by stimulating other control methods and, above all, by teaching on all levels in order to change the attitude towards nature. It is bizarre that public money is used for the development of alternatives for chemical control when, at the same time, their application is often not encouraged by governmental bodies.

## 7. Present crop protection policies seem to stimulate a reduction of pesticide usage

After years of increasing pesticide sales in the developed world, a stabilization and even a reduction occurred at the start of the 1990s, particularly in Western Europe where pesticide sales decreased by 20% in 1992 (expressed as sales value). In the Netherlands the reduction was even 42% in 1993, when compared with average pesticide use in the period 1984–1988 (expressed in volume a.i. used).

Several European governments, e.g. Sweden, Denmark, Germany and the Netherlands, have made programmes to reduce the use, the emission and the dependence on pesticides. The programmes are not merely aimed at banning pesticides, but to create alternative, sustainable means of control. The programmes are at different stages of implementation. The Swedish programme was the first and achieved a 50% reduction in pesticide use between 1987 and 1990, and a further cut of 50% is presently implemented. The German and Danish programme achieved a 50 and 25% reduction, respectively, by the year 1990. The Dutch programme started in 1991 and is well on schedule with an aimed reduction of 65% or more by the year 2000. In all programmes a main goal is to replace environmentally critical pesticides with less harmful ones. In the new pesticide registration guidelines, the Dutch authorities have included the condition that first the usefulness of non-chemical alternatives has to be evaluated before a pesticide can be registered. The Swedish government is working along the same line and hopefully the European Union will also adopt the requirement that pesticides will be registered only if no alternative, non-chemical methods are available.

Long-term views of agricultural developments point at even much stronger reductions in chemical control. A strategy study of the Dutch Scientific Council for Government Policy (see Rabbinge, 1992) indicates that the area of cultivated land will be reduced to about half of its present size in the European Union. As a result, the anticipated reduction in the volume of pesticide used will be tremendous over the next two decades, i.e. in the order of 80–90% Europe-wide, because:

- The technology of pesticide application will be improved (better targeting) and the amount of active ingredient needed for effective control will decrease.
- The application of integrated pest and disease management systems will become standard for all crops.
- The abundant use of pesticides due to narrow crop rotations will be eliminated, which will particularly lead to a reduction of soil fumigants.
- The area on which pesticides are used will be much smaller.

Such developments will strongly support the role that naturally occurring beneficial insects can play, and also allow for increased use of other biologically-based methods.

## 8. Over-specialization in agriculture prevents a reduction of pesticide inputs

Agricultural research has seen a large diversification in this century. In plant breeding, the search for better-yielding varieties has been the central theme for years. Plant production concentrated on finding and overcoming limiting factors in plant growth and crop yields. Plant resistance against pests was not considered to be important, because most pests could easily be controlled with insecticides. This has led to plant varieties becoming less resistant to insects. In addition, over-fertilization resulted in plants which stimulate pest development (e.g. aphids, leafminers, spider mites).

Particularly during the past 50 years, pest control has been seen as an independent area in agriculture. The viewpoint was that those working in this area could easily present solutions to protect the maximally producing crops. This reductionist approach to agricultural research has resulted in negative feedbacks leading to an ever-increasing use of pesticides. A holistic approach, through a strong integration of the research areas of plant breeding/production and crop protection, will prevent such conflicting developments.

## 9. Integrated production approaches make chemical control practically redundant

A first step towards the reduction of environmental problems caused by agriculture is to test new developments within the framework of a holistic farming systems approach. The focal point in this approach is farm economics in the form of *maximizing net income*, which is not synonymous with yield maximization. Top yields are obtained with excessively high inputs of fertilizers and pesticides. Reducing the inputs may lead to somewhat lower yields, but financial inputs are also lower and the net income may be the same or better. In farming systems, environmental effects such as pollution of soil and water by pesticides, can be minimized. In general, integrated farming takes more completely into account the various impacts on ecosystems (preservation of flora and fauna, quality and diversity of landscape, and the conservation of energy and non-renewable resources) as well as sociological considerations (employment, public health and well-being of persons associated with agriculture) than is the case with current farming (Vereijken *et al.*, 1986; Wijnands and Kroonen-Backbier, 1993).

Although the objective of obtaining a better balance between the various influences – based on a growing awareness of problems caused by conventional farming – is often expressed, its realization is difficult to achieve as long as politicians and policy makers merely pay lip service to integrated farming and do not provide the means for its accomplishment. Research in integrated farming is still very limited. In Europe, the farming systems approach is studied e.g. in the Federal Republic of Germany and The Netherlands.

The practices which can be manipulated in integrated farming programmes are crop rotation, cultivation, fertilization, pesticide use, cultural control measures,

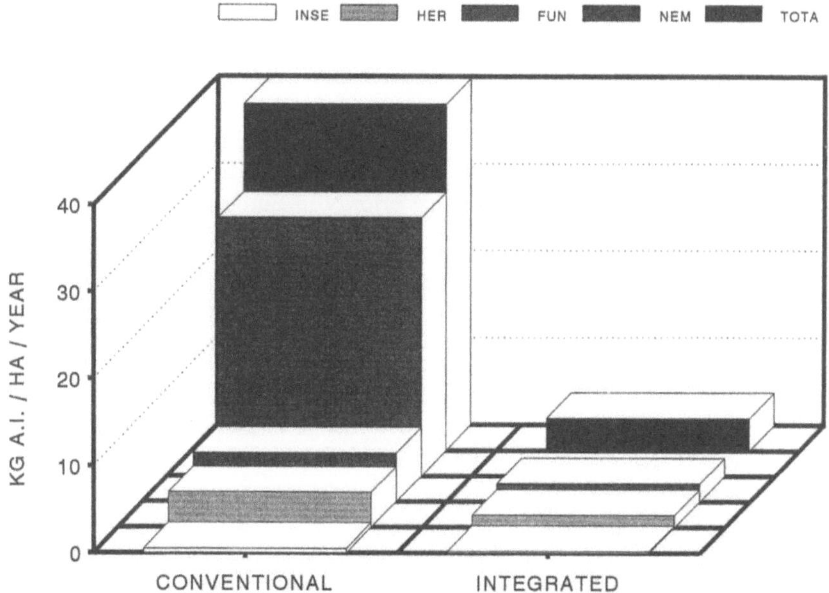


*Figure 2.* Average use of pesticides (kg a.i./ha/year) in conventional and integrated farming (1986–1990; after Wijnands and Kroonen-Backbier, 1993). INSE = insecticides, HER = herbicides, FUN = fungicides, NEM = nematicides, TOTA = total.

biological control and other alternatives to conventional chemical control. The practical results concerning pesticides, obtained in the project in The Netherlands, are summarized in Figure 2. In integrated farming, an important reduction of environmental pollution is realized through a decrease in fertilizer use and the replacement of chemical pesticides by an intensified knowledge of non-chemical measures (crop rotation, use of resistant varieties). In integrated farming, artificial fertilizers tend to be replaced by organic manure, and the total amount of nitrogen is lower to prevent creating a higher sensitivity for pests and diseases. Weed, pest and disease problems are reduced in integrated farming through the use of weed-competitive or disease- and pest-resistant varieties, reduction of N-fertilization, adoption of a specific sowing date and plant spacing, mechanical weed control, natural control, etc. Chemical pest control in integrated farming is based on pest population sampling and use of decision thresholds (supervised control). The more than 90% reduction in pesticide use in integrated farming is for a large part the result of using nematode-resistant plants.

The results on the demonstration farms have been so encouraging that the implementation of integrated farming is strongly supported by the Dutch Ministry of Agriculture, to reduce environmental pollution and to create a firmer basis for the survival of agriculture in the longer term. Integrated farming gives the same economic results as present-day (conventional) farming. The generally lower physical yields for the integrated system were compensated by cost reduction as a result of the lower input of pesticides and fertilizers. Indirect costs of fertilizer and pesticide use are not yet included in this comparison, which would give an even better result for integrated farming. I strongly emphasized the economic aspect of integrated farming because economic figures are more convincing for farmers, policy-makers and politicians in the short term, than pure environmental reasons. In my opinion, development and application of sustainable agriculture, even if it were more costly currently, should have the first priority in Europe.

## 10. The environment should be considered first when choosing pest control methods

Entomology, through chemical pest control, contributes to the environmental crisis. To solve this crisis, simple technological solutions are no longer sufficient. The worldwide environmental problems demand a change in attitude towards nature if policies aimed at improving the environment will lead to success. Governments design and implement environmental policies, but no governmental policy succeeds without broad social support. This means that a certain basic attitude towards nature is essential in the first place to develop ideas for an appropriate environment policy and, second, to have the policy accepted and practised by the general public. Where have developments concerning basic attitudes towards nature gone wrong?

### *10.1 Man as participant*

Historically, man has played a role as a participant in nature. Like other organisms, he used and was used by nature in the form of nutrient and energy cycles. He participates within the ecosystem and is one of the many components of the system. Man is merely part of the system which will also continue to exist when man is absent. In the modern vision of participation, the typical elements of humanity (such as its culture, ethics, science and technology) are not primarily used for, or focused at, the submission of nature to man, but are used to create a situation in which one participates in nature.

### *10.2 Man as despot and ruler*

Since the early days of agriculture man has gradually abandoned the role of participant. Local over-exploitation and pollution interfered with natural cycles. As soon as man developed the idea that he was allowed to manipulate nature, nature was at risk. Cicero (106–43 BC) was one of the first to present a strict anthropocentrical view of nature: '*De Natura Deorom*' – 'everything in nature exists to the benefit of man, nature is created for man'. It was not until the seventeenth century that anthropocentric views were widely accepted and the destruction of nature commenced. Bacon (1561–1626) wrote *The New Atlantis* (1626) in which he designed a technocratic utopia partly realized during the following centuries. Several European scientific societies emerging in the seventeenth century – with the English Royal Society as the first – had as their principal goal to rule nature through science and technology, and referred to Bacon's New Atlantis. The unlimited tendency to control nature, which Bacon described as strongly desirable, has led to our environmental crisis: his utopia has changed into a nightmare (Achterhuis, 1990). Descartes (1596–1650) was reasoning along the same lines as Bacon. He sees man as the lord and master of nature, which, through technical inventions, should be submitted to him.

After 1600, Western man has played the role of despot, of absolute ruler who subjects nature to his own will and sees only its instrumental value. Although this now seems to be an old-fashioned view, essential elements of it can often be found in the vivid optimism of technocrats with a blind trust in technology and for whom growth has no limits. The mere dominance of instrumental thinking (nature is there for man to be used and managed) is now perceived by many as one of the main roots of the environmental crisis.

### *10.3 Man as partner*

In this view man respects the intrinsic value of nature, existing side by side with man. Interests and values of both partners are respected in order to develop jointly in a balanced way. Man interferes with nature in such a way that he does not harm its intrinsic value. He bases his actions on ecological knowledge. Both partners are expected to benefit from this mutualistic relationship. With this attitude, man can play an essential role in order to improve the disturbed relationship which developed as a result of the absolute ruler attitude. The problem with this view is that it implicitly assumes that man knows how to improve the disturbed relationship, that man is willing to sacrifice and not apply technologies which appear economically attractive but environmentally unacceptable.

(The content of this section is based on information presented in Passmore (1974), Rodman (1983) and Achterhuis (1990).)

## 11. Conclusions: from *Homo economicus* to *Homo ecologicus*

The acceptance of and respect for an intrinsic value of nature will result in an attitude which will make possible an acceptance of a new approach for the solution of environmental problems. Such an approach will consist of environmental policies of a radical nature and will demand worldwide stabilization of the human population and reduction of materialistic growth. It is an approach which cannot be realized through negative or positive sanctions alone, but which needs to be based upon a broad social acceptance. Education and information at all levels on the fragile biosphere of Mother Earth will be the crucial elements needed to accomplish this acceptance.

Many of the present-day proposals to improve the deteriorating situation, including the Brundtland report (1987), are aimed at sustainability and management of the environment (not nature). Their main message is: provide the necessities for now but do not risk the possibilities for survival of future generations. They do not speak about an earth which makes life worthwhile, but about a planet with opportunities for survival. Such ideas have been expressed since the seventeenth century (Achterhuis, 1990). They have never led to improvement of nature; on the contrary, the situation has never been as bad as it is today. It is very disappointing to have to conclude that the Brundtland report does not contain a single new idea which has not already been expressed before. Why then should its suggestions for improvement now suddenly be taken up and result in success? Why have all warnings about deterioration of nature since the mid-1650s not resulted in actions? How does one stop the man-caused pollution of nature which has already proceeded at full speed for more than three centuries? Do we wait until the physical limits of nature are reached? The issue here is: is man able to establish and subsequently respect such limits? In my opinion he is far from knowing the latter. Serious studies of ecosystems were initiated some 200 years ago. Due to the great complexity of ecosystems, we are only aware of some of the relationships and feedback mechanisms. We lack the most elementary understanding about the functioning of hundreds of thousands of the participants in ecosystems. It is therefore hazardous and arrogant to manipulate such ecosystems as we do now. When we realize and acknowledge our poor understanding of nature, we should be able to accept our role as partner or participant instead of ruler, and act accordingly.

Finally, let me return to the role of the entomologist with regard to nature. The spin-off of his current activities to control crop pests and vectors of human diseases with chemical pesticides can be classified as very negative for nature. Besides a drastic reduction in the use of conventional chemical pest control based on how we see nature, new cost/benefit analyses, taking indirect costs into account, will lead to a reappraisal of the value of environmentally safe control measures. In order to achieve drastic changes, pest control should not be viewed as an isolated activity, but within an ecosystem approach. In such an approach, many more opportunities for long-term environmentally safe control methods may be achieved.

## References

Achterhuis H J 1990 Van Moeder Aarde tot Ruimteschip: Humanisme en Milieucrisis. Inaugurele Rede, Landbouwuniveriteit Wageningen, 40 pp.

Bellows T S (Ed.) 1997 Principles and Application of Biological Control. University of California Press, Berkeley, CA (in press).

Brundtland G H 1987 Our Common Future. World Commission on Environment and Development. Oxford University Press, 383 pp.

DeBach P and Rosen D 1991 Biological Control by Natural Enemies. Cambridge University Press, 440 pp.

DeFoliart G R 1989 The human use of insects as food and as animal feed. Bull. Entomol. Soc. of Am. 35, 22–35.

Dicke M 1996 Direct and indirect effects of plants on performance of beneficial organisms. In Handbook of Pest Management. Ed. J R Ruberson (in press).

Hogue C L 1987 Cultural entomology. Annu. Rev. Entomol. 32, 181–199.

Huffaker C B and Messenger P S 1976 Theory and Practice of Biological Control. Academic Press, New York, 788 pp.

Huffaker C B, Simmonds F J and Laing J E 1976 The theoretical and empirical basis of biological control. In Theory and Practice of Biological Control. Eds. C B Huffaker and P S Messenger. pp 42–80. Academic Press, New York.

Klassen W 1993 Pest management and biologically based technologies: A look to the future. In Pest Management: Biologically Based Technologies. Eds R D Lumsden and J L Vaughn. pp 410–422. American Chemical Society, Washington, DC.

Lenteren J C van 1987 Environmental manipulation advantageous to natural enemies of pests. In Integrated Pest Management: Quo Vadis? Ed. V Delucchi. pp 123–166. Parasitis 1986 Symposium Book, Geneve, Switzerland.

Lenteren J C van 1993 Parasites and predators play a paramount role in pest management. In Pest Management: Biologically Based Technologies. Eds R D Lumsden and J L Vaughn. pp 68–81. American Chemical Society, Washington, DC.

Lenteren J C, van Minks A K and de Ponti O M B (Eds) 1992 Biological Control and Integrated Crop Protection: Towards environmentally safer agriculture. Pudoc, Wageningen, 239 pp.

Lumsden R D and Vaughn J L (Eds) 1993 Pest Management: Biologically Based Technologies. American Chemical Society, Washington, DC. 435 pp.

Metcalf R L 1980 Changing role of insecticides in crop protection. Annu. Rev. Entomol. *25*, 219–256.

Passmore J 1974 Man's Responsibility for Nature. Scribner, London, 214 pp.

Pimentel D, Andow D, Dyson-Hudson R *et al.* 1980 Environmental and social costs of pesticides: A preliminary assessment. Oikos 34, 126–140.

Rabbinge R 1992 Options for integrated agriculture in Europe. In Biological Control and Integrated Crop Protection: Towards Environmentally Safer Agriculture. Eds J C van Lenteren, A K Minks and O M B de Ponti. pp 211–218. Pudoc, Wageningen.

Rodman J 1983 Four forms of ecological consciousness reconsidered. In Ethics and the Environment. Eds D Scherer and T Attig. Prentice-Hall, Englewood Cliffs, NJ, 236 pp.

Smith R A, Mittler T E and Smith C N (Eds) 1973 History of Entomology. Annual Reviews Inc., Palo Alto, CA, 517 pp.

Southwood T R E 1977 Entomology and Mankind. Proceedings XVth International Congress of Entomology, Washington, DC, 19–27 August 1976, pp. 36–51.

Tisdell C A (1990) Economic impact of biological control of weeds and insects. In Critical Issues in Biological Control. Eds M Mackauer, L E Ehler and J Roland. pp 301–316. Intercept, Andover.

Vereijken P, Edwards C, El Titi A, Fougeroux A and Way M 1986 Proceedings of Workshop on Integrated Farming, Bulletin IOBC/WPRS 1986/IX/2, 34 pp.

Waage J K 1994 Biodiversity, Business and Biotechnology: Shaping the Future of Biological Control. Wageningen Agricultural University, 42 pp.

Wijnands F G and Kroonen-Backbier B M A 1993 Management of farming systems to reduce pesticide inputs: The integrated approach. In Modern Crop Protection: Developments and Perspectives. Ed. J C Zadoks. pp 227–234. Wageningen Pers, Wageningen.

# 3
# Classical biological control: an environmental boon or bane?

DAVID ROSEN

Ours has been the day and age of a rampant population explosion. The human population of the world has been increasing at an alarming, unprecedented rate, and this has confronted mankind with several serious problems, including imminent starvation, depletion of non-renewable energy sources, and the threat of worldwide environmental pollution. Indeed, *Homo sapiens* may soon become an endangered species, and it is imperative that whatever we do to alleviate one problem does not aggravate the others. So, in order to feed the ever-increasing population of the world and avoid starvation, effective programmes should be developed for increasing all agricultural yields; such programmes should however tend to conserve energy and reduce, rather than increase, environmental pollution.

Serious pest problems are a prominent fact of life in modern agriculture. World crop losses in the field have been estimated at about 35% – 12% due to insects and mites, 1% to mammals and birds, 12% to plant pathogens, and 10% to weeds – and are worth about US$400 billion per year (Pimentel, 1986). An additional 15% are lost to pests and diseases after harvest, bringing the total up to about 50% of world crops. Effective pest management is, therefore, an absolute necessity in any modern agro-ecosystem.

Modern chemical pest control, based on the use of toxic synthetic pesticides, has provided us with potent, and usually quite reliable, means for controlling pest populations, and has therefore become the main method of pest management in most world cropping systems. However, the indiscriminate use of chemical pesticides involves some very serious problems and grave hazards. Only very few pesticides are truly selective. Many of them are toxic to humans, plants, domestic animals, wildlife, beneficial natural enemies and various other non-target organisms; all are vulnerable to rapid development of resistance in pest populations; and some contaminate food chains and constitute serious environmental hazards. Their widespread use has become a heavy financial burden on agricultural production, and one of the major components of worldwide environmental pollution. Moreover, unilateral reliance on toxic chemicals, like the use of narcotics, often becomes addictive, because destruction of natural enemies and development of resistance inevitably induce severe target-pest resurgences and secondary-pest upsets, which in their turn call for a vicious cycle of ever more intensive chemical control (DeBach and Rosen, 1991). No wonder that, with rising concern about environmental quality, modern agriculture has been experiencing a severe pesticide crisis, and pressures for

*D. Rosen* et al. *(eds): Modern Agriculture and the Environment, 33–39.*

reducing the amounts of toxic residues in agricultural commodities have been mounting in both domestic and export markets. In view of all these serious drawbacks, attention has been increasingly focusing on the development of safer, and less expensive, alternative methods.

Of the various alternatives available to us – selective chemicals, cultural, mechanical, physical and autocidal controls, etc. – 'classical' biological control, or the importation of exotic natural enemies and their establishment in new habitats, has been by far the most successful to date. Other selective control methods may serve as important components of integrated pest management whenever they are applicable, but classical biological control is the most promising approach for the foreseeable future. When it works, as it often does, classical biological control offers effective, permanent, inexpensive and virtually hazard-free solutions to serious pest problems. Worldwide, by the end of the 1980s, attempts at classical biological control have been made against more than 400 species of *insect* pests belonging to eight orders, and a significant degree of success has been achieved against some two-fifths of them: 75 species have been brought under complete control by imported natural enemies, 74 under substantial control, and 15 under partial control in at least one country. Many of these successes have been repeated in other countries, so that altogether there have been at least 384 successful projects worldwide, including 156 that have been rated as complete successes, 164 as substantial, and 64 as partial successes (DeBach and Rosen, 1991). Achievements with classical biological control of weeds have been as impressive, with 272 exotic invertebrates and fungi released in 272 projects directed against 112 target species in 58 countries (Julien, 1992).

The benefit-to-cost ratio of classical biological control has been calculated as 30:1. Compared with a 3:1 ratio for chemical control, this is indeed remarkable (DeBach and Rosen, 1991). The cost of researching, developing and marketing a new pesticide is now at least US$20 million (Hollingworth, 1987), and the process may take at least ten years. The cost of discovering a new natural enemy, on the other hand, rarely exceeds several thousand dollars. Repeat projects, where a successful natural enemy is simply transferred to another country, may cost no more than the small amount required for a few airmail shipments; but even new projects, involving foreign exploration 'from scratch', may be relatively inexpensive.

Some have cited a low rate of establishment of imported natural enemies as justification for the claim that classical biological control hardly ever works. The rate of establishment is actually very good: out of a total of more than 4200 natural enemy importations that have been recorded to date against insect pests in more than 150 countries, about 30% have resulted in establishment, and the fate of an additional 22% is still unknown. This is an impressive record, considering that, by comparison, only one in 15 000 or 20 000 candidate chemicals ever makes it to the marketing stage as a pesticide. However, the rate of establishment in itself is, of course, irrelevant as a measure of success of biological control. If nine natural enemies fail to become established and the tenth effects complete, permanent control of a serious pest, it is the benefit-to-cost ratio of the entire project that should be taken as the criterion for its success, and we have already seen that this ratio is ten times better than that for chemical control (DeBach and Rosen, 1991).

In view of this outstanding record, it is almost unbelievable that out of about 10 000 insect pest species recorded worldwide, only some 400 species have ever served as targets for classical biological control efforts. In other words, over 95% of the world's insect pests have had no biological control importation project directed against them. Classical biological control is still far from reaching the point of diminishing returns. Not only have the vast majority of *pest* species never served as targets for importation projects, but the vast majority of natural enemy species still remain to be discovered. The fauna of large world areas is very little known, and it has been estimated that between 70 and 90% of all parasitic Hymenoptera – by far the most important group of biological control agents – are still undescribed, and that we have no biological information whatsoever about 97% of all existing species (Kerrich, 1960). Most of the task is still ahead of us, and at the current rate of foreign exploration it is anybody's guess how many potentially powerful weapons may become extinct before we even have a chance to discover them!

For obvious reasons, importation of exotic *phytophagous* organisms for biological control of weeds has always been subject to strict screening, to ascertain the specificity of introduced agents. Indeed, with hundreds of successful projects against weeds, we actually have only one reported case of unintended injury: a tingid bug imported in the early 1960s into Uganda against the weed *Lantana* did switch for a while to sesame, a minor crop, but the effect was temporary, as it was unable to reproduce on it (Greathead, 1971). Even this would not have happened today, with our much tighter procedures.

With classical biological control of arthropod pests, the accepted procedure is to introduce as many primary natural enemies as may be available, for the simple reason that we cannot predict the performance of a natural enemy in a new habitat. It should be stressed, however, that competition among natural enemy species does not present a threat to the success of a biological control project. Multiple importations may result either in complementary action of several natural enemies coexisting in the same habitat, or in competitive displacement of inferior types by ones that are better adapted to that particular habitat, and in either case the outcome should be more effective biological control.

The main advantage of classical biological control lies in its being virtually free of all the grave hazards associated with chemical control. First and foremost, effective natural enemies are *specific* control agents, and pose much less danger to non-target organisms than do toxic chemicals. Even general predators, which may be rather polyphagous, have not been documented to cause any real environmental damage. Biological control is more environmentally friendly than almost any other form of pest management. *We are the environmental good guys*!

Unfortunately, however, some environmentalists have recently raised all kinds of petty concerns about possible detrimental effects that classical biological control may have on the environment. A landmark of sorts has been a paper by Howarth (1991) where, after paying some brief lip service to the beneficial aspects of biological control, he goes on to lambast it with examples of possible negative impacts.

What is Dr Howarth's first and most striking example? It is the tragic demise of the levuana moth in the Fiji Islands. Let us pause for a moment to consider this case, one of the shining landmarks in the annals of classical biological control. Coconut was, and probably still is, *the* major agricultural crop in Fiji and one of the main sources of income there, and it was devastated by a very serious pest, *Levuana iridescens*. So an exotic natural enemy, the tachinid fly *Bessa remota*, was imported in 1925 and, lo and behold, the pest was not only brought under effective biological control, but it seems to have been eradicated and become extinct. *So what?!* All I can say, from the bottom of my bleeding heart, is: 'Tough luck, and good riddance!' Biological control usually does *not* result in complete eradication of the target pest – and this may not have been the case here, as the pest was obviously exotic to Fiji and may have survived elsewhere – but even if it occasionally does, should we really be concerned about protecting serious *pests* from extinction?

Our golden age represents but a tiny blimp in evolutionary time. Extinctions have been as much part and parcel of evolution as have speciation events. Let us not forget that, but for the terrible mass extinction that occurred on this planet some 65 million years ago, *we* would most probably not be here, certainly not in our present form, to discuss the merits of biological control. It goes without saying that Man should not be the cause of yet another mass extinction – but should preservation of the present biodiversity in its exact form be an absolute end in itself? Should extinction of *any* species be prevented at *any* cost? Have we forgotten that we, too, are engaged in a struggle for our own existence? If a pest that poses a direct threat to human welfare is entirely wiped out, I, for one, would not be able to bring myself to complain about it!

Howarth (1991) goes on to cite several other examples, including, for instance, the use of *Bacillus thuringiensis israelensis*, an excellent microbial control agent for mosquitoes that is not toxic to humans, which has apparently affected some mayfly and dragonfly populations – as if broad-spectrum chemical larvicides would not! He then goes on to propose all kinds of restrictions to be imposed on the importation of biological control agents. All this could well have been a tempest in a teapot. Unfortunately, however, his kind of arguments seem to be snowballing. A similar article was soon published by Miller and Aplet (1993), advocating that legislation be enacted to regulate all biological control activities. I have just recently been informed that some high-ranking USDA officials are concerned about the prospects of importing exotic natural enemies against the spirea aphid, *Aphis spiraecola*, an important pest of citrus, because the aphid may be of North American origin. In fact, all evidence points to its being an Oriental species (Starý and Schlinger, 1967), but does it really matter? Is the USDA now in the business of conserving native *pests*? Another concern raised was that importation of an exotic natural enemy against the spirea aphid might displace a native parasitoid – which is evidently ineffective, or we would not have a pest to control in the first place! Has conservation of ineffective natural enemies become a goal of the USDA?

There is a lot of talk nowadays about so-called endangered species. By definition, such species are already poorly adapted, and some of them may be doomed for extinction, no matter what we do. Admittedly, establishment of an effective natural enemy that is not strictly monophagous may exert some extra pressure on them and hasten the process, or at least decimate their ranks – but this has never been clearly documented, and should not serve as a basis for curtailing classical biological control activities. The same is true also for all those unsubstantiated, and largely hypothetical, claims about the possible role of species – including arthropod pests – in food chains and food webs. Possible conflict of interest has long been an important concept in biological control of weeds – one man's weed may well be another man's beneficial plant, and all interests should be taken into consideration. But arthropod pests usually do not have any obvious redeeming features, and if we seriously request that all their obscure interactions in any ecosystem be thoroughly investigated before release of an exotic natural enemy is permitted, we may just as well put an end to all classical biological control projects! In the final analysis, it is all a matter of striking a reasonable balance between perceived benefits and risks. To take an example from another area, as we all know, pests are often inadvertently transferred by aeroplanes. We should certainly strive to strengthen our quarantines – but nobody would seriously suggest that commercial aviation be stopped because of this potential – and much better documented – threat!

Above all, let us not lose sight of the alternatives. Continued severe pest damage to agricultural crops is, of course, unacceptable. Farmers will *have* to go on controlling pests, and if classical biological control activities are drastically curtailed, they will simply continue to resort to intensive, broad-spectrum chemical control – which may not cause extinctions, but is certainly likely to be much more devastating to the environment than the worst scenario that can be ascribed to biological control!

If I sound worried, indeed I am. The FAO recently circulated a draft for a new international code of conduct for the import and release of biological control agents (see Table 1). Five years ago I would have applauded it as an excellent, timely document. Today, however, in this changing climate, I wonder whether such seemingly innocent phrases, requesting importers of biological control agents to

*Table 1.* Excerpts from Code of Conduct for the Import and Release of Biological Control Agents (draft, FAO).

| | |
|---|---|
| 4.2 | The importer of biological control agents ... should prepare dossiers with information on the candidate biological control agent, including: |
| 4.2.2 | A summary of all available information on its origin, distribution, biology, natural enemies and impact in its area of distribution. |
| 4.2.3 | An analysis of the host range of the biological control agent and any potential hazards posed to non-target hosts. |
| 4.3 | The importer ... should also prepare a dossier ... with an analysis of potential risk posed to ... human and animal health following introduction. |

prepare a summary of all available information on their origin, distribution, biology, natural enemies *and impact in their area of distribution*, or an analysis of their host range *and any potential hazards posed to non-target hosts*, may not be misinterpreted to call for endless 'in-depth' studies that would halt all importation activities for years to come. By the same token, when the Code requests importers to prepare *an analysis of potential risks posed to human and animal health following introduction*, it should be made clear that 'animal health' does not include the health of non-target arthropods, or we may as well forget about the importation of any natural enemies!

I am even more worried because I wonder, if this line of reasoning is followed to its logical conclusion, what will the next step be? Perhaps advocating against agriculture itself? After all, are not most of our major agricultural crops exotic plants? Should we eliminate citrus from Florida, for instance, because it has usurped land from native vegetation? Moreover, even those few crop plants that *are* native, are commonly grown in extensive monocultures, at the expense of other native vegetation. Should we eliminate them, too, and return to hunting and gathering? But why stop there? Is not Caucasian man, too, an exotic organism on the American continents and in Australia, who has largely displaced such native organisms as the North American Indian, the Australian Aborigine and the buffalo? Should we, then, send Caucasian man back where he came from? Hopefully, not even our most ardent environmentalists would be willing to go that far at this stage, but who can tell what may come next?

In conclusion, then, *Homo sapiens* is the endangered species we should endeavour to protect and feed, and classical biological control is one of the safest, environmentally benign ways of increasing our agricultural yields. We should do our best to protect the innocent bystanders, but our own interests in the struggle for existence should also be taken into consideration. I would, therefore, like to make a plea to all of us who are concerned about the environment, to support, enhance and encourage classical biological control, in the better interests of Mankind and the environment.

## References

DeBach P and Rosen D 1991 Biological Control by Natural Enemies. 2nd edn. Cambridge University Press, Cambridge, 440 pp.

Greathead D J 1991 A Review of Biological Control in the Ethiopian Region. Technical Communication No. 5, Commonwealth Institute of Biological Control, Commonwealth Agricultural Bureaux, Farnham Royal, 162 pp.

Hollingworth R M 1987 Vulnerability of pests: Study and exploitation for safer chemical control. In Pesticides: Minimizing the Risks. Eds N N Ragsdale and R J Kuhr. pp 54–76. American Chemical Society, Washington, DC.

Howarth F G 1991 Environmental impacts of classical biological control. Annu. Rev. Entomol. 36, 486–509.

Julien M H 1992 Biological Control of Weeds. A World Catalogue of Agents and their Target Weeds. 3rd edn. CAB International, Wallingford, 186 pp.

Kerrich G J 1960 The state of our knowledge of the Hymenoptera Parasitica, with particular reference to the British fauna. Trans. Soc. Brit. Entomol. 14, 1–18.

Miller M and Aplet G 1993 Biological control: A little knowledge is a dangerous thing. Rutgers Law Rev. 45, 285–334.
Pimentel D 1986 Population dynamics and the importance of evolution on successful biological control. In Biological Plant and Health Protection. Ed. J M Franz. pp 3–18. Gustav Fischer Verlag, Stuttgart.
Starý P and Schlinger E I 1967 A Revision of the Far East Asian Aphidiidae (Hymenoptera). Dr. W Junk, Den Haag, 204 pp.

# 4
# Soil disinfestation: environmental problems and solutions

JAACOV KATAN

## 1. Introduction

Soilborne plant pests, be they pathogens, arthropods or weeds, cause heavy losses in all major crops and may even lead to a crop's complete destruction. Farmers tend to plant highly profitable crops in the same land frequently, or even continuously, as in monoculture sytems in greenhouses, thus neglecting crop rotation (Katan, 1984). In the long run, such an attitude leads to the build-up of detrimental biological, chemical and physical agents in the soil. With any crop, the rhizosphere, the plant residues incorporated into the soil, and the routine agricultural practices involved, all tend to have a selective effect on factors leading to poor plant health and reduced yield and quality. The nature of the detrimental agents and the rate of their accumulation in agricultural soils depend on many factors such as the plant species, its susceptibility to detrimental biotic agents, the cropping history of the land, the chemical, physical and biological properties of the soil, agricultural practices, climate, control measures employed, sanitation, and the educational level of the farmer.

Frequent or continuous cropping may have additional negative side-effects. Since such practice also involves frequent use of the same pesticides, the development of pest resistance to them is accelerated. Enhanced degradation of pesticides (with rapid loss of effectiveness) can also occur in soils which have been treated frequently with the same pesticide (Kaufman *et al.*, 1985) because of the build-up of populations of soil organisms that degrade the chemical.

There are many approaches available for the control of soilborne plant pests: the use of soil-applied pesticides, developing resistant cultivars, biological control, and the use of cultural practices. In this chapter, we shall concentrate on soil disinfestation as a method for controlling these pests, the environmental consequences of its use, and approaches for coping with its negative side-effects.

## 2. Soil disinfestation as a tool for control of soilborne pests

Soilborne pests can survive in the soil without a host for long periods, usually at considerable depths (50 cm or deeper). Pathogens usually produce resting structures which are embedded in infected plant tissues and are highly resistant to biotic and abiotic environmental agents. It is difficult, therefore, to reach the pathogen propagules at all sites and depths, and this is reflected in the difficulty involved in

*D. Rosen* et al. *(eds): Modern Agriculture and the Environment, 41–45.*

controlling them. The basic principle of soil disinfestation is to eradicate the pests *before* planting (since the killing agents are also harmful to the crop), using highly effective (and drastic) killing agents. These agents have to penetrate the soil to the desired depth. They are either chemical, e.g. fumigants, or physical, e.g. heating. Therefore, if soil disinfestation is carried out properly, and no subsequent reinfestation occurs, the crop will be protected from soil pathogens for the entire season. A major advantage of soil disinfestation as a control method stems from the fact that its spectrum of control is broad and it is effective against many major, minor and insidious soil pests. However, due to its drastic nature and the technical complications involved in its application, this method has also several disadvantages and limitations:

- It is usually drastic and nonselective, possibly resulting in the creation of a 'microbial vacuum'. In such soils, reinfestation may cause severe outbreaks of disease.
- It is an expensive method, due to the sophisticated equipment involved. It is, therefore, used mainly with crops of high value.
- Chemical disinfestation can be hazardous and requires strict safety precautions.
- Beneficial microorganisms may be adversely affected.
- Soil disinfestation is normally carried out before planting and the soil is left without a crop for a time, the duration of which (one to several weeks) depends on the method employed. This is especially true with soil solarization.

The aim of soil disinfestation is drastically to reduce pest populations while having only a minimal effect on soil biological activity. The term 'soil sterilization', used in certain cases to describe soil disinfestation, is therefore incorrect, and its goal is undesirable.

The main approaches to soil disinfestation are:

- Physical disinfestation, e.g. by steaming the soil.
- Chemical disinfestation using volatile or soluble toxicants, such as methyl bromide, metham sodium or formalin.
- Soil solarization, namely, heating the soil with solar energy while mulching for several weeks with transparent polyethylene, or some other suitable material.

The plant protection and physical aspects of soil disinfestation have been dealt with in many publications (e.g. Goring, 1967; Katan and DeVay, 1991; Munnecke and Van Gundy, 1979; Newhall, 1955) and are not referred to in this chapter.

## 3. Environmental problems encountered with soil disinfestation

The contribution of heat to the environment from steaming is negligible. Thus, the major problems encountered with disinfestation are related to the use of chemical disinfestants. These substances are highly toxic to all components of the agricultural and non-agricultural environment: the farmer who applies the chemical, the inhabitants close to, as well as farther away from the treated agricultural area, the factory workers who produce the toxicants, and the consumers of the plant's edible parts, upon which residues accumulate. The chemical disinfestants may contaminate the groundwater as well as the atmosphere. Methyl bromide, which is the major fumigant used worldwide, was detected in the stratosphere and some of it originates from soil fumigation (UNEP, 1995). It should be mentioned that this substance is also produced by natural processes.

The potential hazards of chemical disinfestants confront us with a dilemma, similar to that encountered with any pesticide. On the one hand, there is public, political and economic pressure to avoid the use of agricultural chemicals. On the other hand, pesticides are still the central core of pest control the world over. A sudden ban, or even a substantial reduction in their use, would result in a dramatic food shortage. This is especially true with intensive agriculture, which is highly dependent on this type of soil disinfestation.

## 4. Potential solutions

### *4.1 Non-chemical alternative methods*

Non-chemical control is the ideal approach. Many non-chemical alternative methods of control were examined in the last decades. However, in only a few cases have non-chemical methods replaced pesticides on a large scale (Katan, 1993). The replacement of chemical disinfestation with alternatives is an even more difficult task, since chemical soil disinfestation is very effective and enables the control of a broad spectrum of pests. The potential non-chemical alternatives are: use of resistant cultivars, biological control, and cultural methods. Soil solarization is the only non-chemical soil *disinfestation* method which has been developed in the last few decades. Since the first publication of this method in 1976, research on this approach has been, or is being carried out, in more than 40 countries (Katan and DeVay, 1991). It is used by farmers in several countries including Israel, Greece, Jordan, Morocco, Japan and Italy. One of the major limitations of this method is its dependence on climate. Thus, many studies are currently focusing on the development of techniques which would enable its use under less favourable climatic conditions.

### *4.2 Reducing disinfestant dosage while maintaining its effectiveness in pest control*

#### 4.2.1 Combining control methods

This is a very promising tool for reducing disinfestant dosage, and makes up part of the integrated pest management (IPM) strategy. Several studies have shown that combining chemical soil disinfestation at *reduced* dosages with another method results in effective pest control. Combining metham sodium at reduced dosage with solarization resulted in effective control of delimited shell spot in peanuts (Frank *et al.*, 1986) and of *Fusarium oxysporum* f. sp. *vasinfectum* (Ben-Yephet *et al.*,

1988). Combining methyl bromide at reduced dosages and solarization resulted in the effective control of yield decline in gypsophila (Gamliel *et al.*, 1993). Improved control by combining methyl bromide with *Trichoderma* application was shown for *Sclerotium rolfsii* and *Rhizoctonia solani* (Elad *et al.*, 1982) and for the control of Fusarium crown and root rot in tomato (Sivan and Chet, 1993).

#### 4.2.2 Improving pesticide application

This is one of the unfortunately neglected areas which has the potential to reduce considerably the amount of pesticides released into the environment. This approach was studied with methyl bromide. During application of this fumigant, the soil is typically covered with polyethylene tarp to delay the escape of this gas into the atmosphere. In the past, the use of plastic tarps which are less permeable to this gas than polyethylene, was found to enable reducing its dosage (de Heer *et al.*, 1984). This approach was followed in Holland on a large scale in the 1980s. Recent studies have again shown that the use of less permeable tarping sheets enables reducing methyl bromide dosage by at least 50%, while effectively controlling a variety of diseases such as Verticillium wilt of potatoes and crown and root rot of tomatoes (Gamliel *et al.*, 1995; Peretz *et al.*, 1995).

#### 4.2.3 Monitoring the pest population to determine the optimal timing and frequency of pesticide applications

This approach is used successfully with arthropods to avoid unnecessary pesticide applications. However, it is rarely followed in soil disinfestation. Developing more sensitive and reliable methods for the assessment of pathogen populations in soil would enable the use of soil disinfestation only when necessary.

### *4.3 Disinfestant replacement*

In the past, in Holland, methyl bromide was used on a large scale for soil fumigation. In the early 1980s, its dosages were reduced and in the early 1990s its use for soil fumigation was totally banned (UNEP, 1995). Its elimination was achieved by using a variety of means: expanding the use of artificial growth media, expanding the use of non-chemical approaches, shifting to other chemicals for soil disinfestation, and others. To date, this is the only example of disinfestant replacement on a large scale.

## 5. Concluding remarks

Soil disinfestation is much needed in modern agriculture, but it is also problematic. There is no single solution to cope with this dilemma. There are many potential, partial solutions to replace, avoid, curb or minimize the potential hazards of soil disinfestation. For each situation, an appropriate solution, or a combination thereof, has to be adopted. IPM, a holistic approach that encompasses pest control and crop management and considers environmental, economic, public and legislative aspects, should be adopted in developing the above strategies. Education and extension tools should also play a role in the implementation of IPM programmes. We have to develop approaches which will enable us to reduce the hazards of chemical soil disinfestation while maintaining crop productivity. This is one of the greatest challenges in crop protection.

## References

Ben-Yephet Y, Melero-Vera J M and DeVay J E 1988 Interaction of soil solarization and metham-sodium in the destruction of *Verticillium dahliae* and *Fusarium oxysporum* f. sp. vasinfectum. Crop Prot. 7, 327–331.

Elad Y, Hadar I, Chet I and Henis Y 1982 Prevention with *Trichoderma harzianum* Rifai aggr. of reinfestation by *Sclerotium rolfsii* Sacc. and *Rhizoctonia solani* Kuhn of soil fumigated with methyl bromide, and improved control in tomatoes and peanuts. Crop Prot. 1, 199–211.

Frank Z R, Ben-Yephet Y and Katan J 1986 Synergistic effect of metham and solarization in controlling delimited shell spots of peanut pods. Crop Prot. 5, 99–202.

Gamliel A, Grinstein A, Klein L, Cohen Y, Ucko O, Austerweil M, Seiner B, Aduel A and Katan J 1995 Methyl bromide for soil fumigation towards the year 2000. Phytoparasitica 23, 250 (Abstract).

Gamliel A, Hadar E and Katan J 1993. Improvement of growth and yield of Gypsophila paniculata by solarization or fumigation of soil or container medium in continuous cropping systems. Plant Dis. 77, 993–938.

Goring C A T 1967 Physical aspects of soil in relation to the action of soil fungicides. Annu. Rev. Phytopathol. 5, 285–318.

Heer de H, Hamaker P H, Tuinstra L G M and Burg A M N van 1984 The use of gas-tight film during fumigation of glasshouse soils with methyl bromide. I. Significance of permeation. Acta Hort. 152, 109–126.

Katan J 1984. The role of soil disinfestation in achieving high production in horticulture crops. Proc. British Crop Prot. Conf. 1189–1196, Brighton.

Katan J 1993 Replacing pesticides with nonchemical tools for the control of soilborne pathogens: a realistic tool? Phytoparasitica 21, 95–99.

Katan J and DeVay J E 1991 Soil Solarization. CRC Press, Boca Raton, FL, 267 pp.

Kaufman D D, Katan J, Edwards D F and Jordan J E 1985 Microbial adaptation and metabolism of pesticides. In Agricultural Chemicals of the Future. Ed. Hilton J L. pp 437–451. Towman & Allanheld, New Jersey.

Munnecke D E, Van Gundy S D 1979 Movement of fumigants in soil, dosage responses and differential effects. Annu. Rev. Phytopathol. 17, 405–429.

Newhall A G 1955 Disinfestation of soil by heat, flooding and fumigation. Bot. Rev. 21, 189–250.

Peretz I, Gamliel A, Grinstein A, Klein L, Tsror L, Nachmias A and Katan J 1995 Control of *Verticillium* wilt with methyl bromide at reduced dosage. Phytoparasitica 23, 249 (Abstract).

Sivan A, and Chet I 1993. Integrated control of fusarium crown and root rot of tomato with *Trichoderma harizianum* in combination with methyl bromide or soil solarization. Crop Prot. 12, 380–386.

UNEP 1995 Montreal protocol on substances that deplete the ozone layer – 1994 report of the methyl bromide technical option committee. Kenya, 304 p.

# 5
# Minimizing environmental damage originating from pesticide utilization: abiotic photochemical control and remediation

NICOLA SENESI AND ELISABETTA LOFFREDO

## 1. Introduction

The extensive use of pesticides in agriculture originates large amounts of residues, the presence of which is the main cause of concern and hazard for the global environment, including soils, surface waters and groundwaters, atmosphere, and organisms.

For technical reasons, pesticides are generally applied in amounts that largely exceed the quantity actually needed to fight the target species. Generally, 0.1 to 1% of a chemical hits the target species, whereas the remainder drifts into the environment, causing contamination of soil, water, air and biota (Pimental and Levitan, 1986). The disadvantages are, therefore, reduced efficiency in pesticide use and potential damage for the environment. Obvious solutions are both to increase, and possibly optimize, the efficiency of pesticide use, and to decrease, and possibly minimize, the hazard deriving from their residues.

Pesticide residues, both the parent molecule and its degradation product(s), are subject to several processes in the soil–water–organism system, which include volatilization, surface runoff, leaching to groundwater, uptake by plants and microorganisms, adsorption by soil colloids, biodegradation, and chemical and photochemical decomposition. These phenomena affect to various extents the behaviour and fate of pesticides and their residues in the soil environment, including their transport and movement, physical and chemical immobilization and bioaccumulation, bioavailability and phytotoxicity.

The most important processes that may lead to minimizing the environmental impact and damage related to pesticide utilization are biotic and abiotic degradation. These processes are expected to lead to the formation of non-contaminant and non-toxic products, and, possibly, to the complete mineralization of the pesticide to carbon dioxide and other simple, innocuous inorganic species.

Biological, chemical and photochemical transformations of pesticides in natural systems often occur simultaneously, and many degradative pathways include both biologically and chemically controlled steps. In general, a net distinction between the two types of processes is difficult. Although biodegradation is often indicated to be the dominant process, the importance of abiotic transformations of pesticides in

*D. Rosen* et al. *(eds): Modern Agriculture and the Environment, 47–73.*

the environment has been widely recognised recently, even in media where intense biological activity occurs, such as soil. Abiotic processes are reasonably expected to play a major role, or even predominate, in systems where biological activity is reduced, such as in flooded, ditch irrigated and submerged soils, on soil surfaces, in arid and semi-arid zones, below the root zone of soil, in surface waters and groundwaters, and where the high concentration of toxicants may inhibit microbial activity. Abiotic transformations appear, therefore, to play an important role in the detoxification of contaminated soil and water systems, especially where biotic conversions are not likely to be significant.

The principal abiotic transformations of pesticides and other organic chemicals in soils and water involve photodecomposition, hydrolysis and oxidation–reduction reactions (Valentine, 1986). Sorption phenomena by soil colloids are known to affect markedly these processes when they occur at the interfaces between liquid, gaseous and solid components. Several natural and artificial catalysts and photosensitizers, including clay minerals, metal oxides and hydroxides and humic substances, are known to play important roles in these processes. Detoxification of contaminated water bodies and soils may be achieved by manipulating the contaminated system to enhance the abiotic transformation of the contaminant. Further, specific carriers with photocatalytic properties can be used in the design of pesticide formulations. Products of abiotic transformations of pesticides may, however, still be toxic, or even more toxic than the parent pesticide molecule.

Abiotic transformations of pesticides have been reviewed recently (Senesi and Miano, 1995; Wolfe *et al.*, 1990). The main objective of this review is to discuss the most important abiotic photochemical processes, both naturally-occurring in soils and waters and artificially-promoted by proper remediation treatments, aiming to minimize the potential hazard of pesticide residues for the environment. The review is meant largely to be illustrative, rather than exhaustive, and will, thus, draw principally on the most relevant experimental results obtained recently on this matter.

## 2. Photochemical transformations

Significant amounts of agricultural and industrial chemicals released into the environment are deposited on sunlight-exposed soil and water. Sunlight-induced photochemical transformations are ascertained to be important pathways in the abiotic decomposition of pesticides and other organic chemicals exposed on soil surface and in uppermost water layers (Cessna and Muir, 1991; Choudry and Webster, 1985; Klöpffer, 1992; Mansour *et al.*, 1993; Marcheterre *et al.*, 1988; Miller *et al.*, 1989b; Senesi and Miano, 1994; Wolfe *et al.*, 1990; Zepp, 1992).

The mean photolysis depth of soil for a chemical, i.e., the depth of light penetration, can be estimated by photolysis depth probes, and results have shown it to be shallow (Miller *et al.*, 1989b). Sunlight has a significant effect on the chemical at the soil–atmosphere interface, which includes approximately the top 0.5 cm of soil. Although photochemical reactions are restricted to the very upper layer of soil, this interface receives a large amount of pollutants and can be a significant transformation site. Cultivated lands are often turned over, and a much deeper layer of soil can be brought occasionally to the surface, and can be potentially involved in photochemical reactions. Therefore, the photic zone of the soil can be exposed to sunlight for widely different durations. Further, pesticides applied to the soil and dissolved subsequently in soil water are not only transported down into deeper layers of soil, but also back to the surface, especially during hot and dry periods. Thus, the movement of these chemicals to and from the irradiated surfaces may assume an important role in photochemical processes.

Photolysis of a pesticide molecule is the process in which visible and/or ultraviolet light causes its transformation to compounds with different, generally reduced toxicity, and initiates the processes that may lead to its complete mineralization. Photolysis rates and products may vary considerably with respect to dependence on the conditions of the system exposed to sunlight. In some systems, photolysis may represent the primary degradative pathway of a pesticide, whereas in others it may not be significant. The significance of photolysis should be determined, however, only in relation to other degradation processes that can affect pesticide stability in soils and waters.

Photolysis of pesticides may occur by two processes: direct and indirect photolysis. Direct photolysis is initiated by direct absorption of sunlight ($\lambda > 290$ nm) by the pesticide, which is energetically excited, and can then be transformed into photoproducts:

$$\text{Pesticide} \xrightarrow{h\nu} \text{Pesticide}^{*} \longrightarrow \text{Photoproducts.} \tag{1}$$

Trifluralin and methylparathion are examples of pesticides which absorb sunlight strongly, whereas methoxychlor absorbs sunlight only weakly, and 3,4-dichloroaniline (3,4-DCA) absorbs at shorter wavelengths (Wolfe *et al.*, 1990). Pesticides which may undergo direct photolysis in aquatic environments include the herbicides 2,4-D and 2,4,5-T, pentachlorophenol, PCBs, DDT, trifluralin, isoxaben, flutriafol, diazinon, phenmedipham, tetrachlorvinphos, parathion, and methylparathion (Mansour *et al.*, 1993). Apparently, most pesticides do not photolyse strongly and rapidly on soil, otherwise their application would not be efficient. For example, the insecticides thiodiazuron and synthetic pyrethroids are rapidly degraded in sunlight and not sufficiently stable for outdoor applications (Klehr *et al.*, 1983; Ruzo, 1982).

In indirect photolysis, light is initially absorbed by a compound other than the pesticide. Subsequently, this compound may either transfer the energy to the pesticide molecule (photosensitization), or produce specific photoreactants which may then react with the target pesticide (photoinduction). The exact identities of naturally occurring organic and inorganic species that are able to absorb the solar radiation and to produce transient photoreactants, i.e., to act as initiators or sensitizers of photoreactions in soils and waters, are generally unknown. However, humic substances (HS) and their main fractions, humic acid (HA) and fulvic acid (FA), clay minerals, metal oxides and oxihydroxides, and transition metal ions, have been ascertained to be involved in indirect photolysis of pesticides and other chemicals in the environment.

The photochemistry of transient photoreactants in aqueous systems, including their formation and decay, lifetimes and quantum yields, and the quenching effects of various additives, have been investigated by laser flash photolysis methods (Zepp, 1992). The phototransient identity and concentration have also been inferred indirectly through steady-state analysis of kinetic results obtained in continuous irradiation experiments with monochromatic light or sunlight, in which a 'chemical probe' is used to trap the transient (Zepp, 1992).

Sunlight-induced formation of reactive species in the presence of dissolved or colloidal organic matter (DOM), especially HS, has been reviewed recently (Cooper *et al.*, 1989; Frimmel, 1994; Hoigné *et al.*, 1989; Senesi, 1993; Zepp, 1992). Photochemical reactions involving HS sensitizers or precursors are unanimously indicated as the primary source for the production of photoreactants such as solvated electrons, singlet oxygen, hydroxyl radicals, superoxide radical ions, organoperoxy radicals, and hydrogen peroxide.

## *2.1 Photolysis in natural water systems*

The solvated electron is a powerful and rapid reductant for electronegative chemicals such as benzenes, phenols and halogenated organics, e.g. dioxins, which are dehalogenated (Cooper *et al.*, 1989; Hoignè *et al.*, 1989). The degradation of the chlorocarbon mirex and lindane in natural waters is controlled by hydrated electrons possibly produced when HS are irradiated in sunlight (Burns and Hassett, 1994). Apparently, no degradation of mirex was observed in experiments using Fenton or photo-Fenton reagents, suggesting that mirex does not react with hydroxyl radicals.

Singlet oxygen in natural waters was apparently responsible for the photooxidation of various substrates including thioether pesticides such as disulfoton (Zepp *et al.*, 1981) and resmethrin (Ruzo, 1982). In contrast, singlet oxygen was unable to oxidize thiophosphates to phosphates at appreciable rates (Wolfe *et al.*, 1990).

The photodegradation rate of several pesticides that were non-degradable by direct photolysis in water solution, increased markedly in the presence of increasing amounts of HA in the system (Table 1) (Korte *et al.*, 1987). A photooxidative reaction pathway involving singlet oxygen species produced by the photosensitizing action of HA was suggested to occur.

The degradation of diazinon, atrazine and fenpropathrin in various natural waters and HA solutions under sunlight was markedly accelerated with respect both to distilled water and in any water tested under dark conditions (Figure 1) (Mansour *et al.*, 1993; Takahashi *et al.*, 1985). These results were ascribed to the photosensitizing effect of naturally occurring substances, especially DOM and HS present in natural waters. The major photoreaction of fenpropathrin, i.e., the ester bond cleavage with subsequent photomineralization of the ciano fragment to carbon dioxide, was found to be accelerated by natural substances in river and sea-water, including HA (Takahashi *et al.*, 1985).

Up to half the triplet state of HS are estimated to have energies sufficiently high to photosensitize reactions of many organic chemicals, including polycyclic aromatic

*Table 1.* Half lives[a] (h) for the photodegradation[b] of selected pesticides at various humic acid concentrations (HA, mg $L^{-1}$) (adapted from Korte *et al.*, 1987).

| | HA (mg $L^{-1}$) | | |
|---|---|---|---|
| Compounds | 0 | 10 | 100 |
| Propoxur | 87.9 | 40.8 | 13.0 |
| Ethiofencarb | 51.7 | 9.9 | 1.3 |
| Sulfotepp | n.d.[c] | 38.4 | 12.4 |
| Parathion | 39.0 | 13.4 | 4.4 |
| Propylenethiourea | 100 | 12.7 | 2.5 |
| Methabenzthiazuron | n.d. | 91.2 | 21.9 |
| Triapenthenol | 128 | 32.2 | 15.4 |
| Amitrole | n.d. | n.m.[d] | 7.5 |
| Cyfluthrin | 16.0 | 4.0 | n.m. |

[a] Half lives are calculated for a pseudo first order reaction.
[b] Irradiation was obtained with simulated sunlight for 8 h. Dark control samples showed degradation only in the case of ethiofencarb at a HA concentration of 100 mg $L^{-1}$.
[c] n.d.: no degradation observed.
[d] n.m.: not measured.

hydrocarbons, nitroaromatic compounds, polyenes, and diketones (Zepp *et al.*, 1985).

Photolysis of atrazine in irradiated ($\lambda = 254$ nm) aqueous solution containing 0.01% FA was initially retarded, and followed first-order reaction kinetics with respect to the herbicide concentration, whereas the resulting degradation was ultimately more extensive than in the absence of FA (Khan and Schnitzer, 1978). The initial reaction appeared to be photochemical solvolysis of atrazine to the 2-hydroxy analogue, which was followed by rapid N-dealkylation, which seemed to be the major reaction when FA was present. Photolysis of prometryn at 254 nm in distilled water and HA suspension at pH 3 followed first-order kinetics, whereas second-order reaction rate was measured in the presence of dissolved HA at pH 6 and 8, and FA at pH 3, 6, and 8 (Khan and Gamble, 1983).

The rate of photodegradation of several s-triazines, including atrazine, ametryne, atraton, simazine, and propazine, increased significantly, yielding dealkylated photoproducts, when irradiated under direct sunlight in aqueous solution in the presence of various natural and synthetic photosensitizers such as acetone, riboflavin, and flavin mononucleotide (Burkhard and Guth, 1976; Rejto *et al.*, 1983). The possible participation of hydroperoxide radicals in the photoreaction mechanism was suggested (Rejto *et al.*, 1983).

Bromacil was very stable towards direct sunlight photolysis, whereas in the presence of photosensitizers such as riboflavin or methylene blue it photoreacted very rapidly (Moilanen and Crosby, 1974).

The rate of photodecomposition of terbacil in dye-sensitized aerated aqueous solutions exposed to direct solar irradiation was highly pH dependent, being rapid at alkaline pH and significantly slower at acidic pH (Acher *et al.*, 1981).

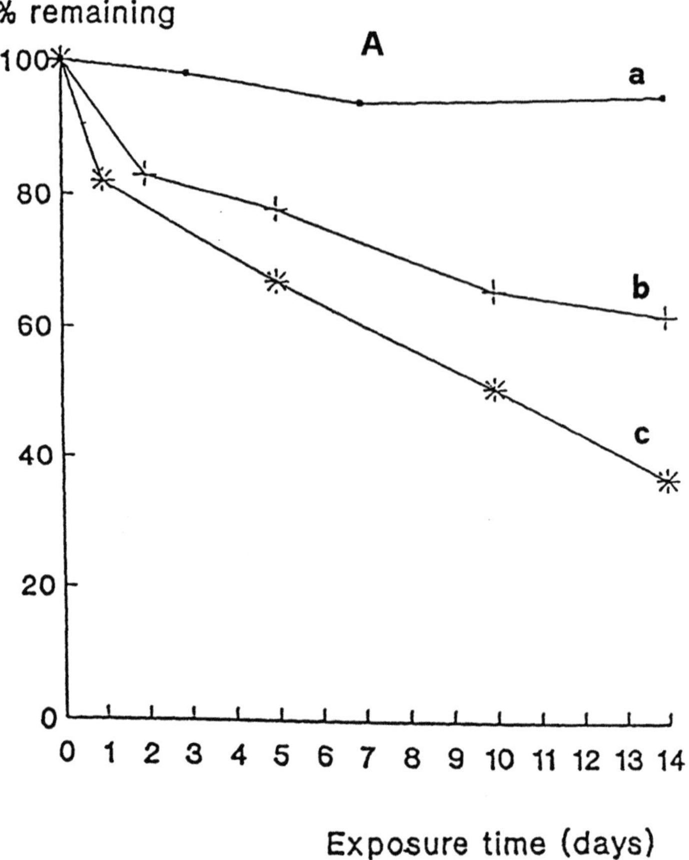


*Figure 1.* Sunlight degradation of atrazine (A) in distilled water (a), Isar river water (b), and Rhein river water (c);

Photodecomposition of bromacil and terbacil was found to occur also during the winter in sensitized aqueous solution below the ice crust (Acher, 1982). The effect of the ice cover was to attenuate the incident light, resulting in decreased rates of photodegradation, which, however, remained high.

The photolysis rate of picloram was only slightly influenced by dissolved substances in natural waters, whereas a FA solution was much more efficient in increasing the photoreaction rate (Skurlatov *et al.*, 1983). Natural substances other than FA appeared somehow to inhibit the FA-mediated photoreaction in natural waters.

The photolysis rate of pentachlorophenol (PCP) irradiated with sunlight in dilute aqueous solution was higher at near neutral pH than at low pH (Wong and Crosby, 1981). PCP yielded photochemical cleavage of the chlorinated ring in sunlight, and was completely photodecomposed after 20 days of outdoor exposure yielding chloride ions, carbon dioxide, and, possibly, small aliphatic fragments as final

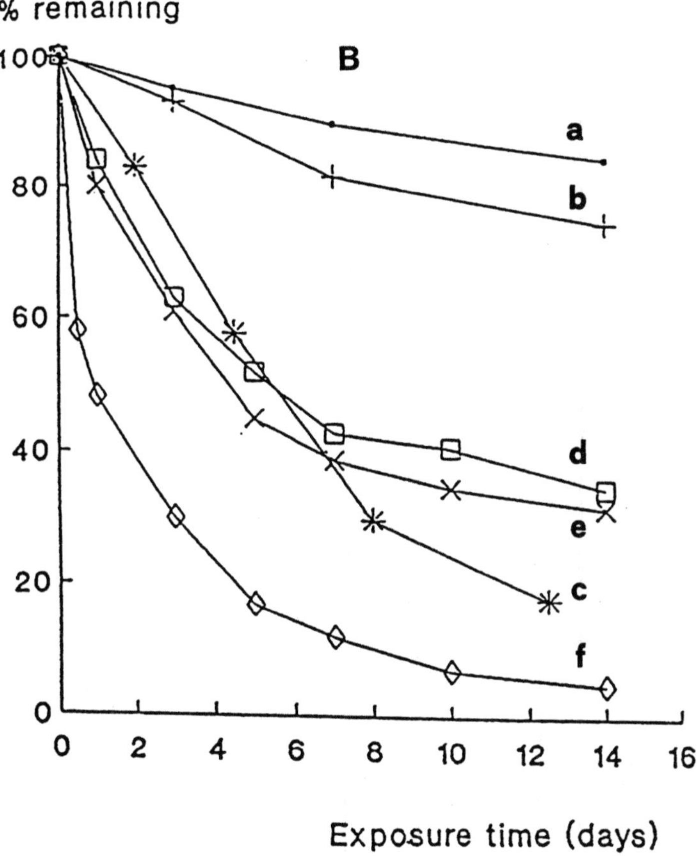


*Figure 1* (cont) and degradation of diazinon (B) in dark conditions in distilled water (a) and river water (b), and under sunlight in Ammer lake water (c), humic acid solution (d), Isar river water (e), and Rhein river water (f), as a function of time in days (adapted from Mansour *et al.*, 1993).

photoproducts. The more extended photoreduction (dechlorination) of PCP measured in seawater, with respect to distilled water and artificial saline water, was possibly ascribed to DOM present in the seawater (Miille and Crosby, 1983).

The non-participation of oxygen in the reductive dechlorination photoreaction of monuron irradiated in aqueous solution containing various amounts of surface-active agents suggested that the aryl moiety of monuron might be either dissolved or adsorbed on the surface of the micellar lipophilic core of the surfactant, away from any oxygen source (Tanaka *et al.*, 1979). Photochemical reactions might thus occur in the organic phase of the micelles rather than in the aqueous phase. These findings are important because the hydrophobic portions of DOM and HS macromolecules present in natural water systems may act in the same way towards several pesticides possessing aryl moieties in their molecules. The presence of surfactant was also shown to increase the photodegradation rate of several carbamate herbicides having low solubilities, including chlorpropham, barban, dichlormate and propham (Tanaka *et al.*, 1981).

The presence of traces of $H_2O_2$, a natural constituent of rice paddy water, was found to increase significantly the photolysis rate of molinate, thiobencarb, and drepamon in aqueous solutions in sunlight (Draper and Crosby, 1984a,b). The involvement and common intermediacy of $H_2O_2$, hydroxyl radicals and molecular oxygen was assessed in the indirect photooxidation of these carbamates to sulphoxides in aqueous solution.

Organic peroxy radicals derived from DOM by sunlight irradiation behave as important transient oxidants for electron-rich phenols, such as alkylphenols, in natural waters (Hoignè *et al.*, 1989).

The sunlight photodegradation of quinoline, a known carcinogen and mutagen, and a common contaminant of environmental waters, was found to be accelerated in solutions containing $NaNO_3$ or DOM (half-lives 3–4 days, in summer), but was not affected appreciably by NaCl or $Na_2SO_4$ solutions, with respect to lake water (half-life 7 days) (Kochany and Maguire, 1994).

### 2.2 *Photolysis in water suspensions and on adsorbed phases*

Natural waters contain a wide range of particulate inorganic materials that may reduce the photolysis rate by either shielding the organic chemical from the incident light or by quenching the excited states of the organic molecules before the photoreaction can occur. Organic photolysis may also be accelerated via photocatalytic processes if light absorption by suspended particles produces excited states or free radicals that can react with the organic chemical (Langford and Carey, 1987; Zafiriou *et al.*, 1984). It is thus possible for organic chemicals that are unable to absorb sunlight directly to be photolysed indirectly. A number of semiconductor solids, such as $TiO_2$, which are common constituents of clays, sediments and soils in the 5–10% range, are able to determine this type of indirect photolysis (Dolcater *et al.*, 1970).

Several studies have indicated that the photochemical behaviour of organic chemicals in the sorbed state differs from that in the dissolved state. Photochemistry of paraquat was greatly altered by sorption on clays suspended in water (Bailey and Karickhoff, 1973). In the presence of suspended bentonite clay, the sensitizing action of methylene blue on bromacil was almost nullified, possibly because the sensitizer strongly adsorbed to colloidal surfaces, thus preventing it from participating in the photooxidation process (Acher and Rosenthal, 1977).

Limited changes in photoreactivity were measured, however, for two hydrophobic chlorinated chemicals common in the aquatic environment, 3,4-dichloroaniline (DCA) and 1,1-bis (*p*-chlorophenyl)-2,2-dichloroethylene (DDE) when they were sorbed on river and pond bottom sediments resuspended in water (Miller and Zepp, 1979). Photoreaction kinetics and photoproducts obtained suggested that the sorbed chemicals were in a less polar microenvironment that was a considerably better hydrogen atom donor than water. The influence of sorption on photoreactivity appeared thus to be distinguished from the effects of light attenuation and scattering by the sorbent.

Suspended river sediments and common clays, including montmorillonite, kaolinite, bentonite, dolomite, and illite, reduced considerably the photodecomposition rate of methoxychlor in irradiated (300 and 350 nm) aqueous slurries with no apparent generation of hydroxyl radicals (Oliver *et al.*, 1979). Thus, even though suspended sediments and clays might contain $TiO_2$ and other semiconductors that absorb sunlight, and produce free radicals, these appeared to be scavenged in the system. As a whole, clays did not appear to photocatalyse the decomposition of methoxychlor in water but reduce its photolysis rate by shielding the chemical from the incident light.

In contrast, the photolysis rate of the pyrethroid insecticide esfenvalerate, which is scarcely soluble in water and has an extremely high soil sorption coefficient, was slightly accelerated in irradiated ($\lambda > 290$ nm) kaolinite and montmorillonite clay suspensions (Katagi, 1993). This effect was ascribed to the presence of $TiO_2$ (0.03–0.2%) in the clays, which might generate hydroxyl radicals photocatalytically; these were mainly responsible for the indirect photolysis of esfenvalerate. Iron oxides (goethite and ematite), present at 0.5–3.5% in the clays tested, were excluded from possibly acting as photocatalysts generating hydroxyl radicals. Further, the clay surface was suggested to play a significant role in the photoproduction of hydroxyl radicals, i.e., the different way of surface adsorption of the chemical to each clay, and the different surface water content would control photoreactivity of the clay surface (Katagi, 1993).

A number of studies have been conducted on the photodegradation of various pesticides adsorbed on different solid supports. The photodegradation rate of propyzamid and phenmedipham on bentonite and alumina was shown to depend on the type of support, and to increase with the presence of water in the alumina support (Figure 2) (Emmelin *et al.*, 1993). The photodegradation of phenmedipham, which consisted of the cleavage of the aromatic ring by a mixed radicalic and ionic mechanism, was immediate on silica and bentonite, and slow on kaolinite (Guittoneau *et al.*, 1994). The photochemical behaviour of carbetamid adsorbed on solid supports such as kaolin, bentonite or silica was shown not to be affected by the nature of the support (Emmelin *et al.*, 1993). However, some physical properties of the support, e.g., specific area and reflexion coefficient, appeared to be important for the photodegradation kinetics of carbetamid, which is different in solution and in adsorbed phase. Butachlor and alachlor were found to phototransform appreciably when desposited as thin films on glass (Fang, 1977).

### 2.3 *Photolysis on soil surfaces*

The photolysis of pesticides on soil surfaces is still poorly understood, and only a few studies are available that have described the phototransformation of pesticides and other organic chemicals in soils and soil surfaces. In contrast to solutions, soils

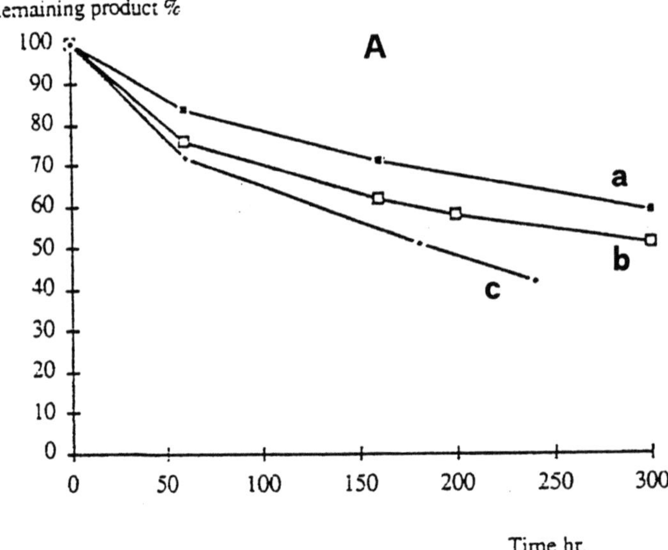


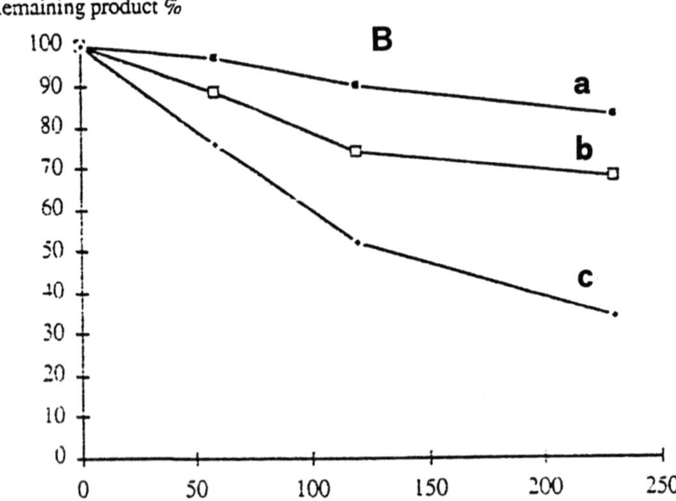


*Figure 2.* Photodegradation of propyzamid (A) on dehydrated alumina (a), hydrated alumina (b), and bentonite (c); and of phenmedipham (B) on bentonite (a), dehydrated alumina (b), and hydrated alumina (c), as a function of time in hours (adapted from Emmelin *et al.*, 1993).

are extremely heterogeneous and irradiation produces light conditions that are difficult to define accurately.

The rate of direct photolysis of organic chemicals is often dramatically decreased on soils. The primary factors responsible for this result are the substantial light attenuating effect on soil, and the photochemical quenching of electronically excited states by intermolecular energy transfer to various soil components occurring on soil surfaces. An additional factor is the inner filter effect, which occurs when a photolabile compound is entrapped inside a soil particle, and is thus shielded from light in the interior of the particle although remaining a few microns from the surface (Yokley *et al.*, 1986). The relative contribution of these effects to the reduction of the photolysis rate on soil is, however, still unclear.

An example of the substantial decrease in photolysis rate occurring on soils, relative to other environmental compartments, is given by the plant growth regulator flumetralin, for which a half-life of one day was measured on soil, with respect to a half-life of less than 30 minutes in solution (Miller and Zepp, 1983). More than 80% of fluchloralin degraded after 60 h irradiation in water, whereas more than 80% remained after 80 h of irradiation on a 0.5 mm deep sandy loam soil (Nilles and Zabik, 1974). Similarly, more than 80% loss was measured for bentazone when irradiated in distilled water for 24 h, while more than 50% remained after 120 h of irradiation on the soil surface (Nilles and Zabik, 1975).

In contrast, indirect photochemical processes are of particular interest as sunlight striking the soil surface can be absorbed by both organic and inorganic chromophores in soil particles, which can thus possibly initiate or sensitize phototransformations of organic chemicals on the soil surface. HS are well-known potent and polyvalent photosensitizers and photoinitiators for these photoreactions.

An important species which may contribute substantially to the photodegradative pathway of susceptible organic chemicals deposited on soil surfaces is the highly reactive singlet oxygen. By using two specific singlet oxygen traps, it has been demonstrated that the photochemical generation of singlet oxygen at environmentally significant rates is a general phenomenon on a variety of sunlight-irradiated soils (Ghore *et al.*, 1986). The ability of irradiated soils to photoinduce the formation of singlet oxygen was independent of the pH and amount of organic matter present, and both organic and inorganic components of the soils appeared to contribute to the photoactivity of the soil. Free radical photoreactions were not observed, whereas non-transition metal oxide (silica and alumina) surfaces were able to produce singlet oxygen, probably involving surface defects that result in band gaps of low enough energy to absorb visible light (Ghore and Miller, 1985).

The photoreactivity of the ricefield soils was found to be quite distinct from that of the other soils. Ricefields are flooded for a considerable length of time during the year, and a common practice is to burn rice plant residues after harvest. The much lower pH of these soils with respect to other soils examined, suggested that aluminium, manganese, and iron ions might contribute to the different photoreactivity of the ricefield soils. Semiconductor type photooxidations promoted by the inorganic fraction of soils were suggested to occur in soils with the lowest organic content (Ghore *et al.*, 1986).

Sulphides are highly reactive to singlet oxygen. Several dialkyl or alkylaryl sulphide-containing insecticides, including disulfoton, fenthion, methiocarb and butocarboxim sorbed on soils were found to undergo rapid photooxidation to the corresponding sulphoxides (Ghore and Miller, 1986). For example, disulfoton exposed to sunlight on soil was converted to sulphoxide with a half-life of two days. Conversely, sulphur in methylthiotriazines, thiocarbamates, or dithiolane insecticides was found not to be oxidized (Ghore and Miller, 1986). By analogy with other thioether compounds, aldicarb and methomyl, which do not absorb sunlight and cannot thus give direct photolysis, would likely react with singlet oxygen to produce sulphoxides and be susceptible to indirect photolysis (Freeman and McCarthy, 1984; Freeman and Nipp, 1984). Thiocarbonyl pesticides were also susceptible to singlet oxygen reactions (Ando, 1981). For example, the fungicides thiophanate and thiram were photooxidized on soil surfaces with a half-life of less than one day (Ghore, 1985).

Little work has been done on the production of photoreactants other than singlet oxygen on irradiated soil surfaces. Parathion was found to be efficiently photooxidized to paraoxon on irradiated soil surfaces in the presence of ozone (Spencer *et al.*, 1980a,b). Hydrolytic transformation pathways possibly involving a photochemical process have been observed for ureas (Crosby and Tang, 1969), organophosphates, such as prothiophos (Takase *et al.*, 1982) and fenitrothion (Ohkawa *et al.*, 1974), and pyrethroids (Ruzo, 1982), exposed to light. Two hydrolytic processes appeared to be uniquely photochemical, i.e., photohydrolysis of DCA (Miller *et al.*, 1979), and photonucleophylic substitution on nitrofen and related compounds to the corresponding phenols (Wolfe *et al.*, 1990).

An experimental procedure was developed for examining the photolysis of pesticides on soil samples, which enabled distinction between the photochemical and non-photochemical transformation of the compound by reducing the penetration of the compound into non-irradiated layers of soil (Klehr *et al.*, 1983). A methanol solution of the defoliant thiodiazuron was sprayed on the surface of sandy loam soil layers, 0.5-mm thick, and irradiated with a filtered xenon arc source providing irradiance qualitatively similar to that of natural sunlight. A rapid disappearance of thiodiazuron was observed, with a photochemical half-life of 0.5 h (Klehr *et al.*, 1983).

The effect of various soil properties, including organic matter and clay contents, particle size distribution, temperature and humidity, on photolysis processes has been examined in few studies. On a very fine, montmorillonitic loam soil, a 15% photolytic loss was measured for 2,3,7,8-TCDD exposed to sunlight for 6 h (Crosby and Wong, 1977). Temperature of the soil was shown to affect photolysis rates of 2,3,7,8-TCDD (Liberti *et al.*, 1978).

The photodecomposition of several dinitroaniline herbicides has been studied on thin-layer, air-dried soil plates with no moisture added during the seven-day exposure to sunlight (Parocchetti and Dec, 1978). Among the ten dinitroanilines

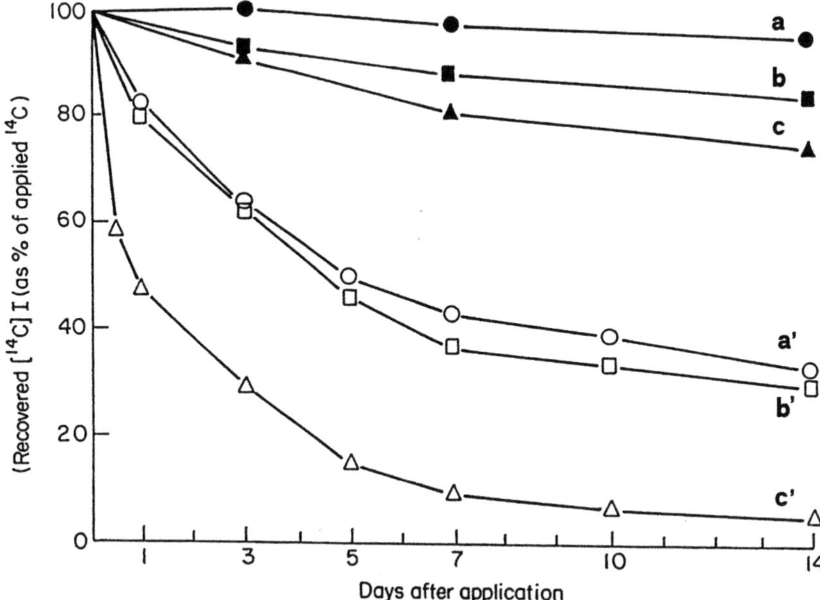


*Figure 3.* Degradation of $^{14}$C-fenpropathrin on 0.5-mm thick layers of sandy loam soil (a), sandy clay loam soil (b), and clay soil (c), in dark conditions and natural sunlight (a′, b′, c′, respectively), as a function of time in days (adapted from Takahashi *et al.*, 1985).

examined, trifluralin, benefin, nitralin, oryzalin, fluchloralin and dinitramine photodegraded appreciably with a percentage of disappearance ranging from 47 to 60%, after the seven-day experiment. Several photoreactions of trifluralin on soil surfaces have been reviewed (Crosby, 1976). Azobenzene and azoxybenzene derivatives were reported to be formed from the photodegradation of trifluralin on soil surfaces (Golab *et al.*, 1979).

The photolysis of the pyrethroid insecticides fenvalerate and fenpropathrin has been studied on 0.5-mm thick layers of three soils on wet plates exposed to natural sunlight (Mikami *et al.*, 1980; Takahashi *et al.*, 1985). The photolysis rate of fenvalerate was highest on a light-clay soil containing 15.3% organic matter (half-life, 2 days), and slower on a sandy loam soil containing 1.9% organic matter (half-life, 6 days), and on a sandy clay loam soil with 2.5% organic matter (half-life, 18 days) (Mikami *et al.*, 1980). The pyrethroid fenpropathrin appeared to photodegrade faster with half-lives of 1, 4, and 5 days for clay, sandy loam, and sandy clay loam soils, respectively (Figure 3) (Takahashi *et al.*, 1985). Dark reaction occurred for fenpropathrin with a half-life greater than 14 days on any of the three soils examined. Among the various photoreactions possibly involved in the photodegradation pathway of fenpropathrin in soil, hydration and ester bond cleavage predominated, and the major photoproducts were biodegradable in nature, and less toxic than the parent molecule. The most rapid photolysis appeared thus to occur on soils with the highest organic matter and clay contents.

The organophosphate fenitrothion, however, was photolysed in approximately 1 day on either a silty loam containing 19% organic matter or a silty loam with 2.5% organic matter (Mikami *et al.*, 1985). In a previous study, photolysis rates of methidathion on six soils with organic matter content between 0.1 and 20% were also found not to be correlated with organic matter content (Smith *et al.*, 1978).

The photoproducts generated on surfaces are often different from those generated in solution, because the chemical environment on dry soil is distinctly different from that in aqueous solution. Further, the sorbent surface of soil can potentially alter the photoproduct distribution by restricting the movement and configuration of excited state molecules (Miller *et al.*, 1989b). For example, substantially different photoproducts of the herbicide basagran have been observed when irradiated on soils relative to those in solution (Nilles and Zabik, 1975). On soils, octachlorodibenzo-*p*-dioxin (OCDD) was photoreduced to a variety of lower chlorinated dioxins, whereas OCDD irradiated in isooctane yielded primarily the laterally photoreduced dioxins, with no low-chlorinated photoproducts (Miller *et al.*, 1989a).

Although standard methods have not yet been developed, six factors have been suggested to be considered when performing photolysis experiments on soil surfaces (Wolfe *et al.*, 1990). These factors are: the compound photochemistry, soil type, pesticide application method, water content, volatilization and advection, and pre-exposure of the surface of the soil. Further, sunlight should be used whenever possible, and the use of simple surface models for soil, such as silica gel and glass surfaces, should be avoided for obtaining quantitative information relevant to photolytic processes in the environment (Wolfe *et al.*, 1990).

## 3. Abiotic remediation of waters and soils

The use of photocatalytic and photochemical degradation processes may represent an efficient and advantageous means for the remediation of waters and soils contaminated with hardly biodegradable and biorecalcitrant chemicals including several pesticides.

The goal of any remediation strategy is the conversion of the targeted chemical to non-toxic, environmentally safe products. In the ideal case, the organic chemical undergoes complete destructive mineralization to simple inorganic species with the use of environmentally-compatible, non-contaminant, or easily degradable reagents and/or catalysts, and the net consumption of light, oxygen and water.

The capability of several organic chemicals to interact with the particulate matter, e.g., by surface adsorption, exposes these species to a different and often enhanced photoreactivity. Several studies have demonstrated that a wide variety of organic chemicals can be mineralized when irradiated in an aerated suspension of a semiconductor oxide such as titanium oxide, $TiO_2$, which acts as a heterogeneous photocatalyst.

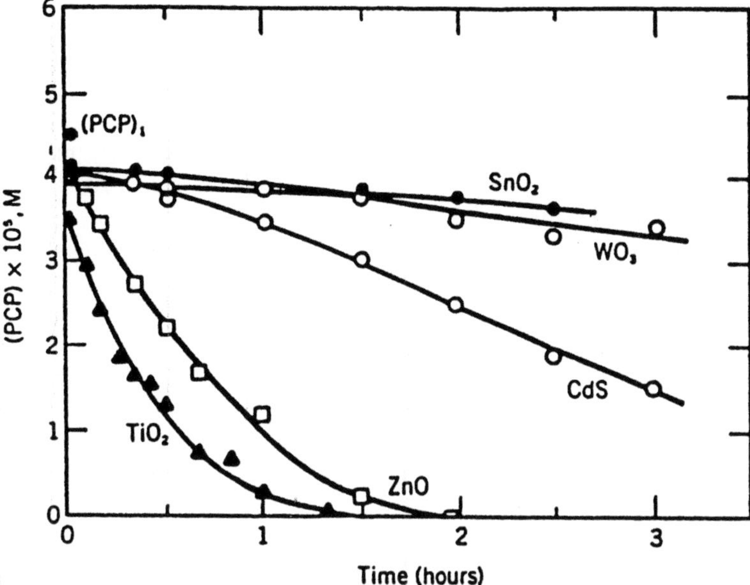


*Figure 4.* Photodegradation of pentachlorophenol in the presence of various semiconductor dispersions irradiated at $\lambda > 310$ nm, as a function of time in hours (adapted from Barbeni *et al.*, 1985).

Ferrous and ferric iron salts in acidic aerated solution, and ferric iron organic complexes at near neutral pH, alone or in the presence of hydrogen peroxide, with or without UV-light irradiation, have been shown to catalyse the production of hydroxyl radicals and other highly reactive radicals, which are able greatly to accelerate the degradation and, possibly, the mineralization of several pesticides *in situ*.

### *3.1 Heterogeneous photocatalysis*

Heterogeneous photocatalysis uses a continuously illuminated, photoexcitable catalyst to convert reactants adsorbed on the photocatalyst surface. Relatively few studies have been made on the photodegradation activity of various semiconductors using the same reactant and reaction conditions. Results obtained generally showed that $TiO_2$, followed by ZnO, are the two most active photocatalysts, whereas other solids, including CdS, $WO_3$, and $SnO_2$ are much less effective as photocatalysts for water purification (Ollis *et al.*, 1989). A revealing example of such studies is shown in Figure 4 for PCP mineralization to $CO_2$ and HCl in water (Barbeni *et al.*, 1985). The surface activity of such catalysts was shown to be pH sensitive.

To achieve the desired goal of the complete and rapid destruction of all intermediates, the occurrence of an extended and strong binding with the catalyst surface is desirable (Ollis *et al.*, 1989). For example, while ZnO showed about the same rate per mole for the photodegradation of PCP adsorbed, the higher concentration of PCP adsorbed on $TiO_2$ led to a correspondingly better photon utilization (Barbeni *et al.*, 1985).

The major events that occur in the irradiation of semiconductor $TiO_2$, and lead to the production of transient photoreactive radical species, which may then photooxidize and, possibly, mineralize the substrate, are summarized by the following scheme (Mamouni *et al.*, 1993):

$$TiO_2 + h\nu \longrightarrow e^-_{cb} + h^+_{vb} \quad (2)$$

$$h^+_{vb} + H_2O \longrightarrow HO^\bullet + H^+ \quad (3)$$

$$e^-_{cb} + H^+ \longrightarrow H^\bullet \quad (4)$$

$$e^-_{cb} + O_2 \longrightarrow O_2^{\bar{\bullet}} \quad (5)$$

$$O_2^{\bar{\bullet}} + H^+ \longrightarrow HO_2^- \quad (6)$$

$$HO_2^- + h^+_{vb} \longrightarrow HO_2^\bullet \quad (7)$$

$$O_2^{\bar{\bullet}} + H^+ \longrightarrow HO_2^\bullet \quad (8)$$

$$2HO_2^\bullet \longrightarrow O_2 + H_2O_2 \quad (9)$$

$$H_2O_2 + O_2^{\bar{\bullet}} \longrightarrow HO^\bullet + HO^- + O_2 \quad (10)$$

$$X + HO^\bullet \text{ (or } HO_2^\bullet) \rightarrow \begin{matrix}\text{photooxidized} \\ \text{products}\end{matrix} \rightarrow \text{mineralization.} \quad (11)$$

Photoredox processes can occur at semiconductor interfaces as a result of the absorption of 'bandgap' radiation, preferably in the solar spectral region. The immediate products of photon absorption are an electron in the conduction band (cb) and an electron vacancy, or positive hole, left in the valence band (vb) of the semiconductor. These reactive species can migrate to the solid/solution interface and produce very reactive hydroxyl radicals, $HO^\bullet$ and $HO_2^\bullet$, which can then react with the organic chemical adsorbed.

The occurrence of photocatalysed dehalogenations and complete mineralization to $CO_2$ and HCl of several chlorohydrocarbon chemicals by the simultaneous presence of illumination and photoexcitable $TiO_2$, suggested that heterogeneous photocatalysis represents a possible tool for water purification and soil decontamination (Ollis, 1985).

A wide range of haloalkanes, haloalkenes, haloaromatics, halocarboxylic acids, and haloaldheydes, and related derivatives, including several pesticides, such as PCP, 2,4,5-T and 4,4′-DDT, have been reported to rapidly photodegrade and mineralize to $CO_2$, HCl and $H_2O$ in oxygenated aqueous suspensions of $TiO_2$ under sunlight (Ollis *et al.*, 1989). Apparently, photocatalysed conversions involved the addition of hydroxyl groups to the reactant (while often displacing chlorine atoms), presumably as hydroxyl radical produced at the $TiO_2$ surface. PCBs sorbed on suspended $TiO_2$ in water were also photodechlorinated rapidly (Carey *et al.*, 1976).

The photolytic degradation of the herbicide bentazon has been demonstrated to occur efficiently and rapidly in the presence of aqueous suspensions of $TiO_2$ as photocatalyst under artificial sunlight (Pelizzetti *et al.*, 1989). The photodegradation process of bentazon led to complete mineralization of the herbicide with half-life of about 30 min at pH 2, and about 4 min at pH 7 in the presence of 0.5 g $L^{-1}$ of $TiO_2$. The process was extremely efficient even at $\mu l^{-1}$ levels of bentazon concentrations (Pelizzetti *et al.*, 1989).

In the presence of particulate $TiO_2$ as photocatalyst under simulated solar light ($\lambda > 340$ nm), several *s*-triazine herbicides, including atrazine, simazine, propazine, trietazine, prometon, and prometryne, were rapidly degraded photocatalytically (Figure 5) (Pelizzetti *et al.*, 1990, 1992). No disappearance of triazines was observed in the presence of either illumination or $TiO_2$ particles alone. Atrazine disappeared rapidly (half-life, 5 to 10 min) with complete removal (to less than 0.1 $\mu l^{-1}$) at 30 min. Only a partial conversion to $CO_2$ occurred but full mineralization of the triazines was not observed. Several intermediates have been identified, which indicated alkyl chain oxidation, and subsequent dealkylation as the main degradation paths, followed by the formation of hydroxy derivatives (Pelizzetti *et al.*, 1990,

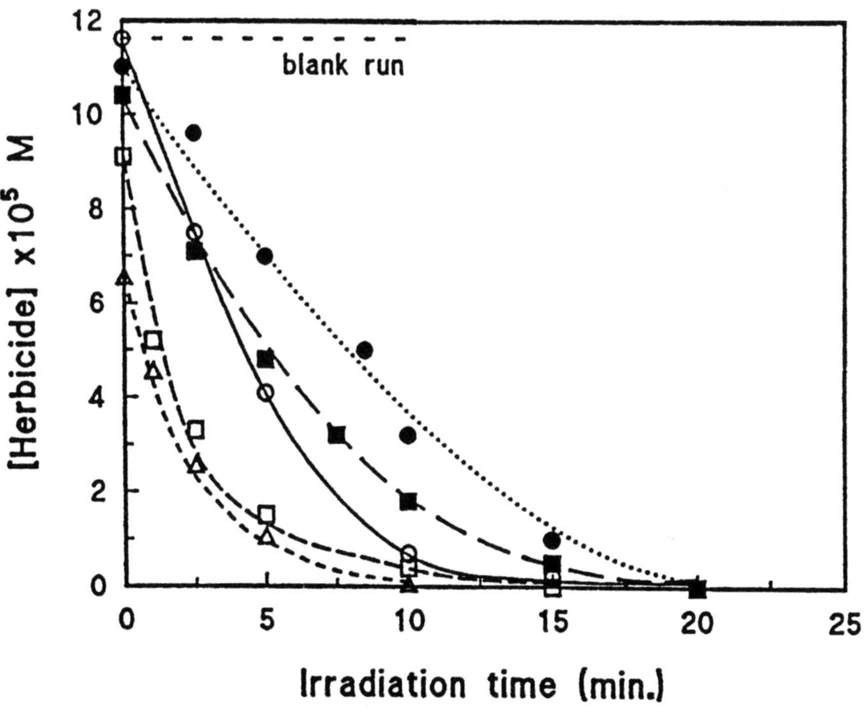


*Figure 5.* Photocatalytic degradation of atrazine ( ), simazine ( ) (scale expanded × 5), trietazine (△), prometon ( ), and prometryne ( ) in the presence of $TiO_2$, as a function of time in minutes (adapted from Pelizzetti *et al.*, 1990).

1992). Cyanuric acid, together with chloride or sulphate, and nitrate, were detected as the major final products of photodegradation for the triazines investigated. Cyanuric acid is a compound reported to have low toxicity and strong stability towards further photooxidation through the attack of hydroxyl radicals (Pelizzetti *et al.*, 1990).

Photocatalytic degradation of the herbicides pendimethalin and alachlor in irradiated aqueous suspensions of $TiO_2$ approximately followed first order kinetics, and involved N-dealkylation and nitro group elimination in the case of pendimethalin, and oxidation on both side chain and aromatic ring in the case of alachlor (Moza *et al.*, 1992). Photoreaction of pendimethalin resulted in 60% decomposition in 3 h in the presence of $TiO_2$, whereas the compound degraded only 10% in the same period of time in the absence of $TiO_2$. Alachlor disappeared much faster (over 95% in 20 min) in the presence of $TiO_2$ than in its absence, producing hydroxyalachlor and ketoalachlor as major photodecomposition products. Complete mineralization of both chemicals was very slow (Moza *et al.*, 1992).

The photodegradation of carbetamide, chloridazon, and metoxuron in water in the presence of suspended $TiO_2$, ozone, or $H_2O_2$ has been investigated under artificial sunlight source ($\lambda > 290$ nm) (Mamouni *et al.*, 1993). The three reactants are known to have in common the capability of producing hydroxyl radical species, but with a different efficiency. $TiO_2$ in the presence of oxygen was more efficient in photodegrading carbetamide and chloridazon than $H_2O_2$, but less efficient than ozone. The isolation and identification of several photoproducts suggested that the main photodegradation pathway of the three pesticides in the presence of either $TiO_2$ or $H_2O_2$ was hydroxylation at the aromatic ring. Ozonation produced rapid photooxidation with opening of the aromatic ring of the three pesticides, and production of low molecular weight carboxylic acids. Further photooxidation possibly converted the acids to full mineralization (Mamouni *et al.*, 1993).

### *3.2 Chemical and photochemical catalysis promoted by ferrous and ferric species*

The photodegradation rates of several *s*-triazines, including atrazine, ametryne, prometon, and prometryne were greatly enhanced by the addition of ferric perchlorate or ferric sulphate in moderately acidic aqueous solution (Larson *et al.*, 1991). The degradation was apparently due to formation of hydroxyl radicals by an electron-transfer photoprocess involving hydrated $Fe^{3+}$ ions. The presence of dissolved materials commonly found in natural waters, however, appeared to inhibit the photoreactions by competing with the triazine for hydroxyl radicals (Larson *et al.*, 1991).

The use of hydrogen peroxide oxidation in the presence of ferrous and ferric iron salts was explored as a promising alternative chemical treatment for the degradation of pesticide residues *in situ*. Such on-site treatment of aqueous or soil-bound pesticides could also avoid costly transport of contamined material to off-site incinerators (Pignatello, 1992).

Hydrogen peroxide in the presence of $Fe^{2+}$ or $Fe^{3+}$ produces reactive hydroxyl radicals, and possibly other strong oxidants. The classical reaction of $Fe^{2+}$ with $H_2O_2$, known as the Fenton reaction, generates hydroxyl radicals, $HO^\bullet$, according to:

$$Fe^{2} + H_2O_2 \longrightarrow Fe^{3+} + HO^\bullet + HO^-. \tag{12}$$

$Fe^{3+}$ ions catalytically decompose $H_2O_2$ yielding the hydroperoxyl radical, $HO_2^\bullet$, and $Fe^{2+}$, according to:

$$Fe^{3+} + H_2O_2 \longrightarrow Fe^{2+} + HO_2^\bullet + H^+ \tag{13}$$

$$Fe^{3+} + HO_2^\bullet \longrightarrow Fe^{2+} + O_2 + H^+. \tag{14}$$

$Fe^{2+}$ may then react with $H_2O_2$ (equation 12) and produce $OH^\bullet$. Hydroxyl radicals may be scavenged or react rapidly and non-selectively with most organic chemicals by H-abstraction and addition to C–C unsaturated bonds. Compared to $HO^\bullet$, $HO_2^\bullet$ radicals are much less reactive.

The herbicides 2,4-D and 3,4,5-T were rapidly degraded in acidic aerated solutions of diluted $H_2O_2$ and $Fe^{2+}$ or $Fe^{3+}$ salts, and complete mineralization could be achieved using $Fe^{3+}/H_2O_2$ in excess peroxide (Pignatello, 1992). Herbicide transformation by $Fe^{3+}/H_2O_2$ was sensitive to pH, with an optimum at about 2.8, and was retarded or inhibited by either chloride or sulphate ions. Dechlorination of the parent herbicide molecule was rapid and quantitative, and completed in less than 1 h. Conversion to $CO_2$ was extensive (about 40 to 70%, depending on $H_2O_2$ concentration and independent of Fe oxidation state), but incomplete in the dark. Degradation was markedly accelerated by irradiation with visible light (greater than 300 nm), and mineralization was complete in less than 2 h (Pignatello, 1992). The narrow effective pH range, however, limits the utility of $Fe/H_2O_2$ systems for treating pesticide contaminated soils, and the required acidification of soil to the proper pH may preclude or severely restrict implementation of $Fe/H_2O_2$ technology for *in situ* treatment.

To circumvent the drawbacks of working under acidic conditions, especially for treating contaminated soils, a number of structurally diverse ferric polydentate chelates that were soluble at near neutral pH (pH 6) were tested for their ability to catalyse the oxidation of 2,4-D, used as a pesticide probe, by $H_2O_2$ in aerated aqueous solution (Sun and Pignatello, 1992). Of 50 chelates tested, about 20 were found to be capable of catalysing the oxidation of 2,4-D in periods ranging from minutes to less than 5 h. The most efficient were 1 m*M* gallate, picolinate, and rhodizonate chelates in 10 m*M* $H_2O_2$, which gave complete transformation of 2,4-D after only 10–30 min, and about 80% mineralization in about 2 h, but further mineralization was difficult (Sun and Pignatello, 1992).

In a successive study carried out in water at pH 6 using $H_2O_2$ and Fe(III) picolinate, gallate, and rhodizonate complexes as catalysts, and without near-UV light, degradation of the herbicides 2,4-D, 3,4,5-T, picloram, and atrazine, and the insecticides baygon and carbaryl, was found to be complete within a few minutes to about 2 h, whereas trifluralin reacted slowly (Figure 6) (Sun and Pignatello, 1993).

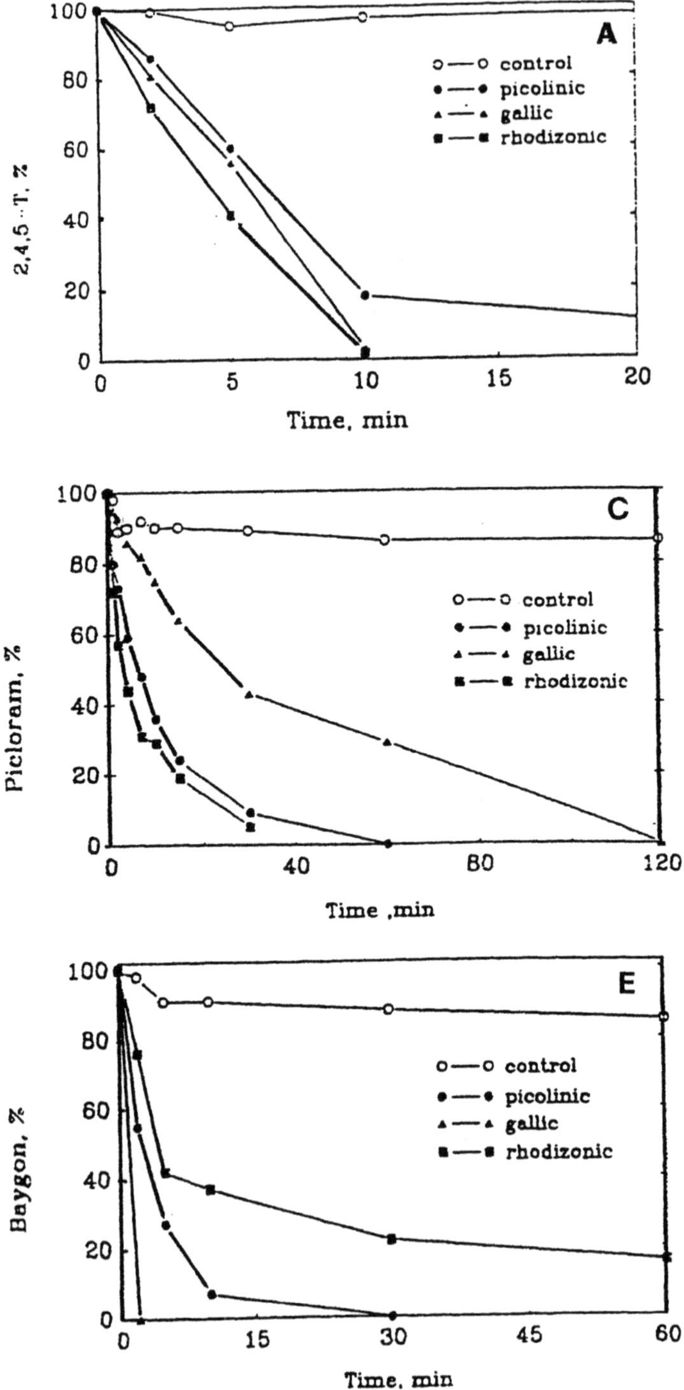


*Figure 6.* Degradation of 2,4,6-T (A), atrazine (B), picloram (C), trifluralin (D), baygon (E), and carbaryl (F) by Fe(III) chelates and hydrogen peroxide at pH 6 without near-UV light, as a function of time in minutes (adapted from Sun and Pignatello, 1993).

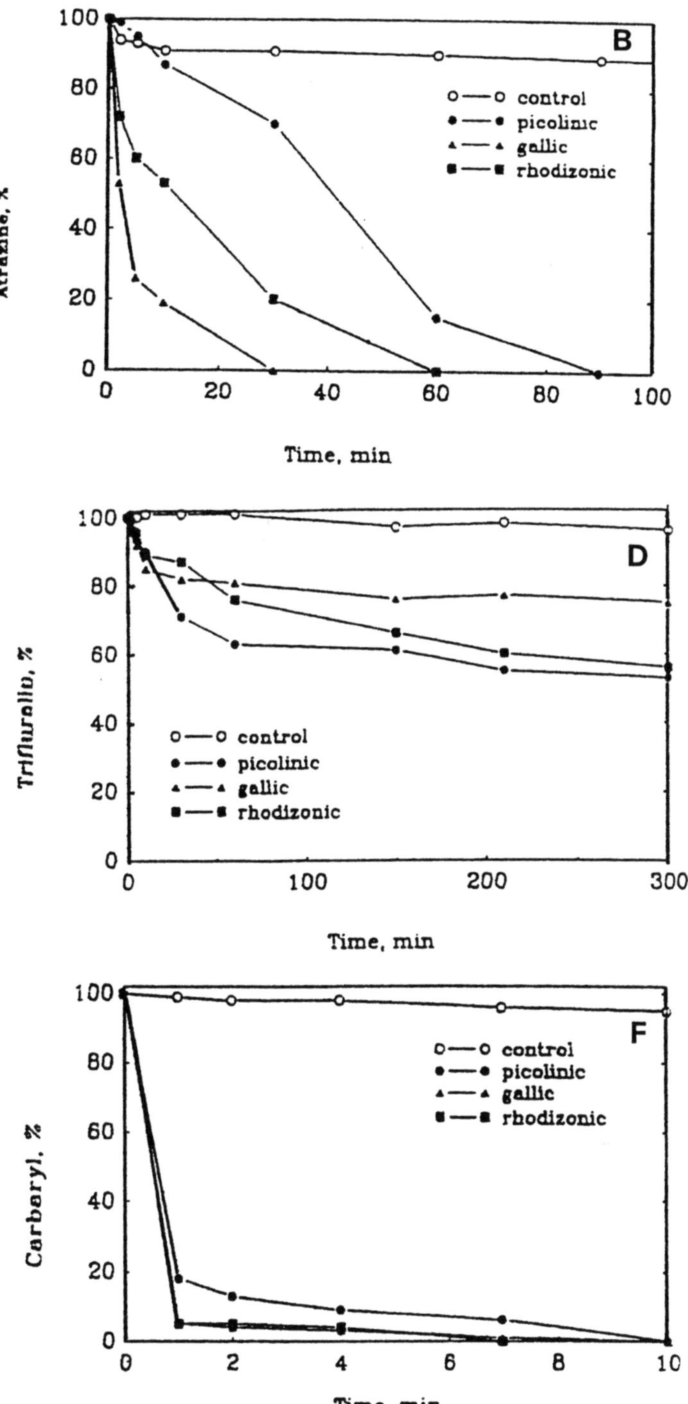


*Figure 6.* (Cont)

Ring- and carboxy-$^{14}C$-labelled 2,4-D and ring-$^{14}C$-labelled 3,4,5-T were 70–80% mineralized to $^{14}CO_2$ in 2–4 h. In the absence of UV light, chemical $H_2O_2$ activation with Fe(III) chelates was suggested to involve formation of $HO^\bullet$ by a mechanism analogous to the uncomplexed $Fe^{3+}/H_2O_2$ system (equations 12–14), with the key step being the Fenton reaction (equation 12) (Sun and Pignatello, 1993). Further, Fe(III) chelates might react chemically with $H_2O_2$ to form ferryl complexes, (L) $Fe^{IV} = O$, or one electron ligand-oxidized ferryl complexes, $(L^{+}_{\bullet})$ $Fe^{IV} = O$, where L is an organic ligand, which are strong and non-selective oxidizing agents, like $HO^\bullet$ (Sun and Pignatello, 1993).

UV light (300–400 nm) irradiation was shown to accelerate degradation and mineralization of 2,4-D by picolinate, gallate and rhodizonate Fe(III) chelates, as well as by chemically inactive oxalate and citrate chelates (Figure 7) (Sun and Pignatello, 1993). UV acceleration was suggested to be most likely due to photodissociation of the Fe(III) chelate by ligand-to-metal charge-transfer excitation:

$$L - Fe(III) \xrightarrow{h\nu} Fe^{2+} + L^\bullet \qquad (15)$$

followed by reaction of the photoreduced iron with $H_2O_2$ in the Fenton reaction (equation 12) to generate $HO^\bullet$. This interpretation explained the high photoactivities of oxalate and citrate chelates, which were practically unreactive without UV-irradiation. The lack of significant reactivity in the absence of $H_2O_2$ indicated that reaction of pesticide with $L^\bullet$ was unimportant (Sun and Pignatello, 1993). Other potential mechanisms for UV enhancement of photoreactivity, including photolysis of $H_2O_2$ itself, and photosensitization by the chelator or the chelate, were shown to be unimportant in this study (Sun and Pignatello, 1993).

## 4. Conclusions and recommendations

Abiotic photochemical and photocatalytic transformations naturally occurring in waters and soils have been shown to be of relevance for pesticide degradation in environmental systems. It has also been shown that technologies based on photochemical processes simulating those naturally occurring in the environment may be successfully applied as efficient abiotic remediation means in several cases. Photochemical remediation technologies are reasonably expected to be particularly promising for application to systems where the efficiency of bioremediation technologies may be compromised by a reduced biological activity, i.e., in submerged, flooded and intensely irrigated soils, in hot and dry soils of arid and semi-arid regions, and where the high concentration of contaminants may inhibit dramatically microbial activity.

As several semiconductor materials are known to be present in soil even if at relatively low concentrations, heterogeneous photocatalysis operated by these compounds can be regarded as an important naturally occurring process controlling the degradation and possible mineralization of organic chemicals, including pesticides, in natural sunlight-illuminated soils. Technologies based on these

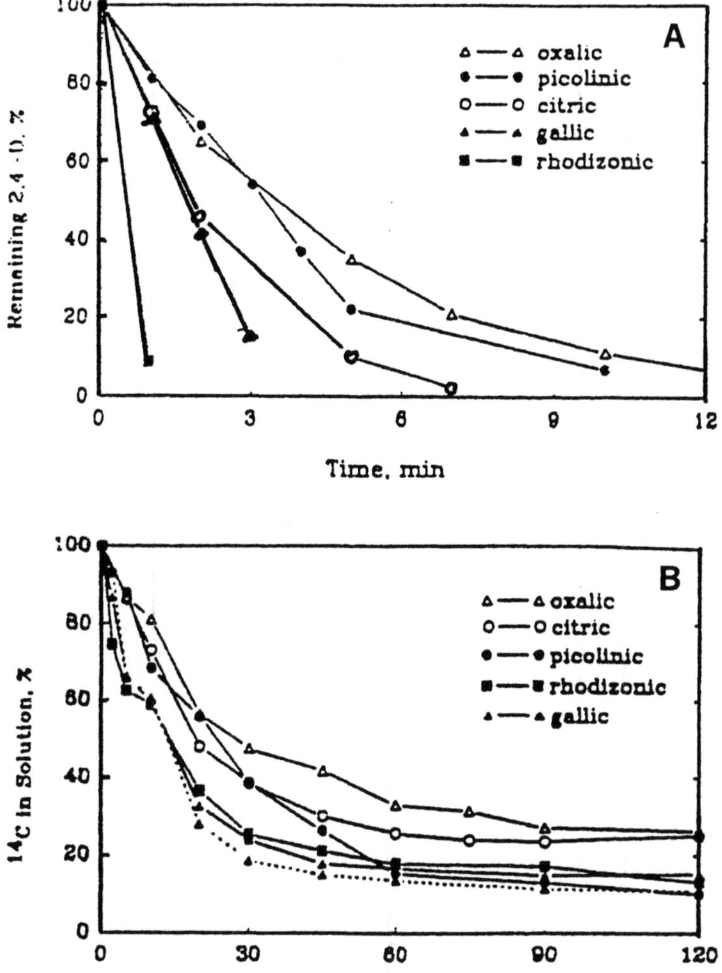


*Figure 7.* UV-assisted transformation (A) and mineralization (B) of 2,4-D by Fe(III) chelates and hydrogen peroxide at pH 6, as a function of time in minutes. Solid lines in B represent ring-labelled and the dashed line represents carboxy-labelled 2,4-D (adapted from Sun and Pignatello, 1993).

materials for the detoxification and/or destruction of hazardous chemicals such as pesticide residues on site are a very attractive and suitable means especially in the case of soil remediation where transport off-site is highly questionable and costly.

Heterogeneous photocatalysis has been shown to be an interesting tool for water and waste water purification. The photocatalytic technology has evolved as a 'soft' technology that may be usefully applied with several advantages also to the remediation of pesticide contaminated soils. In fact, the photocatalytic procedures:

- enable the complete destruction or mineralization of the organic chemical to $CO_2$, $H_2O$, HCl, and other simple inorganic compounds;

- avoid formation of products recalcitrant to further degradation and toxic to the environment;
- require only ambient temperature and ambient pressure conditions, natural solar illumination, and the presence of air oxygen as the oxidant;
- require very short times (minutes to hours), much shorter than in their absence;
- involve the use of inexpensive catalysts with high turnovers; and
- are efficient also at concentrations of the chemical in the $\mu l^{-1}$ range.

Hydrogen peroxide activated by Fe(III) chelates at near neutral pH produces large amounts of reactive hydroxyl radicals, and other reactants, which show attractive potential for the on-site detoxification of pesticide contaminated soils. Further, UV light has been shown to enhance degradation catalysed by Fe(III) chelates. This technology presents several advantages:

- the reagents are inexpensive;
- the reactions are rapid;
- the methodology is simple;
- gallic, citric and oxalic acids are natural products;
- picolinic and rhodizonic chelates are unstable; and
- the use of chelating agents avoids the need for acidification.

Photochemical and photocatalytic technologies applied to real systems such as natural waters and field soils, however, can often be strongly limited and reduced in their expected performance and efficiency due to the presence of several potential inhibitors and scavengers of photoreactants produced. Dissolved, colloidal and solid constituents typically present in soils and waters, including organic matter, clays, and several anions, such as chloride, nitrate, sulphate, and especially, carbonate, bicarbonate, and phosphate, are different scavengers of transient photoreactants, e.g., singlet oxygen and hydroxyl radicals, and have retarding effects and a general unfavourable influence on the process of photoabiotic remediation.

## References


Acher A J 1982 The fate of organic pollutants in frozen waters: Sunlight photodecomposition of uracil herbicides in frozen aqueous solutions. Water Res. 16, 405–410.

Acher A J and Rosenthal I 1977 Dye-sensitized-photo-oxidation. A new approach to the treatment of organic matter in sewage effluents. Water Res. 11, 557–562.

Acher A J, Saltzman S, Brates N and Dunkelblum E 1991 Photosensitized decomposition of terbacil in aqueous solutions. J. Agric. Food Chem. 29, 707–711.

Ando W 1981 Photooxidation of organosulfur compounds. Sulfur Rep. 1, 147–207.

Bailey G W and Karickhoff S W 1973 UV-vis spectroscopy in the characterization of clay mineral surfaces. Anal. Lett. 6, 43–49.

Barbeni M, Pramauro E and Pelizzetti E 1985 Photodegradation of pentachlorophenol catalyzed by semiconductor particles. Chemosphere 14, 195–208.

Burkhard N and Guth J A 1976 Photodegradation of atrazine, atraton and ametryne in aqueous solution with acetone as a sensitizer. Pestic. Sci. 7, 65–71.

Burns S E and Hassett J P 1994 Apparent concentration of hydrated electron inside and outside of the aqueous humic acid matrix. In Proc. 208th ACS National Meeting, Washington, DC, 21–25 August 1994. Vol. 34, No. 2, 64–66.

Carey J H, Lawrence J and Tosine H M 1976 Photodechlorination of PCBs in the presence of titanium dioxide in aqueous suspensions. Bull. Environ. Contam. Toxicol. 16, 697–701.

Cessna A J and Muir C G 1991 Photochemical transformations. In Environmental Chemistry of Herbicides. Eds R Grover and A J Cessna. pp 199–263. CRC Press, Boca Raton, FL.

Choudhry G G and Webster G R B 1985 Protocol guidelines for the investigations of photochemical fate of pesticides in water, air and soils. Residue Rev. 96, 79–136.

Cooper W J, Zika R G, Petasne R G and Fischer A M 1989 Sunlight-induced photochemistry of humic substances in natural waters: Major reactive species. In Aquatic Humic Substances – Influence on Fate and Treatment of Pollutants. Eds I H Suffet and P MacCarthy. Advances in Chemistry Series 219. pp 333–362. American Chemical Society, Washington, DC.

Crosby D G 1976 Herbicide photodecomposition. In Herbicides: Chemistry, Degradation and Mode of Action. Eds P C Kearney and D D Kaufman. Vol. 2, 2nd ed. pp 835–890. Dekker, New York.

Crosby D G and Tang C S 1969 Photodecomposition of 3-(*p*-chlorophenyl)-1,1-dimethylurea (monuron). J. Agric. Food Chem. 17, 1041–1044.

Crosby D G and Wong A S 1977 Environmental degradation of 2,3,7,8-tetrachlorodibenzo-*p*-dioxin (TCDD). Science 195, 1337–1338.

Dolcater D L, Syers J K and Jackson M L 1970 Titanium as free oxide and substituted forms in kaolinite and other soil minerals. Clays Clay Mineral. 18, 71–79.

Draper W M and Crosby D G 1984a Solar photooxidation of pesticides in dilute hydrogen peroxide. J. Agric. Food Chem. 32, 231–237.

Draper W M and Crosby D G 1984b Photochemistry and volatility of drepamon in water. J. Agric. Food Chem. 32, 728–733.

Emmelin C, Guittonneau S, Lamartine R and Méallier P 1993 Photodegradation of pesticides on adsorbed phases. Chemosphere 27, 757–763.

Fang C-H 1977 Studies on the photodecomposition of herbicide alachlor. Chung-Kuo Nung Yeh Hua Hsueh Hiu Chih 15, 14–22.

Freeman P K and MacCarthy K D 1984 Photochemistry of oxime carbamates. 1. Phototransformation of aldicarb. J. Agric. Food Chem. 32, 873–877.

Freeman P K and Nipp E M N 1984 Photochemistry of oxime carbamates. 2. Phototransformation of methomyl. J. Agric. Food Chem. 32, 877–881.

Frimmel F H 1994 Photochemical aspects related to humic substances. Environ. Internat. 20, 373–385.

Ghore K 1985 Singlet oxygen production on metal oxide and soil surfaces: Relation to sulfur containing pesticide photodegradation. PhD thesis. Univ. of Nevada, Reno (Diss. Abstr. 46: 07-B).

Ghore K and Miller G C 1985 Photochemical generation of singlet oxygen on non-transition-metal oxide surfaces. J. Chem. Soc. Faraday Trans. 181, 793–800.

Ghore K and Miller G C 1986 Photooxidation of thioether pesticides on soil surfaces. J. Agric. Food Chem. 34, 709–713.

Ghore K, Scholl R and Miller G C 1986 Singlet oxygen reactions on irradiated soil surfaces. Environ. Sci. Technol. 20, 934–938.

Golab T, Althaus W A and Wooten H L 1979 Fate of $^{14}$C-trifluralin in soil. J. Agric. Food Chem. 27, 163–179.

Guittonneau S, Khelifi F and Meallier P 1994 Photodegradation of pesticides on adsorbed phases. Photodegradation of phenmediphame (in press).

Hoignè J, Faust B C, Haag W R, Scully F E Jr and Zepp R G 1989 Aquatic humic substances as sources and sinks of photochemically produced transient reactants. In Aquatic Humic Substances – Influence on Fate and Treatment of Pollutants. Eds I H Suffet and P MacCarthy. Advances in Chemistry Series 219. pp 363–381. American Chemical Society, Washington, DC.

Katagi T 1993 Photodegradation of esfenvalerate in clay suspensions. J. Agric. Food Chem. 41, 2178–2183.

Khan S U and Gamble D S 1983 Ultraviolet irradiation of an aqueous solution of prometryne in the presence of humic materials. J. Agric. Food Chem. 31, 1099–1104.

Khan S U and Schnitzer M 1978 UV irradiation of atrazine in aqueous fulvic acid solution. J. Environ. Sci. Health B13, 299–310.

Klehr M, Iwan J and Riemann J 1983 An experimental approach to the photolysis of pesticides absorbed on soil: Thiodiazuron. Pestic. Sci. 14, 359–336.

Klöpffer W 1992 Photochemical degradation of pesticides and other chemicals in the environment: A critical assessment of the state of the art. Sci. Total Environ. 123/124, 145–159.

Kochany J and Maguire R J 1994 Sunlight photodegradation of quinoline in water. In Proc. 208th ACS National Meeting, Washington, DC, 21–25 August 1994. Vol. 34, No. 2, 52–55.

Korte U J, Anderson C and Spiteller M 1987 Photodegradation of pesticides in the presence of humic substances. Sci. Total Environ. 62, 335–340.

Langford C H and Carey J H 1987 Photocatalysis by inorganic components of natural water systems. In Photochemistry of Environmental Aquatic Systems. Eds R G Zika and W J Cooper. ACS Symposium Series 327. p 225. Am. Chem. Soc., Washington, DC.

Larson R A, Schlauch M B and Marley K A 1991 Ferric ion promoted photodecomposition of triazines. J. Agric. Food Chem. 39, 2057–2062.

Liberti A, Brocco D, Alligrini I, Cecinato A and Possanzini M 1978 Solar and UV photodecomposition of 2,3,7,8-tetrachlorodibenzo-*p*-dioxin in the environment. Sci. Total Environ. 10, 97–104.

Mamouni A, Mansour M and Schmitt P 1993 Abiotic degradation pathways of selected pesticides in the presence of oxygen species in aqueous solutions. In Fate and Prediction of Environmental Chemicals in Soils, Plants, and Aquatic Systems. Ed. M Mansour. pp 253–274. Lewis, Boca Raton, FL.

Mansour M, Scheunert I and Korte F 1993 Fate of persistent organic compounds in soil and water. In Migration and Fate of Pollutants in Soils and Subsoils. Eds D Petruzzelli and F G Helfferich. NATO-ASI Series, Vol. G32. pp 111–139. Springer, Berlin.

Marcheterre L, Choudhry G G and Webster G R B 1988 Environmental photochemistry of herbicides. In Reviews of Environmental Contamination and Toxicology. Ed. G W Ware. Vol. 103. pp 61–126. Springer, New York.

Miille M J and Crosby D G 1983 Pentachlorophenol and 3,4-dichloroaniline as models for photochemical reactions in water. Marine Chem. 14, 11–120.

Mikami N, Imanishi K, Yamada H and Miyamoto J 1985 Photodegradation of fenitrothion in water and on soil surface, and its hydrolysis in water. J Pestic. Sci. 10, 263–272.

Mikami N, Takahashi N, Haytashi K and Mitamoto J 1980 Photodegradation of fenvalerate (sumicidin) in water and on soil surface. J. Pestic. Sci. 5, 225–236.

Miller G C and Zepp R G 1979 Photoreactivity of aquatic pollutants sorbed on suspended sediments. Environ. Sci. Technol. 13, 860–863.

Miller G C and Zepp R G 1983 Extrapolating photolysis rates from the laboratory to the environment. Residue Rev. 85, 89–110.

Miller G C, Miille M J, Crosby D G, Sontum S and Zepp R G 1979 Photosolvolysis of 3,4-dichloroaniline in water. Evidence for an arylcation intermediate. Tetrahedron 35, 1797–1800.

Miller G C, Hebert V R, Miille M J, Mitzel R and Zepp R G 1989a Photolysis of octachlorodibenzo-*p*-dioxin in soils: Production of 2,3,7,8-TCDD. Chemosphere 18, 1265–1274.

Miller G C, Hebert V R and Miller W W 1989b Effect of sunlight on organic contaminants at the atmosphere–soil interface. In Reactions and Movement of Organic Chemicals in Soils. Eds B L Sawhney and K Brown. SSSA Special Publication 22. pp 99–110. Soil Sci. Soc. Am., Madison, WI.

Moilanen K W and Crosby D G 1974 Photodecomposition of bromacil. Arch. Environ. Contam. Toxicol. 2, 3–8.

Moza P N, Hustert K, Pal S and Sukul P 1992 Photocatalytic decomposition of pendimethalin and alachlor. Chemosphere 25, 1675–1682.

Nilles G P and Zabik M J 1974 Photochemistry of bioactive compounds. Multiphase photodegradation of basalin. J. Agric. Food Chem. 22, 684–688.

Nilles G P and Zabik M J 1975 Photochemistry of bioactive compounds. Multiphase photodegradation and mass spectral analysis of basagran. J. Agric. Food Chem. 23, 410–415.

Ohkawa H, Mikami N and Miyamoto J 1974 Photodecomposition of sumithion (O,O-dimethyl O-(3-methyl-4-nitrophenyl) phosphorothioate). Agric. Biol. Chem. 38, 2247–2255.

Oliver B G, Cosgrove E G and Carey J H 1979 Effect of suspended sediments on the photolysis of organics in water. Environ. Sci. Technol. 13, 1075–1077.

Ollis D F 1985 Contaminant degradation in water. Environ. Sci. Technol. 19, 480–484.

Ollis D F, Pelizzetti E and Serpone N 1989 Heterogeneous photocatalysis in the environment: Application to water purification. In Photocatalysis. Fundamentals and Applications. Eds N Serpone and E Pellizzetti. pp 603–637. Wiley, New York.

Parocchetti J V and Dec G W Jr 1978 Photodecomposition of eleven dinitroaniline herbicides. Weed Sci. 26, 153–156.

Pelizzetti E, Maurino V, Minero C, Zerbinati O and Borgarello E 1989 Photocatalytic degradation of bentazon by $TiO_2$ particles. Chemosphere 18, 1437–1445.

Pelizzetti E, Maurino V, Minero C *et al.* 1990 Photocatalytic degradation of atrazine and other *s*-triazine herbicides. Environ. Sci. Technol. 24, 1559–1565.

Pelizzetti E, Minero C, Carlin V, Vincenti M, Pramauro E and Dolci M 1992 Identification of photocatalytic degradation pathways of 2-Cl-*s*-triazine herbicides and detection of their decomposition intermediates. Chemosphere 24, 891–910.

Pignatello J J 1992 Dark and photoassisted $Fe^{3+}$-catalysed degradation of chlorophenoxy herbicides by hydrogen peroxide. Environ. Sci. Technol. 26, 944–951.

Pimental D and Levitan L 1986 Pesticides: Amounts applied and amounts reaching pests. Bioscience 36, 86–91.

Rejto M, Saltzman S, Acher A J and Muszkat L 1983 Identification of sensitized photooxidation products of *s*-triazine herbicides in water. J. Agric. Food Chem. 31, 138–142.

Ruzo L A 1982 Photochemical reactions of the synthetic pyrethroids. Prog. Pestic. Biochem. 2, 1–33.

Senesi N 1993 Organic pollutant migration in soils as affected by soil organic matter. Molecular and mechanistic aspects. In Migration and Fate of Pollutants in Soils and Subsoils. Eds D Petruzzelli and F G Helfferich. NATO-ASI Series, Vol. G32. pp 47–74. Springer, Berlin.

Senesi N and Miano T M 1995 The role of abiotic interactions with humic substances on the environmental impact of organic pollutants. In Environmental Impact of Soil Component Interactions: Natural and Anthropogenic Organics. Eds P M Huang, J Berthelin, J M Bollag and W B McGill. Vol. I. pp 311-355 CRC Press, Boca Raton, FL.

Skurlatov Y I, Zepp R G and Baughman G L 1983 Photolysis rates of (2,4,5-trichlorophenoxy)acetic acid and 4-amino-3,5,6-trichloropicolinic acid in natural waters. J. Agric. Food Chem. 31, 1065–1071.

Smith C A, Iwata Y and Gunther F A 1978 Conversion and disappearance of methidathion on thin layers of dry soil. J. Agric. Food Chem. 26, 959–962.

Spencer W F, Adams J D, Shoup T D and Spear R C 1980a Conversion of parathion to paraoxon on soil dusts and clay minerals as affected by ozone and UV light. J. Agric. Food Chem. 28, 367–371.

Spencer W F, Shoup T D and Spear R C 1980b Conversion of parathion to paraoxon on soil dusts as related to atmospheric oxidants at three California locations. J. Agric. Food Chem. 28, 1295–1300.

Sun Y and Pignatello J J 1992 Chemical treatment of pesticide wastes. Evaluation of Fe(III) chelates for catalytic hydrogen peroxide oxidation of 2,4-D at circumneutral pH. J. Agric. Food Chem. 40, 322–327.

Sun Y and Pignatello J J 1993 Activation of hydrogen peroxide by iron (III) chelates for abiotic degradation of herbicides and insecticides in water. J. Agric. Food Chem. 41, 308–312.

Takahashi N, Mikami N, Yamada H and Miyamoto J 1985 Photodegradation of the pyrethroid insecticide fenpropathrin in water, on soil and on plant foliage. Pestic. Sci. 16, 119–131.

Takase I, Ohama H and Uelyama I 1982 Photodecomposition of prothiofos (O-2,4-dichloro-phenyl O-ethyl S-propyl phosphorodithioate). Pestic. Chem. 7, 463–471.

Tanaka F S, Wien R G and Mansager E R 1979 Effect of nonionic surfactants on the photochemistry of 3-(4-chlorophenyl)-1,1-dimethylurea in aqueous solution. J. Agric. Food Chem. 27, 774–775.

Tanaka F S, Wien R G and Mansager E R 1981 Survey for surfactant effects on the photodegradation of herbicides in aqueous media. J. Agric. Food Chem. 29, 227–230.

Valentine R L 1986 Nonbiological degradation in soil. In Vadose Zone Modeling of Organic Pollutants. Eds S M Hern and S Meloncon. pp 223–243. Lewis, Chelsea, MI.

Wolfe N L, Mingelgrin U and Miller G G 1990 Abiotic transformation in water, sediments, and soil. In Pesticides in the Soil Environment: Processes, Impacts, and Modeling. Ed. H H Cheng. SSSA Book Series 2. pp 103–168. Soil Sci. Soc. Am., Madison, WI.

Wong A S and Crosby D G 1981 Photodecomposition of pentachlorophenol in water. J. Agric. Food Chem. 29, 125–130.

Yokley R A, Garrison A A, Wehry E L and Mamontov G 1986 Photochemical transformation of pyrene and benzo (a) pyrene vapor-deposited on eight coal stack ashes. Environ. Sci. Technol. 20, 86–90.

Zafiriou O C, Joussot-Dubien J, Zepp R G and Zika R G 1984 Photochemistry of natural waters. Environ. Sci. Technol. 18, 358A–371A.

Zepp R G 1992 Sunlight-induced oxidation and reduction of organic xenobiotics in water. In Fate of Pesticides and Chemicals in the Environment. Ed. J L Schnoor. pp 127–153. Wiley, New York.

Zepp R G, Baughman G L and Schlotzhauer P F 1981 Comparison of photochemical behaviour of various humic substances in water: I. Sunlight induced reactions of aquatic pollutants photosensitized by humic substances. Chemosphere 10, 109–117.

Zepp R G, Schlotzhauer P F and Sink R M 1985 Photosensitized transformations involving electronic energy transfer in natural waters: Role of humic substances. Environ. Sci. Technol. 19, 74–81.

# 6
# Neem, eco-friendly, IPM-compatible plant-derived bioinsecticide: some new findings

K.R.S. ASCHER

## 1. The tree and its potential

The neem tree or Indian lilac, *Azadirachta indica*, a member of the family Meliaceae and a botanical relative of mahogany, is known in Sanskrit as *arishta* (reliever of sickness) and in Kiswahili as *marubaini*, meaning 40 cures. It is a tree held in high esteem in India, a country which boasts at least 18 million neem trees, and in many other tropical countries – not only because of its conspicuous medicinal potential, but, especially in the last few decades, because of the excellent and astonishing pest control properties of its easily accessible products. Apart from the latter, it is a source of fuel, firewood, timber, folkloristic herbal medicines and natural contraceptives.

Some of the ethnobotanical or neem culturing descriptions now following are cited from National Research Council (1992).

Neem was introduced into most African countries in the last century. However, to Ghana neem was brought by Brigadier-General Sir Frederick Guggisberg, who was governor from 1919 to 1927 of what was then known as the Gold Coast, and as a result of his efforts, neem is now found growing throughout Ghana and the Sahelian region. Another outcome of this introduction was the vernacular name of neem in the Dyula language in Mali, which is *goo-gay*, a corruption of Guggisberg, whereas in Ghana the tree is normally called 'king', the local title for governor.

As already intimated, the tree is propagated vigorously in all tropical and many subtropical countries of the world. In our corner of the world, in the plains of Arafat near Mecca (Saudi Arabia), where according to tradition the prophet Muhammad delivered his farewell sermon 1400 years ago, a city of thousands of tents springs up each year to accommodate the pilgrims. In this area, one of the hottest in the world, there is little relief from the intense heat – and therefore the world's largest neem plantation, about 50 000 trees, has recently been planted through the initiative of a Saudi philanthropist. The project is designed to provide shade to the 2 million Muslim pilgrims who camp there annually for the '*haj*'. Shade will hopefully be available within a short time, since the fast-growing neem tree reaches a height of more than 11 metres after six years. Neem, an attractive broad-leaved evergreen, grows to a 30-m height, is 2.5 m in girth and has a 20-m crown. It can take considerable abuse. It regrows rapidly after pollarding (repeated lopping at heights above 1.5 m), as well as coppicing (repeated lopping at near-ground level), because it is served by a root system large enough to feed a full-grown tree.

*D. Rosen* et al. *(eds): Modern Agriculture and the Environment, 75–79.*

At the October 1991 BARD Workshop on 'New Targets for Insect Management in Crop Protection' at Shoresh, Israel I presented an overview on the state of art of neem pesticidal research, but since that time large strides have been taken in this area, and each year hundreds of papers have been published and some spectacular advances have been accomplished. Some very recent overviews (Ascher, 1993; Ley *et al.*, 1993; Mordue (Luntz), 1993; National Research Council, 1992; and especially Schmutterer, 1995) give testimony to these achievements. Some selected conspicuous findings published since 1991 will be described here.

## 2. Advances in industrial neem products

The most prominent insecticidally active ingredient of neem is the tetranortriterpenoid azadirachtin (AZA). Nowadays, the established procedure of preparing AZA formulations consists of first cold-pressing milled neem seeds to remove the oil. Proprietary processes are then used by the industries to extract AZA with solvents from the seed meal; powder concentrates containing about 20–30% AZA are thus produced, which are then formulated to 3–5% AZA emulsifiable concentrates (ECs). Neem industrial formulations produced in the USA, which, until 1992, had Environmental Protection Agency (EPA) registration on non-food and non-feed crops and ornamentals only, have, since mid-July 1993, USEPA registration and ratification for unlimited use on food and feed crops. The most prominent products of this recent upgrading are Margosan-O* (also called Bioneem in the USA), an older 0.3% AZA and 14% neem oil containing preparation of the US firm Grace & Co., as well as its more recent product Neemix 45[†] (also called Superneem), with a 4.5% AZA content; Neemix 45[†] has, *inter alia*, an excellent record in both the USA and Israel for controlling the citrus leaf miner, *Phyllocnistis citrella*. Another US firm, Agridyne Technologies, has a food and feed crop formulation called Align[†] which contains 3% AZA. Neem is essentially a tropical plant and the US firms either import the seeds from countries such as India, Ghana, etc., or have erected plants overseas, e.g. in India, to produce primary, highly concentrated extracts locally. Efforts are also underway to establish large plantations in Florida, California, Puerto Rico, and Hawaii, all of which have a tropical climate. Many countries in Central and South America are vigorously initiating neem plantations, whereas the Dominican Republic, Nicaragua and Haiti already have huge numbers of neem trees. In Haiti, for example, this stemmed from a USAID project in the late 1970s to plant 200 000 neem trees as part of a road beautification programme. Later, neem became a popular species for planting in Haiti and thus a major tree species in that country. The trees have grown so well that neem seed has become a Haitian export; approximately 40 tons of Haitian neem were processed for azadirachtin by an American company in 1990. Since then, other companies have also sought to buy Haiti's neem seed (National Research Council, 1992).

* These Grace products are now distributed by Thermo Trilogy, Columbia, MD, USA.
[†] Product now owned by biosys, Columbia, MD, USA.

Apart from the USA, further neem pesticidal products are proliferating in Germany, Canada, and especially India, where numerous individual firms are offering neem products called, among others, RD9-Repelin, Wellgro, Nimbosol, Biosol, Neemrich, Neemark, and so on. Since 1991 further developments in the use of neem products in many countries have been most impressive. Neem was found excellent in the management of pests affecting rice, sorghum, cowpea, beans, cabbage, citrus crops, etc., as well as of storage pests in many African states. This has motivated European sponsor institutions to establish a sizeable international grant administered from Kenya for the sole purpose of fostering 'neem awareness' in six African countries: Kenya itself, Ethiopia, Malawi, Tanzania, Uganda and Zambia.

## 3. Further achievements of neem research

The last three years have led to important advances in the structure–activity relationships of AZA and closely related compounds, and to a breakthrough towards the synthesis of the parent molecule, AZA (Ley *et al.*, 1993).

Further spectacular biological activities of neem in general, and of AZA in particular, were discovered in addition to those that had been described until 1991 against nearly 200 insect pest species. I will highlight a few of these fascinating new findings. In the past, it was mainly the antifeedant, moult-inhibiting and other IGR effects, oviposition deterrency, and sterilising potential, which led to the practical application of neem products. However, German teams in Germany and Africa have now demonstrated that $N_1$–$N_5$ nymphs of the desert locust, *Schistocerca gregaria*, when sprayed with neem oil, suffered, apart from the classical growth and development inhibition, and the resulting progressive mortality of nymphs and adults, from conspicuous, very peculiar defects: at lower dosages, nymphs and adults surviving the treatments showed a series of hitherto unknown morphogenetic effects. These consisted of:

- Parts of the legs missing, which in the moulting sequence was in the order: loss of tarsus, tibia, and femur; there was no loss of the coxa. With rising concentration, there was an increase in the number of such crippling effects regarding femur and especially tibia, with the last pair of legs being especially affected. Very often such legs lost their function of transportation. As to tarsi, nymphs sprayed with the highest dosage had, on the average, three tarsi less than the control. Another defect, getting more pronounced with rising concentration, was curly, crinkled, reduced and sometimes missing wings, all of which prevented full development of the membranous wings, and led to incapability of flight.
- Perhaps the most important defect was the progressive loss of segments of the antennae, which at the highest dosage left a maximum of two segments, and mostly there were no antennae at all. Also the eyes were affected, showing bright patches; many of such damaged adults no longer responded to optical stimuli. The deformation and/or losses of limbs were the greater, the earlier that treatment with neem oil had taken place in the nymphal development.

- The most fascinating effect was the obvious tendency of treated desert locust nymphs to solitarization, indicated by appearance of green haemolymph, and brown and green intermediate or totally solitary forms of the insect with very little black pigment, whereas the gregarious forms are yellow and black. These results implicated that the oil-treated locusts demonstrated colours characteristic for the solitary phase, viz., greenish, green and brown colours; even at very low dosages, e.g. 2.5 L neem oil/ha, the haemolymph was green. Whereas only a few of the nymphs turned greenish after the first postapplicative moult, the rate of abnormally coloured insects grew successively with development of the nymphs. In cage experiments these changes in colour were independent of the number of locusts in the cage, that is to say, even when normally untreated insects kept under these conditions would have developed the usual gregarious colours.

Another typical phenomenon of the difference between the neem-solitarised and the usual gregarious forms is the occurrence of six nymphal stages in the solitary form, *vs.* five in the gregarious form. Generally, an additional ecdysis occurred between $N_3$ and $N_4$, more rarely between $N_5$ and the moulting to adult. In all instances in which one of the first nymphal instars was treated, there appeared cases of additional moults.

Finally, as an outcome of all the defects, there was a conspicuous reduction in overall fitness of the treated locusts, such as inability of some of the insects to climb. The treated insects' movements were much slower than in the untreated controls; even if the jumping legs were undamaged, the insects were unable to jump. If the legs were damaged badly, the insects lay on their thorax and were unable or hardly able to move. Nymphs treated in early stages lay with the head and the first thoracic segment in contact with the ground, even when they had an intact first pair of legs. Use of the highest neem oil dosage tested in the experiments (10 L/ha) left insects that could either only walk or creep, or could not move at all. The fitness of the insects in field-cage trials was, of course, strongly dependent on the dosage. Whereas untreated animals were nimble and light-footed and could jump high and far, insects treated once with 2.5 L neem oil/ha up to the $N_3$ stage showed a reduced jumping and climbing capability; with 5 or 7.5 L/ha they were reduced to complete lack of jumping and climbing. Application of 10 L/ha produced animals that could only run or creep or even could not move at all. All these effects described in locusts have been reported and discussed by Schmutterer and Freres (1990), Freisewinkel and Schmutterer (1991), Nicol and Schmutterer (1991), Langewald and Schmutterer (1992), Schmutterer (1993), Schmutterer *et al.* (1993), and Langewald and Schmutterer (1995).

Semi-field trials with *S. gregaria* (Schmutterer *et al.*, 1993) and field trials just completed in Africa (Langewald, 1994; Schmutterer, 1995; Schmutterer *et al.*, 1993) with the red locust, *Nomadacris septemfasciata*, a species closely related to *S. gregaria*, have fully corroborated the results of these laboratory and field-cage studies and supplied additional interesting findings. It seems that treatment with neem oil is indeed a measure to prevent formation of swarms. The treated locusts remain solitary, with a maximum of three insects being seen to be aggregated; they are unable to jump. In addition there is strong cannibalism, since due to the damage to the hindlegs, they are unable to protect themselves from their congeners. Moreover, they were observed to be much easier prey to foraging birds.

Very recent results have shown that other insect mass plagues, such as maychafers (Kaethner, 1991), the gipsy moth, or the Colorado potato beetle (Kaethner, 1992) can be combated successfully in the field with neem preparations. Thus, in 1994 field trials in forests in Hassia, Germany, neem products gave excellent control of the maychafers *Melolontha melolontha* and *M. hippocastani*, and the gipsy moth, *Lymantria dispar* (Schmutterer and Nicol, 1994).

To summarize, neem has increasingly emerged, during the last three years, as a promising insecticide exhibiting unexpected non-conventional pesticidal effects.

## References

Ascher K R S 1993 Nonconventional effects of pesticides available from the neem tree, *Azadirachta indica*. Arch. Insect Biochem. Physiol. 22, 433–449.

Freisewinkel D and Schmutterer H 1991 Kontaktwirkungen von Niemoel bei der afrikanischen Wanderheuschrecke *Locusta migratoria migratorioides* (R. & F.). Z. Angew. Zool. 78, 189–203.

Kaethner M 1991 Untersuchungen ueber die Eignung von Niemsamenprodukten zur Bekaempfung des Waldmaikaefers *Melontha hippocastani* F. und des Feldmaikaefers *M. melolontha* L. (Col., Scarabaeidae). J. Appl. Entomol. 112, 345–352.

Kaethner M 1992 Fitness reduction and mortality effects of some neem-based pesticides in the Colorado beetle, *Leptinotarsa decemlineata* Say (Col., Chrysomelidae). J. Appl. Entomol. 113, 456–465.

Langewald J 1994 Zur Wirkung und Anwendung von Produkten des Niembaums [*Azadirachta indica* (A. Juss.)] gegen die Wuestenheuschrecke [*Schistocerca gregaria* (Forskål) und andere Schadheuschrecken in Westafrika und Madagaskar. PhD thesis, University of Giessen, Germany.

Langewald J and Schmutterer H 1992 Effects of neem oil treatment on the phase status of desert locust, *Schistocerca gregaria*. In Biological Control of Locusts and Grasshoppers. Proc. Workshop Int. Inst. Trop. Agric., Cotonou, Benin, 29 April–1 May 1991, pp 142–154.

Langewald J and Schmutterer H 1995, Colour change in the desert locust, *Schistocerca gregaria* (Forskål) (Orth., Cyrtacanthacridinae), induced by topical treatment with azadirachtin-enriched neem oil in semi-field trials in Benin at different population densities. J. Appl. Entomol. 119, 221–226.

Ley S V, Denholm A A and Wood A 1993 The chemistry of azadirachtin. Nat. Prod. Rep., 10, 109–157.

Mordue (Luntz) A J and Blackwell A 1993 Azadirachtin: An update. J. Insect Physiol. 39, 903–924.

National Research Council 1992 Neem: A Tree for Solving Global Problems. National Academy Press, Washington, DC, 141 pp.

Nicol C M Y and Schmutterer H 1991 Kontaktwirkungen des Samenoels des Niembaumes *Azadirachta indica* (A. Juss.) bei gregaeren Larven der Wuestenheuschrecke *Schistocerca gregaria* (Forskål). J. Appl. Entomol. 111, 197–205.

Schmutterer H 1993 Some effects of neem (*Azadirachta indica*) products on locusts and grasshoppers. World Neem Conf., Bangalore, India, Souvenir, pp 44–52.

Schmutterer H, Ed. 1995 The Neem Tree, *Azadirachta indica* A. Juss. and Other Meliaceous Plants – Sources of Unique Ingredients for Integrated Pest Management, Medicinal, Industrial and Other Purposes. VCH Verlagsgesellschaft, Weinheim, Germany, 696 pp.

Schmutterer H and Freres T 1990 Beeinflussung von Metamorphose, Faerbung und Verhalten der Wuestenheuschrecke *Schistocerca gregaria* (Forsk.) und der Afrikanischen Wanderheuschrecke *Locusta migratoria migratorioides* (R. & F.) durch Niemsamenoel. Z. Pflanzenkr. Pflanzenschutz 97, 431–438.

Schmutterer H and Nicol C M Y 1994 Gleichzeitige erfolgreiche Bekaempfung des Waldmaikaefers *Melolontha hippocastani* und des Schwammspinners *Lymantria dispar* mit NeemAzal-T + Telmion im Fruejahr 1994. Mitt. Dtsch. Ges. Allg. Anger. Entomol. 10, 135–138.

Schmutterer H, Baumgart M, Freisewinkel D, Langewald J and Nicol C M Y 1993 The effects of neem oil and other neem products on nymphs and resting adults of *Schistocerca gregaria, Nomadacris septemfasciata, Locusta migratoria migratorioides*, and *Zonocerus variegatus*. J. Appl. Entomol. 116, 178–186.

7

# Utilization of scorpion insecticidal neurotoxins and baculoviruses for the design of novel selective biopesticides

MICHAEL GUREVITZ, NOAM ZILBERBERG, OREN FROY, DANIEL URBACH, ELIAHU ZLOTKIN, BRUCE D. HAMMOCK, RAFAEL HERRMANN, HAIM MOSKOWITZ AND NOR CHEJANOVSKY

## 1. Natural toxins suitable for insect pest control

Public concern over the risks associated with widespread use of insecticidal chemicals has motivated our efforts to develop safer and more effective approaches to pest control. Thus, natural insecticidal compounds derived from venomous animals may serve as rational alternatives. Venoms of arthropods such as scorpions, spiders, braconid and sphecid wasps, reduviid bugs and centipedes (Zlotkin, 1985) which prey on insects, possess anti-insect selective polypeptidic neurotoxins. Such toxins have already shown promise when incorporated into insect pathogens by enhancing their killing efficacy (Chejanovsky *et al.*, 1995; Maeda *et al.*, 1991; Stewart *et al.*, 1991; Tomalski and Miller, 1991). These toxins bind to insect sodium channels (Gordon *et al.*, 1992), modify their properties and disrupt normal neuromuscular functions leading to paralysis and death. Thus, the incorporation of anti-insect selective scorpion toxins into insecticidal microorganisms offers an opportunity to improve and augment current crop protection strategies (Maeda *et al.*, 1991; Stewart *et al.*, 1991; Tomalski and Miller, 1991). In order to minimize biological hazards which may be associated with such a genetic approach and to avoid the possibility of resistance build-up of insect pests, the molecular basis for anti-insect selectivity of these toxins should be clarified. Furthermore, usage of insecticidal toxins should be limited to those revealing the 'animal group specificity' phenomenon where a toxin shows activity against a given group of organisms and is not effective against other groups of organisms. In cases where a venomous organism feeds on a given limited group of animals, the 'animal group specificity' is manifested already on the level of the whole venom. For example, venomous marine Conidae snails are subdivided according to their feeding habits into fish, mollusc and worm eaters. A series of early bioassays (Endean and Rudkin, 1963) demonstrated that this feeding preference was associated with the specific toxicity of their venoms aimed at the respective groups of prey animals. An extreme example of anti-insect selectivity revealed by an entire venom (gland extract) is demonstrated by Braconidae wasps. Their venom is employed primarily for securing hosts (moth larvae) for their offspring. The venom is exclusively effective against moth larvae, in

*D. Rosen* et al. *(eds): Modern Agriculture and the Environment, 81–96.*

which it induces a flaccid paralysis of long duration (weeks) as a consequence of a presynaptic blockage of excitatory transmitter (glutamate) release at the neuromuscular junction (Piek and Spanjer, 1986).

A distinction between arthropod affecting and mammal non-affecting scorpion and spider venoms has been made (Zlotkin, 1985). Clear detection of anti-insect specific toxins has been essentially obtained by chemical fractionation of arthropod venoms. Thus, vertical column electrophoresis of the black widow spider (*Latrodectus*) venom distinguished between several protein fractions (Mr = 60–80 kDa), each causing either fast reversible, or slow progressive irreversible paralysis of flies or toxicity to guinea pigs (Frontali and Grasso, 1964). Similarly, molecular exclusion column chromatography coupled to starch gel electrophoresis enabled the isolation of polypeptide factors from scorpion (Buthidae) venoms which selectively affect insects, vertebrates or crustaceans (Zlotkin, 1985, 1988; Zlotkin *et al.*, 1991).

Among the scorpion toxins that show selectivity for insect sodium channels over their vertebrate counterparts, are the excitatory (Zlotkin, 1983) and the depressant (Zlotkin *et al.*, 1993) toxins. The first excitatory anti-insect toxin, AaIT, was isolated in 1971 (Zlotkin *et al.*, 1971) and its specificity was confirmed by bioassays on intact animals, *in vitro* neuromuscular preparations and binding assays to various neuronal membranes. The excitatory toxins are single chained polypeptides of 70 amino acids cross-linked by four disulphide bridges, three of which appear at homologous positions as compared to other toxins affecting sodium channels. They interact with insect neuronal preparations through a single class of high affinity ($K_D = 1$–$2$ nM) and low capacity (1.2–2.0 pmol $mg^{-1}$ membranal protein) binding sites. The rapid excitatory paralysis of insects promoted by these toxins is due to the induction of repetitive electrical activity of motor neurons. This excitatory effect is attributed to the increase of the activatable action potential sodium conductance and a voltage-dependent slowing of its turning off (Pelhate and Zlotkin, 1981; Zlotkin, 1985, 1988). The depressant toxins constitute 61–64 amino acid long polypeptides with unique primary sequences; however, their four disulphide bridges are positioned homologously to other scorpion toxins. These toxins reveal a dual mode of action, where a short, transient period of excitability (similar to that obtained with the excitatory toxins) precedes the progressive flaccid paralysis induced by them in insects. The depressant toxins perturb binding of the excitatory toxins to their binding site at the insect neuronal membrane (Zlotkin *et al.*, 1991) although their receptor site was shown to be distinct from that of the excitatory toxins (Moskowitz *et al.*, 1994). The end result of their effect, flaccid paralysis, is a consequence of blockage of the evoked action potentials (Zlotkin, 1985) which may result from two mechanisms: first, suppression of the activatable sodium conductance as revealed by voltage clamp data; and second, depolarization of the axonal membrane attributed to the increase of the resting sodium permeability (Zlotkin, 1985, 1988).

Another group of scorpion toxins, the $\alpha$-toxins, shows activity against both insects and mammals. This class constitutes a family of structurally and functionally related polypeptides, each containing 62–65 amino acid residues cross-linked by four disulphide bridges (Rochat *et al.*, 1979; Figure 1). Alpha mammal toxins from the

```
               10        20        30        40        50        60
                :         :         :         :         :         :
LqhαIT  .VRDAYIAKNYNCVYECFRDAYCNELCTKN.GASSGYCQWAGKYGNACWCYALPDNVPIRVP..GKCR
Lqq4    G-------DDK----T-GSNS---TE---D.--E------L---------IK---K----I-..----
Bot1    G.------QPE------AQNS---D----D.--T------L---------KD--------I-..---HF
Bot11   .LK-G--VDDR--T-F-GTN-----E-V-LK-.E------V-R--------K---H-RTVQA..-R--S
BeM10   .---G---DDKD-A-F-G-N---D-E-K-..--E--K-WY--Q--------K---W---KQKVS---N
Lqq5    .LK-G--VDDK--TFF-G-N----DE-K-K.-GE------ASP--------K---R-S-KEK..-R-N
Amm5    .LK-G--I--L--TFF-------DD--K-K.-G-------ASP------------R-S-KEK..-R-N
AaH2    .---G--VDDV--T-F-G-N-----E---LK-.E------ASP-----Y--K---H-RTKG-..-R-H
```

*Figure 1.* Sequence comparison between scorpion alpha neurotoxins. Identical residues with the LqhαIT sequence are designated by dashes. Lqq4 and Lqq5, *Leiurus quinquestriatus quinquestriatus* toxin 4 and toxin 5; Bot1 and Bot11, *Buthus occitanus tunecatus* toxin 1 and toxin 11; BeM10, *Buthus eupeus* toxin 10; LqhαIT, *Leiurus quinquestriatus hebraeus* alpha insect toxin; Amm5, *Androctonus mauretanicus mauretanicus* toxin 5; AaH2, *Androctonus australis* Hector toxin 2.

scorpion *Androctonus australis* Hector (such as AaH1 and AaH2) or from *Leiurus quinquestriatus quinquestriatus* (such as Lqq4 and Lqq5) bind in a voltage dependent manner to the receptor site of vertebrate voltage sensitive sodium channels (Catterral, 1977; Couraud *et al.*, 1978; Ray and Catterral, 1978). Their toxic effect is due to a modification of the channel activity by slowing inactivation and altering its voltage dependence (Catterral, 1980, 1986; Gordon, 1990). As specific modifiers of the inactivation process, they are practically used as tools for the clarification of the molecular mechanism of ion channel gating. The large repertoire of various α-scorpion toxins is of particular significance as it is useful for the identification, characterization and discrimination between sodium channels in different organisms, tissues, and even in different locations within the same tissue (Barchi, 1987).

Recently, a new α-toxin, LqhαIT, with salient similarity in sequence and in mode of action to the anti-mammal alpha scorpion toxins, was characterized in the venom of the scorpion *Leiurus quinquestriatus hebraeus* (Eitan *et al.*, 1990). This toxin is active against both insect and mammalian sodium channels and its binding site does not overlap with those of either the excitatory or depressant anti-insect selective toxins in reciprocal radioligand binding experiments (Eitan *et al.*, 1990; Gordon and Zlotkin, 1993). LqhαIT thus recognizes a binding site on the sodium channel that may be common to insects and mammals, while at the same time having an apparent higher affinity for the insect channel.

On this basis, our goal to clarify the molecular basis for the anti-insect selectivity of scorpion neurotoxins is focused, in its initial stages, on LqhαIT as a representative of the α-toxin class. There are several advantages in pursuing this study with α-toxins:

- Alpha scorpion toxins comprise a large group of the most studied scorpion neurotoxins displaying natural variability with respect to primary sequences, binding sites and displaceability characteristics, potency, and the ratio of their anti-insect/mammal toxicity.

- Biological assays, including radioligand binding and neurophysiological assays for insect and mammalian sodium channel function are available.
- Several cDNAs encoding α-scorpion neurotoxins were cloned by us and their sequences determined. Furthermore, we established a reliable system for *ex vivo* functional expression of LqhαIT (Zilberberg *et al.*, 1996) which is most suitable for site-directed modificational analysis and determination of its three-dimensional structure.
- Alpha scorpion toxins display different binding capabilities to various sodium channels of the same organism or in different animals. For example, AaH2 and Bom3 (*Buthus occitanus mardochei*) do not displace each other from the receptor site in rat brain synaptosomes (Vargas *et al.*, 1987); Lqq4 hardly affects insects when compared to the strong anti-insect toxicity of LqhαIT (Zilberberg *et al.*, 1996); the anti-insect/mammal toxicity ratio can be altered (Zilberberg *et al.*, 1996). These data suggest that the binding ratio of alpha toxins to various channels could be modified by genetic means.
- The three-dimensional structure of three alpha scorpion toxins, AaH2, AaH3 and BeM9 (*Buthus eupeus*), has been solved by X-ray crystallography and $^{1}$H 2D-NMR (Fontecilla-Camps *et al.*, 1982, 1988; Mikou *et al.*, 1992; Pashkov *et al.*, 1988). This enables inclusion of assayable predictions in the genetic approach based on a combination of comparative analyses of alpha toxin sequences and three-dimensional structural models.
- LqhαIT-cDNA was placed under a strong promoter (polyhedrin) in the genome of the insect pathogen *Autographa californica* Nuclear Polyhedrosis Virus (AcNPV) and viral isolates with enhanced insecticidal properties were obtained (Chejanovsky *et al.*, 1995).

## 2. Structure–activity relationship of alpha scorpion toxins

Several approaches have been undertaken in studying structure–function relationship of scorpion toxins: chemical modifications introduced to isolated toxins (Darbon *et al.*, 1983; El Ayeb *et al.*, 1986a; Kharrat *et al.*, 1989); immunochemical studies to identify regions participating in binding to the receptor site in sodium channels (El Ayeb *et al.*, 1986b); a biophysical approach to determine the three-dimensional structure (Fontecilla-Camps *et al.*, 1982, 1988; Mikou *et al.*, 1992; Pashkov *et al.*, 1988); and a genetic approach using isolated cDNA clones and *ex vivo* expression systems (Bougis *et al.*, 1989; Chejanovsky *et al.*, 1995; Gurevitz *et al.*, 1991; Maeda *et al.*, 1991; Martin-Eauclaire *et al.*, 1994; Stewart *et al.*, 1991; Zilberberg and Gurevitz, 1993; Zilberberg *et al.*, 1996).

### *2.1 Chemical modifications*

A number of chemical modifications of various toxins have been carried out for localizing residues involved in the binding to the receptor or in their pharmacological properties. Biotinylation of Lys56 in AaH1 and Lys58 in AaH2 of *A. australis* Hector or in Lqq5 of *L. q. quinquestriatus* reduced the toxicity by 80–99% (Darbon *et al.*, 1983). Arg2 and Arg60 were modified in the toxin AaH1 by *p*-hydroxyphenylglyoxal causing a reduction of 60–70% in toxicity (Kharrat *et al.*, 1989). Arg2 modified by 1,2-cyclohexanedione in AaH3 reduced the binding to the receptor site by 99% (Kharrat *et al.*, 1989). Similar treatment of Lys60 in Lqq5 caused 40% loss of activity (Darbon *et al.*, 1983). Nitration of Tyr residues in AaH3 by tetranitromethane caused reduction of 50% in the activity when either Tyr14 or Tyr60 were affected. Nitration of both residues reduced the activity by 68%. When Tyr5, Tyr14 and Tyr60 were nitrated, the remaining activity was 10% (Kharrat *et al.*, 1989). Sulphenylation of Trp38 in AaH2 decreased the activity to 9% (Kharrat *et al.*, 1989). The same treatment of Trp45 in AaH3 did not affect the toxicity (Kharrat *et al.*, 1989). Proteolytic cleavage of the N-terminal pentapeptide of Bom3 of *Buthus occitanus mardochei*, reduced its activity to 10% (El Ayeb *et al.*, 1986a). All of these residues are located in or near a conserved hydrophobic region proposed by Fontecilla-Camps *et al.* (1982) to carry the 'toxic site'. A conclusion drawn out of these studies was that all important residues are clustered on one face of the $\alpha$ anti-mammal toxin, suggesting a multipoint interaction with residues of the mammalian sodium channel.

### 2.2 *Immunochemical studies*

An immunochemical approach using five antibody populations specific towards various regions of the toxin AaH2, was conducted to probe the interaction of the toxin with its receptor site. The results revealed two antigenic sites around the disulphide bridge 12–63 and one encompassing residues 50–59 whose blocking avoided binding to the receptor site (El Ayeb *et al.*, 1986b). Monoclonal antibodies that recognized an epitope which included Lys58, Val10 and His64, had similar effects. However, antibodies directed against stretches 19–28 and 28–39 did not affect the capability of the toxin to bind to its receptor site.

### 2.3 *Determination of three-dimensional structure*

The three-dimensional structures of the $\alpha$-anti mammal toxins AaH2 and AaH3 from *Androctonus australis* Hector (Fontecilla-Camps *et al.*, 1982, 1988; Mikou *et al.*, 1992), and BeM9, an $\alpha$-anti mammal toxin from *Buthus eupeus* (Pashkov *et al.*, 1988), have been determined by X-ray crystallography and $^{1}$H 2D NMR. These data confirmed the hypothesis that these toxins share a very similar carbon backbone structure. This fact is important when structure–activity relationship of such a toxin is being studied by using a comparative approach.

### 2.4 *Genetic approach*

Several cDNA clones encoding scorpion neurotoxins have been cloned and in some cases expressed either in COS cells (Bougis *et al.*, 1989) or insect cultured cells

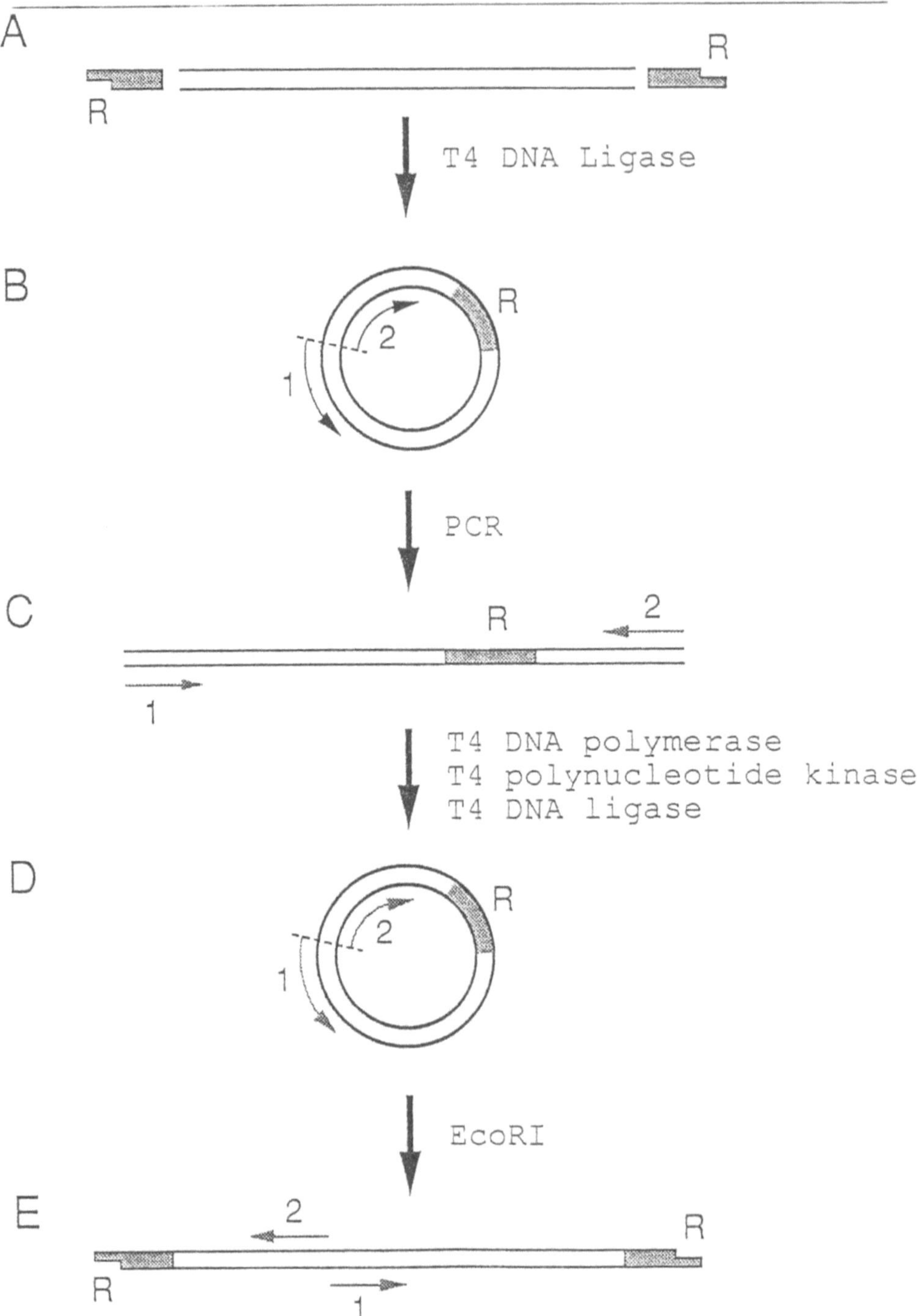


*Figure 2.* Isolation of cDNA clones. A, molecular adaptors containing unique restriction sites, are ligated to the cDNA mixture; B, circularisation of the cDNA by T4 DNA ligase followed by inverse-PCR (the 'back to back' primers are indicated); C, the PCR product (note the central position of the adaptor and former positions of the primers); D, the linear PCR product is circularised by T4 DNA ligase after 'fill-in' and 5′-phosphorylation; E, digestion with EcoRI generates a linear clonable PCR-product. R indicates EcoRI site. 1 and 2 designate primers.

using a baculovirus vector (Chejanovsky *et al.*, 1995; Maeda *et al.*, 1991; Stewart *et al.*, 1991). However, genetic modifications using these expression systems have not been reported. We have established an expression system that is most suitable for genetic alteration studies. We first isolated (Figure 2) and characterized several cDNAs encoding different alpha toxins from the Israeli yellow scorpion (*L. q. hebraeus*) (Gurevitz *et al.*, 1991; Zilberberg and Gurevitz, 1993). The cDNA clone of Lqh$\alpha$IT, a unique alpha toxin with a high ratio of anti-insect/mammal activity, was successfully expressed in *E. coli* by using the pET-llc translational vector bearing a strong T7 promoter (Figure 3). The toxin obtained in large insoluble amounts was completely denatured and then reoxidized under controlled conditions (Zilberberg *et al.*, 1996). The recombinant protein was purified by HPLC and analysed for its activity. Toxicity was estimated by injection to blowfly larvae and mice; electrophysiological properties were determined by current and voltage clamp experiments using a cockroach isolated axon, and binding to the receptor was performed with a neuronal membrane preparation from locust synaptosomes. The recombinant toxin, engineered with an additional methionine residue at its N-terminus, was found to be identical to the authentic toxin purified from the scorpion venom (Eitan *et al.*, 1990). This expression system was used for amino acid substitutions of specific C-terminal and N-terminal residues. Design of the modifications was based on comparative analyses with other alpha toxins of high anti-mammal/insect toxicity ratio (Lqq4 and AaH2; Figure 1). The substitutions of three residues on the C-terminal side had a minute effect on the toxicity to blowfly larvae and on the binding affinity to insect neuronal membranes. However, the ratio of anti-insect/mammal activity (toxicity and binding to the receptor) increased several-fold (Zilberberg *et al.*, 1996). Conversely, the substitution of three amino

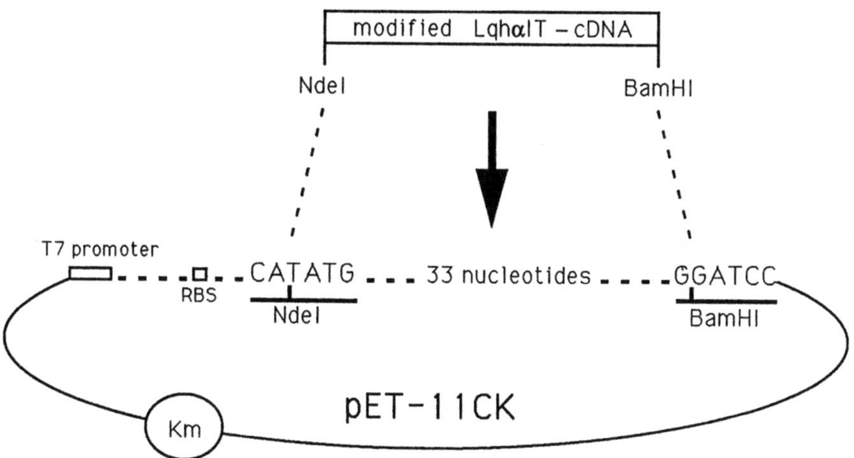


*Figure 3.* Construction of the expression vector. The engineered cDNA was inserted between the NdeI and BamHI sites in the vector pET-11c, under the control of the T7 promoter.

*Table 1.* Biological activity of variant genetically modified LqhαIT toxins.

| Toxins and modificants | $ED_{50}$ (ng 100 $mg^{-1}$ body weight) | $IC_{50}$ (nM) | $LD_{50}$ (ng $g^{-1}$ body weight) |
|---|---|---|---|
| LqhαIT-rec | 14 | 0.38 ± 0.1 | 250 |
| C-terminus | 20 | 0.44 ± 0.1 | 1 000 |
| N-terminus | 3000 | 2500 ± 1500 | 70 000 |
| Lqq4 | 950 | Unknown | 70 |

$ED_{50}$, Effective dose 50% determined on 100–150 mg blowfly larvae.
$IC_{50}$, Inhibition concentration 50% determined on locust synaptosomal membranes.
$LD_{50}$, Lethal dose 50% determined on 20 g ICR mice.
LqhαIT-rec stands for the nonmodified recombinant toxin.
C-terminus and N-terminus stand for the modified toxins.
Lqq4, purified toxin from the venom of *L. q. quinquestriatus*.

acid residues on the N-terminal side of the toxin, reduced the biological activity ($ED_{50}$ and $IC_{50}$) by several orders of magnitude (Table 1). Thus, the *E. coli* expression system followed by reconstitution of active toxins *in vitro* paved the way for a genetic study aiming to clarify the toxic site of scorpion alpha nuerotoxins and the molecular basis for the 'animal group specificity' phenomenon.

## 3. Baculovirus-mediated targeting of scorpion toxins

The nervous system of insects serves as a major target for conventional efforts of insect control which relies currently on hazardous chemicals of four groups: organochlorines, organophosphates, carbamates and pyrethroids (Elliot, 1986). It is our aim to develop new safer means for pest control by utilizing natural anti-insect agents. Insecticidal polypeptidic neurotoxins from scorpion venoms targeted to insect sodium channels of excitable tissues can be used for this purpose. However, in order to be efficiently utilized, they require a delivery system to reach their target sites. In this respect, an insect pathogen which naturally infects and spreads within the insect host and is amenable for genetic engineering, should be an ideal vector.

Baculoviruses, double stranded DNA viruses, infect insects and do not infect mammals (Huber, 1986). More than 500 baculovirus species have been isolated from various insect orders (Martignoni and Iwai, 1986; Matthews, 1982) and almost all serious lepidopterous pest species are susceptible to infection by one or more species of baculovirus. The baculoviruses are divided into three groups: the non-occluded viruses (NOVs) characterized by the absence of an occlusion body; the granulosis viruses (GVs) comprised of a single virion occluded in a polyhedral capsule; and the nuclear polyhedrosis viruses (NPVs) in which many virions are occluded in a polyhedral crystal. The polyhedra increases the stability of the encapsulated viruses, protects the viral particles from environmental effects, and its ingestion by insect larvae initiates the infectious cycle. The polyhedra dissolves in the alkaline conditions of the insect midgut and the released virions fuse with epithelial cells. The virions migrate through the cytoplasm, uncoat in the nucleus (Granados and Lawler, 1981), replicate and then bud through the plasma membrane. This process results in a secondary infection of various tissues of the larval host. It was recently shown that the major pathway for spread of the viral infection is through the tracheal system (Engelhard *et al.*, 1994). At a later stage of infection, occlusion bodies containing mature virions are produced in the infected cell nuclei. The continuous replication of the virus results in the death of the infected larvae. The cadavers full of polyhedra rupture easily and millions of polyhedra are released to the surrounding foliage and soil.

There are several advantages in utilizing baculoviruses as biopesticides:

- They infect only arthropods and are not pathogenic to vertebrates or plants (Carbonell and Miller, 1987; Huber, 1986; Miller, 1988; Miller *et al.*, 1983; Summers *et al.*, 1975).
- Most baculoviruses are specific to a single family or genus of insects (Entwistle and Evans, 1985).
- No undesirable residues are left in the environment after baculovirus use (Huber, 1986).
- They are used successfully with other control agents in integrated insect pest management (IPM) systems (Huber, 1986).
- Eukaryotic proteins expressed by recombinant baculoviruses in insects are properly processed posttranslationally (glycosylation, phosphorylation, amidation and internal cleavage of polypeptides), accumulate to high amounts and are biologically active.

Baculoviruses have been registered in the past as bioinsecticides despite a major disadvantage in their slow killing rates (6–10 days post-infection) of insect pest larvae. During this period of time, extensive damage to the crop may occur (Huber, 1986). This problem illustrates the need to engineer baculoviruses with foreign genes encoding insecticidal agents which impair normal functions in the insect development, metabolism and neural system. The availability of insect cell cultures, used for the engineering of recombinant baculoviruses (for a comprehensive review see Luckow and Summers, 1988), has motivated us and others (Chejanovsky *et al.*, 1995; Hammock *et al.*, 1990; Maeda *et al.*, 1991; Stewart *et al.*, 1991; Tomalski and Miller, 1991) to utilize baculoviruses as suitable vectors for the expression of various anti-insect components in insect pests.

### *3.1 Recombinant baculoviruses expressing polypeptides involved in insect development and metabolism*

Insect neuropeptides and hormones have been expressed by using baculoviral vectors (Bonning and Hammock, 1992; Hammock *et al.*, 1990; Maeda, 1989). However, only the *Bombyx mori* virus (BmNPV) expressing the diuretic hormone from *Manduca sexta* revealed an increased insecticidal effect on *Bombyx mori* larvae by 20% reduction in time to kill as compared with the wild-type virus (Bonning and

Hammock, 1992). The first insect enzyme expressed via a baculovirus vector was the juvenile hormone esterase (JHE), responsible for degradation of the juvenile hormone (Hammock *et al.*, 1990). The recombinant enzyme reduced the amount of diet consumed by first instar *Trichoplusia ni* but no effect was observed in the other instar stages. Two mutant forms of JHE, constructed for improved stability, were found to have an improved effect of reducing the time to kill by 30% when compared to the wild-type baculovirus (Bonning and Hammock, 1992). Another approach was undertaken by O'Reilly and Miller (1991) by deleting the gene encoding ecdysteroid UDP glucosyl transferase (EGT) from the genome of the baculovirus AcNPV (this enzyme blocks the activity of the moulting hormone ecdysone). Viruses lacking the gene coding for EGT reduced by 20% the time to kill *Spodoptera frugiperda* when compared to wild-type.

### *3.2 Recombinant baculoviruses expressing toxin genes*

The expression of a number of different toxins in *Spodoptera frugiperda* cells (SF9 and SF21 cell lines) has been obtained by using the baculovirus vector AcNPV. Among them, BeIT, an anti-insect toxin from the scorpion *Buthus eupeus* (Carbonell *et al.*, 1988), δ-endotoxin from the bacterium *Bacillus thuringiensis* (Merryweather *et al.*, 1990), TxP1 from the mite *Pyemotes tricini* (Tomalski and Miller, 1991) and AaIT from the scorpion *Androctonus australis* (Maeda *et al.*, 1991; McCutchen *et al.*, 1991; Stewart *et al.*, 1991). TxP1 and AaIT reduced the time to kill *Heliothis virescens* larvae by 30% as compared to the wild-type virus. Larvae infected by recombinant baculoviruses expressing the AaIT toxin fell off the plant 5–8 hours before death and were unable to feed or recover (Hoover *et al.*, 1995; Hoover *et al.*, in preparation; Table 2). This resulted in 60% reduction in damage to the plants as compared to larvae infected by the wild-type virus (Figure 4). This result demonstrates that lethal times (LTs) do not necessarily represent the ultimate

*Table 2.* Oral infectivity of wild-type and recombinant AcNPV-AaIT to *Heliothis virescens* larvae.

| Time post infection (h) | Mortality (%) | |
|---|---|---|
| | Wild-type | Recombinant |
| 24 | 0 | 0 |
| 48 | 0 | 0 |
| 72 | 0 | 26 |
| 96 | 7 | 72 |
| 120 | 20 | 94 |
| 144 | 50 | 100 |
| 168 | 72 | 100 |
| 192 | 90 | 100 |

Three replicates (24 larvae each) of third instar *Heliothis virescens* larvae were fed with virus contaminated diet (4000 PIB/plug). Mortality was recorded daily and the data were analysed using a probit analysis program (Russell *et al.*, 1977). $LT_{50}$ = 82 hours post-infection.

*Figure 4.* Reduction in feeding damage by *Heliothis virescens* larvae treated with recombinant and wild-type baculoviruses. Tomato plants were infected with third instar *H. virescens* larvae. Prior to the infection, the larvae were treated with wild-type (represented by the plant on the left) or recombinant AcNPV-AaIT (represented by the plant on the right).

parameter for evaluating repelling efficacy of pest larvae because feeding stops earlier and crop damage is then minimized.

We engineered a recombinant AcNPV with the LqhαIT-cDNA (Chejanovsky *et al.*, 1995) under the control of the potent polyhedrin promoter (Smith *et al.*, 1983) and obtained expression of the toxin in *Spodoptera frugiperda* culture cells. The functional LqhαIT produced was processed (leader peptide removed) and secreted by the cells. Toxicity of the recombinant toxin was demonstrated on blowfly (*Sarcophaga falculata*) and on lepidopterous (*Spodoptera littoralis*) larvae. The bioinsecticidal effect of the recombinant baculovirus was assayed on *Heliothis armigera* larvae. A significant reduction in the time of killing and in the mean weight gained by the *H. armigera* larvae was obtained when compared to wild-type viruses (Table 3).

## 4. Conclusions and future prospects

The potency and pharmacological specificity of the anti-insect neurotoxins prompt us to clarify their structure–function relationship and toxin–receptor interactions at the molecular level. The fact that the selective neurotoxin interacts with a well conserved membranal structure serving a defined neuronal function, enables its utilization as a specific marker of the above structure and a key for clarifying its function. As such, the anti-insect specific scorpion toxins and their modificants serve

*Table 3.* Oral infectivity of wild-type and recombinant AcNPV-LqhαIT to *Heliothis armigera* larvae.

| Time post infection (h) | Mortality (%) | | Weight gain (mg) | |
|---|---|---|---|---|
| | Wild-type | Recombinant | Wild-type | Recombinant |
| 24 | 0 | 0 | 9.5 | 7.8 |
| 48 | 0 | 0 | 13.1 | 12.0 |
| 72 | 0 | 4.5 | 63.2 | 36.1 |
| 96 | 11.4 | 72.7 | 125.0 | 43.0 |

Second instar larvae were fed with virus-contaminated diet (4000 PIB/plug). Mortality and weight gain were determined in population samples of 48 larvae per treatment.

as valuable tools in comparative neurochemistry, neuropharmacology, and developmental neurobiology. The genetic modificational approach may clarify structural elements involved in the 'toxic site' of some toxins which could be further utilized to design non-peptide agents active specifically against insect pests (Gurevitz and Zilberberg, 1994). The results of the chemical, immunochemical and genetic approaches in studying alpha scorpion toxins, can be interpreted according to the three-dimensional structure published for AaH2 and our three-dimensional model of LqhαIT. The 'conserved hydrophobic surface' (CHS), suggested previously to be involved in binding to the receptor site, is in fact located away from the putative binding surface. However, the outstanding impact of the substitutions at the N-terminal region of LqhαIT is probably due to a conformational change in the three-dimensional structure and does not necessarily imply their involvement in binding to the receptor site. Thus, the genetic system has become a powerful tool in helping to determine the structural motifs participating in binding and in conferring insecticidal activity.

The ability to design molecules capable of distinguishing between insect and mammalian nerves is of practical value. The industrial and agricultural significance of the insect-selective peptide neurotoxins can be appreciated on the background of recent advances in developing baculovirus vectors bearing scorpion toxin genes for plant protection (Chejanovsky *et al.*, 1995; Maeda *et al.*, 1991; Stewart *et al.*, 1991; Tomalski and Miller, 1991).

Several approaches can be envisaged to improve further the insecticidal properties of recombinant baculoviruses. The first is by purification of novel proteins with high insecticidal activity from natural venomous parasitoids or predators of lepidopterous larvae, or by generating recombinant toxins with increased potency via genetic alterations. The second is by improving the efficacy of recombinant baculoviruses through the utilization of various promoters, the expression of more than one toxin by each virus, and their combinations. Early expressing promoters would produce most likely smaller amounts of toxin; however, in our experiments we realized that minute amounts of toxin might already be sufficient for the desired effects. Furthermore, stabilization of RNA transcripts or their translated products by genetic means and the utilization of other leader sequences for better secretion, could improve the efficacy of the viruses. A third approach could use classical selection. Virus isolates from cultured cells that express higher amounts of proteins or replicate at increased rates, could be selected. The insecticidal properties of existing viruses expressing anti-insect selective neurotoxins can also be improved. Injecting different combinations of scorpion neurotoxins revealed a 5–10 fold increase in the toxicity of some combinations used. These results indicate that baculovirus-mediated simultaneous expression of two different toxins would produce an improved selective biopesticide (Herrmann *et al.*, 1995). Another improvement can be achieved by the combined application of toxins with classical insecticidal chemicals leading to a synergistic effect. The combination of a recombinant AcNPV expressing AaIT with very low doses of classical insecticides ($LC_{10}$) reduced the time for kill by 30–40% as compared to the recombinant baculoviruses alone (McCutchen *et al.*, 1996).

In summary, genetically engineered viral biopesticides have considerable potential for pest control in the future. Their efficacy as compared with classical chemicals, currently under massive use, depends on their efficiency, speed of kill and reduction in crop damage, safety (selectivity and biodegradativity) and on the environmental risks to other organisms (limited host range and transferability among viral species). Extensive studies are presently underway on aspects of safety. For instance, an ongoing project at the Oxford NERC laboratory is evaluating the effects of engineered viruses on non-target lepidoptera. Our laboratory is examining the interaction of the recombinant viruses with beneficial arthropods. Predators of Lepidoptera, the green lacewing *Chrysopa carnea* and the insidious flower bug *Orius insidious* were not adversely affected by feeding on larvae of *H. virescens* infected with recombinant viruses. In addition, a pollinator, the honey bee *Apis mellifera*, showed no significant differences in survival rates when injected with budded virions of wild-type or recombinant baculoviruses (Heinz *et al.*, 1995). More studies and examination of recombinant baculoviruses and their utilization are required to estimate their future potential in replacing hazardous existing insecticides and prior to their registration and commercial application.

## Acknowledgements

This project is being supported by US-Israel Binational Agricultural Research & Development (BARD) grants IS-1982-91 (M.G., E.Z.), IS-2486-94C (M.G.), IS-2139-92 (N.C., B.D.H.), and a USDA grant 91-37302-6185 (B.D.H.). N.Z. was supported by a Levi Eshcol fellowship from the Israel Ministry of Science. R.H. was supported by a fellowship from Crop Genetics International, Columbia, MD, USA. H.M. was supported by a BARD postdoctoral fellowship award FI-0157-93.

## References

Barchi R L 1987 Sodium channel diversity: Subtle variations on a complex theme. TINS 10, 221–223.

Bonning B C and Hammock B D 1992 Development and potential of genetically engineered viral insecticides. Biotec. Genet. Engineer Rev. 10, 453–487.

Bougis P E, Rochat H and Smith L A 1989 Precursors of *Androctonus australis* scorpion neurotoxins: Structures of precursors, processing outcomes, and expression of functional recombinant toxin II. J. Biol. Chem. 264, 19259–19265.

Carbonell L F and Miller L K 1987 Baculovirus interaction with non-target organisms: A virus-borne reporter gene is not expressed in two mammalian cell lines. Appl. Environ. Microbiol. 53, 1412–1417.

Carbonell L F, Hodge M R, Tomalski M D and Miller L K 1988 Synthesis of a gene coding for an insect specific scorpion neurotoxin and attempts to express it using baculovirus vectors. Gene 73, 409–418.

Catterral W A 1977 Membrane potential dependent binding of scorpion toxin to the action potential sodium ionophore. Studies with a toxin derivative prepared by lactoperoxidase catalyzed iodination. J. Biol. Chem. 252, 8660–8668.

Catterall W A 1980 Neurotoxins that act on voltage-sensitive sodium channels in excitable membranes. Annu. Rev. Pharmacol. Toxicol. 20, 15–43.

Catterall W A 1986 Molecular properties of voltage-sensitive sodium channels. Annu. Rev. Biochem. 55, 953–985.

Chejanovsky N, Zilberberg N, Rivkin H, Zlotkin E and Gurevitz M 1995 Functional expression of an alpha anti-insect scorpion neurotoxin in insect cells and lepidopterous larvae. FEBS Lett. 376, 181–184.

Couraud F, Rochat H and Lissitzky S 1978 Binding of scorpion and sea anemone neurotoxins to a common site related to the action potential $Na^+$ ionophore in neuroblastoma cells. Biochem. Biophys. Res. Commun. 83, 1525–1530.

Darbon H, Jover E, Couraud F and Rochat H 1983 $\alpha$-Scorpion neurotoxin derivatives suitable as potential markers of sodium channels. Int. J. Pept. Protein Res. 22, 179–186.

Eitan M, Fowler E, Herrmann R, Duval A, Pelhate M and Zlotkin E 1990 A scorpion venom neurotoxin paralytic to insects that affects sodium current inactivation: Purification, primary structure, and mode of action. Biochemistry 29, 5941–5947.

El Ayeb M, Darbon H, Bahraoui EM, Vargas O and Rochat H 1986a Differential effect of defined chemical modifications of antigenic and pharmacological activities of scorpion $\alpha$ and $\beta$ toxins. Eur. J. Biochem. 155, 289–294.

El Ayeb M, Bahraoui EM, Granier C and Rochat H (1986b) Use of antibodies specific to define regions of scorpion $\alpha$-toxin to study its interaction with its receptor site on the sodium channels. Biochemistry 25, 6671–6678.

Elliot M 1986 In Neuropharmacology and Pesticide Action. Eds M G Ford, G G Lunt, R C Reay and P N R Usherwood. pp 498–500. Ellis Horwood, UK.

Endean R and Rudkin C 1963 Studies of the venoms of some Conidae. Toxicon 1, 49–64.

Engelhard E K, Kam-Morgan L N W, Washburn O J and Volkman L E 1994 The insect tracheal system: A conduit for the systemic spread of *Autographa californica* M nuclear polyhedrosis virus. Proc. Natl. Acad. Sci. USA 91, 3224–3227.

Entwistle P F and Evans H F 1985 Viral control. In Comprehensive Insect Physiology, Biochemistry and Pharmacology. Eds L I Gilbert and G A Kerkut. Vol. 12, pp 347–412. Pergamon Press, Oxford.

Fontecilla-Camps J C, Almassy R J, Suddath F L and Bugg C E 1982 The three-dimensional structure of scorpion neurotoxins. Toxicon 20, 1–7.

Fontecilla-Camps J C, Habersetzer-Rochat C and Rochat H 1988 Orthorhombic crystals and three-dimensional structure of the potent toxin II from the scorpion *Androctonus australis* Hector. Proc. Natl. Acad. Sci. USA 85, 7443–7447.

Frontali N and Grasso A 1964 Separation of three toxicologically different protein components from the spider *Latrodectus tredecinquttatus*. Arch. Biochem. Biophys. 106, 213–218.

Gordon D 1990 Ion channels in nerve and muscle cells. Curr. Opin. Cell Biol. 2, 695–707.

Gordon D and Zlotkin E 1993 Binding of an $\alpha$ scorpion toxin to insect sodium channels is not dependent on membrane potential. FEBS Lett. 315, 125–128.

Gordon D, Moskowitz H, Eitan M, Warner C, Catterall W A and Zlotkin E 1992 Localization of receptor sites for insect selective toxins on sodium channels by site directed antibodies. Biochemistry 31, 7622–7628.

Granados R R and Lawler K A 1981 In vivo pathway of *Autographa californica* baculovirus invasion and infection. Virology 108, 297–308.

Gurevitz M, Urbach D, Zlotkin E and Zilberberg N 1991 Nucleotide sequence and structure analysis of a cDNA encoding an alpha insect toxin from the scorpion *Leiurus quinquestriatus hebraeus*. Toxicon 29, 1270–1272.

Gurevitz M and Zilberberg N 1994 Advances in molecular genetics of scorpion neurotoxins. J. Toxicol. Toxin Rev. 13, 65–100.

Hammock B D, Bonning B C, Possee R D, Hanzlik T N and Maeda S 1990 Expression and effects of juvenile hormone esterase in a baculovirus vector. Nature 344, 458–461.

Heinz K M, McCutchen B F, Herrmann R, Parrella M P and Hammock B D 1995 Direct effects of recombinant nuclear polyhedrosis viruses on selected non-target organisms. J. Econ. Entomol. 88, 259–264.

Herrmann R, Moskowitz H, Zlotkin E and Hammock B D 1995 Positive cooperativity among insecticidal scorpion neurotoxins. Toxicon 33, 1099–1102.

Hoover K, Schultz C M, Lane S S *et al.*. 1995 Reduction in damage to cotton plants by a recombinant baculovirus that knocks moribund larvae of *Heliothis virescens* off the plant. Biol. Control 5, 419–426.

Huber J 1986 Use of baculoviruses in pest management programs. In The Biology of Baculoviruses. Eds R R Granados and B A Federici. Vol. II, pp 181–202. CRC Press, Boca Raton, FL.

Kharrat R, Darbon H, Rochat H and Granier C 1989 Structure/activity relationships of scorpion $\alpha$-toxins. Eur. J. Biochem. 181, 381–390.

Luckow V A and Summers M D 1988 Trends in the development of baculovirus expression vectors. Bio/Technology 6, 47–55.

Maeda S 1989 Increased insecticidal effect by recombinant baculovirus carrying a synthetic diuretic hormone gene. Biochem. Biophys. Res. Commun. 165, 1177–1183.

Maeda S, Volrath S L, Hanzlik T N *et al.* 1991 Insecticidal effects of an insect-specific neurotoxin expressed by a recombinant baculovirus. Virology 164, 777–780.

Martignoni M E and Iwai P J 1986 Catalog of Viral Diseases of Insects, Mite and Ticks. USDA Forest Service PNW-195. USGPO, Washington, DC.

Martin-Eauclaire M-F, Sogaard M, Ramos C, Cestele S, Bougis P E and Svensson B 1994 Production of active, insect-specific scorpion neurotoxin in yeast. Eur. J. Biochem. 223, 637–645.

Matthews R E F 1982 Classification and nomenclature of viruses. Fourth Report of the International Committee on Taxonomy of Viruses. Intervirology 17, 1–199.

McCutchen B F, Choudary P V, Crenshaw R *et al.* 1991 Development of a recombinant baculovirus expressing an insect-selective neurotoxin: Potential for pest control. Bio/Technology 9, 848–852.

McCutchen B F, Betana M D, Hoover R, Herrmann R and Hammock B D 1996 Interactions of recombinant and wild-type baculoviruses with classical insecticides and pyrethroid-resistant tobacco budworm (Lepidoptera, Noctuidae). J. Econ. Entomol. (in press)

Merryweather A T, Weyer U, Harris M P G, Hirs M, Booth T and Possee R D 1990 Construction of genetically engineered baculovirus insecticides containing the *Bacillus thuringiensis* ssp. Kurstaki HD-73 delta endotoxin. J. Gen. Virol. 71, 1535–1544.

Mikou A, Laplante S R, Guittet E, Lallemand J Y, Martin-Eauclaire M F and Rochat H 1992 Toxin III of the scorpion *Androctonus australis* Hector: Proton nuclear magnetic resonance assignments and secondary structure. J. Biomolec. NMR 2, 57–70.

Miller L K, Lingg A J and Bulla L A 1983 Bacterial, viral and fungal insecticides. Science 219, 715–721.

Miller L K 1988 Baculoviruses as gene expression vectors. Annu. Rev. Microbiol. 42, 177–199.

Moskowitz H, Herrmann R, Zlotkin E and Gordon D 1994 Variability among insect sodium channels revealed by selective neurotoxins. Insect Biochem. Molec. Biol. 24, 13–19.

O'Reilly D R and Miller L K 1991 Improvement of a baculovirus pesticide by deletion of the EGT gene. Bio/Technology 9, 1086–1089.

Pashkov V S, Maiorov V N, Bystrov V F, Hoang A N, Volkova T M and Grishin E V 1988 Solution spatial structure of 'long' neurotoxin $M_9$ from the scorpion *Buthus eupeus* by $^1$H-NMR spectroscopy. Biophys. Chem. 31, 121–133.

Pelhate M and Zlotkin E 1981 Voltage-dependent slowing of the turn off of $Na^+$ current in the cockroach giant axon induced by the scorpion venom 'insect toxin'. J. Physiol. 319, 30–31.

Piek T and Spanjer W 1986 In Venoms of the Hymenoptera. Ed. T Piek. pp 161–308. Academic Press, London.

Ray R and Catterral W A 1978 Membrane potential dependent binding of scorpion toxin to the action potential sodium ionophore. Studies with a 3(4-hydroxy 3-[$^{125}$]iodophenyl) propionyl derivative. J. Neurochem. 31, 397–407.

Rochat H, Bernard P and Couraud F 1979 Scorpion toxins: Chemistry and mode of action. In Advances in Cytopharmacology. Eds B Ceccarelli and F Clementi. Vol. 3, pp 325–334. Raven Press, New York.

Russell R M, Robertson J L and Savin N E 1977 POLO: A new computer program for probit analysis. Bull. Entomol. Soc. Am. 23, 209–213.

Smith G E, Summers M D and Fraser M J 1983 Production of human beta interferon in insect cells infected with a baculovirus expression vector. Mol. Cell. Biol. 3, 2156–2165.

Stewart L M D, Hirst M, Ferber M L, Merryweather A T, Cayley P J and Possee R D 1991 Construction of an improved baculovirus insecticide containing an insect-specific toxin gene. Nature 352, 85–88.

Summers M D, Engler R, Falcon L A and Vail P 1975 Baculoviruses for Insect Pest Control: Safety Considerations. American Society for Microbiology, Washington, DC.

Tomalski M D and Miller L K 1991 Insect paralysis by baculovirus-mediated expression of a mite neurotoxin gene. Nature 352, 82–85.

Vargas O, Martin M F and Rochat H 1987 Characterization of six toxins from the venom of the Moroccan scorpion *Buthus occitanus mardochei*. Eur. J. Biochem. 162, 589–599.

Zilberberg N and Gurevitz M 1993 Rapid isolation of full length cDNA clones by 'inverse PCR'. Purification of a scorpion cDNA family encoding α-neurotoxins. Anal. Biochem. 209, 203–205.

Zilberberg N, Gordon D, Pelhate M *et al.* 1996 Functional expression and genetic modification of an alpha scorpion neurotoxin. Biochemistry 35, 10215–10222.

Zlotkin E 1983 Insect selective toxins derived from scorpion venoms: An approach to insect neuropharmacology. Insect Biochem. 13, 219–236.

Zlotkin E 1985 In Comprehensive Insect Physiology, Biochemistry and Pharmacology. Eds G A Kerkut and L I Gilbert. Vol. 10, pp 499–546. Pergamon, Oxford.

Zlotkin E 1988 In Comparative Invertebrate Neurochemistry. Eds G G Lunt and R Wolsen. pp 256–324. Croom Helm, London.

Zlotkin E, Rochat H, Kupeyan C, Miranda F and Lissitzky S 1971 Purification and properties of the insect toxin from the venom of the scorpion *Androctonus australis* Hector. Biochimie (Paris) 53, 1073–1078.

Zlotkin E, Eitan M, Bindokas V P *et al.* 1991 Functional duality and structural uniqueness of depressant insect-selective neurotoxins. Biochemistry 30, 4814–4821.

Zlotkin E, Gurevitz M, Fowler E and Adams M E 1993 Depressant insect selective neurotoxins from scorpion venom: Chemistry, action and gene cloning. Arch. Insect Biochem. Physiol. 22, 55–73.

8

# Nitrogen fixing cyanobacteria as BTi toxin genes delivery system – a biotechnological approach to control malaria mosquitoes

SAMMY BOUSSIBA, WU XIAOQIANG AND ARIEH ZARITSKY

## 1. Introduction

Control and prevention of epidemic diseases has been consistently one of the major subjects of public concern. The most efficient way to reach this goal is to control effectively the spreading of disease transmitting vectors such as mosquitoes. With the advance of modern science and technology, the impact of epidemic diseases on civil life has been more or less alleviated by the heavy use of pesticides and related chemicals. Their negative effects on environmental and related ecological problems have gradually been causing public awareness. The wide application of toxic chemical pesticides has become a major source of environmental pollution. Therefore some more powerful and more specific alternative measures must be developed for the same purpose, which will be more safe and friendly to human life and the surrounding environment. Biological control of disease transmitting vectors is generally considered to be permanent, inexpensive and free of environmental hazards. One of such innovative approaches will be discussed in this chapter, i.e., use of genetically engineered cyanobacteria as a BTi toxin to gene delivery system for mosquito control of vector borne diseases.

The harmful and beneficial insect species comprise only a few thousand of the millions of species (80% of the animal kingdom). A vast group of insects is classified as neutral, but are essential components of the world's ecosystems, and thereby, contribute to man's long-term well-being. Use of genetically engineered cyanobacteria carrying BTi toxin genes will be very specific for mosquito larvae control and cause no harm to other organisms. Therefore, it is environmentally friendly.

## 2. Environmental impact of chemical insecticides for mosquito control

Mosquitoes and black flies are a constant threat to public health and comfort. Current strategies developed for control of these epidemic vectors through decades

*D. Rosen* et al. *(eds): Modern Agriculture and the Environment, 97–109.*

of struggle by scientists from multidisciplines include pesticides and biological control.

### *2.1 Pesticides*

Pesticides decrease the incidence of mosquito larvae only temporarily, but cause resurgence of resistant strains and have adverse effects on the natural predators and competitors.

The total amount of pesticides being used is decreasing, largely because the newer compounds are often ten times more potent than the ones being replaced. New agrochemicals generally require more steps for their synthesis than in the past and there is an expanding use of fluorine in optimizing structures for activities. Commercial pesticides now comprise 50 000 chemicals; it takes about 15 years and $100 million to develop a new agrochemical.

Biopesticide discovery, development and management depend on understanding of the biochemical targets, secondary effects and their toxicological consequences. There is an increasing number and diversity of new biochemical targets being discovered in natural product and molecular design programmes and, of course, by serendipity. The push for genetic-engineered biopesticides is moving slower than predicted because of the relatively narrow pest species range and potential selection for resistance. Newer insecticides disrupt the acetylcholine and $\gamma$-aminobutyric acid receptors and a number of targets within the sodium channel and the respiratory system. The targets of developmental herbicides include protoporphyrinogen IX oxidase, phytoene desaturase, and P-hydroxyphenyl pyruvate dioxygenase while those of newer fungicides include polymerase II.

Even though newer pesticides may be more efficient for pest control, the residue effect which causes environmental contamination and lack of specificity of toxicity – which can be more or less harmful to the neutral or beneficial organisms – consequently often disturbs the finely balanced ecosystem. The rapid resurgence of pesticide resistant pests or pathogens will devalue the efficiency of even the most efficient chemical pesticide. In the future, with the rapid increase of living standards, preservation of the green environment will be of utmost importance. The wide application of chemical pesticides and the choice of the category of pesticides, will probably soon be restricted.

### *2.2 Biological control*

Biological control uses predators, competitors, pathogens and parasites to reduce the density of target organisms while conserving species diversity. Fish, nematodes, microsporidians, mermithids, as well as viruses, bacteria and fungi are all used to control mosquitoes and blackflies (Lacey and Undeen, 1986). With this approach, ecological balance is finely sustained; biological control of pests is therefore safer to human life than use of chemical pesticides.

### 3. ***Bacillus thuringiensis* subsp. *israelensis*** *(BTi) as a biopesticide against mosquitoes*

One of the main health problems in tropical countries, in which rice is one of the major agricultural crops, is diseases transmitted by mosquitoes – which proliferate in the rice field habitats – such as malaria. Such tropical diseases devastate the economy of less developed countries by debilitating the farmers, thus reducing dramatically their production capacity. The most common treatment for this situation is the use of chemical insecticides to overcome the parasites' invasion. However the emergence and spread of insecticide resistance in many species of vectors, the concern with environmental pollution, and the high cost of the new chemical insecticides, make it apparent that vector control can no longer be safely dependent upon the use of chemicals.

A more sophisticated technology uses the bacterium, *Bacillus thuringiensis* subsp. *israelensis* (BTi), as a biocontrol agent. This bacterium, discovered in the Negev years ago, produces crystals ($\delta$-endotoxin) during sporulation with highly specific toxicity against mosquito larvae.

Over the past decade, the discovery of the unique capacity of the bacterium *Bacillus thuringiensis* subsp. *israelensis*, serotype H-14 (BTi) to synthesize insecticidal protein crystals has spurred intensified research activity into the molecular and genetic make-up of this organism. This has led to a comprehensive understanding of the toxin gene families, its expression and regulation, and the structure and function of the toxin molecule itself. As a result, which in turn has improved efficiency of biocontrol formulations, knowledge of the ecology of the bacteria has been gained and more effective toxins and delivery systems have been constructed. BTi is now the most valuable agent for controlling mosquito and blackfly larvae while causing no harm to other organisms including predators of the vectors (de Barjac and Sutherland, 1990; Entwistle *et al.*, 1993). The high larvicidal activity of BTi is specific to water-dwelling dipteran species. The origin of the toxicity resides in $\delta$-endotoxin contained in parasporal crystals produced during sporulation; it is composed of five major polypeptides (Hofte and Whiteley, 1989) differing in their immunogenicity (Ibarra and Federici, 1986) and mosquito larvicidal potential. Synergism seems to exist between combinations of the polypeptides (Chang *et al.*, 1993; Ibarra and Federici, 1986; Yoshida *et al.*, 1992; Yoshisue *et al.*, 1993; Poncet *et al.*, 1995).

The literature of the last decade is full of descriptions of attempts to clone various genes from *B. thuringiensis* (BT) and *B. sphaericus* into other organisms to raise their field efficacies in controlling respective target insects and to study these genes and their products (de Barjac and Sutherland, 1990; Entwistle *et al.*, 1993). The most exciting, successful examples are the engineering of plants to express BT toxins. It is now commonly accepted that these genes can be expressed in other organisms. In some cases they can reach a high expression level inside cyanobacteria simply with their original promoters (Xudong *et al.*, 1993).

## 4. Potential of cyanobacteria as a BTi toxin delivery system

Classical biological control, i.e. the importation and establishment of exotic natural enemies, is by far the most successful and promising way to replace the role of chemical pesticides for vector control. However some possible detrimental effects of imported natural enemies may cause environmental concerns and lead to serious legislative restrictions on classical biological control projects. The introduction of the engineered indigenous strain of cyanobacteria will avoid this problem of disturbing the fine ecological balance with minimum effects to the surrounding environment.

The use of BTi is limited by economic constraints, due to low persistence under field conditions, in which it does not reproduce (de Barjac and Sutherland, 1990; Entwistle *et al.*, 1993). The major reasons for the low efficacy of current BTi preparation are:

- Consumption by other organisms.
- Sinking to the bottom of the water body (Ohana *et al.*, 1987).
- Absorption to silt particles and organic matter.
- Inactivation by sunlight.

A great deal of effort is being expended around the world to solve these problems with attempts to isolate new mosquitocidal strains or species, to produce formulations of BTi with enhanced persistence in natural habitats of dipteran larvae, as well as to clone the genes responsible for toxicity into organisms which will express them co-ordinately and be ingested by the target organisms. Cyanobacteria which are abundant in diverse environments are suitable candidates for this purpose. The scientific background for this approach is based on the finding that the abundance of these algae species in water bodies, particularly in rice fields, and their ability to float in the upper layer and to resist different environmental conditions during the entire rice growth cycle, are unique features which make them good candidates for carrying the genes for toxicity. Shuttle vectors harbouring cyanobacterial replicons have been constructed and used for gene transfer (Elhai and Wolk, 1988). Their transferability is used for the production of BTi's $\delta$-endotoxin inside the cells of cyanobacteria.

In the past few years, some progress has already been achieved (Angsuthana-sombat and Panyim, 1989; Chungjatupornchai, 1990; De Marsac *et al.*, 1987; Murphy and Stevens, 1992; Xudong *et al.*, 1993). In 1987, pioneer work was done by a French group (De Marsac *et al.*, 1987). They cloned a 3.6 kb HindIII DNA fragment from *B. sphaericus* 1593M and expressed it in *E. coli* and *Bacillus subtilis* using pHV33 as shuttle vector, and in the cyanobacterium *Anacystis nidulans* R2 with PUC303 as shuttle vector. The test of larvicidal toxin activity against *Culex* mosquito larvae of these transformed organisms proved that:

- The cloned DNA fragment contained a larvicidal toxin gene(s).
- The gene with its original promoter can be expressed in other prokaryotics *B. subtilis*, *E. coli* and unicelluar cyanobacteria *A. nidulans* R2, with a similar level of toxin activity (100% mortality/mg of protein per ml in 48 h).

This work has opened new perspectives for genetically manipulated cyanobacteria as potential biocontrol agents with amplified persistence in mosquito habitats.

Two years later, another result demonstrating the potential use of transgenic cyanobacteria for biological control of vectors was published (Angsuthanasombat and Panyim, 1989). In this work, a gene coding for 130 kDa mosquito larvicide isolated from BTi was subcloned from pMU388 into shuttle vector pAQE19LPC Sal and introduced into cyanobacteria *Agmenellum quadruplicatum* PR-6 by plasmid transformation. Test of the larvicidal activity against *Aedes aegypti* showed that the toxic level of transformed *E. coli* with plasmid pMU388 containing the toxin gene with its original promoter is high enough to be comparable with the previous result reported by the French group. When the toxin gene is put under the control of cyanophycocyanin (CPC) promoter in construct of pAqPc388i, the expressed larvicidal activity of transformed *E. coli* with this recombinant plasmid is reduced to only 20% at 24 h and 60% at 48 h of those transformed cells with pMU388, and only 45% mortality of that pMU388 expressed in *E. coli* at 48 h when the shuttle vector expressed in the transformed cyanobacteria *A. quadruplicatum* PR-6. The explanation for this lower level of larval toxicity in this cyanobacterium is that the biosynthesis of 130 kDa polypeptide from pAqPc388i plasmid in *E. coli* was much stronger than in its *A. quadruplicatum* PR-6 counterpart as a result of relatively low copy number of this plasmid in cyanobacteria. The greater degradation of this gene product in *A. quadruplicatum* PR-6 was another apparent reason shown by Western hybridization. The toxin gene cloned in pMU388 was later designated as *cryIVB* coding for a polypeptide with an apparent MW of 126 kDa. This report indicated that greater expression of undegraded $\delta$-endotoxin is necessary for genetically engineering a local cyanobacterium found in mosquito habitats for effective control of tropical disease vectors.

Recently, the gene bt8 from BTi encoding the 128 kDa polypeptide (i.e. 126 kDa) was placed under transcriptional control of $P_{psbA}$, which is derived from higher plant chloroplast and has been patented (Vaeck *et al.*, 1988), and the construct was directed to a specific chromosomal site of *Synechocystis* PCC6803 (Chungjatupornchai, 1990). Total protein from the recombinant organism was toxic to mosquito larvae comparable to that expressed in *E. coli* with an $LC_{50}$ of mean value 1.3 mg $ml^{-1}$. Both Western blot analysis and ELISA revealed that the amount of 126 kDa protein was about 0.2% of total soluble protein in transformed *Synechocystis* PCC6803 cells.

The above mentioned achievements certainly gave us positive data supporting the idea that expression of the BTi $\delta$-endotoxin genes in cyanobacteria, which proliferate near the water surface, may provide an alternative for prolonged delivery of BTi $\delta$-endotoxin to feed mosquito larvae in the upper layer of aquatic habitats. There were however still at least two technical problems that needed to be solved. One is that the intact cells of unicellular cyanobacteria transformed with BTi $\delta$-endotoxin gene did not show much toxicity to vector larvae except when cell extracts or sonicated broken cells were used for toxicity assays. The other problem is that the transformed cells containing only a single toxin gene did not show a satisfactory level of larvicidal activity even under the direct control of the strong promoter $P_{psbA}$. The intracellular concentration of those expressed endotoxins was too low to be effective *in vivo*. The high toxicity of intact living cells is very important for practical application. Most recently some improvements have been reported by Murphy and Stevens (1992). The *cryIVD* gene, coding for 72 kDa protein, a major component of the BTi crystal complex highly toxic to mosquito larvae, was introduced into and expressed in *A. quadruplicatum* PR6. The transformation is based on a derivative of shuttle vector pAQE19 Sal in which the *cryIVD* gene was translationally fused to the highly expressed PR-6 cpcB gene promoter (Murphy and Stevens, 1992). Immunoblot analysis indicated that the cpcB-*cryIVD* gene fusion is expressed in the cyanobacterial cells which were proved to be readily ingested by the freshly hatched *Culex pipiens* mosquito larvae and were toxic to them. This is the first instance of a genetically engineered cyanobacterium that is lethal to mosquito larvae feeding directly on living cyanobacterial cells. So far only one laboratory model species of filamentous nitrogen fixing cyanobacterium *Anabaena* 7120 has been used and also shown certain higher capability for mosquito control after transformation with larvicidal endotoxin genes from *B. sphaericus* 2297 (Xudong *et al.*, 1993). Attempts to introduce the laboratory strain with foreign genes into the field were however unsuccessful because of their low adaptive values and quick disappearance, due to competition with indigenous species. No attempts have been performed yet to clone the BTi genes coding for the mosquito larvicidal polypeptides into a filamentous nitrogen-fixing cyanobacterial species prevailing in rice fields, which is ingested by mosquito larvae.

This latter approach is currently being undertaken by our group. It is the main goal of our study to reintroduce an indigenous strain into its natural habitat. The rice field isolate cyanobacterium *Anabaena siamensis* may thus be considered as an appropriate shuttle organism to carry the BTi endotoxin genes. At the present time, various BTi toxin genes have been cloned in different combinations (BTi toxon) and the most efficient operon has been screened out in *E. coli* and proved to be highly toxic to mosquito larvae (Ben-Dov *et al.*, 1995). These BTi toxons mounted on shuttle vectors have been introduced into *Anabaena* 7120 by conjugational mating, and those exconjugant cells of *Anabaena* 7120 carrying BTi toxons show certain toxic activities (Wu *et al.*, 1995a).

## 5. Potential of *Anabaena siamensis* as a nitrogen biofertiliser and vector for mosquito control

Rice is one of the most important crops in developing countries. About two-thirds of the 143 million hectares of the world's riceland are used as wetland cultures: water from rainfall and irrigation sources is conserved to maintain shallow flooded conditions for almost the entire growth duration of the crop. Such a water body creates ecological conditions favourable to the propagation of vectors of diseases such as malaria, schistosomiasis, and Japanese encephalitis. The vectors (mosquitoes and the blackfly) require an aquatic environment at certain stages of their life cycle. Numerous studies have shown the association of wetland rice culture with mosquito-borne diseases in various tropical areas. Most of the world's riceland farmers in tropical countries, east Asia and Latin America, suffer from a high incidence of diseases carried by mosquito larvae, which breed in water, as well as from lack of adequate nitrogen fertilizers for the rice plants at reasonable cost (Boussiba, 1988, 1991, 1993; Querijero-Palacpac *et al.*, 1990). With application of a bifunctional transgenic cyanobacterial strain which will be released into the rice field for control of mosquito larvae and for excretion of nitrogen-fixed ammonium as a fertilizer, while prevailing in rice fields longer than the commonly used laboratory strains, the life quality of the population in those areas will be improved. The idea has been realized by introducing the genes involved in the specific mosquitocidal $\delta$-endotoxin formation from BTi into ammonium releasing mutants which we have derived from a rapidly growing rice field indigenous strain of *Anabaena siamensis,* either by chemical mutagenesis or by recombinant techniques (Wu *et al.*, 1995b).

This dinitrogen fixing cyanobacterial strain is widespread in the rice fields of east Asian countries where it is already marketed as a substitute for nitrogenous fertilizers. *A. siamensis* is first isolated from a rice paddy, Kaserstat University agricultural farm (Bangkok, Thailand) (Antarikanonda, 1984). Based on its characteristic terminal heterocysts and mode of differentiation (by symmetrical division of the vegetative cell in the middle of the trichome), *A. siamensis* is named as a new species in the family Nostocaceae. We have isolated a mutant (SS1) from this species which releases ammonium continuously during active growth (Thomas *et al.*, 1990).

With the advance of recombinant DNA technology, genes can be freely transferred from one species to another, ignoring the natural species barriers (Sambrook *et al.*, 1989). Some cases are problematic, when the gene products are deleterious to the recipient organisms, but these problems can be overcome by certain sophisticated cloning techniques (Studier and Moffatt, 1986) and relying on basic knowledge in classical genetics and physiology of the recipients.

Some basic questions need to be addressed before embarking on this study:

- Is this filamentous microorganism accessible to mosquito larvae?
- Is it accessible to foreign DNA by transformation?
- What problems are associated with its production in large scales?
- Release of the recombinant DNA into the environment.

Our recent work showed that:

- The size of this algal filament (about 20–40 $\mu$m long) is suitable for ingestion by mosquito larvae. The filaments are readily ingested by all developmental stages of *A. aegypti* larvae. Following passage through the larval digestive system the vegetative cells (but not the heterocysts) were disrupted, as required for access of intracellular synthesized polypeptides. Previous work has also shown that even longer filaments of *Anabaena* 7120 can be easily ingested and digested by vector larvae.

- Methods for transformation of model species *Anabaena* 7120 are well developed (Elhai and Wolk, 1988). For applied purposes, however, field indigenous strains such as *A. siamensis* are much preferred. To develop a transformation system for a new *Anabaena* species, the standard protocols should be modified keeping in mind the following aspects: Does it harbour indigenous restriction enzymes?; What is the effect of its different morphology and physiology?; What is the final goal of research?

In work on molecular biology of filamentous cyanobacteria, transformation is often problematic except for *Anabaena* 7120 and a few other species. Cyanobacterial microbiologists were handicapped due to the lack of a reproducible gene-transfer system until Wolk *et al.* (1984), for the first time, reported successful gene transfer in *Anabaena* using PRL series of shuttle vectors, broad host range conjugal plasmid RP4 and helper plasmid pGJ28 or pDS4101 (a derivative of which, pRL528, carrying with AvaI, AvaII methylase genes, was constructed later and is used currently). It has opened the way for gene manipulation studies in other biotechnologically important filamentous cyanobacteria, providing success for surety of conjugal contact, and avoiding degradation of transferred DNA from the host's (recipient) restriction enzyme digestion; the transferred gene (plasmid) is capable of replication or integration into the recipient cells. The following is a general overview on how to transform unknown strains of filamentous cyanobacteria.

In most cyanobacteria, the problem is that the introduced DNA gets chopped to pieces or otherwise remains useless while conjugation using IncP plasmids such as RP4 can probably get DNA into any Gram negative bacteria as far as we know. A survey of the ability to get DNA into heterotrophic *Nostoc* by conjugation of a shuttle vector produced rather unsatisfactory results: some strains accepted the plasmid, but in no case was conjugal transfer very efficient (Flores and Wolk, 1985). Cyanobacteria in general are loaded with restriction enzymes, up to five different activities per strain. Data have shown that restriction diminishes conjugal efficiency by a factor of 5 to 50 per restriction site (Thiel and Poo, 1989; Wolk *et al.*, 1984). There are at least two general ways to protect the recombinant plasmids from restriction:

- *In vitro* methylation: DNA may be protected against restriction prior to electroporation by incubating it in extracts of the target strain under such conditions that methylases are active but restriction enzymes are not (Swanson *et al.*, 1992).
- Intraspecific DNA transfer: DNA isolated from the target organism is already fully methylated against its own restriction enzymes. Therefore, DNA transfer from one strain to another within a species is unaffected by restriction.

For a successful transformation, persistent propagation of introduced foreign gene (or plasmid) is also necessary, which can be realized by self-replicative shuttle vectors or integrative suicide vectors.

- Shuttle vectors: A few shuttle vectors have been constructed from plasmids isolated from cyanobacteria, and their host range appears to extend at least to members within the genus of the original strain. Furthermore, RSF1010-derived plasmids have been shown to propagate in a wide variety of cyanobacteria.
- Homologous recombination: Site directed integration of DNA via general homologous recombination has been demonstrated in many strains and is probably possible in all cyanobacteria. In some cases, e.g. *Anabaena variabilis* ATCC 29413, introduction of DNA via homologous recombination is far more efficient than introduction of a shuttle vector.

To achieve the final goal of this proposed research, an efficient transformation system must be developed for *A. siamensis*. Several methods known for other cyanobacteria will be tried in parallel to find out the best for this rice-field indigenous species.

As the first step of this project, we have discovered that *A. siamensis* contains restriction endonucleases that behave like BamHI, PstI and HaeIII. Such isoschizomers have also been found in other cyanobacteria (Houmard and Tandeau de Marsac, 1988; Muro-Pastor *et al.*, 1991). Consequently we have deleted BamHI and PstI recognition sites on the recombinant plasmids used for transformation or protected it by sophisticated *in vitro* or *in vivo* methylation procedures. Because of the frequent appearance of four base recognition sites on vectors, it is very difficult to get rid of all the HaeIII sites. It is better to be protected by the methylation method. The methylase genes used for constructing the helper plasmids include BamHI, HaeIII and SssI methylase genes. The SssI methylase is a type of methylase which recognizes two base sites (GC) and shows wide range of methylation activities. The application of SssI methylation has been proved to be effective for transformation of *A. siamensis*.

Considering the stability of introduced plasmid inside host cells distributed in the field under natural conditions without the selection pressure of antibiotics, it is best to approach the final goal by the choice of integrational plasmids, which are believed to be more stable than self-replicating plasmids. Based on this, we have constructed recombinants from pRL435K (and pRL447) with sequences of 5–10 kb fragments of *A. siamensis* chromosome DNA and with fragments of glnA from *Anabaena* 7120 (pAn503). Our recent results of electroporation with this series of integrational plasmids show that several steps are critical to get satisfactory transformation frequencies, including methylation protection against host restriction activities, sonication to single cell or shorter filaments and supplementing the culture with nitrogenous compounds.

- Mass production of this species has been achieved successfully in a tubular reactor (Boussiba, 1993). About half a gram of dry weight per litre was obtained each day during at least two months, while maintained in a steady state without contamination by other microalgae. Very dense cultures are easily obtained in tubular, closed reactors because:

- Maximal output rate is achieved at high cell densities due to the high light intensities inside the relatively thin tubes (5 cm diameter) used.
- Microalgae are not washed off, as occurred in open ponds during the rainy seasons of tropical countries. Optimal production is maintained for longer periods than in open-air reactors due to the high temperatures which persist inside the tubes.

With advanced microalgal biotechnology, the engineered species of *A. siamensis* can be quickly applied into the rice field successfully with several approaches:

- Large amounts of viable cell powder of transgenic *A. siamensis* produced in a blue-green algae factory can be applied directly into the rice fields at any time.
- A closed tubular culture reactor system established at the rice field can continuously supply the rice field with a certain amount of *Anabaena* culture for specific purposes. Our past achievements in complementary fields which are closely related to this project improve the chance of success.

• The main goal of the project is to construct a bifunctional cyanobacterial strain which will control mosquito-borne diseases while facilitating use of nitrogenous compounds in rice fields. There will be special concerns about ecology and safety whenever a genetically engineered strain is released into the environment (Lajoie *et al.*, 1992). There is however much information available proving that BTi $\delta$-endotoxin is very specifically toxic to harmful insects and vector larvae (mainly belonging to lepidoptera and diptera), and is safe to the environment and friendly to human beings. Several special concerns on safety management, tracing and controlling of the released recombinant strains must also be addressed:

- Temporarily release the recombinant strains of *A. siamensis* or *Anabaena* 7120.
- Another innovative way of keeping stringent control of released strains of recombinant is to put some specific genes involved in the catabolism of certain substrates into the recombinant strain which will be under the control of this substrate in the surrounding environment.
- A specific marker is linked with toxin genes for controlling and tracing the recombinant strains, such as luxAB genes as reporter. A reporter gene in a plasmid makes the method of detecting the recombinant strain from the released region much easier. This method seems now to be widely adopted in the applied research (Kaniga *et al.*, 1992). An enhanced amplifying system of luminescence from transformed cells developed by Wolk's group has been adopted for quick screen of true transformant colonies appearing on the agar plate with proper concentration of antibiotics (Wolk *et al.*, 1993). This will also be used as a simple and quick method to trace the recombinant species under the environment conditions.

An efficient way of monitoring the expression of toxin genes inside the host cells is to put the ORF of these genes under the control of temperature sensitive promoters

(Sambrook *et al.*, 1989). The following step would be used to design a way by which the operon can be regulated, so that it is not always expressed, to reduce the chances of development of resistance to the toxin by mosquito larvae (Williams *et al.*, 1992). It has been argued that constitutive expression will increase selective pressure on mosquitoes to evolve toxin resistance (Gibbsons *et al.*, 1992). Strategies to limit such undesired evolution involve the provision of spatial or temporal refugia from exposure to the toxin (McGaughey and Whalon, 1992). The expected lack of absolute fitness to the environment of the engineered strain (in spite of its rice field origin) provides one such temporal refugia, but we shall not rely on this alone. We shall design temporal refugia by ligating the operon downstream of a promoter that will only be expressed when appropriate conditions prevail for developing mosquitoes. We propose to isolate a cold-sensitive promoter that will be inactive at temperatures below 20°C but active at above 25°C.

## 6. Summary

The future of human health may once again face the problem of resurgence of the antibiotic resistant pathogens and the spreading of vector-borne epidemic diseases. Biological control that uses predators, competitors, pathogens and parasites, instead of heavy use of pesticides and chemicals, to reduce density of target organisms while conserving species diversity, will be one of the favourite controls against harmful organisms in the future. By then, humankind will be living in harmony with microorganisms and indeed using them as a symbiotic aid to good health. As a biotechnological approach to control malaria mosquito vectors, we suggest application of nitrogen fixing cyanobacteria as a BTi toxin gene delivery system. Constructing a bifunctional cyanobacterial strain which will control mosquito-borne diseases, while facilitating use of nitrogenous compounds, is the final goal of our research.

## Acknowledgments

We are very grateful to Prof. Peter Wolk and Dr Jeff Elhai for the supply of different plasmids. We thank also UNESCO-MCBN (Project No. 252) and Moriah Fund (Project – Biological Control of Mosquito-borne Diseases) for partial support of this project.

## References

Angsuthanasombat C and Panyim S 1989 Biosysnthesis of 130-kilo-dalton mosquito larvicide in the cyanobacterium *Agmenellum quadruplicatum* PR-6. Appl. Environ. Microbiol. 55, 2428–2430.

Antarikanonda P 1984 Production of extracellular free amino acids by cyanobacterium *Anabaena siamensis*. Curr. Microbiol. 11, 191–196.

Ben-Dov E, Boussiba S and Zaritsky A 1995 Mosquito larvicidal activity of *Escherichia coli* with combinations of genes from *Bacillus thuringiensis* subsp. *israelensis*. J. Bact. 177, 2851–2857.

Boussiba S 1993 Production in tubular reactor of the nitrogen-fixing rice field isolate cyanobacterium *Anabaena siamensis* for rice farming. Microb. Releases 2, 35–39.

Boussiba S 1991 Nitrogen-fixing cyanobacteria: potential uses. Plant and Soil. 137, 177–180

Boussiba S 1988 Cyanobacteria as nitrogen biofertilizers: A study with the isolate *Anabaena azollae*. symbiosis. 6, 129–138.

Casida J E 1994 International Conference on Modern Agriculture and the Environment, 2–6 October 1994, Program and Abstract Book, Rehovot, Israel. P A1.

Chang C, Yu Y M, Dai S M, Law S K and Gill S S 1993 High-level *cryIVD* and *cytA* gene expression in *Bacillus thuringiensis* does not require the 20-kilodalton protein, and the coexpressed gene products are synergistic in their toxicity to mosquitoes. Appl. Environmtl. Microbiol. 59, 815–821.

Chungjatupornchai W 1990 Expression of the mosquitocidal protein genes of *Bacillus thuringiensis* subsp. *israelensis* and the herbicide-resistance gene bar in *Synechocystis* PCC6803. Cur. Microbiol. 21, 283–288.

Clark G A G and Sears M 1991 Mosquito Vector Control and Biology in Latin America – A Symposium. J. Am. Mosquito Control Assoc. 7, 633–645.

de Barjac H and Sutherland D J (Eds) 1990 Bacterial Control of Mosquitoes and Black Flies. Rutgers University Press, New Brunswick.

De Marsac N T, de la Torre, F and Szulmajster J 1987 Expression of the larvicidal gene of *Bacillus sphaericus* 159M in the cyanobacterium *Anacystis nidulans*. Mol. Gen. Genet. 209, 396–398

Delecluse A, Charles J-F, Klier A and Rapoport G 1991 J. Bacteriol. 173, 3374–3381.

Douek J, Ranaweera S, Einav M and Zaritsky A 1990 Proc. 1st Asia-Pacific Conf. Entomol., pp 1–7.

Douek J, Einav M and Zaritsky A 1992 Molec. Gen. Genet. 232, 162–165.

Elhai J and Wolk C P 1988 Conjugal transfer of DNA to cyanobacteria. Methods Enzymol. 167, 747–754.

Entwistle P F, Cory J S, Bailey M J and Higgs S (Eds) 1993 *Bacillus thuringiensis*, an Environmental Pesticide: Theory and Practice. Wiley & Sons, New York.

Flores E and Wolk C P 1985 Identification of facultatively heterotrophic, $N_2$-fixing cyanobacteria able to receive plasmid vectors from *Escherichia coli* by conjugation. J Bacteriol. 162, 1339–1341.

Gibbsons A 1992 Moths take the field against biopesticide. Science 254, 646.

Golop R and Zaritsky A 1987 Annu. Meeting Israel Soc. Genet. p 18.

Hofte H and Whiteley H R 1989 Insecticidal crystal proteins of *Bacillus thuringiensis* Microbiol. Rev. 53, 242–255.

Houmard J and Tandeau de Marsac N 1988 Cyanobacterial genetic tools: Current status. Methods Enzymol. 167, 808–847.

Ibarra J E and Federici B A 1986 Parasporal bodies of *Bacillus thuringiensis* subsp. *morrisoni* (PG14) and *Bacillus thuringiensis* subsp. *israelensis* are similar in protein composition and toxicity. FEMS Microbiol. Lett. 34, 79–84.

Kaniga K, Sory M-P, Delor I, Saegerman C, Limet J N and Cornelis G R, 1992 Monitoring of *Yersinia enterocolitica* in murine and bovine feces on the basis of the chromosomally integrated *lux*AB marker gene. Appl. Environ. Microbiol. 58, 1024–1026.

Lacey L A and Undeen A H 1986 Microbial control of black flies and mosquitoes. Annu. Rev. Entomol. 31, 265–295.

Lahkim-Tsror L, Pascar-Gluzman C, Margalit J and Barak Z 1983 J. Invertebr. Pathol. 41, 104–116.

Lajoie C A, Chen K-C O H and Strom P F 1992 Development and use of field application vectors to express nonadaptive foreign genes in competitive environments. Appl. Environ. Microbiol. 58, 665–663.

van Lenteren J C 1994 International Conference on Modern Agriculture and the Environment, 2–6 October 1994, Program and Abstract Book, Rehovot, Israel. p A4.

McGaughey W H and Whalon M E 1992 Managing insect resistance to *Bacillus thuringiensis* toxins. Science 258, 1451–1455.

Muro-Pastor A M, Herreo A and Flores E 1991 Sequence-specific endonucleases from the cyanobacterium *Nostoc* sp. ATCC 29132. FEMS Microbiol. Lett. 77, 1–4

Murphy R C and Stevens Jr E S 1992 Cloning and expression of the *cryIVD* gene of *Bacillus thuringiensis* subsp. *israelensis* in the cyanobacterium *Agmenellum quadruplicatum* PR-6 and its resulting larvicidal activity. Appl. Environ. Microbiol. 58, 1650–1655.

Ohana B, Margalit J and Barak Z 1987 Fate of *Bacillus thuringiensis* subsp. *israelensis* under simulated field conditions. Appl. Environ. Microbiol. 53, 828–831.

Poncet S, Delecluse A, Klier A and Rapoport G 1995 Evaluation of synergistic interactions among the CryIVA, CryIVB, and CryIVD toxic components of *B. thuringiensis* subsp. *israelensis* crystals. J. Invertebr. Pathol. 66, 131–135.

Querijero-Pslscpsc N M, Martinez M R and Boussiba S 1990 Mass cultivation of the nitrogen fixing cyanobacterium *Gleotrichia natans*, indigenous to rice fields. J. Appl. Phycol. 2, 319–325

Renqiu K, Xudong X and Yuxiang H 1992 Chin. J. Parasit. Dis. Contr. 5, 821–824.

Sambrook J, Fritsch E F and Maniatis T 1989 Molecular Cloning: A Laboratory Manual. 2nd ed. Cold Spring Harbor Laboratory Press, Cold Spring Harbor, New York.

Studier F W and Moffatt B A 1986 Use of bacteriophage T7 RNA polymerase to direct selective high-level expression of cloned genes. J. Mol. Biol. 189, 113–130.

Swanson R V, Zhou J, Leary J A, *et al.* 1992 Characterization of phycocyanin produced by *cpc*E and *cpc*F mutants and identification of an intergenic suppressor of the defect in bilin attachment. J. Biol. Chem. 267, 16146–16154.

Thiel T and Poo H 1989 Transformation of a filamentous cyanobacterium by electroporation. J. of Bacteriol. 171, 5743–5746.

Thomas S P, Zaritsky A and Boussiba S 1990 Ammonium excretion by a methionine sulfoximine resistant mutant of the rice field cyanobacterium *Anabaena siamensis.* Appl. Environ. Microbiol. 56, 3499–3504.

Vaeck M A, Chungjatup W and McIntosh L 1988 World Patent 8 806 631. Assignee: Michigan State University and Plant Genetic Systems.

Visick J E and Whiteley H R 1991 J. Bacteriol. 173, 1748–1756.

Williams S, Friedrich L, Dincher S, Carozzi H, Keddmann H, Ward E and Ryals J 1992 Chemical regulation of *Bacillus thuringiensis.* δ-endotoxin expression in transgenic plants. Biotechnology 10, 540–543.

Wolk C P, Elhai J, Kuritz T and Holland D 1993 Amplified expression of a transcriptional pattern formed during the development of *Anabaena.* Molec. Microbiol. 7, 441–445.

Wolk C P, Vonshak A, Kehoe P and Elhai J 1984 Construction of shuttle vectors capable of conjugative transfer from *Escherichia coli* to nitrogen-fixing filamentous cyanobacteria. Proc. Natl. Acad. Sci. USA 81, 1561–1565.

Wu X Q, Xu X, Ben-Dov E, Dahan E, Zaritsky A and Boussiba S 1995a Expression of mosquitocidal *Bacillus thuringiensis* subsp. *israelensis* δ-endotoxin genes in the filamentous cyanobacterium *Anabaena* 7120. Proceeding of third european workshop on the molecular biology of cyanobacteria. Sevilla, Spain. p 38.

Wu X Q, Shi D J, Liu X L, Li G B, Wang H J, Qiu Z S, Zhang C Q, Yin L P, Zhao W P, Liu H R, and Boussiba S, 1995b Site-directed integration of NPTII gene into *gln*A of the chromosome in *Anabaena azollae* and its effects on morphology, ultrastructure and ammonia secretion of the transformed cells. Chinese J. Bot. (English J.), 7, 43–48.

Xudong X, Renqiu K and Yuxiang H 1993 High larvicidal activity of intact cyanobacterium *Anabaena* sp. PCC 7120 expressing gene 51 and gene 42 of *Bacillus sphaericus* sp. 2297. FEMS Microbiol. Lett. 107, 247–250.

Yoshisue H, Fukada T, Yoshida K-I *et al.* 1993 Transcriptional regulation of *Bacillus thuringiensis* subsp. *israelensis* mosquito larvicidal crystal protein gene *cryIVA*. J. Bacterial. 175, 2750–2753.

Yoshida K-I, Sen K, Sakai H and Komano T 1992 Effects of *Bacillus thuringiensis* var. *israelensis* 20-kDa protein on production of the *Bti* 130-kDa crystal protein in *Escherichia coli.* Biosci. Biotech. Biochem. 56, 1429–1433.

Zaritsky A, Einav M and Douek J 1990 Abstract, EMBO Workshop on Bacterial Growth and Division.

# 9 Structure–function analysis of PBAN/MRCH: a basis for antagonist design

MIRIAM ALTSTEIN, EZRA DUNKELBLUM, YOAV GAZIT,
ORNA BEN AZIZ, TAL GABAY, ZVI VOGEL AND JACOB BARG

## 1. Introduction

Insect neuropeptides play a key role in the regulation of a variety of physiological functions such as embryonic and post-embryonic development, homeostasis, osmoregulation, migration, oviposition and mating (for review, see Menn and Borkovec, 1989; Kelly *et al.*, 1994). In recent years, improvements of biochemical and chemical techniques have facilitated the chemical identification of insect neuropeptides, and over 80 insect neuropeptides have been isolated and sequenced to date (for review, see Kelly *et al.*, 1994). The varied functional capacities of this group of regulatory substances provide new strategies for insect control. Control could be based on interference with the basic processes associated with neuropeptide activity: biosynthesis, release, transport, binding and degradation. Among these processes, the most effective and susceptible to manipulation is the binding to the target receptors. Blocking of the binding may be achieved by selective antagonists, which prevent receptor activation by the neuropeptide. To design potent antagonists, the structure–activity relationship of the neuropeptide must be revealed and the active as well as inactive sequences in the molecule have to be identified. We have used this approach to study the structure–function relationship of an insect neuropeptide, termed pheromone biosynthesis activating neuropeptide (PBAN; Raina and Klun, 1984; Raina *et al.*, 1989), also known as melanization and reddish coloration hormone (MRCH; Matsumoto *et al.*, 1981, 1990), which regulates two functions in moths: sex pheromone biosynthesis and cuticular melanization (for review, see Raina, 1993).

## 2. Materials and methods

*Heliothis peltigera* moths (Lepidoptera: Noctuidae) were reared on an artificial diet as described previously (Dunkelblum and Kehat, 1989). *Spodoptera littoralis* larvae (Lepidoptera: Noctuidae) were fed on castor bean leaves as described elsewhere (Altstein *et al.*, 1994). All insects were kept in a thermostatically regulated room with a light:dark regime of 14:10 h at 25 ± 2°C and 60% RH. Adult moths were held in screen cages and supplied with a 10% sugar solution.

*D. Rosen* et al. *(eds): Modern Agriculture and the Environment, 111–118.*

C-terminal free and amidated fragments derived from the amino acid sequence of *Helicoverpa zea* PBAN (Hez-PBAN) (Raina *et al.*, 1989): PBAN 9–33$NH_2$, PBAN 13–33$NH_2$, PBAN 17–33$NH_2$, PBAN 19–33$NH_2$, PBAN 26–33$NH_2$, PBAN 28–33$NH_2$, PBAN 1–33COOH, were synthesized as described previously (Gazit *et al.*, 1990). Pheromonotropic activity was tested by the ability of the full length synthetic Hez-PBAN (PBAN 1–33$NH_2$) and derived fragments to induce sex pheromone biosynthesis in adult female moths using the *H. peltigera* bioassay as described by Gazit *et al.* (1992). Sex pheromone content was monitored with the major component, (Z)-11-hexadecenal (Z11–16:Ald) (Dunkelblum and Kehat, 1989), 2 h post-injection (unless otherwise indicated) by capillary gas chromatography as described previously (Gazit *et al.*, 1992). Melanotropic activity was determined by the ability of the same fragments to induce cuticular melanisation in fifth-instar *S. littoralis* larvae using the bioassay described by Matsumoto *et al.* (1981). The degree of cuticular melanisation (expressed as the optical density and size of the pigmented and non-pigmented areas) was determined by a quantitative computerised image analysis (Biological Detection Systems, Pittsburgh, PA, USA) as described previously (Litman *et al.*, 1994; Altstein *et al.*, 1996). The effect of Hez-PBAN and its derived fragments on cuticular melanisation was examined by comparing the size of the melanised and non-melanised areas in peptide-versus buffer-injected larvae. Analysis was performed at concentrations where PBAN 1–33$NH_2$ demonstrated maximal activity (10 and 100 pmol/larva) (Altstein *et al.*, 1996).

## 3. Results

### *3.1 Pheromonotropic activity of PBAN: structure–function analysis*

Analysis of the activity of 1 pmol of the PBAN-derived peptides (Table 1), 2 h post-injection, revealed that removal of the first eight N-terminal amino acids did not alter significantly the pheromonotropic activity. The amount of the major sex pheromone component that was formed in response to the injection of PBAN 9–33$NH_2$ was 96 ng/gland, compared with 108 ng/gland obtained with PBAN 1–33$NH_2$. Further removal of four and eight amino acids from the N-terminus resulted in a complete loss of activity, and the amount of pheromone produced in response to the injection of PBAN 13–33$NH_2$ and PBAN 17–33$NH_2$ did not differ from that evoked by the injection of buffer (Table 1). Interestingly, injection of a shorter fragment lacking 18 amino acids from the N-terminus showed a marked activity (74 ng/gland), which did not differ significantly from that of PBAN 9–33$NH_2$ (Table 1). The C-terminal derived octa- and hexapeptides, PBAN 26–33$NH_2$ and PBAN 28–33$NH_2$, as well as the C-terminal free acid peptide, PBAN 1–33COOH, failed to stimulate pheromone biosynthesis, and the amounts of sex pheromone that were obtained in response to their injection were 6, 5 and 3 ng/gland, respectively (Table 1). These data indicated that the region between amino acids 9 and 13 plays an important role in maintaining the activity of the PBAN. The fact that the C-terminal free acid PBAN 1–33COOH was devoid of any activity, and the fact that the potency of the PBAN-derived

*Table 1.* Effect of synthetic Hez-PBAN and derived fragments on sex pheromone biosynthesis in *Heliothis peltigera* at different concentrations and in the presence or absence of protease inhibitors.

| Injected peptide | Z11–16:Ald content (ng/gland) ± SEM | | |
|---|---|---|---|
| | 1 pmol | 1 pmol + inhibitors | 10 pmol |
| *Amidated peptides* | | | |
| PBAN 1–33$NH_2$ | 108 ± 10 ($n = 44$) a | 193 ± 33 ($n = 10$) a* | 177 ± 15 ($n = 37$) a |
| PBAN 9–33$NH_2$ | 96 ± 19 ($n = 14$) ab | n.t. | 211 ± 57 ($n = 10$) a |
| PBAN 13–33$NH_2$ | 4 ± 3 ($n = 8$) c | 12 ± 4 ($n = 8$) c | 180 ± 33 ($n = 5$) a |
| PBAN 17–33$NH_2$ | 33 ± 10 ($n = 10$) c | 51 ± 27 ($n = 9$) c | 197 ± 41 ($n = 10$) a |
| PBAN 19–33$NH_2$ | 74 ± 17 ($n = 18$) b | 123 ± 29 ($n = 10$) b | 141 ± 21 ($n = 27$) a |
| PBAN 26–33$NH_2$ | 6 ± 2 ($n = 9$) c | n.t. | 21 ± 9 ($n = 9$) b |
| PBAN 28–33$NH_2$ | 5 ± 2 ($n = 21$) c | 0.3 ± 0.2($n = 10$) c | 6 ± 2 ($n = 32$) b |
| *Free acid peptide* | | | |
| PBAN 1–33COOH | 3 ± 2 ($n = 9$) c | 3 ± 2 ($n = 10$) c | 1 ± 1 ($n = 10$) b |
| Buffer | 4 ± 1 ($n = 54$) c | 17 ± 12 ($n = 10$) c | 4 ± 1 ($n = 54$) b |

Peptide injection was performed using 3.5-day-old females in photophase. Injection of protease inhibitors (phenylmethylsulphonyl fluoride, N,$\alpha$-p-tosyl-L-lysine chloromethyl ketone, aprotinin and bacitracin, 0.5 $\mu$g each) was performed 30 min prior to the injection of the peptides. The activity of all peptides was tested 2 hours post-injection. Statistical analysis was performed by ANOVA (using the Newman–Keuls test at $P < 0.05$) within columns among the different peptides (indicated by letters) and within rows (for each peptide separately) comparing activity in the presence and absence of inhibitors (indicated by an asterisk). Means with the same letter do not differ significantly (n.t., not tested). An asterisk(*) indicates a significant difference in pheromone content at $p < 0.05$.

fragments did not correlate directly with the length of the peptide (PBAN 19–33$NH_2$ was more potent than PBAN 13–33$NH_2$ or PBAN 17–33$NH_2$), indicated that proteolytic activity may affect the potency of PBAN, and led to a further analysis of the activity in the presence and absence of protease inhibitors, at higher doses and shorter times post-injection.

Injection of a mixture of protease inhibitors, prior to the PBAN peptides, did not affect significantly the pheromonotropic activity of PBAN 13–33$NH_2$, PBAN 17–33$NH_2$, PBAN 19–33$NH_2$, PBAN 28–33$NH_2$ and PBAN 1–33COOH. Only in the case of the complete PBAN 1–33$NH_2$, did the administration of the protease inhibitors cause a significant rise in pheromone levels, from 108 to 193 ng/gland (Table 1). Injection of PBAN 1–33$NH_2$, and all derived shorter peptides at 10 pmol, for 2 h, raised the level of pheromone significantly in all cases with the exception of the C-terminal octa- and hexapeptides as well as the free acid PBAN 1–33COOH (Table 1).

Examination of the pheromonotropic activity of the PBAN-derived fragments that were inactive at a dose of 1 pmol (PBAN 13–33$NH_2$, PBAN 17–33$NH_2$, PBAN 28–33$NH_2$) after their infection for 30 min (instead of the usual 2 h) revealed accumulation of significantly larger amounts of pheromone than the amounts that were obtained with these peptides at 2 h post-injection (Table 2). The activity of PBAN 13–33$NH_2$ was even significantly higher than that of PBAN 1–33$NH_2$. Even

*Table 2.* Comparison of the effect of PBAN 1–33$NH_2$ and derived fragments on sex pheromone biosynthesis at different times post-injection.

| Injected peptide (1 pmol) | Z11–16:Ald content (ng/gland) ± SEM | |
|---|---|---|
| | 30 min post-injection | 120 min post-injection |
| *Amidated peptides* | | |
| PBAN 1–33$NH_2$ | 67 ± 8 ($n = 30$) b | 108 ± 10 ($n = 44$) a* |
| PBAN 13–33$NH_2$ | 91 ± 14 ($n = 10$) a | 4 ± 3 ($n = 8$) c* |
| PBAN 17–33$NH_2$ | 79 ± 8 ($n = 10$) b | 33 ± 10 ($n = 10$) b* |
| PBAN 28–33$NH_2$ | 30 ± 4 ($n = 10$) c | 5 ± 2 ($n = 21$) c* |
| *Free acid peptide* | | |
| PBAN 1–33COOH | 15 ± 7 ($n = 10$) d | 4 ± 1 ($n = 9$) c |
| Buffer | 11 ± 3 ($n = 47$) d | 4 ± 1 ($n = 54$) c |

Statistical analysis was performed by ANOVA (using the Newman–Keuls test at $P < 0.05$) within columns comparing the activity of the different peptides with that of the full length PBAN or buffer (indicated by letters), and within rows (for each peptide) comparing the activity at 30 and 120 min (indicated by an asterisk). Means with the same letter do not differ significantly. An asterisk (*) indicates a significant difference in pheromone content at $P < 0.05$.

PBAN 28–33$NH_2$, which was inactive under all previously tested conditions, stimulated formation of 30 ng Z11–16:Ald/gland when injected for 30 min. The free acid PBAN 1–33COOH remained entirely inactive (Table 2).

Based on these data we have extended the kinetic analysis and performed a time-response study using the C-terminal hexapeptide and the full length PBAN in the presence or absence of protease inhibitors. Comparison of the activity of 10 pmol PBAN 28–33$NH_2$ with that of a similar dose of PBAN 1–33$NH_2$ at 5, 15, 30, 60 and 120 min post-injection (in the absence of protease inhibitors), revealed that the activity of the short fragment does not differ from that of the full length PBAN at 5, 15 and 30 min post-injection (Table 3). In the presence of protease inhibitors, the pheromone content of females injected with PBAN 28–33$NH_2$ (at 15 min post-injection) was even significantly higher than that obtained with PBAN 1–33$NH_2$. The activity of PBAN 28–33$NH_2$, in the presence of the inhibitors, lasted longer, and even after 60 min the pheromone content did not differ significantly from that obtained with the full length peptide (Table 3). Injection of the hexapeptide for longer periods resulted in a marked decrease in its activity as compared with that of PBAN 1–33$NH_2$ (Table 3). Conversely, the amount of pheromone stimulated by the full length peptide increased with post-injection time.

### 3.2 *Melanotropic activity of PBAN: structure–function analysis*

Injection of 10 pmol PBAN 9–33$NH_2$ resulted in a slight, non-significant decrease in the size of the pigmented area compared with the area obtained by the injection of PBAN 1–33$NH_2$ (Table 4). Other PBAN-derived peptides lacking 12, 16 and

*Table 3.* Time–response curve of 10 pmol synthetic PBAN 1–33NH$_2$ and PBAN 28–33NH$_2$ in the presence or absence of a mixture of protease inhibitors.

| Time post-injection (min) | Z11–16:Ald content (ng/gland) ± SEM | |
|---|---|---|
| | PBAN 1–33NH$_2$ (10 pmol) | PBAN 28–33NH$_2$ (10 pmol) |
| *Without protease inhibitors* | | |
| 5 | 24 ± 7 ($n = 9$) c | 40 ± 15 ($n = 10$) a |
| 15 | 30 ± 3 ($n = 19$) c | 37 ± 4 ($n = 17$) a |
| 30 | 53 ± 5 ($n = 29$) c | 42 ± 4 ($n = 35$) a |
| 60 | 89 ± 13 ($n = 19$) b | 26 ± 5 ($n = 18$) a* |
| 120 | 177 ± 15 ($n = 37$) a | 6 ± 2 ($n = 32$) b* |
| *With protease inhibitors* | | |
| 15 | 17 ± 4 ($n = 10$) b | 39 ± 5 ($n = 8$) ab* |
| 30 | 34 ± 6 ($n = 10$) b | 64 ± 8 ($n = 33$) a |
| 60 | 46 ± 3 ($n = 8$) b | 63 ± 11 ($n = 22$) a |
| 120 | 323 ± 67 ($n = 10$) a | 11 ± 4 ($n = 16$) b* |

Statistical analysis was performed within columns (for each peptide) as a function of time (indicated by letters) and within rows (between the two peptides) comparing activity at each time point (indicated by asterisks). Values with the same letter do not differ significantly (at $P < 0.05$). An asterisk (*) indicates a significant difference in pheromone content at $P < 0.05$.

*Table 4.* Effect of synthetic Hez-PBAN and derived fragments on cuticular melanization in *Spodoptera littoralis* larvae.

| Injected peptide | Percentage of melanized area ± SEM | |
|---|---|---|
| | 10 pmol | 100 pmol |
| *Amidated peptides* | | |
| PBAN 1–33NH$_2$ | 93 ± 1 ($n = 59$) † | 94 ± 2 ($n = 15$) † |
| PBAN 9–33NH$_2$ | 88 ± 6 ($n = 5$) † | 89 ± 4 ($n = 10$) † |
| PBAN 13–33NH$_2$ | 75 ± 4 ($n = 5$) * | 82 ± 3 ($n = 10$) * |
| PBAN 17–33NH$_2$ | 82 ± 5 ($n = 5$) * | 75 ± 5 ($n = 10$) * |
| PBAN 19–33NH$_2$ | 80 ± 3 ($n = 10$) * | 88 ± 4 ($n = 10$) † |
| PBAN 26–33NH$_2$ | 85 ± 5 ($n = 4$) † | 77 ± 4 ($n = 10$) * |
| PBAN 28–33NH$_2$ | 86 ± 2 ($n = 14$) *† | 81 ± 4 ($n = 19$) *† |
| *Free acid peptide* | | |
| PBAN 1–33COOH | 79 ± 4 ($n = 9$) * | 77 ± 2 ($n = 10$) * |
| Buffer | 75 ± 1 ($n = 61$) * | 75 ± 1 ($n = 61$) * |

The size of the melanized area is expressed as a percentage of the total area of the cuticle. Values represent the mean ± SEM. Statistical analysis was performed by Hochberg's (1988) multiple comparisons procedure, where family-wise error rate is controlled at $\alpha = 0.05$. The activity of each peptide was compared with that of PBAN 1–33NH$_2$ and buffer-injected larvae using two-sample Student *t*-tests, which provided the corresponding *P*-values. The symbols * and † indicate mean values significantly different from PBAN 1–33NH$_2$ and buffer, respectively.

18 amino acids (PBAN 13–33$NH_2$, PBAN 17–33$NH_2$ and PBAN 19–33$NH_2$, respectively) at 10 pmol/larva, did not stimulate melanisation, and the extent of the melanised areas did not differ significantly from that obtained in response to injection of phosphate buffer. Similarly, the C-terminal free acid PBAN (PBAN 1–33COOH) failed to evoke melanisation. Interestingly, injection of the octapeptide and hexapeptide (PBAN 26–33$NH_2$ and PBAN 28–33$NH_2$, respectively) evoked melanisation and the extent of the melanised area was significantly greater than that obtained by buffer injection (Table 4). The size of the melanised area obtained in response to the injection of PBAN 28–33$NH_2$ was smaller than that obtained by the injection of the full length PBAN. Examination of the extent of melanisation at 100 pmol revealed a similar pattern to that obtained at 10 pmol (Table 4). Most of the inactive peptides did not exhibit any greater activity as a result of their injection at the higher dose. The only peptide that showed an elevated activity (which did not differ significantly from that of PBAN 1–33$NH_2$) was PBAN 19–33$NH_2$. Analysis of the melanisation intensity (= optical density) of cuticles obtained from larvae injected with PBAN 1–33$NH_2$ and its derived fragments revealed a similar structure–activity relationship (data not shown).

## 4. Discussion

A structure–activity relationship study of Hez-PBAN was performed with respect to its pheromonotropic and melanotropic activities. Using a series of Hez-PBAN-derived fragments at various doses (1 and 10) and times post-injection (5–120 min), we were able to demonstrate that peptides lacking 12 and 16 amino acids from their N-terminus exhibit under certain conditions a pheromonotropic activity similar to that of the full length PBAN, and that the C-terminal derived hexapeptide (when analysed at short post-injection times) is capable of stimulating sex pheromone production to a similar extent as PBAN 1–33$NH_2$, suggesting that this sequence constitutes the biologically active site of the neuropeptide. Within the C-terminal sequence, the amide group was found to play a crucial role. In addition, it was observed that the region between amino acids 9–13 is important for the pheromonotropic activity of the full length PBAN. The structure–activity relationship of PBAN with respect to its melanotropic activity was found to be similar to that of its pheromonotropic activity, although the effects on melanin formation were less pronounced. We observed that the C-terminal region is essential for induction of cuticular melanisation, with the amide being most important. The first eight amino acids at the N-terminus of the peptide do not play an important role in the biological activity. The region between amino acids 9–13, however, is important, as indicated above for sex pheromone production. As in the case of the pheromonotropic activity, the melanotropic activity of the PBAN-derived peptides did not correlate with their length, and PBAN 13–33$NH_2$ and PBAN 17–33$NH_2$ were less active than the hexapeptide PBAN 28–33$NH_2$.

Our findings regarding the importance of the C-terminal region in sex pheromone production are in agreement with the studies of Kitamura *et al.* (1989), Kuniyoshi *et al.* (1991) and Raina and Kempe (1990), who have demonstrated (using various *Heliothinae* species and *Bombyx mori*) that the C-terminal pentapeptide of PBAN is the minimal sequence required for pheromonotropic activity.

To the best of our knowledge, this is the first study of the structure–activity relationship of Hez-PBAN with respect to its melanotropic activity. To date, structure–activity studies have been performed only on *Pseudaletia separata* pheromonotropin (Pss-PT, also termed Pss-MRCH), an 18-amino-acid neuropeptide which was also found to induce cuticular melanisation, and on members of the pyrokinin and myotropin neuropeptide family (Lem-PK, Lom-MT-II and Lom-MT-III) (Matsumoto *et al.*, 1992, 1993). The results of our study correspond well with these studies that demonstrated the importance of the C-terminal region, and C-terminal amide and the insignificant role of the first six N-terminal amino acids.

The ability of a variety of peptides to stimulate sex pheromone biosynthesis and cuticular melanisation suggests two possible mechanisms:

- existence of multiple pheromonotropic and melanotropic mechanisms which may be mediated by multiple PBAN/MRCH receptors that are activated at different concentrations and kinetics; and
- existence of only one mechanism mediated by the short C-terminal part of the peptide.

In the first case, the C-terminal derived sequences fulfil the conformational requirement of one class of receptors, and other regions in the PBAN/MRCH molecule (e.g. 9–13) fulfil the conformational requirements of a second (or other) class of receptors. In the second case, the C-terminal derived sequence is the only conformationally important sequence, and other sequences, which have a role in the biological activity, serve for other non-conformational purposes (e.g. protection against proteolytic degradation). The existence of two separate PBAN/MRCH receptors has been suggested also by Raina and Kempe (1992), who based their assumption on the fact that a pentapeptide, derived from the internal part of Hez-PBAN (amino acids 15–19), showed high activity, in *H. zea*, at a dose of 1 pmol. This issue will receive further clarification when the PBAN/MRCH receptor(s) are characterised. Meanwhile, the identification of the active and non-active sequences may serve as a basis for the design of PBAN/MRCH agonists and antagonists which will provide a better insight into the mode of action of the neuropeptide and serve as a basis for the design of efficient inhibitors capable of interfering with the activity of the endogenous peptide.

## Acknowledgments

We thank Mrs Irit Schafler for her excellent assistance in the GC analysis of pheromone production. This research was supported by the Endowment Fund for Basic Research in Life Sciences: Charles H. Revson Foundation, administered by the Israel Academy of Sciences and Humanities.

## References

Altstein M, Ben Aziz O and Gazit Y 1994 Pheromone biosynthesis activating neuropeptide (PBAN) and color polymorphism: An immunochemical study in *Spodoptera littoralis*. J. Insect Physiol. 40, 303–309.

Alstein M, Gazit Y, Ben Aziz O *et al.* 1996 Induction of cuticular melanization in *Spodoptera littoralis* larvae by PBAN/MRCH: Development of a quantitative bioassay and structure function analysis. Arch. Insect Biochem. Physiol. 31, 335–370

Dunkelblum E and Kehat M 1989 Female sex pheromone components of *Heliothis peltigera* (Lepidoptera: Noctuidae): Chemical identification from gland extracts and male response. J. Chem. Ecol. 15, 2233–2245.

Gazit Y, Dunkelblum E, Benichis M and Altstein M 1990 Effect of synthetic PBAN and derived peptides on sex pheromone biosynthesis in *Heliothis peltigera* (Lepidoptera: Noctuidae). Insect Biochem. 20, 853–858.

Gazit Y, Dunkelblum E, Ben-Aziz O and Altstein M 1992 Immunochemical and biological analysis of pheromone biosynthesis activating neuropeptide (PBAN) in *Heliothis peltigera*. Arch. Insect Biochem. Physiol. 19, 247–260.

Hochberg Y 1988 A sharper Bonferroni procedure for multiple tests of significance. Biocetrika, 75, 800–802.

Kelly T J, Masler E P and Menn J J 1994 Insect neuropeptides: Current status and avenues for pest control. In ACS Symposium Series 551. Eds P A Hedin, J J Menn and R M Hollingworth. pp 292–318. American Chemical Society, Washington, DC.

Kitamura A, Nagasawa H, Kataoka H *et al.* 1989 Amino acid sequence of pheromone-biosynthesis-activating neuropeptide (PBAN) of the silkworm *Bombyx mori*. Biochem. Biophys. Res. Commun. 163, 520–526.

Kuniyoshi H, Kitamura A, Nagasawa H *et al.* 1991 Structure–activity relationship of pheromone biosynthesis activating neuropeptide (PBAN) from the silkmoth, *Bombyx mori*. In Peptide Chemistry Ed. Y Shimonishi. pp 251–254. Protein Research Foundation, Osaka, Japan.

Litman P, Barg J and Ginzburg I 1994 Microtubules are involved in the localization of tau-mRNA in primary neuronal cell culture. Neuron 13, 1463–1474.

Matsumoto S, Isogai A, Suzuki A, Ogura N and Sonobe H 1981 Purification and properties of the melanization and reddish colouration hormone (MRCH) in the armyworm, *Leucania separata* (Lepidoptera). Insect Biochem. 11, 725–733.

Matsumoto S, Kitamura A, Nagasawa H *et al.* 1990 Functional diversity of a neurohormone produced by the suboesophageal ganglion: Molecular identity of melanization and reddish colouration hormone and pheromone biosynthesis activating neuropeptide. J. Insect Physiol. 36, 427–432.

Matsumoto S, Yamashita O, Fonagy A *et al.* 1992 Functional diversity of a pheromonotropic neuropeptide: Induction of cuticular melanization and embryonic diapause in lepidopteran insects by *Pseudaletia* pheromonotropin. J. Insect. Physiol. 38, 847–851.

Matsumoto S, Fonagy A, Schoofs L *et al.* 1993 Induction of cuticular melanization in the armyworm larvae *Pseudaletia separata*, by insect myotropic neuropeptide possessing FXPRLamide at the C-terminus. J. Pestic. Sci. 18, 127–129.

Menn J J and Borkovec A B 1989 Insect neuropeptides: Potential new insect control agents. J. Agric. Food Chem. 37, 271–278.

Raina A K 1993 Neuroendocrine control of sex pheromone biosynthesis in Lepidoptera. Annu. Rev. Entomol. 38, 329–349.

Raina A K and Kempe T G 1990 A pentapeptide of the C-terminal sequence of PBAN with pheromonotropic activity. Insect Biochem. 20, 849–851.

Raina A K and Kempe T G 1992 Structural activity studies of PBAN of *Helicoverpa zea* (Lepidoptera: Noctuidae). Insect Biochem. Mol. Biol. 22, 221–225.

Raina A K and Klun J A 1984 Brain factor control of sex pheromone production in the female corn earworm moth. Science 225, 531–533.

Raina A K, Jaffe H, Kempe T G *et al.* 1989 Identification of a neuropeptide hormone that regulates sex pheromone production in female moths. Proc. Natl. Acad. Sci. USA 244, 796–798.

10

# Estimation of metolachlor volatilization from agricultural fields using a micrometeorological approach

J.H. PRUEGER, J.L. HATFIELD AND T.J. SAUER

## 1. Introduction

Studies on the impact of agricultural pesticides on water quality have been expanding in recent years. However, of potentially equal importance is the impact of pesticides on air quality. Pesticide volatilisation is a major dissipation route for a wide variety of pesticides used in agriculture (Caro and Taylor, 1971; Farmer *et al.*, 1974; Spencer and Farmer, 1980). Two modes of pesticide entry into the atmosphere are volatilization from soil and plant surfaces, and the physical transport from wind erosion of soil particles on which pesticides were adsorbed onto the soil particulate surface.

Pesticide volatilisation can be affected by a variety of chemical and environmental factors. Volatility of a pesticide is related to its vapour density, while actual rates of volatilization are dependent upon microclimatic parameters that modify the effective vapour pressure of pesticides (Basile *et al.*, 1986). Parochetti and Warren (1966) found that chlorpropham volatilization from wet soils increased with temperature but declined to almost zero in dry soils. Spencer *et al.* (1969) reported that vapour densities of dieldrin in dieldrin–soil mixtures increased with temperature and dieldrin concentration, but were not affected by soil water content. Relative humidity was found to have an effect on dieldrin volatilization (Spencer and Cliath, 1973). Glotfelty *et al.* (1984) showed in results from field experiments, that pesticides applied to the surface of fallow soil initially volatilise at rates proportional to the vapour density of the pure chemical; this is affected by soil water content and incident solar radiation.

Profile techniques are typically used to estimate pesticide volatilisation flux from agricultural fields and in this study the methods included the aerodynamic and energy balance (Bowen-ratio) approaches. The aerodynamic method involves use of the Thornthwaite–Holzman equation in which pesticide flux is expressed as:

$$P = \frac{k^2(\Delta\bar{c})(\Delta\bar{u})}{\phi_m\phi_p[\ln(z_2/z_1)]^2} \tag{1}$$

where $P$ is the pesticide flux ($\mu$g m$^{-2}$ h$^{-1}$), $k$ is von Karman's constant (dimensionless, $\approx 0.4$), $\Delta\bar{c}$ ($\mu$g m$^{-3}$) and $\Delta\bar{u}$ (m s$^{-1}$) are average pesticide and

*D. Rosen* et al. *(eds): Modern Agriculture and the Environment, 119–125.*

horizontal wind speed differences between heights $z_2$ and $z_1$ (m), and $\phi_m$ and $\phi_p$ are atmospheric stability corrections for momentum and pesticide respectively.

The Bowen-ratio method makes use of a fundamental concept in atmospheric theory relating the transfer of any conservative scaler, i.e., water vapour, momentum, heat, or chemical vapour from a surface to the overlying atmosphere by turbulence produced by wind traversing over a surface. These fluxes can be expressed as flux gradient equations similar to equation 2 for latent heat:

$$L_E = L\frac{\rho_a \epsilon}{P} k \frac{\partial e}{\partial z} \tag{2}$$

where $L_E$ is latent heat flux (W $m^{-2}$), $L$ is latent heat of vaporisation (J $kg^{-1}$), $\rho_a$ is density of air (kg $m^{-3}$), $\epsilon$ is the ratio of the mole weight of water vapour to air (0.622), $P$ is atmospheric pressure (kPa), and $\partial e/\partial z$ is the vapour pressure gradient (kPa). Gradient fluxes expressed in a manner similar to equation 2 have corresponding transfer diffusivities ($k$) which are assumed to be identical under adiabatic conditions. This assumption has been debated in the literature (Raupach and Legg, 1984; Thom *et al.*, 1975; Webb, 1975) but in general has been found to be satisfactory for many different applications. In this study it was assumed that transfer diffusivities for water and pesticide vapour are equal so that pesticide flux can be expressed as:

$$P = -k_e\left(\frac{\partial c}{\partial z}\right) \tag{3}$$

where $k_e$ is the transfer diffusivity for water vapour ($m^2$ $h^{-1}$) and $\partial c/\partial z$ is the vertical pesticide concentration gradient.

The energy balance Bowen-ratio technique for estimating evaporation is derived from the surface energy balance equation expressed as:

$$R_n - G - H - L_E = 0 \tag{4}$$

where $R_n$ is net radiation (W $m^{-2}$), $G$ is soil heat flux (W $m^{-2}$), and $H$ and $L_E$ are sensible and latent heat fluxes (W $m^{-2}$) respectively. The Bowen-ratio is the ratio of sensible to latent heat flux expressed as:

$$\beta = \frac{H}{L_E} = \gamma\frac{\partial T}{\partial e} \tag{5}$$

where $\beta$ is the Bowen-ratio, $H$ and $L_E$ have been previously defined, $\gamma$ is the thermodynamic value of the psychrometric constant, and $\partial T/\partial e$ are vertical gradients of temperature and vapour pressure. Latent heat flux can now be calculated as:

$$L_E = \frac{(R_n - G)}{1 + \beta} \tag{6}$$

where all terms have been previously defined.

Dual, a pre-emergence herbicide (active ingredient metolachlor), has become one of the herbicides of choice among many farmers in the Midwest in recent years. Metolachlor is a relatively non-volatile (vapour pressure $1.3 \times 10^{-5}$ mm Hg at 20°C), clear to amber colour odourless liquid. Parochetti (1978) found that while 0.1% of metolachlor was lost from the soil surface after eight days, approximately 50% was lost from a glass surface, and from 11.5 to 36.6% volatilised from various plant residue surfaces. Burkhard and Guth (1981) calculated that metolachlor at a concentration of 80 $\mu g\ g^{-1}$ on a wet soil weight basis volatilises from moist soil at 20°C at a rate of between 1.5 and 4.5 ng $cm^{-2}\ h^{-1}$ when the air flow rate is 30 L $h^{-1}$. Nations and Hallberg (1992) reported that metolachlor along with 13 other pesticides were found in rainfall samples in northeastern Iowa. The mode of metolachlor transport into the atmosphere is not well understood.

The objective of this study was to develop a quantitative analysis of the volatilisation of metolachlor from bare soil into the atmosphere using turbulence exchange models and transfer coefficients derived from temperature, and wind speed profiles above a bare soil surface.

## 2. Material and methods

The field selected for this study is located in the Walnut Creek watershed approximately 10 kilometres south of Ames, Iowa. The geographic position of the field is 41° 57′ 53″ latitude and 93° 39′ 8″ longitude. The area of the field was approximately 237 hectares which was planted to corn using conventional tillage practices on a Webster soil series (fine-loamy, mixed, mesic Typic Haplaquolls).

An array of micrometeorological instruments were erected 150 metres from the southern edge of the field to measure the surface energy balance components of net radiation, soil, sensible, and latent heat flux densities. Sensible and latent heat flux were calculated using a Bowen-ratio technique originally described by Bowen (1926) as a method for estimating latent heat loss from any water surface. The Bowen-ratio station consisted of two aspirated psychrometers vertically placed at 25 and 125 cm above the soil surface. The psychrometer housing was constructed of schedule 40 poly(vinylchloride) (PVC) pipe which contained a pair of dry and wet bulb thermocouples. The thermocouples were made of 0.4 mm diameter copper-constantan (Type T) wire. The wet bulb thermocouple was a type T junction encased in a porous ceramic wick which was connected to a reservoir of distilled water maintained at a constant level to provide a constant supply of water to the ceramic wick. A Micronel fan was attached to the back of the psychrometer to provide a constant air flow of 3.8 m $s^{-1}$ past the dry and wet bulb thermocouples. The thermocouple accuracy was rated at 0.2°C. The vertical position of the psychrometers was held constant for the duration of the study. Prior to the experiment the two psychrometers were placed at the same height for monitoring dry and wet bulb temperatures continuously for 72 hours to assess potential sensor bias. Temperature values were within the error specifications of the copper-constantan wire. A Q6 REBS net radiometer was used to measure net radiation, while a REBS soil heat flux plate was employed to measure soil heat flux. The scan rate for all sensors was 30 seconds with a 30-min output to a Campbell Scientific 21× datalogger.

### *2.1 Pesticide sampling mast*

The pesticide sampling mast was patterned after Majewski *et al.* (1990) in which two Teflon canisters (Savillex Corporation) were placed at two heights coincident with the heights of the Bowen-ratio psychrometers and were filled with 40 ml XAD-8, a non-ionic polymeric adsorbent made of macroparticulate resin beads. The sampling canisters were connected to a Thomas Industries diaphragm pump capable of total sampling rate of 30 L $min^{-1}$. The flow rate through each of the Teflon canisters was 10 L $min^{-1}$.

The air over the treated field was sampled continuously beginning at 0800 hours and ending at 1600 hours using the same XAD-8 material for the eight-hour period. Analysis for metolachlor concentration was achieved by gas chromatography with N-P detection followed by verification analysis using gas chromatography/mass selective detection (GC/MSD). The recovery and extraction efficiency rates for the analytical procedures ranged between 90 and 97% for metolachlor.

## 3. Results and discussion

### *3.1 Micrometeorological parameters*

A typical illustration of the partitioning of the surface energy balance at the surface is shown in Figure 1 for 6 May, 1992. The partitioning is dominated by net radiation which generally provides the upper limit of energy for the evaporation of water from soil and plant tissue surfaces in the Midwest region of the United States. On this day latent and sensible heat values can be seen to be similar in magnitude from the early morning hours to about midday at which time latent heat values become greater than the sensible heat throughout the afternoon hours. During the morning hours the energy from net radiation is partitioned between evaporating surface soil moisture (latent heat) and heating of the bare soil surface (sensible heat).

### *3.2 Metolachlor flux*

A comparison of the two micrometeorological techniques for estimating metolachlor flux is shown Figure 1a where it can be seen that for both techniques the temporal distributions for metolachlor fluxes were similar. As flux rates increased the aerodynamic method produced higher fluxes than the Bowen-ratio technique; however the order of magnitude for metolachlor flux was consistent. During the first four days after application, metolachlor fluxes ranged between approximately 20 and 60 $\mu g\ m^{-2}\ h^{-1}$ which was followed by a large plume of volatilisation on 12 May (DOY 132).

Fluxes returned to about 40 $\mu$g m$^{-2}$ h$^{-1}$ and lower, then sharply increased again on 18 May (DOY 138). Two large metolachlor plumes occurred on days with a brief precipitation event during the early morning hours (before sunrise) of the two

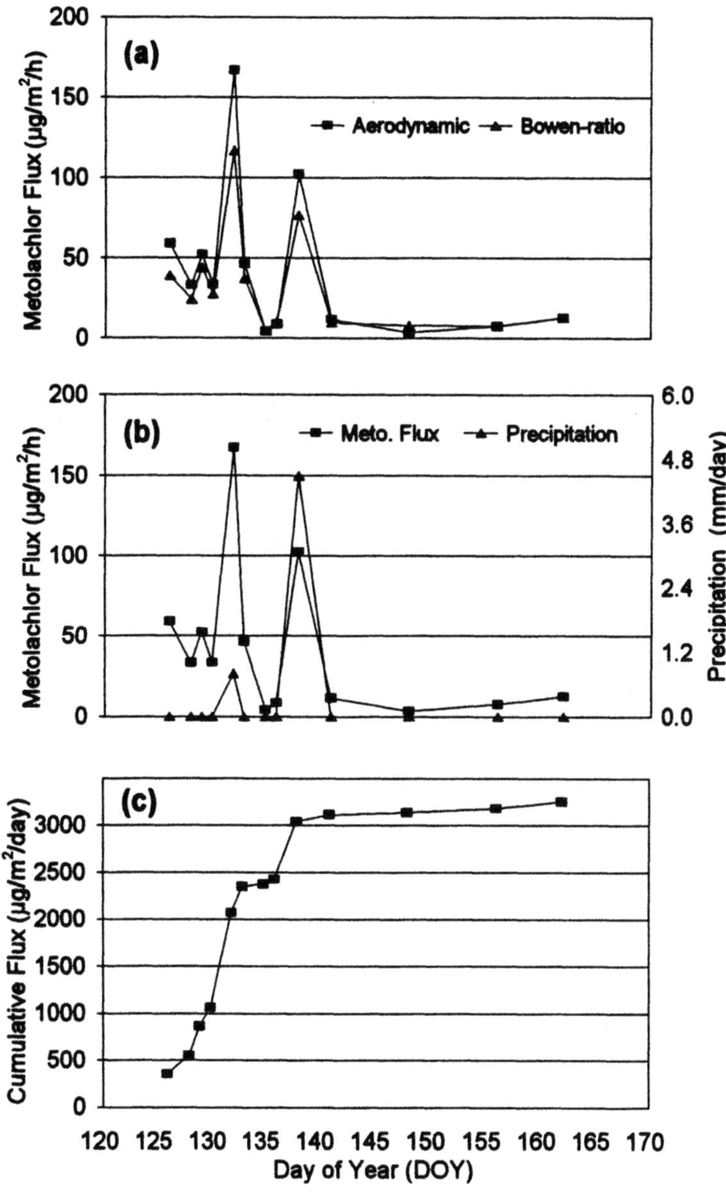


*Figure 1.* (a) Average temporal distribution of metolachlor flux as calculated using the aerodynamic and Bowen-ratio techniques; (b) metolachlor flux using the aerodynamic method and daily precipitation; (c) cumulative metolachlor flux calculated for the period from DOY 126 to DOY 162.

respective days (Figure 1b). Increased metolachlor volatilisation was associated with increased surface soil water content from the light precipitation events. Metolachlor volatilisation in this study was well-coupled to soil water content in the top 1 cm depth of soil. Both precipitation events were relatively light (0.8 and 4.5 mm), providing sufficient moisture to wet the surface of the soil but insufficient moisture to promote infiltration of metolachlor below the soil surface. It is important to understand that when making environmental measurements for pesticide vapour, and in particular pesticides with low vapour pressures, absolute concentrations of pesticide vapour are often very difficult if not impossible to quantify. In general, consistency in the magnitudes of pesticide concentrations is often a good indication that the amounts of metolachlor trapped from the profile sampling masts represent a reliable measurement of pesticide concentration. Differences between the aerodynamic and Bowen-ratio methods are believed to be related to errors associated with the assumption that the transfer coefficients and stability functions calculated for microclimate parameters are equivalent for pesticide vapours when in fact no such equivalence may exist.

The areas under the curves in Figure 1a were integrated to determine cumulative volatilisation from the field in 1992 and are presented in Figure 1c. These results show that 94% of volatilisation losses occurred during approximately the first ten days after application. After the initial period of accumulation, metolachlor volatilisation from the soil surface was minimal. Over the next 30 days the amount lost was 6% of the total lost in the 40-day period. Volatilisation losses were significantly reduced due to source depletion of metolachlor on the soil surface. The results from this study indicate that measurable and significant quantities of metolachlor had volatilised from the soil surface during the first ten days after application using two micrometeorological techniques.

## Note

The USDA-ARS does not endorse any product or company; trade/product names are only for facilitation purposes for the reader.

## References


Basile M, Senesi N and Lamberti F 1986 A study of some factors affecting volatilization losses of 1,3-dichloropropene (1,3-D) from soil. Agric. Ecosys. Environ. 17, 269–279.

Bowen I S 1926 The ratio of heat losses by conduction and by evaporation from any water surface. Phys. Rev. 27, 779.

Burkhard N and Guth J A 1981 Rate of volatilization of pesticides from soil surfaces; comparison of calculated results with those determined in a laboratory model system. Pestic. Sci. 12, 37–44.

Caro J H and Taylor A W 1971 Pathways of loss of dieldrin from soils under field conditions. J. Agric. Food Chem. 19, 379–384.

Farmer W J, Spencer W F, Shepherd R A and Cliath M M 1974 Effect of flooding and organic matter applications on DDT residues in soil. J. Environ. Qual. 3, 343–346.

Glotfelty E D, Taylor A W, Turner B C and Zoller W H 1984 Volatilization surface-applied pesticides from fallow soil. J. Agric. Food Chem. 32, 639–643.

Majewski M S, Glotfelty D E, Paw U K T and Seiber J N 1990 A field comparison of several methods for measuring pesticide evaporation rates from soil. Environ. Sci. Technol. 24, 1490–1502.

Nations B K and Hallberg G R 1992 Pesticides in Iowa Precipitation. J. Environ. Qual. 21, 486–492.

Parochetti J V 1978 Photodecomposition, volatility and leaching of atrazine, simazine, alachlor, and metolachlor from soil and plant material. Weed Sci. Soc. Am. (Abstract No. 17).

Parochetti J V and Warren G F 1966 Vapor losses of IPC and CIPC. Weeds 14, 281–285.

Raupach M R and Legg B J 1984 The uses and limitations of flux-gradient relationships in micrometeorology. In Evaporation from Plant Communities. Ed. M L Sharma. pp 119–132. Elsevier, Amsterdam.

Spencer W F and Cliath M M 1973 Pesticide volatilization as related to water loss from soil. J. Environ. Qual. 2, 284–289.

Spencer W F and Farmer W J 1980 Assessment of the vapor behavior of toxic organic chemical. In Dynamics, Exposure, and Hazardous Assessments of Toxic Chemicals in the Environment. Ed. R Hague. pp 142–162. Ann Arbor Sci., Ann Arbor, MI.

Spencer W F, Cliath M M and Farmer W J 1969 Vapor density of soil-applied dieldrin as related to soil-water content, temperature, and dieldrin concentration. Soil Sci. Soc. Am. Proc. 33, 509–511.

Thom A S, Steward J B, Oliver H R and Gash J H 1975 Comparison of aerodynamic and energy budget estimates of fluxes over a pine forest. Q.J.R. Meteorol. Soc. 101, 93.

Webb E K 1975 Evaporation from catchments. In Prediction in Catchment Hydrology. Eds T G Chapman and F X Dunin. pp 203–236. Australian Academy of Science, Canberra.

# 11
# Titanium dioxide photocatalysis for the treatment of contaminated waters

LEA MUSZKAT

## 1. Introduction

The photocatalytic process takes place by a combination of photochemistry and catalysis, which implies that light and a catalyst are necessary to bring about a chemical reaction. It is well-known that ultraviolet light below about 390 nm excites the titanium dioxide semiconductor to promote photoexcited electrons ($e^-_{cb}$) in the conduction band and form positive holes ($h^+_{vb}$) in the valence band. The holes migrate to the oxide surface, generating highly oxidising species which carry out degradation of organic water contaminants. This process enables reduction to occur at the negative sites.

Photocatalysis is found to be effective for removing heavy metals from water by reduction of metal ions to their metallic form, which plate out onto the photocatalyst. Consequently, two actions can be accomplished simultaneously: organics can be decomposed by photocatalytic oxidation while toxic metals, and also precious metals, are removed by reduction. In addition to removal by electrochemical reduction and oxidation, many pollutants, both organic and inorganic, can be removed by adsorption to the photocatalyst surface. Considerable public attention has been focused on this possibility (Carey *et al.*, 1976; Dieter, 1989; Hafner, 1989; Prairie *et al.*, 1993). Several reviews have recently been published which discuss the reaction mechanism and give examples of promising laboratory and field studies (Fettwell, 1989; Hecht, 1990, Robin, 1989; Thornton, 1989). However, while many of the laboratory studies demonstrated the feasibility of the method for almost all classes of hazardous chemicals, mechanistic studies have not been extensive. Titanium dioxide in its anatase form shows the highest photocatalytic detoxification efficiency. It is a wide gap semiconductor with Eg~3.2 eV. Therefore only light below 390 nm is absorbed and capable of forming the $e^-/h^+$ pairs which allow the process to occur. This amounts to 5% of the total solar energy reaching the soil surface.

## 2. Materials and methods

The experimental details are described in previous studies (Muszkat *et al.*, 1992, 1995). Titanium dioxide was used at 100 mg $L^{-1}$.

*D. Rosen* et al. *(eds): Modern Agriculture and the Environment, 127–132.*

## 3. Results and discussion

### *3.1 Solar photocatalytic decomposition of polluted well water*

This occurs by treatment with $Fe^{3+}$, and titanium dioxide and hydrogen peroxide (Muszkat *et al.*, 1992).

#### 3.1.1 Halocarbon solvents

Results are presented in Table 1. The concentration of the very common solvents 1,2-dichloroethane, trichloroethylene and tetrachloroethylene was decreased by at least one order of magnitude. The results for the very volatile chloroform were erratic (not shown), probably due to evaporation losses. At least 15 hours of sunlight were required in order to reduce the extremely high initial concentrations of some of these solvents (about 2000 ppb) by more than one order of magnitude.

#### 3.1.2 Halobenzenes

The initial concentrations of several halobenzenes in the well water were in the range 0.5–20 ppb. As shown in Table 1, exposure shorter than 3.5 h was sufficient for complete detoxification of most compounds of this group.

#### 3.1.3 Aromatic hydrocarbons

The two xylenes were completely degraded within less than 3.5 h to below 0.1 ppb.

#### 3.1.4 Triazine herbicides

The triazines detected in the well water were atrazine, trietazine, propazine, terbutryn and an isomer of atrazine ($m/z = 215$). These molecules were decomposed to less than 0.1 ppb. It can be seen (Table 1) that 15 h exposure was sufficient to reduce the level of atrazine, of its isomer and of propazine to less than 5 ppb, but trietazine and terbutryn required a longer time.

#### 3.1.5 Acetamides, bromacil and urea herbicides

Three widely used herbicides, alachlor, propachlor and bromacil, were found at high levels in the contaminated well. Exposure of 3.5 h resulted in complete disappearance of the urea derivatives methobromuron and 1-methoxy-1-methyl-3-phenylurea. In the case of alachlor the initial concentration was relatively high and degradation required a longer exposure time.

#### 3.1.6 Chloroxynil, bromoxynil and some benzene derivatives

The herbicides chloroxynil and bromoxynil (dichloro and dibromo hydroxybenzonitrile, respectively) and also dichloro benzaldehyde and di-tert-butyl hydroxy toluene (BHT), which were detected at levels of 0.5–6 ppb, were all completely degraded (to less than 0.1 ppb) within 15 h.

### 3.1.7 Phthalate esters and fatty acid derivatives

The long-chain carboxylic acids, tetradecanoic, hexadecanoic and octadecanoic acid, which appeared in rather minute concentrations (< 12 ppb) in the contaminated well,

*Table 1.* Solar photocatalytic degradation of polluted well water; treatment with $TiO_2$ (1 g $L^{-1}$), $Fe^{3+}$ (0.3 mM) and $H_2O_2$ (0.1 M).

| Compound | Concentration (ppb) | | | |
|---|---|---|---|---|
| | Untreated | Exposure time (h) | | |
| | | 3.5 | 15 | 75 |
| $CHCl_2–CH_2Cl$ | 3 | 1 | 0 | 1 |
| $CH_3–CCl_3$ | 5.5 | 0 | 0 | 0 |
| $CH_2Cl–CH_2Cl$ | 1950 | 750 | 70 | 105 |
| $CCl_4$ | 75 | 55 | 0 | 0 |
| $CHCl = CCl_2$ | 1900 | 160 | 4 | 1 |
| $CCl_2 = CCl_2$ | 2400 | 160 | 4 | 1 |
| Chlorobenzene | 21 | 0 | 0 | 0 |
| Dichlorobenzene I | 13 | 0.6 | 0 | 0 |
| Dichlorobenzene II | 3 | 0 | 0 | 0 |
| Bromoxylene I | 0.5 | 0 | 0 | 0 |
| Bromoxylene II | 0.5 | 0 | 0 | 0 |
| Trichlorobenzene I | 12 | 1 | 0 | 0 |
| Trichlorobenzene II | 8.5 | 0.5 | 0 | 0 |
| Bromo-t-butylbenzene | 5.5 | 0.6 | 0 | 0 |
| Toluene | 7 | 0 | 0 | 0 |
| Xylene-I | 3.5 | 0 | 0 | 0 |
| Xylene-II | 5.5 | 0 | 0 | 0 |
| Atrazine | 43 | 8.5 | 4 | 0 |
| Atrazine isomer | 2.5 | 0 | 0 | 0 |
| Trietazine | 23 | 9.5 | 5 | 0.3 |
| Propazine | 1 | 0 | 0 | 0 |
| Terbutryn | 4 | 3 | 3 | 0 |
| Alachlor | 640 | 3.5 | 0 | 0 |
| Propachlor | 1 | 0 | 0 | 0 |
| Bromacil | 115 | 40 | 6 | 0 |
| Metobromuron | 22 | 0 | 0 | 0 |
| 1-Methoxy-1-methyl-3-phenylurea | 11 | 0 | 0 | 0 |
| Chloroxynil | 6 | 4.5 | 0 | 0 |
| Bromoxynil | 3 | 0 | 0 | 0 |
| Dichlorobenzaldehyde | 0.5 | 0 | 0 | 0 |
| Dibutyl hydroxy toluene | 0.5 | 0 | 0 | 0 |
| Tetradecanoic acid | 4 | 4.5 | 6 | 2 |
| Hexadecanoic acid | 12 | 0 | 0 | 0 |
| Octadecanoic acid | 2 | 1.5 | 3.5 | 4 |
| 9-Octadecenamide | 3.5 | 0 | 0 | 0 |
| Diethyl phthalate | 0.5 | 0 | 0 | 0.2 |
| Dioctyl phthalate | 2.5 | 2 | 2 | 2 |
| Hydroxymethyl coumarin I | 36 | 0 | 0 | 0 |
| Hydroxymethyl coumarin II | 1 | 0 | 0 | 0 |
| Dibutoxymethane | 1.5 | 0.3 | 0 | 0 |

were essentially unaffected by photocatalysis. Octadeceneamide was degraded within 3.5 h to below 0.1 ppb, possibly by photolysis to octadecenoic acid. Diethyl and diisooctyl esters of phthalic acid were not appreciably degraded.

#### 3.1.8 Hydroxymethyl coumarins and dibutoxy methane

These compounds were completely removed during illumination. The two isomers of hydroxymethyl coumarin and butylmethyl pyrimidinedione (butyl methyl uracil) were degraded within 3.5 h but dibutoxymethane (formaldehyde dibutyl acetal) disappeared only after 15 h.

### *3.2 Titanium dioxide – solar photocatalysis in the absence of hydrogen peroxide (Muszkat et al., 1995)*

Field studies of polluted well water point to the possibility of inhibition of the photodegradation by some unknown factors. This inhibitory effect is prevented by hydrogen peroxide (Muszkat *et al.*, 1995). The data in Table 2 present the photodegradation of dilute malathion and atrazine solutions. It can be noted that photodegradation of these solutions is also effective in the absence of hydrogen peroxide. Comparison with the direct excitation, non-catalytic photodecomposition, emphasises the decisive contribution of titanium dioxide. Photocatalysis of malathion formulations (diluted to 70 ppm) was also examined (Table 3) in open and in polyethylene-covered systems. The half-life value for malathion is about 1 h and the reaction follows first order kinetics over several half-life times. The reaction efficiency seems to be satisfactory in closed systems. While the role of hydrogen peroxide in the photodegradation of trace organics at the sub-ppm levels is not clear, we consider that it is indispensable for the decomposition of inhibitory constituents of polluted water. We have encountered one such case when examining the photodegradation of groundwater from the polluted Nir-Galim well (Table 4).

### *3.3 Photocatalytic degradation of surfactants*

The removal of surfactants which reach waste water in increasingly growing amounts can be achieved by biodegradation. However, although this is the most important process in the waste water treatment plants, biodegradation may be problematic when dealing with some cationic and non-ionic surfactants which are not easily biodegraded. Thus alkylbenzenesulphonates and alkylphenol-polyethoxylates are resistant to degradation by bacteria. In the case of nonylphenolethoxylate, biodegradation results in formation of nonyl phenol, a toxic product (Giger *et al.*, 1984; Swisher, 1986). Recently it has been shown that titanium dioxide photocatalysed degradation is one of the more promising methods to mineralise these substances. Hidaka *et al.* (1993), who studied the kinetics of surfactant photodegradation, classified it into fast and slow events. The fast events include the formation of the electron hole pair and of hydroxyl radicals, whereas the slow

*Table 2.* $TiO_2$ Photocatalysed decomposition.

| | Residual concentration (ppb) | | | |
|---|---|---|---|---|
| | $TiO_2$ (100 mg $L^{-1}$) | | Blank (no $TiO_2$) | |
| Exposure (h) | Atrazine | Malathion | Atrazine | Malathion |
| 0 | 90 | 104 | 90 | 104 |
| 4 | 17 | 0.34 | 81 | 83 |
| 12 | 0.1 | 0.2 | 73 | 84 |

*Table 3.* Effect of solar irradiation through polyethylene film on $TiO_2$-photocatalysed decomposition of malathion (70 ppm)[abc].

| | Residual concentration (ppm) | |
|---|---|---|
| Exposure (h) | Open[d] | Polyethylene cover |
| 1 | 34 | 35 |
| 3 | 9 | 13 |
| 6 | 2.5 | 4.5 |
| 12 | 0.2 | 0.3 |
| 18 | > 0.2 ppb | 0.8 ppb |

[a] Polyethylene film, 100 $\mu$m thick.
[b] In the presence of 1.15 mg $L^{-1}$ detergent TPBS. Detergent is decomposed completely after 1 h of irradiation.
[c] Irradiation in crystallizer (diameter, 22.5 cm); solution volume, 2 L; in the absence of $H_2O_2$.
[d] Concentrations corrected for evaporation.

*Table 4.* Solar, $TiO_2$-photocatalysed decomposition of organic contaminants in water of well Nir Galim 1.

| Compound | Concentration (ppb) after exposure for | | | |
|---|---|---|---|---|
| | 0 h | 1 h | 5 h | 10 h |
| Dichloroaniline | 45 | 13 | 12 | 4 |
| Benzopyran (*m/z* 176) | 10 | 4 | 3.5 | 3 |
| Atrazine | 23 | 9 | 6 | 5 |
| Propazine | 6 | 3 | 3 | 3 |
| Alachlor | 273 | 86 | 83 | 83 |
| Prometryn | 25 | 12 | 11 | 11 |
| Bromacil | 27 | 13 | 12 | 10 |
| Cyanobenzoate (*m/z* 253) | 24 | 12 | 10 | 8 |

processes are the benzene ring opening, the intermediate formation of peroxides, aldehydes and carboxylic acids, and finally the mineralisation to carbon dioxide. A significant factor in the photodegradation kinetics is the adsorption of surfactants onto the photocatalyst surface. The following order of photodegradation rate has been noticed: anionic > nonionic > cationic surfactants. In the surfactant groups the rate of photooxidation changes according to the specific structure, in the order, aromatic > oxyethylene > alkyl group.

### *3.4 Intermediate products in the photocatalysed oxidation of organics*

The formation of carboxylic acid intermediates, mainly formic acid, was observed in the photocatalysis of surfactants (Hidaka *et al.*, 1993). Phosphorus containing surfactant was mineralised to form $PO_4^{3-}$ ions, and nitrogen containing surfactant yielded ammonium, $NH_4^+$, which is very stable at acidic pH. These results may indicate the necessity of a post-treatment or the combination with other methods. In the photocatalytic oxidation of 1,2,4-trichlorobenzene, the formation of biphenyl derivatives has been observed (Pelizzetti *et al.*, 1993). However, these potentially detrimental intermediates were completely degraded following an irradiation of two additional hours. In the photocatalysis of aromatic pollutants (Pichat *et al.*, 1993) it has been shown that hydroxylation of the aromatic nucleus is a general pattern which is sometimes accompanied by side chain abstraction and decarboxylation. The photocatalytic degradability of chlorophenols was shown to be correlated with descriptors such as the Hammet coefficients and with the octanol–water partition coefficient. The rate of pollutant disappearance is slowed by chloride, sulphate and phosphate, which interact with the photocatalyst at pH below the zero charge point (6.3 for titanium dioxide) and at concentrations higher than those of natural water.

## References

Carey J H, Lawrence J and Tosine H M 1976 Bull. Environ. Contam. Toxicol., 16, 697.

Dieter H H 1989 Wechselwirkung, 43, 4.

Fettwell J 1989 New Sci. 10 June, 36.

Giger W, Brunner P H and Shaffner C 1984 Science 225, 623.

Hafner M 1989 Nature 10, 20.

Hecht J 1990 J. New Sci. 14 April, 28.

Hidaka H, Zhao J, Nohara K *et al.* Photocatalyzed mineralization of non-ionic, cationic and anionic surfactants at $TiO_2/H_2O$ interfaces 1993 In Photocatalytic Purification and Treatment of Water and Air. Eds D F Ollis and H Al-Ekabi. pp 251–259 Elsevier.

Muszkat L, Hallman M and Raucher D 1992 J. Photochem. Photobiol. Chem. 65, 409.

Muszkat L, Bir L and Feigelson L 1995 J. Photochem. Photobiol. Chem (in press).

Pelizzetti E, Minero C, Sega M and Vincenti M 1993 Formation and disappearance of biphenyl derivatives in photocatalytic transformation of trichlorobenzene In Photocatalytic Purification and Treatment of Water and Air. Eds D F Ollis and H Al-Ekabi. pp 291–300. Elsevier.

Pichat P, Guillard C, Maillard C, Amairic L and D'Oliveira J 1993 $TiO_2$ photocatalytic destruction of water aromatic pollutants, In Photocatalytic Purification and Treatment of Water and Air. Eds D F Ollis and H Al-Ekabi. pp 207–223. Elsevier.

Prairie M R, Stange B M and Evans L R 1993 $TiO_2$ photocatalysis for the destruction of organics and the reduction of heavy metals In Photocatalytic Purification and Treatment of Water and Air. Eds D F Ollis and H Al-Ekabi. p 353. Elsevier.

Robin E 1989 San Francisco Examiner, 20 May, D-3.

Swisher R D Ed. 1986 Surfactant Biodegradation. M Dekker, New York.

Thornton J 1989 SERI S&T in Review, 8.

# SECTION II

## Fertilisers in Agroecosystems

# 12
# Site-oriented ecosystem management: precondition to reducing the contamination of waters and the atmosphere

R.E. HANTSCHEL AND F. BEESE

## 1. Introduction

During the last decades, agricultural land-use practices have changed tremendously. More intensive use and the subsequent changes in cultivation systems in Central Europe have caused the following main effects (Tivy, 1993):

- reduction of farm labour;
- increase of farm and field size;
- increased mechanisation and application of anthropogenic resources (mineral fertilisers, pesticides, diesel);
- discoupling of nutrient cycles by specialised systems (i.e. monocultures);
- separation of plant and animal production;
- simplification and standardisation of crop rotations; and
- neglect of site variability (in soil physical, soil chemical and soil biological properties, as well as in micrometeorological and relief properties).

In large areas, environmental problems are caused by these changed land-use practices (Knauer, 1993):

- destruction of landscape heterogeneity by removing hedges, boundary strips and humid biotopes;
- eutrophication of neighbouring natural ecosystems;
- impoverishment of local flora and fauna elements;
- deterioration of soil structure and therefore increased soil compaction and erosion;
- pollution of ground and surface waters by increased nitrate and pesticide loads;
- pollution of air and atmosphere by increased emissions of trace gases ($CO_2$, $CH_4$, $NH_3$, $N_2O$); and
- reduction of food quality.

One of the endangered areas in Upper Bavaria, Germany, is the region of the tertiary hillslopes north of Munich. It is an important area for agricultural production.

To develop sustainable management strategies the ecosystem research centre 'Forschungsverbund Agrarökosysteme München' (FAM) is running an experimental farm in this region. The strategies will combine the economic benefits of agriculture

*D. Rosen* et al. *(eds): Modern Agriculture and the Environment, 135–145.*

with the preservation and regeneration of natural resources. The main goal is to improve the methodological tools for monitoring and predicting changes in agroecosystems and their neighbouring ecosystems. The available methods are still insufficient to estimate changes:

- representative for single crops;
- representative for fields; and
- representative for different land-use systems.

This also holds true for the prognosis of future effects and their evaluation with respect to sustainable agricultural land use.

## 2. Materials and methods

The methodological approach of the FAM network is based on the integrated investigation of systems at three levels of aggregation: a set of laboratory systems; different model ecosystems; and a section of an agricultural landscape (Figure 1). Important processes and their dependency on selected key parameters (climate, soil type, land use) are investigated under defined laboratory conditions. Results and parameters are used to feed deterministic simulation models. Model ecosystems defined as controlled quasinatural systems are used to improve the understanding of the interaction of single processes as well as a first test for the simulation models and

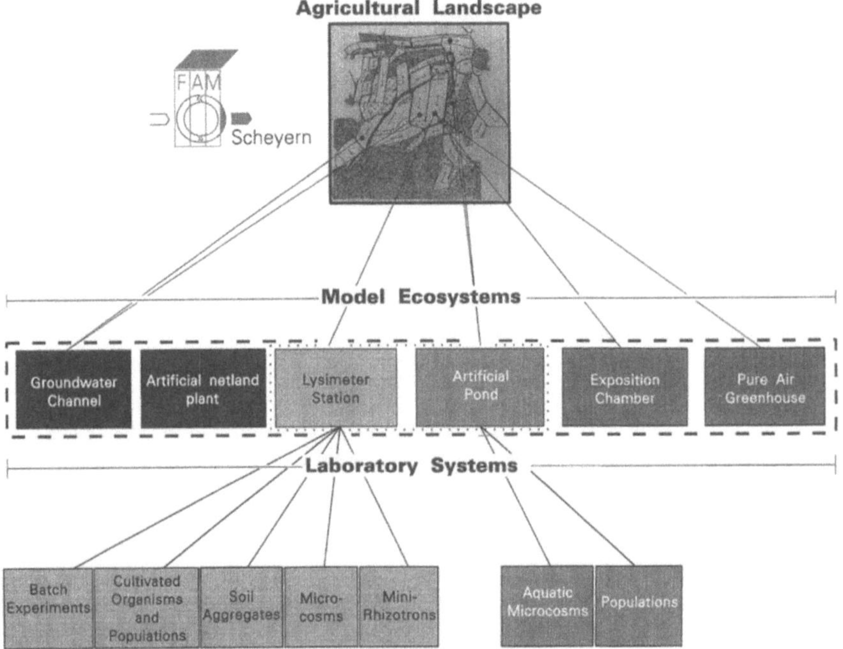


*Figure 1.* Hierarchy of investigated systems and methodological approach of the FAM project.

their adaptation to a larger scale. On the landscape scale, energy, water and element fluxes through the compartments of the single ecosystems and between them are studied. This approach was developed in the forest ecosystem studies during the 1980s (Ulrich, 1987). The dynamics of crops and the indicators of the organism communities are also studied.

To collect and handle all data, prove and use them for different kinds of evaluation, an approach as illustrated in Figure 2 has been developed (Lenz and Schmitt, personal communication, 1994). A geographical information system (GIS) combined with the central database is used to produce area information of ecosystem parameters. Different approaches of an integrated assessment towards critical points of the system have been chosen. One is the definition of environmental standards which are used in an agroecological information system to evaluate ecological and economic risks of different land-use scenarios.

The field study of the FAM, called 'Long-term Study Scheyern', is run on a 150-ha experimental farm north of Munich. The hilly landscape derived from tertiary sediments is covered with loess, which is eroded mainly from the hilltops downwards, forming a thick colluvium in the valley bottoms. The high variability of the landscape morphology and the geological parent material is reflected in the patchy distribution of different soils (sandy, loamy and clay soils with different content of gravel) and their ecological factors. Mean annual precipitation in the area is 833 mm and the mean air temperature is 7.4°C. Part of the precipitation comes down as storms, enhancing erosion of the loess cover.

During the first period of the FAM project (1990–1992), basic parameters (soils, flora, biota) of the landscape were characterised along a 50 × 50 m grid (Figure 3) in order to characterise the spatial variability on about 400 points and to make an inventory of the biotic and abiotic state initially present in the landscape. During the first main phase of the project (1993–1997), the parameters were recorded on 21 permanent measuring points, each representative of certain soil types, landscape morphology and crop rotations. At the end of the initial phase, in autumn 1992, the landscape was restructured based on the ecological information obtained. An integrated and an ecological land-use system were installed on 45–60 ha each. Hedges, boundary strips and biotopes have been set up on about 15–20% of the total area. Steep slopes have been transferred to grassland left for succession.

The dynamics of changes in the development of both systems will be studied during the following years. This will help to quantify the development of the two land-use systems ecologically and economically.

## 3. Results and discussion

Applying geostatistics to the soil data recorded on the nearly 400 sampling points provides good area information to study further developments. Variograms were used as the basis for spatial interpolation and kriging was used for transfer (Webster and Oliver, 1990). As an example, the distribution of plant-available water in the effective rooting zone of cereals is shown in Figure 4. The capacity differs by a

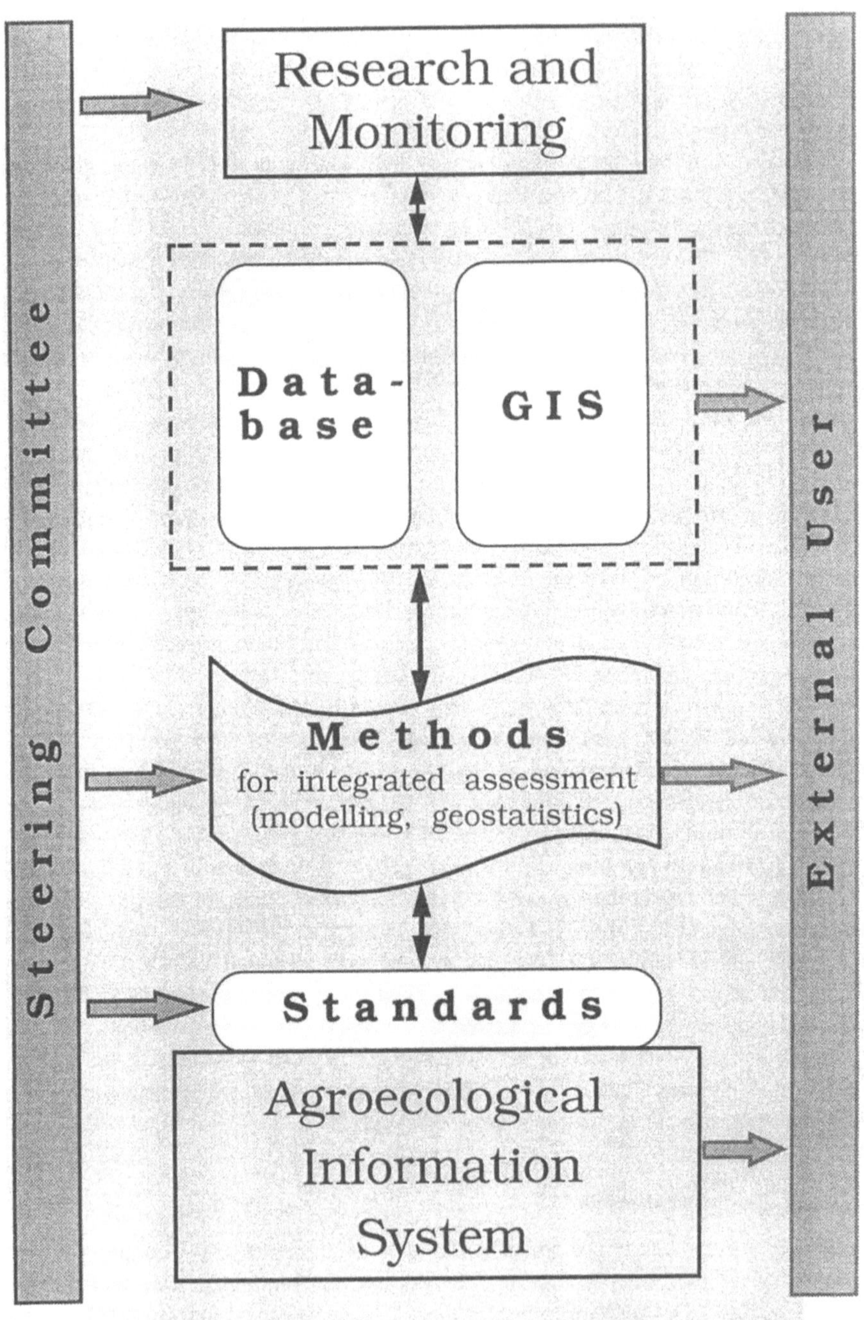


*Figure 2.* Scheme for the data and modelling management in the FAM project.

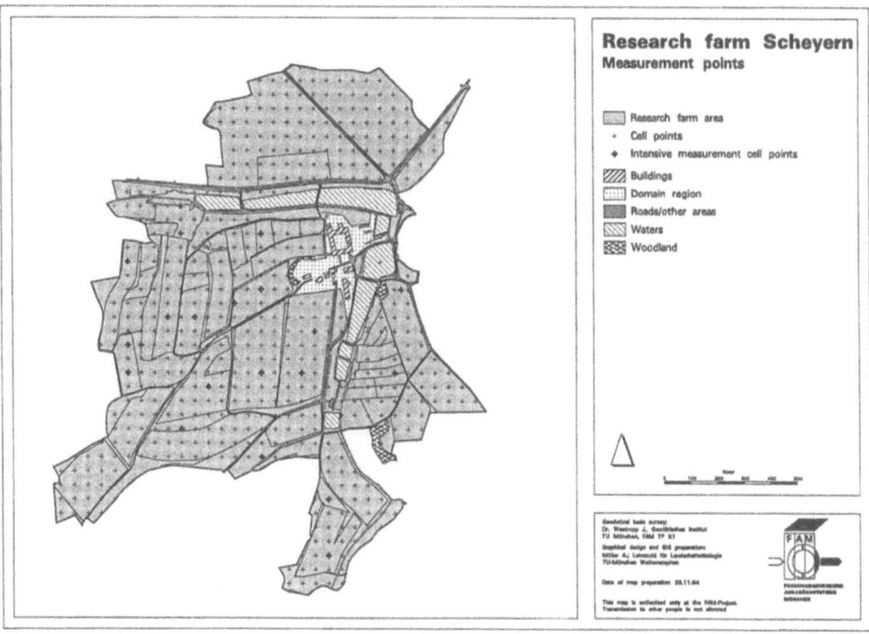


*Figure 3.* Map of the experimental farm in Scheyern with all measurement points.

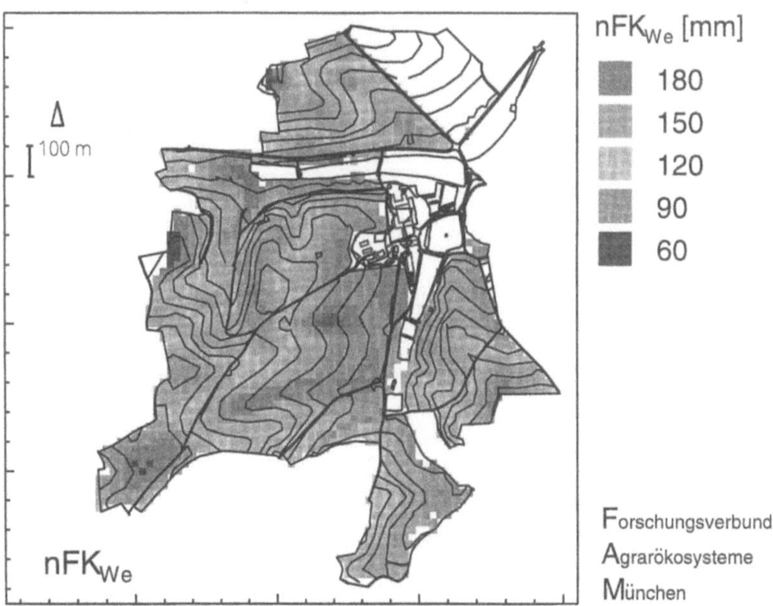


*Figure 4.* Map of plant-available water in the rooting zone ($nFK_{We}$) on the experimental farm in Scheyern.

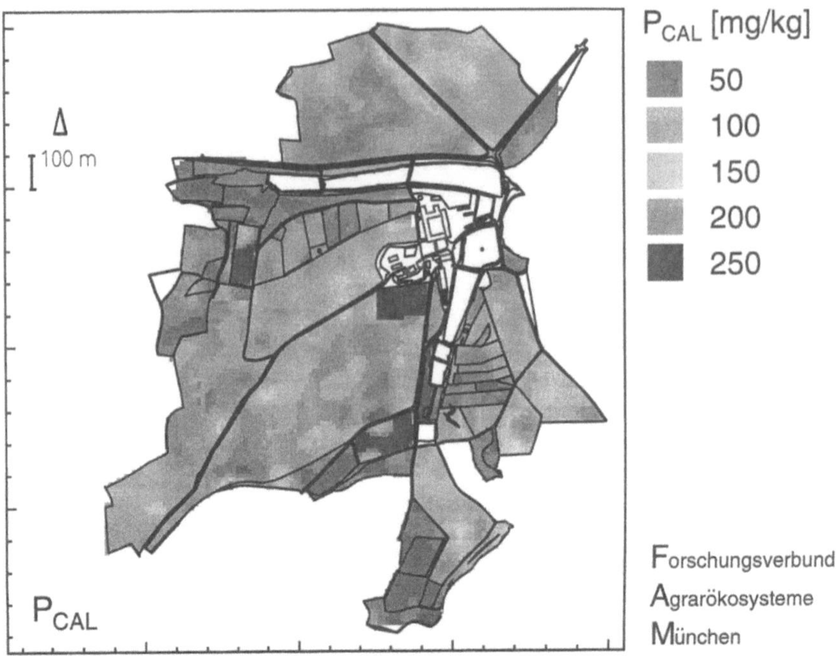


*Figure 5.* Map of plant-available phosphorus on the experimental farm in Scheyern.

factor of 3 within the area. The distribution is closely related to the main soil types (Sinowski, 1995). Two geomorphological processes explain the similarity of both maps (Schwertmann *et al.*, 1993):

- Solifluction during the Pleistocene exposed tertiary sediments on the hilltops. These are mainly characterised by medium or low water capacities. In the valley bottoms, the Quaternary sediments with high water capacities were accumulated.
- Water erosion has transported silty and organic material with high water capacity from the tops and slopes towards the valley bottoms.

Other soil physical parameters, for example grain size distribution, also reflect the spatial variability shown. In contrast to these results, the basic soil chemical parameters do not show site differentiation. Especially in the top horizon (plough layer), quite homogeneous concentrations of organic carbon, organic nitrogen, pH and plant-available potassium were mapped (Sinowski, 1995). Figure 5 shows the similar and very high amounts of plant-available phosphorus on all fields. The distribution of nutrients is strongly influenced by land-use practices; as management was intensive in former times, contents of P and K are very high and fertilisation of these nutrients has now been stopped. Soil biological parameters such as microbial carbon and abundance of Collembola were also equilibrated on the investigated area by the former land-use practices (Fromm *et al.*, 1993). The only exception was

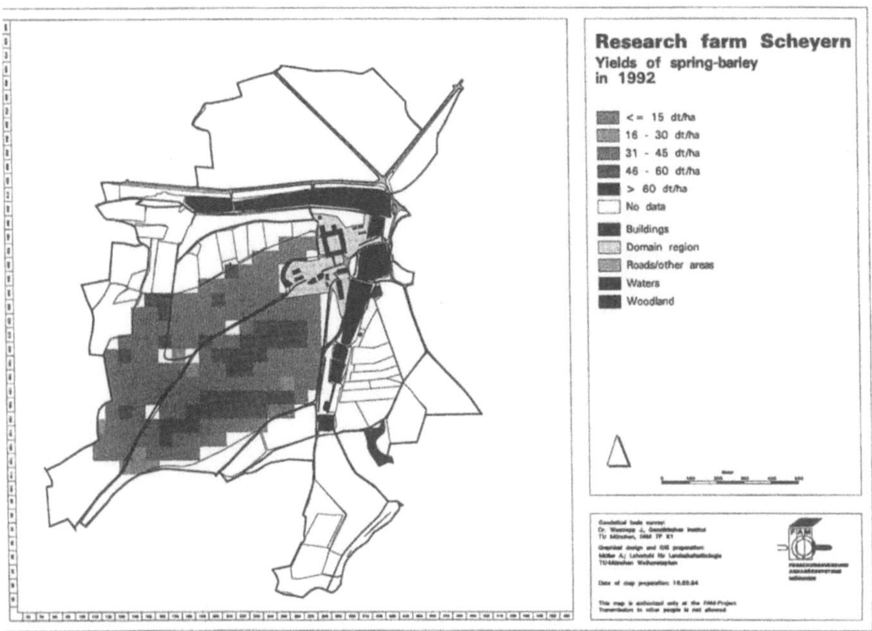


*Figure 6.* Map of yields of spring barley in 1992 on the experimental farm in Scheyern.

detected in the area of former hop fields where soil microbes and animals are reduced by the high available copper content in the soil, which was heavily treated with chemicals during hop monoculture (Filser *et al.*, 1995).

The spatial heterogeneity of soil physical and chemical parameters and processes should influence the yields on the fields. To prove these interactions, the yields of winter wheat in 1991 and spring barley in 1992 were determined by means of a global positioning system which helps to localise the harvesting machine in the field, and a flow-through balance which registers the weight of the grain. Using this system (Auernhammer *et al.*, 1994), yield maps of different scales could be drawn. An example for spring barley is shown in Figure 6. Grain yields of barley and also wheat in 1991 were highly variable in the investigated landscape section and differed by at least a factor of 2.

The mineral nitrogen in the rooting zone showed no pronounced spatial pattern in 1992 (Hantschel and Stenger, 1997). The differences in grain yields – which are well reflected in the differential nitrogen uptake of the barley crop (Hantschel and Stenger, 1997) – could not be explained by the nitrogen availability. Scheurer (1995) developed a GIS-coupled water model to describe the water budget in the landscape depending on relief, land use and climatic conditions. This landscape model produces maps of interflow and transpiration. The spatial distribution of the total loss of water by transpiration during the vegetative period of 1992 (Figure 7) fits very well with the map of grain yields. These results indicate that under the present

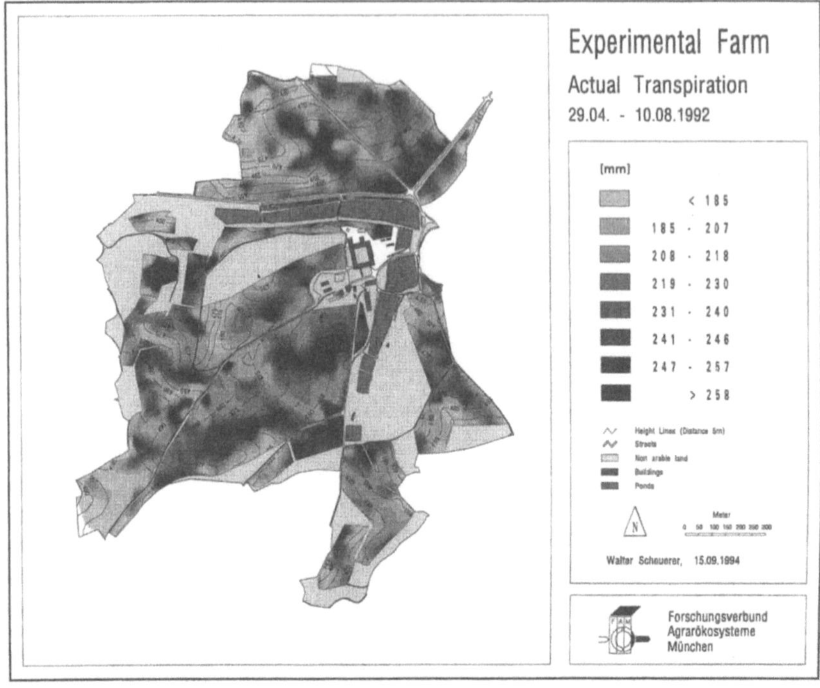


*Figure 7.* Actual transpiration of spring barley during the vegetative period of 1992 on the experimental farm in Scheyern.

management practices, the differentiation of yields in the agricultural landscape studied is mainly caused by atmospheric water availability and demand, because nutrient contents are kept at a high level and there are no limiting factors for growth.

The spatial information on soil parameters, plant uptake of nutrients and landscape water fluxes can also be used to identify areas which are sensitive to losses from nitrogen from pollutants to the ground and surface waters by leaching, interflow or erosion. Using this information on soil parameters, fertilising and cropping concepts can be improved to reduce environmental pollution without negative effects on yields.

However, it is important not only to detect sensitive areas or 'hot spots', but also to regard so-called 'hot periods' (climatic events) or 'hot events' (cultivation events) which contribute substantially to the yearly flux balance of matter. As an example, Table 1 depicts the annual $N_2O$–N fluxes in two crops managed differently. It is surprising that the highest rates of the annual emissions take place during winter time in both systems (Flessa *et al.*, 1995). Hantschel and Stenger (1997) could show in a soil-heating experiment that the $N_2O$ emissions dramatically increase during freezing–thawing cycles. It is not possible to reduce emissions during these events if their relationship to soil factors and cultivation history is not known. Another important period of high emissions, a 'hot event' after mineral or organic

*Table 1.* Emissions of $N_2O$–N from soil in two land-use systems from July 1992 to July 1993.

| | Yearly emission $N_2O$–N (kg/ha) | Emission during freezing–thawing cycles (January 1993) (% of yearly emission) | Emission after fertilisation (June–July 1993) (% of yearly emission) |
|---|---|---|---|
| Integrated system | | | |
| Wheat | 16.8 | 26 | 31 |
| Ecological system | | | |
| Sunflower | 9.6 | 45 | 7 |

fertilisation, can be influenced by the mode of fertiliser application or the schedule of application. Process studies in laboratory systems are necessary for a better understanding of the main factors influencing $N_2O$ emissions after fertilisation.

To define the frame within which sustainable land use is guaranteed, it is necessary to evaluate the matter fluxes and status of the system. One of the goals is the development of a new evaluation scheme which is generally applicable to different land-use strategies. The evaluation scheme used in this project (Figure 8) can be regarded as an extended critical-load approach (Nilsson, 1988). It shows much more flexibility to include specific site conditions than the widely used fixed threshold values and allows inclusion of dynamic developments. According to land-use scenarios, the critical inputs, critical operations, critical states and critical outputs to neighbouring systems have to be determined using the site and specific

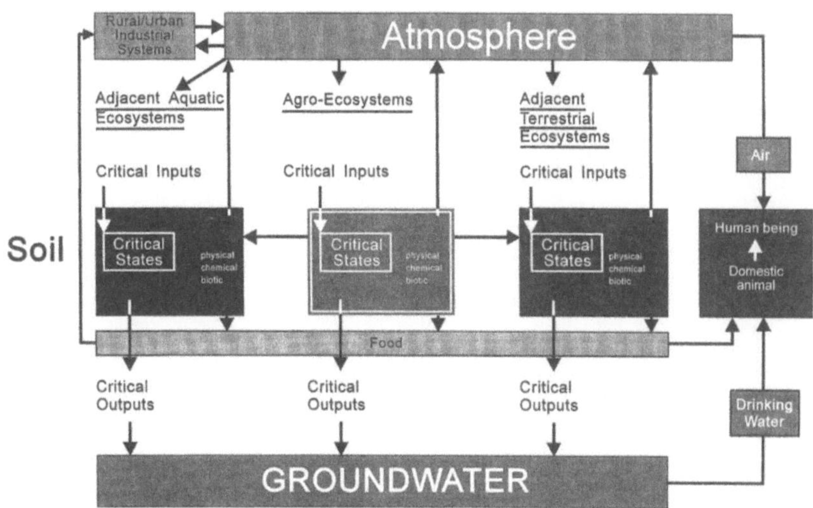


*Figure 8.* Evaluation scheme of the FAM project.

knowledge of the ecosystem processes. In the FAM project this approach will be used for the element nitrogen in different systems in the near future.

## 4. Conclusions

To optimise land-use strategies in the area studied – and this holds true for important parts of the agricultural landscapes in Central Europe – it is necessary to better understand, ecosystems and their spatial interactions in the landscape in terms of the main parameters and processes. For a fine-tuning of the soil management, it is not enough to identify sensible or vulnerable areas (hot spots) where improper land use harms the biotic and abiotic resources the most. It also is necessary to know 'hot events' and 'hot periods' during which important nutrient and pollutant fluxes in the ecosystems take place within short periods of time. The knowledge has to be combined in evaluation schemes or information systems to optimise different land-use regimes with respect to their economic benefits and ecological effects.

## Acknowledgements

The scientific activities of the research network, 'Forschungsverbund Agrarökosysteme München' (FAM), are financially supported by the Federal Ministry of Research and Technology (BMFT 0339370). Rent and operating expenses of the experimental farm in Scheyern are paid by the Bavarian State Ministry for Education and Culture, Science and Art.

## References

Auernhammer H, Demmel M, Muhr T, Rottmeier J and Wild K 1994 Site specific yield measurement in combines and forage harvesting machines. Comput. Elect. Agric. 11, 53–68.

Filser J, Fromm H, Nagel R and Winter K 1995 Effects of previous intensive agricultural management on microorganisms and the biodiversity of soil fauna. Plant and Soil 170, 123–129.

Flessa H, Dörsch P and Beese F 1995 Seasonal variability of $N_2O$ and $CH_4$ fluxes in diffferently managed arable soils in Southern Germany. J. Geophys. Res. 100, 23115–23124.

Fromm H, Winter K, Filser J, Hantschel R E and Beese F 1993 The influence of soil type and cultivation system on the spatial distributions of the soil fauna and microorganisms and their interactions. Geoderma 60, 109–118.

Hantschel R E, Kamp T and Beese F 1995 Increasing the soil temperature to study global warming effects on the soil nitrogen cycle in agroecosystems. J. Biogeog. 22, 375–380.

Hantschel R E and Stenger R 1997 Site effects on the variability of the nitrogen turnover. In Processes in Managed Ecosystems: Spatial and Temporal Variability. Eds R E Hantschel and F Beese. Ecological Studies Series. Springer Verlag, Heidelberg (in press).

Knauer N 1993 Ökologie und Landwirtschaft – Situation, Konflikte, Lösungen. Verlag Eugen Ulmer. 280 p.

Nilsson J 1988 Critical loads of sulfur and nitrogen. In Air Pollution and Ecosystems. Ed P. Mathy. D Reidel Publishing Company. 85–91.

Scheurer W 1995 Fähenhafte Simulation des Bodenwasserhaushaltes in Scheyern unter einheitlicher Bewirtschaftung. PhD work. University of Munich, Institute of Geography.

Schwertmann U, Auerswald K, Scheinost A and Sinowski W 1993 Fähenhafte und punktverdichtete Erfassung von Bodenparametern. In FAM-Bericht 3. Eds R Hantschel and M Kainz. Abschlu*β*bericht Aufbauphase 1990–1992. 279 pp.

Sinowski W 1995 Die dreidimensionale Variabilität von Bodeneigenschaften – Ausma*β*, Ursachen und Interpolation. PhD Work. Technical University of Munich/Weihenstephan, Institute of Soil Science.

Tivy J 1993 Landwirtschaft und Umwelt – Agrarökosysteme in der Biosphäre. Spektrum Akademischer Verlag, Heidelberg.

Ulrich B 1987 Stability, elasticity and resilience of terrestrial ecosystems with respect to matter balance. In Potentials and Limitations of Ecosystem Analysis. Eds E-D Schulze and H Zwölfer. Ecological Studies No. 61, Springer Verlag, Berlin. pp 11–49.

Webster R and Oliver M A 1990 Statistical Methods in Soil and Land Resource Survey. Oxford University Press, Oxford, UK. 328 pp.

13

# Minimising surface and ground-water pollution from fertiliser application

PAOLO SEQUI AND ROBERTO INDIATI

Transport of nutrients to natural waters is largely controlled by the volume and rate of water movement through the soil body and across the soil surface. Precipitation and infiltration are of primary importance in affecting these volumes and these rates, but interception and evapotranspiration processes are also important. Precipitation is the driving force of the hydrologic processes. Infiltration controls the distribution of precipitation between surface and subsurface flow processes. Infiltrated water moves downwards through the soil profile, first replenishing storage and then transporting chemicals dissolved in the soil solution to greater depths (*leaching*). Surface runoff, i.e. the part of rainwater which flows over the land surface, may provide detachment, entrainment and transport mechanisms for the movement of solid soil particles (*erosion*) by the impact of raindrops and the shear stress resulting from overland flow. Soil particles may also be detached and transported by the action of wind.

This paper takes for granted that fertilisation is correctly practised, as outlined in another review (Kafkafi, 1994), and aims to discuss the mechanisms of nutrient transport from soil to waters through the leaching and erosion processes, focusing on nitrogen (N) and phosphorus (P) behaviour. Anthropic factors may influence the action of natural factors, and should be calibrated to manage the situation properly in order to realise a strategy of integrated plant nutrition.

## 1. Leaching

Leaching, i.e. displacement of the soil solution with consequent transport of contained chemicals downwards out of the soil profile, occurs whenever:

$$\mathrm{P} - \mathrm{SR} > \mathrm{WHC} + \mathrm{ET}$$

i.e. precipitation, P, less the fraction of water flowing through surface runoff, SR, overcomes the sum of soil water-holding capacity, WHC, plus evapotranspiration water, ET, lost to the atmosphere during the period of time considered. Often SR water is underestimated, but can be an impressive amount in Mediterranean climates; the average figure for Italy is 51% of the precipitation water (ISTAT 1991). Soil WHC may range from very low values, such as 0.2 mm in sandy soils, to ten times as much in clay soils. By considering the highest potential ET values occurring

*D. Rosen* et al. *(eds): Modern Agriculture and the Environment, 147–158.*

in semi-arid climates, it cannot be surprising to find that no leaching processes take place in many areas.

As P is strongly adsorbed on soil colloids (Barrow, 1978; Sibbesen, 1981), its contribution to pollution of ground waters is generally negligible. On the contrary, N supplied to soil in the nitrate form is repulsed by negatively charged soil surfaces and generally can readily leach with percolating water down the soil profile. Nitrogen applied as ammonium may resist leaching because the $NH_4^+$ concentration in soil solution is comparatively low due to the sorption by the soil surface. Soil nitrate can also derive from mineralisation of soil organic matter and conversion of ammonium to nitrate over time; therefore, whether or not utilised by the crop, it can accumulate and potentially constitute a risk factor for the contamination of ground water.

The amount of leached nitrate depends mainly on climate, soil type, amount of nitrate in the soil, cropping system and irrigation practices.

### *1.1 Climate*

The principal elements which affect the rate of soil-nutrient transport are rainfall (amount, intensity and seasonal distribution) and temperature; both these factors act directly on the balance between precipitation, soil water-holding capacity and evapotranspiration. In addition temperature affects N biological transformations, particularly denitrification by reduction of nitrate to $N_2O$ and $N_2$ and mineralisation of the organic N pool. The last mechanism is very important, because often the largest amount of leached nitrate originates from this process.

A preliminary assessment of the climatic conditions must be taken into account in evaluating the potential hazards of contamination of underground waters by fertiliser application.

Considering, for instance, a climatic situation typical for most Mediterranean areas, the main features are:

- Total yearly rainfall may be comparatively abundant, though irregularly distributed during the year and in each separate month: a common pattern is a cold season with more uniform rains and a dry, hot season with occasional storms.
- Some stresses of crops due to soil dryness are common during the hot season, though they generally do not lead to withering.
- Typically, soil moisture may exceed field capacity only during one or two months of the year. As a consequence only some, if any, of the water brought to the soil by rains during the rainy season may be lost from the soil body, draining freely to the underground water; during the hot season, humidity moves upwards through the soil profile unless irrigation is practised in order to restore soil water storage.

The most important features of this climatic pattern lead to equally important operational principles: first of all, the loss of water from soil by drainage occurs during a single season. For water rich in nitrates, drainage might also be a problem in undisturbed soils (Sequi and Vittori Antisari, 1989). In fact, nitrate content in the soil solution tends to be high only in the cold and rainy season, due to both the limited absorption by plant roots and the strong decrease of denitrification processes. Sometimes the contribution of agriculture may be negligible with respect to natural nitrate sources, as demonstrated experimentally in comparisons between cultivated and forested watersheds (Caporali *et al.*, 1991).

### *1.2 Soil type*

The significance of the soil type for the leaching of nitrate is primarily connected with the pore geometry and the hydraulic conductivity. Sandy soils have large pores, the highest hydraulic conductivity and the smallest water-holding capacity, and therefore leaching from this soil is usually higher than from clay soils. The denitrification capacity is greater in clay soils, and this also contributes to reduced leaching. In addition, experimental evidence shows that biologically induced macropores – formed as a result of biological activities such as soil animal boring and root channels – can play an important role in determining preferential flows of water and solutes (Li and Ghodrati, 1994). Finally, soil pH and the presence of a positively charged surface can be of primary importance, if we consider that each mEq of positive charge per 100 g soil equals the adsorpion of 1 t $NO_3$–N per hectare.

### *1.3 Nitrates in soil*

Nitrogen fertilisation is the basis for a sufficiently large and adequate crop production. In order to minimise the potential risk of contamination of ground water following N application, it is essential to take into account the correct timing of application of N fertilisers, so as to ensure the highest concentration of nitrate in the soil solution over the period of maximum plant growth and therefore to protect N from leaching. In addition, leaching of nitrate may be reduced by: N fertiliser placement near the zone of active root growth (Clay *et al.*, 1993); and use of the appropriate type of N fertiliser. In applying chemical fertilisers along with animal manures and effluents, sewage sludges and slurries, the farmer should consider not only the total N content, but also the distribution of this element on the basis of the main fractions present: i.e. inorganic N; easily mineralisable organic N; and slowly releasable organic N. Also, several chemical components (nitrification inhibitors) that can prevent the conversion of ammonium to nitrate and thus reduce N losses, have been developed over recent years. As $NH_4^+$ is not exposed to leaching to the same extent as $NO_3^-$, they have significance in connection with nitrogen fertilisation, especially for early application, as leaching of N may be reduced.

### *1.4 Cropping system*

The presence of a plant cover prevents the soil from accumulating nitrates by virtue of their uptake by roots, and thereby from N leaching. Crops covering the soil during the whole year can take up nitrogen during the whole period in which nitrogen is mineralised. In order to reduce N losses by leaching, crop sequences should ensure a covering of soil by specific catch crops during the rainy season. In addition, if a crop rotation includes leguminous species, cultures able to utilise the early fixed nitrogen efficiently should be cropped. In any case, rotation of crops should be planned with the object of optimising the utilisation of both residual and mineralised soluble N by plants. Furthermore, ploughing in of vegetal residues presenting high values for the C:N ratio (> 40–50) should promote organisation of soluble N in soil and thus protect it from leaching.

### *1.5 Irrigation practices*

With respect to the influence of irrigation practices on N transport by percolating water, the entity of nitrate leaching is generally seen to decrease with increasing efficiency of water distribution. The most used irrigation systems are furrow and sprinkler. Higher application efficiency for conventional furrow may be achieved through practices such as tailwater reuse and surge irrigation techniques; finally, EPA (low-energy precision application) represents an improvement of the centre pivot sprinkler system, by employing long drop tubes and specially designed emitters to minimise irrigation losses. Whatever the system used, the quantity of water to be supplied should moisten only the soil layer in which the plant roots are mainly active.

It can be seen then that studies regarding crop yields and N movement according to soil type, cropping and irrigation system, soil N fertilisation practices and climatic zone would be the best tools in developing policies that protect the environment while sustaining agricultural productivity. Increased interest has been shown, in recent years, in developing methods of predicting crop N requirements and the correct timing of N application. Results are not yet adequate as N demand and uptake vary considerably and are difficult to predict. There is variation between fields and between years depending on soil type, preceding crop, rooting depth and amount of inorganic N present in the root zone.

The methods for prognosis rest upon the premise that the optimum N application depends on the N pool in the soil, N mineralisation and uptake of applied N fertiliser. The accuracy of the N prognosis thus depends on the certainty with which these three factors are determined.

### *1.6 Environmental issues*

As previously discussed, the contribution to water pollution from the application of N fertilisers may be considerable provided that:

- Precipitation and/or irrigation water is in excess, which causes the presence of surplus water in the soil body, in excess of the soil moisture capacity; this causes percolation of the soil water from the upper layers to underground water.
- Soil solution contains nitrates when a surplus of moisture occurs in the soil.

According to the FAO Fertiliser Handbooks, for instance, in the last 20 years, arable soils and permanent crops of The Netherlands have received about 500 kg $ha^{-1}$ N yearly from chemical fertilisers. When it is considered that in this country many soils have a sandy constitution, it seems reasonable to believe that agriculture substantially contributes to the enrichment in nitrates of ground waters.

In other countries, however, the average amounts applied to the soil are much lower and do not reach 50–100 kg $ha^{-1}$ N. This is the case in many Mediterranean countries, including Italy. It is necessary therefore, first of all, to disaggregate the whole country data for smaller geographic areas, to show the potentially important differences in local additions of fertilisers, and then to take into account the possible utilisation of animal manure in addition to mineral fertilisers. Afterwards, it is imperative to establish a precise hydrologic balance in soils, defining vulnerable areas in function of both precipitation plus irrigation waters, and soil physical and chemical characteristics related to water infiltration and retention parameters (Sequi *et al.*, 1991a). This gives a good picture of the situation concerning possible water loss from soil to surface and ground water.

A difficult step to be estimated is the closure of the N cycle in soil with the return of nitrogen as $N_2$ to the atmosphere; in fact it is hard to model denitrification in the different environmental compartments. Despite the quantitative transformation of nitrates to $N_2$ which occurs rapidly in submerged soils, it may be possible to find models in which nitrate is easily leached together with the percolating water from paddy soils!

The situation can only be outlined by drawing a comprehensive balance of N inputs into the environment from soil and other sources. Strong inputs may arise from sewage plants and sanitary landfills. An often neglected input depends on combustion of fossils. In Italy, $NO_3$–N emitted from fossil combustions is evaluated to be about 0.8–1.0 × $10^6$ t (Ministero dell'Ambiente, 1992), in the order of fertiliser N used in (and not lost from) agriculture.

In Mediterranean countries, therefore, it appears that in some areas agriculture may contribute to the burdening of ground water with nitrate-N, though the relevance of the loadings may be less than expected. In other areas, agriculture purifies the environment and especially rainwater, by intercepting and denitrifying nitrates. This is similar to the conclusions drawn by Spalding and Exner (1993) after a review on the data from 200 000 $NO_3$–N data points of United States ground water.

## 2. Erosion

The movement of nutrients in the solid phase, i.e. associated to soil particles, occurs as an effect of transport of the molecules being more or less strongly sorbed onto the solid phase. Consequently, the movement of soil particles will promote movement of the linked molecules. Erosion in surface run-off is generally assumed to occur from an effective average depth of soil and run-off–water interaction termed EDI (Sharpley *et al.*, 1981). The EDI has been computed to be in the range 0.2–0.3 cm, with the degree of interaction decreasing exponentially with depth; it is affected by soil aggregation, soil slope and kinetic energy (intensity and turbulence) of raindrops. Erosion is a harmful process because it not only decreases the reservoir of soil nutrients but also causes losses of the surface soil mass, generally more fertile in nutrients than deeper soil layers.

Phosphorus, contrary to nitrogen, is transported from terrestrial to aquatic environments principally sorbed to eroded soil particles (Burwell *et al.*, 1977; Sharpley and Syers, 1979). In fact, the movement of dissolved P through the soil profile is highly unlikely, the mechanism of sorption of the small amounts of soluble P by generally P-deficient subsoil being prevalent. Exceptions may occur for soils presenting a permanent or temporary low P retention, i.e. coarse-textured, waterlogged and organic soils. P mobility in coarse-textured, silica-rich soils depends on the low adsorption capacity. In waterlogged soils, reduction of highly insoluble ferric to soluble ferrous ions may cause solubilisation of substantial P amounts linked to the iron compounds. In organic soils, P mobility is often attributed to leachability of organic P, though it is known that the retention mechanism for organic P in soil is similar to that of inorganic P. In addition to the sorption to iron and aluminium by a specific sorption reaction – ligand exchange or anion penetration – retention is due to interaction of the organic portion of the phosphate esters with inorganic soil components. Mobility may therefore depend on the lower insolubility of P sorbed to the organic portion with respect to the inorganic portion of soil constituents.

The degree to which soil erosion takes place on cultivated land is influenced by a complex mixture of both physical (nature of the soil, topography and climate) and anthropic factors (cultivation practices and land treatment).

### 2.1 *Nature of the soil*

Many soil properties in combination affect water infiltration, water retention and erodibility, such as the chemical composition, the extent of weathering, the content of organic matter, the particle size and structural characteristics. For instance, an agronomically good structure may favour infiltration of water, hindering erosion, and also water retention, impeding nutrient leaching. The soil clay fractions are known to favour adhesion of soil particles, but soil organic matter can absorb and store water to a greater extent and is the first fraction responsible for the formation of aggregates. Soils rich in organic matter and clay components are generally considered to be very resistant to erosion.

### 2.2 *Topography*

Relief, or topography, mainly controls water drainage by means of the aspect and the direction of the soil slope, which contributes to the soil microclimate. The slope of the land, that is its gradient, plays an important role in the movement and direction of rainwater, soil erosion and soil nutrient loss. The length of the slope is also important; it can be expected that the longer the slope, and assuming the same rate of rainfall, the greater will be the volume of excess water accumulating upon it, all of which must run down the slope at increasing volume and velocity.

### *2.3 Climate*

The principal climatic factors influencing the rate of erosion are rainfall, wind and temperature. The areas of the world which are most susceptible to rainfall erosion include all lands receiving medium to high rainfall within latitudes approximately 42° north and south. In the temperate regions in excess of these latitudes the hazard of soil erosion tends to be less than in the tropics, because the rainfall is more often relatively gentle and spread more evenly throughout the year. In addition, higher temperatures allow the soil organic matter to decrease rapidly through decomposition, thus resulting in a soil even more susceptible to erosion. The potential damage to the soil can be evaluated from the seasonal distribution of rainfall, the amount of rain falling during each storm, the frequency of storms and the rate at which the rain falls.

### *2.4 Cultivation practices*

The combined influences of soil and slope on the rate of erosion are in turn modified by the cultivation or farming practices employed – the way in which crops are grown, livestock raised and the land treated by tillage and other means. The rate of erosion is reduced to a minimum when the soil is completely covered by vegetation. Pasture plants grow so close together that the soil is very efficiently protected from both water and wind erosion. Well managed pasture and grass-lined waterways and drainage channels thus exibit some of the best anti-erosion measures which can be devised, and are applicable over a wide range of soils, slope and climatic conditions. In the case of annual crops (maize, cotton, etc.), attention must be given to optimising the spacing between and within the rows, by choosing varieties that can be planted as close as possible, without a reduction in yield. In addition, if the rows are planted up and down sloping land, and the soil is erodible, serious erosion may occur from run-off water rushing down the strip of bare soil during heavy rainstorms. On the other hand, if the rows are planted on or nearly on the contour, the rate of erosion will be greatly reduced because the relatively closely spaced plants in the row will tend to check the velocity of run-off and allow more water to percolate into the soil. The erosion hazard from run-off can be reduced if crops are allowed to grow on ridges along the contour. Some additional measures which can provide effective protection to soil are: mulching technique; strip-cropping technique (rotational and buffer strip-cropping); and land treatments such as the setting up of waterways and terraces (bench and channel terraces). A deeper till must be done for the soil-tillage practices. The importance of tillage from the soil conservation point of view is its effect on soil structure and moisture status, the amount of crop residue available, and how it is managed; these factors influence the erosion-control effectiveness of tillage systems.

With conventional tillage systems, generally referred to as mouldboard ploughing followed by a secondary tillage operation such as discing and/or harrowing, water erosion may occur at any time on most soils, but the potential is greatest while the surface is bare after ploughing, during seedbed preparation and at seedling establishment. Conservation tillage systems such as no-tillage, till planting, etc., are however especially effective for controlling water erosion, and thus nutrient transport in flowing water, by retaining most crop residue on the soil surface.

The main advantages of conservation tillage systems include: effective soil-erosion control; less use of fossil fuels normally associated with land preparation; more intensive use of sloping lands; and less risk of environmental pollution (Blevins and Frye, 1993). Disadvantages that may be associated with conservation tillage practices include: higher herbicide costs; more difficulty in controlling weed infestation; and aggravation of the existing wetness limitation for imperfectly drained soils by creating a more anaerobic soil environment. In addition, crop residue mulch left at the soil surface may cause slower warming of soil during the spring and thus delay germination and early growth in temperate regions; in contrast, the surface residual mulch may become a positive factor in the tropics by reducing the high temperatures that adversely affect plant growth. From the results of several studies (e.g. Seta *et al.*, 1993), it can be concluded that as tillage intensity is reduced: run-off water volumes and sediment losses decrease significantly; average concentrations of nutrients decrease with increasing tillage intensity; and total losses of nutrients are smaller from no tillage than from conventional tillage systems. Furthermore, controlling P loss is not as simple as controlling sediment yield; in fact it involves not only the selection of tillage systems and crop rotation but also the selection of proper nutrient management systems (including rate, technique and rate of application). However, regardless of the tillage system, concentrations of chemicals in the run-off water are related to the rainfall events, tending to be greatest in the first rainfall event and decreasing with each consecutive run.

### *2.5 Environmental issues*

Many estimates of P loadings to surface water in developed countries attribute a relevant proportion of the responsibility to agriculture, of the order of 16–38% (Isermann, 1990). For Italy, Chiaudani *et al.* (1978) ascribed to P fertilisers a contribution of 18% of total P loadings. This percentage was calculated by assuming that 3% of the fertiliser P applied is released to surface water as the sum of soluble orthophosphate-P and particulate P. Such an estimate hardly agrees with the mean residence time in soil of about 20 000 years evaluated for P by Sequi (1989). The 3% annual loss of soil-applied mineral fertiliser P to waters, suggested by Chiaudani *et al.* (1978), relies on a series of data collected by Vollenweider (1968), who did not take into account, however, variation due to the different analytical methodologies followed by the different authors. Ciavatta *et al.* (1990), for instance, showed that in analytical methods currently used in environmental studies, interference of silica may be so high as to render experimental data meaningless: overestimates of the

P concentration by up to 400% may occur. Many other errors may combine to misrepresent the true amounts of dissolved inorganic P as determined in soil solutions and in stream or lake water; they have been discussed by Sequi *et al.* (1991b).

Since losses of particulate P by erosion may be large, it is essential to consider the changes between soluble P and particulate P. They can take place during transport in stream flow, the direction of the exchanges depending on the concentration of sediment and soluble P in the stream flow and the equilibrium P concentration (EPCo) of the sediment contacted; this parameter is defined as the soluble P concentration in equilibrium with a solid material at which neither net sorption nor desorption occurs. The degree of these changes depends on the desorbable P content of the sediment contacted and the rate of the stream flow.

It is important to emphasise that if the soluble P concentration of run-off and stream flow is lower than EPCo relative to the suspended material contacted, a P desorption from material will take place, and vice versa, the processes involved in the sorption and desorption of P by sediment materials in a water basin being analogous to those occurring in a stream flow. While dissolved orthophosphate is for the most part immediately available for algal uptake, bioavailable particulate P can be a variable portion of particulate P transported in run-off as a function of agricultural management (Sharpley *et al.*, 1994).

As an alternative to lengthy algal - culture tests (7–100 days), several chemical-extraction procedures have been used to quantify the availability of P to algae; however, because not all the chemically extracted P is algal-available, caution must be exerted in relating P bioavailability of sediment material determined by these chemical extractants (principally NaOH solution and citrate–dithionite–bicarbonate solution) to the potential of the sediment to increase algal growth. Sharpley (1993) developed a method which uses iron oxide-impregnated paper strips to estimate bioavailable P (defined as the amount of inorganic P a P-deficient algal population can utilise over a period of 24 hours or longer) in run-off or sediment samples; the strip method is based on the principle of P sorption onto an iron oxide strip (acting as a sink for P) and subsequent removal of P by dilute acid.

Three main points must be considered in evaluating the significance of particulate P according to Sequi *et al.* (1991b):

(1) *Availability of particulate P to organisms living in waters.* Despite the fact that a part of particulate P is biologically available, availability is only potential: furthermore, apatites were available to acidifying microorganisms, while they formed submarine deposits in geological times. Potentially available P is defined as the proportion of total P an algal population can utilise under standard conditions, e.g. in the absence of protozoa which can hinder algal growth, and in a P-deficient medium (Sonzogni *et al.*, 1982). Most rivers receiving effluents from sewage-treatment plants have high dissolved inorganic P concentrations, so that particulate P can be considered as non-utilisable P.

(2) *Adsorption of soluble P by suspended materials.* Eroded fine soil particles can show a strong purifying action in terms of P. Such action has been proved by

various authors and, for Italy, it is appropriate to recall an exemplary case reported by Candussio and Visintini Romanin (1980): the letting of turbid waters of the stream Corno during a flood in the Stella river, whose P content was very high due to trout breeding, caused the disappearance of dissolved P. Sharpley *et al.* (1981) showed, in different basins, an inverse relationship between the concentration of soluble P and that of suspended solids. The higher the suspended concentration of fine particles in surface run-off, the lower the concentration of dissolved inorganic P in waters, that can become as low as about 0.001 mg $L^{-1}$, i.e. the mean concentration reported for the soil solution.

(3) *Sedimentation of particulate P.* A continuous decrease of the concentration of transported solid materials is to be expected as a function of time and the different hydraulic characteristics of the stream. Sedimentation of particulate matter will cause a decrease of total P transported by water. Such a process may be particularly evident some hundreds of metres beyond sewage plants, especially if sediments have been recently deposited. Sequi *et al.* (1991b) showed that the concentration of dissolved inorganic P of the river Marecchia increased sharply after the sewage-treatment plant of Rimini and decreased 300 m after the plant, whereas the total P content of the sediments of the same river increased in parallel. The process was made more evident because the sediments were recently deposited; in fact, dredging of the river bed is performed regularly every 10–15 years.

A similar synergistic process has been experimentally shown to occur by the US Soil Conservation Service (Sharpley, 1985). Water from two agricultural basins reached rivers after passage through two small impoundments; particulate loading was measured before and after passage. This ‘impoundment effect’ is comparable to that of any action reducing water velocity and increasing permanence time. Sedimentation of particulate material in the two basins resulted not only in a decrease of more than 90% of total P in water, but also of almost 80% of dissolved inorganic P. Similar data were reported by Okereke *et al.* (1983), who studied the effect of a large impoundment along the course of Des Moines river: in this case a decrease of about 90% of suspended solids and 57% of total P was reported.

In conclusion, a correct interpretation and evaluation model of the role of sediments arising from agricultural land must take into account a reduction coefficient. Sedimentation could reasonably involve up to 90% of total particulates and at least 75–80% of total P. According to the findings of US scientists, it seems reasonable to assume an average abatement of at least 70% of particulate inorganic P from agricultural basins. If we consider waters polluted from sewage plants, like those of rivers of all advanced countries, a comprehensive effect of agriculture is a negative contribution, because agriculture efficiently contributes to reduce the P content of surface water.

## 3. Conclusions

Surface and ground waters belong to the entire human community. Their quality therefore must be ensured and defended for the society.

Whenever a policy-maker needs to adopt a strategy to protect water quality, he must establish a priority scale of the measures that have to be taken. In other words, a policy-maker must decide which are the more important causes of pollution in order to take remedial measures without both dissipating financial resources and damaging the course of innocent productive activities.

Of course, when policy-makers attempt to ascribe any kind of responsibility, it is unavoidable that lucrative activities are compelled to protect themselves: as expected, they try to throw their own responsibilities on as many as possible jointly responsible different activities. This should hopefully induce the representatives of agriculture, which are the weakest link of productive chains in most advanced countries, to critically examine any responsibility attribution which is directed to them. In the frequent case of falsity, they must vigorously dissuade the representatives of national policies from taking measures which reflect only the interests of classes of activities that are economically stronger.

Each of the commonly recurrent claims should be carefully evaluated with respect to: actual damage or risk to the environment; actual origin from agriculture; and economic interests involved. This last item could seem inappropriate or excessive. However, some limitations and banning of chemicals in recent years have seemed to depend more on economic factors, i.e. substitution with more expensive products, than on true environmental or sanitary considerations.

## References

Barrow N J 1978 The description of phosphate adsorption curves. J. Soil Sci. 29, 447–462.

Blevins R L and Frye W W 1993 Conservation tillage: An ecological approach to soil management. Adv. Agron. 51, 32–78.

Burwell R E, Schuman G E, Heinemann H G and Spomer R G 1977 Nitrogen and phosphorus movement from agricultural watersheds. J. Soil Water Conserv. 32, 230–266.

Candussio R and Visintini Romanin M 1980 Fosforo e azoto di origine agraria nei corpi idrici della Regione Friuli-Venezia Giulia. Ann. Ist. Sper. Nutriz. Piante 10, 1–52.

Caporali F, Nannipieri P and Pedrazzini F 1991 Nitrogen contents of streams draining an agricultural and a forested watershed in Central Italy. J. Environ. Qual. 10, 72–76.

Chiaudani G, Gerletti M, Marchetti R, Provini A and Vighi M 1978 Il problema dell'eutrofizzazione in Italia. Quaderni IRSA 42, 3–93.

Ciavatta C, Vittori Antisari L and Sequi P 1990 Interferences of soluble silica in the determination of orthophosphate-P. J. Environ. Qual. 19, 761–765.

Clay D E, Clapp C E and Molina J A E 1993 Mineralization of nitrogen in fertilised-acidified lime-amended soils. Biol. Fertil. Soil 15, 249–252.

Isermann K 1990 Share of agriculture in nitrogen and phosphorus emissions into the surface waters of Western Europe against the background of their eutrophication. Fertil. Res. 26, 253–269.

ISTAT 1991 Statistiche Forestali. Istituto Poligrafico dello Stato, Roma.

Kafkafi U 1994 Considerations in fertiliser application to prevent ground water contamination. Proc. Int. Conf. on Modern Agriculture and the Environment. Rehovot, Israel. 2–6 October.

Li Y and Ghodrati M 1994 Preferential transport of nitrate through soil columns containing root channels. Soil Sci. Soc. Am. J. 58, 653–659.

Ministero dell'Ambiente 1992 Relazione sullo stato dell'ambiente. Istituto Poligrafico e Zecca dello Stato, Roma.

Okereke V I, Baumann E R, Al Dustin T, Schulze S and Lutz D 1983 Midwest (USA) reservoir water quality modification, part 1, 2 and 3. Water, Air and Soil Pollution 37, 309–354.

Sequi P 1989 L'importanza del suolo negli equilibri ambientali. In Chimica del suolo. Ed. P Sequi. pp 533–545. Pàtron, Bologna.

Sequi P and Vittori Antisari L 1989 Dinamismo chimico dell'azoto: Aspetti agronomici ed ambientali. Riv. Agron. 23, 30–42.

Sequi P, Vianello G and Ciavatta C 1991a An Italian approach to the determination of areas vulnerable to pollution. In Soil Vulnerability to Pollution in Europe. Eds N H Batjes and E M Bridges, pp 91–97. ISRIC, Wageningen.

Sequi P, Ciavatta C and Vittori Antisari L 1991b Phosphate fertilisers and phosphorus loadings to river and seawater. Agrochimica 35, 200–211.

Seta A K, Blevins R L, Frye W W and Barfield B J 1993 Reducing soil erosion and agricultural chemical losses with conservation tillage. J. Environ. Qual. 22, 661–665.

Sharpley A N 1985 Phosphorus and eutrophication. In Proceedings of Symposium on Plant Nutrient Use and the Environment. pp 247–270. Kansas City, MO.

Sharpley A N 1993 An innovative approach to estimate bioavailable orthophosphate in agricultural runoff using iron oxide-impregnated paper. J. Environ. Qual. 22, 597–601.

Sharpley A N and Syers J K 1979 Phosphorus inputs into a stream draining an agricultural watershed: II. Amounts and relative significance of runoff types. Water Air and Soil Pollution 11, 417–428.

Sharpley A N, Ahuja L R and Menzel R G 1981 The kinetics of phosphorus in runoff sediments. J. Environ. Qual. 10, 386–391.

Sharpley A N, Indiati R, Ciavatta C, Rossi N and Sequi P 1994 Interlaboratory comparison of iron oxide impregnated paper to estimate bioavailable phosphorus. J. Environ. Qual. 23, 14–18.

Sibbesen E 1981 Some new equations to describe phosphate sorption by soils. J. Soil Sci. 32, 67–74.

Sonzogni W C, Chapra S C, Armstrong D E and Logan T J 1982 Bioavailability of phosphorus inputs to lakes. J. Environ. Qual. 11, 555–563.

Spalding R F and Exner M E 1993 Occurrence of nitrate in groundwater. A review. J. Environ. Qual. 22, 392–402.

Vollenweider R A 1968 Water management research. Scientific fundamentals of the eutrophication of lakes and flowing waters, with particular reference to nitrogen and phosphorus as factors in eutrophication. Technical Report, OCDE, Directorate for Scientific Affairs.

14

# Considerations in fertilizer application to prevent groundwater contamination

U. KAFKAFI

## 1. Introduction

Nitrogen fertilizers are essential to provide the major food supply for the increasing world population. It is expected that the world population will reach 8 billion in the year 2025 (Figure 1).

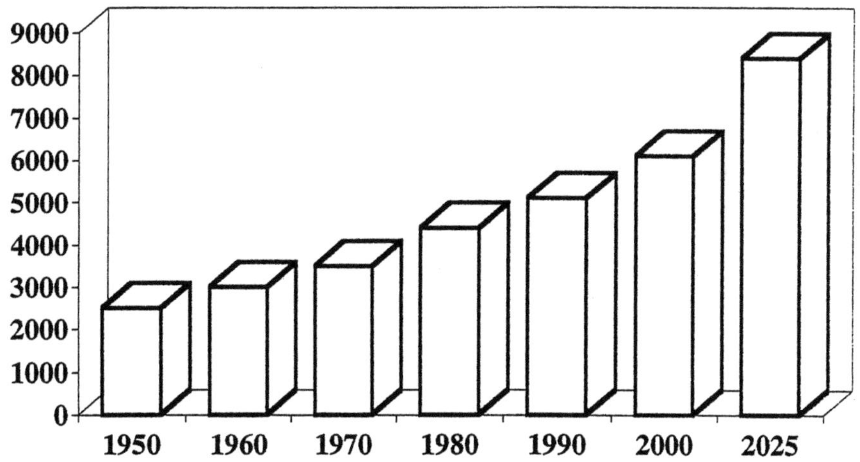


*Figure 1.* Past and future expected world population.

The potential demand for grain yield closely follows the projected population growth. Most of the demand for nutrients to satisfy this population growth will come from fertilizers (Figure 2) of which nitrogen constitutes about 80–85%. It should be noted that nutrients from soils and manure sources will not be enough to support the world's growing population.

The world average cereal yield increased from 1.2 t ha$^{-1}$ in 1945 to 2.5 t ha$^{-1}$ in 1982 (Staple, 1982). Part of that increase is due to the breeding of new cultivars but no yield increase was obtained without N fertilization. Data from the Rothamsted long-term

*D. Rosen* et al. *(eds): Modern Agriculture and the Environment, 159–171.*

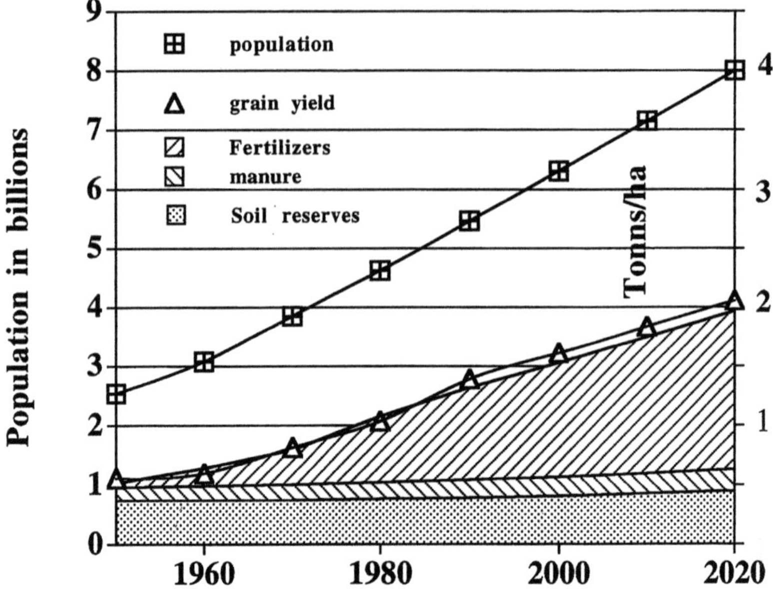


*Figure 2.* Global trends in sources of mineral nutrients, grain yield and population growth. Source: Stapel (1982) with latest data added and projected estimate to year 2020

experiment (Jenkinson, 1982) demonstrate the long-term effect of breeding and nitrogen on grain yields of six wheat varieties that have been dominant in England since 1843 (Figure 3). It is clear that changing cultivars did not increase the wheat yield if it was not supplemented by nitrogen. The cultivars introduced since the 1960s are using nitrogen more efficiently, but it is clear that without nitrogen fertilization, no yield can be expected to support the world's growing demand for cereal grains. Introducing the

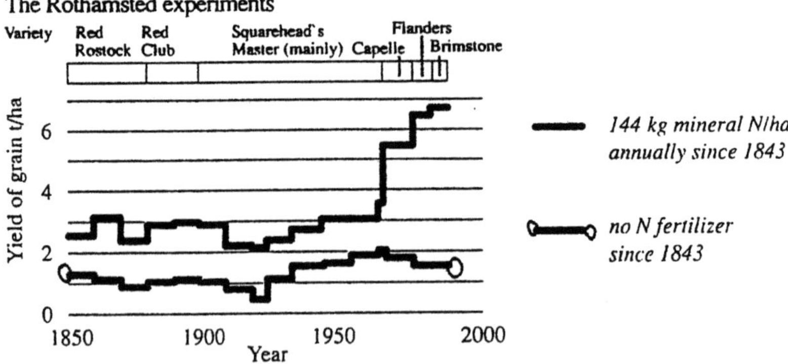


*Figure 3.* One-hundred-and-forty years of grain yield with and without fertilizers for six wheat cultivars in the Rothamsted permanent plots experiment. Source: Jenkinson (1982, 1988/89).

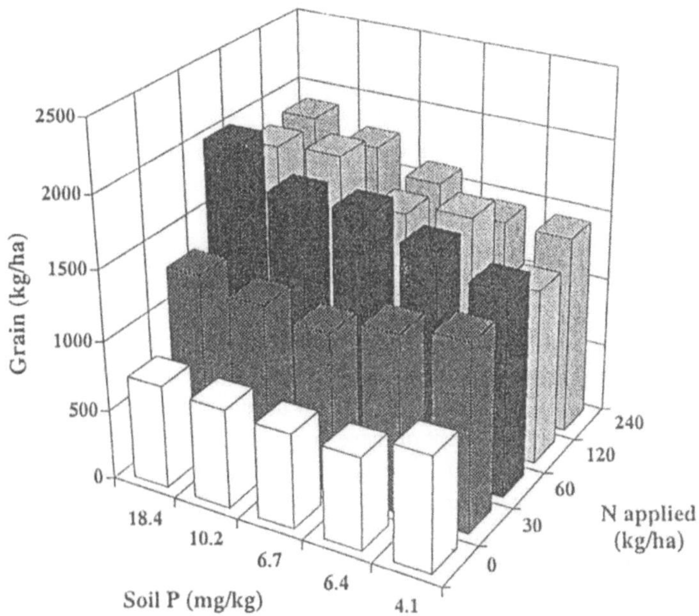


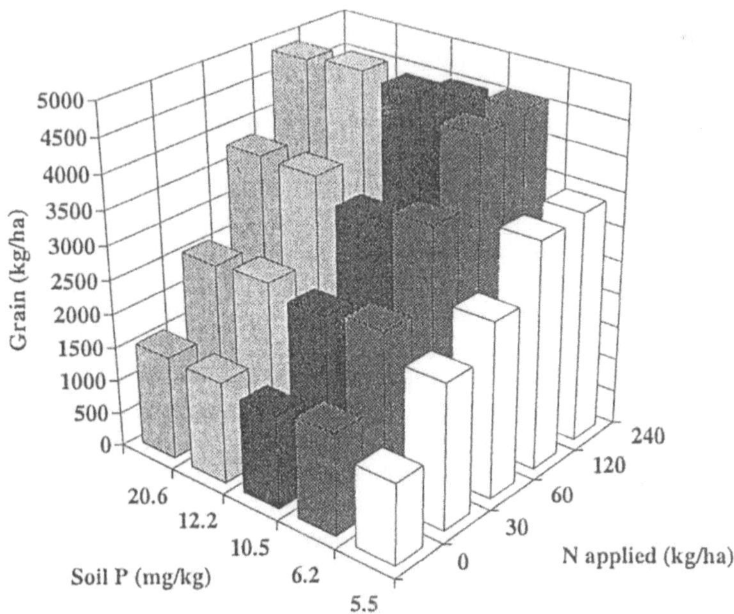


*Figure 4.* Response of two wheat cultivars to N and P in the permanent plot experiment at the Volcani Centre, Bet Dagan, Israel.

semi-dwarf wheat cultivars in Israel increased their response to fertilizers but again no gain is achieved without application of N and P fertilizers (Figure 4). Irrigation and general N management are blamed for transport of large quantities of N to ground water sources (Ayers and Wescot, 1985; Bouwer, 1986; Follett, 1989).

The two main practical methods employed in agriculture to minimise fertilizer N from being leached below the root zone are: split applications of N fertilizers (Mossadeaq and Smith, 1994); and slow-release N fertilizers (Obreza and Rouse, 1992). The slow-release fertilizers are very expensive compared to regular fertilizers, so their usage as a potential solution to the leaching problem is possible only in very expensive crops such as home gardens and glasshouse crops where fertilizers are a minute item in the total production costs, or in orange production in Florida where public pressure is important (see Chapter 15).

Field crops such as rain-fed wheat, where fertilizers constitute about 10–20% of the total production cost (Bøckman *et al.*, 1990) must rely on the cheapest available N source. Since wheat and corn supply about two-thirds of the world consumption of carbohydrate, nitrogen management of these crops is being tested continuously all over the world. Any increase in N-usage efficiency and prevention of losses will be of huge global impact. The fate of applied N to the soil has been extensively studied during the last 50 years. The number of publications for the years 1984–1994 directly relating N (as nitrogen, nitrate or nitrite) to pollution of ground water sources is 418 in the Agricola database.

The purpose of this chapter is to emphasise practical considerations on the effect of N application rates and the effect of balanced fertilization on the recovery, leaching and disappearance of N from cultivated soils on an annual basis and long-term cultivation practices.

The rate of nutrient uptake by annual plants as a function of time forms a typical sigmoid curve. A typical nutrient uptake rate is given in Figure 5 for wheat grown in the permanent plots experiment in Israel. The rates of growth of wheat on three levels of P are illustrated. The bicarbonate soil P extractable values (Kafkafi, 1979) were: $P_0$– 3, $P_2$– 10 and $P_4$– 29 ppm P. Nitrogen was applied at a rate of 240 kg N $ha^{-1}$, split into two doses, one applied just before seeding on 1 December, and the second six weeks after emergence, on 20 January. Just after the second N application, continuous rain for a week supplied 115 mm and the total accumulated precipitation to that point in time was 235 mm. On that soil, with a field capacity of about 30% by volume, any nitrate present on the soil surface could have been washed to a depth of about 400 mm. When the plants were 6 weeks old, on low soil P level the total N uptake was only 20 kg $ha^{-1}$ while in plots with high soil P level it was already 100 kg $ha^{-1}$. The shortage of P during early growth caused very slow plant development (Figure 5), and the main rainfall event leached the nitrate below the effective root depth. Increasing P levels shortened the time to heading and the rate of N uptake was very high. The overall balance of N on the low and high P content in that soil is presented in Table 1. It is clear that when N is limiting the yield ($N_0$), the final total N uptake is less than the amount present in the soil at seeding time. When phosphate was a limiting growth factor, of 420 kg N $ha^{-1}$ in the soil

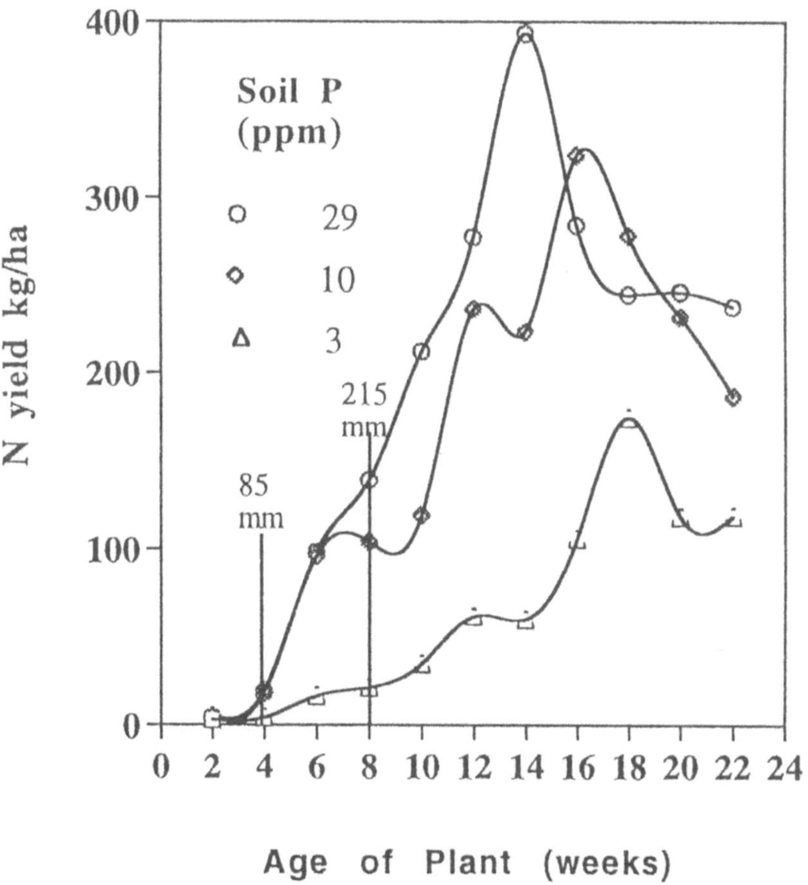


*Figure 5.* Amounts of nitrogen uptake by wheat during the growing season as a function of soil phosphate level. Total N applied was 240 kg N $ha^{-1}$.

only 127 were taken up by the plant, but, when P level was high, all the nitrogen applied was recovered in the plant, leaving no N for downward leaching. This estimate was confirmed at the end of the season (see Table 2). It is clear from Table 2 that by increasing the soil phosphate level, less nitrate N is left at harvest time in the soil and consequently, less N is leached down from that treatment. When soil P is very low, any nitrate that escapes the 60-cm depth is out of the plant's reach. The peak nitrate concentration was found at a depth of about 60–80 cm. The plants on high P treatments were able to recover the nitrate from deeper soil layers or to take it

*Table 1.* Response of maize to split application of 90 kg ha$^{-1}$ N fertilizer during the rainy season in Malawi within the first month after germination. (Plant germination was on 22 December 1988.)

| Application time | Amounts of N doses (kg ha$^{-1}$) | Total rainfall* after last N application (mm) | Yield (kg ha$^{-1}$)† |
|---|---|---|---|
| 19 December | 90 | 295 | 7289 c |
| 19, 29 December | 45 + 45 | 265 | 8789 bc |
| 19, 29 December 12 January | 30 + 30 + 30 | 211 | 10888 ab |
| 19, 29 December 12, 21 January | 22.5 + 22.5 + 22.5 + 22.5 | 0 | 12218 a |

* To 21 January 1989.

†Different letters designated significant difference at the 0.05 level by Tukey – Kramer HSD.

Within one month after germination, split application of the same amount of N was enough to raise the yield from 7 to 12 t ha$^{-1}$ grain. This is the main rainy month and split application can secure N uptake by

up before it had a chance to escape the root zone. It is clear that for wheat the major uptake period for N and P is between three and ten weeks. Increasing the soil P level shortens the period of maximum N uptake by about four weeks. However to build a strong plant that will be able to take up such high quantities of N, a high nutrient concentration in the soil near the root must be present during the first four weeks of wheat growth.

## 2. Split applications of nitrogen fertilizers

The normal soil preparation practices for arable crops usually supply the N fertilisers before seeding. As a result, any uncontrolled rain or excess water supply before and

*Table 2.* Nitrogen balance, N present in soil and N recovery by wheat grown on five levels of N and two levels of P.

| | Bicarbonate extractable soil P (mg kg$^{-1}$) | | | | | | | | | |
|---|---|---|---|---|---|---|---|---|---|---|
| | 3 | | | | | 29 | | | | |
| N treatment | $N_0$ | $N_1$ | $N_2$ | $N_3$ | $N_4$ | $N_0$ | $N_1$ | $N_2$ | $N_3$ | $N_4$ |
| N in soil (kg ha$^{-1}$) at emergence upper 40 cm | 89 | 112 | 124 | 154 | 302 | 76 | 70 | 161 | 139 | 277 |
| N added at top dressing (kg ha$^{-1}$) | 0 | 20 | 40 | 80 | 120 | 0 | 20 | 40 | 80 | 120 |
| Total N available (kg ha$^{-1}$) | 89 | 132 | 164 | 234 | 422 | 76 | 81 | 141 | 255 | 393 |
| Total N uptake (kg ha$^{-1}$) | 38 | 73 | 98 | 155 | 127 | 38 | 81 | 141 | 255 | 393 |

*Table 3.* Nitrate-N distribution with depth at emergence (E) and harvest (H) of wheat.

| | N-$NO_3$ (mg $kg^{-1}$ dry soil) | | | | | |
|---|---|---|---|---|---|---|
| | P[a] (mg P $kg^{-1}$ dry soil) | | | | | |
| | 3 | | 10 | | 29 | |
| Soil depth (cm) | E | H | E | H | E | H |
| 20 | 27.0 | 3.2 | 17.0 | 9.0 | 16.8 | 7.4 |
| 40 | 13.0 | 7.7 | 7.6 | 3.3 | 11.2 | 5.1 |
| 60 | 13.5 | 22.8 | 9.1 | 5.9 | 11.8 | 4.3 |
| 80 | 10.3 | 22.2 | 7.5 | 14.0 | 7.1 | 6.1 |
| 100 | 8.4 | 13.5 | 6.4 | 12.3 | 4.2 | 9.1 |
| 120 | 7.6 | 7.7 | 7.2 | 11.3 | 3.4 | 5.6 |

[a] Na-bicarbonate extractable P.

during the first month after seeding may cause losses of N since no crop is available to take it up in time. An example of corn-yield losses by early-season rains due to N fertilizer losses in a field in Malawi is demonstrated in Table 3. When 90 kg N $ha^{-1}$ were applied in four split applications, the nitrogen fertilizer supplied enough N to produce 12 t $ha^{-1}$ of corn grains, as compared to only 7 t $ha^{-1}$ when the same amount of N was applied in one dose just before seeding. An amount of N sufficient to produce 5 t $ha^{-1}$ corn grains was leached by the rains below the root zone within the first six weeks of growth.

Split application, although advantageous, is not simple to apply and is a costly operation. Sometimes the exact time for application falls within the rainy season; as the soil is not able to carry machinery transport, aerial application is needed – an expensive operation that is not available in many parts of the world.

The cost of fertilizers as a proportion of total production costs of arable crops is 10–20% in western Europe (Bøckman *et al.*, 1990). Until the late 1960s, the consumption of the main three nutrients in Europe was in the same ratio and increased annually at a rate of about 0.118 million tonnes per year (Figure 6). The increase in P and K consumption has stopped since the oil crisis in 1973 but the demand for N has increased at about 0.315 million tonnes per year. The current trend of increasing N without addition of P and K fertilizers might be justified for a short period of several years but in the long run, N recovery will decrease if P application does not follow N fertilizer use. The amount of N recovered by crops varies. Estimations range from 40 to 60% of that applied (Machet, 1987). Loss by denitrification as $N_2O$ and $N_2$ gases to the atmosphere is estimated at 2–30% of all N applied. Leaching is estimated to be only in the range of 2–10%. Even this relatively small fraction of the N applied represents about 0.2 to 1 million tonnes of nitrogen which is leached to the underground water in EEC countries only. The minimum N loss by leaching per kg grain produced is found to be at the optimal rate of N to

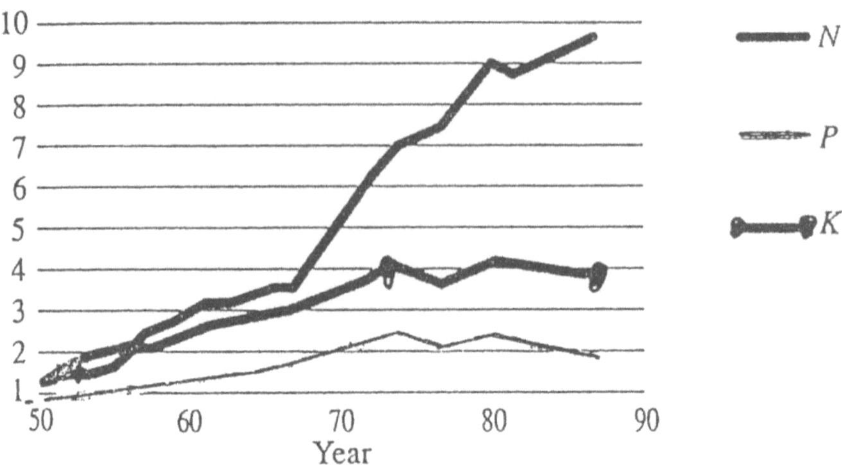


*Figure 6.* Fertilizer nutrient consumption in western Europe in the period 1950–1986.

obtain maximum yield (Figure 7). A growing plant which removes the nutrients from the soil is the main protector of underground water quality.

On permanent grass, nitrate leaching is minimal. Ploughing and drying the soil increase the organic N which is oxidized to nitrate, and hence increase leaching (Bøckman *et al.*, 1990). Losses of N by leaching may occur in the period between cropping, if rain occurs in that period, or during the growing season of annual crops when an excess of nitrogen compared to the plant requirements is present in the soil. In winter crops in the Mediterranean climate, rainfall cannot be controlled in irrigated crops, leaching may occur only if irrigation is given in excess of plant transpiration and surface evaporation.

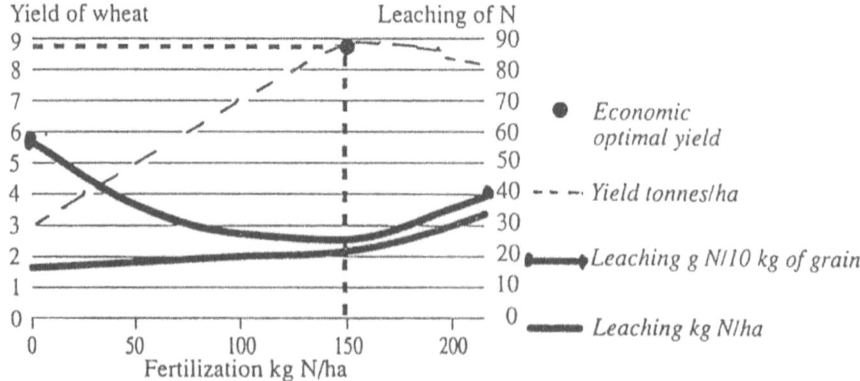


*Figure 7.* Nitrogen fertilization effect on grain yield and N leaching.

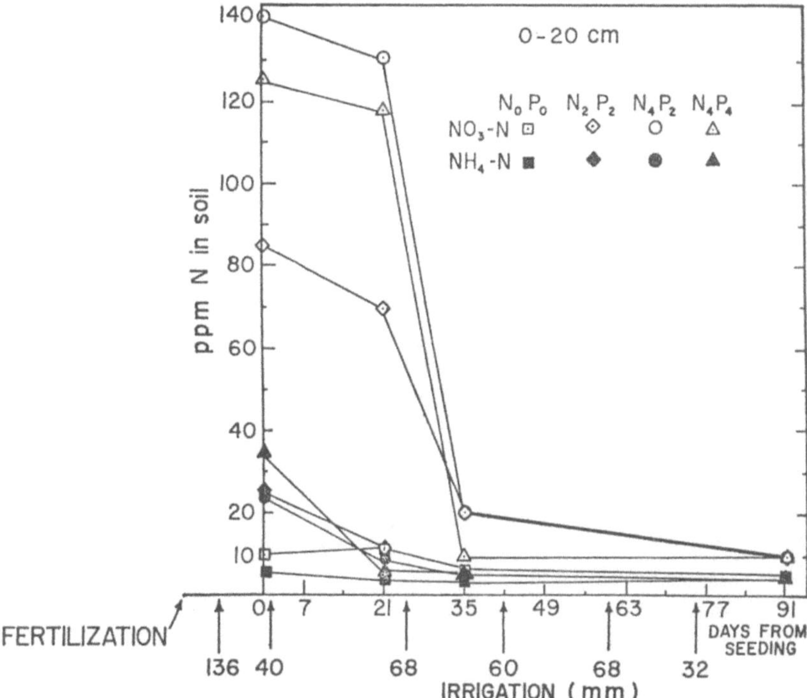


*Figure 8.* Nitrogen concentration in the soil during the growing period of irrigated corn in the permanent plot experiment at the Volcani Centre, Bet Dagan, Israel.

Another way in which nitrogen is lost from the soil and the leaching hazard reduced is by reduction of the $NO_3$ to $N_2$ or $N_2O$ through the biological activity known as denitrification. The rate of irrigation in the presence of a growing corn crop has a significant effect on the fate of N-$NO_3$ in the soil. The quantities of N losses to the atmosphere by denitrification in summer-grown corn, under irrigation on clay soils, were very high when a high concentration of nitrate was present and the irrigation was in excess of the infiltration rate (Figure 8). Fertilizer N nitrification in the soil under July soil temperatures in Israel converted almost all of the 400 kg N $ha^{-1}$ applied ammonium sulphate fertilizer into nitrate form within a week (Figure 8, Bar Yosef and Kafkafi, 1972). Irrigation of 40 mm to ensure plant emergence did not reduce the amount of nitrate in the soil. Irrigation of 68 mm 23 days later, when the plants were at knee-high stage, caused the disappearance of about 250 kg N-$NO_3$ $ha^{-1}$ from the soil. This amount of irrigation was not enough to leach the nitrate below 20 cm. Measurement of nitrate down to 120 cm could not recover any of the N. It was estimated that denitrification due to the simultaneous presence of five conditions was responsible for that process: high levels of nitrate and moisture in the soil; high soil temperatures; high level of available carbon from root exudates of growing plants; high activity of soil microorganisms; and a sudden reduction of

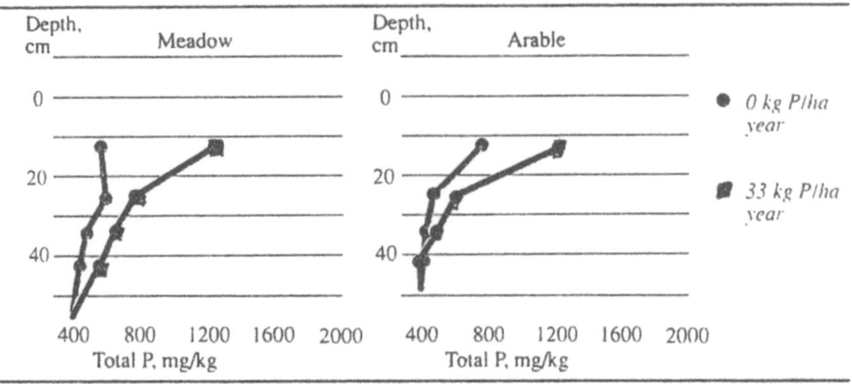


*Figure 9.* Soil phosphate profile after 100 years of P fertilizer application. Source: Cooke and Williams (1970).

oxygen supply due to continuous irrigation which lasted seven hours. On that particular clay soil, such an irrigation caused water saturation of the upper soil layer and the supply of oxygen, as demanded by the total biological activity in the soil, was hampered. In such conditions, nitrate is readily reduced and nitrate is lost from the soil in gaseous form (Bar Yosef and Kafkafi, 1972).

Soil P accumulates in the upper soil layers with and without fertilization. The effect of 100 years of arable cultivation and annual application of 33 kg P $ha^{-1}$ resulted in P accumulation in the upper 20 cm of the soil. Even after 100 years without P application (Cooke and Williams, 1970), the total P in the upper 20 cm is twice that in the upper 40 cm (Figure 9). Phosphate is not leached downward in significant amounts, but the main loss of P from the soil is by soil erosion (Larsen, 1967).

The depth of nitrate movement in the soil depends on its total water-holding capacity, the initial soil moisture conditions and the total amount of effective leaching water (precipitation minus evapotranspiration). Nitrate is very soluble in water and does not sorb to the soil. This characteristic causes the levels of nitrate at the soil surface to change in time and direction. When evaporation exceeds precipitation, as in arid conditions, the soluble salt might accumulate at the soil surface. Such a phenomenon is particularly frequent when trickle irrigation is employed (Figure 10) (Kafkafi and Bar Yosef, 1980).

When applying N, in order to minimise N losses, the following should be considered: soil characteristics including constant and variable parameters (Kafkafi *et al.*, 1978); the rate of growth and nutrient consumption of the plant and the exact seeding time in the case of arable crops; and climatic data such as temperature, and precipitation during the growing season.

When irrigation is applied, careful monitoring of the water balance to a depth of about 1 m is required to prevent over-irrigation or under-irrigation. Knowing the

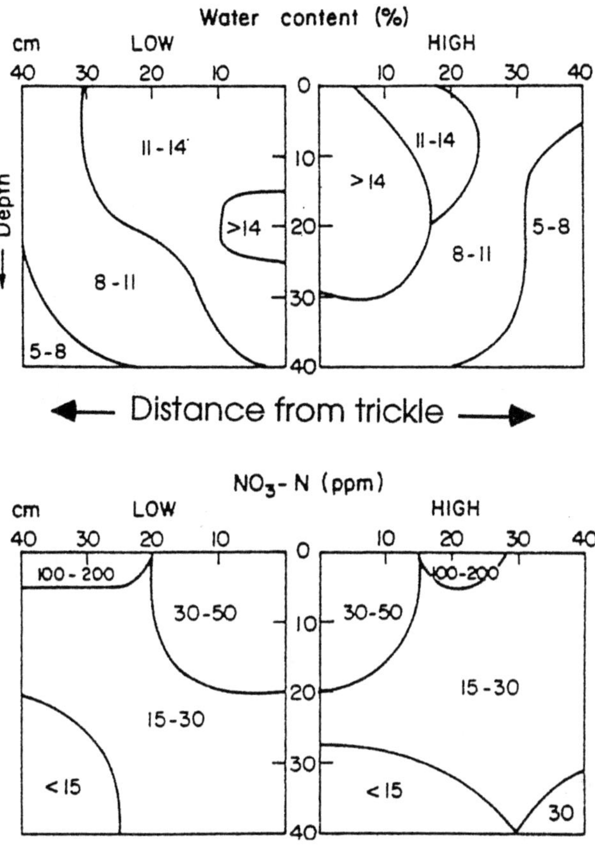


*Figure 10.* Distribution of water and nitrate in a soil when applied from a trickle source.

constant parameters of a specific soil site, the most important decision regarding the amount and time of N application is made according to the crop type grown, its rate of growth and rainfall distribution.

While for yield purposes increasing the soil P levels above 12 mg $kg^{-1}$ does not show any benefit to grain yields (Figure 4, and Kafkafi, 1979), from an ecological point of view there must be less nitrate available for leaching under very high P levels as compared to soils with low P content (Henkin *et al.*, 1994).

Feigin and Halevi (1994) measured the cumulative effect of 30 years of fertilizer application on the same plots as those reported in Table 2. Their results are shown in Figure 11. They measured selected treatments down to 12 m. It can be seen that down to 2 m the minimum nitrate level occurs for zero application of fertilizers. This is not unexpected; however, the yields on these plots were always very low. Adoption of zero N fertilization will not be enough to sustain the farming activity, nor to supply the increasing demand for grains all over the world. On plots with the

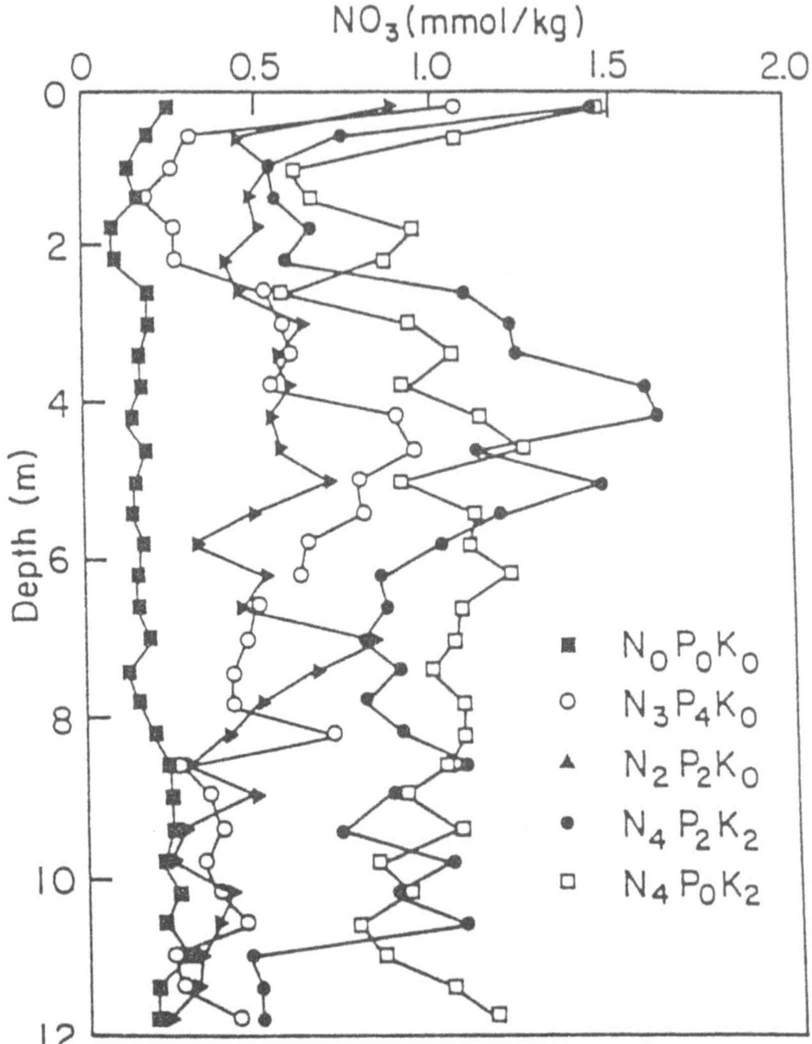


*Figure 11.* Long-term effects of N, P, and K fertilization on the distribution of nitrate in a 12-m deep soil profile, 27 years after continuous application of fertilizers in the permanent plot experiment at the Volcani Centre, Bet Dagan, Israel.

highest P level, the lowest concentration of nitrate is found in the profile down to 12 m. At the $N_4P_0K_2$ combination, the maximum nitrate concentration is found down to a 2-m depth. It is clear therefore that a long history of a plentiful supply of P that takes care of early removal of nitrogen from the soil by annual crops is the best method to prevent downward N leaching in arable crops. In nature, phosphate is the most frequent growth-limiting factor (Larsen, 1967). Application of superphosphate to natural grazing areas increased the total herbage growth without additional N application (Henkin *et al.*, 1994).

Understanding the crop demand for nutrients during its growing period is the main information needed to plan the application rates and timing of N that could prevent nitrate losses. Maintaining a high P level in the soil (provided that precautions are taken against soil erosion) is a safeguard to prevent N losses due to excessive plant growth and utilization of the nitrogen. High soil P is therefore a guard against downward movement of nitrates in agricultural and natural growing habitats.

Balanced fertilization over time is the key to minimizing nitrate pollution of ground water.

## References

Ayers R S and Wescot D W 1985 Water quality for agriculture. Rev. Edn. FAO Irrig. Drain. Paper No. 29. Rome, 174 pp.

Bar-Yosef B and Kafkafi U 1972 Rates of growth and nutrient uptake of irrigated corn as affected by N and P fertilisation. Soil Sci. Soc. Am. Proc. 36, 931–936.

Bøckman O C, Kaarstad O, Lie O H and Richards I 1990 Agriculture and Fertilisers. Agricultural Group, Norsk Hydro, Oslo, Norway.

Bouwer H 1986 Effect of irrigated agriculture on ground water. J. Irrig. Drain. Div. ASCE 113, 4–15.

Cooke G W and Williams R J B 1970 Losses of nitrogen and phosphorus from agricultural land. Water Treat. Exam. 19, 253–276.

Feigin A and Halevi Y 1994 Effect of crops and fertilisation regimes on the leaching of solutes in an irrigated soil. In: DC Adriano, A K Iskandar and I P Murarka (editors) Contamination of Ground Water. Advances in Environmental Science Reviews, Northwood, UK. pp 367–393.

Follett R F (Ed.) 1989 Nitrogen Management and Ground Water Pollution. Developments in Agricultural and Managed Forest Ecology 21. Elsevier, Amsterdam. 395 pp.

Henkin Z, Noy-Meir I, Kafkafi U, Seligman N and Gutman M 1994 Soil phosphorus in a managed Mediterranean woodland ecosystem: Herbage response and cattle grazing effect. Agric. Ecosys. Environ. 47, 299–311.

Jenkinson D S 1982 The nitrogen cycle in long-term field experiments. Phil. Trans. R. Soc. Lond. B 296, 563–571.

Kafkafi U 1979 Measurements of available nutrients in the soil for prediction of fertiliser requirements by crops in the Mediterranean region. In Soils in Mediterranean type Climate and Their Yield Potential. Proc. 14th Colloq. of International Potash Institute, Sevilla, Spain.

Kafkafi U and Bar-Yosef B 1980 Trickle irrigation and fertilisation of tomatoes in high calcareous soils. Agron. J. 72, 893–897.

Kafkafi U, Bar-Yosef B and Hadas A 1978 Fertilisation decision model. Soil Sci. 125, 261–268.

Larsen S 1967 Soil phosphorus. Adv. Agron. 19, 151–185.

Machet J M 1987 Làpport des techniquès à la stratégie de fertilisation. Perspect. Agric. 115, 37–42.

Mossadeaq F and Smith D H 1994 Timing nitrogen application to enhance spring wheat yields in a Mediterranean climate. Agron. J. 86, 221–226.

Nielsen N E 1983 Plant parameters controlling the efficiency of nutrient uptake from the soil. In UN Econ. Comm. Europe. Efficient Use of Fertilisers in Agriculture. Dev. Plant Soil Sci. 10. pp 199–219. Kluwer, The Hague.

Obreza T A and Rouse R E 1992 Controlled release fertiliser use on young Hamelin orange trees. Soil Crop Sci. Soc. Fl. Proc. 51, 64–68.

Shimshi D and Kafkafi U 1978 The effect of supplemental irrigation and nitrogen fertilisation on wheat (*Triticum aestivum* L.) Irrig. Sci. 1, 27–38.

Staple C 1982. Ecological agriculture in global and national context (in Danish). Ugeskrift for Landbrug, 127, 495–500.

UN 1989 Prospects of World Urbanisation Population studies No. 112. Dept. Int. Econ. and Social Affairs, New York.

# 15
# Application of controlled release fertilizers in the Florida citrus industry

J.J. FERGUSON AND F.S. DAVIES

## 1. Introduction

According to recent Florida surveys, 35% of citrus growers applied controlled release fertilizers to young citrus trees after planting in 1992 (Ferguson and Taylor, 1993), up from 19% of growers reported in 1987 (Taylor *et al.*, 1989). Sixty-two per cent of respondents used readily water-soluble fertilizers and 49% used liquid fertilizers applied through irrigation systems. Controlled release fertilizers with different coatings and nitrogen sources are applied as the only nutrient source, are sometimes blended with water soluble, granular fertilizers, and are also spread in fertigated groves to increase root zone coverage of mature trees (Zekri and Koo, 1991). These materials are broadcast, incorporated into the root zone after planting or placed in the planting hole at different depths during planting, both as granules and as briquets (Ferguson *et al.*, 1988). When applied at lower than recommended N rates, controlled release fertilizers can stimulate growth of young trees comparable to that of trees fertilized with readily water-soluble, granular fertilizers applied much more frequently (Boman, 1993; Ferguson *et al.*, 1988; Marler *et al.*, 1987; Obreza and Rouse, 1991), and can reduce nutrient leaching (Alva and Tucker, 1993). The N content of commercial controlled release fertilizers currently used in Florida ranges from approximately 16 to 20%.

Our objective was to compare the growth of young 'Hamlin' orange trees on 'Swingle' citrumelo rootstock and yield of bearing 'Hamlin' orange trees on sour orange rootstock, fertilized with water-soluble, granular fertilizers, commercial controlled release fertilizers containing 17 to 19% N, and an experimental 44.5% N controlled release fertilizer applied at reduced rates.

## 2. Materials and methods

Current recommended N rates for young citrus trees range from 0.14–0.27, 0.25–0.40 and 0.33–0.58 kg N per tree per year for years 1, 2 and 3, respectively; 0.18 kg N per 40 kg box oranges; and 0.14 kg N per box grapefruit (Table 1). Standard, water-soluble fertilizer was applied at recommended rates and controlled release fertilizers – Osmocote, IBDU and an experimental 44.5% urea-N fertilizer, coated with a sulphonated ethylene-propylene-diene polymer (Sherritt, Inc.) – were

*D. Rosen* et al. *(eds): Modern Agriculture and the Environment, 173–179.*

*Table 1.* Recommended fertilization rates for young and mature citrus trees in Florida (Koo *et al.*, 1984).

| Years in grove | Applications per tree per year | Total fertilizer per tree per year (kg) | | |
|---|---|---|---|---|
| | | N | $P_2O_5$ | $K_2O$ |
| 1 | 5–6 | 0.14–0.27 | 0.14–0.27 | 0.14–0.27 |
| 2 | 4–5 | 0.25–0.40 | 0.25–0.40 | 0.25–0.40 |
| 3 | 3–4 | 0.33–0.58 | 0.33–0.58 | 0.33–0.58 |
| 4 | 3–4 | 0.38–0.65 | 0.38–0.65 | 0.38–0.65 |
| 5 | 3–4 | 0.44–0.72 | 0.44–0.72 | 0.44–0.72 |
| 6 | 3–4 | 0.48–0.80 | 0.48–0.80 | 0.48–0.80 |
| 7 | 3–4 | 0.54–0.87 | 0.54–0.87 | 0.54–0.87 |
| 7+ (mature trees) | 2–4 | 0.18/40 kg box oranges or 111–336 kg/ha; 0.14 kg/40 kg box grapefruit or 99–267 kg/ha | | |

applied according to manufacturers' recommendations to young citrus trees for three years, and to ten-year-old bearing trees at 75% the recommended rate for one year (Table 2). Controlled release fertilizers, applied once a year, were either broadcast or incorporated within the dripline, while the standard, granular fertilizer was broadcast three, four and six times a year to young trees. Bearing trees were fertilized once a

*Table 2.* Fertilization of young and mature citrus trees with controlled release fertilizers in Florida.

| Fertilizer | Applications per year | Analysis | Total N per tree per year (kg) | | | |
|---|---|---|---|---|---|---|
| | | | Year 1 | Year 2 | Year 3 | Year 10[b] |
| Standard[a] (water soluble) | 3–6 (broadcast) | 8–2–8 | 0.22 | 0.34 | 0.38 | 0.54 |
| Sherritt[c] (controlled release) | 1 (incorporated) | 44.5–0–45 | 0.043 | 0.067 | 0.087 | 0.40 |
| IBDU (controlled release) | 1 (broadcast) | 19–6–12 | 0.043 | 0.065 | 0.087 | 0.40 |
| Osmocote (controlled release) | 1 (incorporated) | 17–6–9 | 0.039 | 0.065 | 0.087 | 0.40 |

[a] Standard rate based on 0.18 kg N/40 kg box or 133 kg N/acre.
[b] Controlled release rate based on 75% of standard rate.
[c] A sulphonated ethylene propylene diene polymer produced by Exxon technology licensed exclusively to Sherritt Inc., Canada; US Patent No. 4988377.

*Table 3.* Growth of 'Hamlin' orange trees/'Swingle citrumelo' rootstock fertilized in Florida with standard and controlled release fertilizers (12 trees per treatment)[a].

| Fertilizer | Trunk diameter (mm) | | | Tree height (cm) | | | Rating[b] | | |
|---|---|---|---|---|---|---|---|---|---|
| | 1991 | 1992 | 1993 | 1991 | 1992 | 1993 | 1991 | 1992 | 1993 |
| Standard (water soluble) | 17.3 | 36.3 | 60.4 | 94.3 | 153.0 | 231.0 | 2.3 | 4.8 | 4.8 |
| Sherritt (controlled release) | 18.2 | 39.1 | 61.4 | 108.0 | 159.0 | 225.0 | 2.6 | 5.1 | 4.4 |
| IBDU (controlled release) | 18.6 | 37.7 | 57.3 | 109.0 | 165.0 | 219.0 | 2.8 | 5.7 | 5.9 |
| Osmocote (controlled release) | 18.6 | 39.3 | 60.2 | 110.0 | 162.0 | 222.0 | 2.7 | 5.5 | 5.8 |

[a] No significant difference among treatments for any growth parameter.
[b] Rating: 1 = poor; 6 = vigorous.

year with controlled release fertilizers and three times a year with water-soluble fertilizers. Trees were arranged in a completely randomized block design with 12 individual tree replications per treatment. Trunk diameter and tree height were measured during the early spring and late autumn of each growing season. Samples for leaf analysis were taken from four to six-month old leaves during the autumn of each year to determine nutrient levels. Yield was also measured in year 3, the first bearing year for young trees, and in year 1 of the experiment on mature trees.

## 3. Results and discussion

Growth of 1–3 year-old young 'Hamlin' orange trees was similar for trees receiving controlled release fertilizers or standard, granular fertilizers (Table 3). There were no differences in trunk diameter, tree height or tree rating among treatments in any year, although there appeared to be a slight reduction in rating for some trees treated with the Sherritt fertilizer in 1993, which displayed biuret toxicity symptoms.

Young tree yields in year 3 ranged from 25–35 fruit per tree with considerable tree-to-tree variability independent of fertilizer treatment, which is characteristic of young bearing trees. For ten-year-old bearing 'Hamlin' orange trees, yield was similar for trees treated with controlled release and standard fertilizers, even though controlled release nitrogen was applied once a year (0.40 kg/tree) at 75% the rate of water-soluble fertilizer applied three times a year (0.54 kg/tree/yr).

Leaf nitrogen, phosphorus, potassium, calcium, magnesium (Table 4) and micronutrient levels (Table 5) for young citrus trees were acceptable for all treatments in 1991, 1992 and 1993 with the exception of the Osmocote treatment which had low to deficient nitrogen and potassium levels in 1992. After one year no significant differences in yields or leaf nutrient levels were observed in mature trees. However changes in the nutritional programme typically do not affect yields in the first season.

*Table 4.* Effects of controlled-release and granular soluble (GS) fertilizers on leaf macronutrient content of young 'Hamlin' orange trees/'Swingle' citrumelo rootstock from 1991 to 1994 at Gainesville, Florida (mean of five leaves from 12 trees per treatment)[a].

| Fertilizer type | N[y] | P | K | Ca | Mg |
|---|---|---|---|---|---|
| | | | (% dry wt) | | |
| *Pretreatment – April 1991* | | | | | |
| | 2.20 | 0.16 | 1.26 | 2.17 | 0.11 |
| *Year 1 – Post-treatment – October 1991* | | | | | |
| | 2.40 | 0.15 | 1.68 | 2.31 | 0.31 |
| | 3.50 | 0.17 | 1.19 | 3.33 | 0.27 |
| | 2.70 | 0.16 | 1.55 | 2.86 | 0.29 |
| | 2.40 | 0.16 | 1.00 | 3.60 | 0.28 |
| *Year 2 – Post-treatment – August 1992* | | | | | |
| Standard[b] (water soluble) | 3.10 | 0.41 | 1.60 | 3.20 | 0.25 |
| Sherritt[c] (controlled release) | 2.80 | 0.34 | 1.20 | 4.20 | 0.30 |
| IBDU (controlled release) | 2.30 | 0.42 | 0.80 | 5.10 | 0.27 |
| Osmocote (controlled release) | 2.50 | 0.38 | 1.30 | 4.50 | 0.22 |
| *Year 3 – Post-treatment – August 1993* | | | | | |
| Standard[b] (water soluble) | 3.10 | 0.16 | 1.50 | 4.09 | 0.21 |
| Sherritt[c] (controlled release) | 3.20 | 0.14 | 1.11 | 4.34 | 0.25 |
| IBDU (controlled release) | 3.00 | 0.14 | 1.00 | 4.93 | 0.29 |
| Osmocote (controlled release) | 3.20 | 0.14 | 1.07 | 4.58 | 0.28 |
| *Year 4 – Post-treatment – September 1994* | | | | | |
| Standard[b] (water soluble) | 3.38 | 0.18 | 1.14 | 5.07 | 0.34 |
| Sherritt[c] (controlled release) | 3.52 | 0.15 | 1.24 | 4.72 | 0.33 |
| IBDU (controlled release) | 3.42 | 0.15 | 1.07 | 5.10 | 0.36 |
| Osmocote (controlled release) | 3.38 | 0.16 | 0.86 | 5.67 | 0.37 |

[a] No statistically significant differences among treatments.
[b] Controlled release rate based on 75% of standard rate.
[c] A sulphonated ethylene propylene diene polymer produced by Exxon technology licensed exclusively to Sherritt Inc., Canada; US Patent No. 4988377.

*Table 5.* Effect of controlled-release and granular soluble fertilizers on leaf micronutrient content of young 'Hamlin' orange trees/'Swingle' citrumelo rootstock at Gainesville, Florida, 1991 to 1994 (mean of five leaves from 12 trees per treatment)[a].

| Fertilizers | Fe[y] | Mn | Zn | Cu | B[b] | Na[c] |
|---|---|---|---|---|---|---|
| | | | (ppm) | | | |
| *Pretreatment – April 1991* | | | | | | |
| | 338 | 29 | 38 | 23 | 85 | – |
| *Year 1 – Post-treatment – October 1991* | | | | | | |
| Standard (water soluble) | 52 | 35 | 16 | 8 | 49 | – |
| Sherritt (controlled release) | 49 | 21 | 24 | 18 | 43 | – |
| IBDU (controlled release) | 48 | 22 | 17 | 8 | 45 | – |
| Osmocote (controlled release) | 55 | 23 | 16 | 8 | 44 | – |
| *Year 2 – Post-treatment – August 1992* | | | | | | |
| Standard (water soluble) | 225 | 77 | 19 | 10 | 61 | 623 |
| Sherritt (controlled release) | 325 | 34 | 13 | 6 | 68 | 500 |
| IBDU (controlled release) | 339 | 27 | 14 | 4 | 38 | 610 |
| Osmocote (controlled release) | 273 | 20 | 13 | 6 | 47 | 617 |
| *Year 3 – Post-treatment – August 1993* | | | | | | |
| Standard (water soluble) | 64 | 36 | 17 | 3.0 | – | – |
| Sherritt (controlled release) | 62 | 23 | 14 | 3.5 | – | – |
| IBDU (controlled release) | 70 | 23 | 12 | 3.8 | – | – |
| Osmocote (controlled release) | 62 | 27 | 13 | 4.0 | – | – |
| *Year 4 – Post-treatment – September 1994* | | | | | | |
| Standard (water soluble) | 83 | 32 | 33 | 12.0 | – | – |
| Sherritt (controlled release) | 83 | 19 | 37 | 6.2 | – | – |
| IBDU (controlled release) | 86 | 21 | 25 | 5.7 | – | – |
| Osmocote (controlled release) | 83 | 21 | 40 | 11.0 | – | – |

[a] No statistically significant differences among treatments.
[b] No samples were taken in 1991, 1993 and 1994.
[c] No leaf Na samples were taken in 1991.

*Table 6.* Revised fertilization rates for young[a] and mature[b] citrus trees and mature trees[b] (Tucker *et al.*, 1995).

| Years in grove | Applications per tree per year | Total fertilizer per tree per year (kg) | | |
|---|---|---|---|---|
| | | N | $P_2O_5$ | $K_2O$ |
| 1 | 4–6 | 0.07–0.14 | 0.07–0.14 | 0.07–0.14 |
| 2 | 4–6 | 0.14–0.27 | 0.14–0.27 | 0.14–0.27 |
| 3 | 4–6 | 0.20–0.40 | 0.20–0.40 | 0.20–0.40 |
| 4 | 4–6 | 133–168 kg/ha (oranges)<br>133–168 kg/ha (grapefruit) | | |
| 5+<br>(mature trees) | 4–6 | 168–122 kg/ha (oranges)<br>146–180 kg/ha (grapefruit) | | |

[a] Recommended fertilizer ratio: N – 1, $P_2O_5$ – 1, $K_2O$ – 1, Mg – 1/5, Mn – 1/20, Cu – 1/40, B – 1/300 A fertilizer with an analysis of 8–8–8–1.6–0.4–0.2–0.025 will contain these elements in this ratio and order.

[b] Recommended fertilizer ratio: 1–1–1.

Given its 44.5% N analysis, the total amount of Sherritt controlled release fertilizer applied to young citrus trees was 4% that of the standard, water-soluble fertilizer and from 39 to 45% that of the two other controlled release fertilizers in years 1, 2 and 3, suggesting potential savings in fertilizer hauling, storage and application costs. This research (Ferguson and Davies, 1995) and other field experiments with controlled release fertilizers has led to a reduction in recommended N application rates for both young and mature trees (Table 6) (Tucker *et al.*, 1995).

## Acknowledgments

This research was supported by Sherritt Inc., Edmonton, Canada.

## References

Alva A K and Tucker D P H 1993 Evaluation of a resin coated nitrogen fertiliser for young citrus trees on a deep sand. Proc. Fla. State Hort. Soc. 106, 4–8.

Boman B 1993 A comparison of controlled release to conventional fertiliser on mature 'Marsh' grapefruit. Proc. Fla. State Hort. Soc. 106, 1–4

Fergusan J J and Davies FS 1995 Fertilisation of young citrus trees with controlled release fertilisers. Proc. Fla. State Hort Soc. 108, 156–160.

Ferguson J J, and Taylor C L 1993 The 1992 Citrus Management Survey. Univ. Fla. Coop. Ext. Ser. Bull. 288. Gainesville, FL.

Ferguson J J, Davies F S, Matthews C H and Davis R M 1988 Controlled release fertilisers and growth of young 'Hamlin' orange trees. Proc. Fla. State Hort. Soc. 101, 17–20.

Koo R C J, Anderson C A, Stewart I, Tucker D P H, Calvert D V and Wutscher H K 1984 Recommended Fertilisers and Nutritional Sprays for Citrus. Univ. Fla. Ag. Expt. Sta. Bull. 536D. 30 pp Gainesville, FL.

Marler T E, Ferguson J J and Davies F S 1987, Growth of young 'Hamlin' orange trees using standard and controlled release fertilisers. Proc. Fla. State Hort. Soc. 100, 61–64.

Obreza T A and Rouse R E 1991 Controlled release fertiliser use on young 'Hamlin' orange trees. Proc. Soil Crop Sci. Soc. Fla. 51, 64–68.

Taylor C L, Ferguson J J, Israel G D and Summerhill W R 1989 Citrus Young Tree Care. Univ. Fla. Coop. Ext. Ser. Bull. PE-7. Gainesville, FL.

Tucker D P H, Alva A K, Jackson L K and Wheaton A T (Eds) 1995 Nutrition of Florida Citrus Trees. Univ. Fla. Coop. Ext. Ser. Pub. SB 169. Gainesville, FL.

Zekri M and Koo R 1991 Comparative effects of controlled-release and soluble fertilisers on young 'Valencia' orange trees. Proc. Fla. State Hort. Soc. 104, 199–202.

# SECTION III

# Environmental Aspects of Aquaculture

16

# Aquaculture and the environment: change and challenge

JAMES F. MUIR

## 1. Introduction

Many analyses suggest that after significant growth in the last three decades, production from capture fisheries is unlikely to grow much further, and will be increasingly limited by physical and biological capacity, by the costs of exploitation, and by deteriorating environments. Aquaculture production has been widely expected to compensate for shortfalls from traditional fisheries (see for example Pillay, 1990). Here, inputs, production processes and quality of output can be at least partially controlled, and ownership, care and environmental responsibility might be more easily established. By removing natural constraints to survival and productivity, and by husbandry and management, production need be limited only by availability of simple inputs such as land, water, seed, fertilizer and feeds. According to FAO statistics, aquaculture has grown steadily in recent decades (Table 1). On present trends, by the year 2000, farmed production might account for some 20 million t $y^{-1}$, some 16–18% by weight, perhaps as much as 45% by value of total output.

*Table 1.* Fishery sector production 1991, by environment/habitat.

| Environment | Total output $10^6$ t (% total) | Fisheries $10^6$ t (% total) | Aquaculture $10^6$ t (% total) |
|---|---|---|---|
| Inland | 15.2 (15.5%) | 7.5 (7.6%) | 7.7 (7.8%) |
| Coastal/marine | 83.1 (84.5%) | 78.1 (79.5%) | 5.0 (5.1%) |
| Total | 98.3 (100.0%) | 85.6 (88.8%) | 12.7 (12.9%) |

Source: FAO (1993), excluding aquatic plants.

An outline of production by species groups is shown in Figure 1, which demonstrates the increasing role of fish and crustacea, the basis of much of the sector's development and expansion. Based originally on semi-natural 'extensive' systems, these are increasingly being intensified, with a corresponding increase in the use of external inputs. In regional terms, more than 80% of production comes from Asia; Latin America contributes less than 5%, while Africa produced only 60 000 t, 0.2% of global production, recording a decline over the past three decades.

*D. Rosen* et al. *(eds): Modern Agriculture and the Environment, 183–196.*

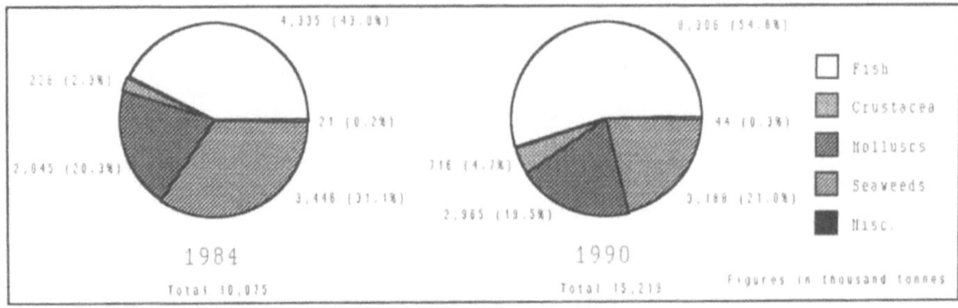


Source: FAO (1992).

*Figure 1.* Aquaculture production by major species category, 1984, 1900.

Although as Table 2 indicates, the rate of growth has been higher in developing countries, much of the newer investment involves higher-value species – these are often the only species which can support capital development and production input costs.

Within a given range of aquatic resource potential, therefore, lower value products such as seaweeds, molluscs and local food fish have not generally developed at the same rate as crops for export or for the more prosperous sectors of the society. In basic nutrition terms, the areas of the world most likely to benefit from aquaculture development have been less able to support it, though many of the physical resources, including good levels of photosynthetic energy, may be equally available. A key issue under current examination is the role of social and institutional factors in adopting aquaculture, and the specific features of technology which may affect this (Harrison, 1994). In the many areas where aquaculture has traditionally offered lower-cost supplies – for example brackish water *tambak* milkfish production in Indonesia and the Philippines, or carp culture of China, SE Asia and India, products are either increasing in price beyond the everyday means of the poorer rural communities, or are being supplanted by other higher-value cash crops.

*Table 2.* Aquaculture production by country group (figures in thousand tonnes).

| Country group | 1984 | 1986 | 1988 | 1990 | % annual growth |
|---|---|---|---|---|---|
| Developed[1] | 2 761 (27.3%) | 3 069 (25.1%) | 3 366 (23.1%) | 3 408 (22.2%) | 3.6 |
| Developing | | | | | |
| All species | 7 357 (72.7%) | 9 141 (74.9%) | 11 190 (76.9%) | 11 915 (77.8%) | 8.4 |
| Fish only | 3 386 (33.5%) | 4 811 (39.4%) | 6 170 (42.4%) | 6 824 (44.5%) | 12.4 |
| Total | 10 118 | 12 210 | 14 556 | 15 323 | 7.2 |

Source: FAO (1992); including seaweeds, excluding mammals.

[1] According to standard UN categories of GDP.

## 2. Aquaculture systems and resource use

Systems and techniques have developed significantly, moving from traditional semi-natural methods towards modern intensive tank and cage-based operations. The pace of technical change has accelerated, with increasing production and yield, but greater environmental consequence. Traditional systems are also evolving, in response to changes in resource availability, economic demand, and technology. It is therefore important to consider the relationships between aquaculture and its resources. As with any economic activity, a range is involved, broadly categorised as:

- primary – i.e. land, water, energy, primary productivity;
- secondary – various developed or manufactured goods, derived from primary inputs; and
- tertiary – such as skill, information, technological expertise.

All of these interact to define the relationship between the activity and the environment, but the role of primary resources, particularly land and water, is particularly crucial. Table 3 estimates the additional water and land resources needed to support the extra production required to maintain average per capita supply at 1989 level to the year 2000. These are based on estimates of the type/intensity of system used, and typical water and land usages (see e.g. Muir, 1992).

For coastal aquaculture, the use of land or coastal zone depends on the intensity of production, typically ranging from 0.01 to 5 ha $t^{-1}$ of output. Here, production or loading per km or $km^2$ of land/water area can be a useful indicator. Csavas (1993) has estimated loading levels in SE Asia to range from 16 t $km^{-1}$ coastline in Malaysia, 18 t $km^{-1}$ in the Philippines, 59 t $km^{-1}$ in the ROK, 80 t $km^{-1}$ in Thailand, 85 t $km^{-1}$ in Taiwan and 210 t $km^{-1}$ in China. As Table 4 shows, indicative ratios can also be developed between aquaculture production and water volume. By combining these overall load rates with factors such as water use/exchange, and with nutrient loading inputs, a general assessment of the overall 'loading' of aquaculture can be developed.

## 3. Key environmental problems

On a global basis, steadily rising demands on natural resources are accompanied by increasing public sensitivities to environmental impact. The 1992 UNCED 'Rio' conference gave significant international focus to these issues, and produced a series of commitments for environmental protection and sustainable development. These have highlighted the need for better environmental management of natural resource-sensitive processes such as aquaculture, and have increasingly attempted to bring externalities of their actual or potential impacts into direct environmental costs (Folke and Kautsky, 1992), and into environmental policy. Together with increasing commercial pressures, these trends have brought about growing need for efficiency, in which inputs are more carefully applied and managed. The main issues and key

*Table 3.* Projected freshwater aquaculture resource use by major species/production system.

| System | Species involved (estimated % species total) | Production $10^3$ t (% total) | Water usage | | Area usage | |
|---|---|---|---|---|---|---|
| | | | Estimated average $10^3$ $m^3$ $t^{-1}$ | Total use $10^6$ $m^3$ $y^{-1}$ (% total) | Estimated average ha $t^{-1}$ $y^{-1}$ | Total use $10^3$ ha (% total) |
| Extensive | Carp[1] (40), tilapia[2] (30) | 1700 (27.4) | 10 | 17 000 (9.2) | 10 | 17 000 (53.5) |
| Semi-intensive | Carp (55), tilapia (55), trout (10), catfish (25), prawns[3] (95), eels (5) | 2540 (40.9) | 5 | 12 700 (6.9) | 3 | 7 620 (24.0) |
| Intensive | Carp (5), tilapia (5), trout (35), catfish (65), prawns (5), eels (50) | 470 (7.6) | 20 | 9 400 (5.1) | 1 | 470 (1.5) |
| Super-intensive | Trout[4] (50), catfish (10), eels (40) | 180 (2.8) | 20 | 36 000 (19.4) | 0.003 | 0.54 (neg) |
| Total inland | | 4890 (78.7) | | 75 100 (40.5) | | 25 090 (78.9) |

Source/notes: table based on projections by New (1991), assuming basis of 1989 per capita supply, with stable capture fishery production at 100 million tonnes. Net increase in production excludes molluscs and aquatic plants, and is based on species distribution in 1989, with author's assumptions for percentages of each species group grown in particular environments or systems.

[1] Includes common, grass, silver and bighead carp.

[2] Tilapia also grown in coastal/ marine areas but assumed negligible here.

[3] Includes *Macrobrachium* and red swamp crawfish.

[4] Salmon also grown in freshwater, but insignificant.

*Table 4.* Outline aquaculture loading per available ecosystem volume.

| Environment | Available volume $km^3$ | Production $10^3$ t | Loading ratio t $km^{-3}$ | Loading ratio g $m^{-3}$ |
|---|---|---|---|---|
| Inland | 132 000 | 7 996 | 60.6 | 0.06 |
| Coastal | 330 000 | 8 551 | 25.9 | 0.03 |
| Total | 462 000 | 16 547 | 35.8 | 0.04 |

Sources: FAO (1993), Dolopsakis, personal communication.
Coastal resources based on an assumed breadth of 5 km.

environmental problems (GESAMP, 1991; Hakanson *et al.*, 1988) associated with these factors are listed in Table 5.

To illustrate the scale of the waste and nutrient factors alone, open water flows as defined by oxygen demand are typically 0.005–0.025 $m^3$ $sec^{-1}$ $t^{-1}$, either ducted through land-based units, or flushed through in water-based systems. Feed loadings in intensive systems are typically 0.1–5 kg $m^{-2}$ $day^{-1}$, or 0.01–0.05 kg $m^{-3}$ water used, producing waste solids at the level of 0.01–1 kg $m^{-2}$ $day^{-1}$ or 0.001–0.015 kg $m^{-3}$ water used. With water-based, e.g. cage systems, benthic loading, assuming 10–100 fold dispersion rates and average wastes outputs, would range from 0.0001–0.1 kg $m^{-2}$ $day^{-1}$. On land-based sites (Bergheim *et al.*, 1991) similar total amounts would be dispersed into streams, lakes or coastal areas, or in many cases may concentrate around the discharge area. Overall, 1000 t production could require the use of 100–200 × $10^6$ $m^3$ of water, adding 200–500 t of solids, 150–400 t BOD, 20–80 t of N, 5–25 t of P, plus micronutrients, treatment chemicals, etc.

For extensive and semi-intensive (partially fed) systems loadings are relatively lower, but may be significant locally, especially during seasonal draining or cleaning. Traditionally, requirements are accommodated by supplying copious

*Table 5.* Environmental problems associated with aquaculture.

| Problem area | Nature of problem |
|---|---|
| Waste and nutrient loadings | Outputs of solids, N, P, vitamins, minerals, husbandry/disease chemicals, antibiotics; impacts of waste materials on the adjacent benthos and the water column; on species/ community diversity, quality indices, possible stimulation of blooms |
| Water exchange | In intensive landbased systems, or flushing through freshwater or marine cage or enclosures; quantities required, effects of abstraction, dilution with 'low grade' wastes, at concentrations sufficient to diminish measured quality, but too low for simple treatment |
| Escaped stocks | From damaged systems, or through flooding, damaged or ineffective discharge screens; risks of competition with/genetic contamination of local stocks, disease transmission, directly or indirectly reduced biodiversity |
| Predation | Causing damage, loss, stress-related disease to farmed stocks, requiring controls without compromising conservation interests |
| Increasing demand for pre-emptive control | Requiring a precautionary or even 'zero-tolerance' approach to existing and intended development, implying anticipatory understanding of processes and risks, and provision for even very low areas of environmental risk |

quantities of water, discharging to receiving systems having a high absorptive capacity. Limitations in either may be tolerable in the short term, but may quickly lead to unacceptable effects (Beveridge, 1984). Consequences include local and general environmental degradation, mixing of stocks, reduced biodiversity, and greater disease management problems, the imposition of increasingly stringent controls and regulations, and direct economic loss. Chua (1993) and Briggs (1993) noted the problems of the Taiwanese shrimp industry in 1988, 1992, and 1993, and the collapse of production in central Thailand in 1989–90, with an estimated loss of some $30 m and the dereliction of some 45 000 ha of land. Problems have also been noted in the Philippines, Indonesia, Sri Lanka, and in China (Chua, 1993), where production has also been heavily affected. The need to supply such demands, the increasing need to 'fallow' sites, i.e. move production to allow sites to recover from loadings, and increasing management restrictions, provide serious constraint for continuation or development (Jensen, 1993). Two approaches are considered in the next sections: improving efficiency, and increasing environmental control.

## 4. Improving efficiency

Intensification is typically accompanied by increased water use. In common carp and tilapia culture in Israel however, a decrease in water requirements has occurred (Hepher, 1985; Sarig, 1988a), primarily because of the high cost of water, and priority being given to conservation. The production of species tolerating poor water quality is also possible with little increased water use. There is an increasing interest in the integration of aquaculture with other industries or with agriculture. There may also be benefits where for example aquaculture can add value to a water resource, e.g. fish production in sewage fed ponds, to improve water quality and generate income from a waste material (Edwards *et al.*, 1987; Little and Muir, 1987). The development of dual-purpose reservoirs in Israel has improved utilisation and reduced costs of water for aquaculture. Water loss from ponds can also be reduced by better siting and management, and by increasing pond depth. Sarig (1988a, b) reported water losses from ponds to be 6–10 times as high (35–60 000 $m^3\ ha^{-1}\ y^{-1}$) as those from deeper dual-purpose reservoirs (5941 $m^3\ ha^{-1}\ y^{-1}$).

It is obviously preferable for aquaculture operations to be set up from the outset to minimize impact, though for various reasons – lack of incentive, absence of awareness of problems and/or site or operational characteristics – many problems may need to be dealt with retrospectively. Table 6 summarises the key issues.

Routine water conservation is practised in many aquaculture systems and can be implemented in various ways. Deepening of ponds increases storage and may widen the potential for other uses (Hepher, 1985). Loss of water during harvesting can be reduced by avoiding draining, although this may result in long-term deterioration of water quality. Integration with agriculture offers scope for efficiency, increasing fertility of irrigation water, to maintain or improve yields with lower levels of fertilization. Pond water has been reported as a fertilizer for maize and for rice and vegetables in the Philippines, and pond sediments and water have been widely used

*Table 6.* Minimizing impact of aquaculture operations.

| Factor | Features/implications/effects |
|---|---|
| *Physical development* | |
| Conceptual approach | Design the system to maximize pay-off between productivity and environmental effects; i.e. get best, most efficient performance from as little as possible environmental cost |
| Careful site selection | Minimize impact on sensitive areas; visually, in terms of cover/erosion, in terms of ecosystems and/or individual species |
| Aquatic system capacity | Provide adequate mixing, dilution and dispersal e.g. of wastes, or e.g. self-purification in river systems |
| Waste and impact management | Careful attention to location, layout, position, visual effects, colour, shading/protection, storage/work areas |
| Water and effluent management | Collection, concentration and disposal – self-cleaning tanks, solids collection points |
| Stock security | Design to minimize releases, contact with external water supplies, transfer to other stock, disease contamination |
| Construction management | Limit construction disturbance, keep top cover as much as possible, do not discharge materials into water courses, etc. |
| *Operational management* | |
| Stock management | Care over incoming stock, also genetic consequences; stock density/system management control |
| Feed management | More digestible and water-stable feeds, better feeding methods, feed control, optimizing efficiency rather than growth |
| Disease/loss management | Predator control, changing ways of disposal of morts (dead stock) and wastes; restricting numbers of different stocks and hence possibility of disease transfer |
| Site management | Using fallow sites; controlling use of chemicals, transfer of resistance |
| Water management | Reduction of seepage losses, intensification of use, and reuse of water. Integration of agricultural water with aquaculture |

as fertilizer in China (Edwards, 1980). Fish culture in rice fields is also returning into practice, as low-pesticide approaches are more widely adopted. Marginal water supply and fertilization costs are very low, but social and economic benefits can be significant, through production of fish and increased rice yields. Fish culture may also improve water resources, through environmental enhancement or rehabilitation; e.g. improving water quality in acidified lakes through selective 'pollution' by cage culture; stocking irrigation canals with macrophyte feeders such as grass carp *(Ctenopharyngodon idella),* java carp *(Puntius gonionotus)* and *Tilapia zilli* to improve water flow.

Cage and pen culture also offer good scope for improved use of water resources as they do not require water abstraction. Cages have been used successfully in irrigation reservoirs in several countries and in a wide range of lakes, reservoirs and running waters (Beveridge, 1987). Freshwater reservoir resources are likely to increase significantly suggesting niches for cage aquaculture to utilize an expanding resource. Greater understanding of interactions between cage culture and environment and development of models to predict stocking levels should assist optimal use. In warm waters, integration of extensive with intensive cage culture

*Table 7.* Aquaculture control and environmental management.

| Operational factor | Implications |
|---|---|
| System integrity | Stock releases, predator control |
| Water exchange | Water usage, quality, health status |
| Behavioural monitoring | Feeding efficiency, stock, health, predator control |
| Feed inputs | Feeding efficiency, water quality, waste output |
| Process management | Water usage, quality, pathogen control |
| Waste management | Waste control, removal, disposal |

offers scope for minimizing eutrophication from intensive cage farming. Pond farming may also be developed to provide more water efficient systems, although land resources may be a constraint. The management problems of large water volumes and low stock densities may be partly overcome by culture using cages or enclosures within large ponds. Finally, though relatively undeveloped since a period in the mid-1970s, the integration of aquaculture with domestic and industrial waste waters is likely to become increasingly important, particularly where low-cost sanitation is to be developed.

## 5. The potential for environmental control

The control of environmental aspects of aquaculture production is set to become an increasingly critical management issue. The objective of environmental management is closely associated with efficient production processes, and may also involve dealing with 'non-process' matters such as predation or accidental events (Muir and Bostock, 1994). Opportunities for environmental monitoring and control clearly vary with the production system, the degree of management control inherent within the system, its scale and operating environment, and external pressures – financial and/or regulatory, which might affect operation. The main areas of control concern water quality, and in close association, ecosystem quality. Other social and/or aesthetic impacts considered undesirable might also be controlled. In many cases, monitoring, management and control approaches for aquaculture are as yet only being conceptualised and developed (GESAMP, 1991; Muir and Baird, 1991; Muir and Beveridge, 1994). At the detailed and specific level, they may also have to be 'tailored' for particular systems and environments. However, given the inefficiencies in many aquaculture processes, and positive consequences of even simple interventions, better environmental control may have a profound effect on the relationship between aquaculture production and environmental impact. The main areas of use are outlined in Table 7. These are detailed further in Table 8, to illustrate the control factors, and the combined objectives of cost, efficiency, security and environmental control which may be involved.

An area of increasing interest in more intensive aquaculture is that of water management and reuse. Such systems commonly involve higher capital costs for water circulation and treatment units, and additional energy cost in circulating water and running treatments. Their record to date has been variable, and commercial

*Table 8.* Monitoring/control factors for aquaculture and their implications.

| Factor | Control variables | Implications/response |
|---|---|---|
| *Input* | | |
| Water | Rate, quality, timing | Security, efficiency, cost |
| Oxygen | Rate, timing | Security, performance, cost |
| Feed | Rate, quality, timing | Efficiency, growth, wastes, cost |
| Energy | Rate, source, timing | Efficiency, cost |
| *System and process* | | |
| System integrity | Presence, continuity, rates | Security, environment, cost |
| System function | Mixing, water quality, rates | Efficiency, safety, cost |
| Stock responses | Size, number, behaviour, distribution | Efficiency, planning, growth, health, safety, cost |
| *Output* | | |
| Harvest | Size, number, quality, timing | Safety, reputation, revenue, |
| Water | Rate, quality, timing | Environment, cost |
| Wastes | Rate, type, timing | Environment, safety, cost |

Source: Beveridge and Muir (1994).

ongrowing in particular has been limited by higher than average production costs (Muir, 1982). However they can offer specific advantages, including:

- the control of key water quality parameters, particularly temperature, dissolved gas and nitrogen (ammonia and nitrite);
- relative independence from external water supplies, hence less locational constraint, and better inbuilt control over discharge wastes;
- opportunities to manipulate the rearing environment, to control/optimize growth rates, production timing, output quality; and
- improved flexibility of supply, overcoming seasonal constraints, meeting stringent market demands.

An important distinction can be drawn between technological and ecological approaches, as outlined in Table 9. The former are the most commonly described,

*Table 9.* Approaches to water reuse.

| Technological | Ecological |
|---|---|
| Water is moved outside culture area | Water is retained within culture area |
| Specific treatments are applied to the process water, with defined water quality objectives | Water use is maximized within the culture area, with little specialized treatment |
| Liquid and gas pumping and energy usually applied; often also heating/cooling | System is driven primarily by photosynthetic energy for water treatment, temperature conditions |
| Specifically designed treatment stages, with at least some degree of control | Based primarily on biological processes, over which there is limited control |
| More commonly used for intensive, usually monoculture systems | More often used for semi-intensive systems often with polyculture |
| Typically capital intensive, with 'high technology' methods | Usually lower capital input, with simpler technology, where used |
| Typically high energy consumption, though can be controlled | Usually lower energy use, but may be wasted within poorly controlled systems |

but it is valuable to recognise recycling functions within conventional or modified aquaculture systems, as these may often be capable of enhancement through design or management to achieve useful gains in water yield. A range of methods may be incorporated, from simple in-tank processes, to complex and carefully controlled treatment systems. Another trend involves pond-based recycle systems, which though relatively simple may be potentially more significant for many of the pressing resource challenges facing the main aquaculture sector. These have evolved from several directions: simpler Japanese eel systems (Gousset, 1992); the early systems for channel catfish; intensive Taiwanese and Israeli pond systems (Hepher and Pruginin, 1981); emergency water shortage systems in salmonid farming; combined intensive/extensive systems for Mediterranean seabass/bream; and modified shrimp farming systems in SE Asia.

These systems usually operate on low-medium levels of recycle (typically 50–85%), are developed from traditional open-flow systems, use ambient temperatures, and use simple treatment methods. Though based around more sophisticated systems, they are usually far more loosely designed, using simple large-scale pond construction techniques. Aeration and solids removal are standard, often based on paddle-wheel aerators, in the Taiwan and Israeli systems, used also to direct water and collect solids. Solids removal is normally by sedimentation, sometimes in production ponds themselves. Biological treatment is used for higher levels of recycle, through lagoons, sometimes operating as semi-intensive fish ponds, simple rock or gravel filter beds, mollusc culture, macro-algal beds and/or designed wetlands (i.e. reed-beds). The choice of system depends on local environmental conditions and preferred management practice. Water is pumped back to the main aquaculture system, on a continuous intermittent, or partial replacement basis. These systems are quite 'ecological' in nature, and are as yet little developed, though broad design and operating parameters have been defined empirically. They offer direct improvements to water quality, and usually also reduce external loadings, thereby benefiting the wider resource system. They may be based on very simple modifications of existing systems; one or more production units may be converted for water treatment, and though productive area decreases, better water quality may result in improved food conversion and higher yields, and some systems produce secondary crops of fish, algae or molluscs. These systems need further assessment and development, as there are obvious areas of opportunity for optimising production, treatment efficiency, and water management.

## 6. Environmental management

The need for a more strategic perspective in the environmental management of aquaculture has been increasingly recognized, and issues have been highlighted in a number of reviews (see e.g. NCC, 1989; Pullin *et al.*, 1993). Coastal aquaculture development has received particular attention, in the context of coastal planning and management (Barg, 1992; Clark, 1992; Muir and Beveridge, 1994). Environmental management of inland aquaculture has also been considered, whether in flow-through systems or for cage culture (NCC, 1989). The need for a wider resource management framework has been noted (Muir, 1992), in which aquaculture is placed into appropriate context, rather than being singled out as a unique and universally damaging activity. In practical terms, Muir and Baird (1991) suggested that environmental management of aquaculture, though stringent if necessary, should not be over-elaborate and burdensome to administrators or industry. The approach used should be:

- simple: easy to operate, clear and understandable to all involved, including the public;
- rational: based on logical and scientific foundations, providing testability and, once operational, offering predictive power;
- equitable: operating fairly on all those 'using' the resources, according to degree of use; this should apply to all users, as well as aquaculturists;
- suitable: to the capacity of the institution(s) involved; with only a modest amount of training and/or equipment, within the operational capability of the system, whether public or private sector, or both; and
- inexpensive: not too burdensome, and not either overload or excessively inhibit activities it is designed to cover, or place too great an imposition on agencies responsible for its operation.

According to these criteria, the main components might be:

- *at installation*: an environmental impact assessment (EIA) procedure appropriate for reviewing new projects, and amendments or supplements to existing projects, prior to their establishment and possibly retrospectively; and
- *during operation*: an environmental management procedure capable of monitoring effects of aquaculture activities, and if necessary providing guidance for modifying installations and/or operations to minimize impacts.

The rise of aquaculture as a new and apparently demanding user of resources has however highlighted in many countries the absence of appropriate planning frameworks, and the difficulty of defining rational policy and development strategy (see e.g. Van Houtte *et al.*, 1989). This is by no means unique to aquaculture, as many sectors are only now being brought into the scope of environmental management. The need for rational and uniform cross-sectoral approaches has increased focus on economic criteria, in terms of resource allocation decisions, and in the use of economic instruments for pollution control, including licensing, loading-based charges, and environmental recovery taxes. Many of these, based around the 'polluter pays' principle, and on maximizing economic rent for natural resources, are coming into planning and regulatory systems. Indices such as net economic output per unit of organic loading, and other comparators, are of increasing interest in making or influencing strategic and allocative decisions.

## 7. Conclusions

The aquaculture sector uses a range of resources, and results in a range of consequences, not all of which may be desirable. It is clear that any means to reduce demands on resources would be potentially beneficial, and various approaches may contribute to lessening impacts in terms of external resource use and undesirable effects. Primary consequences of this could include:

- the potential to increase production, with lower costs, within the existing resource allocated;
- better acceptance of aquaculture within the overall public opinion and regulatory environment, and the location/resource planning context, with improved predictability of impact, and better opportunities for 'fine-tuning';
- the ability to adhere to increasingly rigorous external controls without undue financial penalty or competitive disadvantage; and
- investment and development confidence, and the continued opportunity for sustained longer-term growth.

Three major trends can be distinguished in the context of aquaculture and environment;

- production and market changes, including demands for cleaner and 'greener' products;
- technological change within aquaculture systems; and
- regulatory/legislative changes, increasing control over aquaculture and other sectors.

For a full strategic perspective, these should be explicitly linked, though in most cases they have a rather discontinuous interaction. Although environmental costs are increasing in real terms, and regulation is becoming more stringent, these have not fully acted to force change. Nonetheless, a combination of tightened control, increased public pressure, greater perception of actual benefits, and realisation of potential gains, may provide a positive environment for further development. There may be significant benefits in improving aquaculture system designs, management and control systems. If properly designed and operationally tested, there are opportunities in stock and predator control, metabolite reduction, waste removal, and water quality restoration. Real gains have been recorded in feed management and system monitoring, and these may result in significant improvements in operating efficiency, risk reduction and profitability.

Aquaculture also requires to be seen within the wider views of e.g. aquatic resource use, land-use development, rural, regional, general economic development. Though it has been claimed that aquaculture can profitably use underutilized resources, all forms of aquaculture require the organisation and mobilization of resources, whose use represents an opportunity cost – whether in productive, recreational, aesthetic or other terms. In both generalized planning and specific design, it is important to be aware of resource implications, and consider the capacity of various locations and systems. That said, aquaculture can be expected to feature increasingly in the wider resource context, involving understanding, management and rehabilitation – of individual ecosystems such as lake and river systems, lagoons, mangroves and coastal reef, in terms of characteristics and potential of production systems, and in terms of better characterisation and use of water resources, including domestic, agricultural and industrial 'waste' water. These components would then form part of a set of larger, higher-order multiple-linked systems and management approaches. Such approaches, rather than presenting productive use as a creator of environmental problems, might emerge as ranges of solutions, capable of allowing improved management of aquatic resources.

## References


Barg U C 1992 Guidelines for the promotion of environmental management of coastal aquaculture development. FAO Fisheries Technical Paper 328, FAO, Rome.

Bergheim A, Tyvold T and Seymour E A 1991 Effluent loading and sludge removal from landbased salmon farming tanks. Proceedings from Aquaculture Europe 1991, Dublin 10–12 June.

Beveridge M C M 1984 Cage and pen fish farming. Carrying capacity models and environmental impact. FAO Fisheries Technical Paper 255, 133 pp.

Beveridge M C M 1987 Cage Aquaculture. Fishing News Books, Surrey, England.

Beveridge M C M and Muir J F 1994 Controlling aquaculture systems for environmental management. In Proceedings, EAS Conference, 'Measures for Success' Instrumentation and Control in Aquaculture. Bordeaux 23–25 March 1994. EAS Special Publication No. 21, Oostende.

Briggs M R P 1993 Water quality in sustainable shrimp farming. In The Third Academic and Research Conference for Southern Thailand Technical Colleges. Krabi, Thailand, 23–25 July 1993, 46 pp mimeo.

Chua T E 1993 Environmental management of coastal aquaculture development. In Environment and Aquaculture in Developing Countries. ICLARM Conference Proceedings 31. Eds R S V Pullin, H Rosenthal and J L Maclean. pp 199–212.

Clark J R 1992 Integrated management of coastal zones. FAO Fisheries Technical Paper 327, FAO, Rome.

Csavas I 1993 In IPFC Symposium on Socio-economic issues in Coastal Fisheries Management, Bangkok, Thailand, 23–26 November 1993.

Edwards P 1980 Food potential of aquatic macrophytes. ICLARM Studies Review 5. ICLARM, Manila, Philippines.

Edwards P, Polprasert C and Kok L W 1987 Resource recovery and health aspects of sanitation. Asian Institute of Technology Research Report, Number 205.

FAO 1992 Aquaculture production 1984–1990. FAO Fisheries Circular No. 815, Revision 5, FIDI/C815 (Rev. 5) Rome.

FAO 1993 Aquaculture production 1985–1991. FAO Fisheries Circular No. 815, Revision 6, FIDI/C815 (Rev. 6), Rome.

Folke C and Kautsky N 1992. Aquaculture with its environment: Prospects for sustainability. Ocean and Coastal Management, 17, 5–24.

GESAMP 1991 Reducing Environmental Impacts on Coastal Aquaculture, 35 pp. Reports and Studies No. 47, FAO, Rome.

Gousset B 1992 Eel culture in Japan. Bulletin de l'Institut oceanographique de Monaco. No speciale 10, Monaco 128 pp.

Hakanson L, Ervik A, Makinen T and Moller B 1988 Basic concepts concerning the environmental effects of marine fish farms. Nord 1988: 90, Nordic Council of Ministers, Copenhagen, 103 pp.

Harrison E 1994 Aquaculture in Africa: Socio-economic dimensions. In Recent Advances in Aquaculture, Vol. 5. Eds J F Muir and R J Roberts. pp 240–299. Blackwell Science, Oxford.

Hepher B 1985 Aquaculture intensification under land and water limitations. Geojournal 10, 253–259.

Hepher B and Pruginin Y 1981 Commercial Fishfarming. John Wiley and Sons, New York.

Jensen A 1993 Carrying capacity for nutrient salts in coastal waters. In Fish Farming Technology. Eds H Reinertsen, L A Dahle, L Jorgensen and K Tvinnereim: pp 151–154. A.A. Balkema, Rotterdam, Brookfield.

Little D and Muir J F 1987 A guide to integrated warm water aquaculture. Institute of Aquaculture, University of Stirling. 238 pp.

Muir J F 1982 Recirculated water systems in aquaculture. In: Recent Advances in Aquaculture. Eds J F Muir and R J Roberts. pp 357–446. Croom Helm, London.

Muir J F 1992 Aquaculture and water resource management. In Proceedings of the Conference on Priorities for Water Resources Allocation and Management. ODA Natural Resources and Engineering Advisers' Conference, Southampton, July 1992. pp 127–148.

Muir J F and Baird D J 1991 Environmental Management of Aquaculture Development. Technical Report, TCP/CYP/9152, FAO, Rome, 65 pp plus annexes.

Muir J F and Beveridge M C M 1994 Resources, planning and management in coastal aquaculture. Proc. Conf. Fisheries and Ocean Industrial Development 1994, Pusan, S Korea. pp 209–234.

Muir J F and Bostock J C 1994 Measurement and control in aquaculture processes – new directions for the information revolution? Proceedings, EAS Conference, 'Measures for Success' Instrumentation and Control in Aquaculture. Bordeaux 23–25 March 1994. EAS Special Publication No. 21, Oostende.

NCC 1989 Fishfarming and the Safeguard of the Natural Marine Environment of Scotland. 136 pp. Nature Conservancy Council, Edinburgh.

New M B 1991 Turn of the Millennium Aquaculture: Navigating troubled waters or riding the crest of the wave?. World Aquacult. Soc. 22 28–49.

Pillay T V R 1990 Aquaculture Principles and Practices. 575 pp. Fishing News Books, Oxford, ISBN 0-85238-168-9.

Pullin R S V, Rosenthal H and Maclean J L (Eds) 1993 Environment and Aquaculture in Developing Countries. ICLARM Conf. Proceedings. Vol. 31. ICLARM, Manila.

Sarig S 1988a The development of polyculture in Israel: A model of intensification. In Intensive Fish Farming. Eds C J Shepherd and N R Bromage. pp 302–331. BSP Professional Books, Oxford, England.

Sarig S 1988b Is an increase in yields in intensive commercial fish farming always accompanied by an increase in profitability? Isr. J. Aquacult. 40, 122–124.

Van Houtte A R, Bonucci N and Edeson W 1989 A preliminary review of selected legislation governing aquaculture. 81 pp. ADCP/REP/89/42, UNDP/FAO, Rome.

# 17
# Biological removal of inorganic nitrogen and organic matter in closed, intensive fish culture systems

JAAP VAN RIJN

## 1. Introduction

Over the past decade there has been rapid development of intensive fish culture systems. Especially in Northern and Western Europe (Rosenthal and Black, 1993) and North America (Colt, 1991) relatively much R&D is being conducted in the field. Intensive fish culture systems can be classified into flow-through and recirculating systems. In flow-through systems clean water is often led in a single pass through the production unit and discharged thereafter. A drawback of open systems is that, apart from their site specificity, they often pollute recipient water bodies; it is therefore expected that expansion of these systems will be limited. Recirculating systems incorporate facilities for maintaining adequate water quality. These systems are environmentally friendly as water input is minimised and, consequently, little water is discharged. Adequate water quality control in recirculating systems is essential for their successful operation. Most recirculating systems incorporate biological and mechanical filtration. Growth of nitrifying bacteria is often induced in order to convert ammonia to nitrate, while mechanical filters are used to capture organic material (fish faeces and unutilised feed). Biological digestion of organic matter and biological reduction of nitrate, essential processes in stagnant fish ponds, are little used in recirculating systems. Incorporation of these processes in recirculating systems is predicted to result not only in more stable water quality within the fish culture units but also in considerable reduction of environmental pollution by intensive fish farming. In the present study results are presented on a recirculating system in which water is purified by a combination of nitrification, denitrification and digestion of organic matter. The study describes the various stages of research towards the development of a treatment system for recirculated, intensive fish culture.

## 2. Materials and methods

### *2.1 Pilot-plant treatment system*

#### 2.1.1 Design of the treatment system

The treatment system consisted of an aerobic trickling filter, a sedimentation basin and a fluidised bed reactor (Figure 1). The trickling filter was situated above the

*D. Rosen* et al. *(eds): Modern Agriculture and the Environment, 197–213.*

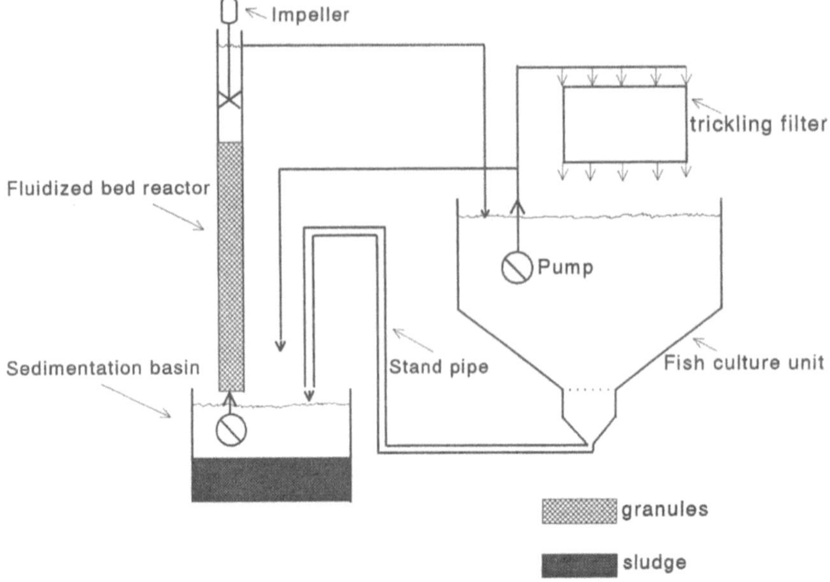


*Figure 1.* Schematic presentation of the pilot-plant fish culture unit and the biofiltration system (not to scale).

experimental fish culture unit, water from the culture unit was sprayed at a rate of 15 $m^3$ $h^{-1}$ on top of the trickling filter (volume: 2 $m^3$; surface area: 400 $m^2$). The fluidized bed column had a height of 4.8 m, a diameter of 19.3 cm and a net volume of 131.5 L. The column was initially filled with clean sand (diameter 0.3–0.9 mm) up to 50 cm of the column height. Water was pumped from the sedimentation basin into the bottom of the column causing dispersion of the sand grains throughout the column. Inoculation and start-up of the column were initiated three months before the experimental period reported in this study. At the start of the experimental period the sand grains were found to be coated with a bacterial biofilm, ranging in thickness from 1.0 to 1.5 mm. The column was checked daily for dispersion of the biofilm-coated sand. If the dispersion of the granules was found to be less than 50% of the column height, clean sand was added (approximately 10 cm of column height). During the experimental period the dispersion of the granules was found to be higher than 50% of the column height on all sampling days and no clean sand was added. The fish culture unit (50 $m^3$) is octagonal in shape with a bottom sloping towards the centre (Zohar *et al.*, 1985). Two 0.35-kW aerators were installed and caused the water to move in a circular motion. Due to the shape of the unit and the circular movement of water, organic matter concentrated at the bottom centre and from here it was drained, by means of a stand pipe, twice a day into the sedimentation basin (3 $m^3$). The water overlying the organic sludge in this sedimentation basin was pumped at a rate of 0.6 $m^3$ $h^{-1}$ into the fluidized bed reactor. The hydraulic retention time of the fluidised bed reactor was 13.1 min while that of the sedimentation basin was 5 h. Retention time of particulate organic matter in the sedimentation basin could not be determined but exceeded by far the hydraulic retention time as an organic sludge layer with a thickness of between 30–70 cm was measured during the experimental period. The water level in the sedimentation basin was kept constant by pumping water from the fish culture unit into the basin. The entry of water from the culture unit into the sedimentation basin was regulated by a floating device in the basin that allowed entry of pond water once a fixed, minimal water level in the basin was reached.

#### 2.1.2 Management of pilot-plant ponds

Two experimental ponds were allocated for this study at the Ginosar experimental station (Israel). During the experimental period (from 26 July until 8 October) both ponds were stocked with common carp (*Cyprinus carpio* L.) at an initial biomass of 20 kg $m^{-3}$. Commercial feed pellets with a protein content of 30% (Central Feedmills Association, Israel) were supplied at a rate of 2.5% of the total fish biomass per day (maximal: 25 kg $day^{-1}$ $pond^{-1}$). Equal portions of feed were supplied every 80 min between 6.00 a.m. and 2.00 p.m. by means of an automatic feeding device. One of the ponds was connected to the treatment system, described before. This pond was operated in a virtually closed fashion as only make-up water was added to compensate for evaporation and seepage losses (3 $m^3$ $day^{-1}$). The second pond served as a control pond and had no treatment system attached. The concentration of inorganic nitrogen in the control pond was examined against different dilution rates using clean, unpolluted dilution water.

### 2.2 *Laboratory-scale anaerobic treatment system*

#### 2.2.1 Design of treatment system

The anaerobic laboratory-scale treatment system consisted of a fluidised bed reactor connected to a digestion basin (540 L) and a nitrate stock solution reservoir (500 L). The flow circuit and flow rates between the various components are depicted in Figure 2. The fluidized bed reactor (height 198 cm, diameter 6.1 cm, volume 6.25 L) was made of Perspex. The upper 18 cm of the reactor was widened (diameter: 15 cm), causing a lower flow rate in this part of the reactor, whereby wash-out of granules was minimised. Silica sand (average diameter: 0.4 mm) was used as carrier material and was added to the reactor at an initial concentration of 50 g $L^{-1}$. The theoretical total surface area in the reactor at full bed expansion was 1.14 $m^2$. Clean sand was periodically added in order to compensate for the sand lost due to wash-out. Different concentrations of nitrate at the inlet of the fluidized bed reactor were obtained by changing the nitrate concentration in the nitrate stock solution reservoir. The flow rate at the inlet of the fluidized bed reactor was 1190 ml $min^{-1}$ (25 m $h^{-1}$), while the retention time of the reactor was 9.05 min. Retention time of the overall system was seven days. The system was operated at a temperature of 25 ± 1°C. Peristaltic pumps (Cole–Parmer, model 7518) were used for water circulation. The digestion basin was supplied daily with 250 g of fish feed pellets (commercial carp pellets, Central Feedmills Association, Israel), containing 25% protein on a dry weight basis.

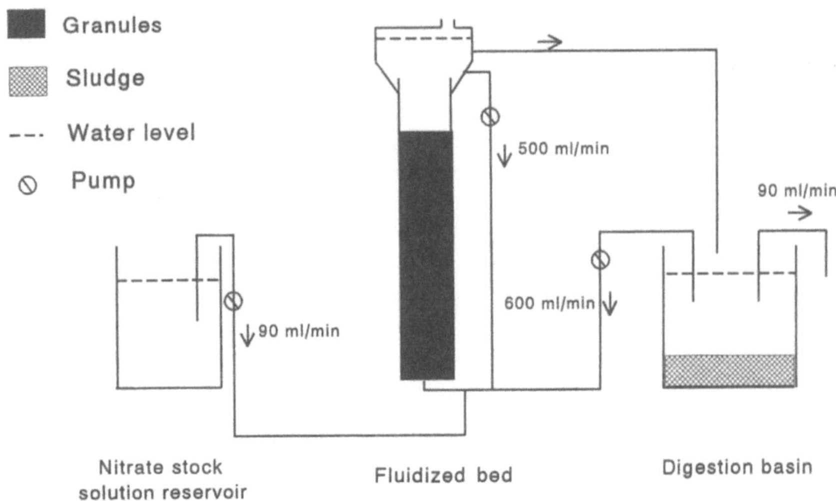


*Figure 2.* Schematic presentation of laboratory-scale anaerobic treatment system (not to scale).

### 2.2.2 Anaerobic digestion of fish feed and sludge in digestion basin

Degradation of the fish feed was examined by incubation of feed in nylon mesh bags which were suspended into the sludge layer of the digestion basin. The nylon bags (mesh size: 40 $\mu$m ) were initially filled with approximately 5.0 g of fish feed. Periodically, 2–4 bags were removed, thoroughly washed under a running tap and dried at 105°C. Degradation was expressed in percentage of the initial dry weight of the feed in the bags. Degradation of sludge, derived from the digestion basin, was determined as follows. Sludge was added to nylon bags and thoroughly washed under a running tap. The bags containing the wet sludge were dried for 24 h at room temperature and the weight of the moist sludge was determined. Sludge of four bags was analyzed for dry weight while the remaining bags were incubated in the sludge layer of the digestion basin. Periodically, two or three bags were removed, rinsed under tap water and dried at 105°C. Degradation was expressed in percent of the initial dry weight of sludge in the bags. The nylon bag method has been applied extensively for measuring litter decomposition in terrestrial ecosystems (Witkamp and van der Drift, 1961) as well as measuring ruminal digestibility of various feeds (Nocek, 1988). First-order kinetics were used to describe the observed feed degradation in the nylon mesh bags. The feed was considered to consist of two fractions, a labile or rapidly decomposing fraction and a recalcitrant or slowly decomposing fraction. Using first-order kinetics (Avnimelech, 1984; Reddy *et al.*, 1986), decomposition rate constants for the two fractions can be determined by the following equation:

$$C = C_o e^{-kt} \tag{1}$$

where $C$ is the concentration of organic matter at the given time ($t$), $C_o$ is the concentration of organic matter at $t = 0$ and $k$ is the decomposition rate constant.

The degradation of the feed can be described by assuming two degradation stages (Hunt, 1977) according to the following equation:

$$D = 100 - 100(Se^{-k_1 t} + (1 - S)e^{-k_2 t}) \tag{2}$$

where $D$ is decomposition rate (as percentage of initial concentration); $S$ is the initial proportion of labile material; $(1 - S)$ is the the initial proportion of recalcitrant material; $k_1$ is the decomposition rate constant for labile material; and $k_2$ is the decomposition rate constant for recalcitrant material.

### 2.2.3 Release of volatile fatty acids during anaerobic digestion of feed

Release of volatile fatty acids (VFA, including acetate, propionate and butyrate) was examined during anaerobic incubation of fresh fish feed in digestion basin medium. Feed (25 g) was incubated in Erlenmeyers containing 5 L water obtained from the digestion basin. The Erlenmeyers were flushed with nitrogen gas and incubated at 25°C. Periodically, samples (5 ml) were withdrawn for determination of VFA. The various incubations were conducted over a pH range of 6.0–6.5 by periodical addition of NaOH. This range was similar to the pH values normally found in the water overlying the sludge in the digestion basin.

### 2.2.4 Accumulation rate of sludge and daily VFA release

Theoretical accumulation rates of sludge and VFA release can be calculated based on equations (1) and (2). For this purpose, a computer program was developed to permit calculation of sludge accumulation and VFA release based on (specified) daily additions of feed. Sludge accumulation and VFA release in the digestion basin were examined at days 243 and 244 of operation. At day 243, the digestion basin was disconnected from the fluidized bed reactor and VFA accumulation in the supernatant of the digestion basin was examined over a 24-hour period. The concentration of sludge in the digestion basin was calculated at day 244 by determination of the dry weight of subsamples taken from the sludge layer and by measuring the total sludge volume in the digestion basin. Values of sludge accumulation and VFA release at days 243 and 244 could be predicted since the daily feed addition (250 g) over this period was known.

## *2.3 Analytical procedures*

Oxygen and temperature in the ponds were measured daily at dawn, while ammonia, nitrite and nitrate concentrations in the ponds were measured every other day at midday. The treatment system was examined weekly for ammonia, nitrite, nitrate and COD at the following sampling points: inlet and outlet trickling filter, inlet and outlet fluidized bed reactor and the sedimentation basin. Total ammonia ($NH_3$ and $NH_4^+$)

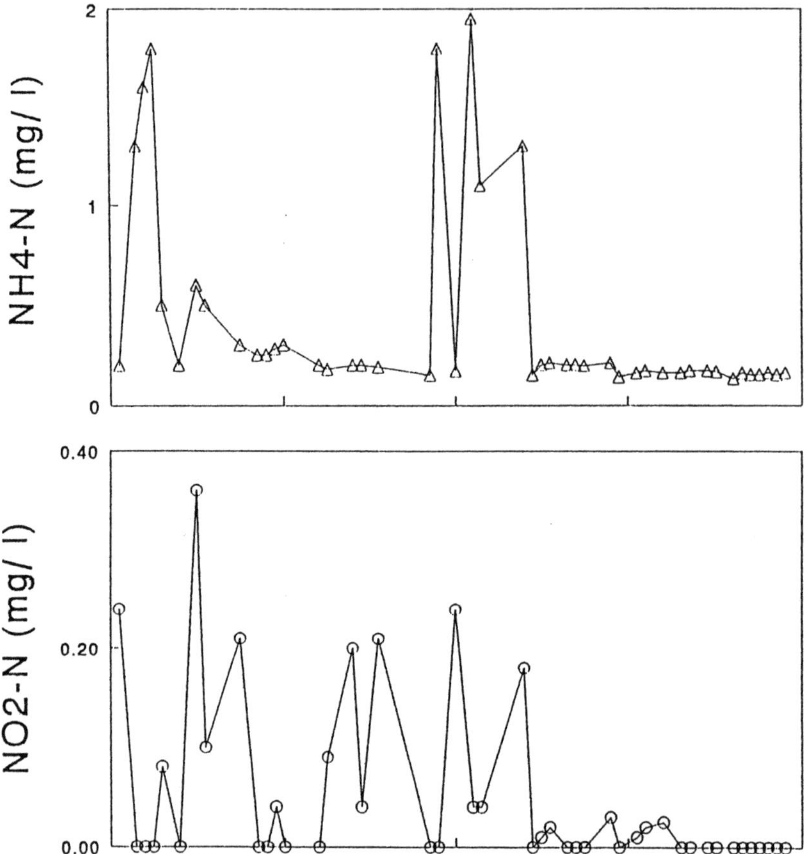


*Figure 3.* Concentrations of ammonia, nitrite, nitrate and particulate organic matter during the experimental period in the fish culture unit equipped with a biofiltration system. (Arbiv and van Rijn, 1995)

was determined as described by Scheiner (1976), nitrite was measured according to Strickland and Parsons (1968) and nitrate was measured with the Szechrome NAS reagent (Ben Gurion University, Applied Research Institute). Chemical oxygen demand (COD) was measured by oxidation of organic matter with potassium dichromate according to standard methods (APHA, 1989). Oxygen and temperature were measured with a YSI (model 57) temperature oxygen probe (Yellow Springs Instruments, Ohio). The pH was measured with an EL Hama pH meter (model PBS 5, El Hama, Mevoh Hama, Israel). Acetate, propionate and butyrate were determined with a Hewlett Packard (model 5890) gas chromatograph equipped with 1/8 inch internal diameter stainless steel column (length: 200 cm) packed with 60–80 mesh Chromosorb W. Injector temperature, oven temperature and FID detector temperature were set at 170, 140 and 175°C, respectively. Nitrogen was used as carrier gas at a flow rate of 20 ml $min^{-1}$. Analytical grade acetate, propionate and butyrate were used

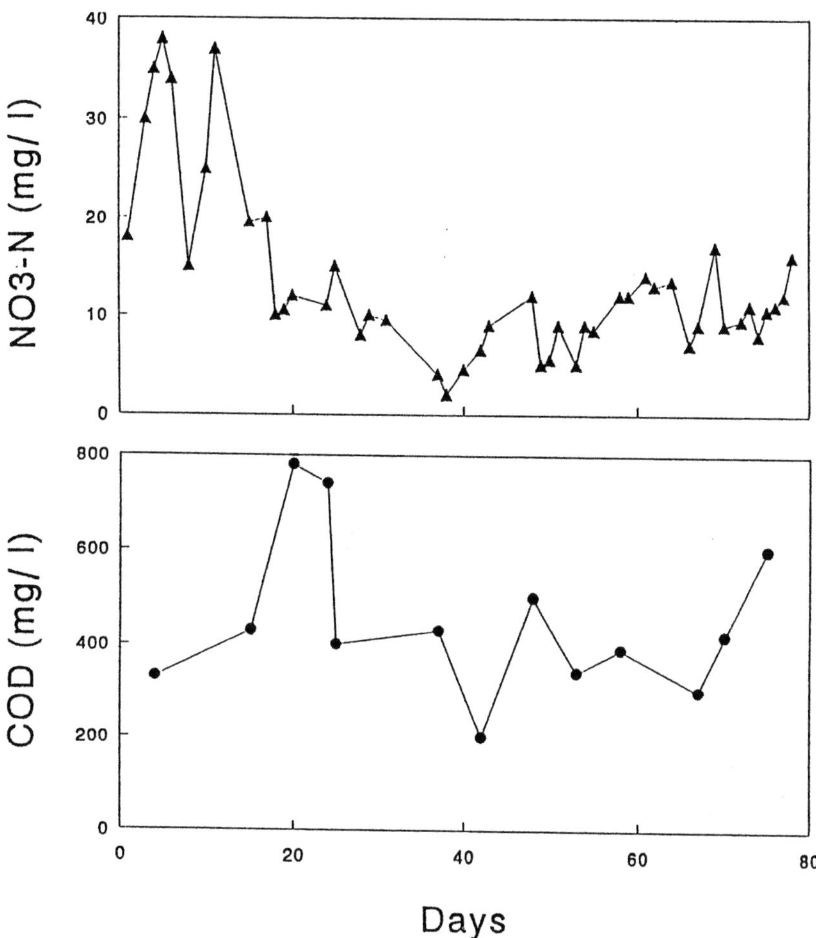


*Figure 3.* (cont).

as standards. Concentrations of VFA were expressed as COD equivalents. The following conversion factors were used: acetate (g/L) = 1.064 COD (g/L); propionate (g/L) = 1.512 COD (g/L); and butyrate (g/L) = 1.816 COD (g/L).

## 3. Results and discussion

### *3.1 Pilot plant treatment system*

The effect of the treatment system on water quality parameters in an experimental pond was examined during the summer season (from 27 July until 8 October). Temperature of the pond water fluctuated between 22.5 and 27°C, pH values between 7.0 and 7.9 and oxygen concentrations between 5.8 and 7.2 mg $L^{-1}$ (not shown). The average ammonia concentration during this period was 0.51 mg $NH_4$-N

$L^{-1}$, the average nitrite concentration was 0.05 mg $NO_2$-N $L^{-1}$ and the average nitrate concentration was 11 mg $NO_3$-N $L^{-1}$; levels of organic matter (expressed as particulate COD) in the water column of the fish culture unit reached values as high as 800 mg $L^{-1}$ (Figure 3).

Around day 38 the carp were found to be heavily infested with ectoparasites (mainly *Argulus* and *Lernea*) and heavy mortality set in. This necessitated treatment of the fish with Bromex (dimethyl 1,2-2,2 dichlorethyl phosphate) and formalin. The various treatment stages were shown to be affected by this chemical treatment as indicated by the relatively high ammonia concentrations for a period of approximately ten days (Figure 3).

After this initial treatment a weekly prophylactic treatment with Bromex (0.2 mg $L^{-1}$) for six hours was included as a routine procedure. During the weekly Bromex treatments no feed was administered and the treatment stages were disconnected from the culture unit. No harmful effect of Bromex on the various treatment stages was evident. The effectiveness of the treatment system in regard to inorganic nitrogen removal was examined by operating a control pond, similar in shape and adjacent to the treated pond, in which the fish biomass, feeding rate and flow rate (3 $m^3$ $day^{-1}$) were similar to those in the treated pond. The control pond was only operated for a fortnight in this fashion as fish mortality, probably due to toxicity of inorganic nitrogen compounds, was apparent from day 9. Concentrations of inorganic nitrogen (Figure 4) reached extremely high levels: ammonia as high as 45 mg $NH_4$-N $L^{-1}$, nitrite, 8 mg $NO_2$-N $L^{-1}$; and nitrate, 65 mg $NO_3$-N $L^{-1}$. In

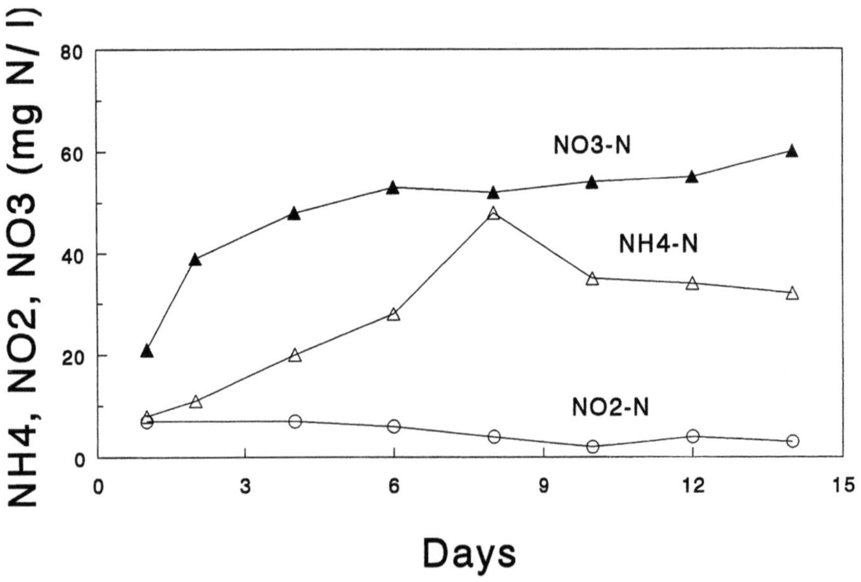


*Figure 4.* Concentrations of ammonia, nitrite and nitrate in the water column of a fish culture unit (without biofiltration) operated without water exchange. (Arbiv and van Rijn, 1995)

*Table 1.* Performance of the pilot-plant fluidised bed reactor over the experimental period[a]. (Arbiv and van Rijn, 1995)

| Day | $NH_4$-N (mg $L^{-1}$) | | $NO_2$-N (mg $L^{-1}$) | | $NO_3$-N (mg $L^{-1}$) | | Dissolved COD (mg $L^{-1}$) | |
|---|---|---|---|---|---|---|---|---|
| | In | Out | In | Out | In | Out | In | Out |
| 1 | 12.5 | 4.6 | 0.2 | 0.1 | 26.6 | 6.7 | 283.0 | 243.0 |
| 11 | 2.2 | 0.6 | 0.5 | bd[b] | 17.6 | 0.3 | nd[c] | nd |
| 15 | 13.3 | 13.2 | 0.3 | bd | 7.9 | bd | nd | nd |
| 25 | 11.1 | 13.5 | 0.4 | bd | 15.9 | bd | nd | nd |
| 31 | 11.3 | 14.4 | 0.1 | bd | 7.5 | bd | nd | nd |
| 39 | 2.0 | 2.8 | 0.3 | bd | 2.7 | bd | nd | nd |
| 43 | 0.7 | 2.1 | 0.1 | bd | 3.3 | bd | 254 | 203 |
| 54 | 0.9 | 1.0 | 0.1 | 0.2 | 10.0 | 4.4 | nd | nd |
| 61 | 0.5 | 0.6 | 0.1 | 0.3 | 6.3 | 3.2 | 163 | 127 |
| 71 | 1.8 | 1.4 | 0.1 | bd | 6.3 | 1.1 | 139 | 79 |
| 75 | 3.1 | 3.1 | 0.3 | bd | 6.0 | bd | 243 | 211 |

[a] Experimental period lasted from 26 July 1992 until 8 October 1992.
[b] Below detection limit (< 0.002 mg $NO_2$-N$L^{-1}$; < 0.03 mg $NO_3$-N$L^{-1}$).
[c] Not determined.

subsequent runs the control pond was operated at various flow rates. It was found (not shown) that even at a flow rate of 150 $m^3$ $day^{-1}$ (300% volume exchange), ambient ammonia concentrations were at levels too high to be tolerated by the fish (around 1.5 mg $NH_4$-N $L^{-1}$).

The performance of the fluidised bed reactor in respect to inorganic nitrogen and dissolved COD removal was measured weekly during the experimental period (Table 1). A complete nitrate and nitrite removal was found at most sampling times while a slight accumulation of nitrite could be measured at sampling times when nitrate was not completely removed. No clear trend was observed in ammonia and dissolved COD removal and no correlation was found between removal of organic matter (expressed as dissolved COD) and removal of nitrate by the fluidised bed reactor (not shown). Due to technical restraints, organic matter in the sedimentation basin could not be quantified accurately. During periodic measurements the height of the organic sludge layer in the sedimentation basin was found to fluctuate between 30 and 70 cm.

It could be concluded that by a combination of nitrification, decomposition of organic matter and denitrification, nitrogen levels can be reduced to acceptable levels for fish growth in a high-density, closed fish culture system. Until day 38 (without prophylactic treatment) fish growth was poor. Hereafter, the fish growth parameters were well within the range commonly experienced with carp of this size (not shown). Maybe contrary to common procedures our approach has been that of testing the treatment system on a semi-commercial scale before conducting more detailed laboratory studies on the different aspects of the treatment system. During operation and examination of the pilot-plant treatment system two questions remained: which of the organic compounds are used by the denitrifying bacteria in the fluidized bed reactor; and what is the rate of accumulation/degradation of organic matter in the sedimentation basin. These questions were addressed by examination of a laboratory-scale treatment system consisting of a degradation basin and a fluidized bed reactor.

### 3.2 *Laboratory-scale anaerobic treatment system*

#### 3.2.1 Decomposition of sludge

After approximately six months of continuous operation of the laboratory-scale treatment system, sludge was derived from the digestion basin and the decomposition of sludge was determined during 40 days using the nylon-mesh bag method (Figure 5). Decomposition, determined over different time intervals, was relatively rapid over the first eight days as compared to sludge degradation beyond this period. In the same figure the degradation coefficients ($k$) of the sludge are depicted (calculated according to equation (1)); $k$ values decreased with time of incubation and reached a value of 0.0063 per day at day 40. Aerobic decomposition rate constants using first order kinetics, between 0.1–0.7 per day have been reported

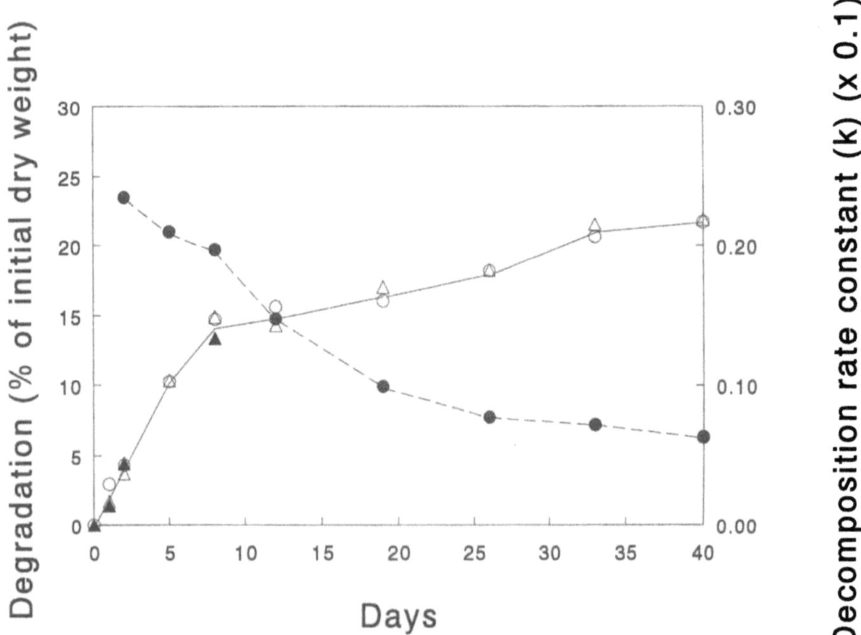


*Figure 5.* Anaerobic decomposition of sludge in the digestion basin (solid line) and corresponding decomposition rate constants (broken line). Each symbol represents the degradation per individual bag harvested at a given day. Decomposition rate constants were calculated from the average decomposition measured at a given day ($n$ = 25). (van Rijn *et al.*, 1995)

for organic material in waste water treatment plants (Benefield and Randall, 1980; Metcalf and Eddy, 1991; Reynolds, 1982). In fish culture systems, aerobic decomposition rate constants of 0.15 per day (Avnimelech, personal communication) and 0.075 per day (Boyd, 1973) were reported. Anaerobic decomposition rate constants for domestic sludge were found to be in the range of 0.02–0.04 per day (Metcalf and Eddy, 1991). The use of decomposition rate constants can be misleading as most often they are obtained by measuring the decay of organic matter during short incubation periods (several hours). The decomposition rate, obtained during these short incubations, will be an indication of the decrease of easily degradable components and might overestimate the actual decomposition of the organic substrate since the decomposition of more recalcitrant components will hardly contribute to the initial decomposition rate. The trend of the anaerobic decay of sludge in this study clearly illustrates this point.

### 3.2.2 Decomposition of feed

Decomposition of the feed incubated in nylon-mesh bags within the digestion basin followed a logarithmic trend (Figure 6). First order kinetics (equation (1)) were applied to model the observed degradation. Similar to that which was observed during incubation of sludge, feed degradation departed from this exponential model

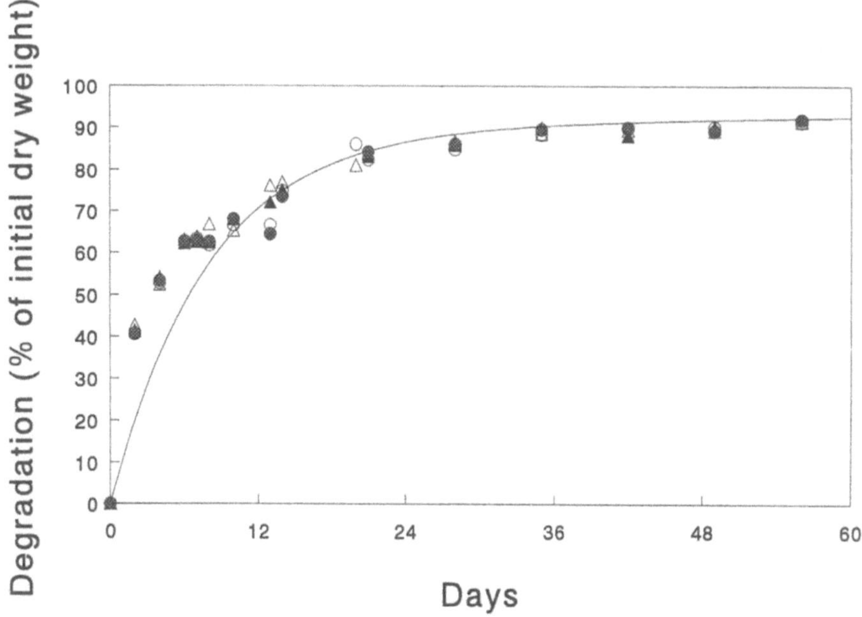


*Figure 6.* Anaerobic degradation of fish feed in the digestion basin. Each symbol represents degradation per individual bag harvested at a given day ($n = 62$). The line fit was obtained by the following equation (see equation (2)): $D = 100 - 100\,(0.89199\, e^{-0.127t} + 0.1081\, e^{-0.00627t})$. (van Rijn *et al.*, 1995)

as the degradation rate constants ($k$), determined over different time intervals, decreased with time. The decrease in degradation rate constants with time clearly indicates that different compounds comprising the feed are broken down at different rates. A line fit through the data (Figure 6) was obtained by considering two degradation stages (Hunt, 1977; Reddy *et al.*, 1986). The degradation of labile components in the feed was considered to take place over the first 34 days, while beyond 34 days the rate of degradation was considered to be governed mainly by recalcitrant compounds. Using equations (1) and (2), the following constants were obtained: $k_1 = 0.127$; $k_2 = 0.00627$; $S = 0.89199$ and $(1 - S) = 0.1081$.

### 3.2.3 Correlation between VFA consumption and denitrification

The laboratory-scale fluidized bed reactor was continuously operated for a period of more than eight months. During this period concentrations of inorganic nitrogen and VFA were determined at the inlet and outlet of the reactor. For the sake of clarity only a selected number of sampling times are presented in Table 2. A strong correlation was found between removal of nitrate and nitrite and removal of VFA. Complete removal of nitrate and nitrite, typical for most sampling dates, coincided with a decrease in VFA (No. 1, No. 2, No. 3, No. 7, No. 8, No. 9). Low concentrations of VFA in the inlet of the column coincided with a partial nitrate and nitrite removal (No. 5) while the absence of VFA at the inlet of the column caused

*Table 2.* Concentrations of oxidized inorganic nitrogen forms and VFA at inlet and outlet of laboratory-scale fluidized bed column.

| No | | $NO_3$-N (mg $L^{-1}$) | $NO_2$-N (mg $L^{-1}$) | Acetate (mg COD $L^{-1}$) | Propionate (mg COD $L^{-1}$) | Butyrate (mg COD $L^{-1}$) | pH |
|---|---|---|---|---|---|---|---|
| 1 | Inlet | 10.36 | 0[a] | 74.69 | 0 | 35.16 | 6.8 |
| | Outlet | 0 | 0 | 43.41 | 0 | 23.97 | 7.1 |
| 2 | Inlet | 11.48 | 0 | 57.46 | 21.23 | 12.78 | 7.3 |
| | Outlet | 0 | 0 | 34.47 | 0 | 3.20 | 7.6 |
| 3 | Inlet | 8.54 | 0 | 49.16 | 13.41 | 11.19 | 6.9 |
| | Outlet | 0 | 0 | 38.30 | 0 | 0 | 7.2 |
| 4 | Inlet | 0 | 0 | 82.35 | 30.17 | 28.77 | 6.9 |
| | Outlet | 0 | 0 | 82.35 | 23.47 | 25.57 | 6.9 |
| 5 | Inlet | 9.94 | 0.168 | 7.66 | 0 | 0 | 7.1 |
| | Outlet | 2.38 | 0.238 | 0 | 0 | 0 | 7.2 |
| 6 | Inlet | 17.50 | 0.56 | 0 | 0 | 0 | 6.7 |
| | Outlet | 14.28 | 0.35 | 0 | 0 | 0 | 6.9 |
| 7 | Inlet | 7.14 | 0.056 | 16.60 | 0 | 44.75 | 6.8 |
| | Outlet | 0 | 0 | 0 | 0 | 0 | 7.1 |
| 8 | Inlet | 6.16 | 0.126 | 54.26 | 32.40 | 13.42 | 7.1 |
| | Outlet | 0 | 0 | 54.90 | 0 | 9.43 | 7.5 |
| 9 | Inlet | 9.52 | 1.26 | 24.90 | 9.83 | 6.87 | 7.1 |
| | Outlet | 0 | 0 | 0 | 0 | 6.87 | 7.3 |

[a] Below detection limit.

only a minor decrease in nitrate and nitrite (No. 6). The absence of nitrate and nitrite in the inlet water coincided with a minor decrease of VFA (No. 4). The pH increase (between 0.1 to 0.4 units) measured at most sampling times is indicative of the denitrification activity in the reactor (Table 2). Ammonia concentrations between the inlet and the outlet of the reactor differed by not more than 1.5 mg $NH_4$-N $L^{-1}$ (not shown). The calculated mean ammonia removal over a period of eight months (ammonia inlet–outlet) was 0.42 mg (± 1.1 mg) $NH_4$-N $L^{-1}$ ($n = 118$).

VFA are major intermediate products during anaerobic degradation of organic matter. In methanogenic reactors, it was found that the production rate of VFA and acetogenic conversion of VFA to acetate are the major factors influencing the overall methanogenic activity in the reactors (Ozturk, 1991; Ten Brummeler *et al.*, 1985; Wiegant *et al.*, 1986). Compared to other organic carbon compounds, VFA are readily available substrates for denitrifying organisms due to the fact that these organic acids are major precursors of the TCA cycle, operating in most denitrifiers (Thauer *et al.*, 1989). Denitrification reactors, however, are usually operated on carbon sources other than VFA and relatively little information is available on the performance of denitrifying reactors in which VFA are used as the main carbon and energy source.

### 3.2.4 Release of volatile fatty acids during feed decomposition

Release of total VFA (acetate, propionate and butyrate) was followed during anaerobic degradation of fish feed. VFA during the three separate runs increased up to days 13–16 and decreased with prolonged incubation (Figure 7). By means of

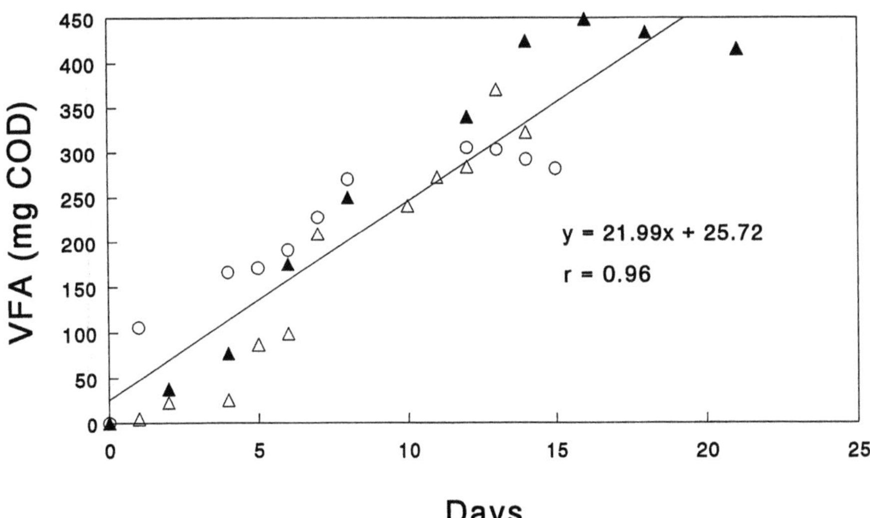


*Figure 7.* Release of volatile fatty acids (acetate, propionate and butyrate) during anaerobic digestion of fish feed examined during three independent incubations ($n = 33$). (van Rijn *et al.*, 1995)

linear regression it was estimated that maximum VFA release (± SE) was 355 (± 34.2) mg COD $g^{-1}$ feed, assuming that VFA release took place only during the first 15 days of feed degradation.

### 3.2.5 Theoretical model for accumulation of sludge and release of volatile fatty acids

Using the information obtained on the breakdown of feed and release of VFA, the theoretical model discussed above was used to predict the accumulation of sludge and daily release of VFA with time. It was assumed that fresh feed released VFA during the first 15 days of degradation. Predicted values for feed accumulation and VFA release, assuming that daily 1 g of feed was added to the digestion basin, are presented in Figure 8.

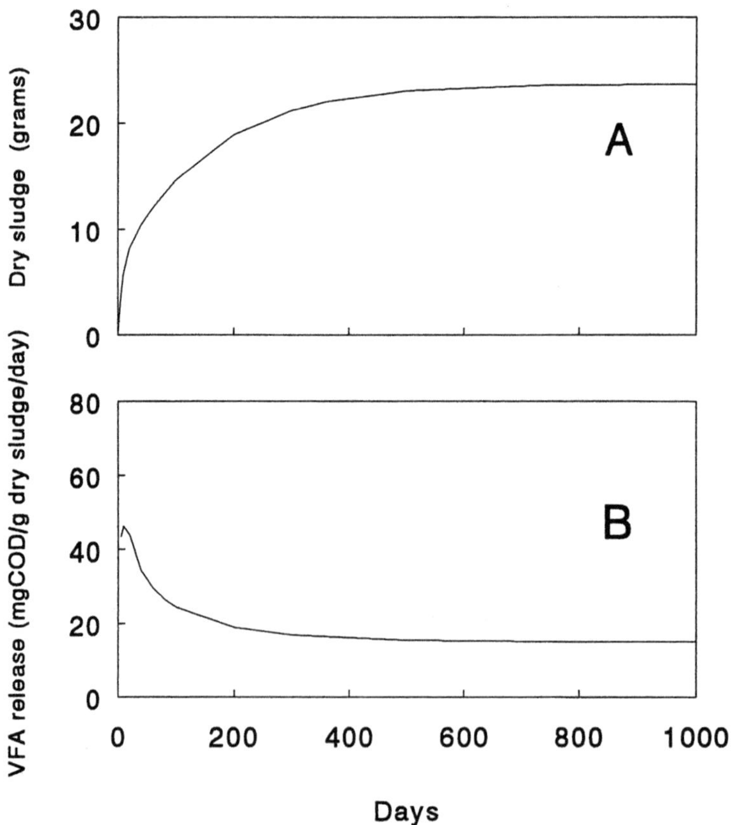


*Figure 8.* Predicted accumulation of sludge (A) and release of volatile fatty acids (B) in an anaerobic digestion basin daily supplied with 1 g of fish feed (25% protein). (van Rijn *et al.*, 1995)

*Table 3.* Daily VFA release during anaerobic digestion of sludge in digestion basin determined 243 days after start of lab-scale anaerobic treatment system[a]. (van Rijn *et al.*, 1995)

| VFA | Concentration (mg COD $L^{-1}$) | | Δ VFA 24 $h^{-1}$ |
|---|---|---|---|
| | $t = 0$ | $t = 24$ h | |
| Acetate | 140.45 (2.10) | 227.91 (3.67) | 87.46 (4.22) |
| Propionate | 98.46 (2.58) | 171.18 (0.12) | 72.72 (2.58) |
| Butyrate | 46.34 (0.58) | 87.89 (2.54) | 41.55 (2.61) |
| Total | 285.25 (3.37) | 486.98 (4.46) | 201.73 (5.59) |

[a] Arithmetic means of three analyses with standard deviation between parentheses.

### 3.2.6 Verification of the theoretical model

After 243 days of operation of the laboratory-scale anaerobic treatment system the digestion basin was disconnected from the fluidized bed reactor and the increase in VFA was determined over 24 h (Table 3). The total increase of VFA (± SD) was 201.73 (± 5.59) mg COD $L^{-1}$ $day^{-1}$. At day 244 the water level in the digestion basin was lowered, the volume of the sludge layer was measured and the dry weight of subsamples taken from the sludge layer was determined. The quantity of sludge present in the digestion basin (arithmetic mean of four subsamples and SD) was 5.48 (± 0.8) kg sludge (dry weight). Multiplying the daily VFA accumulation (mg COD $L^{-1}$ $day^{-1}$) by the total water volume in the digestion basin (540 L) and dividing the outcome by the total amount of dry sludge in the digestion basin results in a daily VFA release of 19.88 (± 15%) mg COD per g dry sludge per day. VFA release as predicted by the model (Figure 8) at day 243 was 20.05 (± 9.6%) mg COD per g dry sludge per day, in agreement with these measurements. Based on a daily feed addition of 250 g, the predicted value for sludge accumulation in the digestion basin was 5.01 kg, in accordance with the actual accumulation of sludge (5.48 ± 0.8 kg).

In a parallel study, VFA requirements of natural consortia of denitrifiers and denitrifying isolates were determined (Aboutboul *et al.*, 1995). Based on this study and based on the model described in this study we were able to predict that VFA production in the sedimentation basin is in excess of the amount required for complete nitrate removal. Hereby assuming that all of the inorganic nitrogen, released in the fish culture system, is oxidised to nitrate.

## 4. Conclusions

Nitrate concentrations as high as 400–500 mg $NO_3$-N $L^{-1}$ have been recorded in intensive, reuse systems (Honda *et al.*, 1993; Otte and Rosenthal, 1979). These concentrations are high when considering that permissible nitrate levels in effluent waters are as low as 11.6 mg $NO_3$-N $L^{-1}$ (European Community Directive). In addition to nitrate, levels of organic matter in fish culture effluents are high. In terms of BOD, production of one ton of fish was estimated to produce organic matter equivalent to the untreated sewage load produced by 20 people (Solbe, 1982). This figure becomes more meaningful when considering that today some 15 million tons of aquatic organisms are produced in aquaculture facilities (Hempel, 1993). In this study we demonstrated that, by means of anaerobic digestion of sludge in combination with nitrification and denitrification, intensive fish culture is possible without the need for effluent-discharge. By means of field experiments it was shown that this kind of treatment system could be successfully applied in a closed aquaculture unit in which fish were kept at a density of 20 kg $m^{-3}$ in a 50 $m^3$ fish basin that comprised a 2 $m^3$ aerobic trickling filter, a 3 $m^3$ digestion basin and a 130 L fluidized bed reactor.

We were able, in this and previous studies (Aboutboul *et al.*, 1995; Arbiv and van Rijn, 1995; van Rijn *et al.*, 1995), to predict the accumulation of sludge and volatile fatty acids under these conditions. Volatile fatty acids, essential in driving the denitrification in the fluidized bed reactor, were produced in quantities large enough to enable complete removal of nitrate.

In most countries it is expected that waste discharge will be legally bound by stringent environmental policies. In the light of this development, I believe that a combination of aerobic and anaerobic treatment of intensive fish culture systems will receive broader attention in the future.

## Acknowledgements

I thank the staff at the experimental station Ginosar for their excellent assistance during the field work. Furthermore, the assistance of Mr Roy Arbiv, Mr Yossi Aboutboul, Dr Nikolai Fonarev and Dr Brian Berkowitz, during the various stages of research, is highly appreciated. This work was supported by the German Ministry for Research and Development (BMFT) in the framework of the joint German–Israeli programme on aquaculture (NCRD project AQ 1 V 1060)

## References

Aboutboul Y, Arbiv R and van Rijn J 1995 Anaerobic treatment of fish culture effluents: Digestion of fish feed and release of volatile fatty acids. Aquaculture (133, 21–32).

APHA 1989 Standard Methods for Examination of Water and Wastewater, 17th ed. American Public Health Association, Washington, DC.

Arbiv R and van Rijn J 1995 Performance of a treatment system for inorganic nitrogen removal in intensive fish culture systems. Aquacult. Engineer (14, 189–203).

Avnimelech Y 1984 Reactions in fish pond sediments as inferred from sediment cores data. In Research on Aquaculture. Special Publication No. 8. Eds H Rosenthal and S Sarig. pp 41–54. European Mariculture Society, Bredene, Belgium.

Benefield L D and Randall C W 1980 Biological process design for wastewater treatment. Prentice-Hall, New York.

Boyd C E 1973 The chemical oxygen demand of waters and biological materials from ponds. Trans Amer. Fish. Soc. 103, 606–611.

Colt J 1991 Aquacultural production systems. J. Anim. Sci. 69, 4183–4192.

Hempel, E 1993 Constraints and possibilities for developing aquaculture. Aquacult. Int. 1, 2–19.

Honda, H, Watanaba Y, Kikuchi K *et al.* 1993 High density rearing of Japanese Flounder, *Paralichthys olivaceus* with a closed seawater recirculation system equipped with a denitrification unit. Suisanzoshoku 41, 19–26.

Hunt H W 1977 A simulation model for decomposition in grasslands. Ecology 58, 469–484.

Metcalf and Eddy Inc. 1991 Wastewater Engineering, Treatment, Disposal, and Reuse, 3rd ed. McGraw-Hill, New York.

Nocek E J 1988 *In situ* and other methods to estimate ruminal protein and energy digestibility: A review. J. Dairy Sci. 71, 2051.

Otte G and Rosenthal H 1979 Management of closed brackish-water system for high density fish culture by biological and chemical water treatment. Aquaculture 18, 169–181.

Ozturk M 1991 Conversion of acetate, propionate and butyrate to methane under thermophilic conditions in batch reactors. Wat. Res. 25, 1509–1513.

Reddy K R, Feijtel T C and Patrick Jr W H 1986 Effect of soil redox conditions on microbial oxidation of organic matter. In The role of organic matter in modern agriculture. Eds Y Chen and Y Avnimelech. pp 117–156. Martinus Nijhoff, Dordrecht, Boston, Lancaster.

Reynolds T D 1982 Unit operations and processes in environmental engineering. Brooks-Cole Engineering Division, Monterey, CA.

Rosenthal H and Black E A 1993 Recirculation systems in aquaculture. In Techniques for Modern Aquaculture. Ed. Jaw-Kai Wang. pp 284–294. ASAE, St Joseph, MI.

Scheiner D 1976 Determinations of ammonia and Kjeldahl nitrogen by the indophenol method. Water Res. 10, 31–36.

Solbe J F de L G 1982 Fish farm effluents: A United Kingdom survey. In Report of the EIFAC workshop on fish-farm effluents. Ed. J S Albaster. pp 29–56. EIFAC Technical paper 41. FAO, Rome, Italy.

Strickland J D and Parsons T R 1968 A practical handbook of seawater analysis. Fish. Res. Board Canada, 167.

Ten Brummeler E, Hulshoff Pol L W, Dolfing J, Lettinga G and Zehnder A B 1985 Methanogenesis in an upflow anaerobic sludge blanket reactor at pH 6 on an acetate-propionate mixture. Appl. Environ. Microbiol. 49, 1472–1477.

Thauer R K, Moller-Zinkhan D and Spormann A M 1989 Biochemistry of acetate catabolism in anaerobic chemotrophic bacteria. Annu. Rev. Microbiol. 43, 43–67.

van Rijn J, Fonarev N and Berkowitz B 1995 Anaerobic treatment of fish culture effluents: Digestion of fish feed and release of volatile fatty acids. Aquaculture 133, 9–20.

Wiegant W M, Hennink M and Lettinga G 1986 Separation of the propionate degradation to improve the efficiency of thermophilic anaerobic treatment of acidified wastewaters. Wat. Res. 20, 517–524.

Witkamp M and van der Drift J 1961 Breakdown of forest litter in relation to environmental factors. Plant and Soil 15, 295–311.

Zohar G, Rappaport U and Sarig S 1985 Intensive culture of tilapia in concrete tanks. Bamidgeh 37, 103–112.

# 18
# Ultrastructure and cellular activity of fish skin in metal-polluted water

Y. IGER, M. ABRAHAM, E. KIMMEL AND S.E. WENDELAAR BONGA

## 1. Introduction

The aquatic environment is subject to deleterious changes, mostly derived from anthropogenic activities. Among these changes, pollution with heavy metals prevails. Heavy metals, including cadmium, lead and copper, exert a wide range of pathological effects on fish and other aquatic organisms (Dethlefsen and Tiews, 1985; Mallatt, 1985).

Short-term exposure to these metals causes changes in several blood parameters, such as decreased haematocrit and leukocrit values, or reduction of plasma electrolytes, e.g. $Na^+$ (Khangarot and Tripathi, 1991; Reid and McDonald, 1988; Tort and Torres, 1988). These changes in blood parameters, or the metals directly, also activate an endocrine stress mechanism. The latter modulates physiological functions via the hypothalamic–pituitary–interrenal axis, resulting in rapid, but transient, elevation of plasma cortisol levels followed by elevation of plasma glucose levels (Fu *et al.*, 1990; Pratap and Wendelaar Bonga, 1990). During long-term exposures, bioaccumulation of heavy metals is observed in internal tissues, in particular the liver or kidneys; the latter display abnormal development or damage after chronic exposure to environmental contaminants (Dethlefsen and Tiews, 1985). Long-term exposure may also induce vertebral lesions or the formation of tumour-like bodies in the skin (Bengtsson *et al.*, 1988; Iger, 1992). As a result of these physiological alterations, heavy metals also have adverse effects on fish populations at the organism level, e.g., by reducing growth and reproduction, and may eventually impair their survival (Dethlefsen and Tiews, 1985; Schreck and Lorz, 1978).

The external epithelia are the first tissues exposed to the water, and are probably also the first tissues affected by aquatic degradation. Indeed, many authors have reported histopathological, or adaptive, changes in, for example, the branchial epithelium after exposure to metals (Schreck and Lorz, 1978; Oronsaye and Brafield, 1984; Karlsson-Norrgren *et al.*, 1985). The skin, which is another external epithelium, has received little attention – albeit in fish this tissue is composed of living cells, e.g., filament cells, mucous cells, sensory cells and club cells, which are directly exposed to the water (Whitear, 1986). Not surprisingly, our previous experiments indicated that minor environmental changes caused prolonged and

*D. Rosen* et al. *(eds): Modern Agriculture and the Environment, 215–228.*

prominent alterations in the skin (Iger and Wendelaar Bonga, 1994; Iger *et al.*, 1994a, b, c). In this chapter we describe, at the ultrastructural level, the effects induced, and the responses evoked, in the skin of fish exposed to heavy metals such as cadmium, lead or copper, for different periods.

## 2. Materials and methods

For exposure with each metal and each concentration tested, 50 to 60 common carps (*Cyprinus carpio*), both males and females of 18–25 g body weight, were kept in two groups for an acclimation period of three weeks. The fish were maintained in either tap water or artificial tap water (demineralised water with the following additives in mmol $L^{-1}$: 3.8 NaCl; 0.8 $CaCl_2$; 0.335 $NaHCO_3$; 0.06 KCl; 0.2 $MgSO_4$, pH 7.5) at 22°C, under continuous aeration and filtration, and with a daily water replacement of about 25%. For each exposure, one group served as control. For the experimental groups, appropriate amounts of solutions of heavy metals were added gradually; within 4 h, Cd (from a stock solution of $Cd(NO_3)_2$) was added to create concentrations of 22 and 560 $\mu$g Cd $L^{-1}$ as well as 1 mg $L^{-1}$; Cu (from a stock solution of $Cu(NO_3)_2$) to create a concentration of 100 $\mu$g $L^{-1}$; and Pb (from a stock solution of $Pb(NO_3)_2$) to create a concentration 1 mg $L^{-1}$. The actual metal concentrations were measured at least once a day, using atomic absorption spectrophotometry (PU 9200X, Philips), and were adjusted when the deviation exceeded 10% of the nominal concentration.

For electron microscopy, small pieces (3 × 3 mm) of skin of three fish from the different groups were excised 1 h, 24 h and 7, 14 and 21 days after reaching the actual metal concentration. Samples were taken from the dorsal areas of the heads of fish that had been anaesthetised lightly with hypno-calmer (Jungle, Texas). The tissues were fixed in 3% glutaraldehyde in sodium cacodylate buffer (0.09 M, pH 7.3), washed in buffer and post-fixed in osmium tetroxide (1%) in the same buffer. After dehydration in ethanol, the tissues were embedded in Spurr's resin. Thin sections were stained with uranyl acetate and lead citrate, and were examined in a Jeol 100 CX transmission electron microscope.

## 3. Results

### *3.1 Controls*

The ultrastructure of the skin of teleost fish, including that of carp, has been described before (Iger and Abraham, 1990; Whitear, 1986). Briefly, the epidermis of our control common carps was composed of filament-containing epithelial cells (filament cells), oval-shaped mucous cells and club cells. Mucous cells were composed of electron-transparent mucosomes. These cells were found from the second innermost cell layer of the epidermis (newly differentiated cells) and up to the skin surface (mature cells). Club cells were located in the mid-epidermal layers.

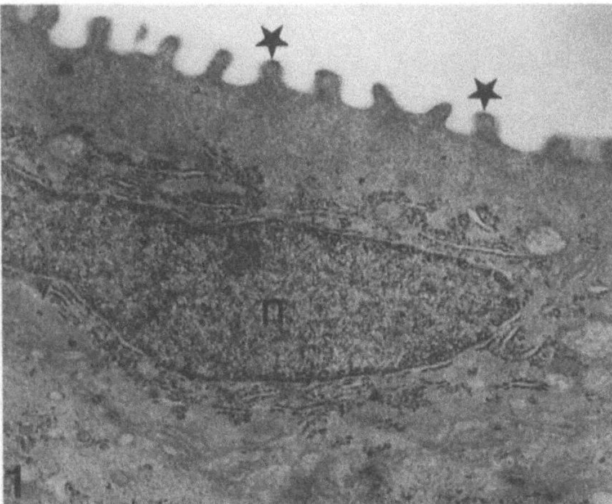

*Figure 1.* Normal flattened pavement cell: n, nucleus; asterisks, ridges at the external surface of the pavement cell. Control, × 9600.

Merkel cells were seen exterior to the club cells. The outer layer of the epidermis was composed of a special form of filament cells, the pavement cells, with few electron-transparent vesicles. Chloride cells and rodlet cells were not found. At the basal pole of the filament cells adjacent to the basal lamina, numerous endocytotic vesicles were found. In the dermis, the outer zone contained loosely arranged collagen fibres, among fibroblasts and pigment cells, in particular melanocytes. Melanosomes were mostly located in the cell bodies of the melanocytes, rather than in the cytoplasmic processes of these cells. Capillaries were found in the inner areas of the dermis. Separate leukocytes, mainly lymphocytes, were occasionally observed in both epidermis and dermis.

### *3.2 Heavy-metal exposed fish*

#### 3.2.1 General

The data represent the changes observed in the skin, when compared with control fish. Unless mentioned otherwise, reported changes were observed under the impact of all metals and concentrations tested.

At 560 $\mu$g Cd $L^{-1}$ and 1 mg Cd $L^{-1}$ mortality had occurred after 3 days and no fish survived for more than 8 days. During the first 24 h, fish of this group appeared darker than the controls. Later on, they were paler than the controls. No mortalities occurred under exposure to any concentration of the other metals. The epidermal surface and the basal lamina of fish exposed to the metals, had a highly wavy appearance throughout the experiment. Single chloride cells and rodlet cells were found, in particular in external epidermal layers.

### 3.2.2 Epidermis

#### 3.2.2.1 PAVEMENT CELLS AND FILAMENT CELLS

Pavement cells, and filament cells of the 3–4 outermost epidermal layers, exhibited increased synthesis and secretory activity: many Golgi systems, well-developed *r*ER, and many vesicles of high electron density. Filament cells also contained lysosome-like bodies as well as phagosomes of cellular debris and autophagosomes.

From day 7 onwards, many pavement cells were cylindrical rather than flattened. The outermost zone of these cells, including the terminal web, the ridges and the glycocalyx around the ridges, was pronounced during the whole experiment. The intercellular spaces of the external epidermal layers were occasionally enlarged.

Mitotic filament cells, not observed in control fish, were detected throughout the epidermis, including in cells close to the skin surface. In fish exposed to copper or lead, mitotic cells were mostly adjacent to club cells, whereas in fish exposed to cadmium, mitosis was also observed in other filament cells. Apoptotic filament cells (showing cellular shrinkage, condensation of cellular components, and loss of junctions) were also found neighbouring club cells. Necrotic pavement cells (swollen cells showing disruption of membranes, increased electron transparency of the cytoplasm and fragmentation of nuclear heterochromatin) were common during the first 24 h, whereas later, apoptotic cells were only occasionally found. Apoptotic pavement cells were rarely found in controls. Necrotic pavement cells were absent in controls.

The endocytotic activity at the inner side of the basally located filament cells was initially (after 1 h) almost completely suppressed. It was resumed later, but remained only 30–40% of that in the controls.

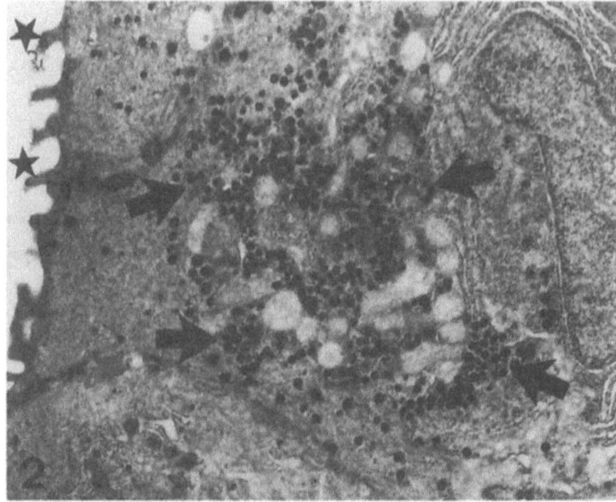


*Figure 2.* Goblet-shaped pavement cell with many electron-dense vesicles (arrows); asterisks, ridges of pavement cells. After 24 h of exposure to copper (100 $\mu g\ L^{-1}$), × 12 600.

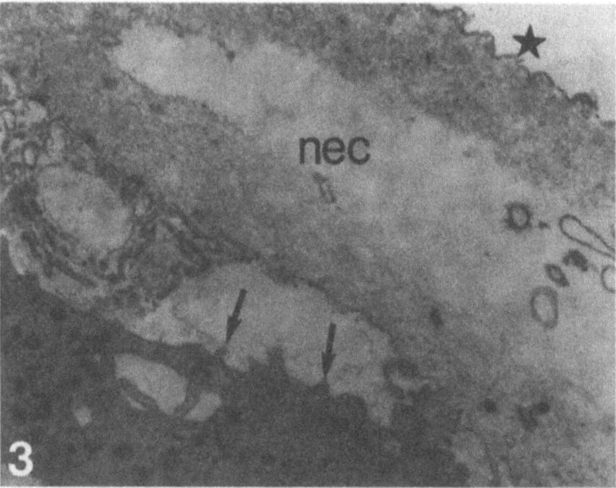


*Figure 3.* Necrotic pavement cell (nec); asterisk, skin surface; arrows, newly formed ridges of deeper filament cell. After 1 h of exposure to lead (1 mg $L^{-1}$), × 12 600.

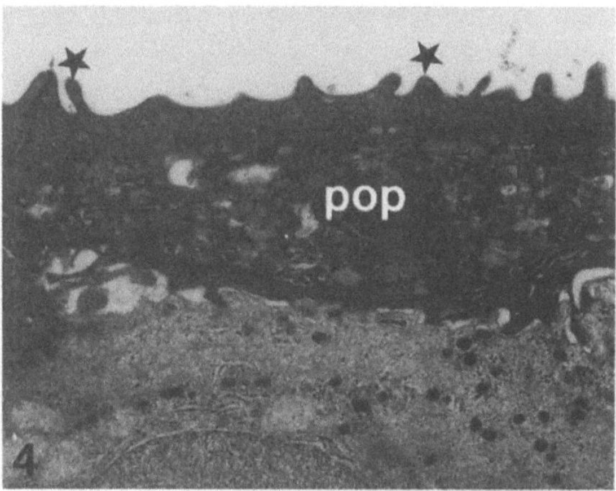


*Figure 4.* Pavement cell at early stages of apoptosis (pop); asterisks, ridges. After 7 days of exposure to copper (100 $\mu g\ L^{-1}$), × 12 600.

In fish exposed to cadmium or lead, filament cells were found with up to 4 osmiophilic vesicles of 0.6–0.8 $\mu$m diameter per cell. Fish from these experiments also exhibited abnormal cell aggregates, at the skin surface, protruding into the water. These structures had a width of 20–60 $\mu$m and a height up to 150 $\mu$m, and were solely composed of filament cells. The latter appeared very active as was indicated by the presence of well developed *r*ER and many Golgi systems as well as many transparent and highly electron-dense vesicles. The nucleus of such cells was enlarged, irregular in shape, and composed of euchromatin. The integrity of the tissue remained in structures of up to 50–60 $\mu$m height, whereas in bigger aggregates the intercellelar spaces were markedly enlarged. The basal lamina, deeper to such irregular aggregates, was mostly missing.

#### 3.2.2.2 MUCOUS CELLS

Enhanced mucus secretion was found after the onset of the exposures. Already after 1 h, most of the mucous cells were located in the external layers of the epidermis. Skin surface became covered by a thick mucus layer attached to the ridges of the pavement cells. Also, there was a dramatic reduction, and an almost complete disappearance of the mucous cells from the epithelium, after 24 h. Later, the number of mucous cells was restored. Most mucous cells were elongated in the direction of the skin surface, rather than oval-shaped. These cells contained numerous Golgi systems, *r*ER and small vesicles at their basal pole, while apparently extruding mucus from their apical pole, at the epidermal surface. Newly differentiated mucous cells were located mainly close to the skin surface, i.e., adjacent to the pavement cells. Several newly differentiated as well as mature mucous cells were necrotic, whereas others were apoptotic. We could also recognise apoptotic mucous cells as apoptotic bodies inside macrophages. Many of the non-degenerated mucous cells contained highly electron dense mucosomes. Other mucous cells contained a fused mass of mucosomes with some remnants of their membranes.

Occasionally, cells were found in which the nucleus was located in the apical part, whereas in controls the nucleus was always basally located.

#### 3.2.2.3 CLUB CELLS

The club cells were also very active. Many of them contained extensive *r*ER, Golgi systems, many free ribosomes and small vesicles (probably primary lysosomes), mainly located around the nucleus. Ribosomes and vesicles appeared also at the cell periphery. Several club cells contained large (about $8 \times 10$ $\mu$m) phagosomes. Phagosome-containing club cells as well as normal club cells lost their

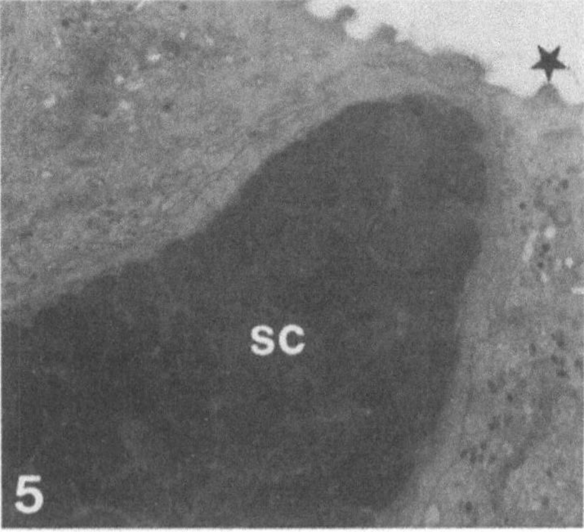


*Figure 5.* Serous mucous cell (sc) located close to the skin surface (asterisk). After 14 days of exposure to copper (100 $\mu$g $L^{-1}$), $\times$ 7800.

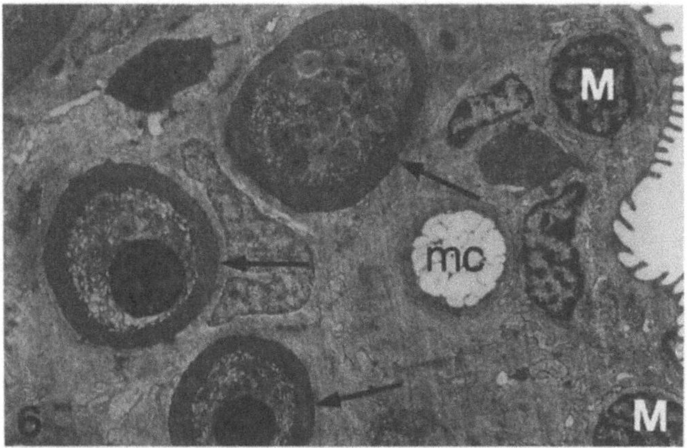


*Figure 6.* Rodlet cells (arrows) in external epidermal layers; mc, mucous cell with electron transparent mucosomes; M, Merkel cells. After 24 h of exposure to cadmium (1 mg $L^{-1}$), × 4400.

interdigitations with neighbouring cells and were elongated and appeared close to the surface. Apparently these cells migrated out of the epidermis, because occasionally only a few club cells remained in the epidermis. Fusion of club cells, indicated by penetration of cytoplasmic extensions of club cells to each other, followed by membrane fusion and appearance of bi-nucleated club cells, occurred during their migration.

#### 3.2.2.4 LEUKOCYTES

Massive extravasation of leukocytes and penetration of these cells into the epidermis were seen. In the epidermis, leukocytes were initially located close to the basal lamina, but they also migrated upward and settled close to the skin surface. Firstly, most intruding leukocytes were lymphocytes and basophilic granulocytes. Later, macrophages and plasma cells were very common, but also eosinophils and neutrophils were seen. Many of the lymphocytes observed were located adjacent to club cells and several leukocytes were found in phagosomes inside the club cells. The leukocytes, in particular macrophages and basophils, appeared to be very active in synthesising of lysosomes as well as phagocytosis. Under the impact of cadmium, mast cells were also seen in the epidermis and associations of two basophils, two eosinophils or of basophil and mast cell were commonly found in both dermis and epidermis. Also, several basophils and mast cells were necrotic. In such cells, many of the granules were found outside the disrupted membrane of the cell. These granules were phagocytosed and found in neighbouring cells, for example, macrophages, fibroblasts or filament cells. As the experiments proceeded, many melanomacrophages were seen in the epidermis and the dermis. The phagosomes in these cells contained cellular remnants, in particular of granulocytes or macrophages.

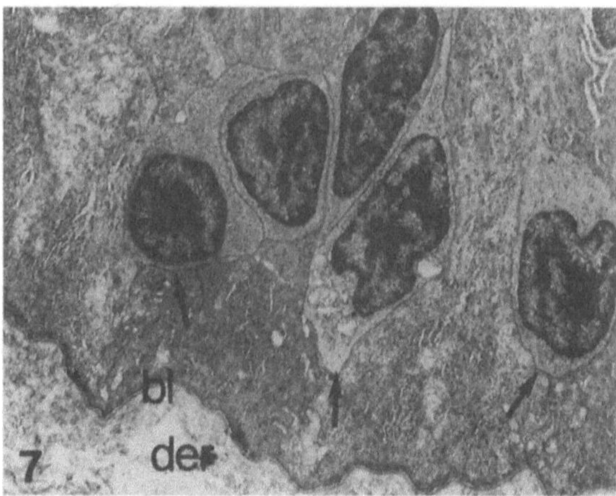


*Figure 7.* Cluster of lymphocytes (arrows) in deeper epidermal layers, close to the basal lamina (bl); der, dermis. After 1 h of exposure to lead (1 mg $L^{-1}$), × 4000.

#### 3.2.2.5 SENSORY COMPONENTS

Merkel cells, normally located in the outer layers of the epidermis, were numerous and were found also in the inner layers of the epidermis. Some of the Merkel cells became necrotic, and in the non-degenerative cells, the number of secretory granules was reduced and remained low during the whole experiment. In the dermis, many axons were found close to the basal lamina. Axon profiles were also common in the epidermis, as were dermal papillae and taste buds.

### 3.2.3 Dermis

#### 3.2.3.1 FIBROBLASTS AND COLLAGEN

In the dermis, fibroblasts with abundant *r*ER and peripheral vesicles were commonly found close to the basal lamina. In fish which survived the whole exposure period, the density of the collagenous matrix of the outer dermis increased markedly by the deposition of new, randomly arranged, collagen fibres. Dermal 'herrings' – bundles of collagen fibres oriented perpendicular to the basal lamina – were also common.

#### 3.2.3.2 CAPILLARIES

The endo- and exocytotic activity of the endothelial cells of the capillaries diminished 1 h after the start of each exposure, and was partially resumed later, after 24 h. Endothelial cells were elongated in perpendicular to the basal lamina, resulting in the appearance of many elongated, rather than rounded, capillary profiles. The distance between capillaries and the basal lamina decreased, and capillaries were even seen adjacent to the basal lamina. Occasionally the tight junctions between

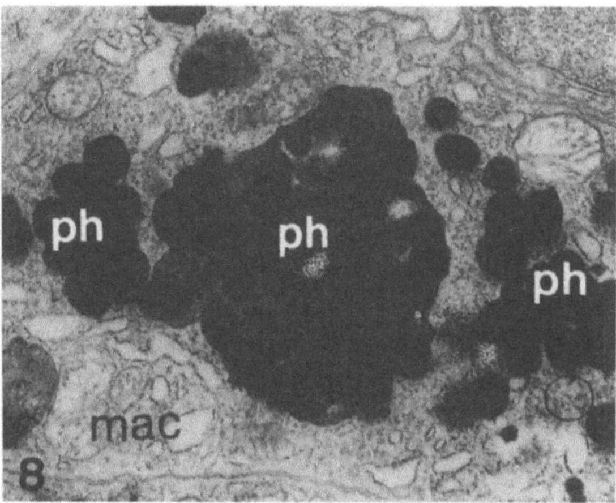


*Figure 8.* Dermal macrophage (mac) with several phagosomes (ph) composed of debris of melanocyte extensions. After 14 days of exposure to cadmium (22 $\mu$g $L^{-1}$), × 16 400.

endothelial cells were open and erythrocytes had moved out of capillaries and were found also in the dermal matrix or in the intercellular spaces of the above-mentioned filament cell aggregates.

#### 3.2.3.3 MELANOCYTES

Melanosomes were located in the cytoplasmic processes of the melanocytes, rather than in the cell bodies. These processes, normally restricted to the dermis, were also found throughout the epidermis. From day 7, aggregates of melanosomes were common and many of the melanocyte processes were apoptotic. These apoptotic processes were confined within macrophages or filament cells. The apoptotic processes were accompanied by re-concentration of melanosomes in the cell bodies of melanocytes.

## 4. Discussion

### *4.1 General*

Exposure of fish to heavy metals resulted in prolonged and profound alterations in skin ultrastructure, reflecting the sensitivity of this tissue to the heavy metals tested. The skin provides some evidence for the damaging effect of the metals (e.g. necrosis or apoptosis of pavement cells), together with evidence for protective and renewal processes (e.g. intense synthesis and secretion of mucus and glycocalyx), probably representing adaptive processes. However, it must be emphasised that complete recovery of this tissue was not found during the exposure period.

## *4.2 Epidermis*

### 4.2.1 Pavement cells and filament cells

Necrotic processes reflect accidental cell death (McConkey and Orrenius, 1994) and thus probably represent direct toxic effects of the metals on the superficial cells, the pavement cells. Necrosis of pavement cells also occurred after exposure to other metals, e.g. zinc (Somasundaram, 1985), as well as after exposure to low-pH water or elevated temperatures (Iger and Wendelaar Bonga, 1994; Iger *et al.*, 1994a). The increased synthesis and secretion of glycocalyx by the non-degenerative pavement cells might eventually lead to the accelerated ageing and exhaustion of these cells, a process reflected by the appearance of apoptotic cells (McConkey and Orrenius, 1994). However, a direct effect of the metals on (the ageing of) cells cannot be excluded. The increased mitotic activity is at least partially ascribed to the mitogenic effect of the metals, in particular cadmium (Eaton, 1974). Nevertheless, the high activity of the external filament cells, the apoptotic figures and the mitosis of inner filament cells indicate that the filament cells had attained a new balance between cell proliferation and cell loss. The results suggest that the turnover of these cells was markedly increased during heavy-metal exposure. Meanwhile the intercellular connections weakened, as was reflected by the increased intercellular spaces. Enlargement of the intercellular spaces is a phenomenon widely observed in the epithelium of stressed fish (Wendelaar Bonga and Lock, 1992; Whitear, 1986). An X-ray microanalysis of the osmiophilic vesicles revealed the presence of the metal to which the fish was exposed (Iger, 1992); thus one of the routes of metal penetration into the fish body is via its skin.

Hyperplasia is commonly seen in external epithelia of, for example, gill or skin of fish exposed to pollutants (Iger *et al.*, 1994a,b,c; Wendelaar Bonga and Lock, 1992). The appearance of cellular aggregates at the skin surface, as observed under the impact of cadmium or lead, also occurs in the initial stage in the development of papillomas (Lamas *et al.*, 1990). These aggregates were solely composed of slightly altered filament cells, indicating that only one cell type was involved in their formation, whereas under exposure to other stressors such as elevated temperatures or acid water, hyperplasia is accompanied by differentiation (Iger and Wendelaar Bonga, 1994; Iger *et al.*, 1994a,b). Proliferation of filament cells was also observed in papillomas of fish from industrially polluted water (Smith and Zaidlik, 1987). The occurrence of neoplasia, however, is an indicator for carcinogenic substances in the water (Couch, 1985).

### 4.2.2 Mucous cells

Mucus secretion was stimulated by heavy metals. Mucus secretion is also stimulated under the impact of a wide range of other pollutants, for example, crude petroleum, mercury pollution or water acidification (Burton *et al.*, 1984; Iger and Wendelaar Bonga, 1994; Lock and Van Overbeeke, 1981). The intense secretion and also the apoptosis of mucous cells, apparently not compensated by enough differentiation of mucous cells, probably contribute to the mucocytopenia commonly reported in the skin of fish stressed by heavy metals and other pollutants (Benedetti *et al.*, 1989; Iger and Wendelaar Bonga, 1994; Iger *et al.*, 1988). Both the mucosomes and the vesicles of pavement cells are released at the skin surface; the appearance of electron-dense mucosomes, termed serous mucous cells (Whitear, 1986), and vesicles of high electron density in the pavement cells, is interpreted as resulting in a modified surface coat. These secretions contain peroxidase activity (Iger and Wendelaar Bonga, 1994). It probably represents a general response of the skin to stressors (Iger *et al.*, 1994a) and might indicate enhanced general protective capacity.

### 4.2.3 Club cells

The migration of the club cells towards the surface of the epidermis, during exposure to heavy metals and other stressors (Iger, 1992), may facilitate the release of their alarm substance (Pfeiffer, 1977). The location of mitotic and inner apoptotic filament cells adjacent to club cells, suggests that the latter are involved in the epidermal cellular regulation, in addition to the secretion of alarm substance. Phagosome-containing club cells have been also found in the skin of carp exposed to the other stressors mentioned (Iger and Wendelaar Bonga, 1994; Iger *et al.*, 1992).

### 4.2.4 Leukocytes

Exposure to heavy metals evoked infiltration of the epidermis by many leukocytes. We have previously described leukocyte infiltration in the skin of carp after administration of manure to the water (Iger *et al.*, 1988) or exposure to low-pH water (Iger and Wendelaar Bonga, 1994), as well as after wounding of the skin (Iger and Abraham, 1990). Massive infiltration of leukocytes has also been described for the epithelium covering the gills under the impact of several water pollutants (Karlsson-Norrgren *et al.*, 1985; Mallatt, 1985; Pratap and Wendelaar Bonga, 1990; Wendelaar Bonga and Lock, 1992). It has been suggested (Iger and Wendelaar Bonga, 1994; Iger *et al.*, 1994a,b,c) that the migration of leukocytes into these epithelia can at least partially contribute to the leukopenia that is characteristic for fish exposed to heavy metals and other stressors (Khangarot and Tripathi, 1991; Pickering and Pottinger, 1987; Tort and Torres, 1988). The high migration and phagocytosis activity of the leukocytes, in particular macrophages and granulocytes, is also reflected by the appearance of apoptotic leukocytes. Beside exhaustion due to enhanced activity, it may reflect the sensitivity of these cells to the metals tested. Another possibility is that the degeneration is a more general phenomenon; cortisol, the major stress hormone in fish, is released in response to a wide range of stressors, including metals (Wendelaar Bonga and Lock, 1992). Cortisol may initiate degeneration of leukocytes (Maule and Schreck, 1990). Degeneration of leukocytes might influence the susceptibility to diseases.

## 4.3 *Dermis*

### 4.3.1 Basal lamina

The higher degree of basal lamina undulation may function to maintain the integrity of the epidermis in the exposed fish. The reduced cytosis along the basal lamina may also have a protective function, by reducing uncontrolled passage of metabolites, if the stressor causes severe epidermal changes. Both responses are controlled by cortisol (Iger, 1992; Iger *et al.*, 1995).

### 4.3.2 Fibroblasts

Migration, rather than mitotic activity, causes the appearance of many fibroblasts in the outer dermal zones (Iger and Abraham, 1990). In fish which survived the whole exposure period, the migration was followed by increased synthesis and secretory activity of fibroblasts, resulting in a higher fibrous dermis. It was also observed in fish exposed to elevated temperatures or acid water (Iger and Wendelaar Bonga, 1994; Iger *et al.*, 1994a), or in fish exposed to water contaminated with sewage sludge (Bucke *et al.*, 1983). Beside providing a structurally rigid area, fibroblast migration and synthesis may further be associated with their ability to synthesise metallothioneins after metal – e.g. copper (George *et al.*, 1992) – exposure. Metallothioneins play a pivotal role in the adaptation to heavy metals (Dixon and Sprague, 1981).

### 4.3.3 Melanocytes

The dispersion of melanosomes into the cytoplasmic extensions of melanocytes was also reported for fish skin after exposure to, for example, manure or acid water (Iger and Wendelaar Bonga, 1994; Iger *et al.*, 1988). Intermediates of the melanin synthetic pathway have bactericidal capacities (Edelstein, 1971). More recent research suggests that melanin penetration might serve as a local modulatory defence mechanism, possibly by scavenging free radicals produced (Bustamante *et al.*, 1993), thereby protecting the tissue from these side effects of the immune function. However, although the role of melanosome penetration is currently not clear, both theories suggest that it represents a defence mechanism.

## 4.4 *Concluding remarks*

In conclusion, the responses observed in the skin seem to be part of a general or mediated response to stressors (e.g. mucus secretion): they were also observed in skin of fish exposed to other challenges and can also be triggered by administration of stress hormones, such as cortisol (Iger *et al.*, 1995). Other responses (e.g. phagocytosis) might be considered as local adjustments of the skin. However, the skin also provides evidence for a direct irritant effect of the metals (e.g. necrotic pavement cells). Several alterations, such as coupling of leukocytes or appearance of cellular aggregates, seem to be specific to the metal tested, in particular to lead and cadmium. However, the wide range of pronounced changes observed in the skin of carp exposed to heavy metals demonstrates the sensitivity of this tissue, as part of the whole organism, to metals. We suggest use of the skin changes as a bio-probe reflecting the water condition and its actual impact on the fish.

## References

Benedetti I, Albano A G and Mola L 1989 Histomorphological changes in some organs of the brown bullhead, *Ictalurus nebulosus* LeSueur, following short- and long-term exposure to copper. J. Fish Biol. 34, 273–280.

Bengtsson A, Bengtsson B E and Lithner G 1988 Vertebral defects in fourhorn sculpin, *Myoxocephalus quadricornis* L., exposed to heavy metal pollution in the Gulf of Bothnia. J. Fish Biol. 33, 517–529.

Bucke D, Feist S W, Norton M G and Rolfe M S 1983 A histopathological report of some epidermal anomalies of Dover sole, *Solea solea* L., and other flatfish species in coastal waters of south east England. J. Fish Biol. 23, 565–578.

Burton D, Burton M P and Idler D R 1984 Epidermal condition in post-spawned winter flounder, *Pseudopleuronectes americanus* (Walbaum), maintained in the laboratory and after exposure to crude petroleum. J. Fish Biol. 25, 593–606.

Bustamante J, Bredeston L, Malaga G and Mordoh J 1993 Role of melanin as a scavenger of active oxygen species. Pigment Cell Res. 6, 348–353.

Couch J A 1985 Prospective study of infectious and non-infectious diseases in oysters and fishes in three Gulf of Mexico estuaries. Dis. Aquat. Org. 1, 59–82.

Dethlefsen V and Tiews K 1985 Review on the effects of pollution on marine fish life and fisheries in the North Sea. J. Appl. Ichthyol. 3, 97–118.

Dixon D G and Sprague J B 1981 Copper bioaccumulation and hepatoprotein synthesis during acclimation to copper by juvenile rainbow trout. Aquat. Toxicol. 1, 69–81.

Eaton J G 1974 Chronic cadmium toxicity to the bluegill, *Lepomis macrochirus*, Rafinsque. Trans. Am. Fish Soc. 103, 729–735.

Edelstein L M 1971 Melanin: A unique biopolymer. Pathobiol. Ann. 1, 309–324.

Fu H, Steinebach O M, van der Hamer C J A, Balm P H M and Lock R A C 1990 Involvement of cortisol and metallothionein-like proteins in the physiological responses of tilapia (*Oreochromis mossambicus*) to sublethal cadmium stress. Aquat. Toxicol. 16, 257–270.

George S, Burgess D, Leaver M and Frerichs N 1992 Metallothionein induction in cultured fibroblasts and liver of a marine flatfish, the turbot, *Scophthalamus maximus*. Fish Physiol. Biochem. 10, 43–54.

Iger Y 1992 Adaptive responses in the skin of carp, *Cyprinus carpio*, under the impact of environmental factors and wounding. Doctoral thesis, Hebrew University of Jerusalem.

Iger Y and Abraham M 1990 The process of skin healing in experimentally wounded carp. J. Fish. Biol. 36, 421–437.

Iger Y and Wendelaar Bonga SE 1994 Cellular responses in the skin of carp (*Cyprinus carpio*) exposed to acidified water. Cell Tissue Res. 275, 481–492.

Iger Y, Abraham M, Dotan A, Fattal B and Rahamim E 1988 Cellular responses in the skin of carp maintained in organically fertilised water. J. Fish Biol. 33, 711–720.

Iger Y, Hilge V and Abraham M 1992 The ultrastructure of fish-skin during stress. In Progress in Aquaculture Research. Eds B Moav, V Hilge and H Rosenthal. pp 205–214. EAS, Oostende, Belgium.

Iger Y, Jenner H A and Wendelaar Bonga S E 1994a Cellular responses in the skin of trout (*Oncorhynchus mykiss*) exposed to thermal elevation. J. Fish Biol. 44, 921–935.

Iger Y, Lock R A, Jenner H A and Wendelaar Bonga S E 1994b Cellular responses in the skin of carp (*Cyprinus carpio*) exposed to copper. Aquat. Toxicol. 29, 49–64.

Iger Y, Lock R A and Wendelaar Bonga S E 1994c Effects of water-borne cadmium on the skin of the common carp, (*Cyprinus carpio*). Arch. Environ. Contam. Toxicol. 26, 342–350.

Iger Y, Balm P H M, Jenner HA and Wendelaar Bonga S E 1995 Cortisol induces stress-related changes in the skin of trout (*Oncorhynchus mykiss*). Gen. Comp. Endocrinol. 97, 188–198.

Karlsson-Norrgren L, Runn P, Haux C and Forlin L 1985 Cadmium induced changes in gill morphology of zebrafish, *Brachydanio rerio* (Hamilton-Buchanan), and rainbow trout, *Salmo gairdneri* Richardson. J. Fish Biol. 27, 81–95.

Khangarot B S and Tripathi D M 1991 Changes in humoral and cell-mediated immune responses and in skin and respiratory surfaces of catfish, *Saccobranchus fossilis*, following copper exposure. Ecotoxicol. Environ. Safety 22, 291–308.

Lamas J, Anadon R, Devesa S and Toranzo A E 1990 Visceral neoplasia and epidermal papillomas in cultured turbot *Scophthalmus maximus*. Dis. Aquat. Org. 8, 179–187.

Lock, R A C and Van Overbeeke A P 1981 Effects of mercuric chloride and methyl mercuric chloride on mucus secretion in rainbow trout, *Salmo gairdneri* R. Comp. Biochem. Physiol. 69C, 67–73.

Mallatt J 1985 Fish gill structural changes induced by toxicants and other irritants: A statistical review. Can. J. Fish Aquat. Sci. 42, 630–648.

Maule A G and Schreck C B 1990 Stress and cortisol affect distribution of leucocytes in juvenile coho salmon (*Oncorhynchus kisutch*). J. Aquat. Animal Health 2, 298–304.

McConkey D J and Orrenius S 1994 Signal transduction pathways to apoptosis. Trends Cell Biol. 4, 370–375.

Oronsaye J A O and Brafield A E 1984 The effect of dissolved cadmium on the chloride cells of the gills of the stickleback, *Gasterosteus aculeatus* L. J. Fish Biol. 25, 253–258.

Pfeiffer W 1977 The distribution of fright reaction and alarm substance cells in fishes. Copeia 653–665.

Pickering A D and Pottinger T G 1987 Crowding causes prolonged leucopenia in salmonid fish, despite interrenal acclimation. J. Fish Biol. 30, 701–712.

Pratap H B and Wendelaar Bonga S E 1990 Effects of water-borne cadmium on plasma cortisol and glucose in the cichlid fish *Oreochromis mossambicus*. Comp. Biochem. Physiol. 95C, 313–317.

Reid S D and McDonald D G 1988 Effects of cadmium, copper, and low pH on ion fluxes in the rainbow trout, *Salmo gairdneri*. Can. J. Fish Aquat. Sci. 45, 244–253.

Schreck C B and Lorz H W 1978 Stress response of coho salmon (*Oncorhynchus kisutch*) elicited by cadmium and copper and potential use of cortisol as an indicator of stress. J. Fish Res. Bd. Can. 35, 1124–1129.

Smith I R and Zaidlik B A 1987 Regression and development of epidermal papillomas effecting white suckers, *Catostomus commersoni* (Lacepede), from Lake Ontario, Canada. J. Fish Dis. 10, 487–494.

Somasundaram B 1985 Effect of zinc on epidermal ultrastructure in the larva of *Clupea harengus*. Mar. Biol. 85, 199–208.

Tort L and Torres P 1988 The effects of sublethal concentrations of cadmium on haematological parameters in the dogfish, *Scyliorhinus canicula*. J. Fish Biol. 32, 277–282.

Wendelaar Bonga S E and Lock R A C 1992 Toxicants and osmoregulation. Neth. J. Zool. 42, 478–493.

Whitear M 1986 The skin of fishes including cyclostomes – epidermis. In Biology of the Integument, Vol. 2 – Vertebrates. Eds J Bereiter Hahn, A G Matoltsy and K S Richards. pp 8–38. Springer, Heidelberg, Germany.

# SECTION IV

# Modelling of Agricultural Pollution

# 19
# Modelling pesticide transport in porous media

R.J. WAGENET, P.M. TILLOTSON AND J.L. HUTSON

## 1. Introduction

The application of pesticides is now a widespread practice in agriculture, and will continue to be for the foreseeable future. However, even when properly applied and managed, pesticides can potentially be displaced beyond the point of application. A key issue in most cases of pesticide displacement is the movement of water, or leaching, of these chemicals through the soil to groundwater, which is emphasised in this chapter. The quantitative description of the interacting processes that determine the degree of displacement has been the focus of numerous studies over the last decade, with a number of predictive tools known as leaching models, being proposed. At this point, it is wise to reflect on the accomplishments of these studies and to identify the next steps that lead us to providing both farmers and the public with the tools necessary to manage pesticides wisely for sustainable agricultural production and minimisation of adverse impact on soil and water quality.

## 2. Contemporary modelling approaches

### *2.1 Deterministic or stochastic as the model structure?*

Pesticide leaching models are a type of solute transport model, and there has been much theoretical and experimental study of the latter, whether related to pesticides or other solutes (Nielsen *et al.*, 1986; Wagenet and Rao, 1990). Solute transport models (and thereby pesticide leaching models) have been categorised in several ways (Addiscott and Wagenet, 1985). The initial separation is between deterministic models which assume that a system or process operates such that the occurrence of a given set of events leads to a uniquely-definable outcome, or stochastic models which recognise that the natural system is variable in space and time, leading to a model structure and output that represents the uncertainty that derives from the variability, usually presented in a statistical manner (Table 1). To date, only a few purely stochastic pesticide leaching models have been published, and as yet these have not been generally used beyond the developers.

In general, uncertainties and variabilities are ignored in the development of deterministic pesticide leaching models. Such models are thereby often criticised as being only one estimate of system behaviour, and probably not a true estimate if only a single model execution is developed to describe a variable space or any length

*D. Rosen* et al. *(eds): Modern Agriculture and the Environment, 231–245.*

*Table 1.* Examples of unsaturated zone pesticide leaching models.

*Model structure*

- (A) Deterministic
  - (1) Mechanistic
    - (a) Analytical (Jury *et al.*, 1983; van Genuchten and Wagenet, 1989; van Genuchten *et al.*, 1974)
    - (b) Numerical (Boesten and van der Linden, 1991; Hutson and Wagenet, 1992)
  - (2) Mixed mechanistic–functional (Hutson and Wagenet, 1993; Mullins *et al.*, 1993)
  - (3) Functional
    - (a) Partly analytical (Chen and Wagenet, 1992)
    - (b) Layer and other simple approaches (Addiscott, 1977; Nicholls *et al.*, 1982; Nofziger and Hornsby, 1986; Rao *et al.*, 1976)
- (B) Mixed deterministic–stochastic (Bellin *et al.*, 1993; Bosma *et al.*, 1993; Jury, 1982; van der Zee and Boesten, 1991; van der Zee and van Riemsdijk, 1987)
- (C) Stochastic (Chen and Wagenet, 1995, 1997; Jury and Gruber, 1989)

*Model user*

Research scientist (A, B, C)

Regulatory personnel (A1b, 2, 3b)

Consultants (A1b, 2, 3b)

Farmers (3b)

Educators (3b)

of time. However, these models are an attractive beginning point for the description of pesticide leaching. First, they allow manipulation of basic input data and process representation so that realistic field cases can be considered. Second, they consist of a clear representation of their component processes, which allows independent studies of biological transformations, soil sorption reactions, volatilisation and internal soil hydraulic properties to be conducted as part of an integrated experimental/modelling programme. Third, the development and use of a properly constructed pesticide leaching model is a learning experience for the scientist who tests his hypotheses, and for the user who takes the time to understand the model, in the process gaining appreciation for the way the soil–water–pesticide system behaves.

## 2.2 *Mechanistic or functional?*

A second method of distinguishing deterministic pesticide leaching models is between those that are relatively mechanistic and those that are more functional. While these have been previously defined as representing distinct categories of leaching models (Addiscott and Wagenet, 1985), it is now wiser to regard them as different ends of a modelling spectrum, with a number of hybrids of each in between. A purely mechanistic model can be envisioned as an attempt to integrate, as best we know them, the fundamental physical, chemical, and biological processes related to leaching. These are also called rate models, or flux/gradient models, because they use instantaneous rates of change of water potential or pesticide concentration to estimate the fluxes of water and pesticide. These models at their most comprehensive also include such rate-dependent processes as sorption, biological degradation, and volatilisation.

By contrast, the purely functional leaching model is one that simplifies the representation of basic process to the point where no claim of fundamentality is made. These are often known as capacity models, because they displace water and pesticide through the profile according to the capacity of a soil layer to retain water and the chemical dissolved in it. While these models are clearly different from the mechanistic, they have recently been presented in mixed mechanistic/functional form, producing models that are intermediate in the spectrum (Carsel *et al.*, 1985; Nicholls *et al.*, 1982; White, 1985). It is in these cases that modelling problems may arise, for the mixture of mechanism with an appropriate level of simplification is difficult if accuracy and sensitivity are to be retained. It is also an opportunity for disciplinary bias to be expressed if the modeller is not sensitive to an issue such as simplification.

Properly constructed functional models usually require less soil characterisation information as input data, execute faster, and provide, perhaps, less quantitative estimates of pesticide leaching than do the highly mechanistic models. Yet, such models can be attractively deceptive when presented to the non-scientific community, who often confuse attractive colour graphics with scientific rigour. However these models can be useful when employed carefully and with full recognition of the assumptions used in their derivation. The ability now to exploit faster computers in repeated executions of multiple ensembles of characterisation sets in both the mechanistic and functional models will allow a test of the degree to which basic mechanism needs to be considered for accurate assessment of variable and uncertain leaching problems.

### *2.3 Who is the user?*

Pesticide leaching models can also be categorised by the type of user (Table 1). They are currently used as research tools, guides in regulation and management, and as educational aids. These models differ in not only their basic formulation, but also in their use, flexibility and data requirements. As research tools, the models differ according to whether they are describing steady-state, laboratory conditions, or whether they are meant to describe the very non-steady, or transient field cases. Exact analytical mathematical models can be derived in the former cases, while more comprehensive numerical models are the tools used to describe field complexity. The deterministic, mechanistic, analytical research models are best used as tools to understand basic transport processes in carefully controlled laboratory soil column experiments, except for the one case (Jury *et al.*, 1983) of an analytical solution developed for placing pesticides into behavioural classes (this model is misused when applied to any other cases, be they laboratory or field). The deterministic, mechanistic, relatively comprehensive numerical models of pesticide leaching (for example LEACHM; Hutson and Wagenet, 1992), previously the domain of the research community for the testing of new hypotheses, are now being extended into the management and regulatory communities. This is resulting in new methodologies for pesticide certification and registration, and also new formulations of the models to serve such purposes better.

The regulatory community makes extensive use of models that fall between the categories. A mixed-mechanistic/functional model of pesticide leaching is PRZM-2 (Pesticide Root Zone Model-release 2; Mullins *et al.*, 1993), a combination of mechanistic basic process and functional simplification produced by and for the US Environmental Protection Agency (USEPA). In the United States, this model is in widespread use by chemical companies that need to satisfy USEPA as to the possible environmental consequences of the leaching of proposed pesticides. It includes a number of possible pesticide fate processes other than leaching, such as foliar interception, run-off, and plant uptake. Another possible choice that treats only pesticide leaching processes is the LEACHA model (Hutson and Wagenet, 1993). Both these models execute quite rapidly on most desktop computers, and serve to screen new chemicals quickly and in a straightforward fashion with a minimum of input data. In the case of such mixed models, issues of internal mass balancing and numerical dispersion become crucial as simplifications are made in the basic structure of the model. For this reason, comparison of these mixed models with analytical solutions that represent one well-defined case of the model simulation are essential as a minimal test of the quality of the numerical code.

Educational models of the pesticide leaching process are functional models of the simplest type (Nofziger and Hornsby, 1986; Rao *et al.*, 1976). They are generally intended for non-scientific audiences and can be very conceptually simple. Such models are intended to demonstrate basic concepts, rather than actually use those concepts as in the mechanistic models. They are at the extreme of functional models, simplifying basic process to the degree that it is usually improper to estimate pesticide concentrations in the profile or in deep drainage. Educational models execute quite rapidly on most computers, and are usually presented with user-friendly interfaces and attractive colour graphics displays. The user is provided with the opportunity to examine and contrast different scenarios of pesticide type and rate of application, soil properties, pesticide properties, rainfall and irrigation. These models are not intended for use outside broad educational purposes, as the nature of their simplifications is such that much of the subtle sensitivity to process interactions is lost. The approximate depth of penetration of the applied mass, or the position in the profile of the centre of mass is often presented. This is entirely appropriate for the broad comparison of scenarios that might result from alternative practices. However, the attempt to take such results further and predict pesticide concentrations as a function of depth and time using superimposed analytical solutions, is underway, but so far not proven accurate. Further refinement of these models by incorporating more basic process is now possible given the power of 386 and 486 computer operating systems. As this occurs, these models will blend with the larger group of functional models described above, while still retaining their educational value.

## 3. Basic processes

During the last 30 years, the theoretical framework for constructing dynamic, deterministic, mechanistic pesticide leaching models has most often been the initial description of water flow using the non-steady state representation of the Richards equation (Richards, 1931), followed by application of the convection–dispersion equation (CDE) to describe the solute (pesticide) leaching process. Extensive reviews of this approach exist elsewhere and in the interests of space constraints the reader is referred to Nielsen *et al.* (1986), Wagenet (1993) or Wagenet and Rao (1990). A few comments important to this chapter are given below.

### *3.1 Transfer processes – water and pesticide*

The first step in accurate prediction of pesticide leaching is accurate description of water movement. In fact, a pesticide leaching model is impossible to develop without a component that accurately estimates water contents and fluxes under unsaturated flow conditions. It is almost always necessary to solve this transient water flow equation by numerical methods to provide estimates of water content with space and time within the soil profile. Once this has been accomplished, estimated time-dependent local water flux between two positions can be calculated. Knowledge of both water content and water flux is required to predict pesticide leaching.

However, the conceptual framework above does not apply directly to flow in non-homogeneous soils, which are characterised by non-matrix water flows, such as water movement down wormholes, structural cracks, and similar such 'preferred pathways' (see, for example, Beven and Germann (1982)). Further discussion of this issue is presented below.

Quantitative description of pesticide leaching is most often described with the convection–dispersion equation (CDE). Since the early 1960s this approach has been widely applied to a variety of solutes in homogeneous soils. Comprehensive pesticide leaching models of deterministic/mechanistic form are based on the CDE and include simultaneous description of the combined physical and chemical processes of pesticide diffusion in response to a concentration gradient in the liquid phase, physical convection (mass flow) in response to the movement of the bulk liquid phase in which the pesticide is dissolved, and the additional transport processes of pesticide diffusion in the gas phase and convection in the gas phase.

The description of steady-state pesticide flux is analogous to that of water flux, with the exception that in this case the driving force is a gradient in pesticide concentration, and a chemical diffusion–hydrodynamic dispersion coefficient is introduced to amend Fickian diffusion for flow in tortuous, partly water-filled porous media. Appropriately formulated to describe non-steady state cases, and for a case of a pesticide that partitions into three phases (sorbed, liquid, gas) a comprehensive mechanistic description of pesticide leaching based on the CDE can be obtained (e.g. Wagenet and Rao, 1990). Linear relationships between solution and sorbed phase concentrations, and between gas and solution phases, allow the expression of pesticide movement and leaching to be only in terms of solution phase pesticide concentrations. This facilitates the solution of the equation for both numerical and analytical deterministic models of leaching. A detailed discussion of the derivation of these equations is presented in Wagenet and Rao (1990).

### *3.2 Degradation*

In most current pesticide-fate models, it is assumed that microbiological processes are predominant in the root zone. Once the pesticide leaches past the biologically active root zone, and migrates well beyond the upper soil layers or enters into the saturated zone (groundwater), chemical degradation pathways such as hydrolysis are likely to be relatively more important.

The degradation process is usually represented in leaching models by first-order kinetics. This allows the derivation of a half-life of pesticide concentration, a widely tabulated value used in many pesticide leaching models, including all those mentioned above. Presently, the assumption of first-order microbiological rate processes in all cases is being challenged by research results of much longer persistence at low concentrations than would be expected with this model. This persistence could be due to diffusion into pore spaces inaccessible to microbes, or binding with organic materials.

The consequences of these proposed mechanisms can be considered as positive due to the non-leaching of ‘bound’, ‘inaccessible’, or ‘sequestered’ pesticide. They can also be considered negative since slow diffusion from micropore spaces or kinetically-controlled sorption surfaces would lead to very small concentrations of pesticide leaching over very long times, an observation that seems to be true in a number of field monitoring programmes. Certainly, the representation of these effects deserves further thought as improvements to pesticide leaching models.

### *3.3 Sorption*

One of the most important non-transport processes that influence chemical leaching is sorption, which is the distribution of the chemical between the solution and solid (sorbed) phases of the soil. Recent sorption studies challenge the current formulation of pesticide leaching models, particularly for miscible organic chemicals such as pesticides, where it is usually assumed that simple equilibrium isotherms adequately describe the partitioning of organic compounds between solution and sorbed phases. This has led to the use of either Freundlich-type sorption isotherms with two coefficients ($K_F$, $n$) describing a non-linear, but equilibrium, partitioning process, or if partitioning is assumed linear, one partitioning coefficient ($K_D$).

One of the assumptions that has made these isotherm approaches attractive is that $K_D$, $K_F$, and $n$ can be determined in batch-type laboratory experiments, and then used in a pesticide leaching model of the form of the CDE. Although this assumption may be adequate for weakly sorbed organics, it does not appear adequate for description of a number of organics that interact with the soil surface to any appreciable degree. Nonetheless, there are many values of $K_D$, $K_F$ and $n$ found in the literature, and the use of these parameters in models of pesticide leaching is widespread.

A possibly better conceptualization of the sorption process is the two-site model, in which sorption is characterised by simultaneous equilibrium and kinetic processes. The soil's sorption capacity is divided into type-1 (equilibrium) 'sites' and type-2 (kinetic) 'sites'. Assuming both equilibrium and kinetic sorption are present, and assuming also that leaching occurs in homogeneous media where dispersion and convection are the primary transport processes, van Genuchten *et al.* (1974) derived the following for steady-state water flow conditions:

$$\left(1+\frac{\rho fK_D}{\theta}\right) = D\frac{\partial^2 c_L}{\partial z^2} - v\frac{\partial c_L}{\partial z} - \frac{\alpha\rho}{\theta}[(1-f)K_D c_L - c_{s2}], \tag{1}$$

$$\frac{\partial c_{s2}}{\partial t} = \alpha[(1-f)K_D c_L - c_{s2}], \tag{2}$$

where $f =$ fraction of type-1 sites, $\alpha =$ first-order kinetic rate coefficient, and $c_{s2} =$ sorbed-phase concentration at type-2 sites. Equations (1) and (2) are of the same general form as equations that describe mobile–immobile porosity. This mathematical correspondence has provided the opportunity to distinguish non-equilibrium physical processes (mobile–immobile water) from non-equilibrium chemical processes (kinetic and equilibrium sorption) through the analysis of soil column data obtained using sorbing (pesticides) and non-sorbing (chloride) solutes.

### 3.4 *Volatilization*

Volatilization in pesticide leaching models is described by relating solution- and gas-phase pesticide concentrations. Recognizing that the liquid–vapour partitioning can be represented by a modified Henry's law, Jury *et al.* (1983) proposed use of a modified Henry's law constant, defined as the ratio of the saturated vapour density and the aqueous solubility. This simplifies the CDE and is useful as shown by Wagenet *et al.* (1989) in modelling the fate of a volatile pesticide, an example being DBCP (dibromochloropropane) transport and transformation.

### 3.5 *Plant uptake*

Pesticide uptake by plants has been ignored in most pesticide leaching models. This is primarily because of an almost total lack of quantitative information available to the modeller on the pesticide uptake process. Recognising this limitation, most modellers have apparently assumed that any inaccuracy in simulation of pesticide fate in the root zone that results from not considering plant uptake is within the noise of inaccuracies produced by other assumptions about the chemical–physical–biological processes operating in the system. It seems prudent to begin to provide estimates of plant uptake as it affects pesticide leaching. Only recently have researchers begun to examine the quantitative relationships between pesticide chemical properties, such as the octanol–water partition coefficient, and plant uptake (Briggs *et al.*, 1982). Further work along these lines will be extremely valuable in modelling pesticide fate.

## 4. 'Revisionist' mechanistic approaches

Field soils are clearly spatially and temporally variable, even in the relatively *homogeneous* soils that the CDE is intended to describe. However, a widespread, and until recently under-researched category of soils, the *heterogeneous* soils, also exist and are proving to be quite important to agriculture and environment. Homogeneous and heterogeneous soils can be distinguished in terms of the size of their representative elementary volume (REV), which is defined in terms of the particular soil property being considered (Bear, 1979; Wagenet, 1985).

Water flow and leaching are primary issues in pesticide leaching, and pore size distribution is a key property. For example, with respect to pesticide leaching, the REV can be defined in terms of pore size distribution as the smallest macroscopic unit of soil that expresses all the microscopic variation in pore size distribution present in the soil. A homogeneous soil is generally composed of a relatively narrow range of pore sizes and the REV that characterises the system is correspondingly small. A heterogeneous soil consists of a larger range of pore sizes, including structural cracks, voids, and macropores, with a commensurately larger REV. A large number of leaching studies have been conducted in homogeneous soils; fewer in heterogeneous soils. New model formulations have been proposed to deal with these more realistic conceptualizations of most soil systems (reviewed by Nielsen *et al.*, 1986).

The two-site/two-region chemical/physical non-equilibrium model (eq. 1,2) has led to improved transport descriptions in laboratory leaching studies (Cameron and Klute, 1977; Gamerdinger *et al.*, 1990; Selim *et al.*, 1976). Basic to this model is the idea that the solid phase of the soil is made up of various constituents (soil minerals, organic matter, iron and aluminium oxides) and that a chemical is likely to react with these constituents at different rates and with different intensities. Unfortunately, the key parameters related to distinguishing the proportions of kinetic and equilibrium sites, and the transfer between the sites, were found in some studies to be functions of the pore water velocity and generally could not be derived independently from batch equilibrium studies. They usually needed to be adjusted for different experiments carried out on the same soil columns, thus suggesting that other than linear first-order kinetic processes dominate the sorption process during water movement.

Only minimal parameterisation of these models has been possible, and no extensive field testing has yet been accomplished. It is proving to be a substantial challenge to consider simultaneously physical non-equilibrium processes (macropore transport, mobile/immobile water), and at the same time to consider chemical non-equilibrium processes that result from combinations of multiple sorption sites, each possibly described as partly at equilibrium in its chemical behaviour, or partly kinetic. Add to this the chemical diffusion of pesticide which redistributes it between mobile and immobile porosity, the swelling and shrinking of organic soils and clay aggregates in response to changes in moisture content and the heterogeneous system becomes conceptually complex and difficult to study.

It is now clear that such systems must be well understood, or at least well described, when developing a pesticide leaching model. The leaching of pesticides is probably dominated more by these non-equilibrium processes than by the more classic forms of conceptually simplified transport models that describe homogeneous systems. Whether it be rapid flow along structural anomalies that bypasses the soil matrix, or very slow kinetic release of sorbed pesticide that creates a slow leakage of chemical over a long time at low concentrations, it is important to understand and relate these processes to field behaviour of applied pesticides. New approaches in this area include multi-region deterministic/mechanistic conceptualizations of the soil in which regions of distinct hydraulic properties are assumed to exist in a unit volume, which are then hydraulically-linked and proportionally weighted to represent soil pore size and aggregation variability (Hutson and Wagenet, 1995). Mixed deterministic/stochastic pesticide models have been employed (Boesten *et al.*, 1989; van der Zee and Boesten, 1991) to represent soil variability and the influence of sorption kinetics and macroporosity on pesticide leaching.

## 5. Future directions

The use of simulation models to address pesticide leaching is now a widespread practice. Yet, there is recognition that these models are limited by the assumptions inherent in their development, the availability and uncertainty of required input data, and the consequent reliability of output predictions. These issues are compounded by the fact that spatial and temporal scales of variability dictate the availability and quality of data and therefore the appropriate modelling choices.

One increasing use of pesticide leaching models is in studies of environmental quality that seek to estimate pesticide leaching within and across the landscape at different spatial and temporal scales (Wagenet *et al.*, 1994; Figure 1). This must be done with care as modelling choices depend on the particular application. One way to conceptualise the effects of scale on modelling choices is to understand the continuum of scales with which we deal. The diagonal line in Figure 1 describes those combinations of time and space scales upon which we have most often focused. At the smallest times and shortest distances are the laboratory studies, with a progression by conceptually similar steps of space and time to ever-increasing scales of soil aggregates, profile horizons, peds, catenas and eventually regional and continental scales. Although not overtly recognised, this has been the classical approach to modelling pesticide leaching, with each new model usually developed to

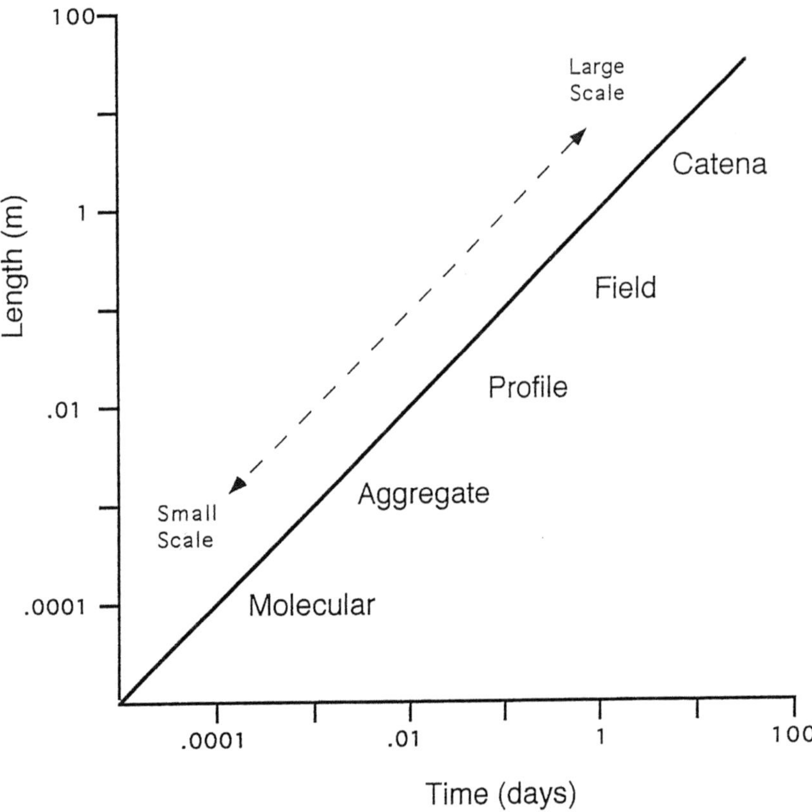


*Figure 1.* Conceptual spatial and temporal scales of application of soil pesticide models.

consider simultaneously a somewhat longer time and commensurately larger distance.

Far less attention has been paid to the off-diagonal areas of Figure 1, which include the processes that occur at very short distances over very long times (above the diagonal), and those that occur at short times over long distances (below the diagonal). For example, pesticide sorption is now recognised as being a combination of equilibrium and kinetic processes for a number of pesticides. While an equilibrium sorption process is appropriate for modelling pesticide–soil interaction on the diagonal, the kinetic process occurs at a short scale of space, but over a very long time. This produces pesticide leaching of very small concentrations over very long time periods, a pattern observed in some field monitoring programmes. Similarly, physical non-equilibrium exists when a physical process occurs at a very short time scale, but over an extended distance. Preferential flow of water and pesticides, sometimes termed macropore flow, is an example. Mechanistic pesticide leaching models do not yet include this process due to the inability to measure and predict accurately the complexity of flow paths that determine the rapid transport event. The consequences of these off-diagonal processes for accurate modelling of pesticide leaching have been almost ignored in most model development, and depend upon further knowledge of time-dependent processes, as discussed above.

Considering further the efforts on the diagonal, example scales can be identified that represent progressively larger units of space and time at which soil and pesticide processes can be considered (Figure 2) (Hoosbeek and Bryant, 1993; Wagenet *et al.*,

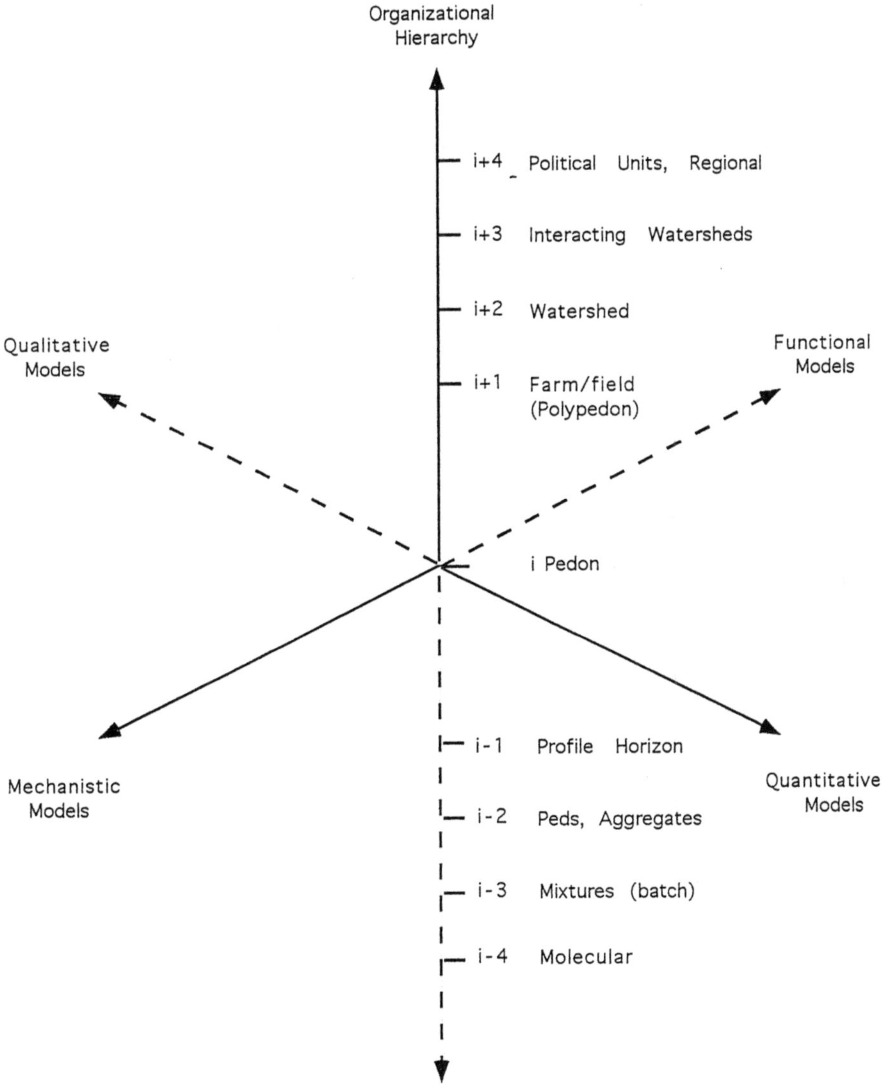


*Figure 2.* Organisation hierarchy of scales at which pesticide models may be applied (from Hoosbeek and Bryant, 1993).

1994). At any particular scale, a plane created by the intersection of two lines related to leaching models can be imagined. One line represents the continuum of mechanistic to functional models. The second line is the continuum of quantitative to qualitative models. Pesticide leaching models are generally developed for scales of soil peds (i−2) to farm/field (i+1), although some of these models are now applied to the larger scales of watersheds and regions generally defined by political boundaries, such as counties or states.

The use of field measurements to test and refine pesticide leaching models has resulted in extensive experimental activity in field soils. However, complicating factors such as spatial and temporal variability of processes and the parameters that describe them has led to a diversity of opinion on several issues: appropriate measurement or estimation methods at successively larger scales (Lauren *et al.*, 1988; Wagenet, 1985; Wagenet *et al.*, 1991, 1994); the merits of one model or another; and even the usefulness of a modelling approach at all. One important direction is increased interaction with the soil morphologist/taxonomist (Wagenet *et al.*, 1991). Methods to estimate soil hydraulic properties from soil survey databases through 'pedotransfer functions' (Bouma and van Lanen, 1987) and to employ a 'building block' approach to summarize and simplify the hydraulic behaviour of soil horizons have been demonstrated as feasible. A soil classification system for use in leaching models has been recently proposed (Quisenberry *et al.*, 1993). There have also been a number of creative applications of new CDE-based approaches intended to address variability of soil parameters and processes related to pesticide leaching (Bellin *et al.*, 1993; Bosma *et al.*, 1993; van der Zee and Boesten, 1991; van der Zee and van Riemsdijk, 1987). All these approaches have promise that should be further explored.

The future also holds new and exciting approaches that exploit evolving technology. We need to move towards models that provide more real-time visualization of transport at various scales. One can even imagine 'virtual reality' presentations of modelling results for better communication with the public. We also need to develop criteria that establish model credibility. We should perhaps establish uncertainty criteria to evaluate fairly model predictions and their 'closeness' of fit to observed data values. We need to establish how well we can predict observed values now and how well we should be able to predict in the future.

There are other issues as well. It is possible that we can create more accurate functional models through response surface analysis of inputs and outputs of existing mechanistic models. Transfer function models appropriate for field scale and watershed scale run-off processes of pesticide fate are a logical approach given the unit-response methods used in hydrology. We need more field sampling at a scale appropriate to model assumptions, so that models can be better tested for precision and accuracy. For mixed deterministic/stochastic models, we need to know the correlation matrix that describes correlations among inputs. Additionally, existing models do not well consider the mass of pesticide actually reaching the soil. Even in the absence of wind-drift, foliar interception, dissipation and wash-off, and the actual spatial distribution of applied pesticide are not well quantified. Finally, we need to improve the ability of leaching models to handle effects of filter strips, no-till versus conventional-till farming, and even broadcast versus banded application methods. These limitations of current models represent exciting opportunities for the future.

## 6. Summary

A wide variety of pesticide leaching models exist, and those that are properly constructed, that are used by the intended user in a professional manner, and that are not stretched beyond their limits, are very good tools. However, the proper construction or use of these tools requires some background in soil science (both in the laboratory and in the field), computer programming, and a modicum of common sense. It is important that any potential model builder, and especially any potential user, appreciate the capabilities and limitations of existing pesticide leaching models, or have access to someone who does. It is easy to misapply pesticide leaching models, which has resulted in erroneous estimations of pesticide transport to groundwater.

First, as models are employed to estimate pesticide leaching, it is important to use measurement and monitoring methods that are relevant at the spatial and temporal domains being modelled. For example, pesticide leaching is very dependent on water flow, and the measurement of water flow at the scale of a farmer's field involves different methods than does measurement of water flow on a watershed basis. Different models would be used to describe the dynamics of leaching at these two scales, and different measurements are also required. This concept is often overlooked as the predictions of a pesticide leaching model formulated on the CDE are evaluated against the 'truth' of whatever field data happen to exist at the particular scale the model is being applied.

Second, it is clear that greater attention has to be given to the non-equilibrium chemical and physical processes that greatly control pesticide partitioning and mobility, and water flow. While other important issues related to pesticide leaching also remain, there seem to be many soil chemists and microbiologists working on them. Study of non-equilibrium cases needs to be enlarged, as results to date indicate they can have significant impact on leaching. Yet, regulatory and educational models, and many research models, are not yet enough advanced to represent these processes.

Third, there should be recognition that modelling of pesticide leaching, and also the leaching of other solutes, is a valuable component of environmental assessments. The models can certainly be improved, and resources should be identified for that purpose. We must gain enough confidence in the use of leaching models that we use them on a prospective, rather than always a retrospective, basis. This seems the only economical way to screen new chemicals, or to evaluate new management practices, either in agricultural or other environmental cases where pesticide is applied. Limited resources will always prevent field studies in every possible case of interest.

Adoption of an integrated, team approach to modelling is always the best choice, and it is particularly wise in the case of modelling pesticide leaching.

Finally, the scientific community must realise that education of the public is critical and that ill-informed public spokesmen can quickly confuse the issues. However, the scientific community is also at fault because they have no clear coordinated efforts to teach the public about modelling and risk issues. The scientific community has failed to present their science in a form that the lay public can understand and appreciate. It is important that we better educate the public on the issues of risk prediction and uncertainty. If we fail to do so, we will not be able to continue our efforts on the public behalf.

## References

Addiscott T M 1977 A simple computer model for leaching in structured soils. J. Soil Sci. 28, 554–563.

Addiscott T M and Wagenet R J 1985 Concepts of solute leaching in soils: A review of modelling approaches. J. Soil Sci. 36, 411–424.

Bear J 1979 Hydraulics of Groundwater. McGraw-Hill, New York.

Bellin A, Rinaldo A, Bosma W J P, van der Zee S E A T M and Rubin Y 1993 Linear equilibrium adsorbing solute transport in physically and chemically heterogeneous porous formations. 1. Analytical solutions. Water Resour. Res. 29, 4019–4030.

Beven K J and Germann P 1982 Macropores and water flow in soil. Water Resour. Res. 18, 1311–1325.

Boesten J J T I and van der Linden A M A 1991 Modelling the influence of sorption and transformation on pesticide leaching and persistence. J. Environ. Qual. 20, 425–435.

Boesten J J T I, van der Pas L J T, and Smelt J H 1989 Field test of mathematical model for non-equilibrium transport of pesticides in soil. Pestic. Sci. 25, 187–203.

Bosma W J P, Bellin A, van der Zee S E A T M, and Rinaldo A 1993 Linear equilibrium adsorbing solute transport in physically and chemically heterogeneous porous formations. 2. Numerical results. Water Resour. Res. 29, 4031–4043.

Bouma J and van Lanen H A J 1987 Transfer functions and threshold values: from soil characteristics to land qualities. In: Proceedings of the International Workshop on Quantified Land Evaluation Procedures, Washington, DC, 27 April–2 May 1986. Eds K J Beek, P A Burrough and D E McCormack. pp 106–110. ITC Publication No. 6, Enschede.

Briggs G G, Bromilow R H and Evans A E 1982 Relationship between lipophilicity and root uptake and translocation of non-ionized chemicals by barley. Pestic. Sci. 13, 495–504.

Cameron D A and Klute A 1977 Convective–dispersive solute transport with a combined equilibrium and kinetic adsorption model. Water Resour. Res. 13, 183–188.

Carsel R F, Mulkey L A, Lorber M N and Baskin L B 1985 The pesticide root zone model (PRZM) : A procedure for evaluating pesticide leaching threats to groundwater. Ecol. Modell. 30, 46–69.

Chen C and Wagenet R J 1992 Simulation of water and chemicals in macropore soils. Part 1. Representation of the equivalent macropore influence and its effect on soilwater flow. J. Hydrol. 130, 105–126.

Chen W and Wagenet R J 1995 Solute transport in porous media with sorption-site heterogeneity. Environ. Sci. Technol. 29, 2725–2734.

Chen W and Wagenet R J 1997 Description of atrazine transport in soil with heterogeneous non-equilibrium sorption. Soil Sci. Soc. Am. Proc. (in press).

Darcy H 1856 Les fontaines publiques de la ville de Dijon. Dalmont, Paris.

Gamerdinger A P, Wagenet R J and van Genuchten M Th 1990 Application of two-site/two-region models for studying simultaneous nonequilibrium transport and degradation of pesticides. Soil Sci. Soc. Am. J. 54, 957–963.

Hoosbeek M R and Bryant R B 1993 Towards the quantitative model of pedogenesis – A review. Geoderma 55, 183–210.

Hutson J L and Wagenet R J 1992 LEACHM, *L*eaching *E*stimation and *CH*emistry *M*odel, Version 3.0, Dept. SCAS Research Report No. 92-3, Cornell University, Ithaca, NY.

Hutson J L and Wagenet R J 1993 A pragmatic field-scale approach for modelling pesticides. J. Env. Qual. 22, 494–499.

Hutson J L and Wagenet R J 1995 A multiregion model describing water flow and solute transport in heterogeneous soils. Soil Sci. Soc. Am. Proc. 59, 743–751.

Jury W A 1982 Simulation of solute transport using a transfer function model. Water Resources Res. 18, 363–368.

Jury W A and Gruber J 1989 A stochastic analysis of the influence of soil and climatic variability on the estimate of pesticide groundwater pollution potential. Water Resour. Res. 25, 2465–2474.

Jury W A, Spencer W F and Farmer W J 1983 Behavior assessment model for trace organics in soil. I. Model description. J. Environ. Qual. 12, 558–564.

Lauren J G, Wagenet R J, Bouma J and Wosten J H M 1988 Variability of saturated hydraulic conductivity in a Glossaquic Hapludalf with macropores. Soil Sci. 145, 20–28.

Mullins J A, Carsel R F, Scarbrough J E and Ivery A M 1993 PRZM-2. A model for predicting pesticide fate in the crop root and unsaturated soil zones: User's manual for release 2.0. EPA/600/R-93/046. Env. Res. Laboratory, USEPA, Athens, GA.

Nicholls P H, Walker A and Baker R J 1982 Measured and simulated behaviour of Fluometuron, Aldoxycarb and chloride ion in a fallow structured soil. Pestic. Sci. 13, 475–483.

Nielsen D R, van Genuchten M Th, and Biggar J W 1986 Water flow and solute transport processes in the unsaturated zone. Water Resour. Res. 22, 895–1085.

Nofziger D L and Hornsby A G 1986 A microcomputer-based management tool for chemical movement in soil. Appl. Agric. Res. 1, 50–56.

Quisenberry V L, Smith B R, Phillips R E, Scott H D, and Nortcliff S 1993 A soil classification system for describing water and chemical transport. Soil Sci. 156, 306–314.

Rao P S C, Davidson J M, and Hammond L C 1976 Estimation of nonreactive and reactive solute front locations in soils. In: Proc. Conf. Hazardous Wastes Res. Symp. pp 235–241. USEPA EPA-600/9-76-015. U.S. Gov. Print. Office, Washington, DC.

Richards L A 1931 Capillary conduction of liquids through porous mediums. Physics 1, 318–333.

Selim H M, Davidson J M and Mansell R S 1976 Evaluation of a two-site adsorption-desorption model for describing solute transport in soils. In Proc. Summer Simul. Conf., Washington, DC. 12–14 July, pp 444–448. Simulation Councils, La Jolla, CA.

van Genuchten M Th, Davidson J M, and Wierenga P J 1974 An evaluation of kinetic and equilibrium equations for the prediction of pesticide movement in porous media. Soil Sci. Soc. Am. Proc. 38, 29–35.

van Genuchten M Th and Wagenet R J 1989 Two-site, two-region models for pesticide transport and degradation: Theoretical development and analytical solutions. Soil Sci. Soc. Am. J. 53, 1303–1310.

van der Zee S E A T M and Boesten J J T I 1991 Effects of soil heterogeneity on pesticide leaching to groundwater. Water Resour. Res. 27, 3051–3063.

van der Zee S E A T M and van Riemsdijk W H 1987 Transport of reactive solutes in spatially variable soil systems. Water Resour. Res. 23, 2059–2069.

Wagenet R J 1985 Measurement and interpretation of spatially variable soil leaching processes. In Soil Spatial Variability. Eds D R Nielsen and J Bouma. pp 209–230. Centre for Agricultural Publishing and Documentation (PUDOC), Wageningen, The Netherlands.

Wagenet R J 1993 A review of pesticide leaching models and their application to field and laboratory data. Proceedings IXth Symposium on Pesticide Chemistry: Degradation and Mobility of Xenobiotics, Instituto di Chimica Agraria ed Ambientale, Piacenza, Italy, October 11–13, 1993. Eds A A M Del Re *et al.* pp 33–62.

Wagenet R J and Rao P S C 1990 Modelling Pesticide Fate in Soil. In Pesticides in the Soil Environment. Eds H H Cheng *et al.* pp 351–399. SSSA Book Series, No. 2, Soil Science Society of America, Madison, WI.

Wagenet R J, Bouma J and Grossman R B 1991 Minimum data sets for use of soil survey information in soil interpretive models. In Spatial Variabilities of Soils and Landforms. Eds pp 161–182. Spec. Publ. No. 28, Soil Science Society of America, Madison, WI.

Wagenet R J, Hutson J L and Biggar J W 1989 Simulating the fate of a volatile pesticide in unsaturated soil: A case study with DBCP. J. Environ. Qual. 18, 78–83.

Wagenet R J, Hutson J L, and Bouma J 1994 Modeling water and chemical fluxes as driving forces in pedogenesis. In Quantitative Modeling of Soil-Forming Processes, pp 17–35. SSSA Special Publication 39. ASA, CSSA, SSSA, Madison, WI.

Wagenet R J and Rao P S C 1990 Modeling Pesticide Fate in Soil. In Pesticides in the Soil Environment. Ed H H Cheng, pp 351–399. SSSA Book Series, No. 2, Soil Science Society of America, Madison, WI.

White R E 1985 A model for nitrate leaching in undisturbed structured clay soil during unsteady flow. J. Hydrol. 79, 37–51.

# 20 Comparative ecosystem research as a basis for agro-ecological modelling

O. FRÄNZLE

## 1. Comparative ecosystem research in the Bornhöved Lake District

The Bornhöved Lake District, selected by means of comparative geostatistics as a core area of the German ecosystem research network (TERN), is situated some 30 km south of Kiel (Schleswig-Holstein) and covers about 50 km$^2$ of interrelated drainage basins. The area is characterised by moraines of the Weichselian Pleniglacial with intercalated outwash sediments including kame deposits, dead-ice depressions with lakes, fens, or wetlands, and Late Saalian moraines in the south. The assemblage thus combines the essential geomorphic features of northern Germany, which ensures a high degree of spatial representativeness of the results obtained from this research.

As the result of a complicated geomorphic history, a great variety of natural ecosystems has evolved during the past 12 000 years. Humans began to have an impact on this environment during the Early Neolithic; this has increased considerably since the Middle Ages. As a consequence, ecosystems have undergone long- and short-term natural and anthropogenic changes which have had to be identified on a comparative basis in terms of basic processes such as energy, water and nutrient cycles in order to improve ecosystem management. This requires manipulative experiments on arable land and pastures in order to define their susceptibility to different management practices and environmental chemical effects, as well as comparable investigations on forest ecosystems and lakes. Agricultural practices must also be evaluated with regard to eutrophication and groundwater management. In this connection, inquiries into the specific role of ecotones are particularly important. More specifically, they relate to exchange and filter processes and the influence of ecotone structures on both the stability and resilience of the adjacent aquatic or terrestrial ecosystems.

In the framework of the socio-economic orientation of the Bornhöved Project, questions of environmental perception and economic expectations among farmers and fishermen, and the resultant private and public investment behaviour in the light of both national and international (EC) marketing policies, are subject to assessment and modelling approaches.

The general aims of this project are:

(1) To define and model structure, dynamics and stability of terrestrial and aquatic ecosystems of the study area in terms of site characteristics and biocenotic

*D. Rosen* et al. *(eds): Modern Agriculture and the Environment, 247–268.*

diversity, natural and anthropogenic fluxes of energy and matter, productivity and land-use patterns.

(2) To determine and model environmental strains and resilience mechanisms of ecosystem components affected by disturbances with particular reference to chemical impacts.

In view of the specific structure of the study area, particular attention is focused on:

(1) Modelling of biotic, energetic and material exchange processes between neighbouring ecosystems of differing land-use and fishery patterns.

(2) Modelling site-dependent relationships between lakes and their drainage basins with a focus on the role of different land/water interfaces.

(3) Modelling of agricultural and pasture ecosystems in light of national and international production and marketing regulations.

(4) Exotoxicological research on the fate and behaviour of environmental chemicals.

(5) Inquiries into the efficiency of environmental protection and conservation measures.

(6) Testing the validity of spatial extrapolations of simulation models by means of comparative site analyses and geographic information systems.

(7) Paleoecological and historical reconstruction of ecosystem evolution in the study area since the end of the Weichselian glaciation.

These objectives are defined in greater detail in a hierarchical system of several hundred working hypotheses organised on five levels, which forms the basis of interdisciplinary co-operation. The project involves some 140 scientists from the following disciplines: agricultural management, bacteriology, chemistry, climatology, computer science, exotoxicology, geobotany, geomorphology, human geography, hydrogeology, ichthyology, limnology, mathematics, meteorology, paleobotany, paleozoology, physical geography, physics, planktology, plant nutrition, plant physiology, soil microbiology, soil science, toxicology and zoology.

## 2. Regionalized hydrological and chemical fate modelling

The process of model construction involves stepwise testing and verification of the hierarchical system of hypotheses. This has the advantage of including a wide variety of different approaches, ranging from empirical models for practical application to abstract ones aimed at general qualitative insights. At one end of this spectrum is a detailed pragmatic description of the interaction of specific systems such as soil horizons or their adsorbents with one or a few chemicals in aqueous solution. This 'tactical' approach, as Holling (1966, 1968) called it, may be of considerable importance for particular resource management projects as well as for risk assessment. At the opposite end of the spectrum are relatively general 'strategic' models, again using Holling's terminology, which sacrifice numerical precision for the sake of general principles. Such models, often presented in the form of comprehensive graphs, need not correspond in detail to any single 'real world' process, but aim to provide a conceptual framework for the discussion of broad classes of phenomena or contentious issues. Rationally handled, tactical and strategic approaches mutually reinforce each other, thus providing insights such as those presented in the following examples.

### *2.1 Nitrogen and carbon fluxes in terrestrial ecosystems*

The fate of chemicals in terrestrial ecosystems is affected by both biotic and abiotic degradation or transformation within the soil, and by transport phenomena within environmental components such as soil, air, water and subsoil. The transformation reactions determine the pattern of exposure and duration, while transport phenomena indicate the target areas likely to be contaminated. Mathematical modelling of these processes is currently a field of intensive research, and as such the variety of models has increased sharply over the past decade. However, while the number of models seems enormous, they are all based on only a very few different modelling concepts. The number of elementary physical and chemical processes described remains limited.

As far as soil–groundwater modelling is concerned, the existing concepts refer to point-source pollution and can be categorized as: unsaturated soil zone or soil; saturated soil and subsoil (groundwater); geochemical; and ranking. The first two categories are designed to describe the behaviour of dissolved organic and inorganic compounds, or immiscible fluids for non-aqueous phase compounds; they follow patterns comparable to the physical-chemical and mathematical approaches. The third category enters into chemistry and speciation modelling, while the fourth is a screening approach (Bonazountas, 1987; Fränzle, 1993; Matthess, 1994).

Models of the above categories, as well as vegetation models, are generally calibrated on the basis of a more or less punctiform set of input data, and the expedient areal extrapolation of the model therefore poses particular problems. An adequate solution involves combining the model with a geographical information system which maintains the topological structure of the environment-specific climatologic, hydrologic and soil data used as input to the model and for validating output. The following examples illustrate this approach on a large scale.

The assessment of acceptable nitrate levels in soil and subsoil of different ecosystems in the Bornhöved study area is based on a combination of the chemical fate model WASMOD/STOMOD (Fränzle *et al.*, 1989; Reiche, 1991) with a geographic information system. The basic structure of WASMOD/STOMOD is summarised in Figures 1 and 2, and in Table 1.

With the aid of a minimum set of input data, submodules (1) to (4) define effective precipitation as stand-specific hydrologic input variable. Submodules (5) to (8) then

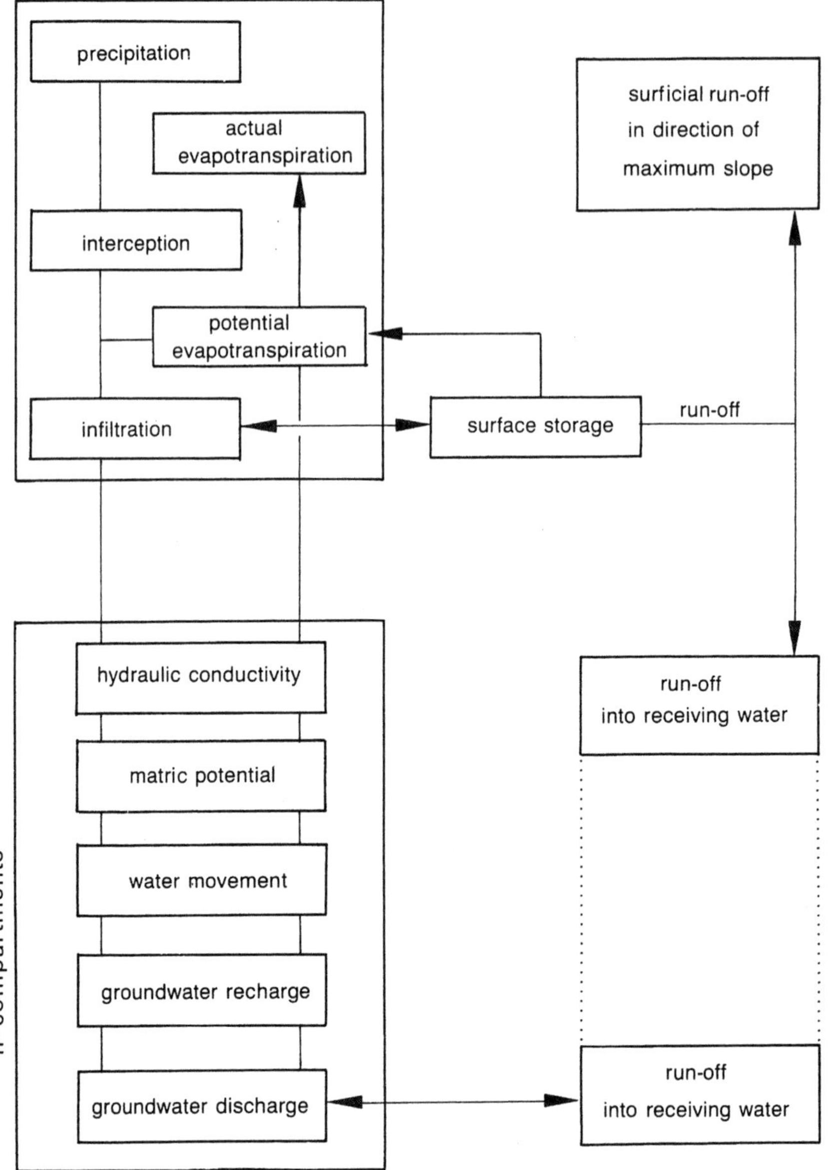


*Figure 1.* Conceptual diagram of the WASMOD submodel (modified from Reiche, 1991).

describe deep drainage through the soil and aeration zones, and submodules (9) and (10), groundwater discharge into a receiving body of water. Based on these input data, the solute module defines the variable proportions of adsorptive and capillary soil moisture relative to the organic matter, clay and silt phases. Thereafter,

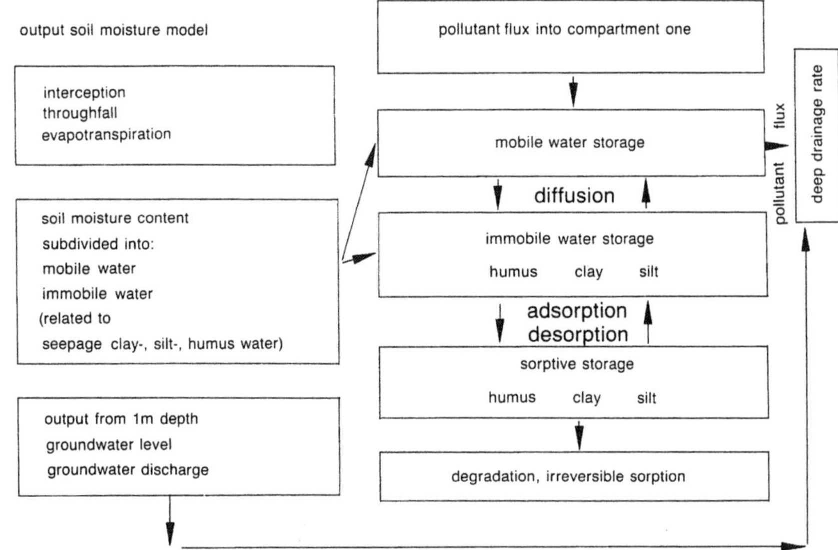


*Figure 2.* Conceptual diagram of the STOMOD submodel (modified from Reiche, 1991).

adsorption/desorption equilibria, irreversible translocation, microbial degradation and water-borne vertical translocation are calculated for each successive horizon of the unsaturated soil and aeration zone. The equilibrium concentrations in the mobile and immobile water fractions and soil matrix are determined on the basis of varying soil-moisture content on the one hand, and time-dependent diffusion and sorption kinetics on the other. The model is free from limiting meso- and macro-scale assumptions about homogeneity (i.e. large-scale equilibria in extensive components).

An extended version of STOMOD including a nitrogen submodule was developed specifically for the purpose of nitrogen modelling (Figure 3) (Reiche, 1991). The extended model was calibrated on the basis of comprehensive data sets from batch, lysimeter and field experiments in 28 plots in different areas of Schleswig-Holstein. A further extension including carbon fluxes is shown in Figure 4.

The application of the model to two different areas of the Bornhöved Lake District was accomplished by means of ARC/INFO (ORACLE), whose software consists of two major subsystems (ESRI, 1992). ARC maintains the topologic structure of the database used to represent mapped features. The database management subsystem INFO (for technical reasons replaced by ORACLE) is a relational databank used to store and process attribute information associated with the geographic features maintained by ARC. In this case, the necessary soil data were derived from large-scale maps of the German Soil Taxation Survey (Reichsbodenschätzung) by means of an automatic translation procedure (Reiche, 1991). Figure 5 illustrates the application of the extended WASMOD/STOMOD version to modelling nitrate translocation into groundwater in a drainage basin of the Bornhöved Lake District.

*Table 1.* Compartments of soil water model.

1. Partial model potential evapotranspiration (ETpot) after Haude
   ETpot $= f$ (temperature, relative humidity)
2. Partial model interception (ETi) after v. Hoyningen-Huene
   ETi $= f$ (leaf area index, precipitation)
3. Partial model suction ($\psi$)
   $\psi = f$ (organic matter, silt, clay)
   Reference data: pF curves
4. Partial model hydraulic conductivity ($K_u$)
   $K_u = f$ (saturated conductivity, suction)
   Reference data: field data of saturated conductivity after Darcy
5. Partial model evapotranspiration (ETa), modified after Braun
   ETa $= f$ (suction, root density, ETp (Haude), ETi)
6. Partial model soil water movement
   $\delta\,\Theta/\delta t = f$ (suction gradient, $K_u$, groundwater level)
   Reference: Darcy's equation
7. Partial model change in water storage
   $\delta$ BF (vol)/$\delta t$ $= f$ ($\delta\Theta/\delta t$, precipitation, ETa)
8. Partial model lowermost compartment
   $\delta$ BF (vol) $u/\delta$t $= f$ ($\delta$ Bf (vol)/$\delta t$, groundwater level)
9. Partial model groundwater discharge
   $\delta$ GW/$\delta t$ $= f$ (Kf, $\alpha$[receiving water])
10. Partial model groundwater level
   $\delta$ GH/$\delta t$ $= f$ ($\delta$ BF (vol) $u/\delta t$, $\delta$ GW/$\delta t$)

Initial data (daily): Temperature, precipitation, relative humidity
Initial data (constant):
- % organic matter, % silt, % clay
- leaf area index
- saturated hydraulic conductivity
- root density
- slope of groundwater surface

Initial data (primary):
- soil moisture
- groundwater level

## 2.2 *Groundwater flows and solute exchange in groundwater/lake ecotones*

An ecologically valid classification of hydrophysical and chemical exchange processes across the numerous lotic and lentic ecotones of the study area combines structural and dynamic characteristics in a multivariate fashion. Dynamic properties can be formulated in terms of hydraulic potential $\phi$ and stream function $\psi$ using the dual formulation. This approach, however, is restricted to saturated hydrogeological systems (FLONET). The equations for two-dimensional, steady-state flow in heterogeneous, anisotropic, porous media, are:

$$\frac{\delta}{\delta x}\left(K_{xx}\frac{\delta\phi}{\delta x}\right)+\frac{\delta}{\delta y}\left(K_{yy}\frac{\delta\phi}{\delta y}\right)=0 \qquad (1)$$

$$\frac{\delta}{\delta x}\left(\frac{1}{K_{yy}}\frac{\delta\psi}{\delta x}\right)^{1}+\frac{\delta}{\delta x}\left(\frac{1}{K_{xx}}\frac{\delta\psi}{\delta y}\right)=0 \qquad (2)$$

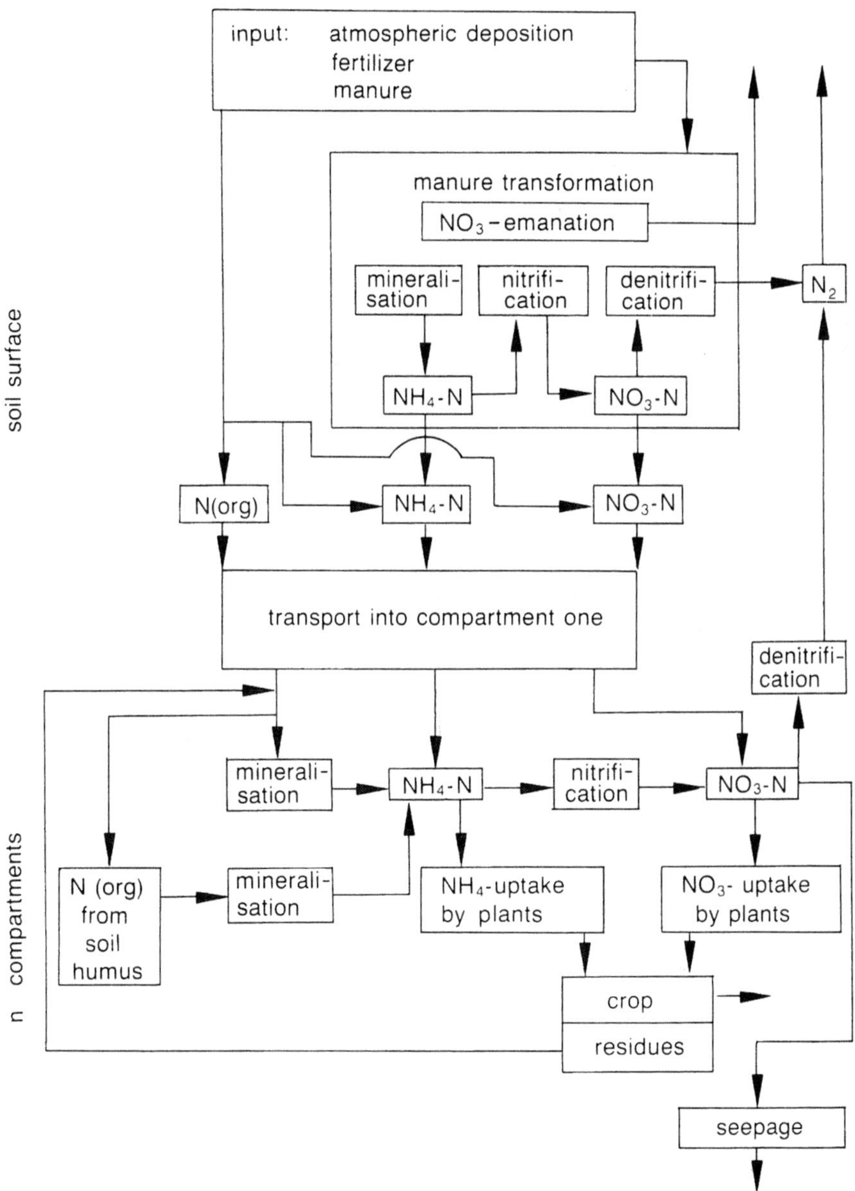


*Figure 3.* Nitrogen submodule of STOMOD (modified from Reiche, 1991).

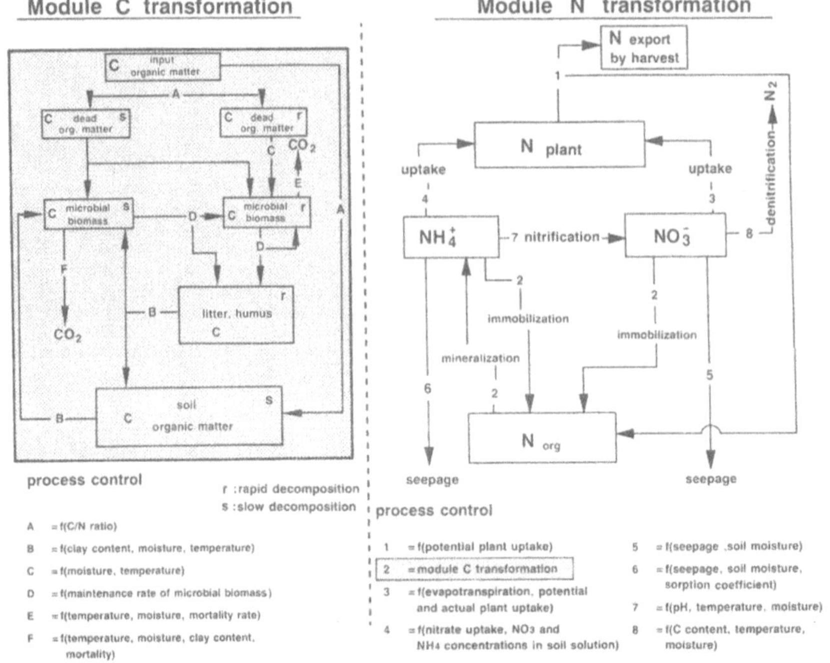


*Figure 4.* C and N submodules of the extended WASMOD/STOMOD model (from Reiche, unpublished).

where:

$K_{xx}$, $K_{yy}$ = principal components of the hydraulic conductivity tensor ($LT^{-1}$)

$\phi$ = hydraulic head (L)

$\psi$ = stream function ($L^2T^{-1}$).

Darcy's law defines the relationship between the properties of the porous medium, the hydraulic gradient and the groundwater flux.

$$q_x = -K_{xx}\frac{\delta\phi}{\delta x} \tag{3}$$

$$q_y = -K_{yy}\frac{\delta\phi}{\delta y}. \tag{4}$$

Written in terms of the stream function, the groundwater flux is given by:

$$q_x = \frac{\delta\psi}{\delta y} \qquad q_y = \frac{\delta\psi}{\delta x}. \tag{5}$$

The Galerkin finite element method is employed to solve the above equations. Its primary advantage over the finite difference method lies in its ability to reproduce

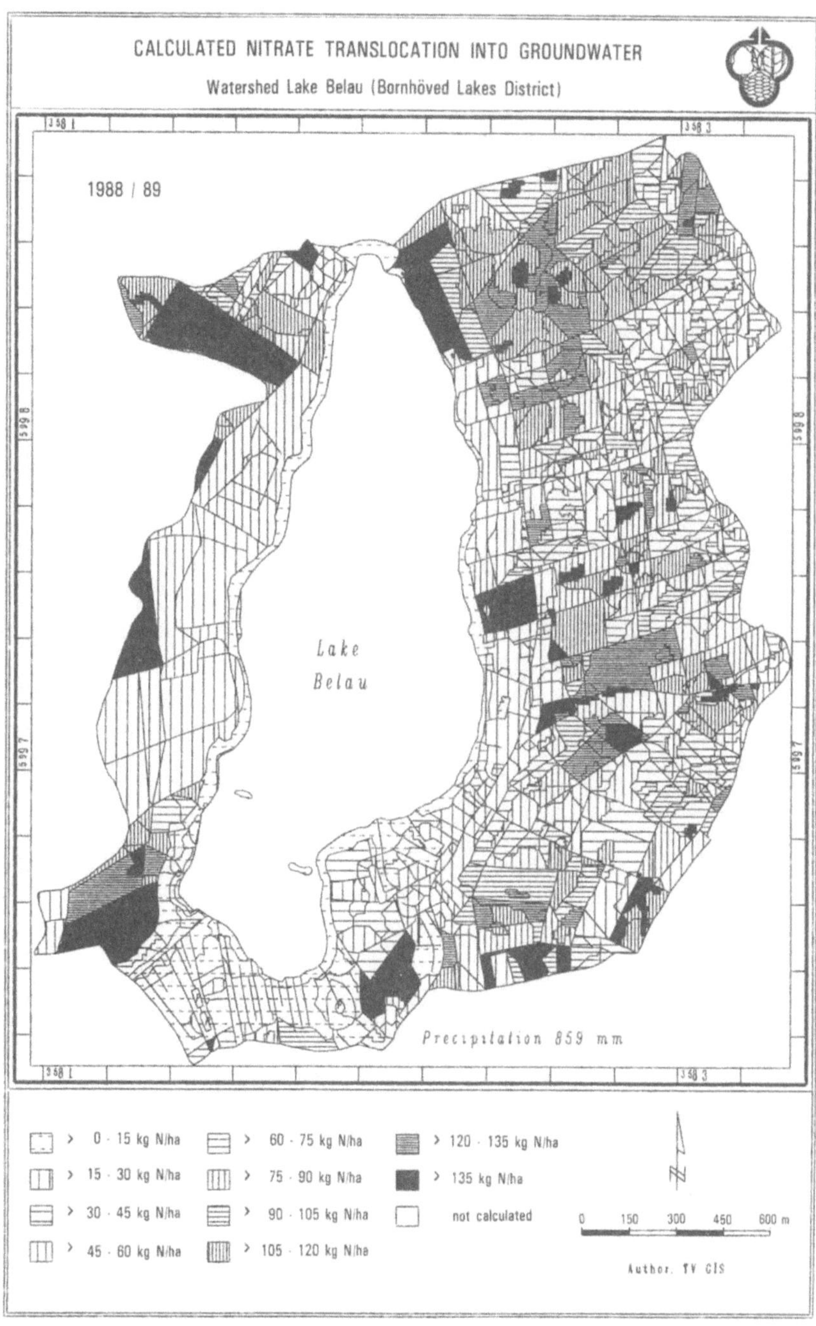


*Figure 5.* Calculated nitrate translocation into groundwater of the Lake Belau catchment (from Reiche, 1991).

*Table 2.* Types of water flow through lentic ecotones.

| |
|---|
| *Water exchange between aquifer and lake* |
| • Direct discharge from phreatic or confined aquifers into the limnetic water body without impeding interface sediments |
| • Free discharge into the hydraulically conductive upper littoral zone, associated with very limited seepage through the basal lake sediments |
| • Preferentially restricted discharge through littoral and profundal zones due to exceedingly high spatial variability of lake bottom sediments |
| • Groundwater recharge from lake to aquifer (exfiltration) in places of very permeable lake sediments |
| • Periodic infiltration and exfiltration with non-stationary groundwater divides or great water-level fluctuations in the lake of adjacent aquifer |
| *Water exchange between wetland and lake* |
| • Natural wetland with or without flooding, with or without groundwater underflow |
| • Wetland with water level regulation by ditches or drainage systems |
| *Hydraulic barriers between lake and surrounding groundwater basin* |
| • Subsurface inflow of perched water or local periodic interflow close to the surface |
| • No subsurface water exchange or three-dimensional flow around the impermeable shore barriers |

irregular geometry with a greater degree of accuracy. The modelling approach allows definition of the following types and subtypes of exchange (Table 2).

Two types are shown in Figures 6 and 7; it should be noted that there are no clear-cut distinctions between these types. In subsurface exchange, the water always selects a path that results in a balance between the spatially varying water levels while covering the shortest distance and with the least hydraulic resistance. Subsurface water exchange is thus based on an energetic principle of minimisation.

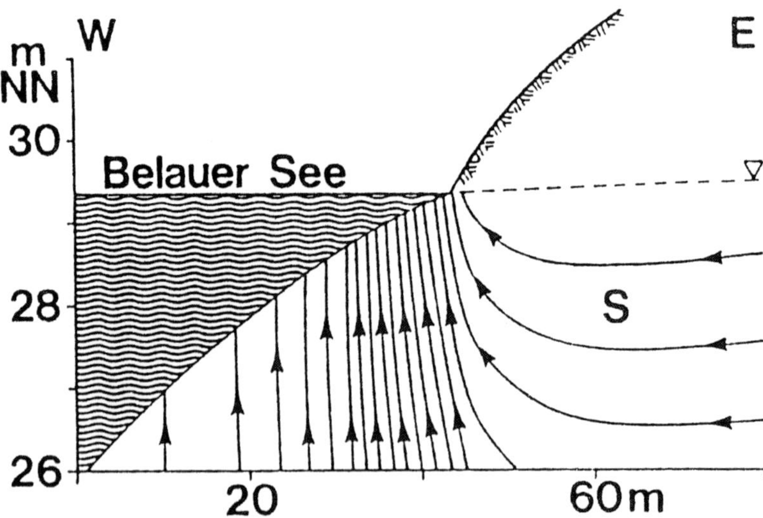


*Figure 6.* Permanent groundwater exchange in lentic ecotone situations of Lake Belau, simulated with FLONET (from Piotrowski and Kluge, 1994); (S = sand).

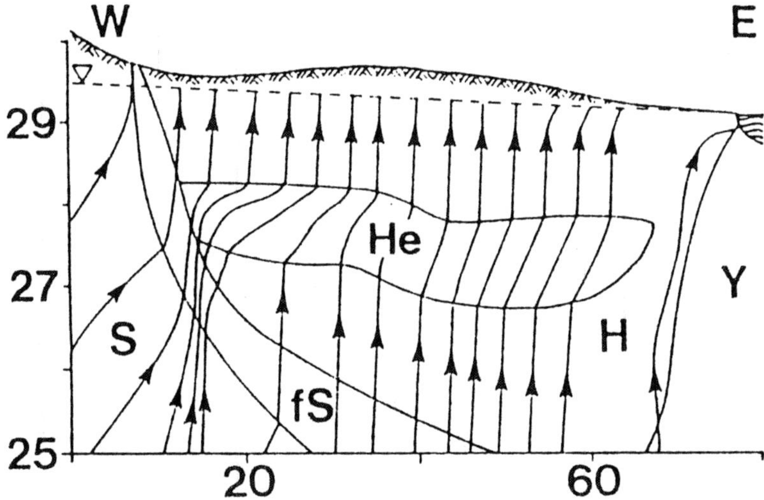


*Figure 7.* Summer-time groundwater exchange between aquifer and wetland with localised inflow into Lake Belau (from Piotrowski and Kluge, 1994); (S = sand; fS = fine sand; He = alder peat; H = moss peat; Y = gyttja).

Given sufficient permeability of the shore deposits, it is concentrated near the water line and the littoral zone. In some lakes, erosion or washing out of fine sediment result in an additional increase in conductivity close to the shoreline.

Proceeding from type a to C type in the above classification, hydraulic resistance increases. The subsurface water avoids those areas with very low permeability, which leads to an extension of the flow lines, and the transition to a three-dimensional heterogeneous flow pattern, which makes it very difficult to define flow conditions precisely.

A frequently occurring increased hydraulic conductivity in the erosion zone, and a reduced hydraulic conductivity in the marginal shore zone and lake sediment (local effects), and differences in the distance to the groundwater divides (long-range effects) give rise to a great deal of spatial variability in the exchange rates along the shoreline. A shift of regional groundwater divides is likely to occur if the aquifer between lakes or streams is continuously developed in alluvial flats and the levels of the receiving waters are headed up or drawn down. In the shore zone, local subsurface divides may even form near the shoreline or in the transitional area between a wider shore wetland and the adjacent aquifer. This may be due to greater fluctuations in the lake water level, or to a rapid rise in the water level in wetland areas following flooding or storms, in comparison to the adjacent aquifer.

The reconstruction of flow lines makes it possible to draw conclusions about the origin of the water, the groundwater recharge areas and the transported substances: with increasing length of flow lines and the residence time or age of the groundwater, the concentrations of nutrients and xenobiotics decrease with depth. In addition to these factors, reduced hydraulic conductivity in the shore zone commonly contributes to preferential groundwater flow along local seepage channels. The average time of contact within the ecotone zone will be reduced under such circumstances, thus decreasing the filter and buffer effects of the shore zone for the transported substances. This influence is caused by the microscale heterogeneity of glacial and lacustrine deposits in the lakeshore area (Piotrowski and Kluge, 1994). The experimental proof of these phenomena was provided by special measurements of the temperature as well as the specific electrical conductivity of Lake Belau sediments during the winter of 1993–94. In some patches varying between 1 and 100 $m^2$, the discharge of groundwater raised the water temperature by up to 5°C and the conductivity by a factor of 2. Comparable results have been reported by Vanek (1991) for lakes and coastal waters in Sweden. In distinctly larger sections of the lake bottom, however, no advective exchange takes place with the groundwater; sedimentation, bioturbation and diffusion predominate in the sediment–water interface.

### 2.2.1 Hydrochemistry of groundwater/lake ecotones

In contrast to groundwater recharge areas, groundwater discharge zones, which serve as the final component of the subsurface water cycle, exhibit a number of distinctive hydrochemical features (Eriksson, 1985). These features are caused by hydromorphic soils, high biological productivity and the accumulation of particulate and dissolved organic matter, the intercalation of permeable and impermeable sediments, and the occurrence of aerobic transition zones and anaerobic stagnation zones. Of particular importance are the hydrolytic and microbiologically catalysed degradation of organic matter, diverse oxidation–reduction reactions, precipitation of calcite, adsorption processes and organic complexation.

Figure 8 shows that the ecotone between the aquifer and Lake Belau is subdivided into several vertical compartments. Just as the age of the water increases with depth, so does the residence time of the subsurface water in the water cycle. The arrows indicate the direction of water and material exchange. Shaded areas correspond to hydrochemical mixing zones between compartments with different hydrochemical conditions. The width of these mixing zones is determined by both periodicity and intensity of water exchange.

A relatively stable hydrochemical environment can only develop in the upper saturated zone and the deeper groundwater zone if the wetland is sufficiently wide. In narrow wetlands, groundwater flow mixes directly with the lake water. Not only the concentration of chemicals, but also their lateral and vertical gradients can exhibit high spatial variability and temporal periodicity in mixing zones. This may be caused by macroscopic dispersion of the advective flow and seasonal variations of the ecotonal water balance.

Figure 9 indicates the relationship between nitrate concentration and redox potential as related to the distance from land to shoreline while the concentration of ammonium increases near the shore. This means that quantitative conclusions about the actual buffer effect of the shore zone on the transport of nitrate from the drainage

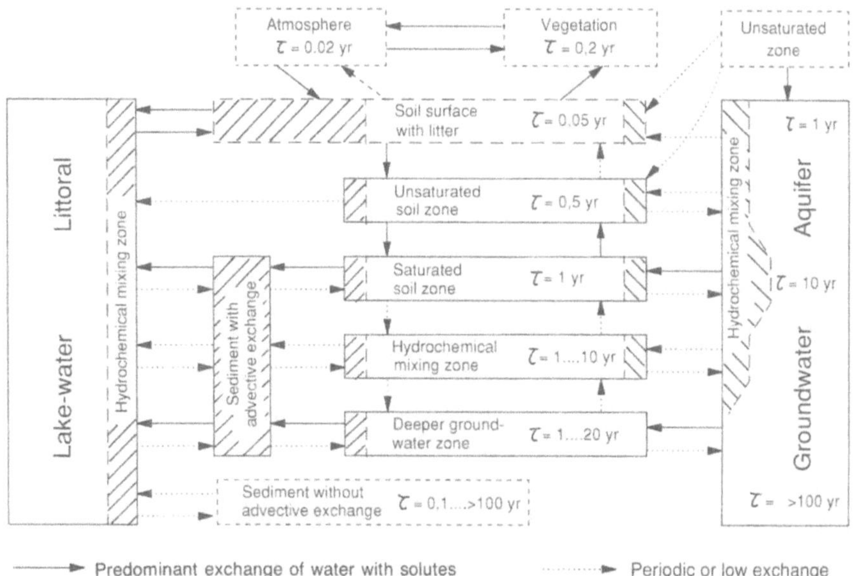


*Figure 8.* Vertical structure of a lentic ecotone with residence times of soil moisture and groundwater and related exchange intensities (from Fränzle and Kluge, in press).

area cannot be deduced from the changes in concentration at just one depth. In the example presented here, the concentration of nitrate in fact increases with depth while the reverse holds true for the concentration of ammonium.

If the groundwater flows in deeper layers through the shore zone, the specific transformation processes are transferred to the lake sediments of the infralittoral or profundal zones. For water exchange rates above 0.05 $m^3/(m^2day)$ or at local springs, the buffer effect of the groundwater/lake ecotone disappears almost completely with the transfer of the groundwater into the lake. The high spatial variability and temporal dynamics of the characteristic hydrochemical conditions make both measurement and modelling difficult. Measurement programmes can supply representative data only if the number of sampling points is greater than 10 at each depth, and if the samples are taken over a long period of time and, preferably, every fortnight.

The nitrogen model applied here is such that the transformations of several nitrogen compounds in the saturated zone can be formulated in the model analogously to the macro-pore concept for aerobic as well as non-aerobic domains between pools of organic matter, dissolved organic nitrogen, dissolved nitrate, and dissolved and adsorbed ammonium.

### 2.2.2 A hierarchical modelling concept of ecotones

The hierarchical modelling system presented in Table 3 summarises a comprehensive concept for application-oriented assessment of lake/groundwater ecotones. Each

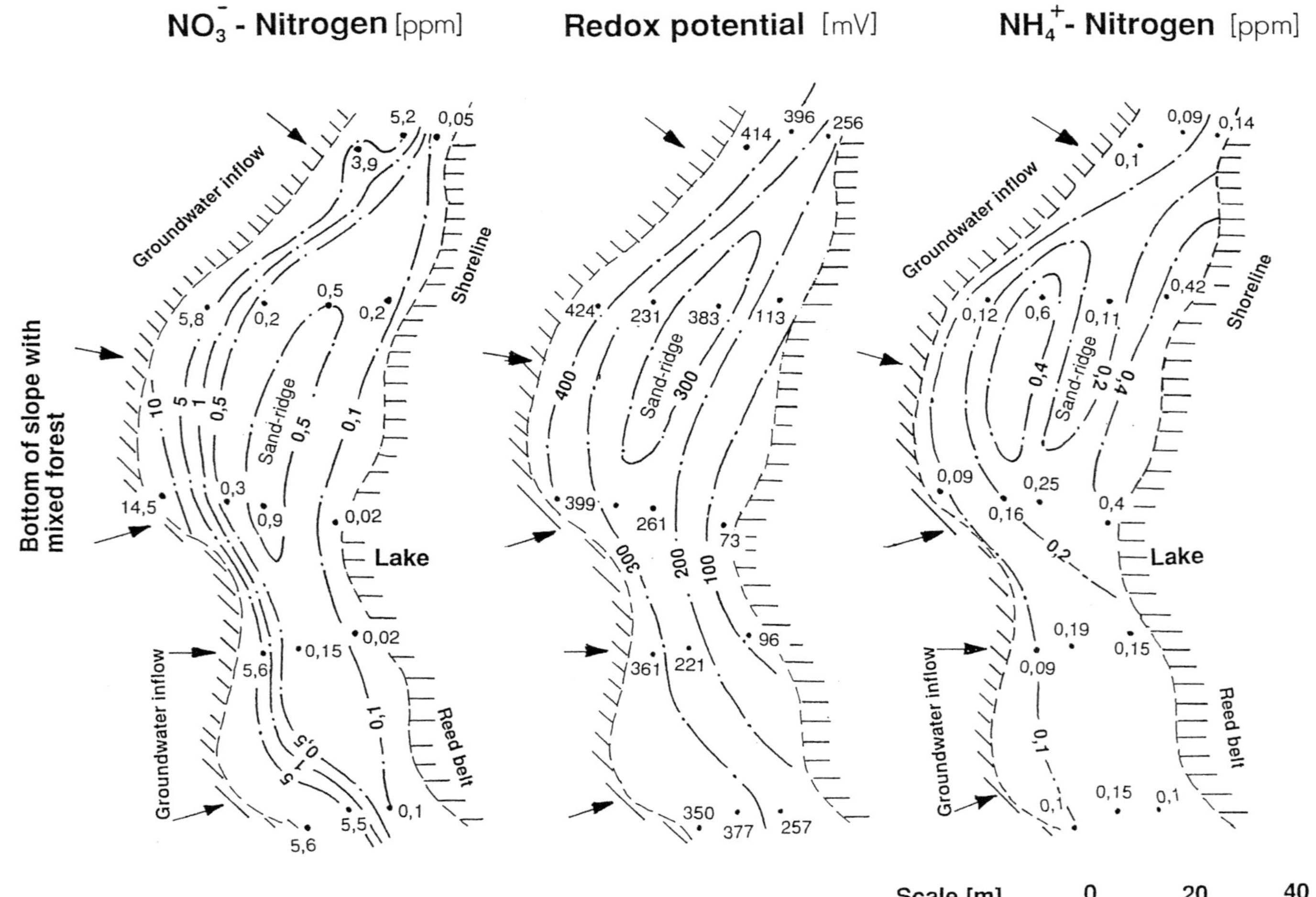


*Figure 9.* Redox potential and related $NO_3^-$-N and $NH_4^+$ concentrations of a lentic ecotone in the Bornhöved Lake District (from Kluge, unpublished).

*Table 3.* Hierarchy of modelling approaches for lentic ecotones.

| Level | Category | Requirements |
|---|---|---|
| A | Expert-system assessment | Fundamental knowledge about system structures, functions, dynamics; database, GIS, simple models, few field observations |
| B | Statistical models (regression, extreme values) | Measurements in space and time |
| C | Process models (deterministic) | Detailed knowledge about processes and structures |
| C1 | Flow and mass transport models with spatially distributed parameters | Physical theory of flow mechanisms and reactions; detailed knowledge of structures and parameter adaptation |
| C2 | Box-balance and reaction models (stirred-tank reactor) | Conceptual framework, balance equations; measurements of calibration and validation |
| D | Stochastic models | Long-term measurement programmes in space and time |
| D1 | Stochastic process models (prognosis) | Complete data sets |
| D2 | System-theoretical operator models (black-box, filtering) | Complete data sets with characteristic events |
| E | Exchangeable modules as intrinsic regulators (e.g. object-oriented models) | Comprehensive knowledge; ecotone typology, ecotone modules; basic data, GIS, simulation language, etc. |

level of generalization is characterized by a specific set of hydrological modelling approaches (i.e. water flow and balance models), coupled with pertinent mass transport and reaction models. Each class of models has a specific realm of application and special requirements as to a-priori knowledge and data. The initial expert assessment of the system utilizes all the available information about the hydrological typology of the lake and individual shore sections. The estimation of current water and mass balances or simple simulations with commercial models form the core of this initial evaluation.

On the following level, two groups of process models may be distinguished: models with spatially distributed parameters (flow and mass transport models) and box balance models. Geohydraulic and geochemical parameters are necessary for every finite volume into which the flow pattern is subdivided. Box models serve as stirred-tank reactor models for larger compartments which possess similar characteristics. Field and laboratory data are required for calibration and validation.

Black-box models describe the transitional hydrological and hydrochemical behaviour of ecotones between a lake and the adjacent aquifer. Weather effects and vegetation give rise to discrete hydrological zones (Figure 8). The transition function can be described in the time domain as well as in the frequency domain by means of time series analysis and systems hydrology, which also permits the application of scale analysis and hierarchical systems theory.

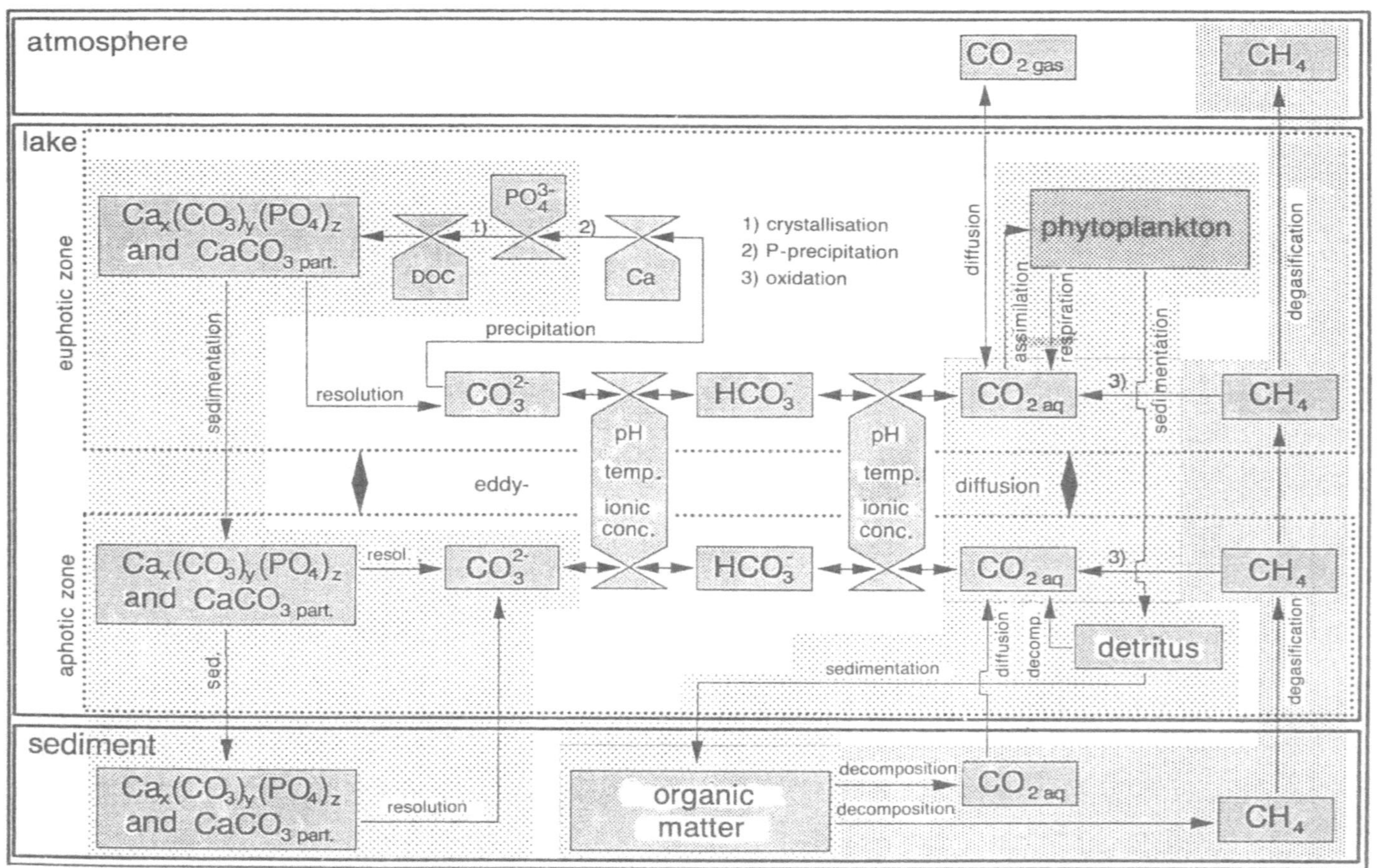


*Figure 10.* Conceptual diagram of carbonate and phosphate dynamics in Lake Belau (modified from Lenz, 1992 and Schernewski *et al.*, 1994).

The future aim is to develop interchangeable simulation models for different ecotone types which adequately describe water and mass exchange between drainage areas and lakes by mean of object-oriented programming approaches (Sekine *et al.*, 1991). This type of model depicts ecotones as black-box modules which are interchangeable, like circuit boards. Prognoses and scenario analyses of mass transfer will be greatly facilitated by the implementation of these modules in regional groundwater transport models, since the system is designed to obtain scale-consistent results for individual groundwater/lake ecotones or for an entire lake, and to optimize experimental programmes (Kluge and Heinrich, 1994). Furthermore, this model approach is intended to fill a gap, hitherto bridged only by non-generalizable statements or impracticable model assumptions.

### *2.3 Calcite/phosphate coprecipitation in Lake Belau*

The nitrogen and phosphorus cycles play a particular role in lake eutrophication (Jørgensen, 1983). In hard-water lakes like Lake Belau, however, the carbon cycle must also be taken into account since calcite precipitation is of decisive importance as a eutrophication-regulating mechanism. The conceptual diagram presented in Figure 10 summarizes the essential relationships between the C and P cycles of Lake Belau as a holomictic–dimictic body of water.

Calcite precipitation as a stabilizing 'self-purification' process has been noted by several authors during the last two decades (Koschel, 1990; Otzuki and Wetzel, 1974; Rossknecht, 1980). It can also function as a scavenger of inorganics by sorption and coprecipitation, and contributes to limiting phosphate and trace elements. This process also leads to a reduction of particulate matter through aggregation and a resulting increase in settling rates.

Rossknecht's model (1977) as adapted to Lake Belau describes carbon dynamics by the following set of equations:

$$CO_{2aq} + H_2O \Leftrightarrow H_2CO_3 \Leftrightarrow HCO_3^- + H^+ \tag{6}$$

$$HCO_3^- \Leftrightarrow CO_3^{2-} + H^+ \tag{7}$$

$$CO_3^{2-} + Ca^2{+} \Leftrightarrow CaCO_3. \tag{8}$$

The resulting activity coefficients are as follows:

$$K_1 = \frac{[HCO_3^-][H^+]f_1^2}{[CO_2aq]}, \tag{9}$$

$$K_2 = \frac{[CO_3^{2-}][H^+]f_2}{[HCO_3^-]}$$

$$K_3 \frac{1}{[CO_3^{2-}][Ca^{2+}]f_2^2},$$

$$K_4 = \frac{[CO_2aq]}{[HCO_3^-]^2[Ca^{2+}]f_1^2 f_2} = \frac{K_2 K_3}{K_1}.$$

The temperature dependence of equilibrium constants ($K_1$, $K_2$, $K_4$) can be expressed as:

$$K_1 = 10^{-(5.9633 + 25.9224/(T + 42.0435))} \tag{10}$$

$$K_2 = 10^{-(9.3864 + 97.6858/(T + 78.5520))} \tag{11}$$

$$K_4 = 10^{(4.0447 - 9.1618/(T - 62.0121))}. \tag{12}$$

The Debye–Hückel theory allows the estimation of activity coefficients ($f_1$,$f_2$) by using the ionic strength ($I$) and ionic radii $a_1$(of $HCO_3^-$) and $a_2$ (of $Ca^{2+}$) (Morel and Hering, 1993):

$$f_1 = \exp\{-(A \times \sqrt{I}/(1 + B \times a_1 \times \sqrt{I})\}, \tag{13}$$

$$f_2 = \exp\{-(A \times 4 \times \sqrt{I}/(1 + B \times a_2 \times \sqrt{I})\}. \tag{14}$$

The constants $A$ and $B$ are temperature-dependent:

$$A = 0.49135 + T \times 0.00077, \tag{15}$$

$$B = 0.32474 + T \times 0.00017. \tag{16}$$

The ionic radii are empirical fitting parameters in the Debye–Hückel theory, whose values vary between 3 Å and 10 Å, depending on the substance and experimental conditions (Morel and Hering, 1993). Here the values suggested by Rossknecht (1977) are used, i.e. $HCO_3^-$: 4 Å; $Ca^{2+}$: 6 Å. The equation for ionic strength ($I$) can be simplified for calcite precipitation in hard-water lakes:

$$I = (x_1 + [HCO_3^-]) \times 0.5 + (x_2 + [Ca^{2+}]) \times 2. \tag{17}$$

The constants $x_1$ and $x_2$ represent the sum of all mono ($x_1$) and bivalent ($x_2$) ions. For Lake Belau, average values are $x_1 = 0.0014$ M $L^{-1}$ and $x_2 = 0.0005$ M $L^{-1}$. Figure 11 illustrates an application of the model to the central part of Lake Belau in the summer of 1991 (Schernewsk *et al.*, 1994). There is very good agreement with the measured values of $CO_2$ and $CaCO_3$ concentrations. The model indicates that calcite formation is limited to the uppermost 7 m of the water, while the water below this level is calcite-aggressive. Calcite is accumulated in the sediment only in parts of the lake with a depth of less than 16 m.

Regarding phosphate coprecipitation, the modelled values of 0.01% to 0.04% during spring are distinctly lower that those reported from Lake Constance (Rossknecht, 1980). Yet about 100 kg of P are thus removed from the epilimnion during the summer months when the average P content amounts to about 150 kg. The coprecipitated phosphorus is partly dissolved in the hypolimnion and partly accumulated in the bottom sediments. Since only a relatively small amount of phosphorus is released back into the epilimnion during the same year, coprecipitation by calcite formation constitutes an efficient mechanism regulating the primary production of Lake Belau. Figures 12 and 13 illustrate the different sources and fluxes of phosphorus relating to the lake and its setting in the landscape in terms of phosphorus pools. It should be noted that these figures were obtained from relatively long-term measurements and thus constitute a framework for better assessment of the relative importance of the coprecipitation phenomena described.

Figure 13, providing a summary cross-section through the ecosystem types studied, is indicative of the fact that phosphorus is unknown in the atmosphere except as a component of dust, since none of its common compounds has any appreciable vapour pressure. It therefore tracks only part of the hydrological cycle, namely from the lithosphere to the hydrosphere, and in terms of modelling this amounts to uncoupling the atmospheric reservoir, except for water.

## References


Bonazountas M 1987 Chemical fate modelling in soil systems: A state-of-the art review. In Scientific Basis for Soil Protection in the European Community. Eds H Barth and P L'Hermite. pp 487–566. Elsevier, London, New York.

Eriksson E 1985 Principles and Applications of Hydrochemistry. pp 79–87. Chapman and Hall, London, New York.

ESRI (Environmental Systems Research Institute) 1992 ARC/INFO Users Guide 6.0. Redlands, USA.

FLONET – Guiguer N, Molson J, Frind E and Franz T FLONET version 1.02. Two-dimensional Steady-State Flownet Generator. University of Waterloo, Canada.

Fränzle O 1993 Contaminants in Terrestrial Environments. pp 208–222. Springer-Verlag, Berlin, Heidelberg, New York.


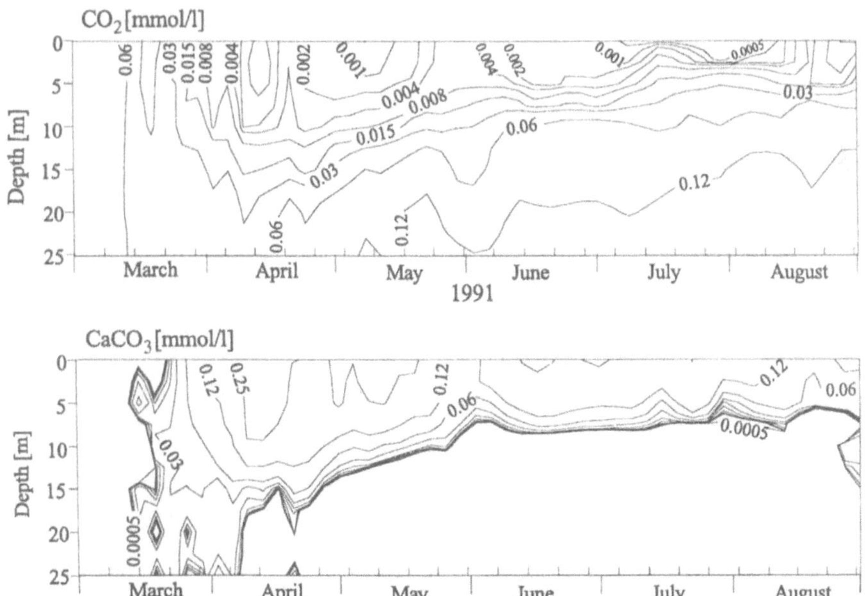


*Figure 11.* Modelled $CO_2$ and $CaCO_3$ concentrations in Lake Belau during spring and summer 1991 (from Schernewski *et al.*, 1994).

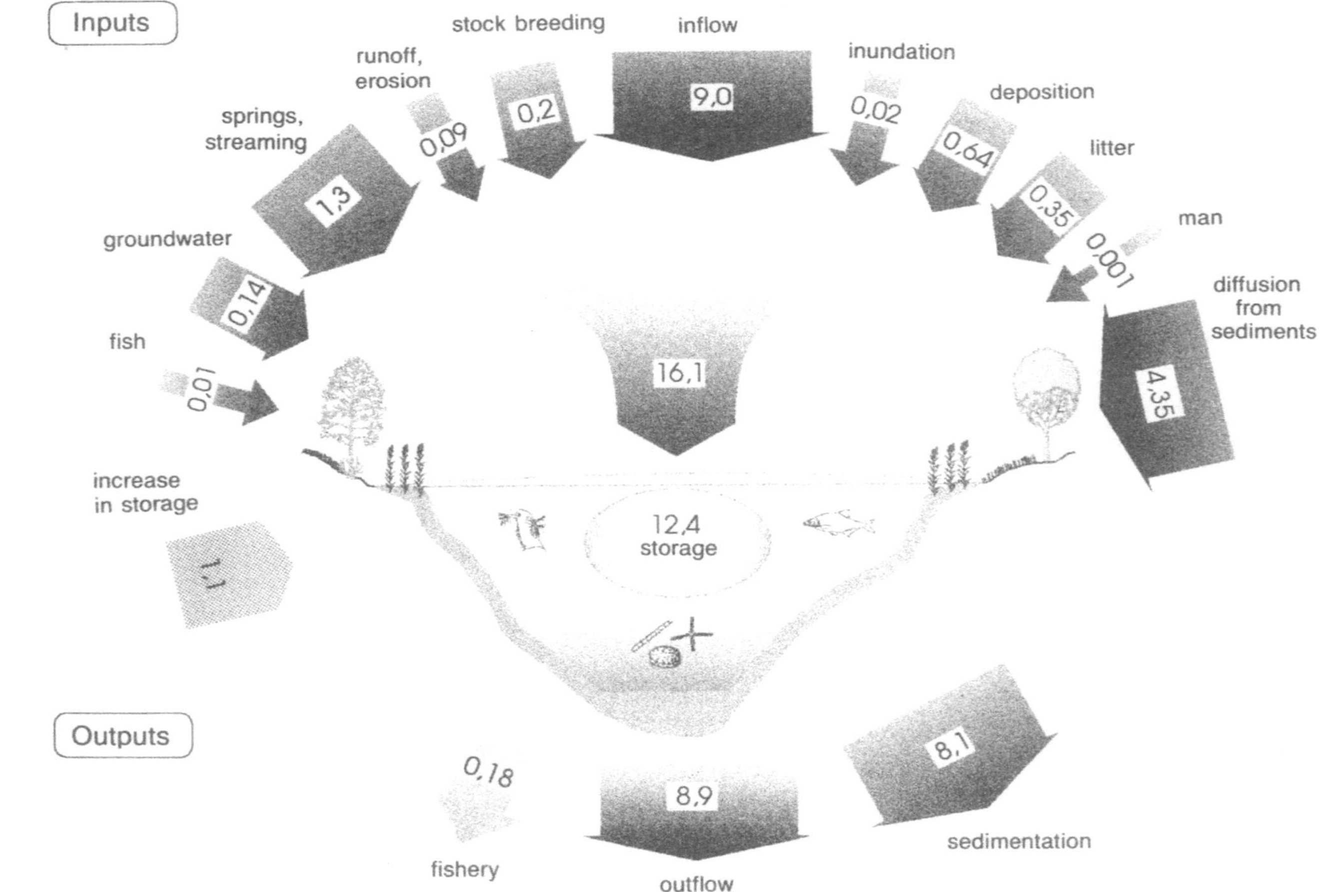


*Figure 12.* Average phosphorus balance of Lake Belau (kg/ha. $y^{-1}$) during the 1989–1992 period (from Schernewski and Niederbröker, unpublished).

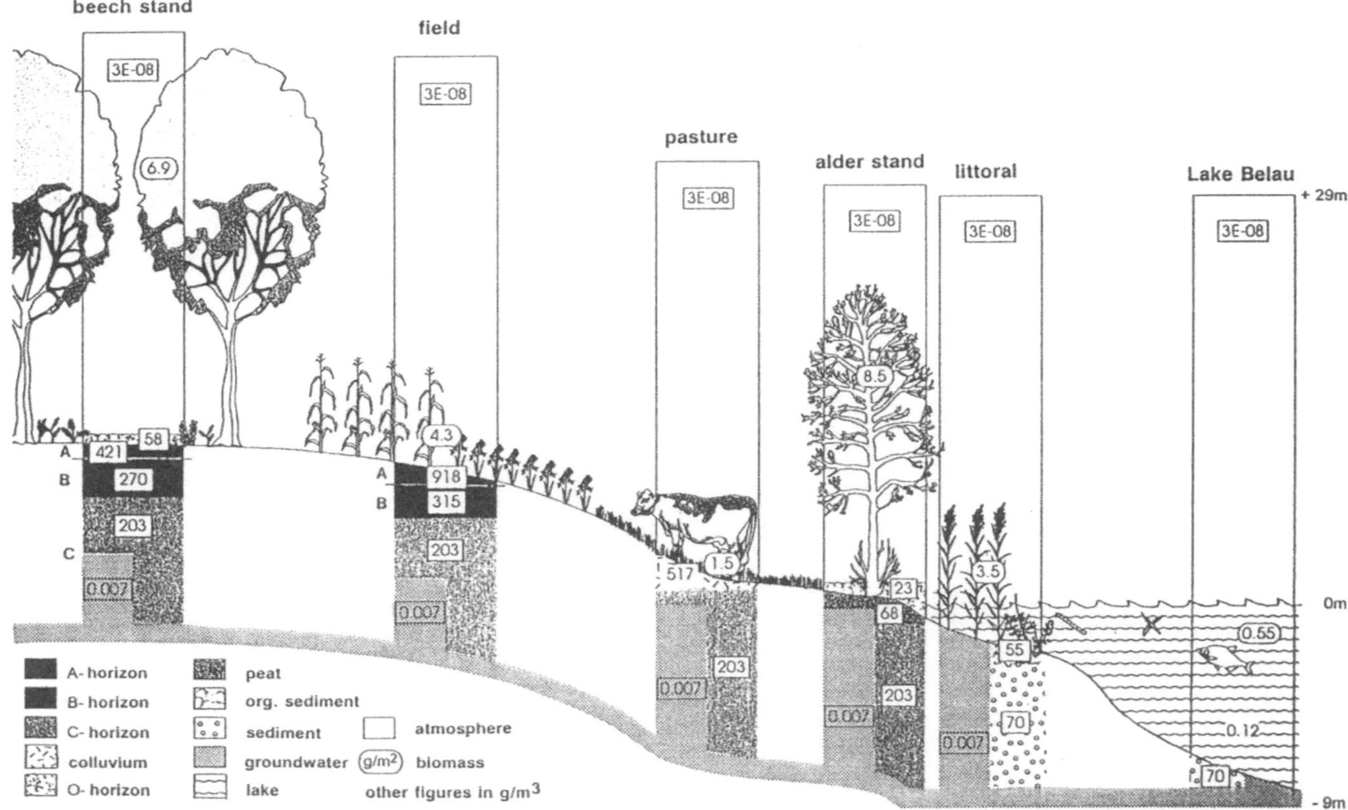


*Figure 13.* Phosphorus contents in an ecosystem catena of Lake Belau catchment (from Dinklage *et al.*, unpublished).

Fränzle O and Kluge W (in press) Typology of Water Transport and Chemical Reactions in Groundwater/ Lake Ecotones. IAHS-Proceedings.

Fränzle O, Bruhm I, Grünberg K-U *et al.* 1989 Darstellung der Vorhersagemöglichkeiten der Bodenbelastung durch Umweltchemikalien. Schriftenreihe 'Texte' des Umweltbundesamtes, Berlin.

Holling C S 1966 The strategy of building models of complex ecological systems. In Systems Analysis in Ecology. Ed. K E F Watt. pp 195–214. New York.

Holling C S 1968 The tactics of a predator. In Insect Abundance. Ed. T R E Southwood. p 47. Oxford.

Jørgensen S E 1983 Eutrophication models of lakes. In Application of Ecological Modelling in Environmental Management. Developments in Environmental Modelling 4A. Ed. S E Jørgensen. pp 227–282. Elsevier, Amsterdam.

Kluge W and Heinrich U 1994 Statistische Sicherung geoökologischer Daten. In Neuere statistische Verfahren und Modellbildungen in der Geoökologie. Eds. W Schröder, L Vetter and O Fränzle. pp 31–67. Vieweg Verlagsges, Wiedbaden.

Koschel R 1990 Pelagic calcite precipitation and trophic state of hardwater lakes. Arch. Hydrobiol. Beih. Ergebn. Limnol. 33, 713–722.

Lenz U 1992 Die Auswirkungen der Frühjahrsblüte auf den Chemismus des Belauer Sees unter besonderer Berücksichtigung der Calcitfällung und ihrer Modellierung. Diplomarbeit am Geographischen Institut der Univ. Kiel.

Matthess G 1994 Die Beschaffenheit des Grundwassers, 3. Aufl. Borntraeger, Stuttgart, Berlin.

Morel F M M and Hering J 1993 Principles and Applications of Aquatic Chemistry. Wiley and Sons, New York.

Otzuki A and Wetzel R G 1974 Calcium and total alkalinity budgets and calcium carbonate precipitation of a small hard-water lake. Arch. Hydrobiol. 731, 14–30.

Piotrowski J A and Kluge W 1994 Die Uferzone als hydrogeologische Schnittstelle zwischen Aquifer und See: Sedimentfazies und Grundwasserdynamik am Belauer See, Schleswig-Holstein. Z. dt. Geol. Ges. 145, 131–142.

Reiche E -W 1991 Entwicklung, Validierung und Anwendung eines Modellsystems zur Beschreibung und flächenhaften Bilanzierung der Wasser-und Stickstoffdynamik in Böden. Kieler Geogr. Schriften 79.

Rossknecht H 1977 Zur autochthonen Calcitfällung im Bodensee-Obersee. Arch. Hydrobiol. 81, 35–64.

Rossknecht H 1980 Phosphatelimination durch autochthone Calcitfällung im Bodensee-Obersee. Arch. Hydrobiol. 88, 328–344.

Schernewski G, Theesen L and Kerger K E 1994 Modelling thermal stratification and calcite precipitation of Lake Belau (Northern Germany). Ecol. Modell. 75/76, 421–433.

Sekine M, Nakanishi H, Ukita M and Murakami S 1991 A shallow-sea ecological model using an object-oriented programming language. Ecol. Modell. 57, 221–236.

Vanek V 1991 Groundwater flow to lakes and coastal waters – Methods and some ecological consequences. Diss., Lund Univ., Sweden.

# 21
# Application of EPIC within an integrated modelling system to evaluate soil erosion in the Canadian Prairies

R.C. IZAURRALDE, P.W. GASSMAN, A. BOUZAHER, J. TAJEK, P.G. LAKSMINARAYAN, J. DUMANSKI AND J.R. KINIRY

## 1. Introduction

There are approximately 30 million hectares of cultivated land in the three Prairie Provinces of Canada: Alberta, Saskatchewan, and Manitoba. Prairie soils are productive but subject to degradation by wind and water erosion, salinisation, and organic matter decline (Coote, 1984; PFRA, 1990). Climatic factors, soil properties, and either direct (e.g. management) or indirect (e.g. farm policy) human activities interact and dictate, in the end, the extent of degradation.

Water stress on plant growth, for example, has been a major determinant of cropping practices in the southern Prairie region. Summerfallow is practised by farmers to alleviate water stress and ensure economic return of major crops such as spring wheat (*Triticum aestivum* L.) and canola (*Brassica* spp.). Frequent summerfallow, especially that relying mostly on tillage for weed control, has been associated with reduced soil organic matter, accelerated wind erosion, and increased dryland salinity (Larney *et al.*, 1993). Water erosion during snowmelt also affects vast cultivated areas of the northern Prairie region (Chanasyk and Woytowich, 1987b; Tajek *et al.*, 1985). In an attempt to conserve soil and water, producers have increased their adoption of conservation methods, especially conservation tillage. While there are relatively few soil and climatic constraints in adopting conservation tillage (Larney *et al.*, 1993), other factors, such as economics, influence adoption decisions. Agricultural production in the Prairies is driven by, but at the same time subject to, export market vagaries. In an effort to protect farmers against inconsistencies in international markets and unfair trade policies, the Canadian government has implemented various agricultural policies. Two recent examples have been GRIP (Gross Revenue Insurance Plan) and NISA (Net Income Stabilization Account), policies designed to protect producers against marketing risks and income volatility. There has been increased recognition, however, of the need to consider the resource neutrality of agricultural policies. It is assumed that resource neutral policies do not distort production decisions, asset values, marketing decisions or environmental stability (Agriculture Canada, 1993a).

The joint evaluation of policy scenarios and status of natural resources requires the availability of consistent and integrated methodologies. In this context, an

*D. Rosen* et al. *(eds): Modern Agriculture and the Environment, 269–285.*

integrated modelling system was developed for Agriculture and Agri-Food Canada Policy Branch capable of determining changes in the status of soil resources and assessing the possible impact of farm policies on degradation (Agriculture Canada, 1993a,b; 1994a,b,c). The modelling system consists of a modified version of the Canadian Agriculture Regional Model (CRAM) (Horner *et al.*, 1992; Webber *et al.*, 1986), the Erosion Productivity Impact Calculator (EPIC) (Sharpley and Williams, 1990), an interface between the two models, and associated databases. Stages of model construction included:

(1) Development of the conceptual framework.

(2) Collection of soil, climate, agronomic, landscape, and economic databases.

(3) Database integration.

(4) Experimental design for EPIC simulation runs.

(5) Development of wind and water erosion metamodels (validated statistical response functions).

(6) Integration of CRAM and EPIC through the use of metamodels.

(7) Formulation and evaluation of policies.

This chapter presents the methodology and discusses the results from EPIC's erosion evaluation of the Canadian Prairies.

## 2. The conceptual framework

A detailed discussion of the conceptual framework used is presented in Agriculture Canada (1993a). Briefly, the integrated modelling system was built around five blocks: agricultural decision; input substitution; policy; environmental; and trade-off (Figure 1). The first three blocks depict a new, resource-sensitive CRAM (RS-CRAM) developed for the integrated modelling system. These blocks are, in turn, linked to an environmental block represented by metamodels and associated data. A metamodel is a validated, empirical, mathematical representation developed through regression analysis that captures all essential information generated by more detailed, process-based models such as EPIC. EPIC was selected because it was specifically developed to estimate the long-term impacts of erosion on productivity (Sharpley and Williams, 1990). The integrated system allows policymakers simultaneously to evaluate the economic and environmental impacts of given policies.

## 3. EPIC and its application in Canada

EPIC consists of ten modules that describe major agro-ecosystem processes such as weather, hydrology, wind and water erosion, nutrient cycling (nitrogen and phosphorus), soil temperature, crop growth, tillage, and pesticide fate as influenced

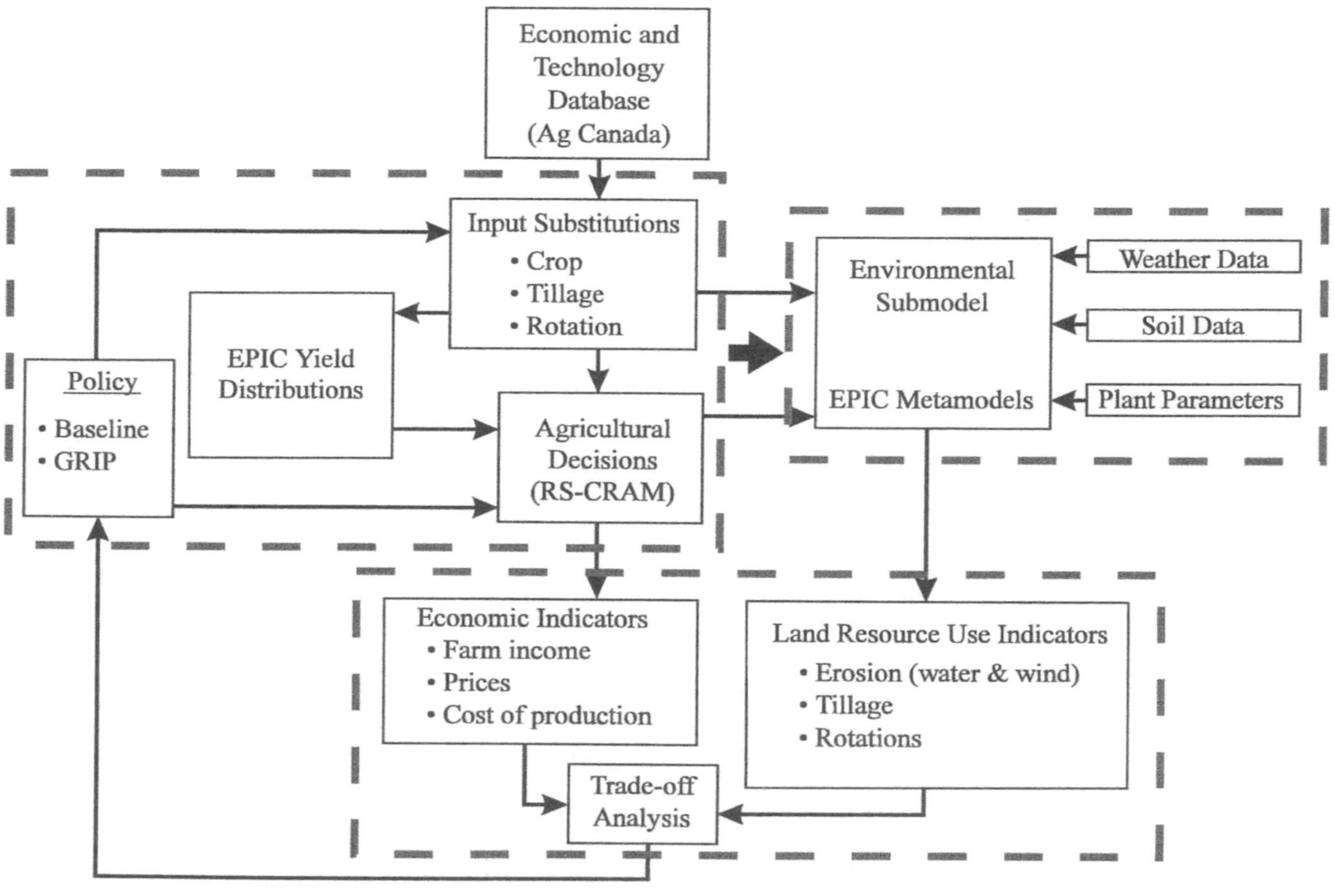


*Figure 1.* Schematic of the general conceptual framework of the integrated modelling system.

by management (e.g., irrigation, fertilization, crop sequence) (Sharpley and Williams, 1990). The model can efficiently simulate complex management over hundreds of years using a daily time step. For examples of EPIC applications see Benson *et al.* (1989), Edwards *et al.* (1994), Cabelguenne *et al.* (1990), Favis-Mortlock *et al.* (1991), and Engelke and Fabrewitz (1991).

Evaluation and application of EPIC in Canada have been more recent. Tests of the crop growth model have been reported by Izaurralde *et al.* (1993), Moulin and Beckie (1993), and Touré *et al.* (1995). Recently, Kiniry *et al.* (1995) developed EPIC model parameters for cereal, oilseed, and forage crops in the northern Great Plains region, that were also used for this study. Izaurralde *et al.* (1994) reported on the ability of EPIC to reproduce changes in soil productivity from artificial erosion. Furthermore, Izaurralde *et al.* (1993) used daily weather data and profile soil properties to simulate changes in soil water over a two-year period at two sites in Alberta. Simulated values compared favourably with observed data. More recently, Puurveen *et al.* (1997) used erosion-plot data to simulate snowmelt runoff and erosion at a site 55°N in Alberta.

Experimental data sets are essential for determining if a model performs satisfactorily under a wide range of conditions. The application of point-based models, such as EPIC, at more regional scales also requires the testing of scale transferability; that is, how reliably model results can be applied at lower resolution scales. Izaurralde *et al.* (1992) evaluated the suitability of EPIC as a tool to estimate erosion from selected landscapes in Alberta. Selected soil polygons within agro-ecological areas were characterized by soil-landscape properties and management. Yield simulations agreed reasonably well with historical records. Erosion predictions were within the range of those predicted either with other simulation models or erosion-risk maps.

## 4. Materials and methods

### *4.1 Environmental databases*

A key aspect of the work was the linking of yield and erosion model outputs with RS-CRAM. Production runs with EPIC across the three provinces required assembling a variety of databases (e.g., soil, landscape, weather, management). While EPIC has rather intense data requirements, the model was built with the premise that most input data had to be readily available; for example, from soil survey maps or weather stations. A detailed discussion of the database construction can be found in Agriculture Canada (1993b).

The two major types of databases assembled for EPIC were soil layer and landscape and weather. The soil layer data contain layered properties of selected cultivated soils down to depths of one to two metres. These data included layer depth (m), bulk density ($Mg/m^3$), sand (%), silt (%), soil pH, organic C (%), sum of bases (cmol(+)/kg), calcium carbonate (%), cation exchange capacity (cmol(+)/kg), and coarse fragments (%). Values of wilting point ($m^3/m^3$), field capacity ($m^3/m^3$), and saturated hydraulic conductivity (mm/h) were estimated internally by EPIC algorithms.

Landform and other information for dominant and subdominant soils were obtained for landscape polygons, which are landscape units characterized by unique combinations of soils, landforms, and parent materials at the 1:1 million scale. Local surface forms, slope class and length data were used to generate slope length (L) and gradient (S) factors, required by EPIC to simulate water erosion with either the Universal Soil Loss Equation (USLE) or the modified USLE (see Wischmeier and Smith (1978) and Williams (1975) for descriptions of the USLE and modified USLE, respectively). The modified USLE was used for this study.

Weather data were assembled for agroecological resource areas (ARAs), defined as natural landscape units that possess relatively uniform agro-climate, landform, soils, and general agricultural potential at the 1:2 million scale. EPIC is driven by historical or generated daily weather inputs, or some combination of both, that include precipitation (mm), maximum and minimum air temperatures (°C), solar radiation (MJ/m$^2$), relative humidity (%), and wind speed (m/s). Historical, 31-year daily records of precipitation and temperature data available for each ARA (Kirkwood *et al.*, 1993) were processed into the proper format and entered into EPIC. Daily values of solar radiation, relative humidity, and wind speed were generated internally in the model using 30-year monthly climate normals obtained for each ARA.

EPIC simulations were performed for specific dominant and subdominant soils in landscape polygons; multiple landscape polygons exist in each ARA. The environmental database was designed so that data generated at the landscape polygon level could be aggregated in a consistent manner to the ARA level or to the RS-CRAM production regions (CRAM regions), the latter of which consist of one or more *Statistics Canada* crop reporting districts. Figure 2 shows the relative size of landscape polygons to CRAM regions for the example of Alberta. An overlay of the 22 CRAM regions on top of the complete set of ARAs is shown for the Prairie Provinces in Figure 3.

### *4.2 Development of crop rotations and management datasets*

Crop distributions are estimated in RS-CRAM using ten-year average census yields for each CRAM region for up to seven crops, that are designated as following stubble or fallow. The crops simulated in RS-CRAM were hard red spring wheat, barley (*Hordeum vulgare* L.), canola, flax (*Linum usitatissimum* L.), field pea (*Pisum sativum* L.), and lentil (*Lens culinaris* L.). In EPIC, a set of 24 crop rotations compiled from land-use information and expert opinion were simulated to capture the more complex cropping systems practised by producers. Examples of these rotations are shown in Table 1. Thirty-year crop distributions estimated by EPIC were incorporated within the risk module developed for RS-CRAM, a separate component from the main RS-CRAM routine that uses the ten-year average census yields.

*Figure 2.* Overlay of RS-CRAM boundaries on landscape polygons in Alberta.

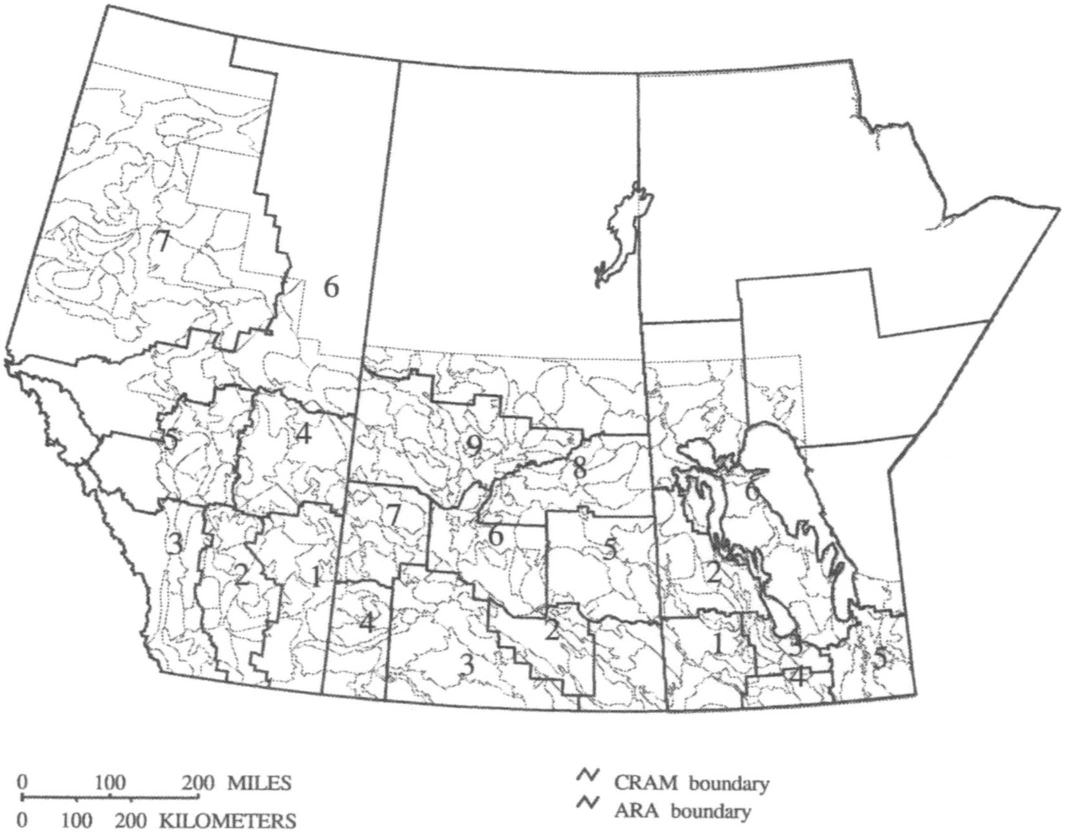

*Figure 3.* Overlay of RS-CRAM production regions (indicated by numbers) on agroecological resource area (ARA) boundaries for Alberta, Saskatchewan, and Manitoba.

*Table 1.* Example rotations included in the EPIC simulation set based on expert opinion.

| Number | Crop rotations in EPIC | Rationale for inclusion |
|---|---|---|
| 1 | Continuous barley | Continuous cereal rotation for northern regions |
| 2 | Barley–barley–hay–hay–hay[a] | Rotation for mixed farming system, short growing season |
| 3 | Canola–wheat–barley | Common oilseed–cereal rotation under continuous cropping |
| 4 | Canola–wheat–fallow | Fallow effect, in comparison with rotation 3 |
| 5 | Continuous wheat | Continuous cereal, longer growing season |
| 6 | Wheat–wheat–fallow | One-third fallow for moisture conservation |
| 7 | Wheat–fallow | Half fallow for moisture conservation |
| 8 | Wheat–wheat–lentils | Alternative crop to replace fallow, adds nitrogen, longer growing season. Lentils require more days from seeding to maturity |
| 9 | Barley–barley–field peas | Incorporates annual legume, adds nitrogen, cooler regions (compares with rotation 1) |
| 10 | Wheat–wheat–flax | Alternative crop to replace fallow |
| 11 | Wheat–wheat–fall rye[b] | Fall rye provides spring cover |
| 12 | Wheat–wheat–sunflower[b] | Rotation with sunflower for Manitoba |

[a]In EPIC, land in hay was represented by alfalfa. Under the current structure of RS-CRAM, hayland is estimated in response to livestock production. Thus, little or no shifts will occur in hayland acreages in response to policy scenarios such as GRIP, resulting in little impact in changes in erosion that can be estimated with the alfalfa metamodels.

[b]Fall rye metamodels were constructed for at least two CRAM regions in each province and sunflower metamodels were built for three CRAM regions in Manitoba. However, fall rye and sunflower are not currently simulated in RS-CRAM due to a lack of cost data. Thus, no links are currently possible between RS-CRAM and the metamodels for these two crops.

Three tillage levels were simulated for annual crops: conventional, reduced, and no tillage. Nearly all conventional tillage systems included at least one fall tillage operation in addition to spring tillage operations. The reduced systems used spring tillage operations while the only soil disturbance in no-till operations was from the planter at sowing. No attempt was made to capture any regional differences in management systems within the three provinces. To further simplify the simulations, standard dates not reflecting year-to-year weather variability were assumed for each operation.

### *4.3 Sampling design and simulation runs*

A detailed description of the sampling design, and the yield and erosion interfaces between EPIC and RS-CRAM, is given in Agriculture Canada (1994a). A schematic diagram of EPIC data inputs and simulation runs performed for the study is presented in Figure 4. EPIC-generated crop yields by soil type, landscape, and climate were aggregated to the CRAM regions for use in the risk component of RS-CRAM, while EPIC wind and water erosion output for each specific crop were interfaced with RS-CRAM via metamodels. A regional analysis with a comprehensive set of EPIC simulations covering all possible soil, climate, management, crop, tillage, and policy options would have required a prohibitive number of simulations. Consequently, a spatial sampling design was used to reduce the number of simulation runs but, at the same time, retain the statistical validity of aggregation and extrapolation into the population of soils, landscapes, weather, and management. In total, nearly 22 000 EPIC simulations were performed for the study. The wind and water erosion metamodels constructed from the EPIC simulation inputs and outputs allow multiple policy scenarios to be performed without having to re-run a new set of EPIC simulations for each policy scenario.

## 5. Results and discussion

### *5.1 Crop yields*

Overall, EPIC predicted greater crop yields, especially for canola and lentil, than the ten-year census data average yields that are used in RS-CRAM. Further, examination of the EPIC predicted crop yields suggested little variation among tillage types (data not shown). This lack of yield variation appears to be consistent with results of tillage experiments in the Canadian Prairie (Larney *et al.*, 1993).

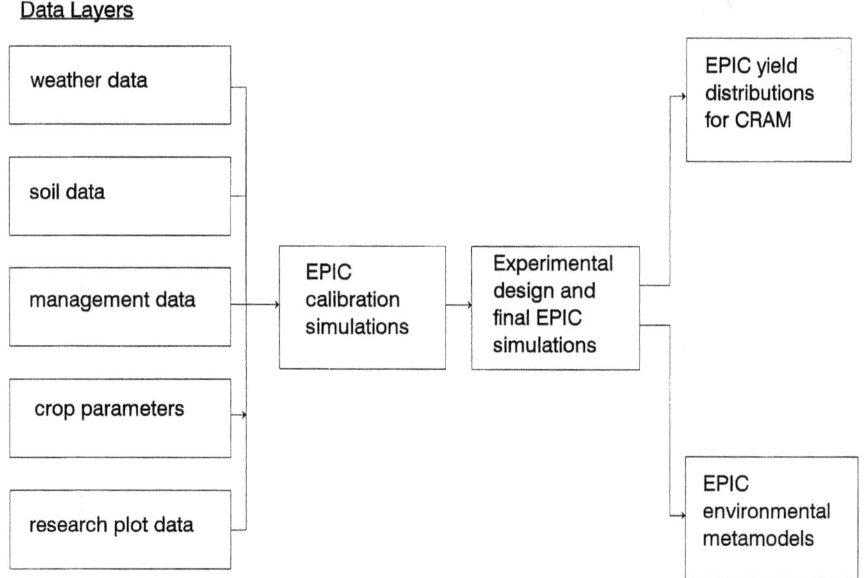


*Figure 4.* Schematic of EPIC inputs and simulation runs.

*Table 2.* Mean and standard deviations of EPIC-simulated crop yields by CRAM regions of Alberta.

| Crop | Tillage | Sequence | Statistic | CR1 | CR2 | CR3 | CR4 | CR5 | CR6 | CR7 |
|---|---|---|---|---|---|---|---|---|---|---|
| | | | | Mg/ha | | | | | | |
| Canola | Conv. | Stubble | Mean | 1.987 | 2.161 | 2.440 | 2.922 | 2.839 | 2.863 | 2.426 |
| Canola | Conv. | Stubble | St. dev. | 0.377 | 0.357 | 0.263 | 0.127 | 0.154 | 0.298 | 0.146 |
| Canola | Conv. | Fallow | Mean | 2.450 | | | 2.757 | 2.663 | 2.227 | 2.162 |
| Canola | Conv. | Fallow | Std. dev. | 0.347 | | | 0.171 | 0.235 | 0.287 | 0.235 |
| Wheat | Conv. | Stubble | Mean | 2.017 | 2.314 | 2.937 | 3.499 | 3.491 | 3.248 | 2.837 |
| Wheat | Conv. | Stubble | Std. dev. | 0.564 | 0.548 | 0.495 | 0.184 | 0.132 | 0.132 | 0.172 |
| Wheat | Conv. | Fallow | Mean | 2.576 | 2.645 | 2.793 | 3.086 | 2.981 | | 2.440 |
| Wheat | Conv. | Fallow | Std. dev. | 0.427 | 0.50 | 0.416 | 0.340 | 0.30 | | 0.261 |

*Table 3.* Mean and standard deviations of EPIC-simulated crop yields by CRAM regions of Saskatchewan.

| Crop | Tillage | Sequence | Statistic | CR1 | CR2 | CR3 | CR4 | CR5 | CR6 | CR7 | CR8 | CR9 |
|---|---|---|---|---|---|---|---|---|---|---|---|---|
| | | | | Mg/ha | | | | | | | | |
| Canola | Conv. | Stubble | Mean | 2.67 | | | | 2.59 | 2.47 | 2.45 | 2.60 | 2.69 |
| Canola | Conv. | Stubble | St. dev. | 0.15 | | | | 0.26 | 0.32 | 0.35 | 0.15 | 0.19 |
| Canola | Conv. | Fallow | Mean | 2.50 | 2.41 | | | 2.33 | 2.46 | 2.61 | 2.36 | 2.39 |
| Canola | Conv. | Fallow | Std. dev. | 0.08 | 0.32 | | | 0.34 | 0.31 | 0.19 | 0.24 | 0.28 |
| Wheat | Conv. | Stubble | Mean | 2.92 | 2.59 | 2.18 | 1.87 | 3.11 | 2.79 | 2.78 | 3.13 | 3.10 |
| Wheat | Conv. | Stubble | Std. dev. | 0.22 | 0.42 | 0.43 | 0.21 | 0.33 | 0.49 | 0.51 | 0.24 | 0.27 |
| Wheat | Conv. | Fallow | Mean | 2.57 | 2.50 | 2.53 | 2.38 | 2.42 | 2.61 | 2.78 | | 2.59 |
| Wheat | Conv. | Fallow | Std. dev. | 0.10 | 0.41 | 0.39 | 0.22 | 0.37 | 0.41 | 0.27 | | 0.37 |

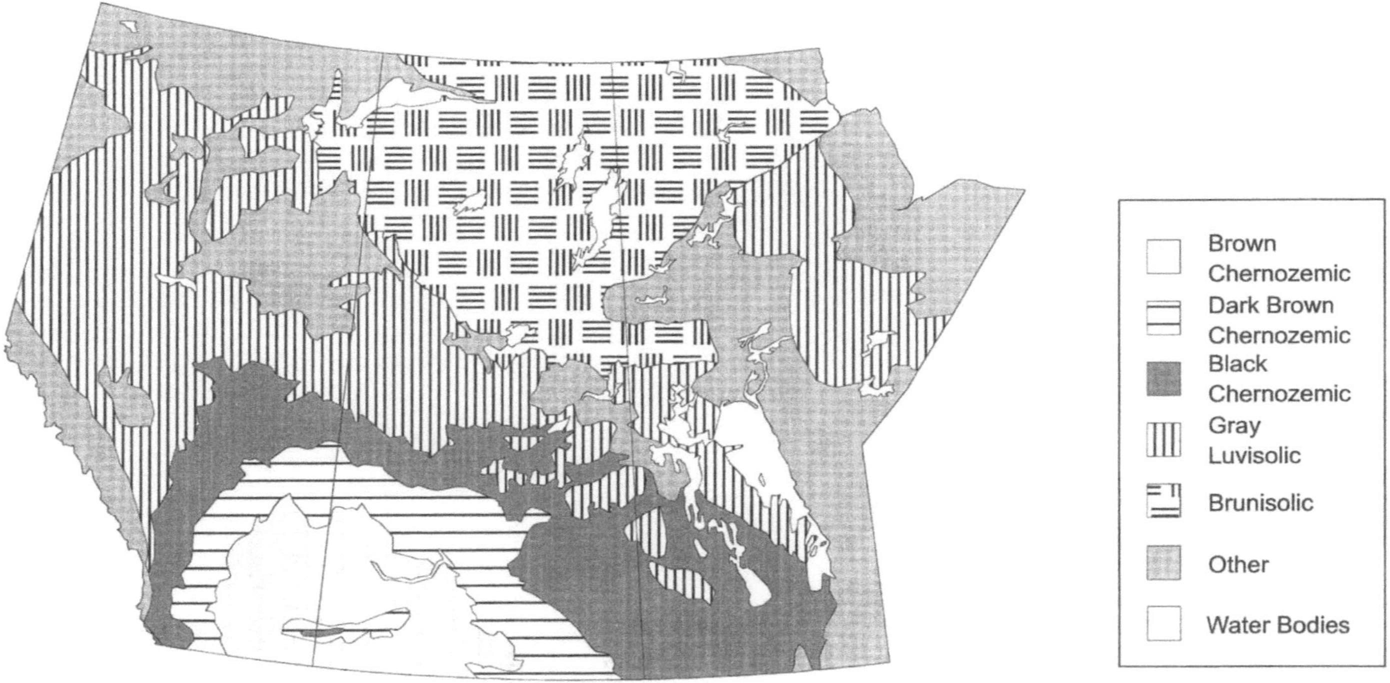

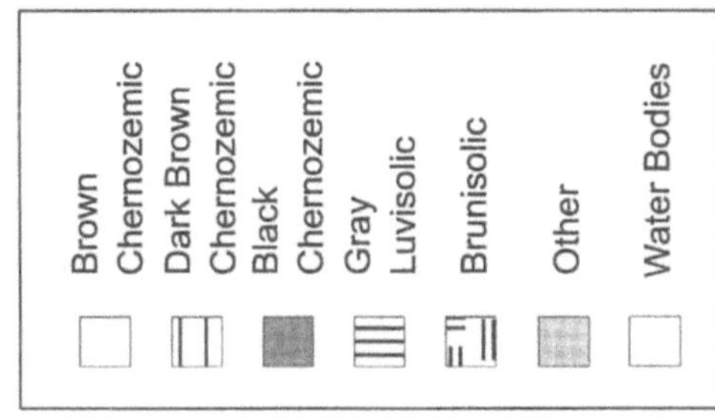


*Figure 5.* Generalised soil zone map of the Canadian Prairies.

Predicted yields and yield variabilities of wheat and canola grown on stubble and fallow land across several RS-CRAM production regions followed expected trends (Tables 2 and 3). Model results suggested that fallowing would provide a definite yield benefit in the Brown soil zone (Figure 5), as shown for Alberta CRAM regions 1 and 2 and Saskatchewan CRAM regions 3 and 4 in Tables 2 and 3. Model results also suggest less yield variability on fallow than on stubble land for the Brown soil zone, an expected result for this region.

Predicted yields on stubble were greater than on fallow for CRAM regions within the Dark Brown, Black, and Grey soil zones (Tables 2 and 3; results not shown for Manitoba). Results for the Dark Brown soil zone (Saskatchewan CRAM regions 2, 6 and 7; Alberta CRAM regions 3 and 4) were inconsistent with those documented by Campbell *et al.* (1990). However, EPIC captured the lack of benefit from fallow practices on crop yield in the more humid black and grey soil zones (although it tended to overpredict stubble yields), results that are consistent with those reported by Campbell *et al.* (1990).

### 5.2 *Wind and water erosion*

EPIC simulated wind erosion to be dominant in Saskatchewan and Manitoba for most crops and CRAM region combinations. Simulation results also suggest water erosion to be more important in Alberta, except in CRAM regions 1 and 2. For example, annual wind erosion rates predicted for fallow land in southeastern Saskatchewan (CRAM region 1) ranged from 5.6 Mg/ha under conventional tillage to 2.5 Mg/ha under no-till. Predicted rates were maximum in southwestern Saskatchewan (CRAM region 4) where the erosion range among tillage varied from 13.5 to 27.7 Mg/ha. Overall, results were consistent with expectations of severe wind erosion in the Brown soil zone under conventional fallow. Predicted water erosion rates were generally less than 10 Mg/ha in Alberta and less than 5 Mg/ha in Saskatchewan and Manitoba. Predicted water erosion was greatest under conventional fallow in Alberta CRAM region 3. Black and Dark Brown soils located on relatively steep slopes are characteristic of this Rocky Mountain foothill region. The simulations also indicate that reduced tillage intensity reduces both wind and water erosion in virtually all cropping sequences.

The scale at which simulated erosion rates are reported here and the lack of direct measurements of erosion in the Prairies prevent direct comparisons between simulated and observed values. In general, EPIC simulated less erosion than the cesium-137 ($^{137}$Cs) estimates reported by de Jong *et al.* (1982, 1983), Kiss *et al.* (1986), and Martz and de Jong (1987) for west-central Saskatchewan and by de Jong and Kachanoski (1988) for south-western Saskatchewan. For example, de Jong *et al.* (1983) calculated annual erosion rates that varied between 9 and 25 Mg/ha over a 20–25-yr period from upper-slope positions. However, erosion estimates from $^{137}$Cs measurements represent the average long-term impact of wind and water erosion and soil movement by tillage. de Jong and Kachanoski (1988) estimated an average annual erosion rate (predominantly wind) of 28 Mg/ha for benchmark cultivated

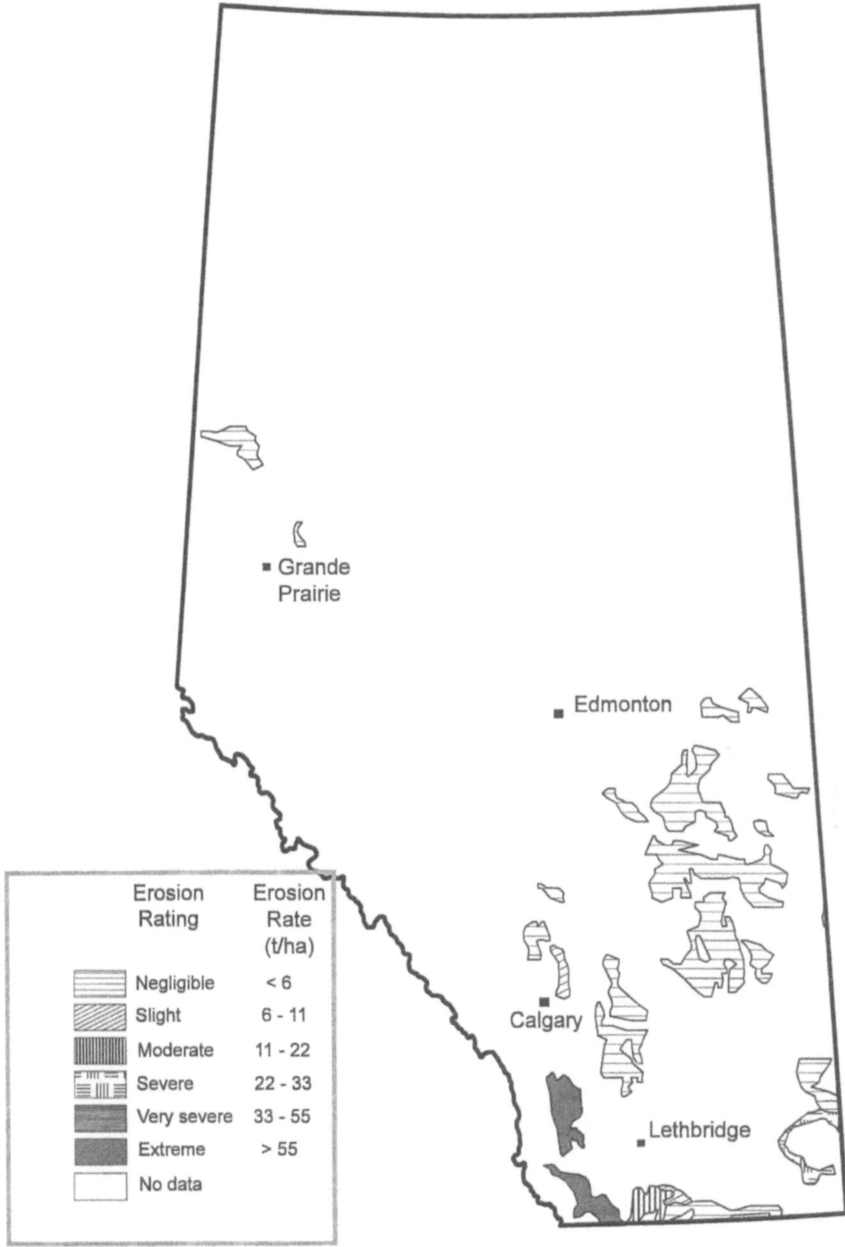


*Figure 6.* EPIC erosion rate predictions for selected landscape polygons in Alberta.

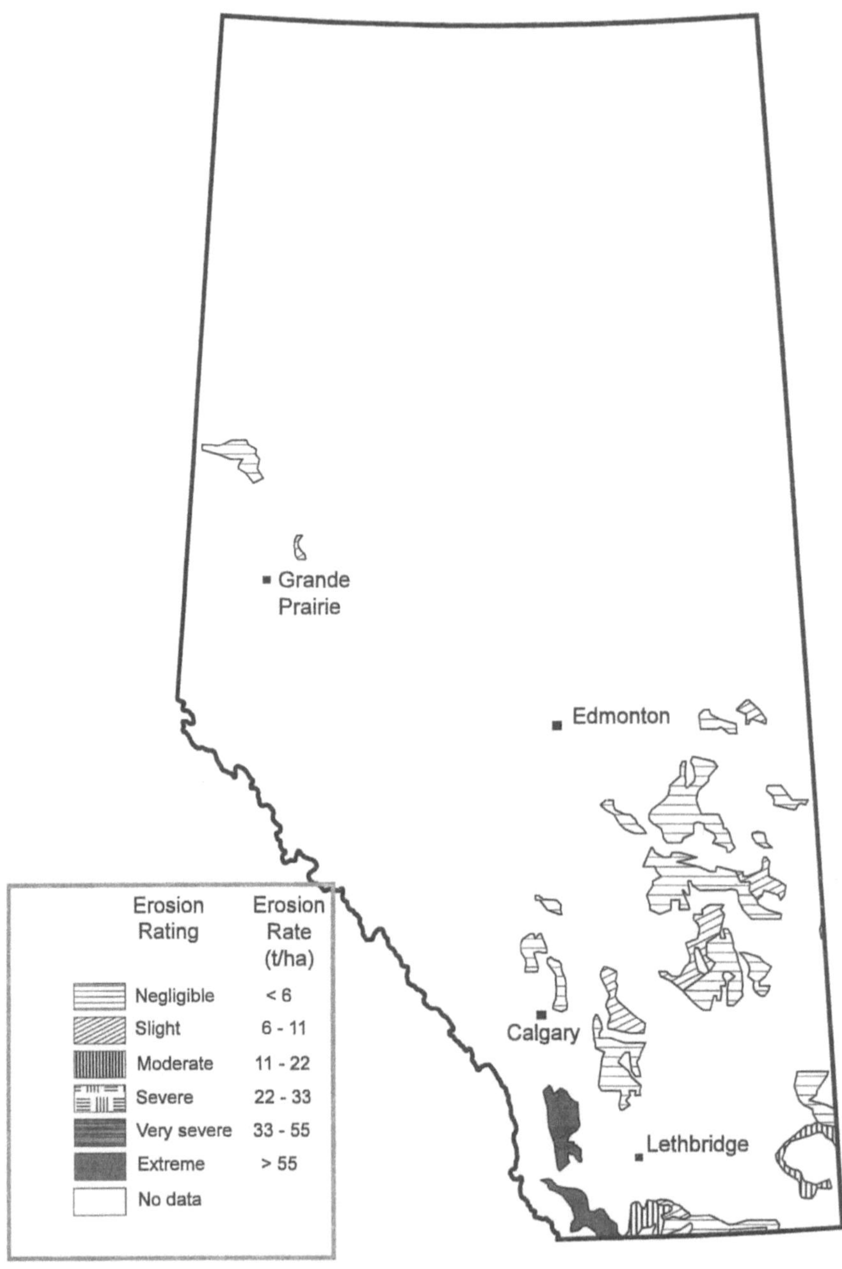


*Figure 7.* Classification of USLE potential erosion rates for selected landscape polygons in Alberta.

soils in the brown zone of Saskatchewan under crop–fallow and crop–crop–fallow rotations. These estimates and those simulated by EPIC for Saskatchewan CRAM region 4 fall within the same order of magnitude.

The highest wind erosion rates in Alberta were simulated for CRAM region 2. Within this region, the maximum annual wind erosion rate predicted by EPIC under conventional summerfallow was 165 Mg/ha. This rate was similar in magnitude to direct measurements totalling 144 Mg/ha during 16 erosion events over 13 months in southern Alberta (Larney *et al.*, 1995).

Other qualitative comparisons suggest an acceptable agreement between EPIC water erosion predictions and experimental erosion data by Toogood (1963), van Vliet and Hall (1991), Chanasyk and Woytowich (1987a) and Nicholaichuk and Read (1978). For example, average erosion rates predicted by EPIC for Alberta CRAM regions 4, 5 and 6 are consistent with Toogood's measured values that showed:

(1) the highest erosion occurring during fallow;

(2) greater erosion on wheat following fallow than following stubble; and

(3) barley and hay being more effective than wheat at controlling erosion.

A final comparison was made on how well EPIC was able to predict erosion rates on a geographic basis. EPIC water erosion rates estimated with the modified USLE for a subset of Alberta landscape polygons were compared with those estimated by Tajek *et al.* (1985) using the USLE (Figures 6 and 7). For most of these polygons, the mapped erosion rates corresponded with those under a wheat–fallow rotation. Erosion rates under continuous barley were used for landscape polygons lying farthest to the north. On a class basis, EPIC rates (Figure 6) agreed well with USLE estimates (Figure 7). For some polygons, the water erosion rates simulated by EPIC were lower in magnitude than those predicted with the USLE. This comparison shows the utility of estimating erosion rates for specific landscapes, crop rotations, tillage levels, and other factors.

## 6. Summary and conclusions

Results suggest that outputs of a process-based model (EPIC) can be efficiently linked to an economic model (CRAM) via regression metamodels. The integrated system assembled allows for repeated execution of the integrated system without having to resort to running EPIC each time a policy is evaluated. This version is a first phase in the development process and is subject to improvement.

Estimates of crop yield and erosion rates were found to be realistic and to conform within bounds of expectations and experimental results. The authors recognise, however, the need to continue the testing and calibration of EPIC to Canadian conditions to improve further the robustness of environmental indicators such as yield and erosion.

## Acknowledgments

Journal Paper No. J-16237 of the Iowa Agriculture and Home Economics Experiment Station, Ames, Iowa, Project No. 2872. Financial support for this study was provided by the Policy Branch of Agriculture and Agri-Food Canada, Ottawa. The authors are indebted to many individuals across Canada and the USA for providing essential expert advice and information. In particular, we thank: J.R. Williams, M. Siemers, V. Kirkwood, R.J. MacGregor, Z. Hassan, R. de Jong, G. Coen, G. Padbury, R. Eilers, G. Coy, A. Johnston, T.L. Jensen, R.H. McKenzie, P. Entz, K. Greer, R. Zentner, and F.J. Larney.

## References

Agriculture Canada 1993a Agricultural policies and soil degradation in western Canada: An agro-ecological economic assessment (Report 1: Conceptual framework). Technical Report 2/93. Agriculture Canada, Policy Branch, Ottawa. 52 pp.

Agriculture Canada 1993b Agricultural policies and soil degradation in western Canada: An agro-ecological economic assessment (Report 2: The Environmental Modelling System). Technical Report 5/93. Agriculture Canada, Policy Branch, Ottawa. 110 pp.

Agriculture Canada 1994a Agricultural policies and soil degradation in western Canada: An agro-ecological economic assessment (Report 3: The Integration of the Environmental and Economic Components). Agriculture Canada, Policy Branch, Ottawa. 130 pp.

Agriculture Canada 1994b Agricultural policies and soil degradation in western Canada: An agro-ecological economic assessment (Report 4: Modifications to CRAM and Policy Evaluation Results). Agriculture Canada, Policy Branch, Ottawa. 131 pp.

Agriculture Canada 1994c Agricultural policies and soil degradation in western Canada: An agro-ecological economic assessment (Report 5: Project Summary). Agriculture Canada, Policy Branch, Ottawa. 31 pp.

Benson V W, Rice O W, Dyke P T, Williams J R and Jones C A 1989 Conservation impacts on crop productivity for the life of a soil. J. Soil Water Conserv. 44, 600–604.

Cabelguenne M, Jones C A, Marty J R, Dyke P T and Williams J R 1990 Calibration and validation of EPIC for crop rotations in southern France. Agric. Syst. 33, 153–171.

Campbell C A, Zentner R P, Janzen H H and Bowren K E 1990 Crop rotation studies on the Canadian Prairies. Publ. 1841/E. Res. Branch, Agric. Canada, Ottawa, ON.

Chanasyk D S and Woytowich C P 1987a Sediment yield as a result of snowmelt runoff in the Peace River Region. Can. J. Agr. Eng. 29, 1–6.

Chanasyk D S and Woytowich C P 1987b A study of water erosion in the Peace River Region. Farming for the Future Proj. No. 83–0145. Dep. Soil Sci., Univ. of Alberta. Edmonton, AB.

Coote D R 1984 The extent of soil erosion in western Canada. In Soil Erosion and Land Degradation, Proc. 2nd Annual Western Provincial Conf. Rationalization of Water and Soil Research and Management. pp 34–38. Inst. of Pedology, Saskatoon, Saskatchewan.

de Jong E and Kachanoski R G 1988 The importance of erosion in the carbon balance of Prairie soils. Can. J. Soil Sci. 68, 111–119.

de Jong E, Villar H and Bettany J R 1982 Preliminary investigations on the use of 137Cs to estimate erosion in Saskatchewan. Can. J. Soil Sci. 62, 673–683.

de Jong E, Begg C B M and Kachanosky R G 1983 Estimates of erosion and deposition for some Saskatchewan soils. Can J. Soil Sci. 63, 607–617.

Edwards D R, Benson V W, Williams J R, Daniel T C, Lemunyon J and Gilbert R G 1994 Use of the EPIC model to predict fertiliser runoff transport of surface-applied inorganic fertiliser and poultry manure constituents. Trans. ASAE 37, 403–409.

Engelke R and Fabrewitz S 1991 Simulation runs with the EPIC model for different data sets. In Soil and Groundwater Research Report II – Nitrate in Soils. Commission of the European Communities Report EUR 13501, Office for Official Publications of the European Communities, Luxembourg.

Favis-Mortlock D T, Evans R, Boardman J and Harris T M 1991 Climate change, winter wheat yield and soil erosion on the English South Downs. Agric. Syst. 37, 415–433.

Horner G L, Corman J, Howitt R E, Carter C A and MacGregor 1992 The Canadian Regional Agriculture Model Structure, Operation and Development. Agriculture Canada, Ottawa.

Izaurralde R C, Tajek J, Larney F J and Dzikowski P 1992 Evaluation of the suitability of the EPIC model as a tool to estimate erosion from selected landscapes in Alberta. Univ. of Alberta, Agriculture Canada, Alberta Agriculture, Edmonton, Alberta. 65 pp.

Izaurralde R C, Chanasyk D S, Grant R G, Quiroga Jakas M C and Puurveen H 1993 Simulation of changes in soil productivity. Final report to National Soil Conservation Program. Univ. of Alberta, Edmonton, Alberta. 57 pp.

Izaurralde R C, Nyborg M, Solberg E D *et al.* 1994 Relationships of simulated erosion and soil amendments to soil productivity. In A Program to Assess and Monitor Soil Quality in Canada. Ed. D F Acton. pp 151–155. Centre for Land and Biological Resources Research. Research Branch, Agriculture and Agri-Food Canada, Ottawa.

Kiniry J R, Major D J, Izaurralde R C *et al.* 1995 EPIC model parameters for cereal, oilseed, and forage crops in the northern Great Plains region. Can. J. Plant Sci. 75, 679–688.

Kirkwood V, Bootsma A, de Jong R *et al.* 1993 Agroecological resource area databases for the Prairies: User's manual. Technical Bulletin 1993-13E. Centre for Land and Biological Resources Research, Research Branch, Agriculture Canada.

Kiss J J, de Jong E and Rostad H P W 1986 An assessment of erosion in west-central Saskatchewan using cesium-137. Can J. Soil Sci. 66, 591–600.

Larney F J, Lindwall C W, Izaurralde R C and Moulin A P 1993 Tillage systems for soil and water conservation in western Canada. In Conservation tillage in temperate regions; development and adaptation to soil, climatic and biological constraints. Ed. M R Carter. pp 305–328. Lewis Publishers, CRC Press, Boca Raton, FLA.

Larney F J, Bullock M S, McGinn S M and Fryrear D W 1995 Quantifying wind erosion on summer fallow in southern Alberta. J. Soil Water Cons. 50, 91–95.

Martz L W and de Jong E 1987 Using cesium-137 to assess the variability of net soil erosion and its association with topography in a Canadian Prairie landscape. Catena 14, 439–451.

Moulin A P and H J Beckie 1993 Evaluation of the CERES and EPIC models for predicting spring wheat grain yield over time. Can. J. Plant Sci. 73, 713–719.

Nicholaichuk W and Read D W L 1978 Nutrient runoff from fertilised and unfertilised fields in western Canada. J. Environ. Qual. 7, 542–544.

PFRA (Prairie Farm Rehabilitation Administration) 1990 Prairie Soils: The Case for Conservation. PFRA Soil and Water Conservation Service, Regina, Saskatchewan.

Puurveen H P, Izaurralde R C, Grant, R F, Chanasyk, D S and Williams J R 1995 Evaluation of EPIC's snowmelt and water erosion submodels using data from the Peace River Region of Alberta. Can. J. Soil Sci. 77, *(in press).*

Sharpley A N and Williams J R (Eds) 1990 EPIC – Erosion/Productivity Impact Calculator: 1. Model Documentation. USDA Techn. Bull. No. 1768, 235 pp.

Tajek J, Pettapiece W W and Toogood K E 1985 Water erosion potential of soils in Alberta: Estimates using modified USLE. Agric. Can. Tech. Bull. No. 1985-29. Ottawa, Canada.

Toogood J A 1963 Water erosion in Alberta. J. Soil Water Cons. Nov–Dec, 238–240.

Touré A, Major D J, and Lindwall C W 1994 Comparison of five wheat simulation models in southern Alberta. Can. J. Plant Sci. 75, 61–68.

van Vliet L J P and Hall J W 1991 Effects of two crop rotations on seasonal runoff and soil loss in the Peace River Region. Can. J. Soil. Sci. 71, 533–544.

Webber R J, Graham J D and Klein K K 1986 The structure of CRAM: A Canadian regional agricultural model. Department of Agricultural Economics, University of British Columbia, Vancouver.

Williams J R 1975 Sediment yield prediction with universal equation using runoff energy factor. In Present and Prospective Technology for Predicting Sediment Yields and Sources. pp 244–252. USDA-ARS Ser. 40. USDA-ARS, New Orleans, LA.

Wischmeier W H and Smith D D 1978 Predicting rainfall erosion losses. US Department of Agriculture, Agriculture Handbook 537.

22

# A model to predict release rate of nitrogen from polymer-coated urea

MINGCHU ZHANG, M. NYBORG AND R.C. IZAURRALDE

## 1. Introduction

Simulation of controlled-release bioactive agents has been developed through a variety of approaches. With controlled-release pharmaceuticals, Higuchi (1963) made a pseudo-steady state assumption in predicting release rate of a drug across a membrane, i.e. the rate of mass transfer through the membrane remains constant while the rate of disappearance of the agent in the core zone is in a transient state. This assumption was widely used in simulating drug release rate (Paul and McSpadden, 1976). Fan *et al.* (1979), Tojo and Fan (1981), and Tojo and Myanami (1983) used a different approach to derive a model in which the release rate of an encapsulated agent could be changed as a function of temperature, decay rate, and dispersion of a coated agent in the core zone. Nitrogen release rate of sulphur-coated urea was predicted at two temperatures by this model. Lu and Lee (1992) developed a model based on constant core concentration–variable extraction medium concentration and variable core concentration–variable extraction medium concentration. They used this model to compare analytical data with the experimental results obtained with latex-coated urea. The diffusivity used in their model was calculated from the experimental data determined in continuously stirred water, and the simulation results were compared with the cumulative urea release from the experiment.

The release rate of coated urea in soil was affected by coating thickness and temperature, but not by soil water content and soil texture (Christianson, 1988; Kochba *et al.*, 1990; Oertli and Lunt, 1962). The release rate of an encapsulated agent in human body is slower than in the stirred water because of the existence of a boundary layer on the coating surface (Baker and Lansdale, 1974). The effect this layer has on the release rate of coated urea in soil has not however been reported. To assess the boundary layer effect on urea flux from coated urea in soil, we introduced a new parameter, $S_b$, the Soil Retardation Factor, which is the ratio of the urea flux in soil and maximum urea flux in flowing water. The objectives of this research were to use $S_b$ to simulate the urea release for polymer-coated urea in soil by determining parameters such as maximum urea flux ($J_{\max}$), urea flux to soil ($J_t$), the permeability of the coating and coating thickness.

*D. Rosen* et al. *(eds): Modern Agriculture and the Environment, 287–296.*

## 2. Analysis

When a coated urea granule is placed in soil, water vapour penetrates the coating and dissolves urea. Because of the dissolution of urea, a concentration gradient is established across the coating, causing diffusion of urea. As the urea solution moves outward, a boundary layer is formed around the outer surface of the coating. The unknown thickness of the boundary layer is indicated by $X$ (Figure 1). Its resistance to urea diffusion is represented by $S_b$. Since urea hydrolysis is a rapid process, it is assumed that urea diffused out of the layer is completely converted to $NH_4^+$, and so the concentration of urea out of the boundary layer is zero, $C = 0$ (Figure 1). The diameters of a polymer-coated urea ball and a urea granule are $a$ and $b$, respectively. The thickness of coating is $L$, i.e. $a-b$. Urea is highly soluble in water; it is assumed that urea concentration in the core zone is in saturation ($C_s$) as long as the solid urea exists. The time required for the depleting solid urea is defined as $t_{\text{sat}}$. When $t > t_{\text{sat}}$, urea concentration in the core zone varies with time. In this research, two release stages were simulated, a linear release stage when solid urea exists in the core zone, and a declining release stage when solid urea is depleted in the core zone.

The governing equation for simulation is:

$$J_t = S_b\, J_{\max} \tag{1}$$

where $J_t$ is urea flux in soil, mg m$^{-2}$ s$^{-1}$; $J_{\max}$ is urea flux in flowing water, mg m$^{-2}$ s$^{-1}$, which is the maximum flux of urea across the coating; $S_b$ is the Soil Retardation Factor, dimensionless, it is the ratio of $J_t/J_{\max}$. Considering the cumulative release of urea, Equation 1 can be written as:

$$M_t = t\, J_{\max}\, S_b$$

where $M_t$ is the total amount of urea released in time $t$. $J_{\max}$ is calculated by:

$$J_{\max} = -PC_s/L.$$

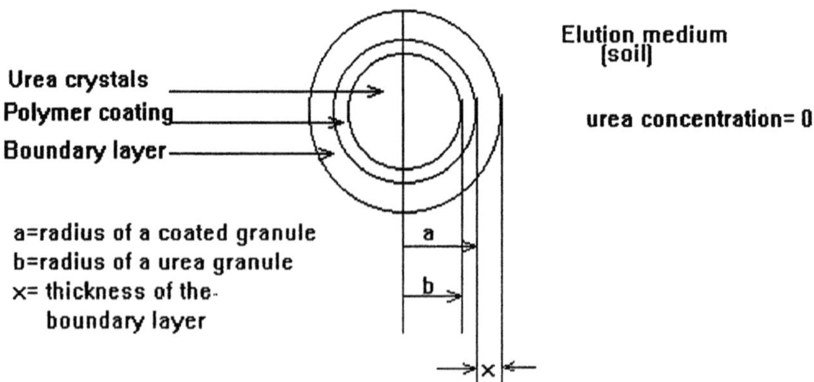


*Figure 1.* The schematic of polymer-coated urea applied in soil with coating thickness of a–b and boundary layer thickness, x.

$P$ is the permeability of coating, $m^2\ s^{-1}$, and it is affected by temperature; $C_s$ is urea concentration at saturation, mg $m^{-3}$; $L$ is the thickness of the coating, m; urea concentration before and after solid urea is depleted in the core zone is:

$$t < t_{sat} \qquad C = C_s$$

$$t > t_{sat} \qquad C = C_i.$$

## 3. Materials and methods

Two batches of a polymer-coated urea with thin and thick coating from Imperial Oil Chemicals Division (Edmonton, Canada) were sieved to a 2- to 3-mm diameter in size. They were used in the simulation research.

### *3.1 Thickness of coating*

The granules were sliced in half and attached to aluminium stubs and examined with the scanning electronic microscope (SEM) (Cambridge S250). The thickness of coating was measured based on the scale in the SEM photographs.

### *3.2 Maximum flux and permeability of coating*

A device was constructed to determine the maximum flux of urea across the coating. The details of the device were reported by Zhang *et al.* (1994). The experiment used urea granules with either of two thicknesses of coating and two water temperatures (12 and 31°C). The coated urea was exposed to flowing water at 12°C and 31°C so that there was no urea accumulation on the outer surface of the membrane. Therefore, a maximum flux was obtained. The permeability ($P$) at each temperature was calculated by:

$$J_{max} = \frac{-P(C_2 - C_1)}{L}$$

where $J_{max}$ is the flux (mg $cm^{-2}\ h^{-1}$), and $C_1$ and $C_2$ are urea concentration at the inner and outer surface of the coating, respectively. In the experiment, $C_1$ is the urea concentration in saturation at a given temperature, and $C_2$ is zero. $L$ is the thickness of the coating. The activation energy of the permeability was calculated by an Arrhenius-type equation (Rogers, 1976), which can be used to calculate any permeability within a given temperature range. The maximum flux is thereafter calculated. Based on the $P$, maximum flux of urea was calculated.

### *3.3 Urea flux to soil ($J_t$)*

Urea flux to soil was determined by incubating thin-coated or thick-coated urea in soil. The incubation was conducted at two temperatures (10 and 23°C) and at the soil water potential −0.1 MPa. Plastic containers (115-mm diameter and 77-mm height) with perforated lids received 400 g soil and 80 mg of fertilizer N (200 mg N $kg^{-1}$ soil) placed 25 mm deep. The number of granules varied from 13 to 14 to supply 80 mg N. Soil in containers at 15, 30 and 60 days was air-dried, sieved (2 mm), and any intact granules were removed. $NH_4$-N and $NO_3$-N in soil samples were extracted by 2 M KCl solution in a 1:5 soil:solution ratio, and their concentrations were determined with a Technicon AutoAnalyzer (Technicon AutoAnalyzer II, 1973a,b). The mineral N ($NH_4$-N and $NO_3$-N) determined at each sampling time was then converted to urea flux (mg urea $cm^{-2}$ $h^{-1}$).

### 3.4 *Experiments for testing the model*

Another incubation experiment was conducted in a different soil to test the simulated results in soil with constant temperatures (10 and 23°C). The nitrogen application rate and experimental conditions were the same as that used for deriving $J_t$.

Field experiments were conducted from autumn 1989 to spring 1990, in three sites in Alberta. Two coated ureas were incorporated into soil at a rate of 127 kg N $ha^{-1}$ on 15 October 1989, 24 October 1989 and 10 October 1989 at Sites 1, 2 and 3, respectively. There were four replicates for each treatment. The experiment was in a complete randomised block design. Soil samples were taken using a 2-cm diameter coring tube when the soil was not frozen. Ten cores were taken from each plot and composited. When the soil was frozen, two blocks of soil (46 × 23 cm area and 15 cm deep) were taken from each plot using heavy chisels. Soil samples were thinly spread and dried at room temperature. Soil was crushed to pass a 2-mm sieve, and 20-g samples were extracted in 100 ml 2M KCl solution. The concentrations of $NH_4$-N and $NO_3$-N in the extract were determined with a Technicon AutoAnalyzer (Technicon AutoAnalyzer II, 1973a,b). Soil bulk density was determined at each site. Daily average soil temperature at the 5-cm depth was recorded at the Ellerslie Meteorological Station (Atmospheric Environment Services); that station was 200 m from Site 3.

## 4. Methods of computation

(1) Saturated urea concentration is varied with temperature and calculated from the regression equation, $y = 65.08 \times 10^{0.101x}$ ($r^2 = 0.996$), based on the data reported by Pratt and Noyes (1965), where $y$ is the solubility of urea (urea g/100 g $H_2O$) and $x$ is temperature. The data were limited to temperatures above 0°C; any urea concentration below zero was an extrapolation from the above equation. When $t > t_{sat}$, urea concentration was calculated by:

$$\frac{W_0 - W_i}{V}$$

where $W_0$ was the initial mass of a granule, $W_i$ was the amount of mass released at time $i$, and $V$ was the volume of the core zone. Since the coating was very thin, the volume of a coated granule was used as that of the core zone.

*Table 1.* Urea flux in flowing water and soil.

| Coated fertilisers | Temperature (°C) | Urea flux ($mg\ m^{-2}\ h^{-1}$) Water | Soil 15[a] | 30[a] | 60[a] |
|---|---|---|---|---|---|
| Thin coating | 10 | 0.15 | 0.024 | 0.018 | 0.011 |
| | 23 | 0.31 | 0.034 | 0.020 | 0.012 |
| Thick coating | 10 | 0.08 | 0.010 | 0.010 | 0.009 |
| | 23 | 0.19 | 0.021 | 0.016 | 0.011 |

[a] Number of days after the coated urea was applied.

(2) The average diameter of the coated urea was 2.5 mm; the surface area and volume were 19.6 $mm^2$ and 8.18 $mm^3$, respectively. Mass per 100 granules for thin or thick coatings was 1.31 and 1.34 g, respectively.

(3) Thickness of coating ($L$) was $8.7 \times 10^{-6}$ m and $14.6 \times 10^{-6}$ m for thin and thick coating, respectively.

(4) Time step for simulation was 24 hours.

(5) Permeability at 31°C was $3.5 \pm 1.5 \times 10^{-14}\ m^2\ s^{-1}$. The details of calculation of permeability from the flowing water device were reported by Zhang *et al.* (1994). Permeability at a given temperature was calculated from:

$$\frac{Ln\ P_2}{Ln\ P_1} = \frac{E_a}{R}\left(\frac{T_2 - T_1}{T_1 T_2}\right)$$

where $P_2$ and $P_1$ are permeabilities at temperature $T_1$ and $T_2$, respectively, $E_a$ is activation energy, 350 kJ $mol^{-1}$, and $R$ is the universal gas constant, 8.314 J $mol^{-1}\ K^{-1}$.

(6) When $t$ was less than $t_{sat}$, urea concentration was $C_s$, and $J_{max}$ was constant. When $t$ was greater than $t_{sat}$, $J_{max}$ was a function of urea concentration. Urea concentration was calculated based on the amount of urea left divided by the volume of the granule after each time step of simulation.

(7) The urea flux in soil, $J_t$, is summarised in Table 1.

(8) Soil retardation factor, $S_b$, was calculated by Equation 1 (Table 2). The $S_b$ varied at each sampling time. After the first 15 days, the value was greater than that found after 30 and 60 days. This may be caused by a quick release of urea from the coating due to the storage or existence of defects (Zhang *et al.*, 1994). The lower $S_b$ after 60 days in comparison with 15 and 30 days was due to depletion of the urea in the core zone. The $S_b$ used in the simulation was 0.1 from the 30-day sample.

*Table 2.* Calculation of Soil Retardation Factor.

| Coated fertilisers | Temperature (°C) | Soil Retardation Factor ($S_b$) | | | |
|---|---|---|---|---|---|
| | | 15[a] | 30[a] | 60[a] | Average |
| Thin coating | 10 | 0.16 | 0.12 | 0.07 | 0.12 |
| | 23 | 0.11 | 0.06 | 0.04 | 0.07 |
| Thick coating | 10 | 0.13 | 0.13 | 0.11 | 0.12 |
| | 23 | 0.11 | 0.08 | 0.06 | 0.08 |
| Average | | 0.13 | 0.10 | 0.07 | |

[a] Number of days after the coated urea was applied.

(9) Simulation was made at two temperature conditions, constant temperature (10 and 23°C) and variable temperature. In the condition of variable temperature, urea flux changed with temperature because polymer permeability and urea solubility are affected by temperature.

(10) All computations were made using Microsoft Excel for Windows, V3.0a.

## 5. Results and discussion

The cumulative urea release over time showed two stages, linear release as $t < t_{sat}$, resulting from constant flux, and a curvature release when $t > t_{sat}$ (Figure 2). The increment of cumulative urea release was proportional to time. After the time reached $t_{sat}$, linear release changed to a curvature release. In this stage, the cumulative release slowed and gradually levelled. This decrease in urea release at each time interval was caused by a declining urea concentration in the core zone. Lu and Lee (1992) described a similar release pattern for latex-coated urea.

The number of days required to reach the $t_{sat}$ varied with temperature and coating thickness (Table 3). For the same temperature, the thick coating had a greater $t_{sat}$ than the thin one. Comparing the same coating thickness, low temperature (10°C) resulted in a longer $t_{sat}$ as compared with high temperature (23°C). Urea release rate was a function of temperature and coating thickness. One coated urea granule contained a limited amount of urea. For the limited resource, urea release rate, in fact, dictated $t_{sat}$. The higher the release rate, the shorter the $t_{sat}$. The higher temperature and thinner coating resulted in a faster release rate, therefore, a shorter $t_{sat}$. Even with the same temperature and thickness of coating, as the elution medium changed, the release rate was proportionally only 10% in soil to that in the flowing water. As a result, the $t_{sat}$ in soil was ten times higher than that in the flowing water (Table 3).

When temperature was variable, urea flux changed with temperature (Figure 3). Before solid urea disappeared in the core zone, the flux changed drastically in accordance with temperature since both urea solubility and polymer permeability

*Table 3.* Comparison of $t_{sat}$ in water and soil.

| | $t_{sat}$ (h) | | | |
|---|---|---|---|---|
| | Soil | | Water | |
| Temperature (°C) | Thin coating | Thick coating | Thin coating | Thick coating |
| 10 | 558 | 1080 | 55 | 108 |
| 23 | 180 | 325 | 18 | 32 |

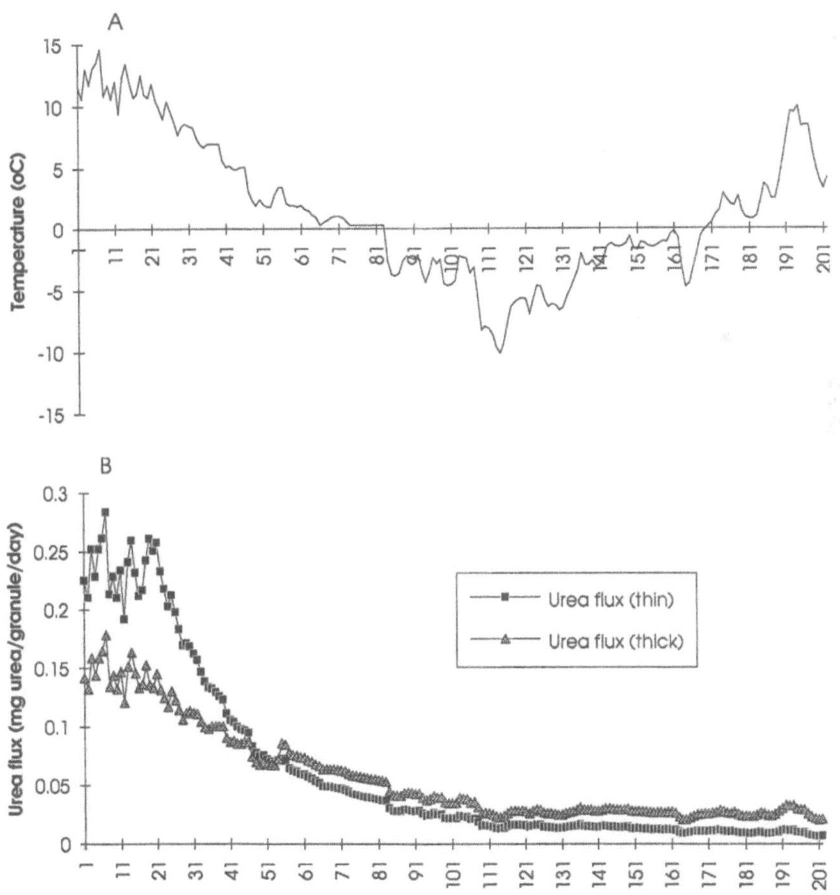


*Figure 2.* Predicted urea release at a constant temperature, A: 10°C; B: 23°C.

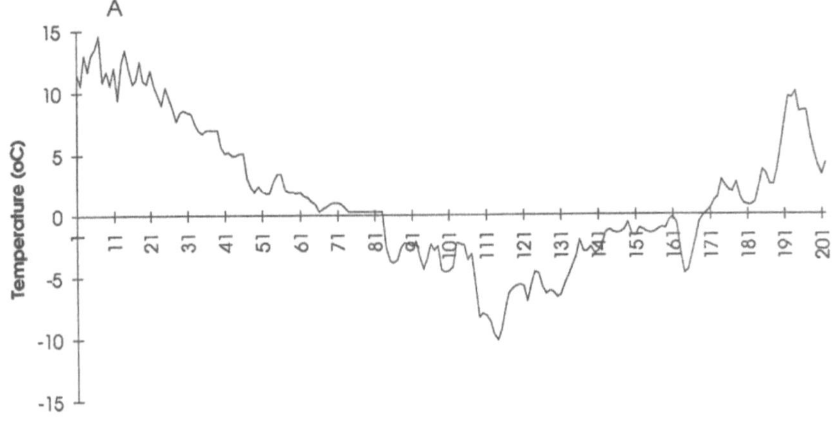


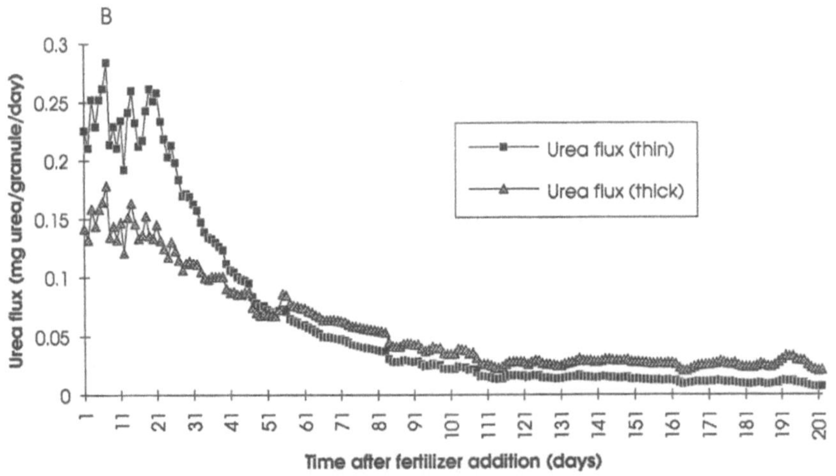


*Figure 3.* Urea flux over time with changing soil temperature, A: soil temperature during 201 days; B: urea flux in soil.

were affected by temperature. When solid urea was depleted, urea flux became less sensitive to temperature at each simulation time step because of a decrease in urea concentration gradient. The thin-coated urea had a faster urea flux compared to thick-coated urea at about 50 days after application (Figure 3). After 50 days, thick-coated urea had a higher flux than thin-coated urea. This trend remained to the end of the simulation. The diffusion path of urea in thin coating was less than in thick coating; therefore, thin-coated urea released its urea faster and had a shorter life compared to the thick-coated urea.

Both urea solubility and membrane permeability were affected by temperature (Pratt and Noyes, 1965; Zhang *et al.*, 1994). To evaluate the sensitivity of these two

*Table 4.* Comparison of calculated cumulative urea release with the amount of urea found in soil on incubation.

| Temperature (°C) | Coating | Calculated urea release (mg cm$^{-2}$) | | | Net urea release measured in soil (mg cm$^{-2}$) | | |
|---|---|---|---|---|---|---|---|
| | | 15[a] | 30[a] | 60[a] | 15[a] | 30[a] | 60[a] |
| 10 | Thin | 20 | 37 | 56 | 28 ± 7[b] | 44 ± 8 | 57 ± 11 |
| | Thick | 10 | 21 | 41 | 11 ± 5 | 22 ± 5 | 44 ± 8 |
| 23 | Thin | 37 | 55 | 65 | 45 ± 4 | 53 ± 4 | 58 ± 7 |
| | Thick | 23 | 41 | 59 | 28 ± 6 | 42 ± 7 | 55 ± 8 |

[a] Number of days after the coated urea was applied.
[b] Standard deviation.

parameters to temperature, i.e. which of the two parameters contributed more to flux as a result of temperature change, we kept one of the parameters constant and calculated urea flux change as affected by temperature. When permeability changed from 31°C to 10°C, flux decreased from 5.43 to 2.16 mg m$^{-2}$ s$^{-1}$ for thin coating; a 60% decrease. For the same coating and temperature range, when urea solubility changed alone, urea flux dropped from 5.43 to 3.47 mg m$^{-2}$ s$^{-1}$, a 36% decrease in flux. The same percentage decrease was found with the thick coating. The analysis clearly indicated that as temperature varied, change in permeability contributed more to flux change.

To test the model, the cumulative N release predicted by Equation (1) at 10 and 23°C was contrasted with the actual amount of N released in an experiment with thin- and thick-coated urea incubated in soil. Two-thirds of the predicted cumulative urea release fell within the standard deviation of urea release found in the soil (Table 4). In the variable temperature condition, simulated data were compared with results of the field experiments, where two coated ureas were incorporated in the three Alberta soils in autumn 1989, with soil samples being taken in October, November, December, February and April. The soil temperature varied from +15 to −10°C. Overall, 52% of the predicted N release matched that actually found in the field soils (Table 5). The percentage match varied with location. At Site 3, where the temperature was recorded and used in this simulation, the predicted release in each instance fell within the standard deviation of the apparent N release as found in the soil samples. But at the other two sites which were 100 to 150 km away from Site 3, the match of simulated to experimental values was poor. Because Site 1 and Site 2 were distant from Site 3, the average temperatures at Site 3 may not have represented the temperature at the other two sites. Urea flux from polymer-coated urea was sensitive to temperature. To make an accurate prediction, temperature should be recorded in the place where the coated urea is applied.

In summary, this study provided a simple simulation approach to predict urea release rate from coating *in situ* by use of $S_b$, the Soil Retardation Factor. The parameters in the model were clearly defined and experimentally determined, yet the

*Table 5.* Comparison of predicted N released from model with the apparent N recovered from soil samples in the field.

| | Days after application | Predicted N release (kg ha$^{-1}$) | | Apparent N release (kg ha$^{-1}$) | |
|---|---|---|---|---|---|
| | | Thin | Thick | Thin | Thick |
| Site 1 | 12 | 29 | 18 | 53 ± 10[a] | 17 ± 5 |
| | 45 | 80 | 50 | 88 ± 40 | 50 ± 24 |
| | 135 | 114 | 90 | 89 ± 13 | 50 ± 16 |
| | 200 | 123 | 108 | 67 ± 14 | 64 ± 22 |
| Site 2 | 17 | 35 | 22 | 46 ± 10 | 31 ± 9 |
| | 126 | 109 | 80 | 82 ± 22 | 47 ± 10 |
| | 179 | 119 | 99 | 78 ± 24 | 55 ± 6 |
| Site 3 | 18 | 43 | 26 | 56 ± 20 | 31 ± 4 |
| | 43 | 82 | 52 | 94 ± 34 | 50 ± 7 |
| | 134 | 116 | 93 | [b] | 87 ± 34 |
| | 190 | 123 | 107 | 138 ± 25 | 139 ± 29 |

[a] Standard deviation.
[b] Missing data.

model could predict most N release either in soil incubated at a constant temperature, or in soil in the field at the changing temperatures.

## References

Atmospheric Environment Services. Monthly Record of Meteorological Observations in Canada. Environment Canada, 4905 Dufferin Street, Downsview, Ontario.

Baker R W and Lansdale H K 1974 Controlled release: Mechanism and rates. In Controlled Release of Biologically Active Agents. Eds A C Tanquary and R E Lacey. pp 15–71. Plenum, New York.

Christianson C B 1988 Factors affecting N release from reactive layer coated urea. Fert. Res. 16, 273–284.

Fan L S, Tojo K and Fan L T 1979 Cyclic dynamics of a nonisothermal controlled release. Math. Biosci. 45, 69–94.

Higuchi T 1963 Mechanism of sustained-action medication. J. Pharm. Sci. 52, 1145–1149.

Kochba M., Gambash L and Avnimelech Y 1990 Studies on slow release fertilizers: I. Effect of temperature, soil moisture and water vapor pressure. Soil Sci 149, 339–343.

Lu S M and Lee S F 1992 Slow release of urea through latex film. J. Contr. Rel. 18, 171–180.

Oertli J J and Lunt O R 1962 Controlled release fertilizer minerals by encapsulating membranes: I Factors influencing the rate of release. Soil. Sci. Soc. Am. Proc. 26, 579–583.

Paul D R and McSpadden S K 1976 Diffusional release of a solute from a polymer matrix. J. Membr. Sci. 1, 33–48.

Pratt C and Noyes R 1965 Nitrogen Fertiliser Chemical Processes. Nyes Development Corp.

Rogers C E 1976 Structure factors governing controlled release. In Controlled Release Polymeric Formulation. Eds D R Paul and F W Harris, pp 15–25. ACS Symposium Series, Washington.

Technicon AutoAnalyzer II 1973a Ammonia in water and waste water. Industrial Method No. 98–70W. October, 1973. Technical Industrial Systems, Tarrytown, NY.

Technicon AutoAnalyzer II 1973b Nitrate and nitrite in water and waste water. Industrial Method No. 100–70WB. Revised January, 1978. Technical Industrial Systems, Tarrytown, NY.

Tojo K and Fan L T 1981 Membrane-moderated controlled release. Math. Biosci. 57, 279–300.

Tojo K and Miyanami K 1983 Mathematical simulation of membrane-moderated controlled release. Powder Techno. 35, 89–96.

Zhang Mingchu, Nyborg M and Ryan J T 1994 Determining permeability of coatings of polymer-coated urea. Fert. Res. 38, 47–51.

# SECTION V

## Wastewater Treatment and Irrigation

23

# Sewage sludge: land utilization and the environment: toxic organic considerations

G.A. O'CONNOR

Toxic organic chemicals (TO) can find their way into soils in a variety of ways, from accidental spills of hazardous wastes to the intentional application of municipal sewage sludges. However, once in the soil, chemicals are subject to the same processes, regardless of the means of addition. Some chemical sources may promote or diminish one process or another, but we will ignore special cases like coal tar or solvent-laden wastes and concentrate on typical reactions and typical application scenarios associated with land application of municipal sludges. This still leaves us with much to consider, including adsorption/desorption, degradation (biotic and abiotic), volatilization, run-off, leaching, and plant and animal (including human) uptake.

Sewage sludges can contain various concentrations of literally thousands of potentially hazardous organics (Table 1). The environmental impact of these chemicals is a function of their concentration in sludge, sludge loading rate, nature of the chemical (especially toxicity), and the rates of various dispersion/dissipation reactions.

Fortunately, data from numerous surveys including the USA 1990 National Sewage Sludge Survey (US EPA, 1990) suggest that the concentrations of TOs in the vast majority of US municipal sewage sludges are low. Sludges typically contain TOs in the 1 to 100 mg $kg^{-1}$ (dry weight) range, with $\geq$ 90% present at concentrations $\leq$ 10 mg $kg^{-1}$. When sludges containing these typical (median) levels are soil-applied at agronomic rates (10–20 Mg $ha^{-1}$), the resulting soil–sludge mixture (15 cm depth) has TO concentrations at least 100-fold less than in the sludge alone. Thus, TO concentrations in sludge-amended soils are similar to (or lower than) routine pesticide-soil concentrations of 0.01 to 1 Mg $kg^{-1}$. Even at sludge rates used to reclaim drastically disturbed lands (100 to 1000 Mg $ha^{-1}$), most TOs will be present in soil–sludge mixtures at concentrations of 1 to 10 mg $kg^{-1}$, or less.

Recognizing these expected low concentrations is important for several reasons:

(1) Low initial soil–sludge concentrations mean low concentrations available to be transferred to plants, groundwater, etc.

(2) The similarity to routine pesticide soil concentrations allows one to utilize a vast literature characterizing pesticide–soil–plant–animal interactions.

(3) It forewarns of analytical difficulties that should be expected in analyzing soils, plants, water, animal tissue, etc., and the need for careful experimental design to demonstrate compound transfer (or lack thereof).

*D. Rosen* et al. *(eds): Modern Agriculture and the Environment, 299–307.*

*Table 1.* Examples of toxic organics in sludge.

| Compound | Typical concentration (mg $kg^{-1}$ dw) |
|---|---|
| *Phthalate esters* | |
| Bis(2-ethylhexyl)phthalate | ≥ 100 |
| *Polynuclear aromatic hydrocarbons (PAHs)* | |
| Benzo(a)pyrene | 1–10 |
| Fluoranthene | 1–10 |
| *Polyhalogenated biphenyls* | |
| Polychlorinated biphenyl (Aroclor 1248,1254) | 0.5–1 |
| Polybrominated biphenyls | ≪ 1 |
| *Chlorinated pesticides and hydrocarbons* | |
| Aldrin/dieldrin | <1 |
| DDT/DDE/DDD | <1 |
| Chlordane | <1 |
| *Acid-extractables* | |
| Pentachlorophenol | 1–5 |
| Dinitrophenol | <1 |
| *Volatile aromatics* | |
| Benzene | 1–10 |
| Trichloroethylene (TCE) | 1–10 |
| *Alkyl amines* | |
| Nitrosamines | ≪ 1 |

Over the last five or so years, the US EPA has undertaken a rigorous risk assessment of sludge-borne toxicant behaviour (Chaney *et al.*, 1991; O'Connor *et al.*, 1991), involving numerous exposure pathways for highly exposed individuals (Table 2). We were able to show, both experimentally and from careful literature reviews, that plant bioavailability of sludge-borne TOs is very low (O'Connor *et al.*, 1991). TOs are accumulated, if at all, at very low concentrations in the edible portion of food-chain crops. Thus, pathways 1, 2 and 8 were assessed as presenting vanishingly small risk.

More important pathways of TO (especially persistent, lipophilic compounds like PCBs) which impact on the environment are:

(1) Direct ingestion of sludge by children (pathway 3).

(2) Human consumption of meat from animals grazing forage contaminated with surface-adhering sludge (pathway 4).

(3) Predators consuming biota living in sludge-amended soil (pathway 10).

(4) Humans living in sludged-areas and exposed to air (pathway 13) or fish and drinking water from surface sources (pathway 12).

Chaney *et al.* (1991) conducted a thorough risk assessment of pathways 3 and 4 with PCBs (the most sensitive TO). They concluded that because sludges contain low levels of PCBs, the estimated cancer risk of a most exposed individual (MEI) was

*Table 2.* Environmental pathways of concern identified for application of sewage sludge to agricultural land.

| Pathway | Description of most exposed individual |
|---|---|
| 1. Sewage sludge → soil → plant → human | Human ingesting plants in sewage sludge-amended soil |
| 2. Sewage sludge → soil → plant → human | Residential home gardener |
| 3. Sewage sludge → human | Children ingesting sewage sludge |
| 4. Sewage sludge → soil → plant → animal → human | Farm households producing a major portion of the animal products they consume. It is assumed that the animals eat plants grown in soil amended with sewage sludge |
| 8. Sludge → soil → plant → human | Plants grown in sewage sludge-amended soil |
| 10. Sewage sludge → soil → soil organism → soil organism predator | Animals eating soil organisms living in sewage sludge-amended soil |
| 12. Sewage sludge → soil → surface water → human | Water quality criteria for receiving water for a person who consumes 0.04 kg/day of fish, 2 litres/day of water |
| 13. Sewage sludge → soil → air → human | Human breathing volatile pollutants from sewage sludge |

much less than 1 in ten thousand. Furthermore, actual low sludge PCB contents and the low probability of simultaneously meeting all the constraining criteria used to define a MEI suggest the lifetime risk for an MEI is actually less than 1 in ten million.

EPA completed the risk assessment (pathways 12 and 13) and concluded that minimal risk existed. Such assessments resulted in organics being largely excluded from the final US EPA sludge rule (40 CFR, Part 503); the exception is incineration (US EPA, 1992).

Officially, the agency used three criteria to deselect organics from the final sludge regulation (US EPA, 1993). To be deleted from the regulation for a particular use or disposal practice, one (and preferably two) of the following criteria had to be satisfied:

(1) Pollutant banned for use, has restricted use, or is not manufactured for use in the US (low concentration expected in sludge now and even lower concentration expected over time).

(2) The pollutant has a low (< 5%) frequency of detection in sludge according to the National Sewage Sludge Survey (NSSS) of 1990 (low concentration actually occurs in sludge).

(3) The limit for the pollutant, as derived from exposure assessment by use or disposal practice, is less than that expected in a '99th percentile' sludge derived from NSSS (minimal risk).

*Table 3.* Toxic organics of concern: organic pollutants by use or disposal practice.

| Pollutant | Use or disposal practice | |
|---|---|---|
| | Land application | Surface disposal |
| Aldrin/dieldrin (total) | X | |
| Benzene | | X |
| Benzo(a)pyrene | X | X |
| Bis(2-ethylhexyl)phthalate | | X |
| Chlordane | X | X |
| DDT/DDE/DDD (total) | X | X |
| Heptachlor | X | |
| Hexachlorobenzene | X | |
| Hexachlorobutadiene | X | |
| Lindane | X | X |
| N-nitrosodimethylamine | X | X |
| Polychlorinated biphenyls | X | X |
| Toxaphene | X | X |
| Trichloroethylene | X | X |

*Table 4.* Organic pollutants that have been banned, have restricted use, or are not manufactured

| Pollutant | Reference |
|---|---|
| Aldrin/dieldrin (total) | [a] |
| Chlordane | [a] |
| DDT/DDE/DDD (total) | [a] |
| Lindane | [a] |
| N-nitrosodimethylamine | [b] |
| Polychlorinated biphenyls | 40 CFR part 761 |
| Toxaphene | [a] |

[a] *Suspended, Cancelled, and Restricted Pesticides*, 20t-1002, US Environmental Protection Agency, February 1990.
[b] *1992 Directory of Chemical Producers*, SRI International, Menlo Park, CA, 1992.

*Table 5.* Percentage of detection for organic pollutants.

| Pollutant | Number of publicly owned treatment works | Percentage detection |
|---|---|---|
| Aldrin/diedrin (total) | 177 | 8 |
| Benzene | 178 | 0 |
| Benzo(a)pyrene | 178 | 3 |
| Bis(2-ethylhexyl)phthalate | 178 | 63 |
| Chlordane | 177 | 0 |
| DDT/DDE/DDD (total) | 177 | 3 |
| Heptachlor | 177 | 0 |
| Hexachlorobenzene | 178 | 0 |
| Hexachlorobutadiene | 178 | 0 |
| Lindane | 177 | 0 |
| N-nitrosodimethylamine | 178 | 0 |
| Polychlorinated biphenyls | 177 | 19 |
| Toxaphene | 177 | 0 |
| Trichloroethylene | 178 | 1 |

*Table 6.* Application to agricultural land, forest, a public contact site, or reclamation site – pollutants deleted

| Pollutant | Criteria met |
|---|---|
| Aldrin/dieldrin (total) | 1 and 3 |
| Benzo(a)pyrene | 2 |
| Chlordane | 1, 2 and 3 |
| DDT/DDE/DDD (total) | 1, 2 and 3 |
| Heptachlor | 1, 2 and 3 |
| Hexachlorobenzene | 2 |
| Hexachlorobutadiene | 2 and 3 |
| Lindane | 1, 2 and 3 |
| N-nitrosodimethylamine | 1 and 2 |
| Polychlorinated biphenyls | 1 |
| Toxaphene | 1 and 2 |
| Trichloroethylene | 2 and 3 |

Organic pollutant remaining: none

A list of the TOs of original concern is given in Table 3. It is important to realise that they are also representative of numerous other organics, so the risk considerations are more generally applicable than for just this group of chemicals. Eight of the TOs satisfy the first criterion for deletion (Table 4).

All but three: aldrin/dieldrin (total), bis (2-ethylhexyl) phthalate, and PCBs satisfy criterion 2 for deletion (Table 5). DEHP is a component of plastics and is ubiquitous – fortunately, it is non-toxic and relatively short lived (O'Connor *et al.*, 1991).

Several chemicals failed criterion 3 depending upon the product use scenario. A summary of criteria met by each chemical for application to agricultural land, forests, a public contact site, or a reclamation site is shown in Table 6. Similarly, the summary for sludge sold or given away in a bag for land application is given in Table 7.

*Table 7.* Sewage sludge sold or given away in a bag or other container for application to the land – pollutant deleted.

| Pollutant | Criteria met |
|---|---|
| Aldrin/dieldrin (total) | 1 and 3 |
| Benzo(a)pyrene | 2 |
| Chlordane | 1, 2 and 3 |
| DDT/DDE/DDD (total) | 1, 2 and 3 |
| Heptachlor | 1, 2 and 3 |
| Hexachlorobenzene | 2 |
| Hexachlorobutadiene | 2 and 3 |
| Lindane | 1, 2 and 3 |
| Polychlorinated biphenyls | 1 |
| Toxaphene | 1 and 2 |

Organic pollutant remaining: none

The only chemical consistently to fail criteria 2 and 3 was PCB. It is scheduled for re-evaluation in Round II review of the regulations (in 1998 or 1999). In fact, however, if a more reasonable percentile concentration (e.g. 90th vs. 99th) is used, even PCBs meet criterion 3 for land application – based on the Chaney *et al.* (1991) evaluation.

Thus, for various reasons, in addition to the EPA criteria tests, it is reasonable (and safe) to delete the original list (and many other) TOs from the regulation.

There are other organics deserving attention in Round II *or* at least greater research attention. I believe they include the following:

(1) Chlorinated dibenzo-*p*-dioxins (CDDs) and dibenzo furamo (CDFs).

(2) Aromatic surfactants.

(3) Others; unknown anthropogenic or natural compounds.

Of the various CDDs and CDFs, 2,3,7,8-tetrachlorodibenzo-*p*-dioxin (TCDD) is the most notorious. It is a persistent xenobiotic that is widely dispersed in the environment, and is acutely and chronically toxic to laboratory animals. Epidemiological studies have failed to document unequivocally significant TCDD-related health effects in humans, but concern persists about its transfer to humans. The cancer potency factor ($q_1^*$) that EPA uses for TCDD is four orders of magnitude larger than any other TO of original concern. Thus, any risk assessment involving TCDD will likely allow extremely low loading rates. Such considerations will also dictate very sophisticated and expensive analytical techniques to confirm risks.

Few reports of CDDs and CDFs in municipal sludges are available (e.g. O'Connor *et al.*, 1991). If detected (often in specialty sludges like wood processing waste), the compounds occur at low $\mu g\ kg^{-1}$ (rather than $mg\ kg^{-1}$) concentrations. Remember that these very low initial (sludge) concentrations would be diluted at least 100-fold when the sludge is land applied and mixed. Further, the OM added by sludge would increase soil sorption capacity for the compounds and likely reduce volatilization and leaching losses. The bioavailability to plants and other organisms of the sludge-borne compounds is expected to be vanishingly small. Given the widespread presence of CDDs and CDFs in the environment, it is likely that no significant difference in plant (or animal) contamination between sludge-amended and control soil grown crops will be detected.

Given these arguments, why do I believe compounds deserve research attention? Animal carcinogens, even if they are only *suspected* human carcinogens, frighten people and are easy prey for well-intentioned, but scientifically naive, regulators. We must generate solid scientific data to guide sound regulatory decisions. We learned this lesson repeatedly in our considerations of the original '503' regulation, even for 'toxics of concern' with much less notoriety (PRC, 1989).

At the other end of the concentration scale of TOs in sludge from CDDs and CDFs are the aromatic surfactants. The substances are mainly detergent-derived and occur in sludges at concentrations orders of magnitude higher than the originally targeted TOs. Chemicals such as linear alkylbenzesulphonates, nonyl phenol, nonyl phenol monoethoxylate and diethyloxylate are often present in sludges in the g $kg^{-1}$ range (rather than $\mu g\ kg^{-1}$ or mg $kg^{-1}$). Some of the compounds are apparently refractory, and have been reported in sludge-amended soils at the mg $kg^{-1}$ level.

There are conflicting data in the literature regarding surfactant degradation rates (O'Connor *et al.*, 1991). Some laboratory studies suggest rapid degradation (half-life as low as 5 days), whereas some field studies suggest that significant residues persist well beyond the time expected for complete degradation. Further, some plant studies suggest uncharacteristically high plant accumulations from soil.

Modern homemakers seem to like the superconcentrated detergents recently introduced, and this has increased demand for surfactants as ingredients. Thus, it is not unreasonable to expect even greater concentrations of the compounds in municipal sewage sludges. Manufacturers are reportedly switching to more biodegradable surfactants, but little is known of the surfactants' environmental behaviour – especially in sludge systems. Work is underway to characterise sludge-borne surfactant behaviour (Overcash, personal communication, 1994). The data are welcomed and could significantly impact future considerations of risk exposure from sludge.

There may also be other organics – both natural and man-made – that we do not know of because we have not looked for them. 40 CFR 503 may have deregulated organics, but municipalities should not completely curtail regular organic screenings.

Table 8 is a summary table of the limiting result for each pathway for TOs from the 503 Technical Support Document. It is a complex table, but highlights several points:

(1) Risk-based assessment has been completed for several toxic organics, involving a large array of environmental possibilities – each pathway with very conservative assumptions.

(2) Although not as complete as we would prefer, the database for TOs is good enough that – combined with common sense about loading rates – concern over some pretty 'nasty' compounds can be largely dismissed.

(3) The process is not perfect, nor complete. Data gaps exist in making some calculations (in my opinion, especially for pathways 12–14) overly simplistic.

This methodology will be used in other scenarios with other solutes and materials. We must make sure we understand the processes involved, that we have appropriate and accurate data, and that we use reliable and validated models. All of these will impact the regulations governing land application of wastes, and we must be vigilant.

Today, the news is good! Sludges in the US and in many countries worldwide are relatively 'clean' with respect to toxic organics. All of our efforts to date – even risk assessment by regulations – suggest that when sludge is appropriately managed to take advantage of the soil's assimilative capacity, land application is efficient, cost-effective, and safe.

*Table 8.* Limiting results for each pathway for organic pollutants.

| Pollutant | 1 | | 2 | | 3 | 4 | | 5–6 | 7 | 8 | 9 | 10 | 11 | 12 | 13 | 14 |
|---|---|---|---|---|---|---|---|---|---|---|---|---|---|---|---|---|
| | RPa | RPc | RPa | RPc | RSC | RPa | RPc | RSC | | | | RPa | RPc | RPa | RPa | RPa |
| Aldrin/dieldrin | | 280 | | 64 | 7.0 | | 17 | 2.7 | | | | | 30 000 | | | |
| Benzo(a)pyrene | 230 | | 54 | | 15 | | | | | | | | | 1.3 | 3 500 | Unlimited |
| Chlordane | | 3 400 | | 790 | 86 | | 13 000 | 2 300 | | | | | | 5.3 | 3.9 | Unlimited |
| DDT | 560 | | 130 | | 320 | 46 | | 150 | | | | | 100 000 | 1.2 | 45 | Unlimited |
| Heptachlor | 990 | | 220 | | 24 | 65 | | 7.4 | | | | | | | | |
| Hexachlorobenzene | 320 | | 75 | 70 | 25 | | | 29 | | | | | | | | |
| Hexachlorobutadiene | 43 000 | | 10 000 | | 1 400 | | | 600 | | | | | | | | |
| Lindane | 2 300 | | 540 | | 84 | 600 | | 140 | | | | | | 2 100 | 10 | Unlimited |
| N-nitrosodimethylamine | 87 | | 20 | | 2.1 | | | | | | | | | 29 000 | 22 | |
| PCBs | 37 | | 8.5 | | 14 | 2.4 | | 4.6 | | | | | 200 | | | Unlimited |
| Toxaphene | 2 800 | | 650 | | 100 | 43 | | 10 | | | | | | 5.0 | 120 | Unlimited |
| Trichloroethylene | 220 000 | | | 10 000 | | | | | | | | | | Unlimited | 420 | Unlimited |

Note: All results rounded down to two significant figures.
RPa = reference annual application rate of pollutant (kg pollutant $ha^{-1}$ $yr^{-1}$).
RPc = reference cumulative application rate of pollutant (kg pollutant $ha^{-1}$).
RSC = reference concentration of pollutant in sewage ($\mu$g pollutant per g sewage sludge, dry weight).

## References

Chaney R L, Ryan J A and O'Connor G A 1991 Risk assessment for organic micropollutants: US point of view. Treatment and Use of Sewage Sludge and Liquid Agricultural Wastes. Ed. P L'Hemite. pp 141–158. Elsevier Applied Science, New York.

O'Connor G A, Chaney R L and Ryan J A 1991 Bioavailibility to plants of sludge-borne toxic organics. Rev. Environ. Contamin. Toxicol. 121, 129–155.

PRC 1989 Peer Review Committee. Peer review of standards for the disposal of sewage sludge (US EPA Proposed Rule 40 CFR Parts 257 and 503). USDA-CSRS. W-170 Regional Research Committee. Washington, DC.

US EPA 1990 National Sewage Sludge Survey: Availability of information and data and anticipated impacts on proposed regulations; Proposal Rule to Cfr 503. Federal Register 55, 47210–47283.

US EPA 1992 Technical Support Document for Land Application of Sewage Sludge, Volume 1. Office of Waste, Washington, DC. Available from NTIS-PB 93-110575. Springfield, VA.

US EPA 1993 Standards for the use or disposal of sewage sludge. (Final rule) Rule 40 CFR Parts 257, 403 and 503. Federal Register 58, 9248–9414.

# 24
# New technologies for disinfection of domestic effluents for agricultural reuse

A. ACHER, E. FISCHER AND Y. MANOR

## 1. Disinfection methods

The reuse of treated wastewater (effluent) for crop irrigation to replace some of the shortfall in natural potable water resources, has become a necessity especially in arid and semi-arid regions.

To avoid health and ecological hazards caused by using effluents in crop irrigation, the treatment of domestic sewage must involve in its last stage a disinfection treatment which inactivates pathogens.

Disinfection of effluents can be achieved by a variety of methods, mostly using chlorine, hypochlorites, chlorine dioxide, ozone and UV-light; other means used to a lesser extent are treatments by: chloramines, bromine, bromine–chloride, iodine, hydrogen peroxide, metal ions, heat, ultrasound, electrostatic processes, gamma- and X-rays, etc. All these methods were developed primarily for use with potable water. Their use in wastewater effluents which contain organic and inorganic materials, suspended solids, dyes, etc., brings about a sharp decrease in the efficiency of the disinfection process, resulting in incomplete disinfection and possible production of toxic and non-desirable materials.

### *1.1 Chlorination*

Chlorination of domestic effluents may produce chlorinated hydrocarbons with toxic (Kott, 1973), mutagenic or suspected carcinogenic properties (Cumming *et al.*, 1982; Zoeteman *et al.*, 1982). Chlorine residuals are toxic to aquatic life (Brungs, 1973; Pieuch, 1974). Even after dechlorination, effluents may still contain carcinogenic and toxic compounds (Legan, 1982). Survival (Dabis and Spruce, 1975) and regrowth of coliforms and faecal coliforms in chlorinated wastewater effluent has been reported (Kelner, 1961; Shuval *et al.*, 1973).

Chlorine dioxide ($ClO_2$) was found to be more effective than chlorine in destroying coliforms, viruses and pathogenic amoebae (Leong, 1983) and does not produce potentially toxic chemical by-products (Narkis and Kott, 1992). However, chlorine dioxide is expensive and difficult to generate. $ClO_2$ does not produce, like $Cl_2$, tri-halo-methanes, but may produce chlorite ($ClO_2^-$) and chlorate ($ClO_3^-$) which are toxic compounds:

$$2ClO_2 + 2OH^- \rightarrow ClO_2^- + ClO_3^- + 2H_2O.$$

*D. Rosen* et al. *(eds): Modern Agriculture and the Environment, 309–315.*

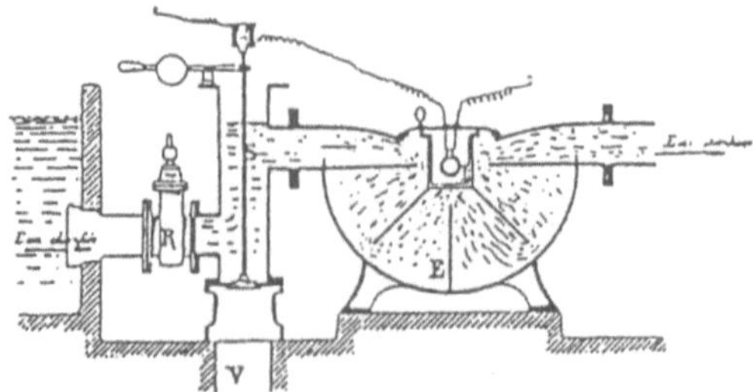

*Figure 1.* A 1910 design of a UV-water disinfection plant (France).

### *1.2 Ozone*

Ozone appears to be a most powerful disinfection agent which can be used in domestic effluents. However, it has some drawbacks that make the ozonation process less attractive: ozone cannot be stored; its residual effect disappears after 7–8 min; its oxidation of organic materials in water brings about formation of biodegradable compounds which can serve as nutrients for regrowth of microorganisms surviving the disinfection process (Venosa *et al.*, 1984); its release in air provokes detrimental effects at 0.3–0.5 ppm for the plant operators (Erb, 1982); and it has a high capital investment and operating cost. Effluents containing bromides ($Br^-$) may produce, during chlorination or ozonation processes, bromates ($BrO_3^-$) which are toxic compounds.

### *1.3 Ultraviolet light*

Ultraviolet light disinfection of effluents could be an attractive method (Fluegge *et al.*, 1981), provided that the treatment processes of domestic sewage result in effluents with low optical density and high transmittance of UV-light (at 254 nm) into the effluent (a rapid sand filtration is often required, before submitting effluents to UV-light). This ecological friendly disinfection method is by no means a new idea (Figure 1). In fact, it was first used for drinking water (large scale) in France at the beginning of this century (von Sonntag, 1988). However, its lack of residual disinfection effect is a drawback which prevents the development of this method until the quality of the network of water supplying pipes has reached a higher reliability against penetration of environmental pollutants into the water pipes.

The disinfection effect of UV-light is caused by the high absorption of the bacterial DNA at 254 nm (von Sonntag, 1988), resulting in formation of pyrimidine dimers which inhibit the reproduction functions of bacterial cells. If the UV-light dose does not reach the efficient killing value, all bacterial cells are able to repair the defective DNA by an enzymatic process (photoreactivation).

An economic comparison of the above disinfection methods is given in Table 1 by one of the leading companies in UV-disinfection technology (Trojan Comp., Canada, 1992).

*Table 1.* Cost comparison of different disinfection methods ($ per 100 m$^3$).

| Disinfection method[a] | | | | Plant size | |
|---|---|---|---|---|---|
| 1 | 2 | 3 | 4 | mm$^3$ y$^{-1}$ | MGD |
| 0.36 | 0.50 | 0.70 | 2.4 | 7.3 | 5.0 |
| 0.33 | 0.42 | 0.50 | 2.1 | 14.6 | 10.0 |
| 0.30 | 0.35 | 0.42 | 1.6 | 146.0 | 100.0 |

[a] 1, UV-light; 2, chlorination; 3, 2 + dechlorination; 4, ozonation.

## 2. New effluent disinfection technologies

New, efficient and ecological friendly methods, based on the use of UV or of visible (sun)light for disinfection of effluents aimed at unrestricted crop irrigation, were developed in our laboratory. These methods are briefly presented below.

### 2.1 *UV-light disinfection of effluents*

To overcome the shortcomings of present UV-disinfection installations in turbid effluents (NTU > 2), a specially designed UV-irradiation device – the multi-tubular-photoreactor (MTP) – was built for this purpose (Acher and Mazur, 1992). The MTP-device has an unique optical design that concentrates the UV-light sources on a transparent pipe through which effluents are passed at up to 10 m$^3$ h$^{-1}$ (Figure 2). Preliminary disinfection experiments with this MTP-device were performed on effluents supplemented with poliovirus (vaccine strain) and infested waters. Although the turbidities of the water exposed to irradiation were high (from 8 to 30 NTU), the results obtained were very promising as can be seen from Table 2.

### 2.2 *Sensitized-sunlight disinfection of effluents*

The use of a sensitized photochemical method for water disinfection was studied previously on a laboratory scale (Acher and Juven, 1977; Hobbs *et al.*, 1977; Martin

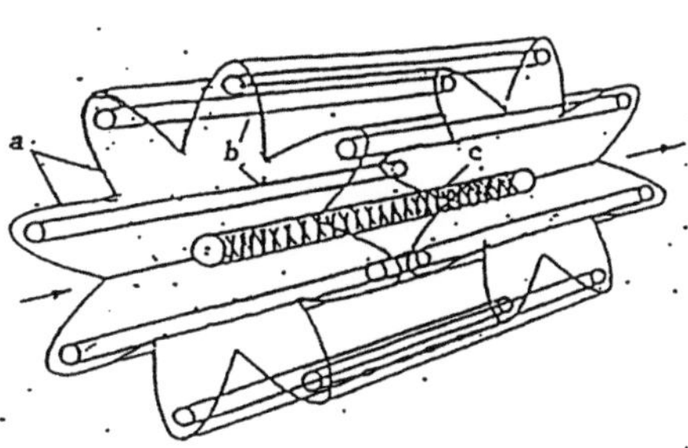


*Figure 2.* The multi-tubular-photoreactor (MTP); a, elliptical reflectors; b, UV lamps; c, transparent pipe.

*Table 2.* Preliminary disinfection results of MTP device.

| Microorganisms (UV-dose needed) | Counts per 100 ml | | Exposure time (s) at 5 $m^3$ $h^{-1}$ |
|---|---|---|---|
| | Initial | Final | |
| *Aspergillus flavus* (99 000 $\mu Wcm^{-2}s^{-1}$) | 250 | 0 | 2 |
| *A. ochraceus* | 250 | 0 | 2 |
| *A. niger* (330 000 $\mu Wcm^{-2}s^{-1}$) | 150 | 0 | 4 |
| *Escherichia coli* | $1.5 \times 10^8$ | 0 | 2 |
| *Streptococcus faecalis* (10 000 $\mu Wcm^{-2}s^{-1}$) | $7.6 \times 10^6$ | 0 | 2 |
| Enteroccoci | $9.3 \times 10^5$ | 0 | 2 |
| *Chlorela vulgaris* (22 000 $\mu Wcm^{-2}s^{-1}$) | 300 | 0 | 2 |
| Poliovirus (21 000 $\mu Wcm^{-2}s^{-1}$) | $2 \times 10^5$ | 0 | 2 |

and Perez-Cruet, 1987; Melnick *et al.*, 1976); and on an experimental pilot scale (Acher, 1984; Acher *et al.*, 1980, 1990, 1994; Eisenberg *et al.*, 1987; Middlebrooks *et al.*, 1987).

The sensitized-sunlight disinfection method, initiated in our laboratory, was specially developed for disinfection of effluents for unrestricted crop irrigation. It is based on the adsorption and transfer of visible sunlight (or artificial light) energy to the dissolved oxygen (DO) present in the effluent, producing active oxidative species ($^1O_2$, $H_2O_2$, $OH^\bullet$, $O_2^-$, etc.). These oxidative species may irreversibly alter the vital cell constituents of the microorganisms, causing lethal damage (Spikes and Livingston, 1969). The material (sensitizer) which absorbs the light energy and transfers it to DO is a dye, methylene blue (MB), used at a low concentration (< 1.0

*Figure 3.* The experimental pilot plant for effluent disinfection.

ppm). The turbidity of the entering effluent should be < 20 NTU; the sunlight intensity (global irradiation) should be > 750 $\mu$E m$^{-2}$ s$^{-1}$ or 0.280 kW m$^{-2}$ (the usually measured blue sky sunlight intensity is about 2400 $\mu$E m$^{-2}$ s$^{-1}$ or 0.840 kW m$^{-2}$ (Manes *et al.*, 1971)). In such conditions the detention time of the effluent, from an activated sludge domestic sewage treatment plant, in the disinfection plant was about 35 min. No photoreactivation of microorganisms was detected during seven days.

The experimental pilot plant (Figure 3) was operated for more than a year, with an effluent flow rate up to 50 m$^3$ h$^{-1}$. The method is simple, effective, reliable, does not use or produce toxic compounds, is economically competitive and it can be used on an industrial scale.

The effluent disinfection cost in a 'sunlight disinfection plant' producing 200 m$^3$ h$^{-1}$ disinfected effluent is estimated to be US$3.95 per 100 m$^3$.

### 2.3 *Concentrate sensitized-sunlight disinfection of effluents*

The use of the energy of concentrated sunlight in a sensitized photochemical process for disinfection of effluents was carried out. The concentrated sunlight was provided by the Solar Tower of the Weizmann Institute of Science, Israel, which is provided with 64 computer controlled mirrors (heliostats) which concentrate and reflect the received sunlight (0.8 kW m$^{-2}$) to the target situated above the plane of heliostats field. The water exposed to disinfection was effluents supplemented with vaccine strain culture of poliovirus and sensitizer MB). The effect of 2 seconds irradiation of concentrated sunlight (up to 200 kW m$^{-2}$) on *E. coli* (+MB and –MB) and on poliovirus is shown in Figure 4. The upper curve represents *E. coli* control

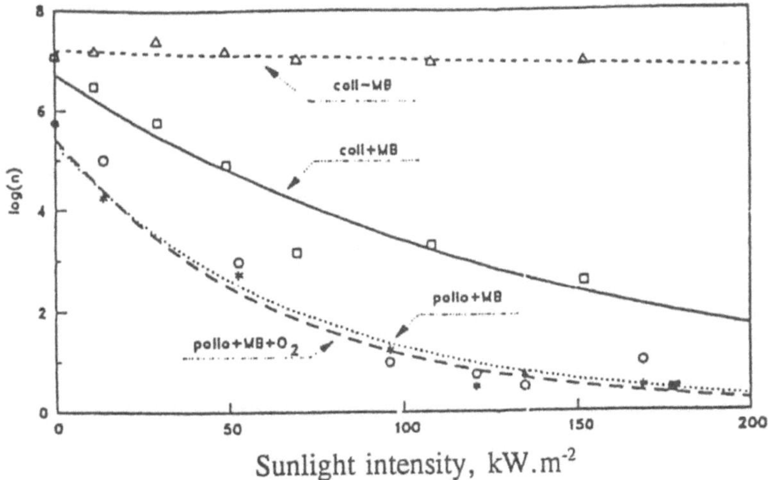


*Figure 4.* The effect of concentrated sunlight on microorganisms.

experiment (exposed without MB); the dotted curves are the poliovirus experiments performed at two different concentrations of $O_2$ (8.0 and about 35 mg $O_2$ $L^{-1}$).

Despite the very spectacular efficiency of killing microorganisms by concentrated sunlight (5 logs decrease in 2 s at about 150 kW $m^{-2}$ as compared with 2100 s–35 min – at 0.8 kW $m^{-2}$), the concentrated sunlight disinfection method has limited practical use. This limitation is due to the fact that the direct sunlight availability (clear blue sky) is not continuously possible. However, concentrated sunlight can be used in biological, chemical and industrial processes research, where high energies are needed for short time periods.

## 3. Conclusions

The disinfection of the effluents has become a compulsory process at domestic sewage treatment plants and efficient and ecological friendly methods are required.

The photochemical disinfection methods are the only ecologically friendly procedures for effluents aimed at reuse or recharge. These methods are efficient, reliable and economically competitive.

## References

Acher A J 1984 Sunlight photooxidation of organic pollutants in wastewater. Wat. Sci. Technol. 17, 623–632.

Acher A J and Juven B I 1977 Destruction of fecal coliforms in sewage water by dye-sensitised photooxidation. J. Appl. Envir. Microbiol. 33, 1019–1023.

Acher A J and Mazur Y 1992 A new apparatus for ultraviolet light disinfection of turbid waters. Israel Patent Appl., 102.135 from 09.06.92.

Acher A J, Marzouk J and Manor J 1980 Disinfection of effluents by photochemical treatment. Progress Report to the Israel Council for R & D, Jerusalem (in Hebrew).

Acher A J, Fischer E, Zellingher R and Manor Y 1990 Photochemical disinfection of effluents – Pilot plant studies. Wat. Res. 24, 837–845.

Acher A J, Fischer E and Manor Y 1994 Sunlight disinfection of domestic effluents for agricultural use. Wat. Res. 28, 1153–1160.

Brungs W A 1973 Effects of residual chlorine on aquatic life. J. Wat. Pollut. Control Fed. 45, 2180–2193.

Cumming R B, Jolley R L, Lewis L R, Lee N E and Thompson J E 1982 Micropollutants produced by disinfection of wastewater effluents. Wat. Sci. Technol. 14, 45–55.

Dabis E M and Spruce R K 1975 Municipal wastewater bacteria capable of surviving chlorination. Hlth Lab. Sci. 11, 132–140.

Eisenberg T N, Middlebrooks E J and Adams V D 1987 Sensitizer photooxidation for wastewater disinfection and detoxification. Wat. Sci. Technol. 19, 1255–1258.

Erb F 1982 Toxicity of Ozon. In Ozonation Manual for Water and Wastewater Treatment. Ed. W J Mawwchelein. pp 197–199. John Wiley & Sons, New York.

Fluegge R A, Metcalf T G and Wallis C 1981 Virus inactivation in wastewater effluents by chlorine, ozone and ultraviolet light. Project Summary. EPA-600/S2-81-008.

Hobbs M F, Gerba C P, Wallis C, Melnick J L and Lennon J S 1977 Photodynamic inactivation of infectious agents. J. Envir. Eng. Div. ASCE 103, 459–472.

Kelner A 1961 Historical background to the study of photoreactivation. In Progress in Photobiology. Proc. of the Third Int. Congress on Photobiology (Copenhagen, 1960). pp 276–278. Elsevier, Amsterdam.

Kott Y 1973 Hazards associated with the use of chlorinated oxidation pond effluents for irrigation. J. Wat. Res. 7, 853–862.

Legan R W 1982 Ultraviolet light takes on CPI role. Chem. Eng. 89, 95–100.

Leong L Y C 1983 Removal and inactivation of viruses by treatment processes for potable water and wastewater – a review. Wat. Sci. Technol. 15, 91–101.

Manes A, Teitelman A and Fruehling I 1971 Solar radiation and radiation balance. Meteorological Notes, Israel, Series A, No. 25, pp 27–28.

Martin D F and Perez-Cruet M I 1987 Preparation of sterile seawater through photodynamic action. Preliminary screening studies. Fla. Sci. 50, 168–176.

Melnick J L, Gerba C P, Wallis C and Hobbs M F 1976 Photodynamic inactivation of virus in sewage. In Virus Aspects of Applying Municipal Wastes to Land. Eds L B Baldwin, J M Davidson and J F Gerber. pp 25–36. Univ. of Florida Press, Gainesville, FL.

Middlebrooks E J, Eisenberg T N, Adams V D, Acher A J and Saltzman S 1987 Uses of solar energy for wastewater disinfection for crop irrigation. Final Report. BARD, Research Project No. US-613-83.

Narkis N and Kott Y 1992 Comparison between chlorine dioxide and chlorine for use as a disinfectant of wastewater effluents. The 16th IAWPRC Biennial Int. Conf., Washington, DC, May.

Pieuch P J 1974 The chlorination controversy. J. Wat. Pollut. Control Fed. 46, 26–37.

Shuval H L, Cohen J and Kolodney R 1973 Regrowth of coliforms and fecal coliforms in chlorinated wastewater effluent. Wat. Res. 7, 537–546.

Spikes J D and Livingston R 1969 The molecular biology of photodynamic action: Sensitized photoautoxidation in biologic systems. Adv. Radiat. Biol. 3, 29–108.

Venosa A D, Petrasek A C, Brown D, Sparks H L and Allen D M 1984 Disinfection of secondary effluent with ozone/uv. J. Wat. Pollut. Control Fed. 56, 137–142.

von Sonntag C 1988 Disinfection by UV-radiation. In Processes for Water Treatment. Ed. S Strucki. Plenum Publishing Corporation.

Zoeteman B C J, Hrubec J, deGreef E and Kool H J 1982 Mutagenic activity associated with by-products of drinking water disinfection by chlorine dioxide, ozone and uv-irradiation. Envir. Hlth. Perspec. 46, 197–205.

# 25
# Effect of irrigation with secondary sewage effluent on the transport of soil-borne pesticides

ELLEN R. GRABER

## 1. Introduction

Efforts to minimize pollution and alleviate chronic water shortages in many parts of the world have resulted in the increased use of treated sewage effluent in irrigated agriculture. One of the problems associated with this practice is the potential for enhanced transport of agrochemicals such as pesticides as a result of complexation with relatively mobile effluent-borne macromolecules and colloids. Interaction of macromolecules and colloids with soil organic carbon may also change the nature of soil sorption sites and reduce the surface area available for pesticide sorption. Enhanced transport of nearly water insoluble pesticides at dissolved organic carbon (DOC) concentrations as low as 5 mg $L^{-1}$ has been demonstrated in laboratory studies (Dunnivant *et al.*, 1992a). The use of effluents in irrigated agriculture with relatively higher levels of organic carbon (15–150 mg DOC $L^{-1}$) therefore may result in deep soil and groundwater contamination by pesticides due to enhanced downward transport.

## 2. Background

Organic matter in soils has been consistently implicated as the major factor controlling the sorption, and hence transport, of pesticides and other organic micropollutants in soils. Strong positive correlations of pesticide retention in soil with soil organic matter content have been reported (e.g., Jury *et al.*, 1987; Karickhoff, 1981); a similar strong sorption of pesticides to suspended and dissolved organic matter colloids and macromolecules in DOC-bearing waters such as effluents can also be expected. The presence of organic macromolecules which can form complexes with pesticides or partition between soil solids and water is expected to reduce pesticide sorption to soil solids. If the DOC–pesticide complex is more mobile in the soil column than the solute pesticide, the overall movement of the pesticide downward into the soil profile will be enhanced.

The sorption of non-polar organic solutes in soils and sediments is commonly described as an equilibrium partitioning of the solute between the aqueous phase and soil organic matter (e.g., Chiou, 1989). A similar mechanism has been proposed for the binding of hydrophobic organic compounds to DOC, such that partitioning into

*D. Rosen* et al. *(eds): Modern Agriculture and the Environment, 317–322.*

the aqueous phase as compared with the solid phase is increased. The higher the DOC concentration, the greater its relative impact on sorption and transport of organic molecules (Magee *et al.*, 1991; Chin *et al.*, 1991). For very hydrophobic pesticides, transport can be enhanced at DOC concentrations as low as 5 mg $L^{-1}$ (Dunnivant *et al.*, 1992a). In common environmental scenarios, DOC levels are lower than 5 mg $L^{-1}$, but end-line effluents used in agriculture usually have much higher DOC concentrations (15–150 mg $L^{-1}$).

Rebhun and Manka (1971) and Manka *et al.* (1974) reported that of the total organic carbon in effluents, approximately 25% is made up of fulvic acids, 12% humic acids, 8% hymatomelanic acids, and 12% anionic detergents, the remainder being represented by lower molecular weight and rapidly degradable sugars, oligosaccharides, proteins, and lipids. Humic substances and surfactants can form complexes with pesticides (Conroy *et al.*, 1981; Stevenson, 1985), and, if mobile, can play an important role in transporting solute species (Ballard, 1971; Ogner and Schnitzer, 1970).

Dunnivant *et al.* (1992b) found that hydrophilic DOC components were rapidly transported through columns of aquifer material with little adsorption to aquifer solids. In contrast, hydrophobic components, and in particular, hydrophobic neutral components, were more strongly adsorbed to aquifer solids. Suspended materials from sewage effluents were also found to be mobile in sandy soil columns (Vinten *et al.*, 1983) and capable of transporting with them sorbed solute species with very high adsorption coefficients. Magee *et al.* (1991) found that the retardation factor of phenanthrene in a sand column was reduced by a factor of 1.8 in the presence of DOC. Dunnivant *et al.* (1992a) found that the breakthrough of both PCB-153 and Cd from aquifer columns was accelerated in the presence of DOC at concentrations as low as 5.2 mg $L^{-1}$.

Enhanced transport of solutes such as pesticides has until recently been virtually undocumented in field studies. Muszkat *et al.* (1993) reported that subsurface contamination by many organic species was significantly greater in effluent-irrigated citrus groves than in higher quality water-irrigated groves. The study examined groves which had been irrigated for over 20 years by waters of differing qualities. The source of pollutants detected in the soil profile under the effluent- and higher quality water-irrigated groves, whether agricultural practices or the effluents themselves, is unknown. Increased downward transport of organic species under effluent-irrigated citrus groves may have been due to changes in soil hydraulic conductivity or the development of preferential flow paths as a result of irrigation with marginal waters.

A recent study by Graber *et al.* (1995) demonstrates the enhanced transport of the herbicide atrazine in plots irrigated with secondary effluent as compared with those irrigated with high quality water. This work is discussed in some detail below.

## 3. Experimental methods

The field experiment was conducted on a 5000 $m^2$ plot of relatively homogeneous Vertisol (fine clayey, montmorillonitic, thermic Typic Haploxerert) in central Israel, which was divided into 40 subplots randomly assigned to irrigation with high quality water or secondary effluent. In two consecutive springs (1990 and 1991), the field was seeded in feed corn and treated with atrazine (6-chloro-$N^2$-ethyl-$N^4$-isopropyl-1, 3,5-triazine-2,4-diamine; 400 g a.i. $ha^{-1}$). This was followed by an irrigated growing season ($\approx$ 400 mm water), crop harvest, and winter rainy season. Biological oxidation demand (BOD), total suspended solids (TSS), and chemical oxidation demand (COD) of the secondary effluent ranged from 60–133 mg BOD $L^{-1}$, 60–226 mg TSS $L^{-1}$, and 181–394 mg COD $L^{-1}$.

Ten high quality water- and ten effluent-irrigated plots were cored at the end of the second rainy season to a depth of 4 m, and each profile was subdivided into 30 cm segments which were analyzed for atrazine content (see Graber *et al.*, 1995, for details).

## 4. Results and discussion

In eight of ten high quality water-irrigated profiles, no movement of atrazine below 145 cm was observed while in nine of ten effluent-irrigated profiles atrazine was distributed below 145 cm, and in most of the profiles, to depths greater than 3 m. The average depth to the centre of the atrazine mass in the ten high quality water-irrigated profiles was found to be 63 $\pm$ 64 cm below the ground surface, while the average depth to the centre of the atrazine mass in the ten effluent-irrigated profiles was found at 115.5 $\pm$ 39 cm, or nearly twice the mean depth of the high quality water plots (Figure 1).

The variance in atrazine distribution around the centre of mass in the effluent-irrigated plots was found to be nearly three times that in the high quality water-irrigated plots. The mean variance in the effluent-irrigated plots was 142 $\pm$ 112 cm around the centre of mass as compared with a mean variance of 58 $\pm$ 73 cm in the control profiles. In other words, atrazine was more widely distributed with depth in effluent-irrigated than in high quality water-irrigated plots.

In addition, it was shown that on average in all ten control profiles, a full 70% of the atrazine content was concentrated in the upper 30 cm, while only 45% of the atrazine in the effluent-irrigated profiles remained in the same uppermost layer. These results were interpreted to indicate that under irrigation with effluent, the transport of atrazine downward through the soil profile was enhanced.

A higher total atrazine content in effluent-irrigated profiles compared with high quality water-irrigated profiles indicated that irrigation with effluent did not accelerate atrazine degradation. This may have been due to faster atrazine transport out of the biologically active root zone. Complexation of atrazine with organic carbon macromolecules and colloids may also have reduced its bioavailability.

Sorption and desorption isotherms for atrazine on soil from an uncultivated field adjacent to that of the experiment were obtained in the presence of 0.005 M $CaCl_2$ and in the effluent used in the field trial (Figure 2) (Graber *et al.*, 1995; Graber, unpublished). The isotherm for atrazine sorption from $CaCl_2$ was linear while that for atrazine sorption from effluent can be described by the Langmuir model with an

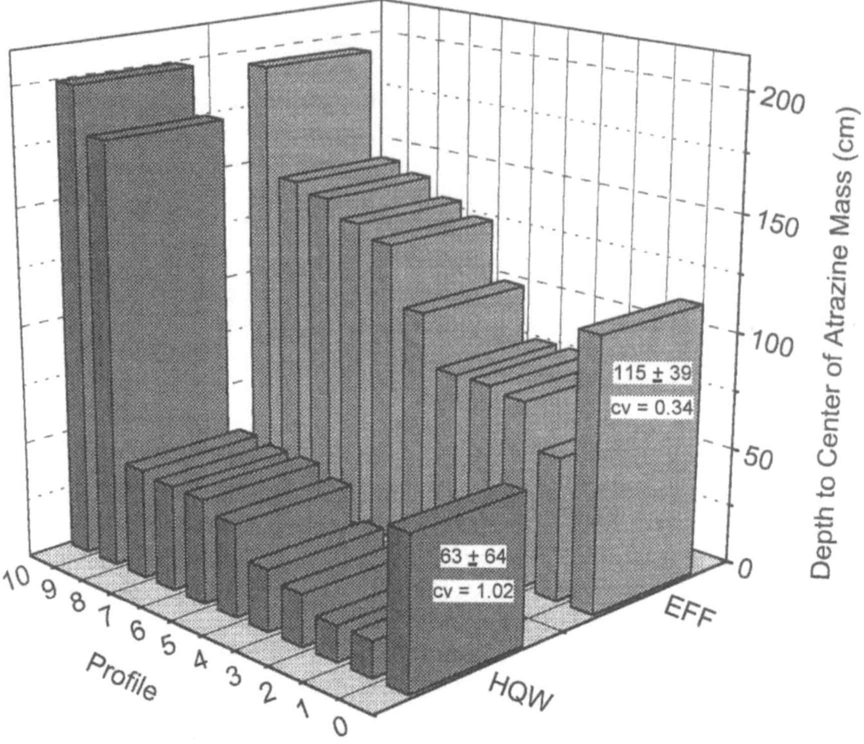


*Figure 1.* Mean (0) centre of atrazine mass and individual centres of mass for ten high quality water (HQW) and ten effluent (EFF) irrigated profiles (From Graber *et al.*, 1995).

estimated surface area of 17.5 mg $kg^{-1}$ and equilibrium constant of 0.06 (standard error of 10% on surface area estimation), or the Freundlich model with an equilibrium constant ($K_f$) of 1.05 and exponent of 0.76 ($r^2 = 0.98$). The slope of the linear portion of the isotherm ($K_d$) is smaller in the presence of effluent than in $CaCl_2$, indicating a reduction in atrazine sorption to soil solids. The good fit for the Langmuir model indicates that the solid phase surface area available for atrazine sorption in the presence of effluent is reduced. This is suggestive that interactions of effluent-derived organic macromolecules and colloids with the soil irreversibly blocked pesticide sorption sites or changed the sorptive nature of some sites.

Desorption of atrazine into $CaCl_2$, distilled water, and effluent was identical (Figure 2). The desorption isotherm obtained from the atrazine–$CaCl_2$ sorption point displayed some hysteresis while the desorption isotherm from the atrazine–effluent point was equivalent to the sorption isotherm. In other words, atrazine desorption was fully reversible when it was sorbed to soil solids in the presence of effluent, while it was only partially reversible when it was sorbed in the presence of $CaCl_2$.

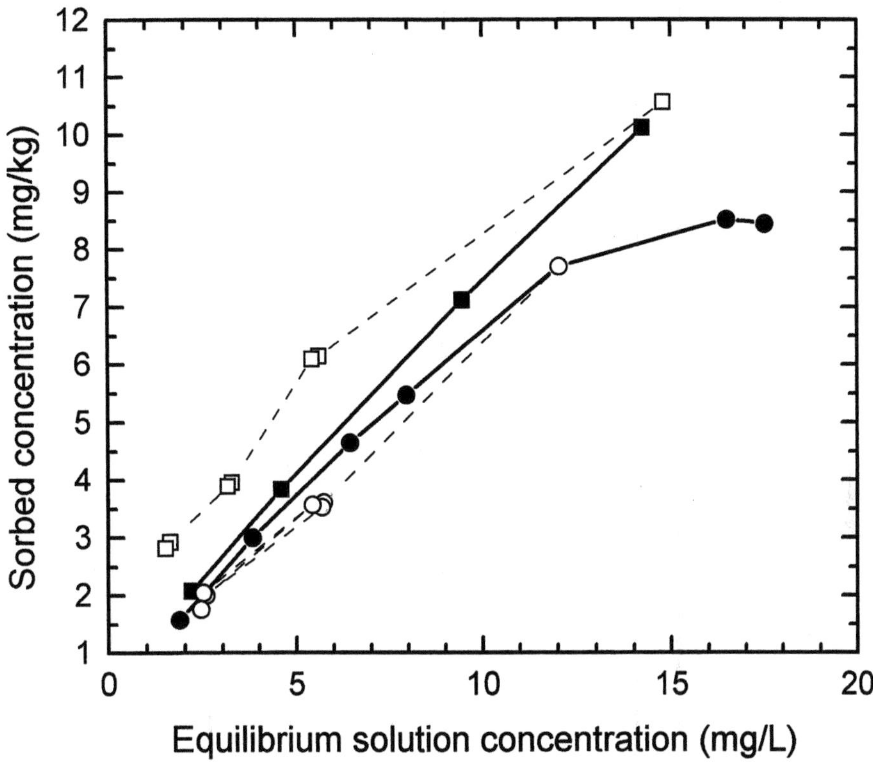


*Figure 2.* Equilibrium (24 h) sorption and desorption isotherms for atrazine in effluent. Desorption into 0.005 M $CaCl_2$, effluent, or distilled water (DW) was identical for both isotherms. Filled circles represent atrazine sorption from effluent; open circles represent desorption from effluent-sorbed isotherm. Filled squares represent atrazine sorption isotherm from 0.005 M $CaCl_2$; open squares represent desorption from $CaCl_2$-sorbed isotherm.

This may represent an additional mechanism which can result in enhanced transport of atrazine under irrigation with effluent.

## 5. Conclusions

Laboratory studies have shown that colloidal, soluble, and suspended organic macromolecules can be mobile through aquifer materials and soils. These macromolecules can form complexes with pesticides and other organic compounds. Interactions of such species with soil organic matter may also block pesticide sorption sites. Reversibility of atrazine sorption to soil solids in the presence of effluent is apparently increased compared with water. Each of these mechanisms can contribute to enhanced transport of pesticides through the soil–aquifer system. The agricultural use of marginal waters high in dissolved organic matter such as sewage effluents can thus pose a risk of deep subsurface and groundwater contamination by soil-borne pollutants. This may have important implications for the continued use of effluent in irrigated agriculture.

Further studies are required to determine how effluent quality, application regime, soil properties, and transported species properties affect the extent of enhanced transport. The possibility of enhanced downward transport of land-borne toxic organic materials due to land application of effluents should be appraised before applying effluents in groundwater recharge zones.

## References

Ballard T M 1971 Role of humic carrier substances in DDT movement through forest soils. Soil Sci. Soc. Am. Proc. 35, 145–147.

Chin Y P, Weber W J and Chiou C T 1991 A thermodynamic partition model for binding of nonpolar organic compounds by organic colloids and implications for their sorption to soils and sediments. In Organic Substances and Sediments in Water, Volume I. Humics and Soils. Ed. R A Baker. pp 251–273. Lewis Publishers, MI, USA.

Chiou C T 1989 Theoretical considerations of the partition uptake of nonionic organic compounds by soil organic matter. In Reaction and Movement of Organic Chemicals in Soils, SSSA Special Publication Number 22. Ed. B L Sawhney and K Brown. pp 1–30. Soil Science Soc. of America, Madison, WI, USA.

Conroy L E, Maier W J and Young-Tzung S 1981 Determination of carbohydrates and primary amines in river water. In Chemistry in Water Reuse. Ed. W J Cooper. pp 65–84. Ann Arbor Science, MI, USA.

Dunnivant F M, Jardine P M, Taylor D J and McCarthy J F 1992a Cotransport of cadmium and hexachlorobiphenyl by dissolved organic carbon through columns containing aquifer material. Environ. Sci. Technol. 26, 360–367.

Dunnivant F M, Jardine P M, Taylor D J and McCarthy J F 1992b Transport of naturally occurring dissolved organic carbon in laboratory columns containing aquifer material. Soil Sci. Soc. Am. Proc. 56, 437–444.

Graber E R, Gerstl Z, Fischer E and Mingelgrin U 1995 Enhanced transport of atrazine under irrigation with effluent. Soil Sci. Soc. Am. J. 59, 1513–1519.

Jury W A, Focht D D and Farmer W J 1987 Evaluation of pesticide groundwater pollution potential from standard indices of soil-chemical adsorption and degradation. J. Environ. Qual. 16, 422–428.

Karickhoff S W 1981 Semi-empirical estimation of sorption of hydrophobic pollutants on natural sediments and soils. Chemosphere 10, 833–846.

Magee B R, Lion L W and Lemley A T 1991 Transport of dissolved organic molecules and their effect on the transport of phenanthrene in porous media. Environ. Sci. Technol. 25, 323–331.

Manka J, Rebhun M, Mandelbaum A and Bortinger A 1974 Characterization of organics in secondary effluents. Environ. Sci. Technol. 8, 1017–1020.

Muszkat L, Lahav L, Ronen D and Magaritz M 1993 Penetration of pesticides and industrial organics deep into soil and into groundwater. Arch. Insect. Biochem. Physiol. 22, 487–499.

Ogner G and Schnitzer M 1970 Humic substances: Fulvic acid-dialkyl phthalate complexes and their role in pollution. Science 170, 317–318.

Rebhun M and Manka J 1971 Classification of organics in secondary effluents. Environ. Sci. Technol. 5, 606–609.

Stevenson F J 1985 Geochemistry of soil humic substances. In Humic Substances in Soils, Sediments, and Water. Ed. G R Aiken, D M McKnight, and R L Wershaw. pp 13–52. Wiley-Interscience, New York, USA.

Vinten A J A, Yaron B and Nye P H 1983 Vertical transport of pesticides into soil when adsorbed on suspended particles. J. Agric. Food Chem. 31, 662–664.

# 26
# Early detection of irrigation environmental risks using monitoring manholes: a case study of the Southern Ukraine

M. ZILBERBRAND

## 1. Introduction

Environmental damage caused by irrigation includes numerous soil problems such as salinization, alkalization, humus flushing, superfluous wetting, and groundwater problems (contamination and water-table rise). Due to inadequate monitoring methods, these problems usually tend to be detected only after considerable delay.

The method of monitoring soil processes is based on repeated surveys using soil sampling. Due to the high spatial variability of soil properties, such surveys provide the possibility of detecting dangerous trends only after they achieve relatively high levels, often damaging agriculture fields.

Groundwater pollution is usually detected after the pollutants have already entered the aquifer. In the best case, monitoring the contamination may be performed in the upper part of the saturated zone using multi-layer samplers (Ronen *et al.*, 1986; Schirmer *et al.*, 1995). Groundwater monitoring without monitoring the unsaturated zone is illogical (Cullen *et al.*, 1992; Durant *et al.*, 1993).

For monitoring contamination of the unsaturated zone, the US Environmental Protection Agency recommends the study of pore fluids of unsaturated sediments using shallow boreholes or trenches installed with vacuum lysimeters and tensiometers (Everett and Wilson, 1986). However, this makes it difficult both to perform the monitoring in dry periods and to obtain information from the wetter lower part of the deep unsaturated zone.

Special methods for monitoring water and solute transport throughout the deep unsaturated zone were developed by Ukrainian hydrogeologists in the early 1970s (Sitnikov, 1986; Zilberbrand and Gvirtzman, 1996). These methods are based on the use of special large-diameter boreholes (manholes). Such manholes, with a diameter of about 1 m, enable installation of monitoring equipment into their insulated walls and control of the functioning of this equipment by a person entering the manhole. This ensures the high quality of sampling and measurements and the possibility of both equipment replacement and addition. It was shown that the monitoring devices reliably describe the natural water and solute transport if put at distance of more than 0.5 m from the manhole wall and from each other (Sitnikov, 1986; Zilberbrand and Gvirtzman, 1996). Capillary pressures in sediments behind the walls and in surrounding sediments equilibrated within one month after manhole installation. The

*D. Rosen* et al. *(eds): Modern Agriculture and the Environment, 323–328.*

head of the manhole was covered with a funnel-like lid for collecting precipitation into a closed bottle. At the bottom of the manhole there was a small borehole for monitoring the water-table zone.

Special manholes were used for long-term monitoring of water-solute transport in the unsaturated zone at irrigated and drained lands (Golovchenko, 1984; Shvets, 1988; Sitnikov, 1986; Zilberbrand, 1987, 1993, 1995; Zilberbrand and Gvirtzman, 1996).

## 2. Site description and methods

The early detection of environmental risks, both to soil and groundwater, is presented here using data from the monitoring manhole near the village of Krestovka, Southern Ukraine, where the Kahovskaya irrigation system functioned.

The unsaturated zone was 10–24 m thick and was built of chestnut soil and loams of loessial formation underlain by clays. These sediments contain soluble salts, mostly sulphates, in quantities less than 0.5–1% of mass.

Groundwater of the Pliocence sand aquifer was of good quality, with chloride concentration less than 0.0015 *N*. Groundwater replenishment resulted from infiltration and amounted to 2–20 mm $yr^{-1}$ at non-irrigated sites and 40–110 mm $yr^{-1}$ under irrigation (Sitnikov, 1986).

Average annual precipitation was about 350–400 mm $yr^{-1}$ of which three-quarters fell in summer. The average temperature of the hottest month (July) was 22–24°C, while the average value for the coldest one (January) was −4°C. The growing of grain and forage crops was supplemented by sprinkler irrigation with intensity of about 400 mm $yr^{-1}$.

Previous observations using tensiometers have shown that in the loamy unsaturated zone preferential water penetration existed to depths of 2–3 m only (Sitnikov, 1986). At greater depths the wetting was much more uniform. From a depth of 3–4 m down to the capillary zone, volumetric water content changed in narrow limits of 30–32%.

Detailed construction of the special manhole has been described by Zilberbrand and Gvirtzman (1996). Using vacuum lysimeters, pore fluid samples (10–15 ml) were collected every 0.5–2 m all over the depth. Sampling was carried out for 2–4 days, once per month by applying a vacuum of 40–80 kPa.

Anion concentrations were determined by automatic titration: $Cl^-$ using the Mohr method, $SO_4^{2-}$ by the complexonometric method, and $HCO_3^-$ by potentiometry. Cation concentrations were determined using an ISP-51 flame photometer. Analytical errors were estimated to be less than 5%.

## 3. Results

The profiles of ion concentrations in pore fluids were obtained (Figure 1) and their changes over time were studied (Figure 2). The upper part of the unsaturated zone was usually flushed from salts. Below the flushed sediments there was an active

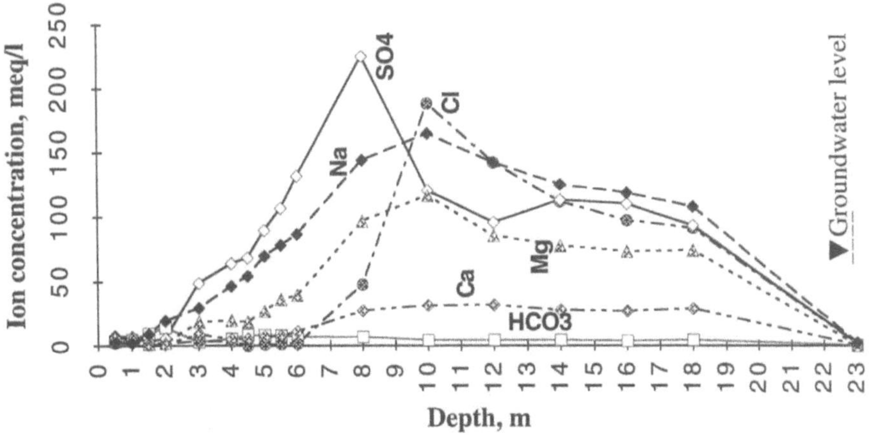


*Figure 1.* Profiles of ion concentration in pore fluids of the unsaturated loess loams under the irrigated field near the village of Krestovka, Southern Ukraine.

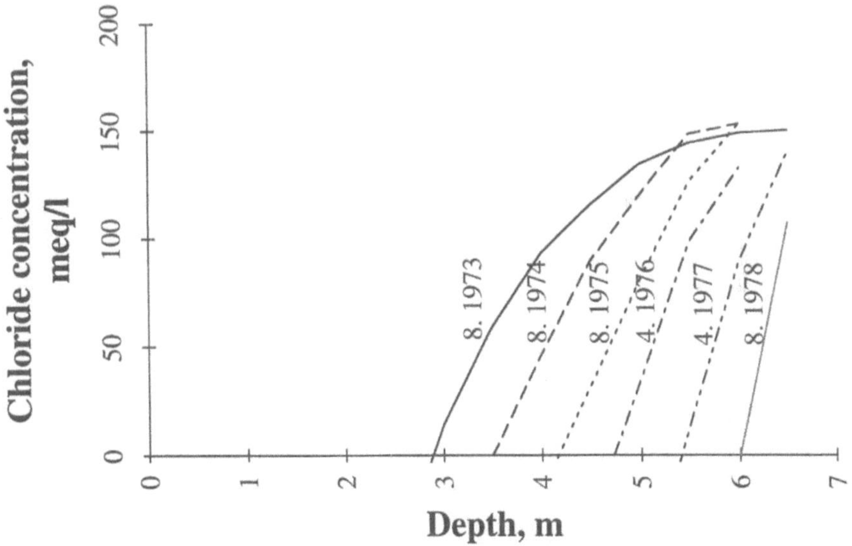


*Figure 2.* Displacement of chloride-concentration profiles under the irrigated field. Numbers show the sampling date.

desalinization zone of 4–6 m. The lower boundary of this zone moved downwards with continued irrigation. The highest ion concentrations were observed near this boundary. In the lower part of the unsaturated zone, ion concentrations decreased gradually with depth and achieved values typical for groundwater.

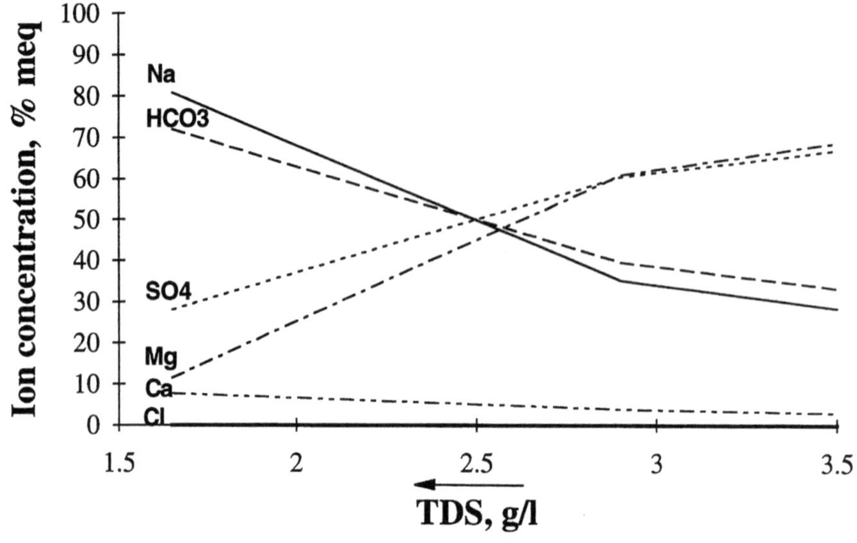


*Figure 3.* Changes in pore-fluid composition at a depth of 1.5 m during soil flushing. An arrow shows a direction of the TDS change.

The measured displacement rate of a moving downwards front of high chloride concentration was 0.5–0.7 m $yr^{-1}$ (Figure 2). For areas with especially intensive irrigation this rate reached 1–1.2 m $yr^{-1}$ (Zilberbrand,1993).

Changes of pore fluid composition during soil flushing and the total dissolved salts (TDS) decrease were studied at constant points all over the sediment profile. Figure 3 presents the trend at a depth of 1.5 m, near the lower boundary of the root layer. A decrease in the TDS during soil flushing caused an increase of relative concentrations (in % of total mEq) of sodium and bicarbonate.

## 4. Discussion and conclusions

Despite performing irrigation using fresh $HCO_3^-$–$Ca^{2+}$ surface water of excellent quality, and the presence of low groundwater TDS, the unsaturated zone contained a ‘time bomb’ consisting of high-salinised pore fluids (up to 0.2 *N* $Cl^-$ and 0.25 *N* $SO_4^{2-}$). This bomb was slowly drifting towards the groundwater and would cause maximum damage in about 20 years. Monitoring groundwater composition did not reveal any dangerous trend.

Monitoring pore fluids in the unsaturated zone using special manholes provided the possibility to predict groundwater hazard and to incorporate this information in planning the water economy of the region. The obtained danger signal has raised the problem of the irrigation regime revision.

The risk of soil alkalization under irrigation using $HCO_3^-$–$Ca^{2+}$ fresh water has never been considered. However, monitoring pore fluids revealed this process. The mechanism of such alkalization has been described in detail by Zilberbrand (1995). This process was found to occur not only under the temperate continental climate of the Southern Ukraine, but also under the arid climatic conditions of the Northern Negev, Israel.

In both regions the root zone acts as a geochemical barrier which retains the calcium predominating in rain and irrigation water, due to calcite and gypsum precipitation. These processes are accompanied by $HCO_3^-$ and $SO_4^{2-}$ extraction from the pore fluid and by an increase in the TDS as a result of evapotranspiration. Outflowing from the root zone, the pore fluids impoverished in calcium but enriched in sodium and magnesium react with the exchangeable complex of loess, composed mostly (80%) of calcium. The cation exchange results in an alkalization of the sediments below the layer of maximum carbonate precipitation.

Preferential water flow in the root zone, after precipitation and irrigation, supplies $HCO_3^-$–$Ca^{2+}$ surface water into the alkalised soil layer. As a result of the following stage of cation exchange, the soda appears and becomes predominant in pore fluids (Figure 3). In principle, it may be transported upwards by an unsaturated water flux occurring in the root zone in hot periods. The described processes, studied using pore fluid data, were not detected during repeated soil surveys based on soil sampling.

The above examples support the use of manhole monitoring pore fluids in the unsaturated zone for the early detection and quantitative estimation of both soil and groundwater environmental problems caused by irrigation.

## Acknowledgements

The author is grateful to Dr Haim Gvirtzman and Dr Rivka Amit, both of the Institute of Earth Sciences, the Hebrew University of Jerusalem, for their comments on the manuscript.

## References

Cullen S J, Kramer J H, Everett L G and Eccles L A 1992 Is our ground water monitoring strategy illogical? GWMR 12, 103–107.

Durant N D, Myers V B and Eccles L A 1993. EPA's approach to vadose zone monitoring at RCRA facilities. GWMR 13, 151–158.

Everett L G and Wilson L G 1986 Permit guidance manual on unsaturated zone monitoring for hazardous waste land treatment units. Office of Solid Waste Rep. No. EPA 530-SW-86–040.

Golovchenko Y 1984 Regularities of water-salt regime of the unsaturated zone and ground water replenishment estimation at the irrigated lands of the Southern Ukraine. PhD Thesis, Institute of Geological Sciences of the Ukrainian Academy of Sciences, Kiev, Ukraine. 203 pp (in Russian).

Ronen D, Magaritz M and Levy I 1986 A multi-layer sampler for the study of detailed hydrochemical profiles in groundwater. Wat. Resour. Res. 20, 311–315.

Schirmer M, Jones I, Teutsch G and Lerner D N 1995. Development and testing of multiport sock samplers for groundwater. J. Hydrol. 171, 239–257.

Shvets A 1988 Water and salt transfer regularities in the unsaturated zone sediments of the Crimea Plain. PhD Thesis, Institute of Geological Sciences of the Ukrainian Academy of Sciences, Kiev, Ukraine. 197 pp (in Russian).

Sitnikov A 1986 Water and salt dynamics in unsaturated soils and rocks. Naukova Dumka, Kiev, Ukraine. 152 pp (in Russian).

Zilberbrand M 1987 Water and salt transfer regularities in unsaturated zone and principles of water-salt regime management at irrigated lands in Southern Ukraine. PhD Thesis, Institute of Geological Sciences of the Ukrainian Academy of Sciences, Kiev, Ukraine. 248 pp (in Russian).
Zilberbrand M 1993 A new method for estimation of contaminant travel rate in unsaturated zone under irrigated soils. Wat. Sci. Technol. 27, 173–178.
Zilberbrand M 1995 The effect of carbonates and gypsum precipitation in the root zone on the chemical composition of groundwater. J. Hydrol. 171, 5–22.
Zilberbrand M and Gvirtzman H 1995 Monitoring of water and solute transport in unsaturated zone using large-diameter borehole. Groundwater, 34, 57–65.

# 27
# Use of wastewater from livestock farms for irrigation in Yugoslavia

J. MILIVOJEVIĆ, M. LJEŠEVIĆ AND M. NEDIĆ

## 1. Introduction

Yugoslavia ranks among the countries rather rich in water and soil resources: 6.3 million ha of agricultural soil, 4.9 million ha cultivated and surface water flow of 1.019 $m^3$ $s^{-1}$; from that, during the summer period 16.03 billion $m^3$ could be used for irrigation of 2.5 million ha without hydroccumulation buildings and use of underground waters.

From the total area of agricultural soil, 96% or 4.6 million ha is private property and the remainder of the 220 000 ha belongs to the state, where a few hundred large livestock farms on relatively small areas are organised on 500 up to 10 000 ha.

The socialist type of economy planning prevailing during the last 50 years has privileged the development programmes of large livestock farms (5000 up to 130 000 heads), with the use of liquid manuring systems; this is now a considerable ecological problem.

In such a situation and in a view of the recently passed Law for Protection of the Environment, and the European regulations, 'Nitrates', related to the limit of nitrogen of 170 units per hectare, farms are facing big problems which they are not able to solve. The concentration of a large number of domestic animals in a rather limited area is coming back to both farmers and the government, like a boomerang.

With the aim of setting the problem in the framework of the national research project, 'Wastewaters and Environmental Protection', parameters for model sizing of farms and the use of liquid manure in irrigation have been worked out, to protect the environment and produce healthy food.

## 2. Production and evacuation technology of liquid manure on a farm

On 500 large livestock farms, production cycles of liquid manure in Yugoslavia are present all the year around. Its evacuation and application on agricultural fields with irrigation could last a maximum of 168 days, depending on the season and vegetative cycles.

The designed accumulation storage for liquid manure cannot accept excrements in washing water under proscribed and inadequate estimates of agricultural field size.

*D. Rosen* et al. *(eds): Modern Agriculture and the Environment, 329–338.*

Problems could be solved if consideration is given to production, placement, transport, use of liquid manure in irrigation, and environmental protection.

### 2.1 Production

Many farms are designed to use 2 L of water for 1 kg of dry organic mass of animal excrement and for its distribution to the storage place. But due to farmers' non-discipline and changed conditions in practical work regarding design, this condition is not adhered to on farms.

### 2.2 Storage

Liquid manure collection is based on free release through special canals from production area to land or concrete pools for 'maturing' and evacuation on the fields (Figures 1 and 2).

*Figure 1.* Concrete pool for liquid manure collection.

### 2.3 Transport

Liquid manure evacuation from lagoons to agricultural fields could be done with typhoons and a centre pivot with irrigation water, with aggregate pumps or tractor operation pumps (Figure 3) or with special tractor cisterns and distribution runs with the help of special plant (Figure 4).

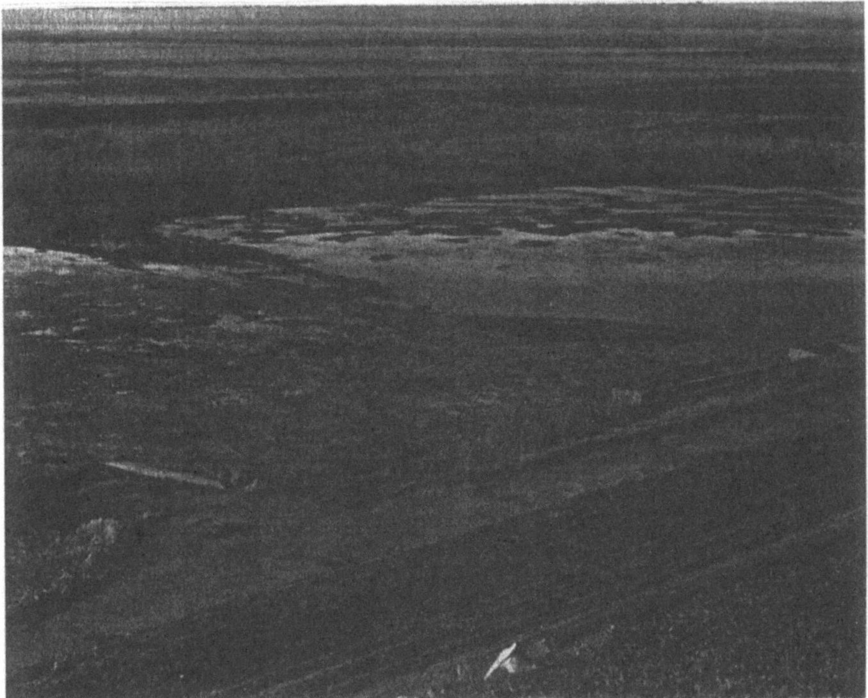

*Figure 2.* Land lagoon for liquid manure collection.

### 2.4 *Using liquid manure in irrigation*

Possibilities for application and some effects of liquid manure use on one correct leaded farm in Yugoslavia are presented in Table 1 (Todorovic, 1993).

### 2.5 *Environmental pollution*

With the aim of solving the problem of wastewater as presented in the example of 'Halas Jozef' farm in Vojvodina, a small number of farms follow in their steps in solving their problems.

Many farms have no storing facilities at all, while some others use water for cattle excrement wash in a ratio of 1:10 or even 1:20 instead of the anticipated 1:2, or they increase the number of cattle above that projected; therefore they soon exceed their capacities.

From lagoons, or direct from stables, farmers release wastewater into rivers, canals, watercourses, pools, lakes and reeds, where nearby farms are usually located. In this way, surface water contains a large volume of organic materials, mineral salts

*Table 1.* Transport intervals from lagoons to production fields, parameters and effects of wastewater (liquid manure) use on 'Halas Josef' farm in Vojvodina.

| Time intervals of liquid manure transport from lagoons to the fields | Crops | Application doses ($m^3$ $h^{-1}$) Irrigation: Water $1200\,m^3 + 400\,m^3$ liquid manure | Non irrigation: – | Irrigation: $400\ m^3$ liquid manure | Possible number of transport days | Accumulation and in pools storage – no. of days |
|---|---|---|---|---|---|---|
| | | Yields (t $ha^{-1}$)[a] | | | | |
| 1 | 2 | 3 | 4 | 5 | 6 | 7 |
| | Wheat (grain) | 7.9 | 6.4 | 6.5 | 30 | – |
| 1.04–30.04 | Maize (grain) | 12.5 | 9.5 | 10.0 | | |
| | Maize (grain) | | | | | |
| 1.05–31.05 | Lucerne (hay) | 18.5 | 14.2 | 15.3 | 31 | – |
| | Peas (consume) | 8.2 | 5.7 | 6.5 | | |
| | Bean pods (consume) | 11.3 | 7.4 | 10.6 | | |
| 1.06–30.06 | Paper (consume) | 45.7 | 22.3 | 25.9 | 30 | – |
| | Paper (industrial) | 28.8 | 19.5 | 20.1 | | |
| | Lucerne | | | | | |
| | Buckwheat | 1.5 | 0.8 | 0.9 | | |
| | Soybean | 4.3 | 3.4 | 3.1 | | |
| 1.07–15.09 | Maize | | | | 76 | – |
| | Sugarbeet | 71.3 | 68.1 | 65.4 | | |
| | Lucerne | | | | | |
| | Fields after reaping | | | | | |
| | – wheat | | | | | |
| 16.09–15.11 | – soybean | – | – | – | 30 | – |
| | – maize | | | | | |
| | – sugar beet | | | | | |
| | – lucerne | | | | | |
| 16.11–31.03 | – | – | – | – | – | 197 |
| Total number of transport days and number of accumulation and storage days of liquid manure: | | | | | 168 | 197 |

[a] Contributions are from production facilities (not experimental) in the frame of irrigation system (F = 1.500 ha), soil type 'chernozem' (one of the most fertile soils in the world). Annual rainfall 630 mm (in the period March–October, 180–420 mm).

*Figure 3.* Transport of liquid manure through pipeline to irrigated field.

and other kind of pollution. One $m^3$ liquid manure makes 10–50 $m^3$ surface water unusable (MAFF, 1979; MZV CSR, 1979; Stehlik, 1988).

On the other hand some farms, designed with production control facilities, storing and liquid manure transport, mostly use these waters without respect for crops' requirements, physical soil quality, neighbouring underground water, or the European 'Nitrates' regulation, 170 units N (RNED, 1991). In many cases the high concentration of $K_2O$ in soil is recognised, that in irrigation conditions become movable (Stehlik, 1988), and washed together with $NO_2$–N and $NO_3$–N, pollutes the underground water; this was intended in Yugoslavia to be used as drinking water and in the nutrition industry.

## 3. Programming model of production and application of liquid manure in irrigation

The model starts with the theoretical point that it is possible to define the characteristics of liquid manure from cattle (block scheme of model is given in

*Figure 4.* Liquid manure transport with tractor cistern–irrigation on field under husbandry crops.

Figure 5). First, the specific production (S) of manure (by one animal) must be known, then the chemical, microbiological, and other characteristics which are specific to different cattle species:

$$S = f(g, t)$$

where g = animal growth and t = animal type.

The chemical characteristics of liquid manure show that it can be used as a substitution for artificial fertilizers in the process of crop production. That means that a certain quantity of artificial fertilizers with their specific nutrition characteristics can be substituted with another, usually much larger quantity of liquid manure, if it is assumed that a larger quantity of liquid manure has an equal quantity of nutritive elements. By this procedure the new model relation that shows that relation of a specific mass of a standardized type of artificial fertilizer and an equivalent mass of liquid manure has been set.

The next step in developing the model is a consideration of economic factors. Limiting factors for using liquid manure must be defined. In this case, the crucial factor is the economically justified distance for transporting liquid manure ($L_{max}$). For its definition the relation between the price of transporting and application of artificial fertilizers, and the equivalent mass of liquid manure, must be established.

This parameter can be crudely defined by supposing that the equivalent transport expenses of artificial fertilizers are equal to the transport expenses of the same mass of liquid manure:

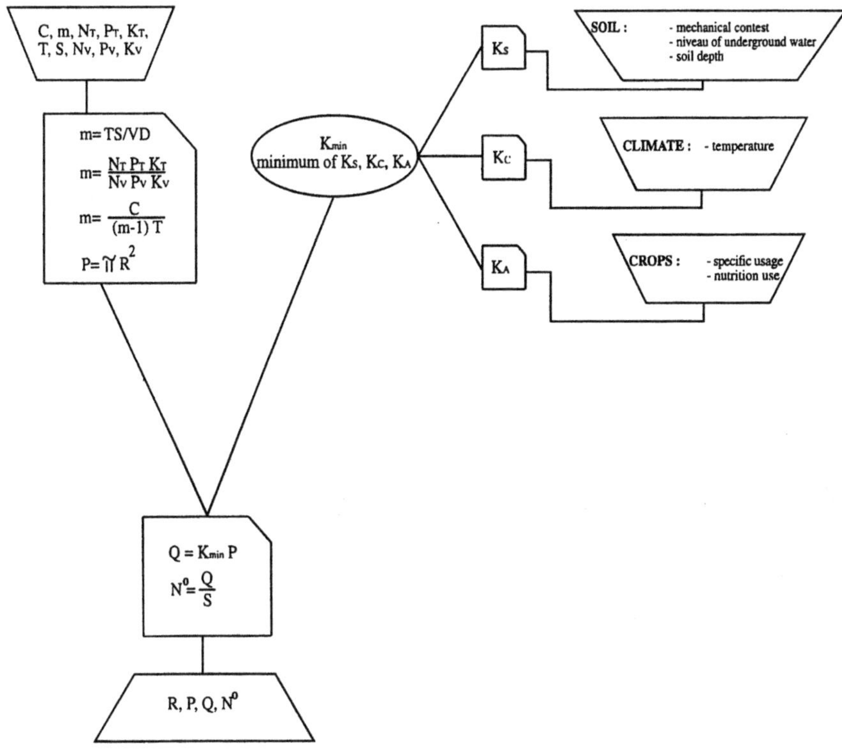


*Figure 5.* Block diagram of liquid manure use programme.

$$L_{max} = \frac{C_f}{(m-1)T}$$

where:

Cf = fertilizer procurement expenses per unit;
T = fertilizer transport expenses per unit;
m = coefficient of relation between mass of liquid manure and mass of artificial fertilizer; and
$L_{max}$ = economically justified distance for manure transport.

By this procedure, in the next phase, can be defined the value of the economically approved field size (F) around the farm where the application of liquid manure is going to be done; the minimal approved distance between farms can also be explained.

The next phase in developing a model is to establish a mass of liquid manure that can be applied on an equivalent field size during the year. For the precise definition of this parameter, the quantity of liquid manure, which the soil can absorb ($K_S$),

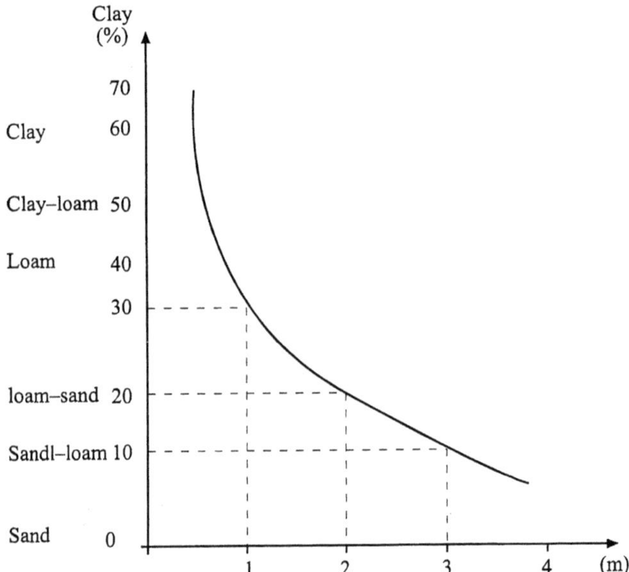


*Figure 6.* Permitted profundity of underground waters in cumulation with clay fraction contents during irrigation with liquid manure (Novotny *et al.*, 1989)

must be established. This parameter depends on soil texture (absorptive soil complex), and on the equivalent clay fraction. It can be three times larger in soils with a heavy mechanical texture (heavy clays), than in sandy soils. Underground water depth is also very important. Sandy soils must not have a shallow underground water level, because in this case the rinsing of liquid manure is extremely large. This problem does not exist in clays, but if the underground water is shallow, the anaerobic process of decomposing organic matter in liquid manure can occur, and the yield can be reduced:

$$K_s = f_s(C_y, D_{gw}, D_s)$$

where

$C_y$ = clay content (< 0.002 mm);
$D_{gw}$ = underground water depth; and
$D_s$ = soil profundity.

The next limiting parameter is the arable crop. If we presume that with liquid manure we are going to apply an equivalent effect on the crop as with an artificial fertilizer, the maximum limit for its usage will be determined by the specific consumption of the arable crop. Because this parameter is different for each arable crop, there is a need for determining the main crop, whose specific usage is going to be used for determining the requirement for mass liquid manure ($K_C$).

$$K_C = f_p\,(S_p)$$

where $S_p$ = basic crop in row.

The final limiting parameters are climatic ones. An increase in temperature accelerates the liquid manure decomposing process. There is a possibility of percolation in underground water; that is why small amounts of liquid manure must be applied on an equivalent field surface. Climatic parameters therefore define the maximal mass of liquid manure, in some areas, which soil can transform ($K_A$).

In the next phase of the developing model, all the parameters which laminate the mass of liquid manure ($K_S$, $K_C$, $K_A$) must be compared. The main limiting parameter which presents the optimal mass of liquid manure that can be applied on an equivalent field surface, without any risks to the soil or environment, during the one year ($K_{min}$), must be established. The total mass of liquid manure (Q) which can be produced during one year, can be presented by multiplication of this parameter (K) with the economically approved field size (F):

$$K_{min} = \min\,(K_s, K_p, K_a)$$

$$Q = K_{min} \times P.$$

In the last phase, the maximum number of cattle, which can be raised during one year, must be established by dividing total mass of liquid manure (Q) and specific production of liquid manure per cattle (S).

$$S = f(g, t)$$

where g is animal growth and
t is animal type

$$N^0 = \frac{Q}{S}$$

where g is animal growth and
t is animal type

## 4. Conclusion

In the ecological conditions of Yugoslavia, wastewaters from livestock farms (liquid manure) are treated more as environmental pollution and an alternative for fertilizers, than as additional water sources for irrigation.

Transport and distribution of liquid manure with irrigation water gives 10–104% higher contribution in agriculture, and by tractor cisterns transport and distribution without irrigation parameters are 1–43%. Because of inappropriate doses for the soil, climate and economic conditions, farmers often pollute soil, underground waters and crops.

A mathematical model for production programming and application of liquid manure in irrigation, could be used during new farm design, during reconstruction of storage capacities of existing farms for liquid manure, by economical effects analyzes regarding liquid manure use in irrigation, or analyzes of environmental protection from livestock liquid manure pollution.

## References

1. RNED 1991 Lisier de porc et fertilisation azotee, RNED Porc–ITP, Techni-Porc. 1991.
2. Ministry of Agriculture, Food and Fisheries (1979) Agriculture and Pollution, 7th Report, Royal Commission on Agricultural Pollution, Cmnd 7644, HMSO, London.
3. Ministarstvo zemedelstvia a vyzivy CSR 1979 Techicke a agronomicke zasady zavlah otpadnimi vodami pp 115–117.
4. Novotny M, and Šanta M and Olbrecht V (1989) Využivanie exkrementov v sulade s ochramou životneho prostredia. p 15. Bratislava, Vzuh.
5. Stehlik K 1988 Zavlahove využivi otpadnich vod iv. Tekuta statkova hnojiva a zemedelske otpadni vody, Ministarstvo zemedelstvi a vyžiny CSR, Prahe, p 43, 239, 374–375.
6. Todorovic N 1993 Proizvodnja i upotreba tecnog stajnjaka u biljnoj proizvodnji.
7. Tivy J 1990 Agricultural Ecology, Longman, Essex.

# SECTION VI

# Recycling of Municipal and Agricultural Wastes

## 28
# Composting and recycling of organic wastes

YONA CHEN, YOSSI INBAR, BENNY CHEFETZ AND YITZHAK HADAR

### 1. Introduction

Both quality and quantity of organic matter (OM) in soils affect plant growth and health. Crop residues and root exudates affect the activity of beneficial microorganisms, as well as pathogens (Chang *et al.*, 1986). The effects of OM decomposition level in the soil on interactions of beneficial microorganisms and pathogens, and on plant growth have been largely overlooked (Grebus *et al.*, 1994; Hoitink and Fahy, 1986). Both the mineral fraction and the yearly introduced organic residues, as well as their resulting transformations in soils, play a role in this complex. In addition, humic substances (HS), the most stable organic components in soil, may have both direct and indirect effects on plant growth (Chen and Aviad, 1990; Chen *et al.*, 1994; Stevenson, 1985).

The composting of organic wastes has been defined as a controlled microbial aerobic decomposition process. Recently composting has gained interest in western countries as a result of the accumulation of large amounts of agricultural, industrial and municipal wastes (de Bertoldi, 1993; de Bertoldi *et al.*, 1987; Golsdtein and Steuteville, 1995; Steuteville, 1994a,b). These wastes create a severe disposal problem involving a number of environmental aspects related to air, soil, ground water and surface water pollution. During the course of the composting process, transformations of various OM fractions can be followed separately without analytical interference from OM already present in the soil. The composting process therefore provides a unique opportunity to evaluate effects of specific OM transformations on plant growth and health and may serve as a model for field soil systems. The main products of aerobic composting are $CO_2$, water, minerals and stabilised OM, often called humus. Well-composted OM is traditionally used as a source of nutritional elements and/or soil conditioner directly in the field. In the last decade, the demand for peat as a potting medium in horticulture has increased while its availability has decreased (Chen *et al.*, 1992). A number of organic wastes such as bark, leaf mould, municipal solid waste (MSW), sewage sludge, sawdust, and treated animal excreta have been introduced as peat substitutes in growth media after proper composting (Bik, 1983; Chen *et al.*, 1988, 1992; Cull, 1981; Inbar *et al.*, 1985; Lohr *et al.*, 1984; Raviv *et al.*, 1986, Verdonck, 1984). Some of the well-matured composts have been found to be suppressive of soilborne plant pathogens when utilised as container media (Chen *et al.*, 1988; Hoitink and Fahy, 1986; Hoitink and Kuter, 1986; Hoitink *et al.*, 1993; Mandelbaum *et al.*, 1988).

*D. Rosen* et al. *(eds): Modern Agriculture and the Environment, 341–362.*

## 2. Principles of composting

Composting is a form of waste stabilization that requires special conditions, particularly of moisture and aeration, to yield temperatures conducive to thermophilic microorganisms (Hoitink and Fahy, 1986). Zucconi and de Bertoldi (1987) proposed the following definitions: composting is a controlled biooxidative process that: involves a heterogeneous organic substrate in its solid state; evolves by passing through a thermophilic phase and a temporary release of phytotoxins; and leads to production of $CO_2$, water, minerals and stabilised OM (compost). Compost is the stabilised and sanitised product of composting which is beneficial to plant growth. The OM has undergone an initial, rapid stage of decomposition and is in the process of humification.

Controlled environmental conditions distinguish composting from natural rotting or putrefaction, which often occurs in open dumps, manure heaps, or field soil. A simplified flowchart of the overall composting process is shown in Figure 1. This flowchart illustrates the interactions between the organic waste, microorganisms, moisture and oxygen. The organic wastes, comprising several components, normally have an indigenous population of microorganisms derived from the atmosphere, water or soil. Once the moisture content of the waste is brought to an appropriate level and the mass is aerated, microbial metabolism speeds up. Energy is obtained by biological oxidation of part of the carbon; some of this energy is used in metabolism, the rest is given off as heat. The end product (compost) is made up of the more resistant residues of the OM, breakdown products, dead and living microorganisms, together with products from further chemical reactions between these materials (Gray and Biddlestone, 1981).

The temperature trend in most composting processes is presented in Figure 2. Basically, the process can be divided into four stages:

(1) *Mesophilic stage* – an initial phase of 1–2 days, during which the mesophilic strains of microorganisms start to decompose the readily degradable compounds, heat is given off and the temperature rises. The pH falls slightly as organic acids are produced.

(2) *Thermophilic stage* – the second stage, lasting 2–4 weeks, when above 40°C the thermophilic strains take over. If the temperature rises above 60°C, the fungi become deactivated, and the reaction is continued by thermophilic bacteria. In this phase the more readily degradable substances such as sugars, fats, starch and proteins are rapidly consumed and most of the human and plant pathogens are destroyed; the pH becomes alkaline as ammonia is liberated from the proteins. The reaction rate decreases as the more resistant materials are attacked; the compost enters stage 3.

(3) *Cooling-down stage* – as the temperature falls, the thermophilic fungi reinvade the composted material from the cooler extremities and start to attack the cellulose, hemicellulose and other polymers. These first three stages of

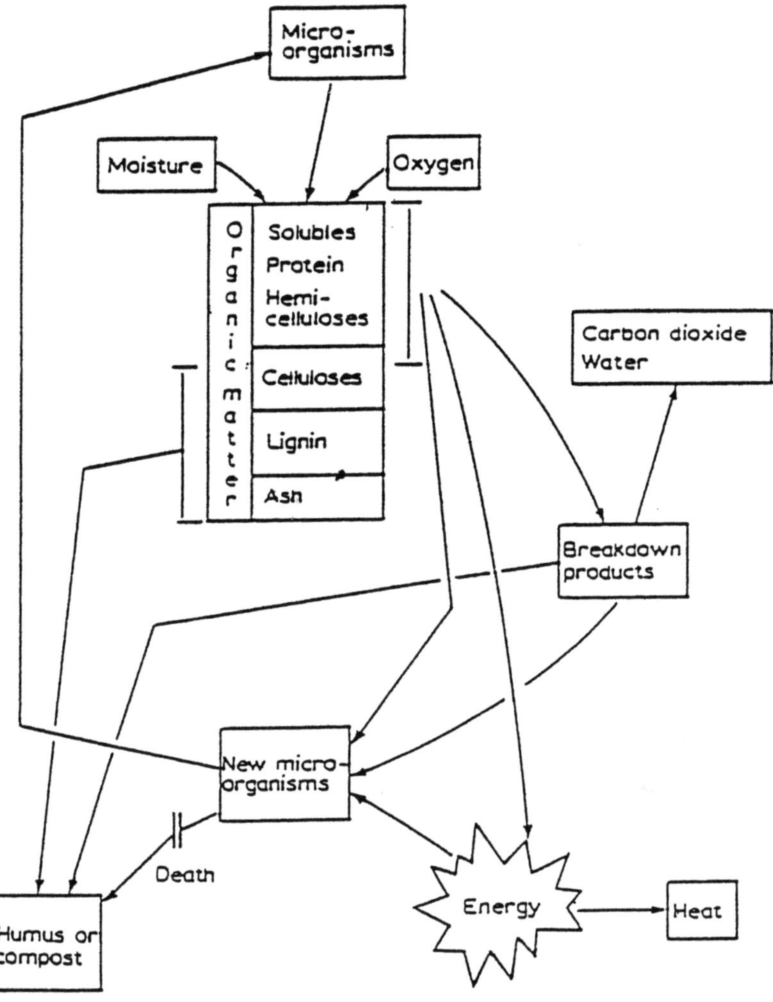


*Figure 1.* Simplified flowchart of the composting process (from Gray and Biddlestone, 1981).

composting last from a few weeks to two months, depending on the composted material.

(4) *Stability and maturity* – requires several months, where little heat is generated and the final pH is normally slightly alkaline. During this phase, mesophilic microorganisms as well as macrofauna colonise the compost. Intense competition for food takes place between the microorganisms, involving antagonism and the formation of antibiotics. Humification occurs in the residual OM to produce the stable composted product (Adani *et al.*, 1995; Chen and

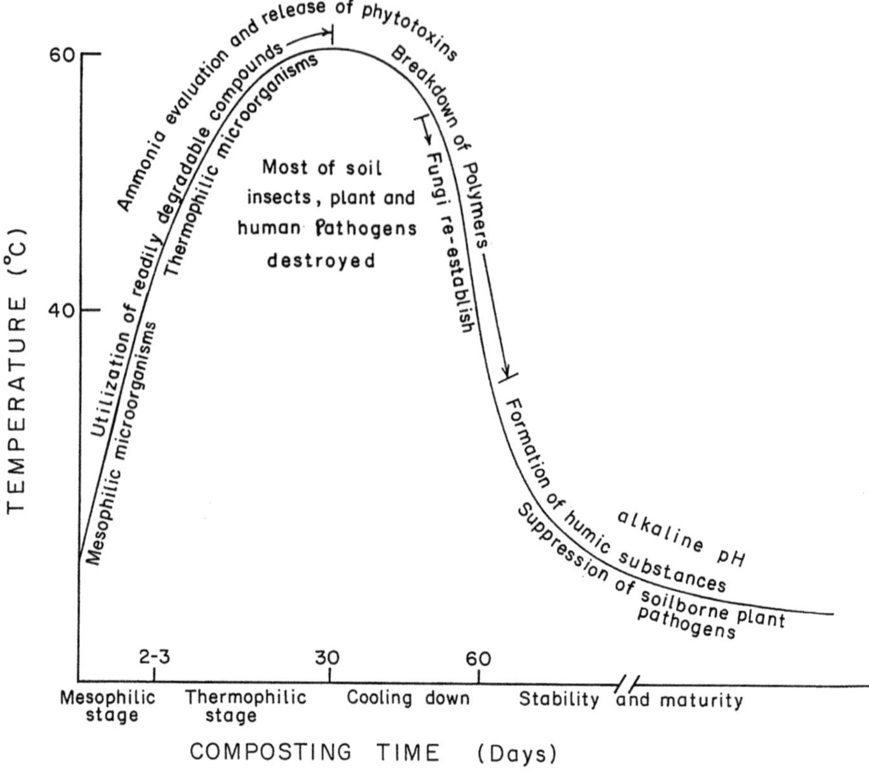


*Figure 2.* Temperature and processes vs. time of composting.

Inbar, 1993; Ciavatta *et al.*, 1993; Diaz *et al.*, 1982; Gray and Biddlestone, 1981; He *et al.*, 1992; Hoitink and Fahy, 1986; Inbar *et al.*, 1993; Jimenez and Garcia, 1992; Zucconi and de Bertoldi, 1987).

The composting of organic wastes is a dynamic and extremely complicated ecological process in which temperature, pH and nutrient availability are constantly changing. As a consequence, the numbers and species of organisms responsible also change markedly. Thermophilic bacteria, particularly *Bacillus* spp. and *B. cereus* (Finstein and Morris, 1975; Kwok *et al.*, 1987), appear to dominate the early phase of high activity, but thermophilic actinomycetes predominate thereafter (Nakasaki *et al.*, 1985). The rate of decomposition depends on the rates of microbial activity, the latter being greater when composts are maintained at low (38–55°C) rather than high temperatures (60–70°C) (Kuter *et al.*, 1985). In addition to temperature, optimum starting material values for composting are: C/N ratio 30–35; C/P ratio 75–150; moisture content 50–60%; particle size 15–35 mm; pile size 1.5 m high and 2.5 m wide.

## 3. Compost maturity

Compost maturity is defined as the degree of OM decomposition during composting. Any definition of maturity must be based on the potential use of the compost. For example, compost used in potting mixes (container media) must be mature enough so that no plant growth inhibition occurs within days after formulating of the mix. On the other hand, composts utilised in field agriculture may be applied weeks or months before planting, giving them time to 'stabilise' in the soil. Thus, the following definitions are proposed for mature composts:

(1) Greenhouse utilization: OM composted to the degree of decomposition which has no adverse effects on container grown plants.

(2) Field application: OM composted to the degree of decomposition that has no adverse effects on the growth of various crops when applied at annual rates of up to 50 tons per hectare, at least six weeks before sowing or planting (during the warm season, in which decomposition can take place).

### *3.1 Chemical parameters for compost maturity*

The chemical nature of the decomposition process and the evaluation of compost maturity are a difficult problem, since several parameters are required to provide a definite evaluation for the readiness of the material for use in horticulture (Morel *et al.*, 1985; Zucconi and de Bertoldi, 1987). Some of the main changes taking place in the solid phase of the OM during decomposition in soils and/or in partially controlled environments such as a compost pile are: an increase in ash content, a decrease in cellulose and hemicellulose, and generally, as a result of concentration, a relative increase in lignin (Harada and Inoko, 1980b; Inoko, 1982).

Several procedures for testing stability or maturity of composts have been proposed. Some examples are: starch content, cellulose content, C/N ratio, C/N ratio in water extracts of composts, concentration of HS, presence of indicator microorganisms, respiration rate, and others (Chanyasak and Kubota, 1981; Chanyasak *et al.*, 1982; Harada and Inoko, 1980a,b; Harada *et al.*, 1981; Inbar and Chen, 1993; Inbar *et al.*, 1989, 1990b; Lax *et al.*, 1987).

The C/N ratio represents one of the major guidelines for the composting process. However often the C/N ratio of bulk composts does not provide useful information; in wood wastes for example, much of the carbon is not readily biodegradable. For such wastes, availability must be considered. High 'available' C/N ratios of immature or inadequately composted materials cause immobilization of nitrogen and nitrogen deficiency in plants. High microbial activity existing in immature compost requires $O_2$ and consequently may cause oxygen deficiency to roots. During the decomposition of fresh or partially decomposed OM, adverse effects on plant growth result from toxic fermentation by-products produced in the early stages of composting. This phenomenon is well-known in relation to the incorporation of crop residues into soils. Lynch *et al.* (1980) observed a reduction in root growth of young barley plants cultured in different soils with added straw. They attributed the reduction to an accumulation of acetic acid produced by decomposition of the straw. Toussan *et al.* (1968) identified benzoic acid and phenylacetic acid as phytotoxins present in the decomposition products of barley sampled in the field. Acetic acid has also been identified as a phytotoxin in immature composts by De Vleeschaawer *et al.* (1981). Anid (1986) identified high levels of acetic, propionic and butyric acids in immature compost, whereas those volatile acids disappeared when stabilisation was achieved. Wong and Chu (1985) noted severe suppression of germination when seeds of *Brassica chinesis* were treated with extracts of fresh refuse compost. Fewer adverse effects were found with 6 week-old compost. Differences from the control were not found with the 12 month-old compost. These investigators speculated that many factors could be influencing germination and root elongation including pH, salt and ammonia contents, ethylene oxide concentration, and heavy metals.

Zucconi *et al.* (1981a,b) tested the response of a wide range of plant species in various bioassays to fresh, immature and mature composts. In general, all plants were inhibited by immature compost. Growth usually improved upon contact with mature compost. These authors concluded that phytotoxicity during composting appears to be strictly associated with the initial stage of decomposition and that it is a transient condition, possibly connected to the presence of readily metabolisable material. Low molecular weight organic acids (acetic, propionic, and butyric) were the most toxic compounds found in the early stages of the composting process (Chanyasak *et al.*, 1983; De Vleeschauwer *et al.*, 1981; Hirai *et al.*, 1986; Zucconi *et al.*, 1981a,b). Finally, allelopathic toxins in wastes can cause problems, but they are generally destroyed within a few weeks of composting. Hoitink and Kuter (1986) concluded that the negative responses of plants grown in bark composts were due to high cellulose content resulting in nitrogen immobilisation, allelopathic chemicals, toxic levels of metabolites of anaerobic metabolism, high salt content and, possibly, other factors. They concluded that appropriate curing is essential to stabilise compost and eliminate or reduce negative plant responses (Hoitink and Fahy, 1986). Inbar *et al.* (1988) measured $O_2$ consumption by fresh and composted separated manure (CSM) and grape marc (CGM) at various moisture levels, with and without the addition of nitrogen, as a function of composting time in a simulation system. For both materials, cumulative $O_2$ consumption increased rapidly at first, then levelled off. Maximum $O_2$ uptake was obtained at moisture levels of 60 to 70% (on a net weight basis). The addition of nitrogen increased the $O_2$ uptake rate in some instances, more markedly in the CGM which is low in nitrogen content. The $O_2$ uptake rate vs. time approached levelling off after a few days of composting in the pilot-scale system, exhibiting a slow approach to a steady state in microbial activity. As opposed to the fresh OM for both materials, the composted products had low $O_2$ uptake rates, indicating that $O_2$ consumption provides a good measure of the rate of decomposition and the degree of stability. A similar relationship between $O_2$ uptake and MSW composting age was observed by Iannotti *et al.* (1994). This measure, as well as $CO_2$ release, has been used by microbiologists for many years for this purpose. However, levelling off or a lower rate of $O_2$ uptake is often achieved before

*Table 1.* Effects of composting of the solid fraction of a slurry produced from methanogenic fermentation of cattle manure on tomato seedling response (from Hadar *et al.*, 1985).

| Treatment | Unfertilised | Fertilised |
|---|---|---|
| | Dry weight (mg/plant)* | |
| Raw material | 166b | 30c |
| Immature compost | 319a | 486b |
| Mature compost | 355a | 612a |

*Values followed by the same letter do not differ significantly at R = 0.05 according to Duncan's multiple range test.

compost can be considered mature for greenhouse utilization (Inbar *et al.*, 1988). This may not necessarily be the case for compost applied to soil well ahead of planting, where only some stability is required.

To avoid problems caused by carbon availability, the ratio of water-extractable organic C/organic N was proposed as an indicator of compost maturity by researchers in Japan (Chanyasak and Kobuta, 1981; Chanyasak *et al.*, 1982, 1983). These researchers found values of 5 to 6 for the soluble organic C/organic N ratio of mature compost of MSW. A highly significant negative correlation was established between the cation exchange capacity (CEC) and the soluble C/N ratio. However, this was not confirmed for CSM by Inbar *et al.* (1989), nor for CGM by Inbar *et al.*, (1990a). It also seems, as proposed by Hoitink and Kuter (1986), that this procedure is not effective for low C/N ratio wastes such as sewage sludge which, after the mixing of raw sludge and bulking agents, already has low ratios in the extract.

### 3.2 *Biological parameters of compost maturity*

The ultimate, and probably most attractive evaluation of compost maturity is based on bioassays. Most are based on plant growth over a few days or weeks, and this limits their usefulness. Data given in Table 1 (Hadar *et al.*, 1985) show that extended composting was needed to obtain maximum growth of plants in separated manure. Growth of unfertilized plants was retarded but fertilization alone could not sufficiently enhance growth unless the compost was mature. This suggests that nitrogen deficiency was not the only growth-limiting factor in this immature compost. An experiment conducted using water extracts of CSM (Inbar *et al.*, 1989) showed that tomato seedlings were inhibited in water extracts obtained from composts less than 40 days old. An increase in inhibition resulting from phytotoxic substances was observed in compost extracts prepared during the thermophilic stages of composting (20–40 days). From day 40 onward, growth in compost extract was significantly better than that in a standard nutrient solution. It was concluded that phytotoxic compounds and competition for $O_2$ in the immature compost were responsible for the inhibition in plant growth. Grebus *et al.* (1994) reported that 60–70 days of composting were required to destroy the inhibitory growth effect of a yard-trimmings mixture on radish. Plant growth bioassays revealed that ryegrass, cucumber and radish differ in their responses to MSW compost age (Iannotti *et al.*, 1994). These bioassays revealed that ryegrass and radish treated with slow-release fertilizer are sensitive to compost maturity, but cucumber is not. Different results were observed by Chefetz *et al.* (1996) who reported that a bioassay based on cucumber growth is a sensitive tool for determining MSW compost maturity.

## 4. Humic substances in composts

HS have attracted the interest of soil scientists due to their importance to soil ecology, fertility and structure and their effects on plant growth (Chen and Avaid, 1990; Chen *et al.*, 1994; Vaughan and Malcolm, 1985). As more agricultural lands throughout the world are treated with composted organic materials such as composts from sewage sludge, municipal refuse, crop residues, and animal excreta, elaboration of the nature of the organic components (especially the HS) present in such composts, as well as of their formation during the composting process, has become essential. HS are known to be involved in a number of interrelated, chemical processes in soil and to influence plant growth directly as well as indirectly. Indirectly, humic acid (HA) may improve soil fertility by increasing the availability of plant nutrients or by sequestering heavy metals such as iron, zinc, and manganese in the soil environment, thereby solubilising these elements for the benefit of plants (Chen and Aviad, 1990). Some reports also suggest that plants roots can take up HS, thereby allowing them to have direct effects on plant growth and metabolism (Vaughan and Malcolm, 1985).

A variety of techniques have been employed to extract HS from soils. The methods vary among researchers, depending upon the nature of the material to be examined, the discipline of the investigators and other arbitrary factors. The common extraction procedures have been reviewed by Hayes *et al.* (1975), Hayes (1985), Stevenson (1982), Watanabe *et al.* (1994) and Adani *et al.* (1995). They stated that aqueous solutions of NaOH are capable of extracting appreciable fractions of OM, including the more complex, high-molecular-weight and less soluble components of humus. Aqueous solutions of NaOH are known to induce slight oxidation of the extracted OM. The alkali solution also dissolves protoplasmic and structural components from fresh organic tissues and these become mixed with the humified OM (Stevenson, 1982). When fresh plant residues and young peat material were extracted with alkali, substances such as hemicelluloses, proteins, alkali-soluble lignin, and tannins entered into the humic fraction. However, as the decomposition proceeded, the fraction of these substances decreased while that of the HS increased (Inbar *et al.*, 1993; Adani *et al.*, 1995). In the case of strongly humified materials, the danger of extracting the admixtures with the HA is practically negligible. To minimise the chemical changes caused by autooxidation during the alkali extraction, all steps should be conducted in the presence of an inert gas (e.g. $N_2$) (Stevenson, 1982).

Among the chemical criteria required for the characterization of a stabilized organic waste product (e.g. compost), humus content and quality are highly ranked

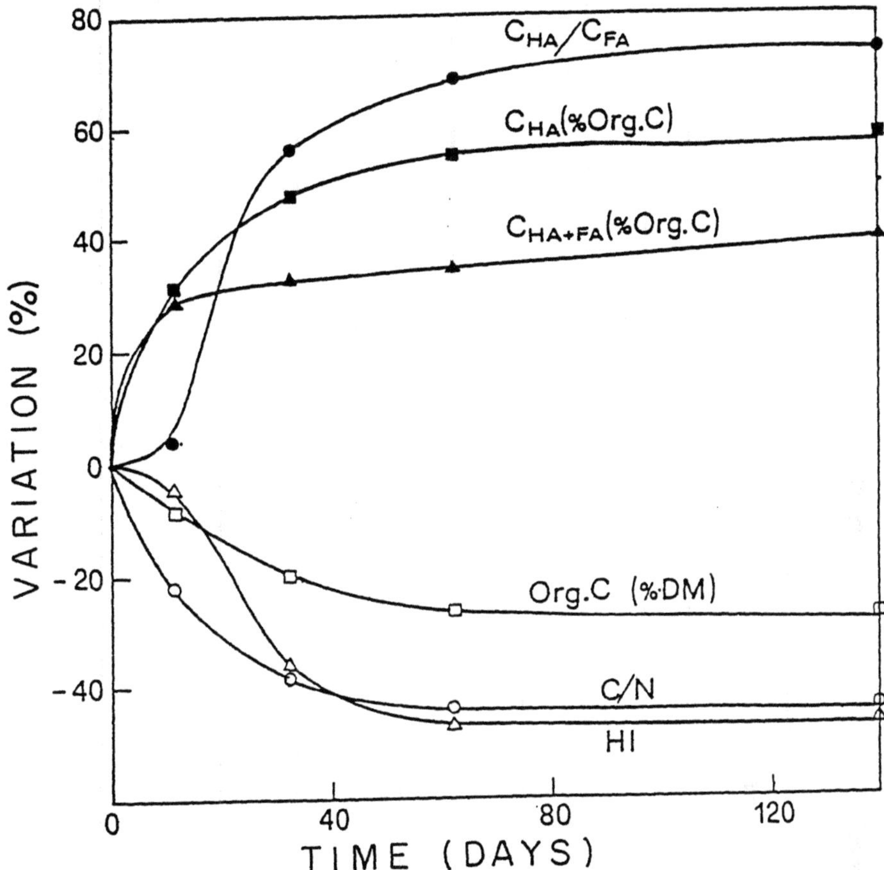


*Figure 3.* Changes and trends in the humification parameters (expressed as percentage of variation) during the composting of wastewater sludge and straw mixtures (from Saviozzi *et al.*, 1988).

(Chen and Inbar, 1993; Jimenez and Garcia, 1992; Roletto and Ottino, 1984; Roletto *et al.*, 1982, 1985; Saviozzi *et al.*, 1988; Sequi *et al.*, 1986; Sugahara and Inoko, 1981). The main trends of several humification parameters (expressed as percentage of variation) over a period of 140 days during the stabilization of waste water sludge and straw mixtures are shown in Figure 3 (Saviozzi *et al.*, 1988). The following humification indexes were suggested:

- C/N ratio;
- CHA – humic acid carbon (percentage of organic C);
- CFA – fulvic acid carbon (percentage of organic C);
- CHA/CFA ratio;
- CHA + CFA (percentage of organic C);
- HI (index of humification) – the ratio between the organic C of the non-humified fraction (NHF) and that of the HA + FA fraction (after Sequi *et al.*, 1986);

- organic C (% DM); and
- stability index (SI) – the ratio between HA after apolar, polar and acid hydrolysis treatments and $NaOH–Na_4P_2O_7$ HA extraction (after Adani *et al.*, 1995).

During the first 30 to 60 days changes in HI occurred rapidly; thereafter the rate of change became progressively slower. All the parameters were reported to be linearly and significantly correlated to the C/N ratio. In their work, Saviozzi *et al.* (1988) suggested minimum values for some humification parameters indicating maturity of farmyard manure (FYM). FYM containing less than 37.1% organic C, and exhibiting an HI value lower than 0.4, a CHA + CFA higher than 16, and a CHA higher than 9.7, could be considered mature. Time-course studies of the composting of various wastes showed an increase in the relative amount of HA (from 18% to 45% of OM in CSM compost and from 5% to 20% in MSW compost), whereas the formation rate of fulvic acid (FA) was inconsistent. Humification ratio (HR = HA/FA) values of 0.9 to 3.4 and 3.0 to 9.2 were typical for immature and mature composts, respectively. Although the values of the HR for various composts differ, they generally increase by more than 80% during composting, indicating a high degree of OM polymerization, since more polymerized components are found in mature compost. HI decrease during composting represents the formation of HS. Elemental and functional-group analysis of the HA originating from various composts indicated only minor differences between composting stages. Moreover, these values fell into a wide range, similar to that of soil HA. In general, even though HA amounts increased during composting, the elemental and functional groups were similar to 'young forms' of HA reported in the literature. The $E_4/E_6$ ratio was higher in compost HA, indicating a lower molecular weight than that of 'model' HA (Chen *et al.*, 1977). Elemental and functional-group analysis of compost FA have been reported in only a few works (Pandeya, 1992; Roletto and Luda, 1987). The compost FA had lower carbon and nitrogen contents but higher amounts of oxygen and carboxyl groups than HA, and the overall elemental and functional groups extracted from compost were similar to 'model' FA (Pandeya, 1992; Roletto and Luda, 1987; Schnitzer, 1978).

## 5. CPMAS $^{13}C$-NMR, FTIR and DRIFT spectroscopies

During the process of compost production, organic materials that are chemically complex and difficult to fractionate are formed. Extraction procedures that are commonly applied to soils and composts remove only a fraction of the C in the system. To profile the organic material of composts during the course of their formation, a method that can be applied to the intact sample is preferable. The only physical method holding some promise for the reliable characterisation of such organic materials appears to be nuclear magnetic resonance (NMR) (Wershaw, 1985; Wilson, 1987). Recently, the application of NMR spectroscopy in studies on HA has gained considerable attention. Both $^{1}H$- and $^{13}C$-NMR, in the solid and liquid state, have been used. Wershaw (1985) indicated that although NMR analysis has been applied for a much shorter period of time than most other techniques, it has provided more definitive structural information than other methods. Recent studies have shown that the technique of solid-state CPMAS (cross polarization magic-angle spinning) $^{13}C$-NMR can provide carbon 'fingerprints' of diverse solid samples such as peats (Preston *et al.*, 1987; Wilson *et al.*, 1983), whole soils (Preston and Ripmeester, 1983; Preston *et al.*, 1994; Wilson *et al.*, 1983), composts (Chefetz *et al.*, 1996; Gerasimowicz and Byler, 1985; Inbar *et al.*, 1990a, Piotrowski *et al.*, 1984; Preston *et al.*, 1986), forages (Eflosen *et al.*, 1984), wood (Haw *et al.*, 1984, Kolodzieski *et al.*, 1982), and fossil fuels (Packer *et al.*, 1983); HS isolated from soils, peat, river water, river sediments, lake water, marine waters, and marine sediments as reviewed by Malcolm (1989). An in-depth review of the state of the art of NMR spectroscopy theory, techniques, and applications to a wide range of organic and inorganic geochemistry and soil chemistry studies are discussed by Wilson (1987). Plant and microbial remains are the substrate for OM transformations in the geosphere. The myriad of chemicals which compose plants and microbes can be grouped into six major types: lignins, polysaccharides, proteins, lipids, resins and pigments (Wilson, 1987). The contribution of CPMAS $^{13}C$-NMR assigned chemical shifts for each of these groups is described by Wilson (1987). Wershaw *et al.* (1986, 1988) compared spectra of HA fractions from several different environments with published spectra of plant components; they showed that these components, when incorporated into HA, retain many of the chemical structural features of the original plant material. They attributed some of the observed bands to lignin-like or melanin-like structures. Sharp and well-defined carbohydrate bands were observed in the spectra of some of the fractions. Lignins are composed of three basic building blocks, namely syringyl, guaiacyl, and *p*-hydroxyphenolic units. Resonances from these structures can be observed in all lignins. Typical $^{13}C$-NMR assignments to these structures are presented in Figure 4. A number of structural differences between lignins from various sources have been reported, but in general, many investigators agree that the basic units are those shown in Figure 4. The

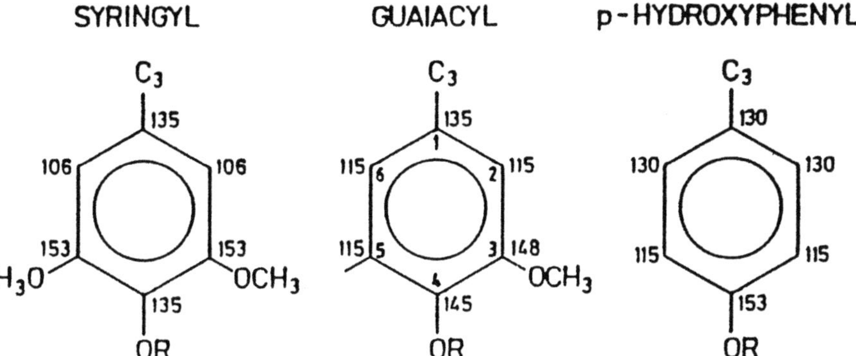


*Figure 4.* Structural components of lignin (the numbers indicate typical $^{13}C$-NMR assignments).

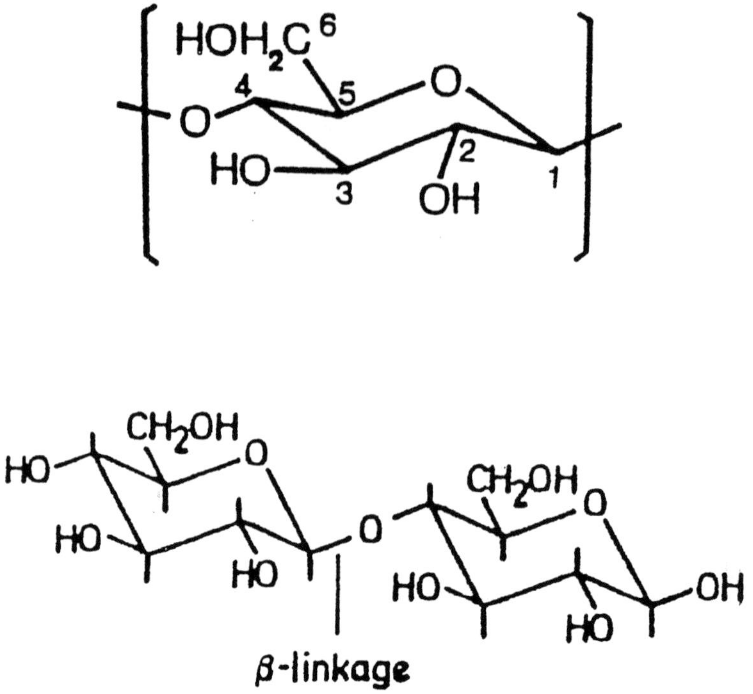

*Figure 5.* Structures of anhydroglucose repeating unit and cellobiose (4-O-($\beta$-D-glucosepyranosyl) $\beta$-D-glucopyranose).

intensity in the $^{13}$C-NMR spectral region from 160 ppm to 110 ppm is due to the aromatic ring carbons of lignin. In addition, the signal around 56 ppm is usually attributed to methoxy carbons, principally of lignin origin (Haw *et al.*, 1984; Himmelsbach and Barton, 1980; Wilson, 1987).

The second most important organic precursors to HS are polysaccharides, of which cellulose is the most important. Cellulose consists of long chains of glucose units, each of which is combined by a $\beta$-glucoside link to the C4 hydroxyl or another glucose unit (Figure 5). Enzymatic hydrolysis of cellulose can vary in nature, as can the number of glucose units that are present. It is not uncommon to find 2–4000 glucose units joined together (Wilson, 1987). In the solid state, cellulose exists in at least four forms – the most common of which are cellulose I and II – that may have slightly different chemical shifts. Moreover, these cellulose units may exist in various degrees of crystallinity (Haw *et al.*, 1984; Wilson, 1987). The features of cellulose I (the native or natural form of cellulose) are: a peak at 105 ppm attributed to the $C_1$ carbon of the anhydroglucose repeating unit (Figure 5); signals at 89 and 84 ppm, resulting from the $C_4$ carbon of cellulose; the highest absorption intensities for $C_2$, $C_3$ and $C_5$ carbons in the 70 to 80 ppm region; and peaks at 63 to 66 ppm assigned to the $C_6$ carbon of cellulose (Haw *et al.*, 1984; Wilson, 1987). Hemicellulose exhibits spectral features which are similar to cellulose. In addition, it exhibits a peak at around 103 ppm due to carbons that are structurally analogous to the $C_1$ carbon of cellulose, and signals at 22 and 174 ppm which result from acetate groups present in hemicellulose. Some of the absorption intensity of the 174 ppm peak could be due to other carbonyl species present in hemicellulose (e.g. $CO_2H$ groups of uronic acid) (Kolodzieski *et al.*, 1982).

The $^{13}$C-NMR spectra can be divided into five regions (Hatcher *et al.*, 1983). Region I (0–50 ppm) is assigned to paraffinic carbon or carbons that are bonded only to other carbons. Region II (50–112 ppm) is assigned to C–O, C–N bonds as in carbohydrates, alcohols, esters and amines. Region III (112–163 ppm) is assigned to aromatic and phenolic carbons. Region IV (163–190 ppm) represents carboxyl, ester and amide carbons groups. Region V (190–215 ppm) represents the carbonyl carbons. The 0 to 112 ppm region was calculated as aliphatic carbon, the 60 to 112 ppm region as polysaccharide carbon and the 112 to 163 ppm region as aromatic carbon. Total aromaticity was calculated by expressing aromatic C (112–163 ppm) as percentage of aliphatic C (0–112 ppm) plus aromatic C (Hatcher *et al.*, 1981; Kögel-Knabner *et al.*, 1991).

To date, many studies have been conducted on HA compost using CPMAS $^{13}$C-NMR spectroscopy, but only a few used this method to study bulk compost (Chefetz *et al.*, 1996; Inbar *et al.*, 1989). Inbar *et al.* (1989) reported only minor differences between the spectra of HA from fresh and composted manure (12% increase in carboxyl content; 8% increase in aromatic carbon and 8% decrease in alkyl carbon). During composting of MSW, HA aromatic carbon increased by 39% while the aliphatic carbon content decreased by 19% (Garcia *et al.*, 1992). During composting of MSW, the following distinct changes in the CPMAS $^{13}$C-NMR spectra of the bulk compost resulting from the composting process were noted:

- Total aliphatic carbon decreased by 14.4% from 74.8% in the raw material to 64.0% in the mature compost; the polysaccharides decreased by 17.0%, from 42.3% to 35.1% in the mature compost; and the alkyl groups (0–60 ppm) decreased by 11.0% during the composting process.
- The aromatic components, including aromatic and phenolic carbons, increased by 54%, from 15.9% in the raw material up to 24.5% of the total carbon in the mature compost. Total aromaticity increased by 58.2%, from 17.5% in the raw material to 27.7% in the mature compost.
- Carboxyl carbon increased by only 9.5%.
- The C=O carbonyl carbon increased by 73.7%, from 1.9% in the raw material to 3.3% in the mature compost (Chefetz *et al.*, 1996).

Infrared (IR) spectroscopy and a wide variety of chemical methods are being used to characterise HS. Fourier transform infrared (FTIR) spectroscopy serves as a qualitative tool to characterise the principal classes of chemical groups that compose the HA non-destructively. Negative feedback as to the usefulness of IR spectral analysis is probably due, in part, to the complexity of the IR spectra. However, IR analysis has contributed substantially to the knowledge of HA chemistry (MacCarthy and Rice, 1985; Schnitzer, 1978; Stevenson, 1982) and complements the information obtained from solid-state CPMAS $^{13}$C-NMR spectroscopy (Gerasimowicz and Byler, 1985; Lobartini and Tan, 1988). The structural complexity of humic materials and the abundance of strongly hydrogen-bonded groups have, until recently, severely limited the usefulness of infrared spectroscopy for structural and compositional studies. With dispersive instruments only broad, ill-defined bands have been observed (Gerasimowicz *et al.*, 1986). FTIR spectroscopy, which provides a much higher signal-to-noise ratio and improved accuracy in the frequency of the instrument, increased the information attainable from the spectra (Griffiths, 1983). Gerasimowicz and Byler (1985) and Gerasimowicz *et al.* (1986) obtained FTIR spectra on solid HA extracted from materials sampled at four selected stages of a municipal wastewater treatment process. From the FTIR spectra, several important differences between the decomposition stages were observed and discussed.

Sugahara and Inoko (1981) investigated HA extracted from city refuse at different stages of decomposition by IR spectroscopy. They observed decreases in aliphatic C–H stretching, increases in COO and decreases in C–O stretching as composting proceeded. On the whole, the shape of the IR spectra of city refuse HA became featureless during composting. Similar changes were observed by Zech *et al.* (1987) for IR spectra of fresh and decomposed spruce and pine litter. In many instances, CPMAS $^{13}$C-NMR spectroscopy data lend themselves somewhat better to quantitative analysis than do the FTIR results. In particular, the relative aromatic nature of the HA was more evident from the NMR spectra and could not be seen as readily from FTIR data (Gerasimowicz and Byler, 1985).

The development of an attachment mounted onto FTIR spectrophotometers has allowed the determination of diffuse reflectance infrared Fourier transform (DRIFT) spectra. Sample preparation for DRIFT spectroscopy is much simpler than for FTIR spectroscopy, interferences due to water adsorption are reduced, and resolution is improved. The spectra obtained using DRIFT spectroscopy had a higher degree of resolution as compared with FTIR spectroscopy. Moreover, the bands indicative of aliphatic C–H, carboxyl and carboxylate functional groups, aromatic C=C, and the C–O stretch of polysaccharides were prominent and very well resolved. The DRIFT spectra obtained can also be used to fingerprint OM acquired from various sources. Relative concentrations of functional groups were found to be fairly constant regardless of sample concentration. The DRIFT spectra peak intensity ratio can be used to determine the humification process. The application of DRIFT spectroscopy to OM research will prove especially useful for characterizing heterogeneous bulk samples such as peat and composts (Niemeyer *et al.*, 1992). A combination of quantitative elemental analysis supplemented by modern, non-destructive NMR, FTIR and DRIFT characterizations could have significant biochemical implications in the study of the nature and properties of HS and other complex heterogeneous matrices.

## 6. Raw material used for composting

### *6.1 Municipal solid waste*

Problems caused by MSW in modern society have become more severe over the last decade due to increasing amounts of waste and decreasing availability of landfill space (Alter, 1991; Finstein, 1992). The amount of MSW generated in the US rose to 323 million tons in 1994, an increase of 5%, whereas during the period between 1988 and 1994, the number of landfills decreased by more than half (from 8000 to 3558 in 1994). The remaining capacity of the landfills has also decreased and in some states such as Ohio, Missouri, and Vermont, the landfills can last another ten years or less. Together with the decreasing landfill capacity, the tipping fee has increased and can reach up to $75 per ton (Steuteville, 1994a,b, 1995). The European community produces more than 2500 million tons of garbage a year of which the MSW fraction can reach up to 10%. Thus recycling has become an attractive solution for MSW treatment. The organic fraction in the MSW varies between 30 and 50%. Thus recycling this fraction can reduce the cost of transportation by 50%. For the organic fraction, composting seems to be a desirable option, having the capacity of reducing the volume and weight by approximately 50% and resulting in a product which can be useful for agriculture (He *et al.*, 1992). An alternative solution for organic waste management is the waste to energy programme, but its high tipping fee and its minor daily capacity make composting the ideal solution for the organic fraction. The main problems facing MSW compost production and utilization stem from the heterogeneous nature of the raw material. Its composition can change with climate, standard of living, type of garbage collection system and season.

### *6.2 Yard waste*

Yard waste can be defined as the amount of leaf fall, grass clippings and wood waste diverted from the MSW stream. In 1988, the EPA estimated that 18% of MSW is yard waste. Yard waste is not homogenized waste and its characteristics vary with weather, area and collection policy. The characteristics of the material affect the collection treatment, the processing equipment and the properties of the end product. Yard waste humidity varies from 30 to 75%, with that of leaves varying between 19 and 28%, of grass from 54 to 64% and of the wood fraction from 17 to 18%. The C/N ratio of yard waste is usually high (90–125) due to the large levels of lignin in woody waste. To reduce the initial C/N ratio to a level suitable for composting (30–35), yard waste is usually mixed with animal manure, sewage sludge, poultry manure or any other waste that can provide additional nitrogen. Recovery of yard waste by composting is an increasingly popular method of recycling solid organic waste (Steuteville, 1995). The common systems used for composting yard waste are static piles and turned windrows. Yard waste compost is useful as a substrate in container media and unlike composts produced from other wastes (MSW and sewage sludge) this compost seems to be 'cleaner'.

### 6.3 *Sewage sludge*

The level of sewage sludge, the by-product of normal wastewater treatment operations, is increasing worldwide. Utilization of sewage sludge in agriculture as an organic fertiliser is regarded as a beneficial and economic process. Chemical properties of sewage sludge indicate high rates of nitrogen and phosphorus. Sewage sludge can be applied to soil as a liquid or solid amendment. An increasing concern over human pathogens in the sludge has enhanced the interest in using composting as a process that can destroy plant and human pathogens and produce useful agricultural products. The main problem facing composting of sewage sludge is its moisture content. Thus before composting, sewage sludge humidity must be brought to less than 60% by drying or by adding a dewatering agent (lime). Usually sewage sludge cannot be used as the only raw material for composting, so yard waste or straw litter are used as bulking agents.

### 6.4 *Separated cattle manure*

The total amount of organic waste products in the US from livestock and poultry production has been estimated at over 1.5 billion tons per annum (on a dry matter basis). Of this amount, about 1 billion tons is cattle manure (Stevenson, 1985). Because of technological and economic considerations, the monetary value of manures and other wastes has significantly decreased in recent years, with the result that these materials are now regarded as a liability rather than an asset to the total farm operation. Farming has become highly specialized and livestock and poultry production is being concentrated into large-scale confinement-type enterprises. These changes imply that large volumes of wastes have to be disposed of in a relatively small area (Stevenson, 1985). Stevenson (1985) summarized the limitations regarding the use of farmland for the disposal of animal manure as follows:

- Animal manure as a source of plant nutrients is a bulky, low-grade fertiliser of variable composition.
- Concentration of nutrients, soluble salts, trace elements, and water vary tremendously and are seldom known. Optimum rates of application are thus difficult to predict.
- In many soils, the leaching of nutrients present in high concentrations in manures (especially $NO_3^-$) into the ground water may limit application rates.
- On-farm management problems are created by the physical properties of manure. Application techniques are inefficient and time consuming.
- Production of animal manures is continuous, while the need for fertilizers is seasonal.
- Odour and associated nuisances.
- The presence of some human and plant pathogens and weeds restricts the use of manures by farmers.

Development of new management systems which will increase the quality of the products, decrease salt concentration, and improve the physical structure and ease of handling is needed. One of the proposed systems now used on farms is based on separating of the manure into solids and liquid (Bohnhoff and Converse, 1987; Huijsmans and Lindley, 1984). This material can be successfully used, after proper composting, as a peat substitute or complementary product for peat mixtures in container media, while the effluent is used as organic liquid fertilizer (Chen and Hadar, 1987; Chen *et al.*, 1984a,b; Hadar *et al.*, 1985; Levanon *et al.*, 1984; Raviv *et al.*, 1984).

### *6.5 Grape marc*

The primary by-product of grape juice processing and wine production is grape marc (GM; pomace). It consists primarily of processed skins, seeds and stems. GM represents as much as 20% of the wet weight of the original fruit and its moisture content is about 65% (Hang, 1988). Annual world production of dried GM is estimated at 7 500 000 metric tons (Aguilera, 1987). Direct incorporation of GM, without any pretreatment, into vineyard soils is a universal practice. However, this method has become a serious problem since degradation materials that are highly incompatible with plant roots are released. Composting GM has the potential to solve the environmental problems related to this by-product and to create a useful product for greenhouse agriculture.

## 7. Conclusions

Composting seems to be a desirable solution for organic waste management. Readily available wastes such as cattle manure, MSW, sewage sludge, winery waste and yard waste have been found to be suitable materials for composting. The composting process has the capacity to reduce the volume and weight of these wastes by approximately 50% and to produce a product which can be useful in agriculture. Composts produced from various wastes have been found to be suitable materials for container media, having the ability to reduce peat and pesticide use, thereby solving environmental problems. The positive effects of the composts on plant growth can be explained by their physical, chemical and biological properties. Their physical properties fit well with optimal requirements. In greenhouse set-ups, where fertilization is continuously applied, the nutrient content is not as important. However, the slow release of nutrients between irrigations, the high content of HS and the bioactivity of the medium improves plant growth. Mature composts have also been found to be suppressive to several soilborne plant pathogens such as *P. aphanidermatum*, *R. solani* and *S. rolfsi*. As a result, application rates of hazardous pesticides can be reduced or eliminated. The addition of either compost or mineral fertilizers to soils is known to enhance crop yield. However, adding both compost and mineral fertilizer often results in a synergistic enhancement of crop yield. The positive effects of mature compost on plant growth are connected to its high HS content. The HS formed during composting seem to have the same characteristics as HS that are formed in soils over long periods. Spectroscopic methods (CPMAS $^{13}$C-NMR, DRIFT and FTIR) are useful tools to determine the transformations of the organic fraction during composting, and can help to define compost maturity. During composting, many parameters must be monitored (temperature, humidity, pH, C/N ratio, aeration), although a number of the parameters stabilise at the maturation stage providing no tools for the actual maturity stage. Only a few parameters can be used to define compost maturity. Many authors have reported that HI, C/N ratio, respiration and microbial activity can be used as maturity indexes. We hypothesise that plant growth may serve as an integrative signal for compost maturity.

## References


Adani F, Genevini P L and Tambone F 1995 A new index of OM stability. Compost Sci. Util. 3, 25–37.

Aguilera J F 1987 Improvement of olive cake and grape by-products for animal nutrition. In Degradation of Lignocellulosics in Ruminants and in Industrial Processes. Eds J M van der Meer, B A Rijkens and M P Ferranti. pp 45–54. Elsevier Applied Science, London.

Alter H 1991 The future course of solid waste management in the US. Waste Manage. Res. 9, 3–9.

Anid P J 1986 Evaluating maturity and metal transfer of MSW compost. BioCycle 46–47.

Bik A R 1983 Substrates in floriculture. Proc. XXI Intern. Hortic. Cong. Hamburg 11, 811–822.

Bohnhoff D R and Converse J C 1987 Engineering properties of separated manure solids. Biol. Wastes 19, 91–106.

Chang Y C, Maker R Kleifield O and Chet I 1986 Increasing growth of plants in the presence of biological control agent *Trichoderma harzianum*. Plant Dis. 70, 145–148.

Chanyasak V and Kubota H 1981 Carbon/organic nitrogen ratio in water extract as a measure of composting degradation. J. Ferment. Technol. 59, 215–219.

Chanyasak V, Hirai M and Kubota H 1982 Changes of chemical components and nitrogen transformation in water extracts during composting of garbage. J. Ferment. Technol. 60, 439–446.

Chanyasak V, Katayama A, Hirai M F, Mori S and Kubota H 1983 Effects of compost maturity on growth of komatsuna (*Brasica rapa* var. *peridis*) in Neubauer's pot. II. Growth inhibitory factors and assessment of degree of maturity by org.-C/org.-N ratio of water extract. Soil Sci. Plant Nutr. 29, 251–259.

Chefetz B, Hatcher P G, Hadar Y and Chen Y 1996 Chemical and biological characterization of OM during composting of municipal solid waste. J. Environ. Qual. 25, 776–785.

Chen Y and Aviad T 1990 Effects of humic substances on plant growth. In Humic Substances in Soil and Crop Science. Eds P MacCarthy, C E Clapp, R L Malcolm and P R Bloom. Am. Soc. Agron., Soil Sci. Am., Madison, Wisconsin pp 161–186.

Chen Y and Hadar Y 1987 Composting and use of agricultural wastes in container media. In Compost: Production, Quality and Use. Eds M. de Bertoldi, M P Ferranti, P L Hermite and F Zucconi. pp 71–77. Elsevier Applied Science, London.

Chen Y and Inbar Y 1993 Chemical and spectroscopical analyses of OM transformations during composting in relation to compost maturity. In Science and Engineering of Composting. Eds H A J Hoitink and H M Keener. pp 551–600. Renaissance Publications, Worthington, OH.

Chen Y, Senesi N and Schnitzer M 1977 Information provided on humic substances by E4/E6 ratios. Soil Sci. Soc. Am. J. 41, 352–358.

Chen Y, Inbar Y, Raviv M and Dovrat A 1984a The use of slurry produced by methanogenic fermentation of cow manure as a peat substitute in horticulture – physical and chemical characteristics. Acta Hort. 150, 553–561.

Chen Y, Inbar Y and Raviv M 1984b Slurry produced by methanogenic fermentation of cow manure as a peat substitute in horticulture. Proc. 2nd. Int. Peat in Agriculture and Horticulture, The Hebrew University, Jerusalem. pp 297–317.

Chen Y, Inbar Y and Hadar Y 1988 Composted agricultural wastes as potting media for ornamental plants. Soil. Sci. 145, 298–303.

Chen Y, Inbar Y and Hadar Y 1992 Composted residues reduce peat and pesticide use. BioCycle 6, 48–51.

Chen Y, Magen H and Riov J 1994 Humic substances originating from rapidly decomposing OM: Properties and effects on plant growth. Humic Substances in the Global Environment and Implication on Human Health. Eds N Senesi and T M Miano. pp 427–443. Elsevier, Amsterdam.

Ciavatta C, Govi M, Pasotti L and Sequi P 1993 Changes in OM during stabilization of compost from municipal solid wastes. Bio. Tech. 43, 141–145.

Cull D C 1981 Alternatives to peat as container media. Organic resources in the UK. Acta Hort. 126, 69–81.

de Bertoldi M, 1993 MSW composting challenges in Europe. BioCycle 34, 75–76.

de Bertoldi M, Ferranti M P, L'Hermite P and Zucconi F (Eds) 1987 Compost: Production, Quality and Use. Elsevier, London.

De Vleeschauwer D, Verdonck O and VanAssche P 1981 Phytotoxicity of refuse compost. BioCycle 22, 44–46.

Diaz L F, Savage M S and Golueke C G 1982 Composting. In Resource Recovery from Municipal Solid Wastes. Volume II – Final processing. pp 95–120. CRC Press, Boca Raton, FL.

Eflosen R M, Ripmeester J A, Cyr N, Milligan L P and G Mathison 1984 Nutritional evaluation of forages by high-resolution solid-state 13C-NMR. Can. J. Anim. Sci. 64, 93–102.

Finstein M S 1992 Composting in the context of municipal waste management. Environmental Microbiology. Ed. R Mitchell. pp 355–374. Wiley-Liss, New York.

Finstein M S and Morris M L 1975 Microbiology of municipal solid waste composting. Adv. Appl. Microbiol. 19, 113–151.

Garcia C, Hernandez T and Costa F 1992 Comparison of humic acids derived from city refuse with more developed humic acids. Soil Sci. Plant Nutr. 38, 339–346.

Gerasimowicz W V and Byler D M 1985 Carbon-13 CPMAS NMR and FTIR spectroscopic studies of humic acids. Soil Sci. 139, 270–278.

Gerasimowicz W V, Byler D M and Susi H 1986 Resolution-enhanced FT-IR spectra of soil constituents: Humic acid. Appl. Spectrosc. 4, 504–507.

Goldstein N and Steuteville R 1995 Solid waste composting plants in a steady state. BioCycle 36, 48–57.

Gray K R and Biddlestone A J 1981 The composting of agricultural wastes. Biological Husbandry. Ed. B. Stonehouse. pp 99–111. Butterworths, London.

Grebus M E, Watson M E and Hoitink H A J 1994 Biological, chemical and physical properties of composted yard trimmings as indicators of maturity and plant disease suppression. Compost Sci. Util. 2, 57–71.

Griffiths P R 1983 Fourier transform infrared spectroscopy. Science 222, 297–302.

Hadar Y, Inbar Y and Chen Y 1985 Effect of compost maturity on tomato seedling growth. Sci. Horti. 27, 199–208.

Hang Y D 1988 Recovery of food ingredients from grape pomace. Process Biochem. February, 2–4.

Harada Y, and Inoko A 1980a The measurement of cation exchange capacity of composts for the estimation of the degree of maturity. Soil. Sci. Plant Nutr. 26, 127–134.

Harada Y and Inoko A 1980b Relationship between cation-exchange capacity and degree of maturity of city refuse composts. Soil Sci. Plant Nutr. 26, 353–362.

Harada Y, Inoko A, Tadaki M and Izawa T 1981 Maturing process of city refuse compost during piling. Soil Sci. Plant Nutr. 27, 357–364.

Hatcher P G, Schnitzer M, Dennis L W and Maciel G E 1981 Aromaticity of humic substances in soils. Soil Sci. Soc. Am. J. 45, 1089–1094.

Hatcher P G, Breger I A, Dennis L W and Maciel G E 1983 Solid-state $^{13}$C-NMR of sedimentary humic substances: new revelations on their chemical composition. In Aquatic and Terrestrial Humic Materials. Eds R F Christman and E T Gjessing. pp 37–81. Ann Arbor Science, Ann Arbor, MI.

Haw J F, Maciel G A and Schroder H A 1984 Carbon-13 nuclear magnetic resonance spectrometric study of wood and wood pulping with cross polarization and magic-angle spinning. Anal. Chem. 56, 1323–1329.

Hayes M H B 1985 Extraction of humic substances from soil. In Humic Substances in Soil, Sediment, and Water. Eds G R Aiken, D M McKnight, R L Wershaw and P MacCarthy. pp 329–362. John Wiley & Sons, New York.

Hayes M H B, Swift R S, Wardle R E and Brown J K 1975 Humic materials from an organic soil: A comparison of extractants and of properties of extracts. Geoderma 13, 231–245.

He X T, Traina S J and Logan T J 1992 Chemical properties of municipal solid waste composts. J. Environ. Qual. 21, 318–329.

Himmelsbach D S and Barton II F E 1980 $^{13}$C-nuclear magnetic resonance of grass lignins. J. Agric. Food Chem. 28, 1203–1208.

Hirai M F, Katayama A and Kubota H 1986 Effect of compost maturity on plant growth. BioCycle 27, 58–61.

Hoitink H A J and Fahy P C 1986 Basis for the control of soilborne plant pathogens with composts. Ann. Rev. Phytopathol. 24, 93–114.

Hoitink H A J and Kuter G A 1986 Effects of composts in growth media on soilborne pathogens. In The Role of OM in Modern Agriculture. Eds Y Chen and Y Avnimelech. pp 289–306. Martinus Nijhoff Publishers, Dordrecht.

Hoitink H A J, Boehm M J and Hadar Y 1993 Mechanisms of suppression of soilborne plant pathogens in compost-amended substrates. Science and Engineering of Composting. Eds H A J Hoitink and H M Keener. pp 601–622. Renaissance Publications, Worthington, OH.

Huijsmans J and Lindley J A 1984 Evaluation of a solid-liquid separator. Trans. ASAE 1854–1858.

Iannotti D A, Grebus M E, Toth B L, Madden L V and Hoitink H A J 1994 Oxygen respirometry to assess stability and maturity of composted municipal solid waste. J. Environ. Qual. 23, 1177–1183.

Inbar Y, Chen Y and Hadar Y 1985 The use of composted slurry produced by methanogenic fermentation of cow manure as growth media. Acta Hort. 172, 75–85.

Inbar Y, Chen Y and Hadar Y 1988 Composting of agricultural wastes for their use as container media: Simulation of the composting process. Bio. Wastes 26, 247–259.

Inbar Y, Chen Y and Hadar Y 1989 Solid-state carbon-13 nuclear magnetic resonance and infrared spectroscopy of composted organic matter. Soil Sci. Soc. Am. J. 53, 1695–1701.

Inbar Y, Chen Y and Hadar Y 1990a Humic substances formed during the composting of OM. Soil Sci. Soc. Am. J. 54, 1316–1323.

Inbar Y, Chen Y, Hadar Y and Hoitink H A J 1990b New approaches to compost maturity. BioCycle 31, 64–69.

Inbar Y, Hadar Y and Chen Y 1993 Recycling of cattle manure: The composting process and characterization of maturity. J. Environ. Qual. 22, 857–863.

Inoko A 1982 The Composting of Organic Materials and Associated Maturity Problems. ASPAC, Food & Fertiliser Technology Center, Technical Bulletin No. 71.

Jimenez E I and Garcia V P 1992 Determination of maturity indices for city refuse composts. Agric. Ecosyst. Environ. 38, 331–343.

Kögel-Knabner I, Hatcher P G and Zech W 1991 Chemical structural studies of forest soil humic acids: Aromatic carbon fraction. Soil Sci. Soc. Am. J. 55, 241–247.

Kolodzieski W, Frye J S and Maciel G E 1982 Carbon-13 nuclear magnetic resonance spectrometry with cross polarization and magic-angle spinning for analysis of lodgepole pine wood. Anal. Chem. 54, 1419–1424.

Kuter G A, Hoitink, H A J and Rossman L A 1985 Effect of aeration and temperature on composting of municipal sludge in full-scale vessel system. J. WPCF 57, 309–315.

Kwok O C H, Fahy P C, Hoitink H A J and Kuter G A 1987 Interactions between bacteria and *Trichoderma hamatum* in suppression of Rhizoctonia damping-off in bark compost media. Phytopathology 77, 1206–1212.

Lax A, Costa, F, Roig A, Hernandez M T and Cegarra J 1987. Indexes of humification grade in manures. Sci. Tot. Environ. 62, 457–458.

Levanon D, Dosoretz C, Motro B and Kahn I 1984 Recycling agricultural waste for mushroom casing. Mushroom J. 133, 13–17.

Lobartini J C and Tan K H 1988 Differences in humic acid characteristics as determined by carbon-13 nuclear magnetic resonance, scanning electron microscopy, and infrared analysis. Soil Sci. Soc. Am. J. 52, 125–130.

Lohr V I, O'Brien R G and Coeffey D L 1984 Spent mushroom compost in soilless media and its effects on the yield and quality of transplants. J. Am. Soc. Hort. Sci. 109, 693–697.

Lynch J M, Gunn K B and Panting L M 1980 On the concentration of acetic acid in straw and soil. Plant and Soil 56, 93–98.

MacCarthy P and Rice J A 1985 Spectroscopic methods (other than NMR) for determining functionality in humic substances. In Humic Substances in Soil, Sediment, and Water. Eds G R Aiken, D M McKnight, R L Wershaw and P MacCarthy. pp 527–543. John Wiley & Sons, New York.

Malcolm R L 1989 Applications of solid-state $^{13}$C-NMR to geochemical studies. In Humic Substances in Soil, Sediments and Water: II. In Search of Structure. Eds M H B Hayes, P MacCarthy, R Swift and R L Malcolm. John Wiley, New York (in press).

Mandelbaum R, Hadar Y and Chen Y 1988 Composting of agricultural wastes for their use as container media: Effect of heat treatments on suppression of *Pythium aphanidermatum* and microbial activities in substrates containing compost. Biol. Wastes 26, 261–274.

Morel J L, Colin F, Germon J C, Godin P and Juste C 1985 Methods for the evaluation of the maturity of municipal refuse. In Composting of Agricultural and Other Wastes. Ed. J K R Gasser. pp. 56–72. Elsevier Applied Science Publishers, London.

Nakasaki K, Sasaki M, Shoda M and Kubota H 1985 Change in microbial numbers during thermophilic composting of sewage sludge with reference to $CO_2$ evolution rate. Appl. Environ. Microbiol. 49, 37–41.

Niemeyer J, Chen Y and Bollag J M 1992 Characterization of humic acids, composts, and peat by diffuse reflectance Fourier transform infrared spectroscopy. Soil Sci. Soc. Am. J. 56, 135–140.

Packer K J, Harris R K, Kenwright A M and Snape C E 1983 Quantitative aspects of solid state $^{13}$C-NMR of coals and related materials. Fuel 62, 999–1002.

Pandeya S B 1992 Characterization of fulvic acids extracted from some organic manures and wastes by potentiometric titration. Biol. Tech. 39, 77–83.

Piotrowski E G, Valentine K M and Pfeffer P E 1984 Solid-state, C-13, cross-polarization, 'magic-angle' spinning, NMR spectroscopy studies of sewage sludge. Soil Sci. 137, 194–203

Preston C M and Ripmeester J A 1983 $^{13}$C-labelling for NMR studies of soils: CPMAS observation of $^{13}$C-acetate transformation in a mineral soil. Can. J. Soil Sci. 63, 495–500.

Preston C M, Ripmeester J A, Mathur S P and Levesque M 1986 Application of solution and solid-state multinuclear NMR to a peat-based composting system for fish and crab scrap. Can. J. Soil Sci. 31, 63–69.

Preston C M, Shipitalo S E, Dudley R L, Fyfe C A, Mathur S P and Levesque M 1987 Comparison of C-13 CPMAS NMR and chemical techniques for measuring the degree of decomposition in virgin and cultivated peat profiles. Can. J. Soil Sci. 67, 187–198.

Preston C M, Hewman R H and Rother P 1994. Using $^{13}$C CPMAS NMR to assess effects of cultivation on the OM of particle size fractions in a grassland soil. Soil Sci. 157, 26–35.

Raviv M, Chen Y, Geler Z, Medina S, Putievski E and Inbar Y 1984 Slurry produced by methanogenic fermentation of cow manure as a growth medium for some horticultural crops. Acta Hort. 150, 563–573.

Raviv M, Chen Y and Inbar Y 1986 Peat and peat substitutes as growth media for container-growth plants. In The Role of OM in Modern Agriculture. Eds Y Chen and Y Avnimelech. pp 257–287. Martinus Nijhoff, The Hague, The Netherlands.

Roletto E and Luda M P 1987 Chemical and infra-red spectroscopic study of humic and fulvic acids from composted poplar bark. Biol. Wastes 19, 205–214.

Roletto E and Ottino P 1984 Analytical characterization of humic substances from composted lignocellulosic residues. Agric. Wastes 11, 181–195.

Roletto E, Barberis R and Zelano V 1982 Gel filtration and absorption spectroscopic investigations on humic substances from organic fertilisers. Plant and Soil 66, 383–390.

Roletto E, Chiono R and Barberis E 1985 Investigation on humic matter from decomposing poplar bark. Agric. Wastes 12, 261–272.

Saviozzi A, Levi-Minzi R and Riffaldi R 1988 Maturity evaluation of organic wastes. BioCycle 29, 54–56.

Schnitzer M 1978 Humic substances: Chemistry and reactions. In Soil Organic Matter. Eds M Schnitzer and S U Khan. pp 1–64. Elsevier, Amsterdam.

Sequi P, De Nobili M, Leita L and Cercignani G 1986 A new index of humification. Agrochimica 30, 175–179.

Steuteville R 1994a The state of garbage in America (II). BioCycle 35, 30–36.

Steuteville R 1994b The stage of garbage in America (I). BioCycle 35, 46–52.

Steuteville R 1995 The state of garbage in America (I). BioCycle 36, 54–58.

Stevenson F J 1982 Humus Chemistry. John Wiley & Sons, New York.

Stevenson F J 1985 Geochemistry of soil humic substances. In Humic Substances in Soil, Sediments and Water. Eds G R Aiben, M D McKnight, R L Wershaw and P MacCarthy. pp 13–52. John Wiley & Sons, New York.

Sugahara K and Inoko A 1981 Composition analysis of humus and characterization of humic acid obtained from city refuse compost. Soil Sci. Plant Nutr. 27, 213–224.

Toussan T A, Weinhold A R, Linderman R G and Patrick Z A 1968. Nature of phytotoxic substances produced during plant residue decompositon in soil. Phytopathology 58, 41–45.

Vaughan D, and Malcolm R E 1985 Influence of humic substances on growth and physiological processes. In Soil Organic Matter and Biological Activity. Eds D Vaughan and R E Malcolm. pp 37–75. Martinus Nijhoff/Dr Junk Publishers, Dordrecht.

Verdonck O 1984 Reviewing and evaluation of new materials used as substrates. Acta Hort. 150, 467–473.
Watanabe A, Itoh K, Arai S and KuWatsuka S 1994 Comparison of the composition of humic and fulvic acids prepared by the IHSS method and NAGOYA method. Soil Sci. Plant Nutr. 40, 601–608.
Wershaw R L 1985 Application of nuclear magnetic resonance for determining functionality in humic substances. In Humic Substance in Soil, Sediments and Water. Eds G R Aiken, D M McKnight, R L Wershaw and P MacCarthy. pp 561–584. John Wiley & Sons, New York.
Wershaw R L, Thorn K A, Pinckney D J, MacCarthy P, Rice J A and Hemond H F 1986 Application of membrane model to the secondary structure of humic materials in peat. In Peat and Water: Aspects of Water Retention and Dewatering in Peat. Ed. C Fuchsman. pp 133–157. Elsevier Applied Science Publishers, London.
Wershaw R L, Thorn K A and Pinckney D J 1988 Characterization of humic acid fractions by C-13 nuclear magnetic resonance spectroscopy. Environ. Tech. Lett. 9, 53–62.
Wilson M A 1987 NMR Techniques and Applications in Geochemistry and Soil Chemistry. Pergamon Press, Oxford, England.
Wilson M A, Heng S, Goh K M, Pugmire R J and Grant D M 1983 Studies of litter and acid-insoluble soil OM fractions using 13C cross polarization nuclear magnetic resonance spectroscopy with magic angle spinning. J. Soil Sci. 34, 83–97.
Wong M H and Chu L M 1985 Changes in properties of a fresh refuse compost in relation to root growth of *Brassica chinensis*. Agric. Wastes 14, 115–125.
Zech W J M B, Hanmaier L and Malcolm R L 1987 CPMAS $^{13}C$ NMR and IR spectra of spruce and pine litter and of Kalson lignin fraction at different stages of decomposition. Z. Pflanzenernahr. Bodenk. 150, 262–265.
Zucconi F, and de Bertoldi M 1987 Compost specifications for the production and characterization of compost from municipal solid waste. In Compost: Production, Quality and use. Eds M. de Bertoldi, M P Ferranti, P L'Hermite and F Zucconi. pp 30–50. Elsevier Applied Science, London.
Zucconi F, Forte M, Monaco A and de Bertoldi M 1981a Biological evaluation of compost maturity. BioCycle 22, 27–29.
Zucconi F, Pera A, Forte M and de Bertoldi M 1981b Evaluating toxicity of immature compost. BioCycle 22, 54–57.

# 29
# Impacts of compost quality on plant disease severity

H.A.J. HOITINK, M.E. GREBUS AND A.G. STONE

## 1. Introduction

Soil organic matter decomposition level plays an important role in soilborne disease severity (Cook and Baker, 1983). Even though this has been recognized for centuries, management of diseases with amendments remains an art rather than a science. Lack of progress in this field in part has been due to our inability to characterize adequately biological energy availability in soil or the amendments. Recent breakthroughs in instrumentation, coupled with advances made in the analysis of microbial populations in soil, promise to solve this problem. Procedures are available for monitoring the stability of composts based on respirometry (Iannotti *et al.*, 1993). Furthermore non-destructive direct spectroscopic procedures are available that in principle allow characterisation of stability or biological energy availability in organic matter (Chen and Inbar, 1993; Inbar *et al.*, 1989). This review emphasises this literature on composting and how it relates to plant disease suppression.

During the 1960s, nurserymen across the United States explored the possibility of using tree bark as a peat substitute to reduce potting mix costs. Since that time, procedures for composting of various types of tree bark have been developed which avoid nitrogen immobilization and allelopathy toxin problems associated with fresh bark (Hoitink and Kuter 1986; Hoitink *et al.*, 1991). During the past decade, similar procedures have been developed for composts produced from food industry and agricultural wastes, as well as from municipal wastes (Hoitink and Keener, 1993).

Early during the utilization of bark composts, improved plant growth and decreased losses caused by phytophthora root rots were observed as side benefits by the nursery industry. In practice, control of such root rots with composts can be as effective as that obtained with fungicides (Hardy and Sivasithamparam, 1991; Hoitink *et al.*, 1991; Ownley and Benson, 1991). Therefore, the nursery industry relies heavily on composts for control of diseases caused by these soilborne plant pathogens.

Composts must be of consistent quality and maturity to be used successfully in biological control of diseases of horticultural crops, particularly if used in container media (Inbar *et al.*, 1993a). Variability in this quality parameter is one of the principal factors limiting its utilization. Maturity is less important in ground bed or field agriculture as long as composts are applied well ahead of planting to allow for additional stabilization. Lack of maturity frequently causes problems here as well, however.

*D. Rosen* et al. *(eds): Modern Agriculture and the Environment, 363–371.*

## 2. Fate of biocontrol agents during composting

The composting process is often divided into three phases. The first phase occurs during the first 24–48 h as temperatures gradually rise to 40–50°C and sugars and other easily biodegradable substances are destroyed. During the second phase, when temperatures of 40–65°C prevail, cellulosic substances that are less biodegradable are destroyed. Lignins, the darker components in plant tissues, break down even more slowly but the rate varies among plant species. Thermophilic microorganisms predominate during this part of the process. Plant pathogens and seeds are killed by the heat generated during this high temperature phase (Bollen, 1993; Farrell, 1993). Compost piles must be turned frequently to expose all parts to high temperature and produce a homogeneous product free of pathogens and weed seeds.

The third or curing phase of composting begins as the concentration of readily biodegradable components in wastes declines. As a result, rates of decomposition and heat output and also temperatures decline. At this time, mesophilic microorganisms recolonize the compost from the outer low temperature layer into the pile. Humic substances accumulate in increasing quantities by this time. Mature composts consist largely of lignins, humic substances and microbial biomass, and have a dark colour (Chen and Inbar, 1993).

Unfortunately, beneficial as well as detrimental microorganisms are killed during the high temperature phase of composting. Therefore, suppression of pathogens and/or disease is largely induced during curing as biocontrol agents recolonize composts after peak heating. *Bacillus* spp., *Enterobacter* spp., *Flavobacterium balustinum, Pseudomonas* spp., other bacterial genera and *Streptomyces* spp., as well as *Penicillium* spp., a host of *Trichoderma* spp., *Gliocladium virens* and other fungi have been identified as biocontrol agents in composts (Chung and Hoitink, 1990; Hadar and Gorodecki, 1991; Hardy and Sivasithamparam, 1991; Hoitink and Fahy, 1986; Nelson *et al.*, 1983; Phae *et al.*, 1990).

The moisture content of compost critically affects the potential for mesophiles to colonize the substrate after peak heating. Dry composts (< 34% moisture, w/w) become colonized by fungi and are conducive to Pythium diseases, for example. The moisture content must be high enough (40–50%, w/w), so that bacteria can colonize the substrate after peak heating. Water must often be added to composts during curing to avoid this problem.

Variability in suppression of Rhizoctonia damping-off encountered in substrates amended with mature composts in part is due to random recolonization after peak heating of compost by a microflora with efficacy against *R. solani.* Compost produced in the open near a forest, an environment that is high in microbial species diversity, is more consistently suppressive to Rhizoctonia diseases than the same produced in a partially enclosed facility where few microbial species survive heat treatment (Kuter *et al.*, 1983). Composts prepared from municipal sewage sludges and MSW are consistently conducive to *R. solani* because care is taken to kill faecal pathogens and parasites with heat exposure. As mentioned earlier, this process kills most beneficial microorganisms as well. These composts, although naturally suppressive to Pythium diseases, have to be incubated for a month or more before they become naturally colonized due to chance by the right-specific microflora and, thus, become suppressive to *R. solani* (Kuter *et al.*, 1988). In field soil, several months may pass before suppression is induced (Lumsden *et al.*, 1983).

To solve the problem of variability in suppressiveness of compost to *R. solani* and other plant pathogens, specific fungal and bacterial inoculants have been developed, which when introduced into compost after peak heating, but before significant levels of recolonization have occurred, induce consistent levels of suppression. Patents have been issued to the Ohio State University for this process (Hoitink, 1990). In Japan, Phae *et al.* (1990) isolated a *Bacillus* strain that induces predictable biological control in composts. It survives peak heating.

## 3. Mechanisms of biological control in composts

Two mechanisms of biological control, based on competition, antibiosis, hyperparasitism, and induced systemic resistance in plants, have been described for compost-amended substrates. Propagules of plant pathogens, including *Pythium* and *Phytophthora* spp., are suppressed through a mechanism known as 'general suppression' (Boehm *et al.*, 1993; Chen *et al.*, 1988a,b; Cook and Baker, 1983; Hardy and Sivasithamparam, 1991; Mandelbaum and Hadar, 1990). Many types of microorganisms present in compost-amended container media function as biocontrol agents for diseases caused by *Phytophthora* and *Pythium* spp. (Boehm *et al.*, 1993; Hardy and Sivasithamparam, 1991). Propagules of these pathogens, if inadvertently introduced into composts, do not germinate in response to nutrients released in the form of seed or root exudates. The high microbial activity and biomass caused by the 'general soil microflora' in compost-amended substrates, prevents germination of spores of the pathogen and infection of the host, presumably through microbiostasis (Chen *et al.*, 1988a; Mandelbaum and Hadar, 1990). Propagules of these pathogens remain dormant and are typically not killed if introduced in compost-amended soil (Chen *et al.*, 1988a; Mandelbaum and Hadar, 1990). An enzyme assay, that determines microbial activity based on the rate of hydrolysis of fluorescein diacetate, predicts suppressiveness of potting mixes to Pythium diseases (Boehm and Hoitink, 1992; Chen *et al.*, 1988a; Mandelbaum and Hadar, 1990). Similar information has been developed for soils on 'organic farms' where soilborne diseases are less prevalent (Workneh *et al.*, 1993).

The mechanism of biological control for *Rhizoctonia solani* in compost-amended substrates is different from that of *Pythium* and *Phytophthora* spp. A narrow group of microorganisms is capable of eradicating sclerotia of *R. solani*. This type of suppression we refer to as 'specific suppression' (Hoitink *et al.*, 1991). *Trichoderma* spp., including *T. hamatum* and *T. harzianum*, are the predominant hyperparasites recovered from composts prepared of lignocellulosic wastes (Kuter *et al.*, 1983; Nelson *et al.*, 1983). These fungi interact with various bacterial strains in biological control of Rhizoctonia damping-off (Kwok *et al.*, 1987). It is of interest that

*Penicillium* spp. are the predominant hyperparasites recovered from sclerotia of *Slerotium rolfsii* in composted grape pomace, a waste high in sugar content (Hadar and Gorodecki, 1991). *Trichoderma* spp. were not recovered from this compost. The composition of the parent product, as expected, thus appears to have an impact on the microflora in composts active in biological control.

## 4. Role of organic matter decomposition level

The decomposition level of organic matter in compost-amended substrates has a major impact on disease suppression. For example, *R. solani* is highly competitive as a saprophyte (Garrett, 1962). It can utilise cellulose and colonize fresh wastes but not the low in cellulose mature compost (Chung *et al.*, 1988). *Trichoderma* isolates, on the other hand, function as biocontrol agents of *R. solani*, and are capable of colonizing fresh as well as mature compost (Nelson *et al.*, 1983). In fresh, undecomposed organic matter, biological control does not occur because both fungi grow as saprophytes and *R. solani* (the pathogen) remains capable of causing disease. Presumably, synthesis of lytic enzymes involved in hyperparasitism of pathogens by *Trichoderma* (de la Cruz *et al.*, 1993) is repressed here due to high glucose concentrations in fresh wastes. The same may apply to antibiotic production. In mature compost, where concentrations of free nutrients are low (Chen *et al.*, 1988a), sclerotia of *R. solani* are killed by the hyperparasite and biological control prevails (Nelson *et al.*, 1983).

Because organic matter decomposition level is so important, composts must be stabilized adequately to reach that decomposition level where biological control is feasible. In practice, this occurs in composts (tree barks, yard wastes, etc.) that have been stabilized far enough not to induce phytotoxicity (due to lack of stability) nor immobilize nitrogen during plant growth, but are colonized by the appropriate specific microflora. Practical guidelines that define this critical stage of decomposition in terms of biological control are not yet available. Industry presently controls decomposition level by maintaining constant conditions during the entire process and adhering to a given time schedule.

Excessively stabilized organic matter, the opposite end of the decomposition scale, does not support adequate activity of biocontrol agents. As a result suppression is lacking and soilborne diseases are severe in highly mineralized soils, where humic substances are the predominant forms of organic matter (Workneh *et al.*, 1993).

The length of time that composts incorporated in soil support adequate levels of activity for biocontrol has not yet been determined. Presumably, it varies with soil temperature and the type of organic matter from which the compost was prepared. Loading rates and farming practices of course also play a role.

We have studied this 'carrying capacity' of soil organic matter in potting mixes prepared with sphagnum peat to bring a partial solution to this problem (Boehm and Hoitink, 1992; Boehm *et al.*, 1993). Sphagnum peat typically competes with compost as a source of organic matter in horticulture. Both the microflora and the organic matter in peat itself can affect suppression of soilborne diseases. The literature on that effect is therefore briefly reviewed here.

Because of its stability, organic matter in sphagnum peat (resistance to decomposition) generally does not support high microbial activity. Dark, more decomposed sphagnum peat, harvested from a foot or greater depths in peat bogs, is low in activity and consistently conducive to Pythium and Phytophthora root rots (Boehm and Hoitink, 1992). On the other hand, light sources of sphagnum peat harvested from the surface of peat bogs, are less decomposed, have a higher microbial activity (FDA activity) and suppress disease. The suppressive effect of light peat to Pythium root rots is of short duration, however (Boehm and Hoitink, 1992; Tahvonen, 1982; Wolffhechel, 1988). Light peats, therefore, can be used most effectively for short production cycles, such as in plug and flat mixes used in the ornamentals industry. Composts have longer lasting effects (Boehm and Hoitink, 1992; Boehm *et al.*, 1993).

The rate of hydrolysis of fluorescein diacetate (FDA) predicts suppressiveness of compost-amended substrates and of peat mixes to Pythium root rot (Boehm and Hoitink, 1992). As FDA activity in suppressive light sphagnum peat substrates declines to less than 3.2 $\mu$g FDA hydrolysed $min^{-1}$ $g^{-1}$ dry weight mix, the population of *Pythium ultimum* increases and root rot develops. During this collapse in suppressiveness, bacterial species diversity in the rhizosphere or the edaphic substrate does not change. However, species composition is affected (Boehm *et al.*, 1993). A microflora typical of suppressive soils, which includes *Pseudomonas* spp. and other rod-shaped Gram negative bacteria as the predominant rhizosphere colonizers, is replaced by pleomorphic Gram-positive bacteria and putative oligotrophs (Wu *et al.*, 1993). Thus, the microflora of the conducive substrate resembles that of highly mineralized niches in soil (Kanazawa and Filip, 1986).

Non-destructive analysis of soil organic matter, utilising Fourier Transform infrared spectroscopy (FT-IR) and cross polarisation magic angle spinning – $^{13}C$ nuclear magnetic resonance spectroscopy (CPMAS–$^{13}C$ NMR), allows characterisation of biodegradable components of soil organic fractions (Chen and Inbar, 1993; Inbar *et al.*, 1989). CPMAS–$^{13}C$NMR allows quantitative analysis of concentrations of readily biodegradable substances such as 'carbohydrates' (hemicellulose, cellulose, etc.) versus lignins and humic substances in soil organic matter (reviewed in Chen and Inbar, 1993). In a preliminary report, Wu *et al.* (1993) have shown that the concentration of 'carbohydrates' in a suppressive light peat mix declines as suppressiveness is lost. During the same time, bacterial genera capable of causing biological control are replaced by those that cannot. The concentration of 'carbohydrates' in the 'worn out' light peat has reached a final concentration not different from that in the conducive dark more decomposed peat. They also have shown that biocontrol agents inoculated into the more decomposed peat are not able to induce sustained biological control of Pythium root rot. Thus, biocontrol of this disease is determined by the 'carrying capacity' of the substrate which regulates species composition and activity and thus, in turn, the potential for sustenance of biological control.

## 5. Compost steepages for disease control

During the past decade, a series of projects have been published on the control of plant diseases of above ground plant parts with compost water extracts (Weltzhien, 1992). The steepages are prepared by soaking mature composts in water (still culture; 1:1, w/w) for 7–10 days. The steepage is filtered and then sprayed on plants. Efficacy varies with composts, crops and the disease under question. Sackenheim (1993), utilising plate counting procedures, has shown that aerobic microorganisms predominate in these steepages. The microflora includes strains of bacteria and isolates of fungi known as biocontrol agents. A number of enrichment strategies, that included nutrients as well as microorganisms, were developed to improve efficacy of the steepages. Disease control induced by compost steepages has in the past been attributed in part to systemic acquired resistance induced in plants by microbes present in the extracts. The recent work by Sackenheim (1993) on grape does not support this assumption, however.

A factor that has been entirely overlooked but could play a role in efficacy of steepages is the condition of organic matter and the associated microflora in the soil in which treated plants are produced. Soils naturally suppressive to soilborne plant pathogens (e.g. compost-amended soils) harbour active populations of biocontrol agents (Boehm *et al.*, 1993) which could induce protection to foliar pathogens in the leaves of plants (Maurhofer *et al.*, 1994; Wei *et al.*, 1991). Thus, further research may show that plants produced in suppressive soils not only have less root rot but could also be less prone to attack by vascular wilt and foliar pathogens and respond better to compost steepages.

## 6. Disease suppression – future outlook

Success in biological control of diseases caused by *Fusarium, Pythium* and *Phytophthora* spp. is possible only if all factors involved in the production of composts are defined and kept constant. Most composts cannot meet these criteria because emphasis on quality control in general is still inadequate. Therefore, composted bark and light sphagnum peat remain the principal organic components used for the preparation of potting mixes or soils naturally suppressive to soilborne plant pathogens. Composted manures, yard and food wastes, are steadily gaining in popularity, and offer the same potential (Gorodecki and Hadar, 1990; Grebus *et al.*, 1994; Inbar *et al.*, 1993b; Marugg *et al.*, 1993; Schüler *et al.*, 1993).

The general suppression phenomenon, based on present concepts, at best covers those diseases caused by pathogens suppressed through microbiostasis. Microbial activity in a mix, based on the rate of hydrolysis of fluorescein diacetate, is one procedure that now can be used effectively to determine suppressiveness to Pythium and Phytophthora root rots. This procedure by itself, however, does not predict how long the effect will last. The potential for biodegradable carbon in a mix to support an active and effective microbial biomass determines that phenomenon. CPMAS–$^{13}$CNMR spectroscopy predicts this property for peat. The same basic information will have to be developed for composts and other sources of organic amendments. Predictive utilization of composts for disease control will remain an art until more quantitative guidelines are developed.

Controlled inoculation of composts with biocontrol agents is a procedure that must be developed on a commercial scale to induce consistent levels of suppression to pathogens such as *R. solani* (Grebus *et al.*, 1993; Hoitink *et al.*, 1991; Phae *et al.*, 1990). This new field of biotechnology is in its infancy, however. Major research and development efforts will need to be directed into this approach to disease control. Recycling through composting is increasingly being chosen as the preferred strategy for waste treatment. This also applies to farm manures (Rynk, 1992). For this reason, composts are becoming available in greater quantities. Peat, on the other hand, is a limited resource that cannot be recycled. Future opportunities for both natural and controlled-induced suppression of soilborne plant pathogens, using composts as the food stuff for biocontrol agents, therefore appear bright.

## Acknowledgements

Salaries and research support provided by State and Federal Funds appropriated to the Ohio Agricultural Research and Development Center, The Ohio State University and by Grant No. US-2196-92 from BARD, the United States–Israel Binational Agricultural Research and Development Fund.

## References

Boehm M J and Hoitink H A J 1992 Sustenance of microbial activity and severity of Pythium root rot of poinsettia. Phytopathology 82, 259–264.

Boehm M J, Madden L V and Hoitink H A J 1993 Effect of organic matter decomposition level on bacterial species diversity and composition in relationship to Pythium damping-off severity. Appl. Environ. Microbiol. 59, 4171–4179.

Bollen G J 1993 Factors involved in inactivation of plant pathogens during composting of crop residues. In Science and Engineering of Composting: Design, Environmental, Microbiological and Utilization Aspects. Eds H A J Hoitink and H M Keener. pp 301–318. Renaissance Publications, Worthington, OH.

Chen Y and Inbar Y 1993 Chemical and spectroscopical analyses of organic matter transformations during composting in relation to compost maturity. In Science and Engineering of Composting: Design, Environmental, Microbiological and Utilization Aspect. Eds H A J Hoitink and H M Keener. pp 551–600. Renaissance Publications, Worthington, OH.

Chen W, Hoitink H A J, Schmitthenner A F and Tuovinen O H 1988a The role of microbial activity in suppression of damping-off caused by *Pythium ultimum*. Phytopathology 78, 314–322.

Chen Y, Hoitink H A J and Madden L V 1988b Microbial activity and biomass in container media predicting suppressiveness to damping-off caused by *Pythium ultimum*. Phytopathology 78, 1447–1450.

Chung Y R and Hoitink H A J 1990 Interactions between thermophilic fungi and *Trichoderma hamatum* in suppression of Rhizoctonia damping-off in a bark compost-amended container medium. Phytopathology 80, 73–77.

Chung Y R, Hoitink H A J and Lipps P E 1988 Interactions between organic-matter decomposition level and soilborne disease severity. Agric. Ecosys. Environ. 24, 183–193.

Cook R J and Baker K F 1983 The Nature and Practice of Biological Control of Plant Pathogens. Am. Phytopathol. Soc., St Paul, MN.

de la Cruz J, Rey M, Lora J M *et al.* 1993 Carbon source control on $\beta$-glucanases, chitobiase and chitinase from *Trichoderma harzianum*. Microbiology 159, 316–322.

Farrell J B 1993 Fecal pathogen control during composting. In Science and Engineering of Composting: Design, Environmental, Microbiological and Utilization Aspects. Eds H A J Hoitink and H M Keener. pp 282–300. Renaissance Publications, Worthington, OH.

Garrett, S D 1962 Decomposition of cellulose in soil by *Rhizoctonia solani* Kuhn. Trans. Br. Mycol. Soc. 45, 114–120.

Gorodecki B and Hadar Y 1990 Suppression of *Rhizoctonia solani* and *Sclerotium rolfsii* in container media containing composted separated cattle manure and composted grape marc. Crop Protection 9, 271–274.

Grebus M E, Feldman K A, Musselman C A and Hoitink H A J 1993 Production of biocontrol agent fortified compost-amended potting mixes for predictable disease suppression. Phytopathology 83, 1406 (Abstract).

Grebus M E, Watson M E and Hoitink H A J 1994 Biological, chemical and physical properties of composted yard trimmings as indicators of maturity and plant disease suppression. Compost Sci. Util. 1, 57–71.

Hadar Y and Gorodecki B 1991 Suppression of germination of sclerotia of *Sclerotium rolfsii* in compost. Soil. Biol. Biochem. 23, 303–306.

Hardy G E St J and Sivasithamparam K 1991 Suppression of *Phytophthora* root rot by a composted *Eucalyptus* bark mix. Aust. J. Bot. 39, 153–159.

Hoitink H A J 1990 Production of disease suppressive compost and container media, and microorganism culture for use therein. US Patent 4960348. Feb. 13, 1990.

Hoitink H A J and Fahy P C 1986 Basis for the control of soilborne plant pathogens with composts. Ann. Rev. Phytopathol. 24, 93–114.

Hoitink H A J and Keener H M (Eds) 1993 Science and Engineering of Composting: Design, Environmental, Microbiological and Utilization Aspects. Renaissance Publications, Worthington, OH.

Hoitink H A J and Kuter G A 1986 Effects of composts in growth media on soilborne plant pathogens. In The Role of Organic Matter in Modern Agriculture. Eds Y Chen and Y Avnimelich. pp 289–306. Martinus Nijoff Publishers, Dordrecht, The Netherlands.

Hoitink H A J, Inhar Y and Boehm M J 1991 Status of composted-amended potting mixes naturally suppressive to soilborne diseases of floricultural crops. Plant Dis. 75, 869–873.

Hoitink H A J, Boehm M J and Hadar Y 1993 Mechanisms of suppression of soilborne plant pathogens in compost-amended substrates. In Science and Engineering of Composting: Design, Environmental, Microbiological and Utilization Aspects. Eds H A J Hoitink and H M Keener. pp 601–621. Renaissance Publications, Worthington, OH.

Iannotti D A, Pang T, Toth B L, Elwell D L, Keener H M and Hoitink H A J 1993 A quantitative respirometric method for monitoring compost stability. Compost Sci. Util. 1, 52–65.

Inbar Y, Chen Y and Hadar Y 1989 Solid-state carbon-13 nuclear magnetic resonance and infrared spectroscopy of composted organic matter. Soil Sci. Soc. Am. J. 53, 1695–1701.

Inbar Y, Chen Y and Hoitink H A J 1993a Properties for establishing standards for utilization of composts in container media. In Science and Engineering of Composting: Design, Environmental, Microbiological and Utilization Aspects. Eds H A J Hoitink and H M Keener. pp 668–694. Renaissance Publications, Worthington, OH.

Inbar Y, Hadar Y and Chen Y 1993b Recycling of cattle manure: The composting process and characterization of maturity. J. Environ. Qual. 22, 857–863.

Kanazawa S and Filip Z 1986 Distribution of microorganisms, total biomass, and enzyme activities in different particles of brown soils. Microb. Ecol. 12, 205–215.

Kuter G A, Nelson E B, Hoitink H A J and Madden L V 1983 Fungal populations in container media amended with composted hardwood bark suppressive and conductive to Rhizoctonia damping-off. Phytopathology 73, 1450–1456.

Kuter G A, Hoitink H A J and Chen W 1988 Effects of municipal sludge compost curing time on suppression of Pythium and Rhizoctonia diseases of ornamental plants. Plant Dis. 72, 751–756.

Kwok O C H, Fahy P C, Hoitink H A J and Kuter G A 1987 Interactions between bacteria and *Trichoderma hamatum* in suppression of Rhizoctonia damping-off in bark compost media. Phytopathology 77, 1206–1212.

Lumsden R D, Lewis J A and Millner P D 1983 Effect of composted sewage sludge on several soilborne pathogens and diseases. Phytopathology 73, 1543–1548.

Mandelbaum R and Hadar Y 1990 Effects of available carbon source on microbial activity and suppression of *Pythium aphanidermatum* in compost and peat container media. Phytopathology 80, 794–804.

Marugg C, Grebus M E, Hansen R C, Keener H M and Hoitink H A J 1993 A kinetic model of the yard waste composting process. Compost Sci. Util. 1, 38–51.

Maurhofer M, Hase C, Meuwly P, Métraux, J-P, and Défago G 1994 Induction of systemic resistance of tobacco to tobacco necrosis virus by the root-colonizing *Pseudomonas fluorescens* strain CHAO: Influence of the *gac*, a gene and of pyoverdine production. Phytopathology 84, 139–146.

Nelson E B, Kuter G A and Hoitink H A J 1983 Effects of fungal antagonists and compost age on suppression of Rhizoctonia damping-off in container media amended with composted hardwood bark. Phytopathology 3, 1457–1462.

Ownley B H and Benson D M 1991 Relationship of matric water potential and air-filled porosity of container media to development of Phytophthora root rot of rhododendron. Phytopathology 81, 936–941.

Phae C G, Saski M, Shoda M and Kubota H 1990 Characteristics of *Bacillus subtilis* isolated from composts suppressing phytopathogenic microorganisms. Soil Sci. Plant Nutr. 36, 575–586.

Rynk R (Ed.) 1992 On-Farm Composting Handbook. Coop. Ext., NE Reg. Agric. Eng. Serv., Ithaca, NY 14853-5701. NRAES-54.

Sackenheim R 1993 Untersuchungen über Wirkungen von wässerigen, mikrobiologisch aktiven Extracten aus kompostierten Substraten auf den Befall der Weinrebe (*Vitis vinifera*) mit *Plasmopora viticola, Uncinula necator, Botrytis cenerea* und *Pseudopezicula tracheiphila.* PhD thesis. Rheinische Friedrich-Wilhelms Universität, Bonn, Germany.

Schüler C, Pikny J, Nasir M and Vogtmann H 1993 Effects of composted organic kitchen and garden waste on *Mycosphaerella pinodes* (Berk, et Blox) Vestergr., causal organism of foot rot on peas (*Pisum sativum* L.) Biol. Agric. and Hort. 9, 353–360.

Tahvonen R 1982 The suppressiveness of Finnish light coloured Sphagnum peat. J. Sci. Agric. Soc. Finl. 54, 345–356.

Wei G, Kloepper J W and Tuzun S 1991 Induction of systemic resistance of cucumber to *Colletotrichum orbiculare* by select strains of plant growth-promoting rhizobacteria. Phytopathology 81, 1508–1512.

Weltzhien H C 1992 Biocontrol of foliar fungal diseases with compost extracts. In Microbial Ecology of Leaves. Eds J H Andrews and S Hirano. pp 430–450. Brock Springer Series in Contemporary Bioscience. BSBN 0387–97579–9.

Wolffhechel H 1988 The suppressiveness of Sphagnum peat to *Pythium* spp. Acta Hort. 221, 217–222.

Workneh F, Van Bruggen A H C, Drinkwater L E and Sherman C 1993 Variables associated with a reduction in corky root and Phytophthora root rot of tomatoes in organic compared to conventional farms. Phytopathology 83, 581–589.

Wu T, Boehm M J, Madden, L V and Hoitink H A J 1993 Sustained suppression of Pythium root rot: A function of microbial carrying capacity and bacterial species composition. Phytopathology 83, 1406 (Abstract).

# 30
# Suppressiveness of municipal solid waste composts to plant diseases induced by soilborne pathogens

C. SERRA-WITTLING, C. ALABOUVETTE, S. HOUOT AND F. ROUXEL

## 1. Introduction

Composts produced from municipal solid wastes (MSW) or other organic wastes have been demonstrated to present suppressive effects towards numerous soilborne plant pathogens (Hoitink and Grebus, 1994) such as *Rhizoctonia* sp. (Mathot, 1987), *Rhizoctonia solani*, *Pythium ultimum* (Schüler *et al.*, 1989), and *Mycosphaerella pinodes* (Schüler *et al.*, 1993). In the experiments presented here, we studied the suppressiveness of MSW-composts to four diseases induced by soilborne pathogens: soft and dry rots of potato tubers, clubroot of Crucifers and Fusarium wilt. Soft rot of potato tubers caused by *Erwinia carotovora* subsp. *atroseptica,* as well as dry rot caused by *Fusarium roseum* var. *sambucinum*, appear during tuber storage. They originate from propagules present at the tuber surface contaminated in soil during the growing period. Suppressiveness of a soil (or a compost) affects rot development during tuber storage (Tivoli *et al.*, 1987). Since the main way of pathogen transmission for soft rot to the tubers is soil water, we studied the suppressiveness of compost water extracts. Clubroot disease of Crucifers causes root galls on colza, cauliflower and various cabbages. Its causal agent, *Plasmodiophora brassicae*, can survive in soil as resting spores for many years. Attempts have been reported to control this disease with composted bark or wool waste (Kinoshita *et al.*, 1988) or with a cattle manure lumbricompost (Szczech *et al.*, 1993). Fusarium wilts, caused by *formae speciales* of *Fusarium oxysporum*, affect a large range of cultivated plants. A variety of bark composts as well as pig slurry compost appeared to be suppressive to these diseases (Filippi and Pera, 1989; Reisinger *et al.*, 1992; Trillas-Gay *et al.*, 1986). The purpose of our work was to compare the suppressiveness of an MSW-compost and an MSW-lumbricompost to the four diseases described previously.

## 2. Materials and methods

### 2.1 Soils and composts

The experiments were performed with soil–compost mixtures. The surface soil (0–30 cm) of a typic Eutrochrept was sampled in a fallow experimental plot in Grignon (Paris area, France). In the experiment on dry rot of potato tubers, three other soils were used for which suppressiveness levels to this disease have been

*D. Rosen* et al. *(eds): Modern Agriculture and the Environment, 373–381.*

*Table 1.* Main characteristics of soils and composts (results expressed on a dry weight basis); TC, traditional compost; LC: lumbricompost.

| | Soils | | | | Composts | |
|---|---|---|---|---|---|---|
| | Grignon | Saint-Thurial | Pontivy | Le Rheu | TC | LC |
| pH – water | 7.3 | 3.8 | 5.6 | 5.4 | 8.6 | 8.7 |
| Organic C (%) | 1.08 | 4.42 | 2.67 | 1.69 | 13.85 | 13.72 |
| Texture | Loamy | Sandy | Sandy | Sandy–Loam | | |

reported previously (Tivoli *et al.*, 1987). The soils, from Saint-Thurial, Pontivy and Le Rheu (Bretagne area, France), were cultivated with potato and sampled 1 month after harvesting.

The composts came from the separately collected organic fraction of MSW of Bapaume (France). Windrow composting and curing lasted 10 weeks and 3 months, respectively, for this 'traditional' compost (TC). The lumbricompost (LC) was produced, after the 10 weeks of composting, through activity of earthworms (*Eisenia andrei* Bouché) during 1 month, followed by a 2-month curing process in windrows. The main characteristics of the soils and the composts are presented in Table 1.

### 2.2 *Assessment of suppressiveness of compost extracts to soft rot of potato tubers*

A strain of *Erwinia carotovora* subsp. *atroseptica* (Eca 88–45 Brettor) was grown on King's B medium. A water suspension was prepared from a 4 day-old culture. Its concentration was determined from its turbidity. Soil and compost extracts were obtained by suspending 50 g samples (sieved through a 3 mm sieve) in 100 ml sterile water, followed by shaking of the mixture for 1 h and centrifugation at 1700 *g* (10 min). The bacterial suspension was diluted in the recovered supernatant to reach final inoculum concentrations of $10^5$, $10^7$ or $10^9$ cfu $ml^{-1}$. This inoculum was preincubated during 0, 4 or 8 days on a rotary shaker at 20°C, before tuber inoculation.

Potato tubers (cultivar 'Bintje'), dipped in 0.5% sodium hypochlorite solution for 15 min and rinsed with water, were inoculated according to the modified method of Colyer and Mount (1984). Two holes (depth and diameter 5 mm) were made on each half tuber and partially filled with 100 $\mu l$ of inoculum. Sterile water was used as a control. Five half-tubers per treatment were inoculated and incubated for 3 days at 20°C in a water saturated atmosphere. The weight of rotted tissue was determined after incubation.

### 2.3 *Assessment of suppressiveness to dry rot of potato tubers*

A highly aggressive strain of *Fusarium roseum* var. *sambucinum* used (90–02–1) was grown on oatmeal agar (40 g $L^{-1}$). The surface of a 2-week-old culture

(25°C) in a Petri dish was covered with 12 ml of sterile water and scraped. The obtained macroconidial suspension was filtered through three layers of gauze to eliminate mycelial fragments. Soils and composts were air dried and sieved (0.5 mm). They were infested ($1.6 \times 10^5$ spores $ml^{-1}$ soil or compost) according to the method of Tivoli *et al.* (1987) and spread on half-tubers (150 mg on each tuber) immediately after infestation or after 4 weeks of preincubation at 15°C. Ten half-tubers per substrate were inoculated. After 15 days of incubation at 15°C in a humid atmosphere, scar tissue on the tubers was excised. Tubers were rated into six classes according to the following rotting scale: class 1 (symptomless tubers); class 2 (0 to 25% of the surface rotted); class 3 (25 to 50%); class 4 (50 to 75%); class 5 (75 to 95%); and class 6 (95 to 100%). Initial (without preincubation of the infested soil or compost) and final (after 4 weeks preincubation) disease indices were calculated: $DI = [(N_1 \times 0 + N_2 \times 12.5 + N_3 \times 37.5 + N_4 \times 62.5 + N_5 \times 87.5 + N_6 \times 100) / 10] \times 100$, where $N_i$ is the number of plants in class $i$ per replicate.

### 2.4 *Assessment of suppressiveness to clubroot disease*

Infected cauliflower roots (stored at −18°C after harvest) were crushed and passed through a cheesecloth. The resulting juice was sieved to 0.1 mm. The concentration of recovered resting spores of *Plasmodiophora brassicae* in suspension was microscopically determined with a haemacytometer.

Suppressiveness was assessed according to the method described by Rouxel and Regnault (1985), using Chinese cabbage (*Brassica campestris pekinensis* 'Granaat') as the test plant. The soil was amended with peat and sand (1:1:1, v/v) and heat-treated (100°C, 1 h). For each soil–compost mixture (0, 10 and 30% of compost), five pots were filled with 300 ml of mixture and seeded with five seeds. Each pot was inoculated with a 30 ml spore suspension to establish final concentrations of 0, $10^3$, $10^4$, $10^5$, $10^6$ or $10^7$ spores $ml^{-1}$ mixture. Plants were produced under controlled conditions: 15 h photoperiod at 7000 Lux, temperature between 20 and 24°C. After six weeks, plants were uprooted and distributed into four classes according to symptom expression on the root system: class 0, no gall; class 1, one or two small galls; class 2, more than two galls and some roots still symptomless; class 3, the entire root system attacked. The disease index was then calculated for each pot: $DI = [(N_0 \times 0 + N_1 \times 0.25 + N_2 \times 0.5 + N_3 \times 1.0) / 5] \times 100$, where $N_i$ is the number of plants per pot in class $i$.

### 2.5 *Assessment of suppressiveness to Fusarium wilt*

The conduciveness bioassay with flax, as proposed by Alabouvette *et al.* (1982), which is representative of all Fusarium wilts, was performed. A benomyl-resistant strain of *Fusarium oxysporum* f. sp. *lini*, designated Foln 35, was used. The method for inoculum production was as described by Amir and Alabouvette (1993). The fungus was grown in liquid malt extract medium (10 g $L^{-1}$) for 5 days (25°C). Conidia were recovered by centrifugation (3000 *g*; 15 min) and mixed with talc (1/3 biomass, 2/3 talc). The number of conidia in the talc was determined. The talc was dried at 20°C by forced air and stored at 4°C.

The soil and the composts were sieved (4 mm) and air-dried. Soil–compost mixtures (0, 10 and 30% of compost) were inoculated with the talc inoculum at initial concentrations of 5000, 10 000 or 20 000 cfu $g^{-1}$ mixture. For each treatment, six pots were filled with 300 ml mixture and seeded with flax seeds (*Linum usitatissimum* 'Opaline'). After emergence, the seedlings were thinned to 10 per pot. Plants were grown in a growth chamber under controlled conditions: 15 h photoperiod at 10 000 Lux; 17°C and 15°C as day and night temperatures, respectively, during the first two weeks; then 25°C and 22°C as day and night temperatures, respectively, until the end of the experiment. The number of wilted plants was recorded weekly from day 28 through day 66. Experiments with the two composts were performed separately.

## 3. Results and discussion

Symptoms of disease did not develop in the non inoculated control treatments (Figures 1, 3 and 4; data not shown for dry rot of potato). This reveals that natural inocula in the soils and composts were below effective levels. It was necessary to assess this natural inoculum potential, especially for *Plasmodiophora brassicae*, because its highly resistant resting spores can survive during composting if adequate precautions are not taken (Bruns *et al.*, 1993).

### *3.1 Soft rot of potato tubers (Figure 1)*

Without preincubation, soft rot increased with inoculum density and there was no highly significant difference between the water control and the soil and compost extracts. Preincubation decreased disease severity, especially at the high inoculum densities. In addition, after preincubation, disease level was highly significantly reduced in the soil and compost extracts treatments compared to the water control. The decrease of rot development after preincubation in the water control could have been due to a reduction of the pathogen population and/or to a loss in nutritional status. In the soil and compost extracts, the influence of preincubation on disease development was more pronounced, although more nutrients were available. Therefore, mechanisms other than nutrient starvation may be involved, such as bacterial antagonism (Chard *et al.*, 1992), especially by *Pseudomonas* sp. (Colyer and Mount, 1984). After preincubation, the disease did not develop on the tubers inoculated with compost extracts, but the results were not significantly different from those in soil. Finally, no significant difference was found between TC and LC.

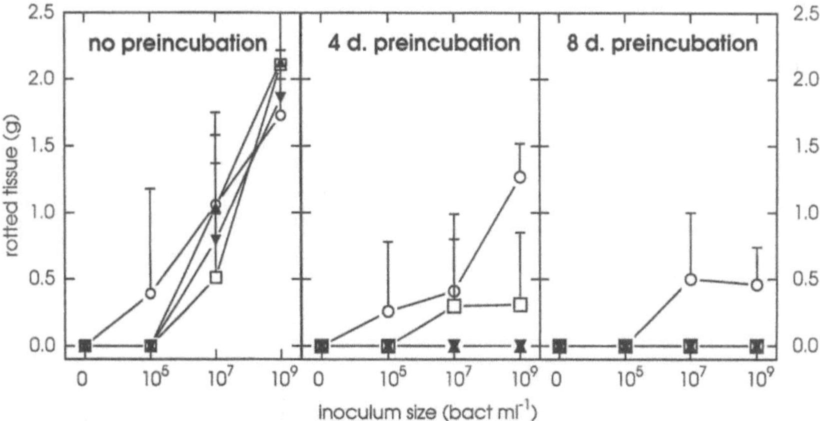


*Figure 1.* Effect of soil and compost extracts on soft rot of potato tubers caused by *Erwinia carotovora* subsp. *atroseptica*, depending upon inoculum preincubation time. d., days; water control; Grignon soil; ▲ traditional compost, TC; ▼, lumbricompost, LC.

## 3.2 *Dry rot of potato tubers (Figure 2)*

Initial disease severity in the soils decreased in the order of Saint-Thurial, Le Rheu and Pontivy. These results are in agreement with those obtained by Tivoli *et al.* (1987), who showed that Pontivy soil is suppressive, whereas Saint-Thurial soil is conducive and Le Rheu soil intermediate in response to the disease. Grignon soil permitted only a low level of disease development and could be considered as suppressive. Initial and final disease indices for the composts were lower than those

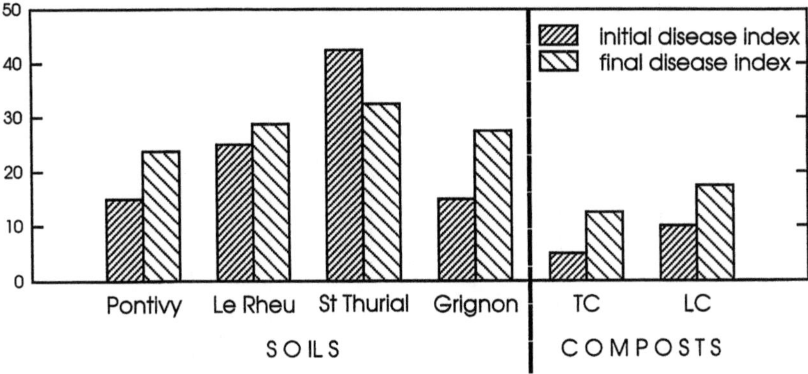


*Figure 2.* Effect of soils and composts (TC, traditional compost; LC, lumbricompost) on dry rot of potato tubers caused by *Fusarium roseum* var. *sambucinum*. Tubers were rated into six classes according to the following rotting scale: class 1 (symptomless tubers), class 2 (0 to 25% of the surface rotted), class 3 (25 to 50%), class 4 (50 to 75%), class 5 (75 to 95%) and class 6 (95 to 100%). Initial (without preincubation of the infested soil or compost) and final (after 4 weeks preincubation) disease indices were calculated: $DI = [(N_1 \times 0 + N_2 \times 12.5 + N_3 \times 37.5 + N_4 \times 62.5 + N_5 \times 87.5 + N_6 \times 100)/10] \times 100$, where $N_i$ is the number of plants in class $i$ per replicate.

of the suppressive soils; both composts were suppressive. Initial disease indices allow one to estimate instantaneous interactions between the soil or the compost and the pathogen population. Preincubation, during which time dependent reactions occur, is more representative of the soil or compost's ability to suppress the disease under natural conditions. Since suppressiveness of the composts was maintained during preincubation, results without preincubation allow one to predict and classify suppressiveness levels. TC and LC did not exhibit different behaviours. As soil suppressiveness is independent of pH (Tivoli *et al.*, 1990), the high pH values of the composts may not account for their suppressiveness. Microbial interactions probably play a role in suppressiveness.

### *3.3 Clubroot of Crucifers (Figure 3)*

In the conducive soil, the disease index increased with the inoculum density until $10^4$ spores $ml^{-1}$, then stabilised around 80%. Addition of 10% of compost (TC or LC) did not modify the conduciveness of the soil, as the results were not

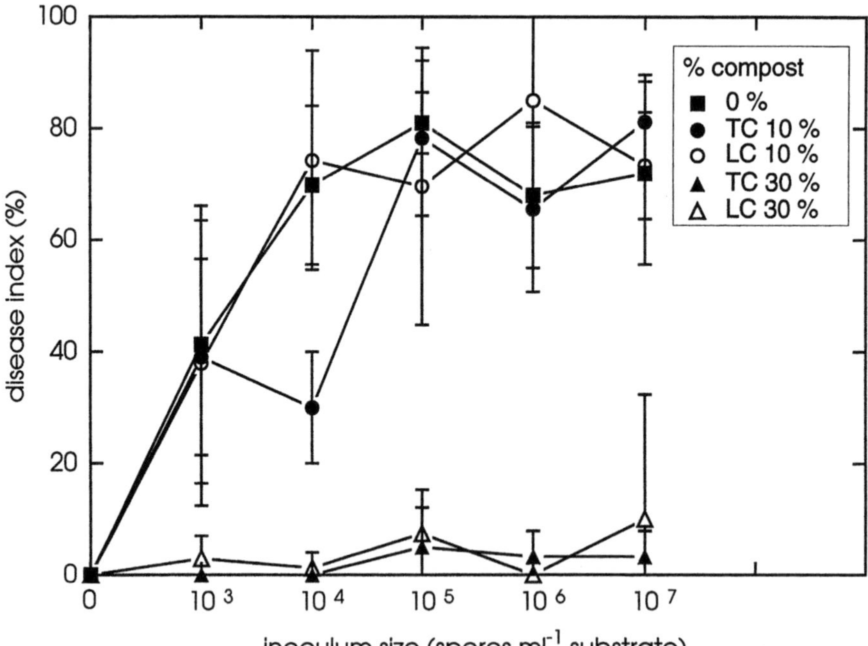


*Figure 3.* Effect of compost addition (TC, traditional compost; LC, lumbricompost) on clubroot of Chinese cabbage caused by *Plasmodiophora brassicae*. Plants were distributed into four classes according to symptom expression on the root system: class 0, no gall; class 1, one or two small galls; class 2, more than two galls and some roots still symptomless; class 3, the entire root system attacked. The disease index was then calculated for each pot: $DI = [(N_0 \times 0 + N_1 \times 0.25 + N_2 \times 0.5 + N_3 \times 1.0)/5] \times 100$, where $N_i$ is the number of plants per pot in class $i$.

*Table 2.* pH of the soil–compost mixtures used in the experiment with *Plasmodiophora brassicae* (TC, traditional compost; LC, lumbricompost).

| Compost proportion | TC | | LC |
|---|---|---|---|
| 0%[a] | | 6.7 | |
| 10% | 7.1 | | 7.2 |
| 30% | 7.5 | | 7.5 |

[a] Soil was added with peat and sand (1:1:1, v/v).

significantly different from those of the soil. In the 30% mixtures, the disease index increased somewhat with the inoculum size, but did not exceed 10%. This indicated that, according to the experimental conditions favourable to the pathogen development, the disease would not have appeared under agronomic conditions. Thus 30% compost made the soil suppressive to the clubroot disease. No significant difference was observed between TC and LC. High pH value of the composts could explain their suppressiveness, as disease development has been related to acid soils, with limit pH of 7.0 (Rouxel and Regnault, 1985). Compost addition increased the pH, as shown in Table 2. But according to this pH criterion, 10% mixtures should have been suppressive, as were the 30% mixtures. Other factors than pH may therefore be responsible for the compost suppressiveness, such as phenolic compounds (Young *et al.*, 1991) and antagonistic bacteria (Einhorn *et al.*, 1991) of composts.

### *3.4 Fusarium wilt of flax (Figure 4)*

Disease incidence increased with the inoculum size in the unamended soil after 66 days of culture, but the results differed from one experiment (with TC) to the other (with LC). According to the disease indices obtained, the soil could be considered as conducive to the disease (Alabouvette *et al.*, 1982). A 10% compost amendment reduced the number of wilted plants, but not significantly. In the 30% amendment, disease incidence was significantly decreased compared to the unamended soil, even at the highest inoculum doses (20 000 cfu $g^{-1}$ substrate for TC and 10 000 and 20 000 cfu $g^{-1}$ substrate for LC). Therefore, both composts could be considered as suppressive to the disease. As results in the unamended soil differed, it was not possible to compare TC and LC. The role of microorganisms in compost suppression to Fusarium wilts has been reported before (Phae *et al.*, 1990; Reisinger *et al.*, 1992; Trillas-Gay *et al.*, 1986), and further experiments have been started to investigate the nature of suppressiveness of the composts.

## 4. Conclusion

In summary, suppressive effects of the composts on soft rot of potato tubers were not different from those of the soil, but both composts were effective in controlling dry

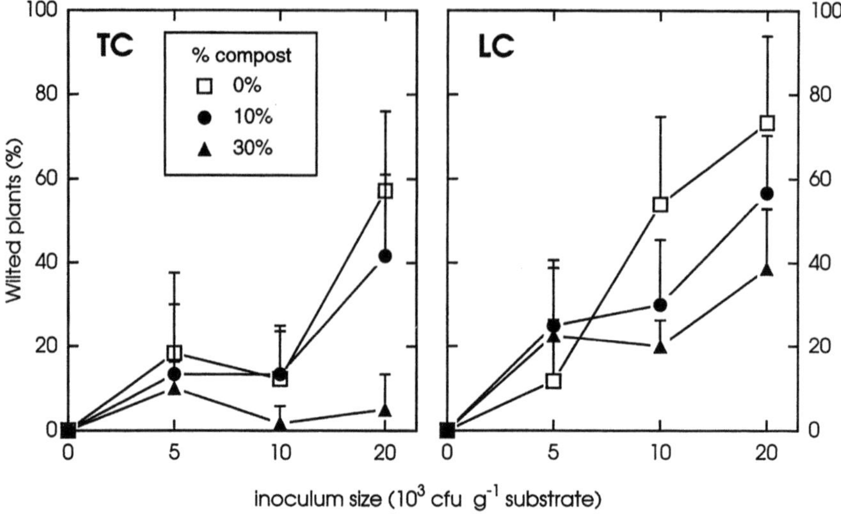


*Figure 4.* Effect of compost addition (TC, traditional compost; LC, lumbricompost) on Fusarium wilt of flax caused by *Fusarium oxysporum* f. sp. *lini*: number of wilted plants after 66 days of culture.

rot of tubers. When incorporated into the soil at a 30% rate, composts appeared to be suppressive to clubroot of Crucifers and Fusarium wilt of flax. Concerning their effects on soft and dry rots of tubers and on clubroot, no significant differences were shown between the two types of composts, indicating that lumbricomposting did not modify the effects of the composts on these diseases.

## Acknowledgments

This research was supported by grants from Procter & Gamble France and the ADEME (Agence de l'Environnement et de la Maîtrise de l'Energie). We specially thank the Research Laboratory on Soilborne Plant Pathogens (INRA Dijon) and the Phytopathology Station (INRA Le Rheu) for supplying us with the pathogen strains.

## References

Alabouvette C, Couteaudier Y and Louvet J 1982 Comparaison de la réceptivité de différents sols et substrats de culture aux fusarioses vasculaires. Agronomie 2, 1–6.

Amir H and Alabouvette C 1993 Involvement of soil abiotic factors in the mechanisms of soil suppressiveness to Fusarium wilts. Soil Biol. Biochem. 25, 157–164.

Bruns C, Gottschall R, Zeller W, Schüler C and Vogtmann H 1993 Survival rates of plant pathogens during composting of biogenic wastes in commercial composting plants under different decomposition conditions. In Soil biota, nutrient cycling and farming systems. Eds Paoletti, Foissner and Coleman. pp 41–51. Lewis publishers/CRC-Press, Boca Raton, FL.

Chard J M, Van Gardingen J and Graham D C 1992 Biological control of *Erwinia* diseases of potato. Bull. OILB/SROP 15, 133–135.

Colyer P D and Mount M S 1984 Bacterization of potatoes with *Pseudomonas putida* and its influence on postharvest soft rot diseases. Plant Dis. 68, 703–706.

Einhorn G, Bochow H, Huber J and Krebs B 1991 Methodische Untersuchungen zum Auffinden von Antagonisten des Kohlhernieerregers, *Plasmodiophora brassicae* Wor. Arch. Phytopathol. Pflanzensch. 27, 205–208.

Filippi C and Pera A 1989 The role of telluric microflora in the control of Fusarium wilt in carnations grown in soils with bark compost. Biol. Wastes 27, 271–279.

Hoitink H A J and Grebus M E 1994 Status of biological control of plant diseases with composts. Compost Sci. Util. 2, 6–12.

Kinoshita T, Arisawa M, Kato K *et al.* 1988 Effects of soil amendment with organic materials on the suppression of clubroot disease in Chinese cabbage. Rev. Plant Pathol. 67, 104 (Abstract).

Mathot P 1987 Emploi de composts dans la lutte contre *Rhizoctonia* sp. Med. Fac. Landbouww. Rijksuniv. Gent 52, 1127–1132.

Phae C, Sasaki M, Shoda M and Kubota H 1990 Characteristics of *Bacillus subtilis* isolated from composts suppressing phytopathogenic microorganisms. Soil Sci. Plant Nutr. 36, 575–586.

Reisinger O, Durecu S and Toutain F 1992 Effect of natural substances on plants: Biological control of telluric phytopathogenic fungi by an antifungal compost. In Humus, its Structure and Role in Agriculture and Environment. Ed. J Kubat. pp 145–153. Elsevier Science Publishers, London/New York.

Rouxel F and Regnault Y 1985 Comparaison de la réceptivité des sols à la hernie des Crucifères: Application à l'èvaluation des risques sur quelques sols à la culture du colza oléagineux. CR Journées ANPP Versailles 2, 375–382.

Schüler C, Biala J and Vogtmann H 1989 Antiphytopathogenic properties of biogenic waste compost. Agric. Ecosyst. Environ. 27, 477–482.

Schüler C, Pikny J, Nasir M and Vogtmann H 1993 Effects of composted organic kitchen and garden waste on *Mycosphaerella pinodes* (Berk. et Blox) Vestergr., causal organism of foot rot on peas (*Pisum sativum* L.). Biol. Agric. Hort. 9, 353–360.

Szczech M, Rondomanski W, Brzeski M W, Smolinska U and Kotowski J F 1993 Suppressive effect of a commercial earthworm compost on some root infecting pathogens of cabbage and tomato. Biol. Agric. Hort. 10, 47–52.

Tivoli B, Tika N and Lemarchand E 1987 Comparaison de la réceptivité des sols aux agents de la pourriture des tubercules de pomme de terre: *Fusarium* spp. et *Phoma* sp. Agronomie 7, 531–538.

Tivoli B, Corbière R and Lemarchand E 1990 Relation entre le pH des sols et leur niveau de réceptivité à *Fusarium solani* var *coeruleum* et *Fusarium roseum* var *sambucinum* agents de la pourriture sèche des tubercules de pomme de terre. Agronomie 10, 63–68.

Trillas-Gay M I, Hoitink H A J and Madden L V 1986 Nature of suppression of Fusarium wilt of radish in a container medium amended with composted hardwood bark. Plant Dis. 70, 1023–1027.

Young C C, Cheng K T and Waller G R 1991 Phenolic compounds in conducive and suppressive soils on clubroot disease of crucifers. Soil Biol. Biochem. 23, 1183–1189.

# 31
# On the agronomic use of municipal solid waste compost: principles and applications

YORAM AVNIMELECH AND MALKA KOCHVA

## 1. Introduction

The use of composts or bio-solids is widely advocated, both as an environmentally sound solution to recycle the increasing amounts of these materials (or wastes) as well as an economic means to raise soil productivity. Soil compaction, soil structure deterioration and an accompanying decrease of soil fertility are worldwide problems (Barnes *et al.*, 1971; Wood, 1985). These are related to the decrease of the organic matter in cultivated soils (Mann, 1986). Thus, a call for the extensive use of composts, manures, and other organic amendments seems obvious. However, the farming experience as well as reported and non-reported research results are not conclusive (e.g. Avnimelech, 1986; Cooke, 1979). In cases, the addition of composts and manures led to a considerable improvement of soil structure and fertility, yet, in other cases the application of the organic amendment was not effective and the expenses involved were not paid back. This irreproducibility leads to a reluctance by farmers to use composts and manures.

A prerequisite for a widespread municipal waste compost (MSW) utilization is the ability to predict its efficiency and the ability to issue recommendations as to the optimal technology that will lead to a consistent positive response in the field.

The use of bio-solids is widely promoted (e.g. O'Dotte, 1994). However, the major obstacle towards a widespread use of these materials is the lack of data on the agronomic and economic value of bio-solids or other organic amendments application.

We present here a review of a series of laboratory and field studies investigating and developing a technology that can consistently support an improvement of soil physical properties and an increase in yields. The consistency of this approach was tested and approved under variable conditions in Israel semi-arid soils. The more basic aspects of these findings and some ecological implications are discussed. Detailed reports of the different experiments have been given elsewhere (Avnimelech and Cohen, 1988, 1989, Avnimelech *et al.*, 1990, 1992, 1993, 1994).

## 2. Working hypotheses

Two major factors were assumed to affect the efficiency of organic amendments to soil. One of the most important factors affecting organic matter metabolism in the

*D. Rosen* et al. *(eds): Modern Agriculture and the Environment, 383–393.*

soil is aeration. The prevalence of either aerobic or anaerobic conditions is a factor that determines the rate and products of organic matter metabolism (e.g. Reddy *et al.*, 1986). Municipal solid waste degradation products were found to be determined by the aeration conditions during the initial stages of the process (Avnimelech and Raveh, 1979). The processes prevailing under anaerobic conditions led to the formation of organic acids, some of them known to be phytotoxic (Lynch and Elliott, 1983) while the aerobic processes led to the production of humic compounds, known to be soil structure stabilizers (e.g. Hayes, 1986).

Another factor expected to affect the efficiency of compost addition is the C/N ratio of the added material. Martin (1945) showed, five decades ago, that polysaccharides are capable of stabilizing and improving the soil structure. Since then, a large number of works have been published, demonstrating the effects of polysaccharides on soil structure (e.g. Hayes, 1986; Martin, 1971; Tisdall and Oades, 1982).

The production of polysaccharides by microorganisms is encouraged when organic carbon supply exceeds the availability of nitrogen or other nutrients needed for protein production (e.g. Harris and Mitchell, 1973; Kronen and Rayburn, 1984; Salle, 1974). Thus, it is expected that the C/N ratio of the compost will be a decisive factor as to its efficiency as a soil conditioner.

## 3. Laboratory experiments

Clay loam soil samples were mixed with MSW compost at a rate of 3% (dry matter) and incubated under aerobic conditions (field capacity moisture in vessels open to the atmosphere), anaerobic conditions (field capacity moisture under $N_2$ atmosphere) or under flooding conditions. Physical and chemical properties of the soil were tested following an incubation at 30°C.

One of the structural problems of the tested soil was its compactness, i.e. its high density and low porosity. Soil aggregates density was clearly reduced following the addition of compost and aerobic incubation (Table 1). Yet, no, or hardly no, effect was found when the soil was incubated anaerobically or under water.

The turbidity of a soil suspension following a mild shaking is a measure of its structural instability (Lynch and Elliott, 1983). The turbidity of the soil suspension was clearly reduced following the aerobic incubation, not affected following the anaerobic incubation and clearly raised following the flooded incubation conditions.

Saccharides, polyuronides and humic acids were reported in the past to be related to soil structure. It was clearly seen that the levels of these components were raised following the addition of the MSW compost and the aerobic incubation, and lowered following the two anaerobic incubations.

The results reported here were repeated with an outstanding reproducibility in a number of other laboratory experiments. It can be clearly concluded that the maintenance of aerobic conditions is an essential condition in order to ensure a positive effect of compost.

*Table 1.* Effects of compost addition and incubation conditions on physical and chemical properties of clay soil.

| Treatment | Incubation conditions | | | |
|---|---|---|---|---|
| | Initial | Aerobic | Anaerobic | Flooded |
| *Soil density (g cm$^{-3}$)* | | | | |
| Control | 1.60 ± 0.03 | 1.58 ± 0.02 | 1.62 ± 0.03 | 1.66 ± 0.02 |
| Compost | 1.60 ± 0.05 | 1.47 ± 0.01 | 1.57 ± 0.03 | 1.56 ± 0.01 |
| Manure | 1.43 ± 0.01 | 1.36 ± 0.02 | 1.40 ± 0.03 | 1.44 ± 0.05 |
| *Residual turbidity (optical density)* | | | | |
| Control | 0.538 ± 0.003 | 0.315 ± 0.01 | 0.416 ± 0.05 | 0.734 ± 0.02 |
| Compost | 0.363 ± 0.006 | 0.254 ± 0.01 | 0.337 ± 0.006 | 0.546 ± 0.01 |
| Manure | 0.305 ± 0.001 | 0.183 ± 0.01 | 1.40 ± 0.03 | 0.308 ± 0.06 |
| *Saccharides extracted by boiling water (mg g$^{-1}$ soil)* | | | | |
| Control | 0.089 ± 0.006 | 0.094 ± 0.001 | 0.080 ± 0.003 | 0.078 ± 0.003 |
| Compost | 0.155 ± 0.012 | 0.200 ± 0.003 | 0.121 ± 0.003 | 0.119 ± 0.004 |
| Manure | 0.644 ± 0.018 | 0.771 ± 0.002 | 0.305 ± 0.016 | 0.393 ± 0.02 |
| *Uronic acid concentrations (mg g$^{-1}$ soil)* | | | | |
| Control | 0.168 ± 0.003 | 0.260 ± 0.01 | 0.126 ± 0.004 | 0.129 ± 0.01 |
| Compost | 0.283 ± 0.01 | 0.725 ± 0.01 | 0.252 ± 0.02 | 0.216 ± 0.01 |
| Manure | 0.798 ± 0.03 | 1.546 ± 0.06 | 0.983 ± 0.2 | 0.895 ± 0.02 |

The effect of the C/N ratio of the added compost was tested in a series of laboratory experiments (Avnimelech and Cohen, 1989). The lowest density and the best distribution of water stable aggregates was found for the C/N range of 10–30. These findings are in line with theoretical considerations mentioned above.

## 4. Surface application of MSW compost

The major conclusion drawn from the laboratory work was the essential need to maintain aerobic conditions in the compost affected soil. A tradition followed by many farmers is to plough the compost into the soil soon after its application. The reason for this tradition seems to be to minimise ammonia volatilisation from nitrogen rich manure, when the manure is the main, or sole, source of plant available nitrogen. This situation does not exist in most modern farming operations and certainly not in MSW compost, a material that is well buffered at slightly acidic pH and not having a high ammoniacal nitrogen level (Table 2). The ploughing down of the compost increases the probability of restricted oxygen supply at the site of the compost interaction with the soil. It seems that a surface application, or a very shallow mixing with the soil, should ensure a better response to the compost application.

The effects of surface applied MSW compost were measured in a series of field experiments. The efficiency of this method was evaluated through the measurement of the induced changes on soil properties and on plant yield. One of the problems that must be resolved is whether the organic matter added onto the surface of the soil has any effect on the underlying soil strata.

*Table 2.* Average composition of MSW compost.

| | |
|---|---|
| Moisture | 34 (%, w/w) |
| Ash | 44 (%, w/w) |
| Nitrogen | 13 (mg $g^{-1}$) |
| Phosphorus | 4 (mg $g^{-1}$) |
| Humic acids | 90 (mg $g^{-1}$) |
| Fulvic acids | 30 (mg $g^{-1}$) |

Compost was applied, at a uniform rate of 100 $m^3$ $ha^{-1}$ on a grumosolic clay loam. The compost was surface applied, mixed with the topmost 5 cm of the soil (using a roto-tiller) or ploughed down to a depth of 25 cm. Control plots were cultivated in a parallel way without addition of compost. The resistance to cone penetration (Bradford, 1986), measured two months after the application of the MSW compost was used as an index to evaluate the effect of the compost on the soil physical properties. Twenty replicated measurements were taken in each triplicated plot.

The results of the cone penetration resistance are given in Figure 1. It is clearly seen that the effects of the surface applied treatments were higher than those of the ploughed down compost.

Ten replicated field experiments, in different soils and with different crops, were conducted to test the effect of surface applied compost. Compost was surface applied and mixed with the soil down to a maximal depth of 5 cm, using a roto-tiller.

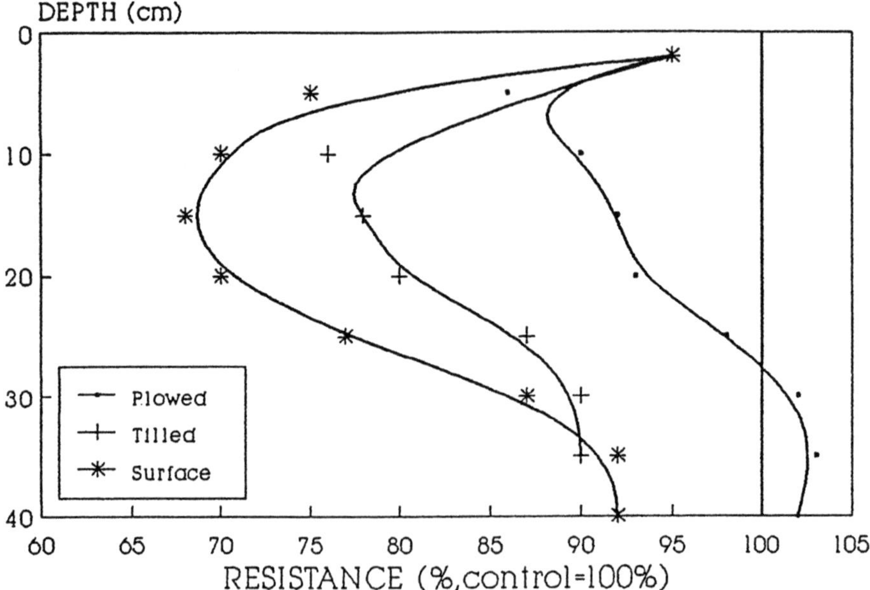


*Figure 1.* Cone penetration resistance.

*Table 3.* Effects of MSW compost application on soil chemical properties.

| Treatment depth (cm) | Organic C (mg $g^{-1}$) | Total N (mg $g^{-1}$) | $NH_4$-N (mg $kg^{-1}$) | $NO_3$-N (mg $kg^{-1}$) | Soluble P (mg $kg^{-1}$) | pH | Electrical conductivity (dS $m^{-1}$) |
|---|---|---|---|---|---|---|---|
| *Control* | | | | | | | |
| 0–10 | 6.1 | 0.1 | 54.9 | 11.5 | 2.2 | 7.5 | 0.21 |
| 10–30 | 7.6 | 1.1 | 20.3 | 8.8 | 2.5 | 7.5 | 0.24 |
| 30–60 | 3.9 | 0.9 | 9.0 | 10.1 | 0.6 | 7.6 | 0.35 |
| *80 $m^3$ $ha^{-1}$* | | | | | | | |
| 0–10 | 10.1 | 1.3 | 51.3 | 5.0 | 0.6 | 7.3 | 0.24 |
| 10–30 | 9.2 | 1.4 | 18.3 | 4.3 | 1.1 | 7.3 | 0.27 |
| 30–60 | 4.6 | 1.0 | 7.5 | 10.2 | 1.7 | 7.6 | 0.36 |
| *160 $m^3$ $ha^{-1}$* | | | | | | | |
| 0–10 | 15.6 | 1.7 | 58.5 | 1.6 | 0.3 | 7.1 | 0.27 |
| 10–30 | 11.2 | 1.4 | 24.2 | 2.3 | 0.9 | 7.3 | 0.28 |
| 30–60 | 4.1 | 1.1 | 7.9 | 10.5 | 0.3 | 7.6 | 0.37 |
| *400 $m^3$ $ha^{-1}$* | | | | | | | |
| 0–10 | 22.6 | 2.3 | 72.0 | 1.2 | 0.3 | 7.1 | 0.34 |
| 10–30 | 16.2 | 1.9 | 38.2 | 1.1 | 0.4 | 7.3 | 0.36 |
| 30–60 | 4.6 | 1.1 | 8.5 | 11.1 | 0.2 | 7.3 | 0.40 |

Leaching of soluble salts in an experiment conducted in a grumosol, following about 200 mm rain, is expressed in the electrical conductivity profile of soil extracts (Table 3). It can be seen that salts were leached down to the 30–60 cm layer. Organic nitrogen and carbon leaching was slower, and their peak concentrations were found between 10–30 cm. Nitrate concentrations were low in the compost applied plots as compared to the control, possibly due to the occurrence of denitrification. The soil pH was reduced due to the addition of the compost, even down to the 30–60 cm layer, way below the depth of compost application. Soluble phosphorus was lowered due to the addition of the compost, most probably due to its immobilization into the growing biomass.

Water content of the soil was determined a few days following rain, thus it can be regarded as a measure of the field capacity. The water content ($H_2O$, %) rose, in the two topmost layers, as a function of the organic carbon contents (OC, mg $g^{-1}$), as shown in Figure 2. The relationship between the two variables was linear as given by Equations (1) and (2), for the 0–10 and 10–30 cm layers, respectively:

$$\text{Water content } (H_2O, \%) = 25.4 + 0.341 \times \text{OC } (0-10 \text{ cm}) \quad (1)$$

$$\text{Water content } (H_2O, \%) = 26.7 + 0.825 \times \text{OC } (10-20 \text{ cm}). \quad (2)$$

The water holding capacity of the added organic carbon is 341% and 825% in the top and the 10–30 cm layers respectively. The difference between the two layers is significant and interesting. The organic carbon in the 0–10 cm layer is mostly made of the native organic matter of the compost. The organic matter in the lower layer, having a water holding capacity almost three times higher, is made mainly of a

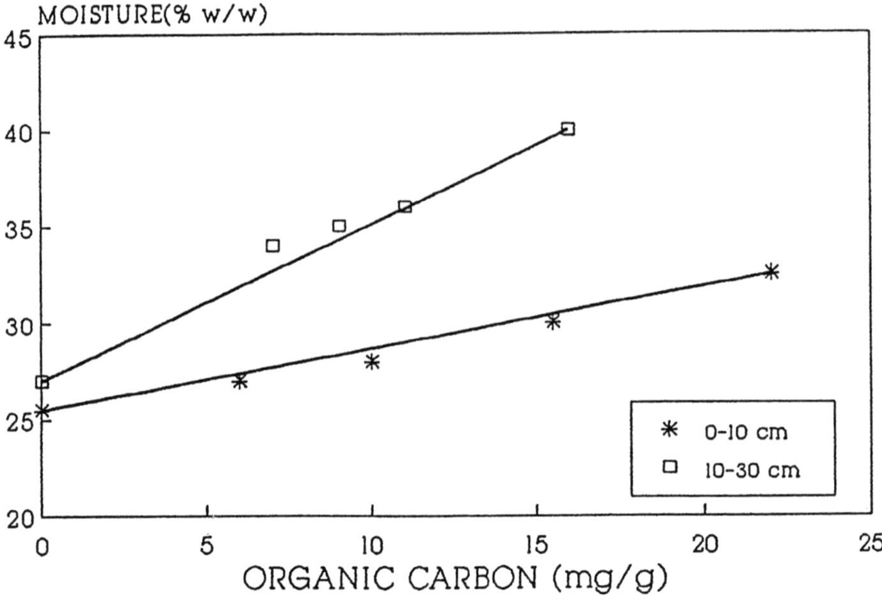


*Figure 2.* Relationship between water and organic carbon content.

fraction of the organic carbon that was leached down from the compost. The high specific activity of that fraction is interesting.

Additional and clear evidence of the effects of surface applied compost on reactions occurring in the underlying soil strata was obtained in an experiment conducted in a highly sodic and saline region at Kfar Baruch, in the Jesreel Valley (Avnimelech *et al.*, 1994).

The addition of compost to the soil in the course of its reclamation was expected to be useful due to the following considerations:

- The addition of compost improves and stabilises the structure of the soil. Thus, the addition of compost is expected to ensure a relatively high infiltration rate.
- Composts, especially MSW composts, contain chelating agents and release acids during their degradation process. This in turn can lead to the solubilization of $CaCO_3$ present in the soil and lead to the replacement of exchangeable sodium by calcium.
- The enhancement of plant growth leads to a better leaching due to the soil mulching effect, soil channelling made by decaying roots, and the resulting addition of organic matter.

These hypotheses were tested in a large scale field experiment in an area that was inflicted with a severe sodium dominated salinity.

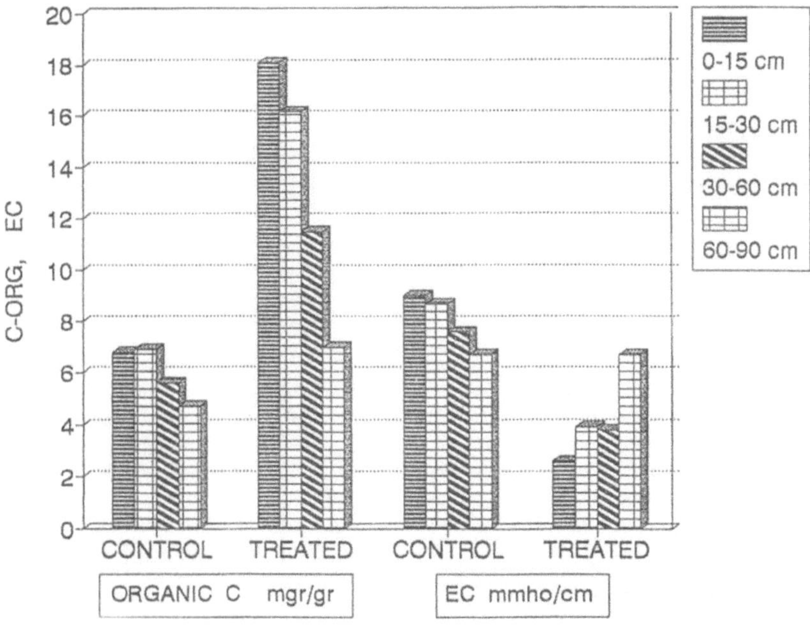


*Figure 3.* Profile of organic matter and electrical conductivity.

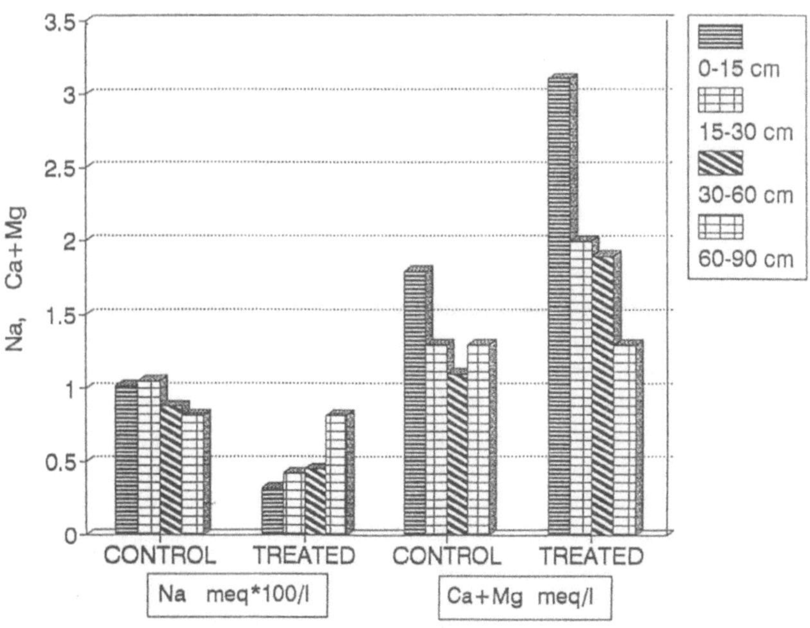


*Figure 4.* Profile of sodium and divalent ions concentration.

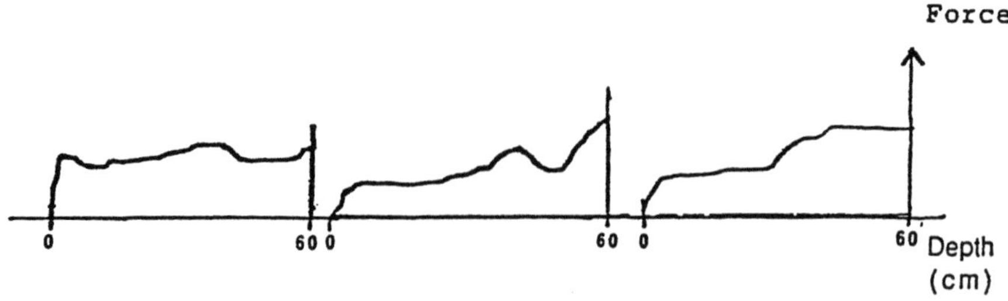


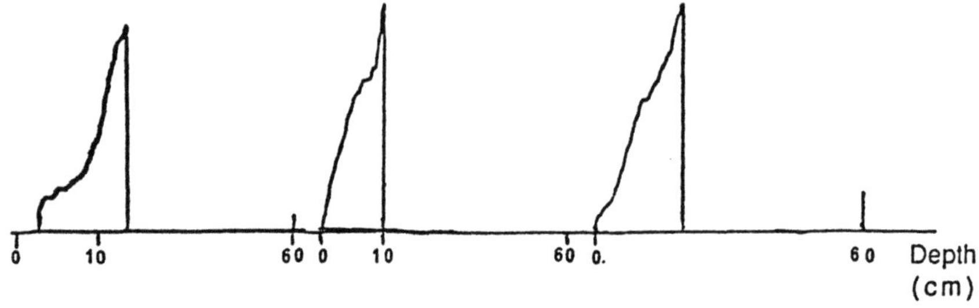


*Figure 5.* Soil structure measured by force needed to introduce a standard cone into the soil.

Profiles of organic matter, total salinity, as expressed through the electrical conductivity of the solution, sodium and divalent ions concentration at the end of a two-year period (total precipitation 900 mm) are given in Figures 3 and 4.

High salt content and sodium percentage in the soil is demonstrated in the control plots. The electrical conductivity of the saturated soil paste was about 9 dS $m^{-1}$ in the top 30 cm layer and about 7 dS $m^{-1}$ in the deeper layers. Values of 4–8 dS $m^{-1}$ are considered to be limiting for most agricultural crops (Richards, 1954). The sodium concentration in the saturated paste was about 100 mEq $L^{-1}$, while the divalent ions, Ca and Mg, were in the range of 1–2 mEq $L^{-1}$, leading to a very high SAR, in the range of 100–200 (sodium adsorption ratio, SAR, above 10, is considered to be indicative for conditions leading to soil alkalisation, soil structure deterioration and instability (e.g. Richards, 1954). Chloride concentrations were in the range of 30–50 mEq $L^{-1}$. Bicarbonate 5–19 mEq $L^{-1}$ and sulphate, at a total level of about 50 mEq $L^{-1}$ made up the ionic balance. The $CaCO_3$ content of the soil was rather high, in the range of 2–5%, yet its solubility was low since the pH of the control soil was high, in the range of 8.5–8.9. The organic carbon in the soil was 5–7 mg C $g^{-1}$, i.e. about 1% organic matter, a typical value for these semi-arid soils.

The addition of compost (applied annually at a level of 300 $m^3$ $ha^{-1}$) led to an expected increase in the organic matter contents that penetrated during the two years of the experiment down to 90 cm (Figure 3). The addition of compost raised the

Ca + Mg level. As seen in Figure 4, the effect of the compost on the divalent ion is detected even in the 60–90 cm layer. The compost applications led to a significant reduction of Na concentration in the surface down to 60 cm layers of the soil, as seen in Figure 4.

The compost application led to an enhanced salt leaching, as reflected in the electrical conductivity profile given in Figure 3, due to the improvement of the structure and the hydraulic conductivity of the soil.

The soil structure was evaluated *in situ* using a recording penetrometer (Bradford, 1986), an instrument that measures the force that is needed to introduce a standard cone into the soil. Original records made one month after the start of the experiment (following 213 mm rain), in the control and compost amended plots are given in Figure 5. It can be seen that the cone penetrometer could be inserted in the control plots only to a depth of about 10 cm, where it was blocked by an impermeable hardpan. In the compost treated plots, the cone penetrometer could be easily inserted down to a depth of 40 cm, the depth of water penetration.

## 5. Effects on soil fertility

A consistent effect on soil fertility, as reflected in yield increase and economical returns, is a prerequisite for a wide scale application of bio-solids and MSW composts. Such an effect was obtained in a series of ten replicated field experiments conducted recently, using surface application of MSW composts. As shown in Table 4, a yield increase of 10% and more was obtained in 9 out of 10 experiments.

It can be concluded that at least some of the inconsistency reported in the literature in regard to the response to compost application is due to the fact that often, when the compost is ploughed down, oxygen consumption exceeds oxygen

*Table 4.* Crop yield in field experiments.

| Location | Soil | Crop | Application rate ($m^3$ $ha^{-1}$) | Yield response |
|---|---|---|---|---|
| Yagur | Grumosol, clay loam | Wheat | 20 | +10% |
| Yagur | Grumosol, clay loam | Corn | | +15% |
| Yagur | Grumosol, clay loam | Corn | 20 | +8.3% |
| Yagur | Grumosol, clay loam | Cotton | | NSD[a] |
| Jordan Valley | Saline clay loam, calcic xerosol | Corn (kernels) | 40 | +6%[b] +18%[c] |
| Nahal oz | Loess, alkaline | Wheat | 30 | +10% |
| Nahal oz | Loess, alkaline | Pea | 30 | +30% |
| Kfar baruch (Jesreel Valley) | Grumosol, highly saline and alkaline | Oat | 400 | +1090% |
| Kfar baruch (Jesreel Valley) | Grumosol, highly saline and alkaline | Wheat | 400 | +687% |
| Gitit (Jordan Valley) | Saline clay loam | Grapes | 50 | +22% |

[a] NSD, highly infested.
[b] Total yield.
[c] Grade ‘A’.

supply and anaerobic conditions prevail. A surface application of compost ensures aerobic conditions and thus supports the positive effect of the added compost.

The consistent effects on soil physical properties, soil chemical properties and yield, as demonstrated in different parts of the country, led to an increase in the demand for MSW compost. This material that until recently was considered to be an inferior soil amendment is now requested and bought by farmers in Israel. In the last years, the two large producers of MSW compost are selling all their products at a price in the range of US$ 5–15 per ton.

## 6. Conclusions

(1) Addition of MSW compost to compacted structurally deficient soils led to a reproducible and consistent improvement of the soil structural properties only if the incubation of the soil–compost mixture was held under aerobic conditions.

(2) The optimal C/N ratio of composts is in the range of 10–30.

(3) A consistent and reproducible yield increase of about 10–20% above that obtained for a highly fertilized crop, was found with the application of MSW compost in heavy structurally deficient soils.

(4) Crop response to compost application may be restricted, at high application rates, due to an induced oxygen deficit. Development of a model and relevant laboratory diagnostic measurements, relating compost level, maturity and application geometry to expected anaerobiosis is underway.

(5) The optimal MSW compost application rate, under the conditions studied in this work, appears to be around 50 $m^3$ $ha^{-1}$.

(6) Compost added onto the surface of the soil affects underlying soil strata. It seems that a fraction of the added organic matter is relatively mobile and migrates readily with the percolating water.

## References

Allison F E 1973 Soil Organic Matter and its Role in Crop Production. Elsevier Science, Amsterdam, 673 pp.

Avnimelech Y 1986 Organic residues in modern agriculture. In The Role of Organic Matter in Intensive Agriculture. Eds Y Chen and Y Avnimelech. pp 1–10. Martinus Nijhoff, Dordrecht, The Netherlands.

Avnimelech Y and Cohen A 1988 On the use of organic manures for amendment of compacted clay soils: Effects of aerobic and anaerobic conditions. Biol. Wastes 26, 331–339.

Avnimelech Y and Cohen A 1989 Use of organic manures for amendment of compacted clay soils – II. Effect of carbon to nitrogen ratio. Common. Soil Sci. Plant Anal., 20, 1635–1644.

Avnimelech Y and Raveh A 1979 Decomposition of organic matter in sanitary landfill models. Agron. Abstr. p 25.

Avnimelech Y, Cohen A and Shkedi D 1990 The effect of municipal solid waste compost on the fertility of clay soils. Soil Technol. 3, 275–284.

Avnimelech Y, Kochva M, Yotal Y and Shkedy D 1992 The use of compost as a soil amendment. Acta Hortic. p 217–236.

Avnimelech Y, Cohen A and Shkedy D 1993 Can we except a consistent efficiency of MSW compost application? Compost Sci. Util. Autumn, 7–14.

Avnimelech Y, Kochva M and Yotal Y 1994 The use of MSW compost for the reclamation of saline and alkaline soils. Compost Sci. Util. Summer, 6–11.

Barnes K K, Carlton W M, Taylor H M, Throckmorton R I and Vanden Berg G E 1971 Compaction of Agricultural Soils. Amer. Soc. Agric. Eng., St Joseph, MI.

Bradford J M 1986 Pentrability. In Methods of Soil Analysis. 2nd ed. Ed. A Klute. Amer. Soc. Agron. Madison, WI.

Cooke G W 1979 Some properties for British soil science. J. Soil Sci. 30, 187–213.

Harris R M and Mitchell R 1973 The role of polymers in microbial aggregation. Annu. Rev. Microbiol. 27, 27–50.

Hayes M H B 1986 Soil organic matter extraction, fractionation, structure and effect on soil structure. In The Role of Soil Organic Matter in Modern Agriculture. Eds Y Chen and Y Avnimelech. p 306. Martinus Nijhoff, Dordrecht.

Kronen W K and Rayburn W R 1984 Influence of growth status and nutrients on extracellular polysaccharides synthesis by soil algae *Chlamidomonas Mexicana* (Chlorophyceae). J. Phycol. 20, 252–257.

Lynch J M and Elliott L F 1983 Aggregate stability of volcanic ash and soil during microbial decomposition of straw. Appl. Environ. Microbiol. 45, 1398–1401.

Mann L K 1986 Changes in soil carbon storage after cultivation. Soil Sci. 142, 279–288.

Martin J P 1945 Microorganisms and soil aggregation I. Origin and nature of some of the aggregating substances. Soil Sci. 59, 163–174.

Martin J P 1971 Decomposition and binding action of polysaccharides in soil. Soil Biol. Biochem. 33, 141–163.

O'Dotte R G 1994 The beneficial use of biosolids: We're getting there! WEF Specialty Conference Series Proceeding: The Management of Water and Wastewater Solids for the 21st Century. pp 9.1–9.13. Water Environment Federation, Alexandria, VA.

Reddy K R, Feijtel T C and Patrick W H 1986 Effect of soil redox conditions on microbial oxidation of organic matter. In The Role of Organic Matter in Intensive Agriculture. Eds Y Chen and Y Avnimelech. pp 117–156. Martinus Nijhoff, Dordrecht, The Netherlands.

Richards L A (Ed) 1954 Diagnosis and Improvement of Saline and Alkaline Soils. US. Dept. of Agriculture, Handbook No. 60, Wahington, DC, 160 pp.

Salle A J 1974 Fundamental Principles of Bacteriology 7th ed. Tata McGraw-Hill, New Delhi, India. pp 110–111.

Tisdall J M and Oades J M 1982 Organic matter and water stable aggregates in soil. J. Soil Sci. 33, 141–163.

Wood A W 1985 Soil degradation and management under intensive sugarcane cultivation in north Queensland. Soil Use Manage. 1, 120–124.

# 32
# Chemical and electron spin resonance properties of municipal solid waste composts

J. DROZD, A. JEZIERSKI AND Y. CHEN

## 1. Introduction

The stage of compost maturity is an important criterion for its utilization in agriculture (Iglesias Jimenez and Perez Garcia, 1992; Inbar *et al.*, 1990; Sophon, 1978). When applied to soils, fresh (immature) compost can raise microbiological activity and introduce phytotoxic compounds (Viel *et al.*, 1987; Wong, 1985). During composting the organic matter is decomposed and new products are created. The transition forms of these organic substances have different properties (Iglesias Jimenez and Perez Garcia, 1992; Inbar *et al.*, 1990; Sugahara and Inoko, 1981). The humic substances formed during composting carry various functional groups, as well as semiquinone free radicals (SFRs) (Almendroz *et al.*, 1987; Senesi, 1992). The latter are very sensitive to pH, ionic strength, hydrolysis, redox reactions and interaction with metal ions (Senesi *et al.*, 1986, 1992). These radicals play an important role in the polymerisation and depolymerisation reactions involved in humification. The signal intensity of electron paramagnetic resonance can change during the transformation of organic material.

The aim of this work was to study relationships between composting time, humification indexes and free radical signals, as a way of determining compost maturity and evaluating it for agricultural use.

## 2. Materials and methods

Studies were conducted on municipal solid wastes (MSW) from the greater metropolian area in Upper Silesia (Poland). The municipal refuse was initially prepared according to Dano technology. Composting was performed in a closed fermentative chamber, which also served as a grinding machine. The material was ground by crushing and blowing and its temperature was raised to about 45°C. After 24 hours a product having some physical properties resembling compost was obtained.

This fresh material, which contained a high level of organic matter, was composted in 100 dm$^3$ plastic containers. Its temperature was measured daily for 90 days and water was added to maintain moisture level of about 50% of the compost weight. Every 10–14 days the compost was turned over and samples were taken.

*D. Rosen* et al. *(eds): Modern Agriculture and the Environment, 395–400.*

*Table 1.* Values of investigated parameters at different stages of maturity of compost from municipal wastes.

| | Composting time (days) | | | | | | | |
|---|---|---|---|---|---|---|---|---|
| Parameter | 1 | 14 | 28 | 38 | 52 | 64 | 78 | 90 |
| $C_t$ (% of d.m.) | 22.11 | 21.69 | 18.78 | 17.97 | 17.46 | 16.81 | 16.34 | 15.60 |
| $N_t$ (% of d.m.) | 0.85 | 0.86 | 0.87 | 0.88 | 0.93 | 1.02 | 1.04 | 1.23 |
| $C_t/N_t$ | 28.6 | 25.1 | 21.6 | 20.5 | 18.6 | 16.4 | 15.7 | 12.7 |
| CHA (% of d.m.) | 3.4 | 3.9 | 5.5 | 5.9 | 5.7 | 5.5 | 5.1 | 6.2 |
| $HR_1$ (CHA/CFA) | 0.58 | 0.69 | 0.84 | 0.82 | 0.70 | 0.82 | 0.69 | 0.89 |
| CEC (dmol $kg^{-1}$) | 51.4 | 52.7 | 52.2 | 57.0 | 58.5 | 58.8 | 76.1 | 82.1 |
| ESRI ($10^{16}$ spins $g^{-1}$) | 19.2 | 16.3 | 14.1 | 18.4 | 16.2 | 13.2 | 20.2 | 23.5 |

*Table 2.* Correlation matrix.

| Parameter | Days | $C_t$ | $N_t$ | $C_t/N_t$ | CEC | CHA | $HR_1$ |
|---|---|---|---|---|---|---|---|
| $C_t$ | –0.9556* | 1.0000 | | | | | |
| $N_t$ | 0.9029* | –0.7822* | 1.0000 | | | | |
| $C_t/N_t$ | –0.9860* | 0.9761* | –0.8647* | 1.0000 | | | |
| CEC | 0.8916* | –0.7644* | 0.9261* | –0.8263* | 1.0000 | | |
| CHA | 0.7473* | –0.8847* | 0.5620 | –0.8327* | 0.5036 | 1.000 | |
| $HR_1$ | 0.5562 | –0.6593 | 0.5196 | –0.6674 | 0.3640 | 0.8371* | 1.0000 |
| ESRI | 0.3829 | –0.2386 | 0.5748 | –0.2802 | 0.7295* | 0.0836 | 0.0090 |

* Significant correlation coefficient at $\alpha = 0.05$; $n = 8$.

These samples were air-dried and the following determinations were made: $C_t$ – total oxidisable carbon; $N_t$ – total nitrogen according to Kjeldahl; cation exchange capacity (CEC) according to Harada and Inoko (1980); humic and fulvic-like carbon (CHA and CFA respectively); humification ratio $HR_1$ (CHA/CFA); and electron spin resonance (ESR) spectra. The ESR spectra were obtained with either SE Radiopan or an ESP 300E Bruker Spectrometer operating at X-band (ca. 9.2 GHz) frequencies. Measurements were performed on solid air-dried samples of composts in different maturities at ambient temperature. TEMPO and DPPH free radicals were used as standards for quantitative measurements of SFRs concentration.

## 3. Results and discussion

During the composting of MSW, marked qualitative and quantitative changes in organic matter were observed (Table 1). $C_t$ content decreased and $N_t$ increased, effecting a reduction in the C/N ratio. Humification increased the amount of CHA with compost maturity. The interdependence between investigated parameters and composting time is presented in a correlation matrix (Table 2). The most significant correlation coefficients with degree of compost maturity (composting time) were found with $C_t$, $N_t$, C/N ratio, CEC and CHA. This observation confirms the

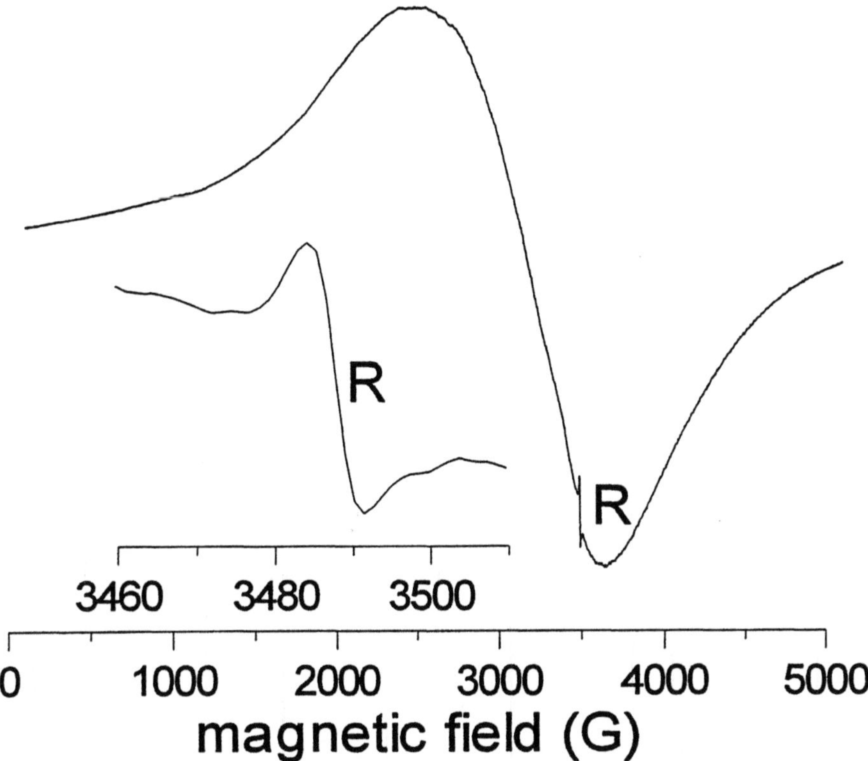


*Figure 1.* ESR spectrum of a compost sample after 90 days of maturation. R – semiquinone free radical signal. Based on the R signal intensity, the ESRI parameter was calculated.

suggestion that organic matter transformation is very closely connected with the degree of compost maturity. The most visible changes in these parameters were observed in the first 4 to 6 weeks. After this time, transformation rate was distinctly slower. ESR spectra consisted of a broad-line signal due to the presence of Fe (III) bound to organic and mineral substances, and a narrow-line signal (R) at G = 2.0035, attributed to SFRs (Figure 1).

The above mentioned changes were influenced by microbiological activity and strongly characterised an exothermic process, as evidenced by the pattern of temperature evolution during composting (Figure 2). The temperature of the compost markedly increased during the first 3 weeks of transformation. It decreased gradually till about 6 weeks, after which it was dependent upon ambient conditions.

The initial, rapid stage of composting (lasting about 20 days and during which temperature increased markedly) was accompanied by a decrease in SFRs concentration (Figure 2). Temperature stabilisation (30–50 days) was connected to additional production of SFRs and an increase in ESRI (intensity of the ESR radical signal). The next local temperature maximum (at about 65 days) was also

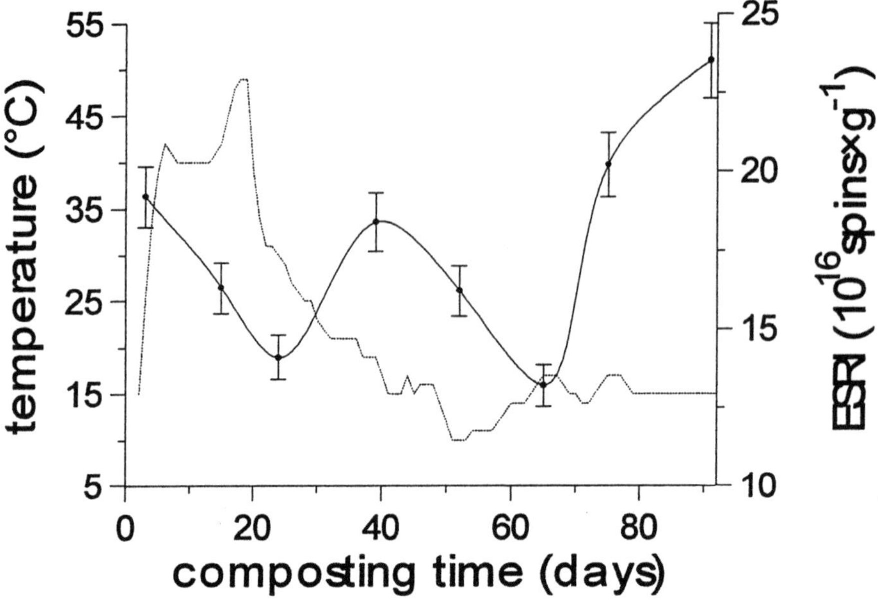


*Figure 2.* Temperature evolution (dashed line) and ESRI (solid line) during composting of the municipal wastes.

accompanied by a decrease in SFRs concentration. Temperature stabilisation in the mature compost (70–90 days) paralleled stabilisation of the SFRs concentration at a higher level (Figure 2).

These phenomena contrast those found with isolated humic substances: in the case of isolated fulvic and humic acids, an increase in temperature resulted in an increase in SFRs concentration (Senesi, 1992). In our system the production of SFRs was connected with the intensity of the composting processes (a humification process).

The ESRI was significantly correlated ($r = 0.73$) with CEC (Figure 3). This last parameter was correlated with composting time (Figure 4). Free radicals may therefore be an important index in evaluating compost maturity. A final estimation of this parameter requires further investigations to confirm this hypothesis.

## Acknowledgements

The authors are grateful to the Agency for International Development (USA) for financial support of this work (Grant No TA-MOU-C13-112).

## References

Almendroz G, Fründ R, Gonzales-Vila F J, Lüdemann H D and Martin F 1987 NMR and ESR investigation of the humification process in defined vegetable starting materials. Z. Pflanzenernähr. Bodenk. 150, 201–207.

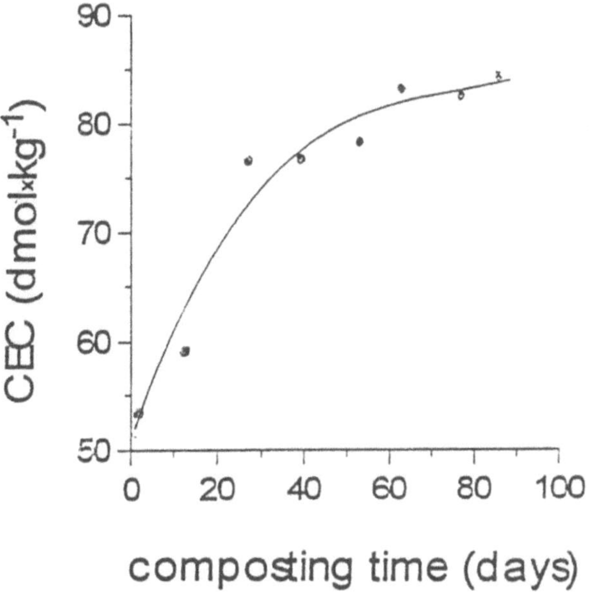


*Figure 3.* Regression between CEC and ESRI.

$y = 0.0000615x^3 - 0.00277x^2 + 0.106x + 51.7$
$R = 0.95$

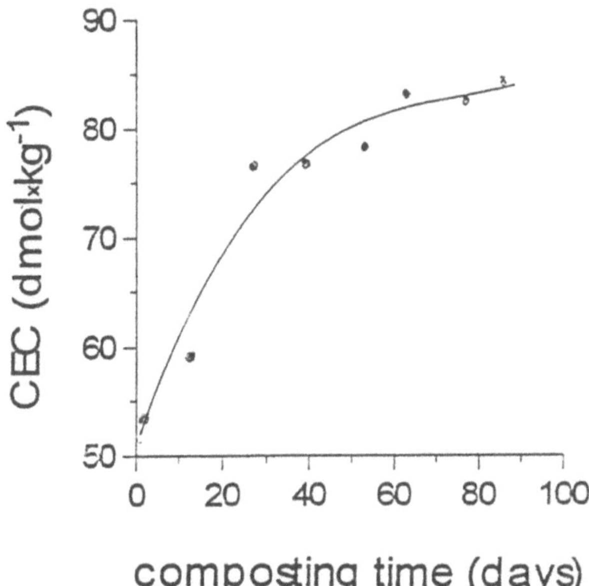


*Figure 4.* Regression between CEC and composting time.

Harada Y and Inoko A 1980 The measurement of the cation exchange capacity of compost for the estimation of the degree of maturity. Soil Sci. Plant Nutr. 26, 127–134.
Iglesias Jimenez E and Perez Garcia V P 1992 Determination of maturity indices for city refuse composts. Agric. Ecosyst Environ. 38, 331–343.
Inbar Y, Chen Y, Hadar Y and Hoitink H A J 1990 New approaches to compost maturity. Biocycle December, 64–69.
Senesi N 1992 Application of electron spin resonance and fluorescence spectroscopies to the study of soil humic substances. In Humus, its Structure and Role in Agriculture and Environment. Ed. J Kubat. pp 11–26. Elsevier Science, Amsterdam.
Senesi N, Sposito G and Martin J P 1986 Copper (II) and iron (III) complexation by soil humic acids: An IR and ESR study. Sci. Total Environ. 55, 351–362.
Senesi N, Saiz-Jimenez C and Miano T M 1992 Spectroscopic characterization of metal-humic acid-like complexes of earthworm-composted organic wastes. Sci. Total Environ. 117/118, 111–120.
Sophon E 1978 Determination of compost maturity. Compost Sci. 19, 26–27.
Sugahara K and Inoko A 1981 Composition analysis of humus and characterization of humic acid obtained from city refuse compost. Soil Sci. Plant Nutr. 27, 213–224.
Viel M, Sayag D, Peyre A and Andre L 1987 Optimization of in-vessel co-composting through heat recovery. Biol. Wastes 20, 167–185.
Wong M H 1985 Phytotoxicity of refuse compost during process of maturation. Environ. Pollut. Ser. A 37, 159–174.

# 33
# Variations with pedoclimatic conditions of an anaerobically digested liquid sewage sludge C mineralisation and its N-$NO_3^-$ availability

S. HOUOT, L. CADOT AND V. GNANALINGAM

## 1. Introduction

Land applications of sewage sludges on agricultural soils are frequently used to recycle nutrients such as N through crop production. The rates of application have to be adjusted to the crop needs to avoid excess concentrations of $NO_3^-$ in soils likely to be leached. The organic matter mineralization in soils or of organic amendments are often characterized via laboratory incubations, and first order kinetics used to describe C and N mineralization, allowing the determination of potentially mineralisable C and N (Lindeman and Cardenas, 1984; Stanford and Smith, 1972). Sewage sludge degradation in soils and the amount of N available for crop use vary with the pedoclimatic conditions including soil characteristics (texture, structure, pH, organic matter, microbial activity, etc.), temperature, soil water content and aeration.

In the work presented here, the impact of soil amendment with a liquid sewage sludge on C and N mineralization and nitrification was studied in two soils, under laboratory controlled conditions. First order kinetics were used to describe the C and N mineralization. The variations of the kinetic parameters with soil temperature, water content and compaction were characterized. Two field experiments were also started on the same soils and with the same sludge. Surface soils were sampled four months after sludge application in control and amended plots. Again, C and N mineralization were characterized during laboratory incubations in order to know how much N was still potentially available after four months of field degradation. To verify that incubations with sieved soils reliably determine the potentially mineralizable C and N, the C and N mineralization in different aggregates sizes was compared to that in sieved soils.

## 2. Materials and methods

### 2.1 Soils and sludge

The sludge came from the Achères sewage plant, near Paris. It was made from domestic–industrial sources and anaerobically digested. Its dry matter content was

*D. Rosen* et al. *(eds): Modern Agriculture and the Environment, 401–411.*

*Table 1.* Main analytical characteristics of the upper layers (0–30 cm) of the two studied soils.

| Soil | C ($g\ kg^{-1}$) | N ($g\ kg^{-1}$) | C/N | pH (in water) | Clay ($g\ kg^{-1}$) | Silt ($g\ kg^{-1}$) | Sand ($g\ kg^{-1}$) |
|---|---|---|---|---|---|---|---|
| Grignon | 11.1 | 1.2 | 9.3 | 7.3 | 220 | 730 | 50 |
| Rambouillet | 8.9 | 0.9 | 9.8 | 6.7 | 215 | 351 | 134 |

28.5 g $L^{-1}$. Its main characteristics were, on a dry matter basis: organic C, 28.0%; organic N, 3.0%; total N, 7.1%; organic C/organic N, 9.5; organic C/total N, 4.0. The soil materials were sampled in November 1989, before sludge application, in the surface horizons (0–30 cm) of a loamy soil (Typic Eutrochrept) and an hydromorphic clay soil (Aquic Eutrochrept) at Grignon and Rambouillet, respectively, both located 40 km west of Paris (France). The main analytical characteristics of the soil surface horizon are presented in Table 1.

## 2.2 *Incubations*

Dry soils were passed through a 5-mm sieve. Fifty-gram soil samples were incubated in sealed 500 ml jars for 84 days, with or without sludge. Soil moisture equivalent to 85% of the water holding capacity (WHC), or 0.1 MPa moisture tension, and temperature of 28°C were considered as the optimal conditions for soil microbial activity. To study the temperature impact on sludge degradation, incubations were also run at 4 and 15°C with soil moisture equivalent to 85% WHC. Incubations at 28°C with soil water content equivalent to 1 MPa and 0.01 MPa moisture tension allowed determination of the soil humidity impact. In the treatments with sludge, 5 ml of liquid sludge were added, corresponding to the 400 Mg $ha^{-1}$ spread in field experiments. Four and five replicate jars for control soils and sludge treated soils, respectively, were sampled after 3, 7, 10, 14, 21, 28, 42, 56, 70 and 84 days of incubation.

The impact of soil compaction was studied by incubating compacted aggregates realised from soil slurries in water dried under pressure and then broken into 1 to 3 cm pieces. The sludge was added either to the soil slurry before or spread over the compacted aggregates at the beginning of the incubation. The aggregates were incubated under the optimal conditions described previously. The same amount of water was added per gram of compacted or sieved soil. Those incubations were intended to simulate agronomic situations where a seed bed is prepared in a compacted soil, with a very low intra-aggregate porosity.

Four months after sludge application in the field experiments, the surface soils were sampled again. Different sizes of natural aggregates were incubated as previously (0–5 mm, 5–20 mm, 20–50 mm), and results were compared to those obtained from incubations of the same total soils sieved to 5 mm.

Mineral N and mineralized C-$CO_2$ were measured at each sampling date. Mineral N was extracted by 1 hour extraction in 200 ml 1M KCl, then analysed using the indophenol blue method for $NH_4^+$ and the Griess-Ilosvay method for $NO_3^-$ on a continuous flow analyzer (SKALAR, The Netherlands). The $CO_2$ evolved between sampling dates was trapped in 10 ml 0.25N NaOH. Residual NaOH, after precipitation of $BaCO_3$ by addition of 1.0 ml saturated $BaCl_2$, was titrated to pH 8.8 with 0.25N HCl. The NaOH traps were replaced at each sampling.

## 3. Results and discussion

### *3.1 Optimal conditions*

In the control soils, the C and N kinetics of mineralization were described using first order kinetics (Table 2). $C_0$ and $N_0$, potentially mineralizable C and N, represented 3% of total C and N in both soils. The kinetics of C mineralization from sludge were obtained by subtracting the C-$CO_2$ produced in the control soil from the C-$CO_2$ produced in the incubations of soils amended with sludge (Figure 1). Similar calculations were realized for N-$NO_3^-$. No $NH_4^+$ was detected in the control incubations, and the $NH_4^+$ measured in the incubations of soil with sludge was considered as coming from the sludge. Under optimal conditions for microbial activity, the nitrification was very rapid and after 14 days of incubation, N-$NH_4^+$ could not be detected. The kinetics of C-$CO_2$ and N-$NO_3^-$ coming from sludge were identical in the two soils. Such a result was expected, since both soils had the same clay contents and pH. Different results have been found concerning the influence of soil characteristics on sludge evolution. Net N mineralization was not dependent on soil type when a sewage sludge was incubated in a clayey and a sandy soil (Lindeman and Cardenas, 1984). Sludge mineralization increased with high clay content (Fine *et al.*, 1989) and was more important in neutral soils than in acidic soils (Garau *et al.*, 1986). Carbon mineralization and N-$NO_3^-$ evolution were also adjusted to first order kinetics, with $C_0$ representing 47% of total sludge C and $N_0$ representing 73% of total sludge N, that is to say 100% of mineral N and 33% of organic N. These results confirmed the model of N availability in anaerobically digested sludges proposed by Hutchings (1984): Available N = 100% N-$NH_4^+$ + 27% organic N.

*Table 2.* $C_0$ and $N_0$ values in the first order kinetics used to describe C mineralisation and net N-$NO_3^-$ evolution from soils and sludge under optimal conditions for microbial activity (28°C and 85% of the water holding capacity).

| | $C_0$ | | $N_0$ | |
|---|---|---|---|---|
| | mg C $kg^{-1}$ | % total org. C | mg N $kg^{-1}$ | % total N |
| *Control soils* | | | | |
| Grignon | 377 ± 27 | 3% | 32 ± 4 | 3% |
| Rambouillet | 278 ± 23 | 3% | 25 ± 2 | 3% |
| *Sludge in soils* | | | | |
| Grignon | 368 ± 13 | 46% | 144 ± 4 | 71% |
| Rambouillet | 388 ± 13 | 48% | 152 ± 6 | 75% |

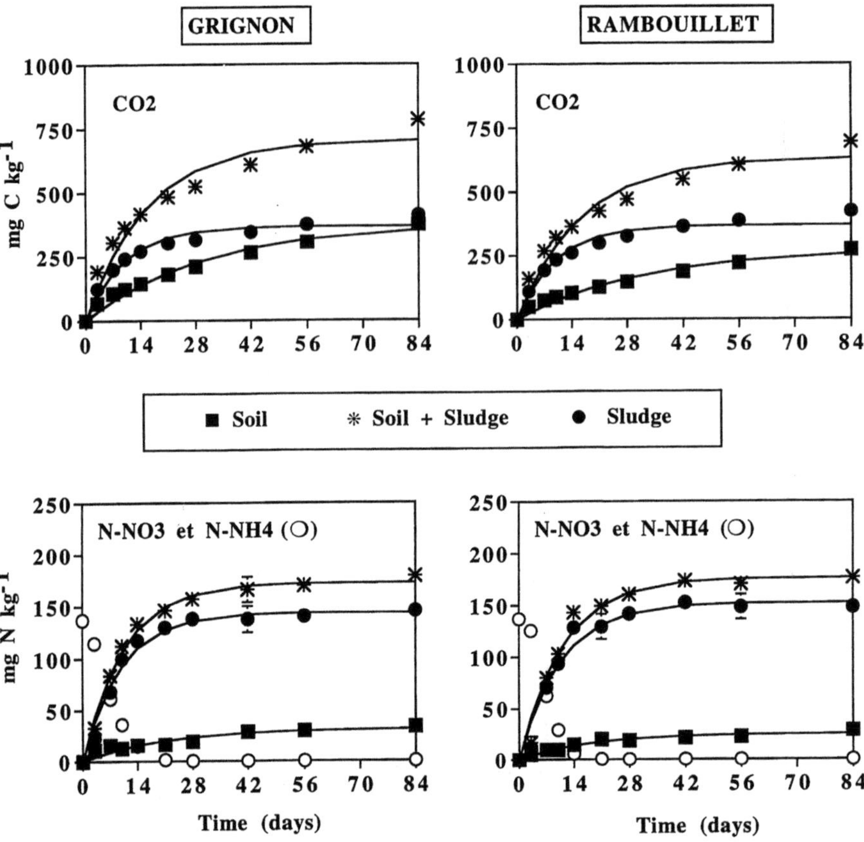


*Figure 1.* Carbon mineralization, N-$NH_4^+$ and N-$NO_3^-$ evolution during soil incubation with or without sludge addition in the two studied soils, Grignon and Rambouillet. The sludge C mineralization was calculated as the difference between C-$CO_2$ mineralization in the soil amended with sludge and in the control soil. The N-$NO_3^-$ evolution from sludge was calculated similarly. Ammonium was only detected in the soil incubations with sludge, and was considered as coming only from sludge. The results are expressed in mg C or N per kg of soil. The symbols correspond to the experimental results and the lines to the first order kinetics. Standard errors are shown when larger than the symbols.

### 3.2 *Impact of temperature*

Temperature impacts on sludge mineralization were similar in the two soils. Sludge C mineralization decreased with temperature, as did N-$NO_3^-$ formation (Figure 2). A lag phase in the N-$NO_3^-$ kinetics appeared at 15°C and was more developed at 4°C, corresponding to the slower nitrification of $NH_4^+$ at low temperatures. Similarly, the N-$NH_4^+$ concentration decrease was slower at 15 and 4°C than at 28°C. We considered that the temperature did not modify the potentially mineralizable C and the potentially available N, that is to say the $C_0$ and $N_0$ values, but did affect the rate

*Table 3.* Variation with temperature of the constant rates of the first order kinetics of C-$CO_2$ ($k_c$) and N-$NO_3^-$ ($k_n$) evolution: results of the control soils and of the sludge in the two soils.

| Temperature (°C) | $k_c$ ($10^{-2}$ $day^{-1}$) | $k_n$ ($10^{-2}$ $day^{-1}$) |
|---|---|---|
| *Grignon control* | | |
| 28 | 3.27 ± 0.50 | 4.66 ± 1.28 |
| 15 | 1.61 ± 0.17 | 3.17 ± 0.37 |
| 4 | 0.72 ± 0.05 | 1.38 ± 0.14 |
| *Rambouillet control* | | |
| 28 | 3.04 ± 0.20 | 5.68 ± 0.92 |
| 15 | 1.64 ± 0.18 | 2.65 ± 0.26 |
| 4 | 0.64 ± 0.06 | 1.21 ± 0.14 |
| *Sludge in Grignon soil* | | |
| 28 | 10.10 ± 0.86 | 10.30 ± 0.70 |
| 15 | 3.20 ± 0.47 | 2.90 ± 0.39 |
| 4 | 1.90 ± 0.29 | 1.03 ± 0.16 |
| Sludge in Rambouillet soil | | |
| 28 | 8.39 ± 0.91 | 9.36 ± 1.29 |
| 15 | 2.71 ± 0.43 | 3.55 ± 0.56 |
| 4 | 1.68 ± 0.29 | 0.99 ± 0.18 |

*Table 4.* Arrhenius linear regression of ln($k_c$) and ln($k_n$) on 1/T in the control soils and for the sludge in the two soils.

| | $k_c$ | | | $k_n$ | | |
|---|---|---|---|---|---|---|
| Treatment | Intercept | Slope (°K) | *r* | Intercept | Slope (°K) | *r* |
| *Control soils* | | | | | | |
| Grignon | 13.9 | −5201 | 0.999** | 10.9 | −4181 | 0.974 |
| Rambouillet | 14.4 | −5365 | 0.990 | 14.9 | −5342 | 0.999** |
| *Sludge in soils* | | | | | | |
| Grignon | 16.8 | −5781 | 0.999** | 24.1 | −7971 | 0.999** |
| Rambouillet | 16.1 | −5610 | 0.978 | 23.4 | −7730 | 0.995 |

**Significant at the level of 1%

constants $k_c$ and $k_n$ (Table 3). The experimental kinetics were still rather well simulated with this hypothesis, although the lag phases were not described. The rate constants, $k_c$ and $k_n$, varied with temperature following a modified Arrhenius equation (Stanford *et al.*, 1973): $\ln k = \ln A - B/T$ where A is a constant, B the slope of the regression and T the absolute temperature (Table 4). The temperature impact on mineral C and N evolutions was assessed by comparing the slopes. The effect was greater on sludge N evolution, probably because of the high sensitivity of the nitrifiers to low temperatures (Sabey *et al.*, 1956; Seifert, 1980). $Q_{10}$ values were calculated between 10 and 20°C: 1.9 for C and N mineralizations in soils, 2.0 for sludge C mineralization and 2.6 for $N\text{-}NO_3^-$ evolution from sludge. Similar $Q_{10}$ have been reported for C mineralization from digested sludge (Hsieh *et al.*, 1981).

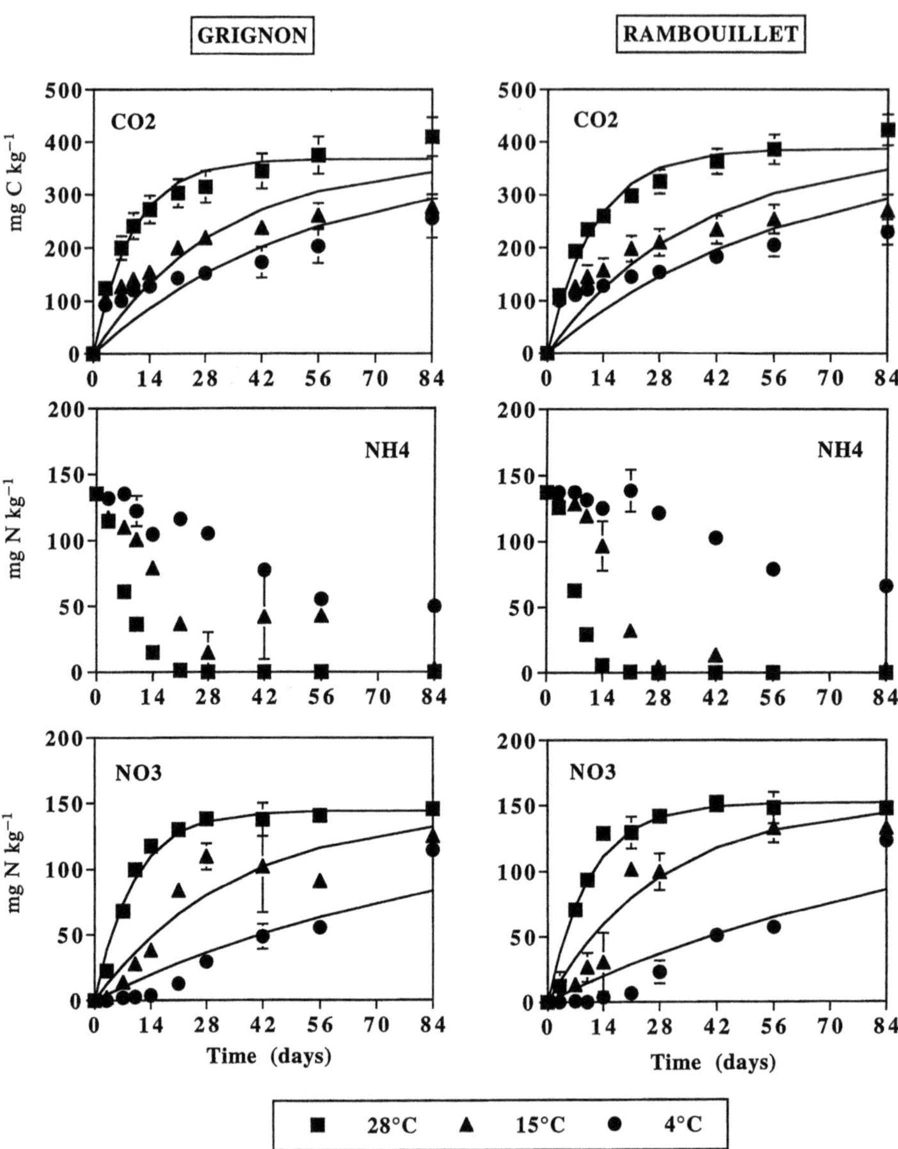


*Figure 2.* Variation with temperature of the sludge C mineralization, $N\text{-}NH_4^+$ and $N\text{-}NO_3^-$ evolution. The data were calculated as explained in Figure 1. The results are expressed in mg C or N per kg of soil. The symbols correspond to the experimental results and the lines to the first order kinetics. Standard errors are shown when larger than the symbols.

### *3.3 Impact of soil water content*

Within the range of soil water contents considered, the soil humidity hardly modified the sludge evolution (Figure 3). The $NO_3^-$ production was only slightly decreased at

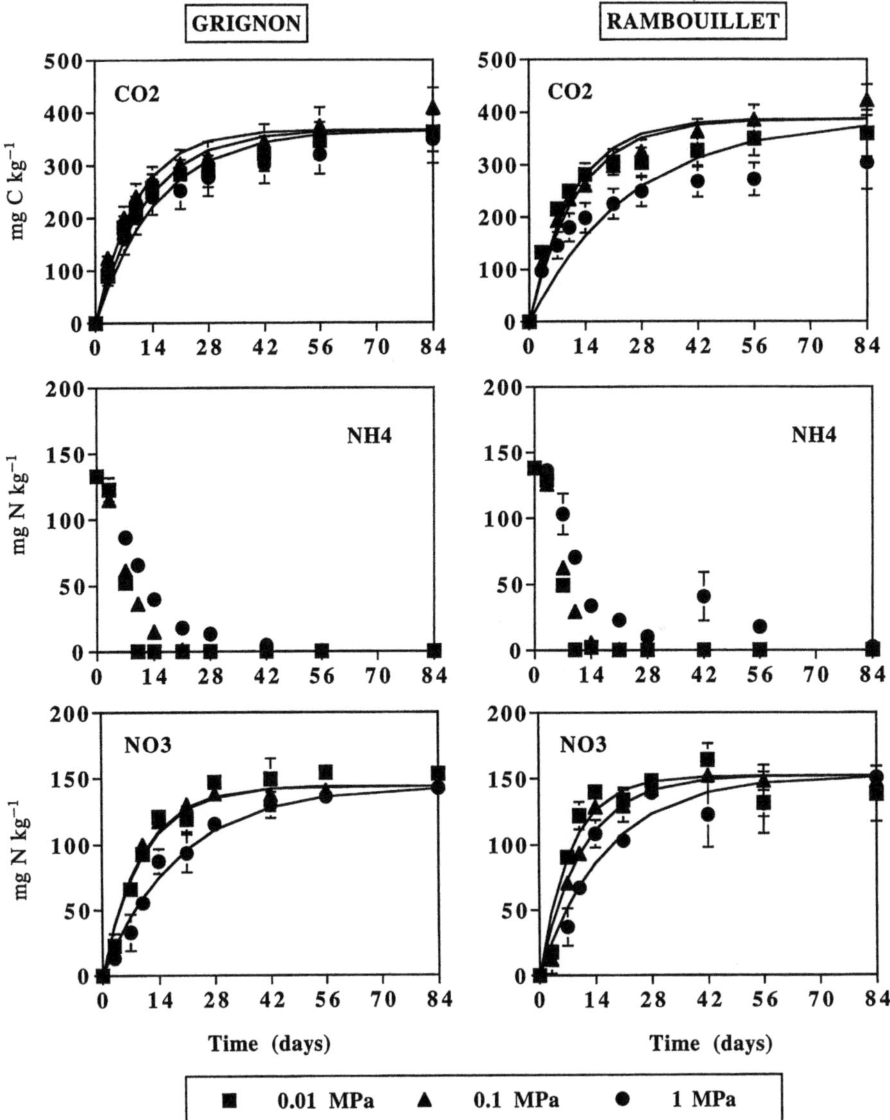


*Figure 3.* Variation with soil water content of the sludge C mineralization, N-$NH_4^+$ and N-$NO_3^-$ evolution. The data were calculated as explained in Figure 1. The results are expressed in mg C or N per kg of soil. The symbols correspond to the experimental results and the lines to the first order kinetics. Standard errors are shown when larger than the symbols.

*Table 5.* Variation with soil water content of the constant rates of the first order kinetics of C-$CO_2$ ($k_c$) and N-$NO_3^-$ (kn) evolution: results of the sludge in the two soils.

| Moisture tension (MPa) | $\Theta_{wt}$ (g $H_2O$ $kg^{-1}$) | $k_c$ ($10^{-2}$ $day^{-1}$) | $k_n$ ($10^{-2}$ $day^{-1}$) |
|---|---|---|---|
| *Sludge in Grignon soil* | | | |
| 0.01 | 35.3 | 7.99 ± 0.63 | 6.41 ± 0.61 |
| 0.1 | 17.8 | 10.10 ± 1.24 | 6.34 ± 0.86 |
| 1 | 10.5 | 6.54 ± 0.68 | 3.43 ± 0.29 |
| *Sludge in Rambouillet soil* | | | |
| 0.01 | 29.4 | 9.25 ± 1.15 | 12.70 ± 1.90 |
| 0.1 | 16.9 | 8.39 ± 0.91 | 9.36 ± 1.29 |
| 1 | 9.9 | 3.92 ± 0.61 | 6.00 ± 0.66 |

the lowest soil humidity, corresponding to permanent wilt. Under field conditions, the limiting factor for N use by plants would not have been the amount of N-$NO_3^-$ present, but the soil water content. At permanent wilt, plants cannot use the water, creating an indirect N stress. Soil humidity equivalent to 0.01 MPa moisture tension gave results similar to the optimal conditions. Metzger and Yaron (1987) also reported that sludge evolution was independent of soil water content in the potential water range 0.025 to 0.1 MPa. Again, the soil water content effect was studied on $k_c$ and $k_n$ constant rates of the first order kinetics, considering $C_0$ and $N_0$ as constant (Table 5). Within the range of soil water contents considered, it seemed more important to consider the temperature impact on C and N evolutions. Nevertheless, saturated conditions which could modify the $NO_3^-$ evolution by enhancing denitrification have not yet been studied.

### 3.4 *Impact of soil compaction*

Both sludge C mineralization and N-$NO_3^-$ evolution followed first order kinetics in the compacted soils (Figure 4). Soil compaction did not modify the sludge C mineralization and N-$NO_3^-$ evolution when sludge was sprayed over compacted soils, but $C_0$ and $N_0$ decreased by 30% when sludge was applied before soil compaction (Table 6). The constant rates were also modified: $k_c$ decreased, but $k_n$ increased. Houot *et al.* (1990) previously found that soil compaction decreased potentially mineralisable N in soils, but simultaneously increased the rates of mineralization. Some of the mineral N was lost during the soil compaction, but it was not enough to explain the $N_0$ decrease. It was mainly due to the low soil aeration in the compacted soils which decreased mineralization and nitrification, and enhanced denitrification (Torbert and Wood, 1992). Again, the results were similar in Grignon and Rambouillet soils.

### 3.5 *Impact of aggregation*

The soils were sampled four months after sludge application. In Grignon, the kinetics of C mineralization were similar in the control and the amended soils, with

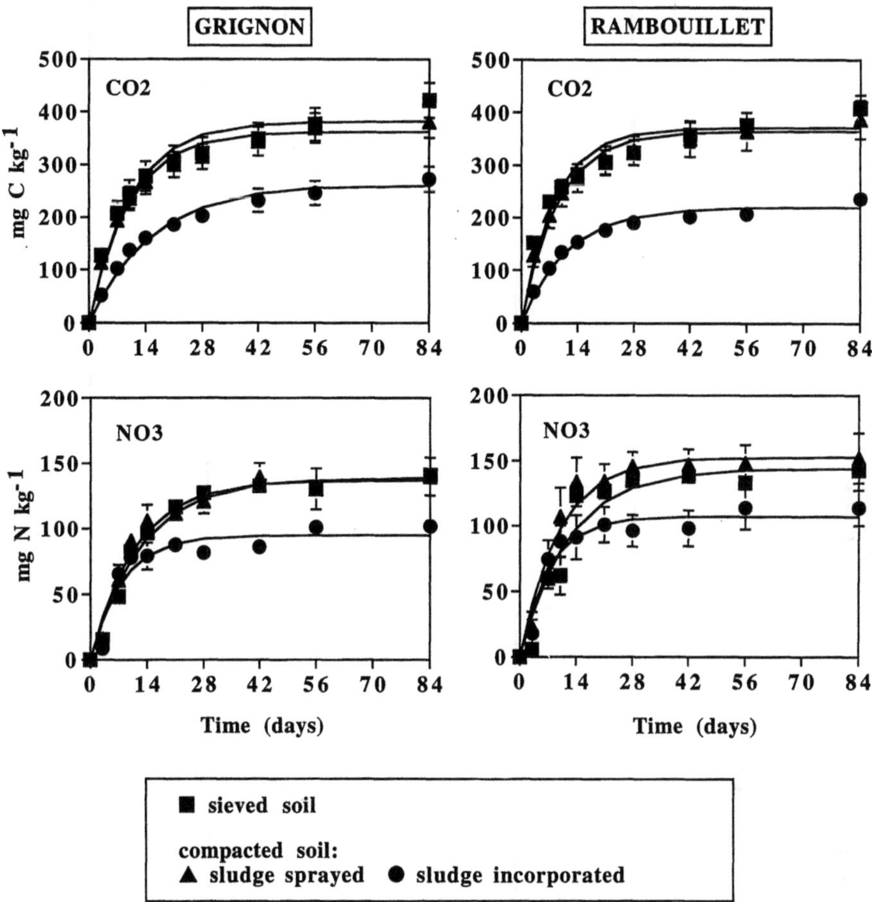


*Figure 4.* Variation with soil compaction of the sludge C mineralization and N-$NO_3^-$ evolution: results of the sludge in the two soils when sludge was added to sieved soils (sieved), sprayed over compacted aggregates (sprayed) or incorporated to the compacted soil aggregates (incorporated). The results are expressed in mg C or N per kg of soil. The symbols correspond to the experimental results and the lines to the first order kinetics. Standard errors are shown when larger than the symbols.

only a mean $C_0$ increase of 8% in the amended soil. The N kinetics were also similar in the control and the amended soils, except for the largest aggregate size (20–50 mm) where $N_0$ increased by 45% (Table 7). Nevertheless, the initial mineral N content was three times larger in the incubations with total amended soil and aggregates than in those with the control soil and aggregates (data not shown). In Rambouillet, the differences between the control and the amended soils were more important. $C_0$ increased by 25 to 50% in the different aggregate sizes, and $N_0$ increased by 27 to 71%. The initial mineral N was only 1.6 times larger in the incubations with total amended soil and aggregates. Under controlled laboratory

*Table 6.* Variation with soil compaction of $C_0$, $N_0$ and of the constant rates ($k_c$ and $k_n$) of the first order kinetics of C-$CO_2$ and N-$NO_3^-$ evolution: results of the sludge in the two soils when sludge is added to sieved soils (sieved), sprayed over compacted aggregates (sprayed) or incorporated to the compacted soil aggregates (incorporated).

| | $C = C_0 \times (1 - \exp(-k_c t))$ | | $N = N_0 \times (1-\exp(-k_n t))$ | |
|---|---|---|---|---|
| | $C_0$ (mg C $kg^{-1}$) | $k_c$ ($10^{-2}$ $day^{-1}$) | $N_0$ (mg N $kg^{-1}$) | $k_n$ ($10^{-2}$ $day^{-1}$) |
| *Sludge in Grignon soil* | | | | |
| Sieved | 382 ± 13 | 9.61 ± 1.20 | 139 ± 5 | 7.78 ± 0.80 |
| Sprayed | 362 ± 8 | 9.96 ± 0.78 | 137 ± 5 | 8.91 ± 1.01 |
| Incorporated | 264 ± 7 | 6.30 ± 0.53 | 95 ± 6 | 12.70 ± 2.74 |
| *Sludge in Rambouillet soil* | | | | |
| Sieved | 372 ± 13 | 11.91 ± 1.63 | 144 ± 9 | 8.13 ± 1.64 |
| Sprayed | 364 ± 8 | 10.94 ± 0.90 | 153 ± 6 | 10.21 ± 1.38 |
| Incorporated | 221 ± 6 | 8.57 ± 0.74 | 107 ± 5 | 13.57 ± 2.29 |

*Table 7.* Variation with soil aggregate size of $C_0$, $N_0$ and the constant rates $k_c$ and $k_n$ of the first order kinetics of C and N mineralization in soils sampled four months after sludge application in the control and the amended plots. Three aggregate sizes were considered: 0–5 mm, 5–20 mm, 20–50 mm; and the total soil sieved to 5 mm.

| | $C = C_0 \times (1 - \exp(-k_c t))$ | | $N = N_0 \times (1 - \exp(-k_n t))$ | |
|---|---|---|---|---|
| | $C_0$ (mg C $kg^{-1}$) | $k_c$ ($10^{-2}$ $day^{-1}$) | $N_0$ (mg N $kg^{-1}$) | $k_n$ ($10^{-2}$ $day^{-1}$) |
| *Grignon: control soil* | | | | |
| Total soil | 504 ± 27 | 4.06 ± 0.58 | 53 ± 4 | 6.00 ± 1.46 |
| 0–5 mm | 565 ± 32 | 3.51 ± 0.50 | 53 ± 3 | 5.37 ± 0.72 |
| 5–20 mm | 444 ± 25 | 4.55 ± 0.74 | 49 ± 3 | 6.75 ± 1.18 |
| 20–50 mm | 395± 22 | 4.10 ± 0.63 | 49 ± 3 | 5.55 ± 1.10 |
| *Grignon: amended soil* | | | | |
| Total soil | 564 ± 31 | 3.74 ± 0.52 | 56 ± 3 | 11.30 ± 2.67 |
| 0–5 mm | 588 ± 34 | 6.60 ± 0.53 | 59 ± 4 | 5.90 ± 1.15 |
| 5–20 mm | 470 ± 26 | 4.37 ± 0.68 | 50 ± 3 | 23.20 ± 9.08 |
| 20–50 mm | 427 ± 22 | 4.72 ± 0.72 | 71 ± 12 | 5.31 ± 2.54 |
| *Rambouillet: control soil* | | | | |
| Total soil | 337 ± 20 | 4.03 ± 0.62 | 38 ± 4 | 7.25 ± 2.34 |
| 0–5 mm | 357 ± 22 | 4.00 ± 0.66 | 35 ± 3 | 18.73 ± 8.50 |
| 5–20 mm | 272 ± 19 | 3.76 ± 0.67 | 33 ± 4 | 7.48 ± 3.31 |
| 20–50 mm | 271 ± 17 | 4.14 ± 0.72 | 80 ± 28 | 1.95 ± 1.23 |
| *Rambouillet: amended soil* | | | | |
| Total soil | 463 ± 28 | 3.46 ± 0.52 | 60 ± 7 | 3.55 ± 0.94 |
| 0–5 mm | 531 ± 33 | 3.27 ± 0.48 | 60 ± 9 | 1.74 ± 0.43 |
| 5–20 mm | 368 ± 26 | 3.49 ± 0.59 | 42 ± 6 | 5.80 ± 2.30 |
| 20–50 mm | 339 ± 20 | 4.17 ± 0.68 | 77 ± 7 | 2.92 ± 0.59 |

conditions, the sludge evolution was identical in the two soils, as the incubated surface layers had similar clay contents and pH. However, it was different in the field experiment, where global soil conditions had to be considered. The more important clay content in the deeper horizons of Rambouillet created saturated conditions which modified the sludge evolution as compared to Grignon. The aggregate size did not seem to influence the sludge evolution, and incubations with sieved soil gave average results.

In summary, the sludge evolution was identical in Grignon and Rambouillet soils in the incubation studies, with a more important impact of temperature than of the soil water content. The temperature impact was more important on N-$NO_3^-$ evolution from sludge, related to the important decrease of the nitrifier activity at low temperatures. This could decrease the sludge N available for plant nutrition at the beginning of spring when temperatures remain rather low. N availability from sludge could also be decreased by soil compaction if it occurred after sludge application. Even with potentially similar sludge evolution in the surface soil where sludge is applied, the global soil conditions must be considered to understand the actual sludge evolution under field conditions. Thus, saturated conditions in the deep soil horizons apparently decreased the sludge mineralization and $NH_4^+$ nitrification in Rambouillet soil.

## References

Fine P, Mingelgrin U and Feigin A 1989 Incubations studies of the fate of organic nitrogen in soils amended with activated sludge. Soil Sci. Soc. Am. J. 53, 444–450.

Garau M A, Felipo M T and Ruiz de Villa M C 1986 Nitrogen mineralization of sewage sludge in soils. J. Environ. Qual. 15, 225–228.

Houot S, Mordelet P, Tardieu F and Molina J A E 1990 Effets du tassement et de la teneur en eau du sol sur la biomasse microbienne et la libération d'azote minéral. In Nitrate-Agriculture-Eau. Ed. R Calvet. pp 201–207. INRA, Paris, France.

Hsieh Y P, Douglas L A and Motto H L 1981 Modelling sewage sludge decomposition in soil: I Organic carbon transformation. J. Environ. Qual. 10, 54–59.

Hutchings N J 1984 The availability of nitrogen in liquid sewage sludges applied to grassland. J. Agric. Sci. Camb. 102, 703–709.

Lindeman W C and Cardenas M 1984 Nitrogen mineralization potential and nitrogen transformations of sludge-amended soil. Soil Sci. Soc. Am. J. 48, 1072–1076.

Metzger L and Yaron B 1987 Influence of sludge organic matter on soil physical properties. Adv. Soil Sci. 7, 141–163.

Sabey B R, Bartholomew W V, Shaw R and Pesek J 1956 Influence of temperature on nitrification in soils. Soil Sci. Soc. Am. Proc. 20, 357–360.

Seifert J 1980 Effect of temperature on nitrification intensity in soil. Folia Microbiol. 25, 144–147.

Stanford G, Frere M H and Schwaninger D H 1973 Temperature coefficient of soil nitrogen mineralization. Soil Sci. 115, 321–323.

Stanford G and Smith S J 1972 Nitrogen mineralization potentials of soils. Soil Sci. Soc. Am. Proc. 36, 465–472.

Torbert H A and Wood C W 1992 Effects of soil compaction and water-filled pore space on soil microbial activity and N losses. Commun. Soil Sci. Plant Anal. 23, 1321–1331.

# 34
# Solid waste management in Israel

YOSSI INBAR, ILAN NISSIM AND DEKEL AMIR-SHAPIRA

Population growth, rising standards of living and changes in consumption patterns have resulted in the discharge of increasing quantities of waste into the environment worldwide. Each person in Israel produces some 1.6 kilograms of solid waste a day. The total quantity of waste produced in the country annually by a population of about 5.3 million is equal to 3.1 million tons – with quantities increasing at an average rate of 2% yearly. In a country with meagre land resources, on the one hand, and ever-increasing quantities of refuse, on the other hand, sound management of solid waste is imperative.

## 1. Solid waste disposal: a historical perspective

While plans for more effective disposal of solid waste date back some 20 years, solutions to the ever-growing problem have not been forthcoming. Until recently, the basis for government policy was rooted in the National Outline Scheme for Solid Waste Disposal, approved in 1989.

The rationale behind the masterplan was to minimize environmental pollution by operating a few large landfills serving as large a population as possible. However, the masterplan set no timetable for establishing new landfills nor for shutting down unauthorised dumps; it allowed for their continued use until an alternative was found. Moreover, the long and difficult statutory approval process for new landfills (requiring approval on the national, district and local levels) was further delayed due to local opposition. In the case of two nationally-approved central landfill sites (Ein Hashofet on the southern part of the Carmel mountains and Beit Guvrin on the southern coastal plain), the detailed approval process was halted at the local and district levels by the NIMBY (not in my back yard) syndrome. As a result, detailed plans for authorised sanitary landfills were not approved in most areas, and over two-thirds of Israel's population remained without a comprehensive solution to the problem of solid waste disposal.

In 1993, some 96% of Israel's domestic waste was landfilled in about 400 waste dumps – 29 of which received over 25 tons of waste per day. Most of the sites were poorly designed and managed; many had reached or are soon to reach full capacity – with no alternative in sight. Some local authorities (e.g. Herzliya, Netanya, Hadera, Haifa, Petah Tikvah, Ra'anana, Kfar Sava and others) have already exhausted their landfill space and will soon have no solution for disposal. The most difficult problem

*D. Rosen* et al. *(eds): Modern Agriculture and the Environment, 413–427.*

lies in Hiriya, a landfill which receives about 2500 tons of refuse a day from the Greater Tel Aviv metropolitan area. This landfill, which will soon reach full capacity, is plagued by the entire gamut of problems associated with improper waste disposal: potential contamination of water sources, stench, air pollution, aesthetic blight, and threats to flight safety due to the congregation of birds in the flight paths leading to Ben-Gurion International Airport.

## 2. Towards environmentally sound waste management

Today, the outlook for solid waste management is no longer grim. In June 1993, the government took a landmark decision designed to expedite the establishment of central landfills, shut down hundreds of illegal waste dumps and create an infrastructure for environmentally-safe solid waste disposal both in the short and long term. Specifically, the decision calls for closure of most of the country's small garbage dumps within the next three years and for their replacement by a few authorised sanitary landfills, as follows:

- Two existing landfills – Ashdod (in the southern coastal area) and Evron (in the Western Galilee) – will be prepared and equipped to serve as regional sites on a temporary basis (about three years).
- Two existing landfills – Duda'im (northwest of Beersheba) and Talya (in the northern Jordan Valley) – will be improved according to strict environmental requirements and expanded within one year to serve as central sites for a significant part of the country's waste, with the exception of hazardous waste.
- The Hiriya landfill, near Tel Aviv, will be closed by the end of 1995 and will subsequently be rehabilitated.
- Oron, a phosphate quarry (still in operation), in the centre of the Negev, will be considered and checked as a central landfill for the long term.

In line with the government decision, the National Planning and Building Board approved a far-reaching amendment to the National Outline Scheme for Solid Waste Disposal in May 1994. The amendment gives expression to the government decision as follows: conversion of the existing landfills at Duda'im, Talya, Ashdod and Evron into central landfills and the addition of a central landfill at Oron, as a long-term alternative to the waste generated by the Tel Aviv metropolitan region. At the end of three years, the Ashdod landfill will be closed and Evron will be transformed once more into a local landfill. To accelerate the implementation of the amended masterplan, all of the new central sites will be subject to national planning at a detailed level, and will be accompanied by environmental impact statements.

The National Board also included the Kalanit site at Beit Guvrin, which met staunch opposition at the regional level of planning, among the sites destined for national planning at a detailed level. Another central landfill in Ein Hashofet is currently undergoing deliberations at the regional level of planning. All landfill sites which appear in the original masterplan will be required to adapt themselves to new

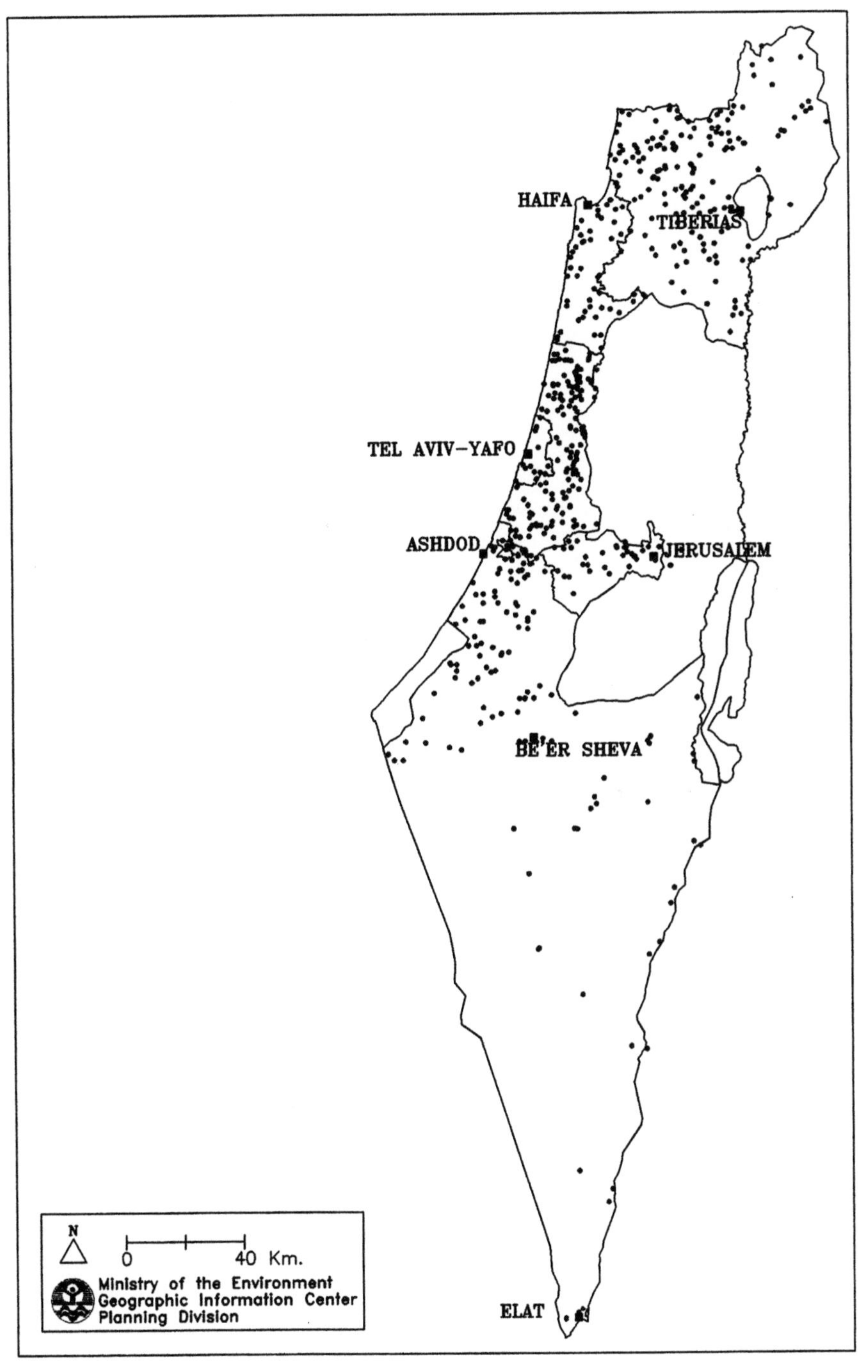


*Figure 1.* Existing solid waste sites (dumps).

state-of-the art standards in accordance with conditions stipulated by the Ministry of the Environment in their business licences. Otherwise they will be slated for closure.

The original masterplan for solid waste disposal was based on the geographic division of the country into waste collection areas, each served by a solid waste disposal site. However, due to problems such as land scarcity and potential risks to groundwater, several geographical areas were not assigned landfills at all. The amended plan views the collection areas as guidelines only. Today, waste generated in any area will be able to be transported to any other area in order to ensure environmentally-safe disposal. Thus, with the closure of the Hiriya landfill (by 31 December 1995), the waste of the Tel Aviv metropolitan area will be transported nearly 90 kilometres southward to the Duda'im site just north of Beersheba.

## 3. The implementation process

An interministerial committee, headed by the director-general of the Ministry of the Environment, has already published tenders for two central sanitary landfills as a first step in implementing the revolutionary plan. The tenders relate to the planning, establishment, operation, closure and rehabilitation of the Duda'im and Talya landfills, based on state-of-the-art standards.

In the case of Talya, vigorous opposition to the landfill was originally voiced by nearby settlements based on fears that the landfill will generate odours, pollution and landscape damage. To deal with the problem, a collaborative conflict management approach was adopted by all parties. Discussions led to the adoption of a workable compromise whereby the Ministry of the Environment agreed to reduce the amount of waste transported to the Talya landfill to a maximum of 1200 tons per day and to include adjacent communities in the entire process of environmental impact statement review and supervision over both waste transport and landfill operation.

All of the regional and central sites, along with other landfills included in the masterplan, will be established and operated according to professional guidelines prepared by the Ministry of the Environment so as to prevent environmental and health risks. They will include state-of-the-art technologies for every stage of landfilling from siting to post-closure, including sealing, leachate detection, collection, treatment and disposal, methane gas collection and use, proper covering of the waste during operation, closure procedures (landfill capping), and monitoring of possible contamination of groundwater during and after closure (up to 30 years).

The Ministries of the Environment and the Interior are responsible for closing and rehabilitating illegal dump sites and for ensuring that all of Israel's solid waste will be discarded at authorised landfills. In 1993, 83 illegal dumps were closed; another 70 major sites are scheduled for closure over the next two years. It is estimated that by 1996, some 80% of the country's waste will be safely discarded at environmentally-sound landfills.

A joint team including representatives of the Ministries of the Interior, Environment, Finance and the Israel Lands Authority will classify sites due for closure according to their economic value. Certain sites, such as Hiriya, may be transformed into parks; others will be rehabilitated and reused for other purposes. In sites where development is not economically feasible, the government will cover rehabilitation costs.

The Ministry of Finance has agreed to offset the increased cost of solid waste disposal which will result from longer transport distances and higher tipping fees. Financing by the Treasury will be based on a formula which will be determined according to financial calculations of the cost of transport plus tipping fee.

The Ministry of the Environment is currently determining priorities for closing down and rehabilitating unauthorised dumps, taking into account, *inter alia*, environmental and logistic considerations. Concomitantly, recycling projects are being promoted, and possibilities for incineration are being examined by an interministerial committee chaired by the Minister of Energy and Infrastructure.

## 4. The recycling path

The impending closure of hundreds of waste dumps and their replacement by a few regional and central landfills, coupled with ever-increasing requirements for high environmental standards in landfilling, will significantly increase the cost of solid waste disposal. Calculation of the true economic and environmental costs of landfilling is expected to expedite the move to low- or non-waste technologies and to encourage the implementation of waste reduction, reuse and recycling options. To facilitate these developments, Israel is redoubling its efforts to promote recycling by every possible means – research, legislation, pilot projects, to name but a few. A recently-established interministerial steering committee on recycling will also be instrumental in further promoting the subject.

On the research front, four feasibility studies have been undertaken in recent years to investigate the economic, environmental and technological feasibility of recycling paper and cardboard, plastic containers, glass and tyres. The studies conclude that collection and recycling make economic as well as environmental sense – saving foreign currency, diminishing investments in the import of raw materials, reducing costs for collection and disposal of waste, increasing the life of existing landfills and minimising environmental nuisances. The studies reveal that increased production in existing recycling plants is possible, but that the bottleneck lies mainly in collection, transportation and marketing.

Several initiatives are currently focusing on recycling projects in local authorities. Pilot facilities for separation at source are being set up in commercial centres, school premises and other locations in local authorities throughout the country. Kiryat Tivon, with 13 500 inhabitants, constitutes a national model for recycling since its pioneer programme was accompanied by a wide-based educational component from the very start. The results so far have been encouraging. Recent data reveal that out of 450 tons of waste produced each month, over 13% (weight) and 30% (volume) are collected for recycling, through six drop-off recycling centres and a curbside commingled collection programme for paper, plastics, glass and textiles. It is estimated that over 60% of the population takes part in the project.

Another project, accompanied by wide-scale education and information activities, has been running successfully in the municipality of Rishon Lezion since 1988. In this municipality of nearly 160 000, over half the population participates in the recycling initiative. Recent estimates point to the collection of some 1200 tons of cardboard, 250 tons of paper and 36 tons of textiles per year, using 70 drop-off centres for paper and textiles (106 bins) and 60 bins for corrugated cardboard. The municipality has recently initiated a pioneer project for chopping yard waste. The chopped waste is currently used for mulching with future plans to sell the trunks to a plywood factory.

Several more pilot projects have been inaugurated this year, with the aid of numerous organisations. These include: separation at source in various neighbourhoods in Jerusalem, Tel Aviv, Ra'anana, Kfar Sava, Hod Hasharon, Petah Tikvah, Kiryat Ono, and the Hadera and Ashdod town associations, under the responsibility of the municipalities and local environmental units; recycling centres in nine local authorities in the south of the country with the co-operation of the Society for the Protection of Nature in Israel; separation at source of plastics in Ramat Hasharon; and purchase of collection and compaction equipment for paper and cardboard and of yard waste choppers in several local authorities. These projects, serving tens of thousands of households throughout the country in more than 30 local authorities, constitute a significant breakthrough in the road towards wide-scale recycling.

Commercial recycling is continuing in Amnir's Afula plant (NAAM) in northern Israel which collects waste from several local authorities with a population of 150 000. The $2.5 million facility began operating in 1989 for the purpose of separating and recycling compost, cardboard, paper, organic materials, plastics and metals. The plant currently recycles some 50% of the 110 000 tons of waste it collects yearly; this figure is expected to increase substantially (to 76%) if tentative plans to use refuse derived fuel are implemented.

In yet another development, a new transfer station including a material recovery facility has recently begun operating in Petah Tikvah. This facility, located at the recently-closed Segula landfill site, provides for sorting of the waste, transfer of paper, carton and plastic to the Amnir recycling plant, and transport of the rest of the waste, after compaction, to the Modi'in landfill. The station is the first such facility in the central part of Israel.

## 5. Recycling facts and figures

Today, 96% of Israel's domestic waste is buried in landfills, but it is expected that within a decade or two, recycling and incineration will increase significantly for the following reasons:

- Even the most well-planned and maintained sites are designed for use for only 20–30 years.
- Inadequate land reserves will further diminish the availability of new landfill sites, thereby increasing the distance between sanitary landfills and the source of the waste.

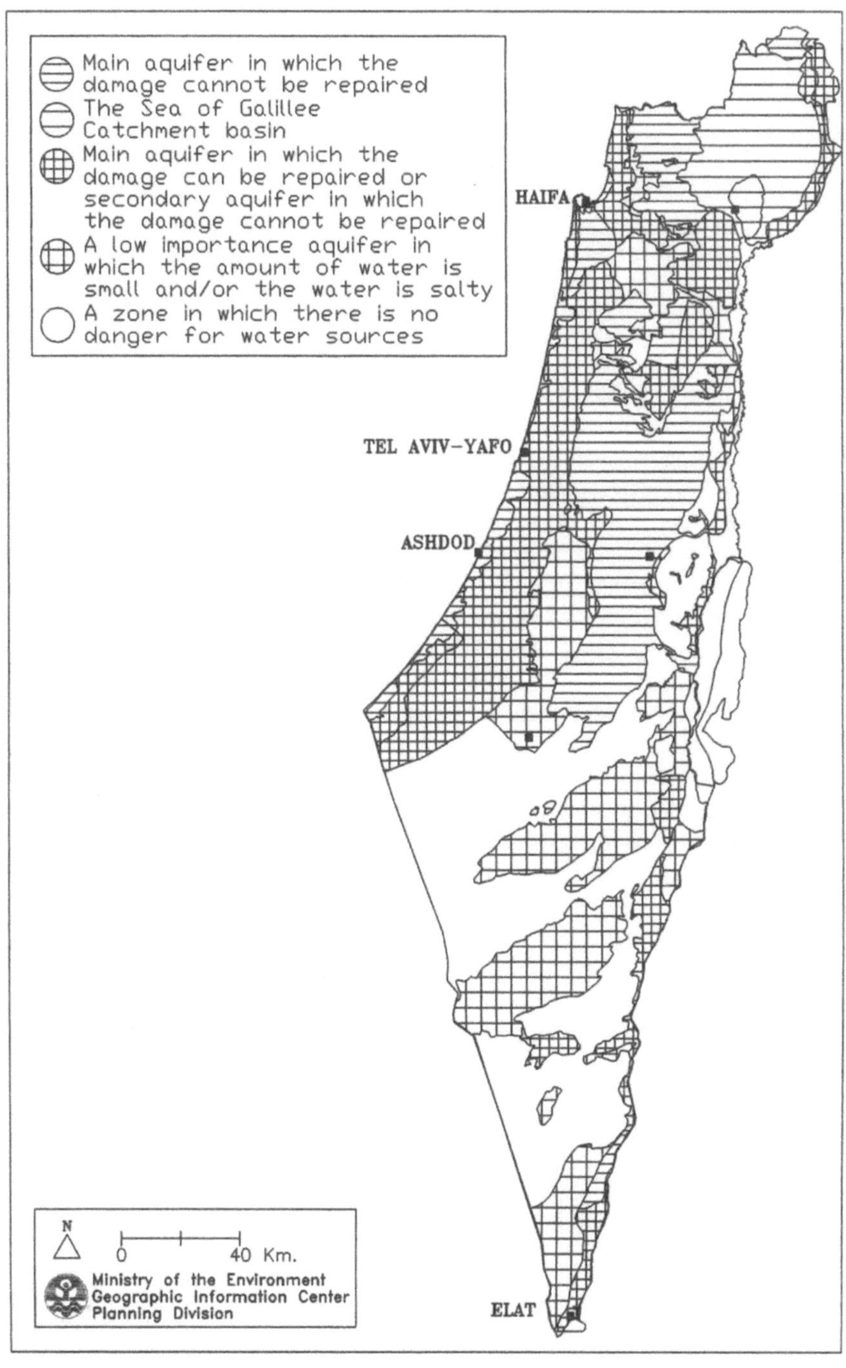


*Figure 2.* Hydrogeological sensitivity zones.

- The opening of new sites will become much more costly as technological means are undertaken to avoid contamination risks.

Comprehensive surveys on the composition of solid waste in Israel have not been conducted in recent years, but the estimated composition of domestic waste in Israel, according to weight, is as follows: 50–54% organic material, 16–21% paper and cardboard, 10–12% plastic and synthetic material; 3–5% metals; 3–4% textiles; 0–15% miscellaneous. Theoretically, all components of solid waste can be recycled. Practically, the current goal is to reach a 10% recycling rate in 1995 and approximately 25% by 2000.

## 6. Paper

One of the most successfully-recycled components of the waste stream worldwide – and one of the major components of municipal rubbish – is paper. Even prior to the passage of recycling legislation, paper was the only component of waste separated at source by Israeli households using special disposal containers. In recent years, some 505 000 tons of paper and cardboard were used in Israel, of which 300 000 tons were imported. The quantity collected for recycling by Amnir Recycling Industries, a subsidiary of American Israel Paper Mills Ltd, is 140 000 – 28% of total consumption. An additional 25 000 tons of office waste paper and cardboard waste are imported into Israel. In all, some 70% of the country's total paper and cardboard production originates in recycled fibres.

Since paper and cardboard comprise some 20% of the total weight and about 30% of the total volume of Israel's solid waste, efforts are being focused on encouraging paper collection, on the one hand, and purchase of recycled paper, on the other hand. A paper collection project geared at government ministries, accompanied by a guidebook on wastepaper recycling, has been in effect for over a year. In addition, an experimental wastepaper collection project for households has been initiated in the municipalities of Ra'anana and Givataim, with the financial support of the Ministry of the Environment.

Special attention is now concentrated on encouraging the production and marketing of recycled products, in order to boost both consumption and production of recycled writing and printing paper. Government offices and the school system are the initial targets of this new campaign, in the short term. In view of the fact that the current school year has been officially designated the Year of the Environment in the educational system, the Minister of the Environment has called upon school principals and teachers to aid in advancing wastepaper recycling by educating students to purchase notebooks from recycled paper and by encouraging separation at source. This will increase the demand for recycled goods, bring about increased collection and reduce the overall volume of the country's waste.

## 7. Plastic

Total plastic production in Israel is estimated at 320 000 tons/year, of which only 4.5% derives from recycled material. Some 35% of the total plastic production is designated for packaging, 28% for agriculture, 14% for construction, 10% for home utensils and the rest (13%) for other uses. The total quantity of plastic waste is estimated at 8–12% of the total weight of waste in Israel, divided almost equally between agricultural, industrial and domestic waste. Plastic consumption in Israel remains relatively low, some 64 kilograms per capita, as compared to 100 kilos in the USA and 130 kilos in Germany. However, in light of the development of the industry and the accelerated rate of population growth, this figure is expected to increase significantly in the near future.

The plastics industry now recycles about 15 000 tons of thermo-plastic materials. In the agricultural sector about 15% (some 5000 tons/year) of the polyethylene sheets and pipes are recycled by Amnir – a quantity expected to increase with the coming into effect of a regulation obligating the collection of plastic sheets from agriculture for disposal at landfills or recycling plants.

The total quantity of plastic containers produced is 25 500 tons per year, of which 15 000 are PET – a material not yet recycled in Israel. In light of the fact that some 430 million litres of light beverages and water (out of a total of 650 million litres) are bottled in PET containers, the country is now investing efforts in expanding the collection and recycling of these containers – which constitute some 5–6% of the total volume of waste in Israel. Industries are encouraged to manufacture each product from the least number of components, and legislation has been proposed calling for the marking of plastic containers with the appropriate codes and emblems to facilitate recycling.

In 1991, Amnir invested in a plastics recycling plant in Hadera. The plant currently processes some 7000 tons of plastic a year and expects to reach 10 000 tons within four years. Larger-scale recycling is conditional upon the expansion of the recycled goods market. At present, the market is restricted to such uses as garbage bags, irrigation pipelines, detergent containers, injection materials, toys and boxes. One industrial plant in the Ramat Hovav area recycles mixed plastic as a substitute for wood (plastic wood), at a scale of some 2000 tons per year. Other plants are expected to follow suit.

## 8. Scrap metal

Israel's steel mills produce about 250 000 tons of different grades of steel annually, of which 177 000 tons are recycled. Additional recycling of metal should be facilitated in the near future as a result of a plan to collect, compact and shred wrecked vehicles. A comprehensive collection and recycling programme will make it possible for Israel to supply all of its consumption needs while helping to rid the country of the visual blight of some 50 000 wrecked vehicles annually.

In a related area, progress in recycling used beverage cans is expected as a result of the recent turnover to production of all-aluminium cans by Caniel, Israel's sole producer of beverage cans. Previously, the cans contained different types of aluminium for the body, base and top, making them virtually non-recyclable. The new cans which entered the market in 1993 also feature push-in tabs, instead of the previously used pull-off rings.

Caniel produces about 250 million beverage cans a year, constituting some 1.5% of the total volume of waste in Israel. Since the economic value of used beverage cans is estimated at $3.5 million a year, collection and recycling are deemed to be economically viable.

## 9. Glass

Annual consumption of glass stands at about 100 000 tons – with a recycling potential of some 60 000 tons per year. In reality, however, less than 1% of the country's glass is recycled. Despite the high cost of transportation to the south of the country, where Israel's only plant capable of glass recycling is located, feasibility studies have shown that recycling can be profitable, given the fact that the bulk of glass containers are concentrated in entertainment centres, thus facilitating collection.

A pilot programme to study the collection of glass, mainly from entertainment centres, was initiated in the summer of 1993. One hundred 2 $m^3$ 'igloo' containers were distributed in various locations in the central part of the country. The six-month experiment (which ended in December 1993) has provided valuable information on the feasibility of the project. Recommendations are to continue the experiment, to transfer the containers to areas in which both commercial centres and residential areas coexist and to increase the scope of education and information campaigns.

Local experience with returnable beer bottles shows that a glass container can last 10–15 years before reaching the landfill. Today, about 30 million litres of beer in 62 million bottles use the deposit system with a 90% return rate.

## 10. Tyres

According to a 1992 survey, the total number of cars in Israel in 1990 was about 1 100 000. These cars produced over 2 300 000 scrap tyres with a total weight of about 38 500 metric tons or 32 000 tons/year, after deducting the renewed tyres. Based on the results of the survey, it is estimated that scrap tyres currently total about 40 000 tons/year and will reach over 65 000 tons by the year 2000. About 38% of the tyres are generated in the greater Tel Aviv area, 13% in the Haifa region, 8% in Jerusalem, 9% in Beersheba and the southern region, and 32% in the rest of the country.

The readiness of several companies to collect scrap tyres from central locations in municipalities for recycling purposes has led the Ministry of the Environment to initiate a national bid for the collection and environmentally-safe disposal of scrap tyres. Local authorities will be required to establish sites for scrap tyre collection, and used tyre producers will convey their waste to these central sites for collection. The results of the tender offer are currently being assessed.

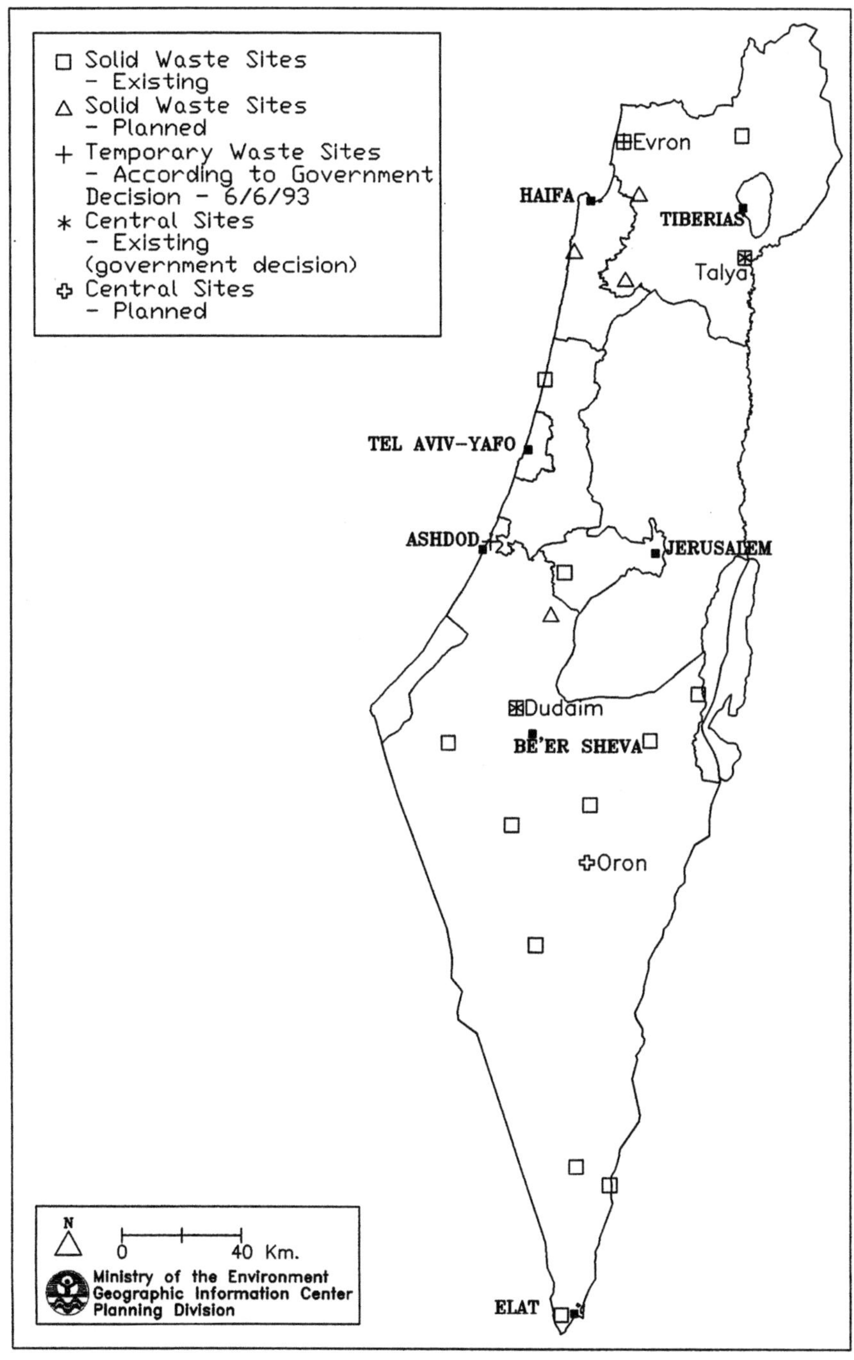


*Figure 3.* National master plan of landfills.

## 11. Oil

Used oil is a hazardous substance; its unregulated use and discharge threatens water and air quality. The quantity of motor oil used in Israel, mostly by garages and industrial plants, is estimated at 50 000 tons, of which over 15 000 tons are recyclable. Today, only 7000 tons are recycled, largely due to the high cost of buying and transporting used oil, tax requirements, and until recently, the lack of legislation requiring safe disposal and treatment of used oil. As a result, some 3000 tons are collected illegally for unregulated fuel use or incineration, and additional quantities find their way to the sewage system, to the soil and to water sources.

It is possible to treat used oil for clean burning or for oil products or to recycle it for reuse as a base oil through a process of separation, purification and rerefining. The renewed oil then serves as a substitute for imported crude oil products – saving foreign currency. A recently-promulgated regulation, prohibiting the pouring, burning or disposal of waste oil by garages and other producers and consumers, should bring an end to the environmental contamination caused by unsafe disposal of used oil. The regulations require the collection of the oil for disposal at a recycling facility or at the hazardous waste site at Ramat Hovav.

## 12. Yard waste

About 550 000 tons of yard waste, including brush, leaves, grass clippings and small trunks, are produced in Israel yearly. Chopping of this waste to reduce volume, and research on alternative uses (energy production, plywood, substitute for sawdust and straw for animal bedding, compost, and mulching) are currently on the agenda. To promote the idea, several local authorities have been granted financial aid to purchase chopping equipment (e.g. Rishon Lezion, Karmiel, Hadera, Lake Kinneret Authority, and the regional councils of Jezreel, Menashe, and Sha'ar Hanegev). It is hoped that their accumulated experience will aid in the promulgation of regulations prohibiting the burning of yard waste and requiring it to be chopped.

The use of yard waste for mulching is especially attractive because of its water-saving capacity, savings in herbicide use and soil cultivation, and contribution to vigorous growth, moderate temperature changes and prevention of soil erosion. The Jewish National Fund has begun experimenting with mulching using chopped yard waste in a four-hectare area in its forests to check the efficiency and benefit of this use.

## 13. Battery collection

It is estimated that each Israeli uses about 17 batteries per year, or 1.4 kilograms of batteries per family on average. The annual quantity is estimated at 1500–1800 tons. About 50 million batteries find their way to waste dumps, threatening groundwater quality, or in the case of incineration, polluting the air. In Israel, only car batteries are recycled; the relatively small quantity of household batteries has thus far

*Table 1.* Estimates of recycling in Israel (1993).

| | Total | Post-consumer |
|---|---|---|
| | (metric tons) | |
| Paper and cardboard | 130 000 | 70 000 |
| Compost | 20 000 | 20 000 |
| Plastic | 20 000 | 5 000 |
| Miscellaneous (metal, glass textile etc.) | 20 000 | 10 000 |
| Total | 190 000 | 105 000 |
| % recycling | 7.0% | 3.9% |

prevented the development of a viable industry for the collection and recycling of batteries. Therefore, present efforts are being focused on the separation of household batteries from ordinary rubbish, prevention of their disposal in landfills, and promotion of disposal at the Ramat Hovav hazardous waste site.

A national battery-collection campaign, organised by the Ministry of the Environment, in cooperation with local authorities and environmental units, was launched in 1993, to coincide with the inauguration of the Year of the Environment. The Ministry of the Environment has already provided local authorities with over 43 000 specially-designed battery-collection containers. Local authorities are then responsible for distributing the containers in convenient locations and collecting them when full for disposal at the central hazardous waste site at Ramat Hovav. The project is accompanied by a national information campaign featuring posters, stickers, media coverage and special service announcements on television. A follow-up survey will be funded by the Ministry of the Environment in order to draw conclusions from results during the first year.

## 14. Cleanliness in the public sector

Since the Maintenance of Cleanliness Law was enacted in 1984, special efforts have centred on involving the general public in implementing the law. An innovative – and effective – enforcement feature of the law enables the appointment of cleanliness inspectors from government and public agencies and cleanliness trustees from the general public, empowered to report on littering violations. Special training seminars are organised for these volunteer law-enforcers to familiarize them with the provisions of the law. Over the past decade, the number of volunteers who have joined the ranks of Israel's cleanliness trustees has risen from several hundred to 100 000. The goal by the end of 1995 is to increase their numbers further to 250 000. Enforcement efforts, both by police and volunteers, have increased littering reports from hundreds of tickets per year to about a thousand per month.

The Ministry of the Environment, in conjunction with other organisations, also organizes annual anti-litter campaigns, which feature press releases, radio programmes, televised special service announcements, beach clean-ups, and distribution of cleanliness stickers and posters in both Arabic and Hebrew.

## 15. Legal framework for solid waste management

The storage, collection and disposal of solid waste are primarily within the jurisdiction of municipalities and are largely governed by municipal by-laws which determine the legal and administrative arrangements for collection and disposal. Regulations promulgated under the Planning and Building Law set out requirements for the size and type of waste containers as well as the size and type of structures for housing these containers.

Siting of landfills and other waste disposal locations is within the authority of municipalities and subject to the Planning and Building Law and its regulations as well as the National Outline Scheme for Solid Waste Disposal.

The Maintenance of Cleanliness Law, 1984, forbids the throwing of waste, construction debris and vehicle scrap into the public domain, or from the public domain into the private domain. It obligates local authorities, either individually or jointly, to establish sites for the disposal of building debris and vehicle scrap. Owners and drivers of public and commercial vehicles are required to affix signs indicating that discarding waste from the vehicle is prohibited. Producers and importers of beverage containers must print or affix a prohibition on improper disposal on the containers. A 1991 amendment to the law broadens its applicability to waste thrown from a vessel and illegal pasting or hanging of notices and announcements.

The law establishes a Cleanliness Maintenance Fund, administered by the Ministry of the Environment, aimed at concentrating the financial means for maintenance of cleanliness, prevention of littering and sponsoring of educational and enforcement programmes. Funds derive from the imposition of a cleanliness fee on manufacturers and importers of disposable beverage containers and from fines paid by violators of the law. In 1993, the Cleanliness Fund budget totalled some $365 000, of which $233 000 were derived from the imposition of the cleanliness fee and the rest from fines.

Within the framework of the Abatement of Nuisances Law, 1961, regulations for the prevention of odours and air pollution from solid waste disposal sites were promulgated in 1990. The regulations prohibit the burning of waste in solid waste disposal sites and require operators to take all necessary means to prevent burning.

Public Health Regulations, in effect since March 1993, prohibit the burning of plastic films used in greenhouses and as plastic covering in agricultural fields and require their collection at the edge of the field for recycling or their transport by the owner to a landfill.

Abatement of Nuisances regulations which prohibit the burning and improper disposal of used oil were promulgated in 1993. Sellers, users and consumers of motor oil must collect used oil in specially-designated receptacles and must transport it for final disposal to the hazardous waste site in Ramat Hovav or to a recycling facility.

The Collection and Disposal of Waste for Recycling Law was passed in June 1993. The law provides the principles and framework for recycling. It authorizes local authorities, and obliges them when so required by the Minister of the Environment, to allocate sites for recycling centres and to install recycling facilities and containers. Municipalities are authorised to pass by-laws specifying procedures for the collection and disposal of waste for recycling, and business and home owners within a municipality operating recycling centres are required to install and maintain recycling containers in accordance with municipal directives. In order to facilitate the adoption of relevant by-laws by local authorities, the Ministry of the Environment has drafted a model by-law on the collection and disposal of waste for recycling. The model by-law was distributed to all local authorities in Israel.

The Minister of the Environment is authorized to promulgate regulations on implementation of the law, including types of waste for recycling, types of containers, etc. Regulations, at various stages of preparation or consideration, include: a waste recycling obligation requiring local authorities to prepare at least 10% of their waste for recycling by the beginning of 1995, rising gradually to 25% by the year 2000; a requirement to have manufacturers mark products with the accepted universal code indicating type of material as well as the appropriate emblem for recyclable or recycled material; bans on the introduction of scrap tyres and unchopped yard waste into sanitary landfills; and a deposit programme for returnable containers and bottles.

# SECTION VII

# Pollution by Heavy Metals

35

# Biofiltration of heavy metals by the aquatic fern *Azolla*

ELISHA TEL-OR, MOTI SELA AND SHAUL RAVID

## 1. Introduction

Traditional and modern industrial activities involved in metal mining, metal-plating and electroplating, the production of batteries, microelectronics, dyes, catalysts and ornaments, discharge toxic heavy metals which contaminate soil and water resources. The process of detoxification and removal of accumulated and discharged Hg, Pb, Cd and other toxic heavy metals are limited and unsatisfactory. These pollutants remain hazardous to humans, and damaging to living systems, food chains and animal welfare.

Industries and water authorities have relied on methods and technologies for detoxification and removal of the heavy metals proposed by the chemical industry. However, these methods have been limited in their efficiency and/or competence, and failed to reduce the content caused by heavy metal ions below the permitted ppm concentration. The re-siting of polluting industries from urban centres to remote areas does not eliminate the damage caused by wastes to land and water resources, and the discharge of toxic metals in both developed and developing countries continues, wherever heavy metal commodities are produced.

The heightened awareness of human welfare, environmental protection and the irreversible damage to soil and water resources, has raised interest in alternative biological solutions for heavy metal detoxification using bacterial, fungal, algal and plant-based technologies. Since 1980, we have focused on the characterization and application of active uptake and passive absorption of heavy metals by the aquatic fern *Azolla* (Sela *et al.*, 1988, 1989, 1990). This technology has been patented in the US (Patent No. 5 000 852, 1991) and in Europe (Patent No. 03332182, 1994) (Tel-Or and Sela, 1991, 1994). This review provides a schematic description of the evolution of the heavy metal biofiltration process by *Azolla*, and the major framework in which its high affinity to bivalent metal ions, and its metal-binding and storing capabilities are developed and applied to the treatment and detoxification of industrial wastes.

## 2. Results and discussion

### *2.1* Azolla *biomass production*

The aquatic fern *Azolla*, characterised as a nitrogen biofertiliser for rice, has been widely distributed and applied as green manure for centuries (Lumpkin and

*D. Rosen* et al. *(eds): Modern Agriculture and the Environment, 431–442.*

Pluncknett, 1982). The symbiosis between the prokaryotic $N_2$-fixer *Anabaena* and the host aquatic fern *Azolla* continues to be of great interest (Peters and Meeks, 1989; Tel-Or *et al.*, 1991). *Azolla filiculoides* (Lamm.) was introduced to the botanical gardens in Tel Aviv University (Abu Kabir, Tel Aviv) where it was propagated in a few artificial ponds. We collected *Azolla* biomass, determined its productivity outdoors, under natural, non-fertilized conditions, and introduced it to controlled growth conditions in the phytotron at the faculty of Agriculture of the Hebrew University of Jerusalem in Rehovot. *Azolla filiculoides*, grown in dilute Hoagland medium or in Peters and Mayne medium (Sela, 1990), was used for heavy metal uptake studies, and for the year-round production of *Azolla* biomass.

Larger-scale production of *Azolla* biomass was established outdoors, in shallow tunnels lined with polyethylene and protected in the summer from excess solar radiation by plastic nets, which eliminated 50% of the incident light (Ravid, 1991). During the cooler season, from November to March, the tunnels were covered with additional 0.1-mm thick polyethylene layers. The yield of *Azolla* biomass in these tunnels is shown in Figure 1, which demonstrates its continuous growth throughout the year, yielding 300–400 g *Azolla* $m^{-2}$ $month^{-1}$, equivalent to 36–48 tons of dried *Azolla* $ha^{-1}$. Cultivation has been recently expanded, and yield on a larger scale – 0.1-ha farm – was found to be similar to that found in the tunnels. In addition to the climatic control of growth, pH needs to be kept at 5.5–7.0. The growth medium is supplemented with phosphorus, potassium and iron to compensate for consumption by the harvested biomass. Fresh *Azolla* can be sun-dried or oven-dried at 80–140°C, and the dry biomass comprises 5–7% of the wet weight, depending on the growth conditions and drying procedure. Relative root-to-shoot weights may change with growth regime and season, and influence the dry weight percentage.

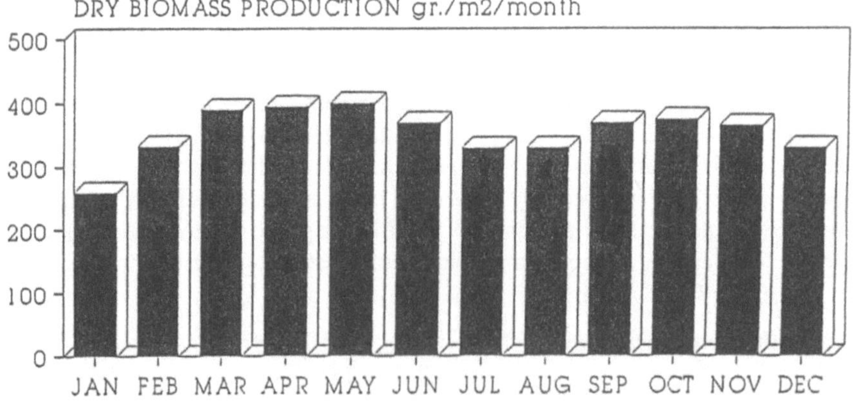


*Figure 1.* *Azolla* biomass production: annual profile of the growth and production of dry biomass. Data obtained from 10 $m^2$ pond growth, during 1991 and 1992, and biomass harvested once a week.

## 2.2 *Heavy metal uptake by* Azolla filiculoides

*Azolla filiculoides* tolerance to heavy-metals was reported by Sela *et al.* (1989), who studied the uptake and accumulation of five metals: Cu as nutritional metal at low concentration but toxic at excessive amounts; Ni and Cd, which are effluents discharged by battery producing industries; and Cr and Zn, which are wastes from metal-plating industries. The pattern of uptake of these metals from solutions containing 9–13 ppm $Cd^{2+}$, $Cr^{3+}$, $Cu^{2+}$, $Ni^{2+}$ and $Zn^{2+}$, was characterized in a 123-h experiment. The analysis of metal taken up by *Azolla* from the solution and accumulated in its roots and shoots is shown in Figure 2. Cu and Cr reached saturation in roots after 35–76 h, peaking at 2.5% (w/w) of root dry weight. Cd, Zn and Ni concentration in *Azolla* roots reached saturation at 0.5–1% metal per root dry weight (Figure 2A). *Azolla* shoot concentration of Ni and Cr reached saturation of 0.1% of the shoot dry weight whereas Cu, Cd and Zn were more mobile in the shoot, reaching concentrations of 0.3–0.5% of shoot dry weight (Figure 2B). These results suggested that some of the Cu, Cd and Zn ions in the shoot could break through osmotic cell-wall barriers of the Casparian strip by symplastic movement. In a short-term uptake experiment (terminated after three days) incubation of *Azolla*, in the presence of 10–12 ppm of each of the five metal ions, demonstrated slow mobility of Cr and Cu from root to shoot and faster mobility of Zn, Cd and Ni, as reflected by the relative content of metal between the roots and shoots (Figure 3). Some of the metal ions stored in the shoot may have been taken up by the leaf ventral lobes, submerged in the metal containing solution, as suggested by experiments with *Azolla* fronds detached from the roots (Sela, unpublished).

*Azolla* was found to remove much of the heavy metal supplemented to the medium, in the experiment described in Figure 3: 66%, 82%, 85%, 96% and 99% of $Cd^{2+}$, $Ni^{2+}$, $Zn^{2+}$, $Cu^{2+}$ and $Cr^{3+}$, respectively was recovered in the biomass at the end of the experiment. The uptake of uranyl ions by *Azolla*, demonstrates a

*Table 1.* Scaled-up removal of Cd and Ni from industrial effluents by *Azolla* biofilter.

| Effluent (L) | Cd in eluted solution (ppm) | Cd removal (%) | Ni in eluted solution (ppm) | Ni removal (%) |
|---|---|---|---|---|
| 20 | 0 | 100 | 0.03 | 98 |
| 40 | 0 | 100 | 0.03 | 98 |
| 60 | 0.03 | 98 | 0.05 | 96 |
| 80 | 0.29 | 81 | 0.12 | 91 |
| 100 | 0.65 | 58 | 0.29 | 78 |
| 120 | 0.82 | 47 | 0.39 | 71 |

Filter size – 150 g dried *Azolla*.
Effluent – industrial effluent from the production batteries containing 1.6 ppm Cd, 1.3 ppm Ni, 3450 ppm Na and 110 ppm K.
Flow – 1 L $min^{-1}$, 4 atm.
Further treatment: the filter which had absorbed Cd + Ni was tested for its capacity to further bind Cd from a concentrated solution. It was treated with Cd solution containing 980 ppm Cd, and was found to absorb 2.2%. Hence, the high binding capacity is retained after the passage of the diluted industrial effluent.

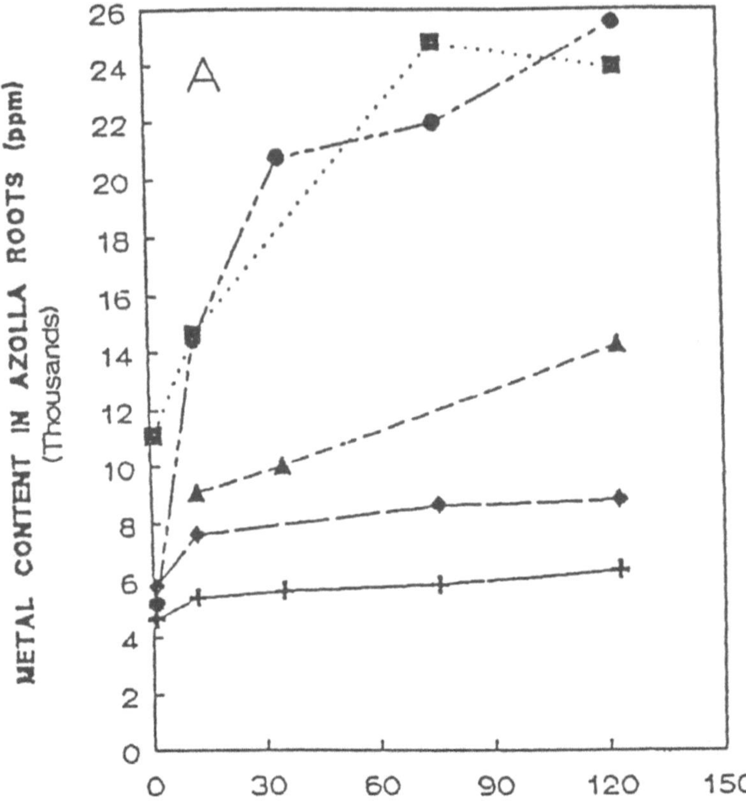


*Figure 2.* Uptake and accumulation of heavy metals by *Azolla* and the distribution of these metals in roots (A) and shoots (B). *Azolla* was grown in the phytotron, at 22/17°C day/night and 16/8 h light/dark regime. Metal salts were added to the growth medium.

uranyl concentration factor which was 390 to 520-fold higher in *Azolla* than in solutions containing 1–17 ppm uranyl. The storage of uranyl ions did not cause any noticeable damage to *Azolla* growth in this experiment, whereas the five aforementioned metals were harmful to both the host fern and the *Anabaena* cyanobiont (Sela *et al.*, 1990).

## 2.3 *Cellular localization and binding of the stored heavy metals in* Azolla

*Azolla* leaves are composed of few layers of parenchymal cells. Water extraction of *Azolla* fronds grown in the presence of Cd, Cr, Ni, Zn and Cu demonstrated that 98% of the metal taken up by *Azolla* was bound to its insoluble cell components (Sela *et al.*, 1989). Electron microscopy analysis of *Azolla* sections by X-ray emission (EDAX), localised Cd, Cu and U to polyphosphate bodies (Sela *et al.*,

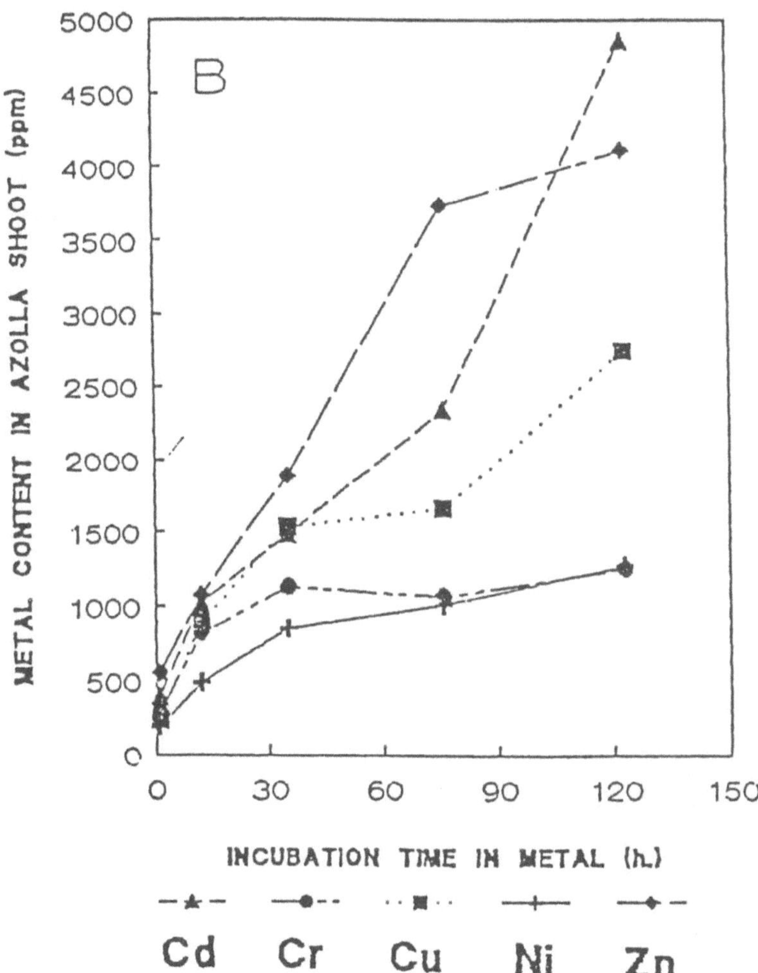


*Figure 2.* (continued)

1988). These results were supported by the fact that *Azolla* accumulated a high amount of phosphate, up to 3% of its weight, and heavy-metal stressed *Azolla* stored larger polyphosphate bodies (0.2–0.5 $\mu$). Membrane structures of *Azolla* grown in the presence of Cd contained more sulphur, suggesting an increase in the sulphydryl residues of metal-binding peptides (Sela *et al.*, 1990). *Azolla* biomass contains little cellulose, and its fronds are very fragile. Pectin extraction revealed that fronds contain 5.4% pectin per *Azolla* biomass (Ravid, 1991). *Azolla* grown in the presence of heavy metals, was chemically treated to dissolve components of the *Azolla* matrix in order to identify the role of pectin, protopectin, hemicellulose and polysaccharides in Cd binding (Figure 4). Pectin was dissolved by 2% ammonium oxalate and

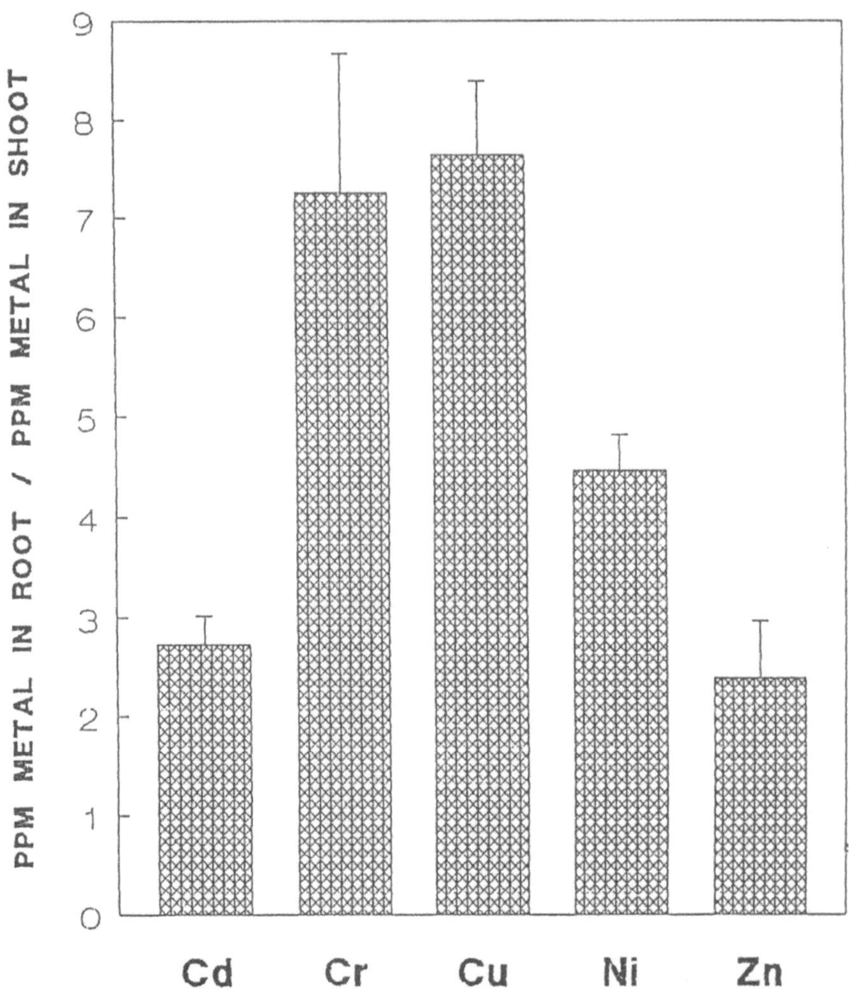


*Figure 3.* Mobility factor of the heavy metals in *Azolla*, expressed by the metal content in shoots and roots. Plants grown in the presence of 10–12 ppm metal ions were harvested after 3–7 days.

protopectin with additional hydrochloric acid (0.05 N). Hemicellulose was dissolved in a separate sample by 0.1 M NaOH, followed by higher NaOH concentration (0.44 M) for polysaccharide removal. As shown in Figure 4, the treatment for pectin removal dissociated 60–70% of the Cu, Ni and Zn bound to the *Azolla* matrix. Almost all the Cd bound to the *Azolla* matrix was solubilized by the treatment dissociating pectin and protopectin. Cr-bound *Azolla* was not greatly affected by the extraction treatments, and hence its binding seems to be different from that of Cd,

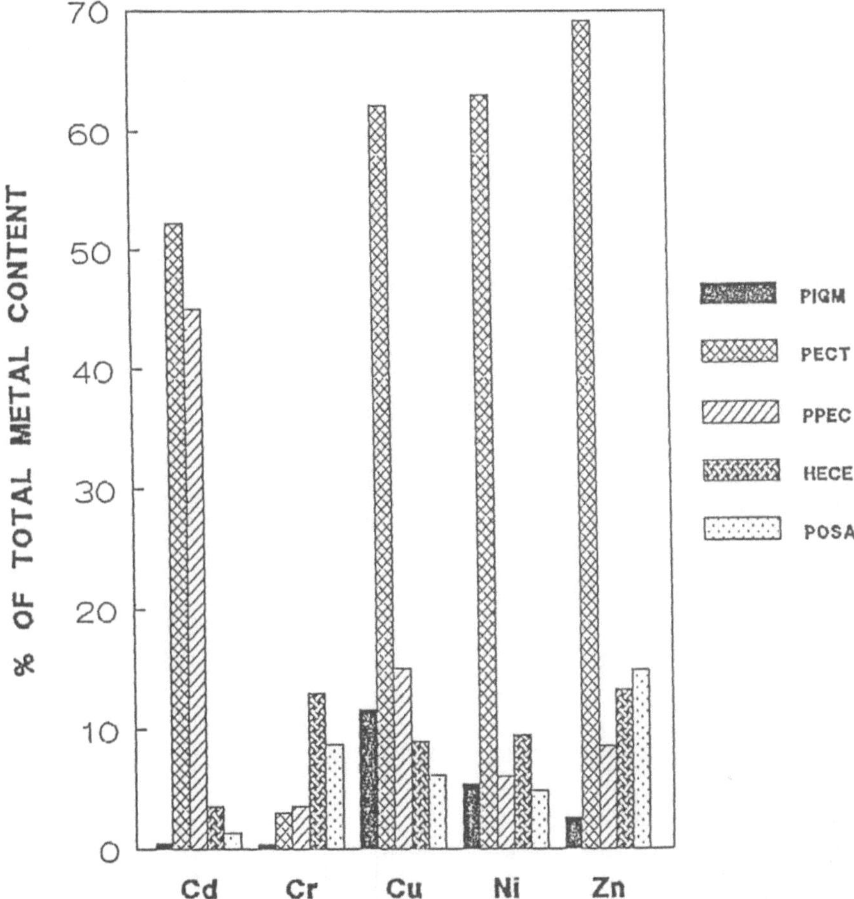


*Figure 4.* Distribution of heavy metals in *Azolla* treated chemically for the selective removal of the different fractions. *Azolla* was grown for 3–5 days in the presence of 30 ppm metal ions. PIGM = pigments; PECT = pectin; PPEC = propectin; HECE = hemicellulose; POSA = polysaccharides.

Cu, Ni and Zn. Cell-wall pectin is complexed with Mg and Ca ions, and recent experiments demonstrated the equivalent exchange of Pb ions bound to *Azolla* biomass with Ca and Mg ions eluted from the biomass (Epstein, unpublished).

We therefore propose that pectin, polyphosphates and metal-binding peptides are the major heavy metal-binding groups for Cd, Cu, Ni, Zn and Pb in *Azolla.* Figure 5 presents a scanning electron micrograph of a broken and freeze-dried *Azolla* leaf, where metal-binding groups are projected schematically, exchanging heavy metals (HM) with $Ca^{2+}$ and $Mg^{2+}$ ions, and bound by the three specific HM binding groups.

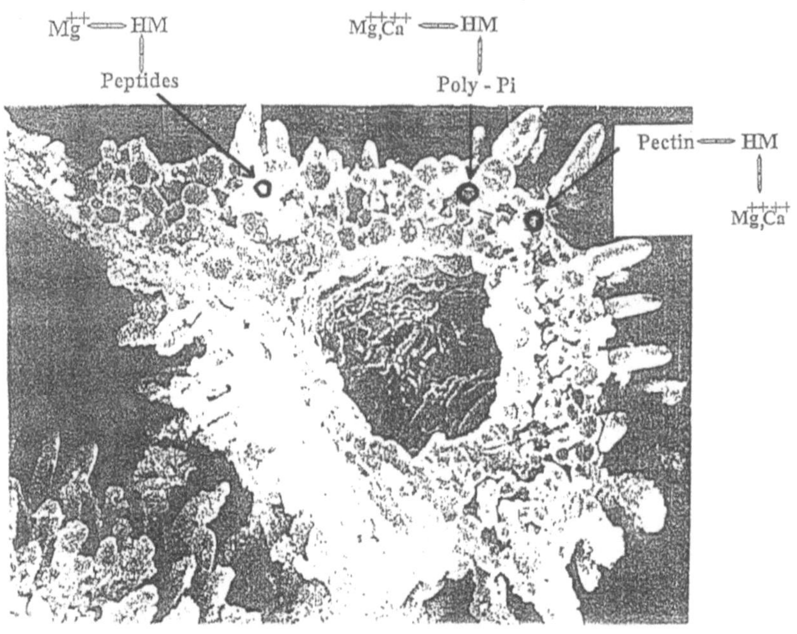


*Figure 5.* Localization of heavy metals in *Azolla*, presented schematically in an electron-micrograph section of *Azolla* leaf. The metal-binding groups are polyphosphates, pectin and metal-binding peptides.

### 2.4 *Binding and adsorption of heavy metals to dried* Azolla *biomass, and biofiltration of industrial wastes*

The binding of heavy metals to insoluble pectin and insoluble polyphosphates and peptides in *Azolla* accumulating Cd, Cr, Cu, Zn, Ni and U during growth, and the limited solubility of the metal ions in homogenised *Azolla* fronds, suggest the high affinity of these cations to the negative charges of phosphate, glucuronic acid and cysteine residues. Some of the bound metal ions may form precipitates in the *Azolla* matrix as well. We tested the binding of soluble heavy metal ions to dry *Azolla* biomass by filtering a contaminated solution through columns made up of dried *Azolla*: the solution was effectively purified. The filtration procedure demonstrated the instant binding of heavy metals, which is more useful for decontamination of industrial effluents than the alternative slow uptake of toxic metals by *Azolla* growing in waste ponds. The biological uptake process is limited to low concentrations of metals (1–20 ppm), whereas the biofiltration procedure was optimised and found useful for purifying industrial effluents containing as much as 17 000 ppm metal (for $Cr^{3+}$). The dried *Azolla* can easily be packed in various configurations, or be mixed with the polluted wastes, binding the heavy metal cation

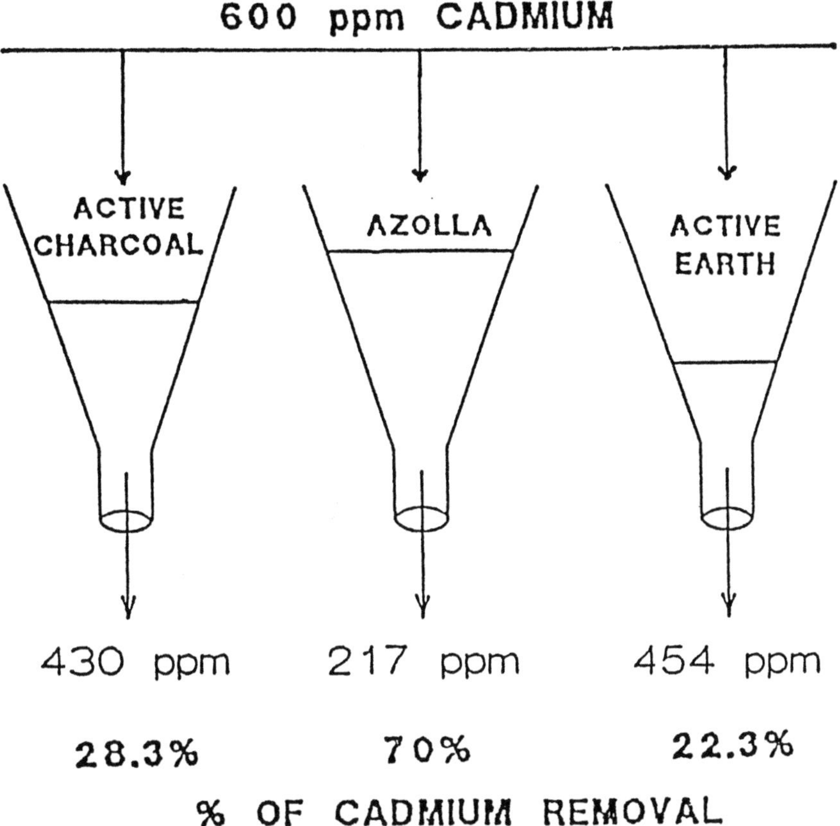


*Figure 6.* Comparison between purification of Cd waste solution by *Azolla*, by active charcoal and active earth; 500 ml of 600-ppm Cd solution was passed through 10 g of each filter. *Azolla* adsorbed 220 mg, active charcoal 84 mg and active earth 65 mg Cd.

to the phosphate and carboxyl groups of the *Azolla* matrix. The use of activated charcoal is a common and effective industrial decontaminating technique. We compared the binding of Cd from a 600 ppm solution to active charcoal, active earth (Paz Chemicals, Israel) and dried *Azolla* biomass, in funnels with 50 g filter of the filter material. As shown in Figure 6, Cd adsorption by *Azolla* was three times more effective than that by active earth and twice as effective as that by active charcoal.

This experiment (conducted in collaboration with Dr Etan Yanir and Shaul Adel of Paz chemicals), enhanced the efforts to treat industrial wastes from a number of industrial heavy metal effluents, which contained Cr, Cd, Ni, Hg and Pb. Extended research was conducted with Cd and Ni wastes of Tadiran battery production, in cooperation with Dr Pnina Dan. Different sizes of *Azolla* filters (3–150 g biomass) were loaded with Cd and Ni mixtures from the concentrated discharge solutions containing 2000 ppm Ni and 1400 ppm Cd. Alternatively we have attempted to

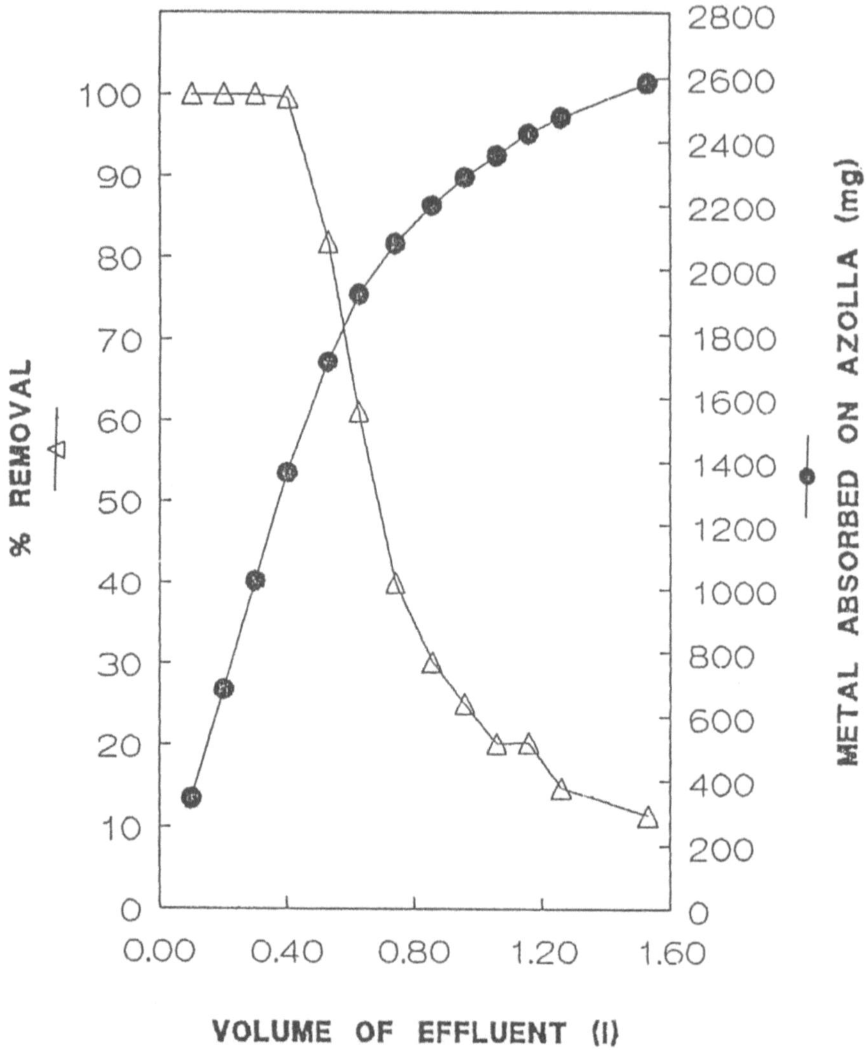

*Figure 7.* Adsorption profile of a Cd and Ni mixture solution containing 2026 ppm Ni and 1383 ppm Cd. *Azolla* filter (90 g) demonstrated a typical solution adsorption profile. 1.6 litre of flow rate 25 ml $min^{-1}$.

polish tails of these metals after base treatment, precipitation and floculation, at the industrial site, containing 1–4 ppm of each metal. A typical biofiltration experiment with the concentrated mixed Cd and Ni original solution is shown in Figure 7. A column of 90 g dry *Azolla*, was used to treat 1.6 litres of waste at a slow 25 ml $min^{-1}$ flow rate. The initial phase of Cd and Ni binding demonstrated almost complete removal of the heavy metals. In the second phase a gradual decrease in decontamination efficiency was observed while heavy metal content accumulated in *Azolla* dry biomass was up to 3% Cd+Ni (w/w). This pattern of Cd and Ni binding suggests an ion exchange mechanism in which heavy metal ions compete with and replace $Ca^{2+}$ and $Mg^{2+}$ bound to polyphosphates and pectin in *Azolla* matrix.

A complementary application of the *Azolla* biofilter to polish Cd and Ni wastes after metal base precipitation was conducted on a larger operation scale at the battery production industrial site (Table 1). The treated solution of Cd+Ni contained about 3 ppm metals plus excess of 3560 ppm Na+K ions, and the solution was introduced at a pressure of 3 atm through the *Azolla* biofilter. The initial phase of the industrial effluent was completely decontaminated by the *Azolla* filter and purified to 0–30 ppb Cd and 30–50 ppb Ni. The high pressure applied in this experiment created tunnels in the *Azolla* biofilter and reduced the binding efficiency.

Recently, controlled biofiltration studies were conducted to optimize flow rates and pH in the metal-binding process. This is an important step to optimize the treatment of specific industrial wastes, taking into consideration the heavy metal ion content and accompanying solutes.

### *2.5 Disposal of the biofilter loaded with the toxic heavy metals*

The final product of the biofiltration process in *Azolla* biomass contains as much as 3–5% Cu, Cd and U or 5–11% Cr, or 5–12% Pb. Thus, the metal-bound *Azolla* cake is in fact toxic waste, which needs to be disposed of or stored to eliminate its contact with soil and water resources.

Two strategies were tested to minimise the volume of biofilter wastes; one was the incineration of metal-bound *Azolla* to inorganic ash, and the second was regeneration of the filter by acid wash for its reuse with other polluted solutions. The incineration of Zn-bound *Azolla* yielded ashes containing 27% Zn (w/w) in the inorganic powder. This metal is potentially reclaimable by melting. The alternative recycling of the metal-bound biofilter by acid wash was demonstrated with Cd-bound *Azolla*; the bound Cd was eluted with acid solution (pH 2.0) and Cd-waste solution was reabsorbed, demonstrating a retention of 80% of the binding capacity of *Azolla*. Such a recycling experiment was repeated for four cycles and demonstrated that the *Azolla* biofilter can be regenerated by acid wash, similarly to Amberlite ion exchangers used as water deionisers.

## Acknowledgment

We thank Mr Ami Avital for his assistance. We appreciate the cooperation and support of Dr Etan Yanir and Mr Shaul Adel of Paz Ltd, and of Dr Pnina Dan of Tadiran Ltd. We wish to dedicate this contribution to the memory of Mr Yaacov Bornstein.

## References

Lumpkin T A and Plucknett D 1982 *Azolla* as a green manure: Use and management in crop production. Westview Tropical Agriculture Series No. 5, Boulder, CO., 230 pp.

Peters G A and Meeks J C 1989 The *Azolla–Anabaena* symbiosis: Basic biology. Ann. Rev. Plant Physiol. 40, 192–210.

Ravid S 1991 Absorption and recovery of precious metals by the water fern *Azolla*. MSc Thesis, The Hebrew University (in Hebrew).

Sela M, 1990 Uptake of heavy-metals by the water fern *Azolla*. PhD Thesis, The Hebrew University of Jerusalem (in Hebrew).

Sela M, Tel-Or E, Fritz E and Huttermann A 1988 The microscopic localization of heavy-metals in *Azolla*. Plant Physiol. 88, 30–36.

Sela M, Garty J and Tel-Or E 1989 The accumulation and the effect of heavy-metals on the water fern *Azolla filiculoides*. New Phytol. 112, 7–12.

Sela M, Fritz E, Huttermann A and Tel-Or E 1990 Studies on Cd localization in the water fern *Azolla*. Physiol. Plant 79, 547–553.

Tel-Or E and Sela M 1991 US Patent No. 5,000,852, Process for the removal of metal ions from solutions.

Tel-Or E and Sela M 1994 European Patent No. 03332128, Process for the removal of metal ions from solutions.

Tel-Or E, Rosen A, Ofir Y, Kobiler D and Schonfeld M 1991 Metabolic relations and intracellular signals in the *Anabaena–Azolla* association. Isr. J. Bot. 40, 171–181.

36

# Modelling the mobilizing effect of olive mill wastewater on heavy metals adsorbed by a soil

L. MADRID AND E. DÍAZ-BARRIENTOS

## 1. Introduction

Modelling those processes which control environmental pollution is a tool that is being increasingly used in environmental chemistry. In many cases, however, models used in the literature are little more than a way of presenting the data, and do not give any real, additional information about the nature of the processes underlying the data. A typical example of such models is the popular use in soil science of the van Bemmelen–Freundlich equation when studying data of equilibrium adsorption (metals, pesticides). The so-called 'Freundlich constant' is estimated for a set of soils by least-square fitting and related to the contents in various soil components (e.g., Roldán *et al.*, 1993). It is well-known that the chemical meaning of this constant is, at the best, doubtful, so that any comparison between values obtained for different soils says little more than the adsorption data themselves. Sposito (1980) showed that the Freundlich equation can be derived rigorously for the trace adsorption of an ion through ion exchange, but only if such an ion is at trace levels in the exchanger, and the Freundlich exponent should remain between 0 and 1. If these conditions are not fulfilled, the use of this equation is questionable.

Moreover, care should be taken to limit the model to as few adjustable parameters as possible, as a large number of them can cause an increase in their uncertainty. Models such as that proposed by Bowden *et al.* (1980) for ion adsorption need many parameters for a good 'theoretical' description of the process; an equilibrium should be sought between a mechanistic representation of all the possible components contributing to the process and the search for a reasonable, 'average' explanation of the data.

We have studied the time-dependence of metal adsorption by a clayey, carbonate-rich soil which has been used as a dumping site for olive mill wastewater (OMW), and observed the changes caused by the residue on the power of the soil to immobilize metals, as it has been observed (Bejarano and Madrid, 1992; Madrid and Díaz-Barrientos, 1994) that OMW can have a mobilizing effect on heavy metals in some conditions.

The comparison between the original and treated soil samples was carried out by a multireaction model developed by Selim (1992), which considers the metal retention to

*D. Rosen* et al. *(eds): Modern Agriculture and the Environment, 443–447.*

be due to several fractions, $S_E$, $S_1$, $S_2$, $S_3$ and $S_{irr}$. Fractions $S_1$ and $S_2$ react kinetically and reversibly with the soil solution, through the forward and backward rate coefficients $k_1$ and $k_2$ for the former and $k_3$ and $k_4$ for the latter, with reaction orders $n$ and $m$ respectively. Fraction $S_E$ reacts instantaneously with the solution, so that it is in equilibrium with the solution *at all times*; the reaction is governed by a distribution coefficient $K_d$ through a Freundlich equation of exponent $b$. $S_{irr}$ reacts kinetically and irreversibly with the solution according to an irreversible, first order reaction, with rate coefficient $k_s$. Finally, $S_3$ is assumed to form from $S_2$ by a slow, reversible, first order reaction, governed by the rate coefficients $k_5$ and $k_6$. The model allows the exclusion of one or more of these fractions if they are not necessary for the explanation of the data. A full mathematical description of the model can be seen elsewhere (Selim, 1992).

## 2. Materials and methods

Duplicates of the original soil, an anfisol from SE Spain containing 55% clay fraction and 42% carbonate, and of the sample after receiving large doses of OMW during three years were shaken with solutions with initial concentrations of 10–60 mg $L^{-1}$ Cd, Cu or Zn with 0.01 M NaCl as background electrolyte, at a soil:solution ratio 1:100, for 1–180 hours. After each reaction time the suspensions were centrifuged for 20 min at 20 000 $g$ and the final metal concentrations were determined in the supernatants by AAS.

## 3. Results and discussion

Figure 1 gives examples of the behaviour of Cd and Zn with both soil samples for 20 mg $L^{-1}$ of initial metal concentration. Other concentrations behave in the same way and are not shown. The original soil shows an initial fast decrease in concentration and a tendency to slow down and stabilise at longer times, with a low value of the solution concentration near equilibrium. The main difference between Cd and Zn is that the latter gives considerably lower concentrations. For the soil receiving OMW, a clearly defined minimum occurs after the first day or a short time later, with a strong increase in concentration afterwards. In many cases the increase occurs after 50–60 hours, so that the behaviour up to 50 hours is similar to that observed for the original soil. The behaviour of Cu (not shown) is also similar, except that the concentrations are somewhat lower than those found for Zn. In some of the Cu experiments, the low concentrations found were affected by large errors and did not allow us to process the data in order to fit them within the model.

When using the model, we interpolated in the graphs the ordinate values up to 50 hours, as it was frequently after this time when the increase in concentration occurred. Therefore, those cases of the three metals in which the increase after the minimum occurred before 50 hours, were considered too limited for their inclusion in the model.

For solving the system of equations describing the model we used the computer program given by Selim *et al.* (1990). In order to fit the best value of the parameters,

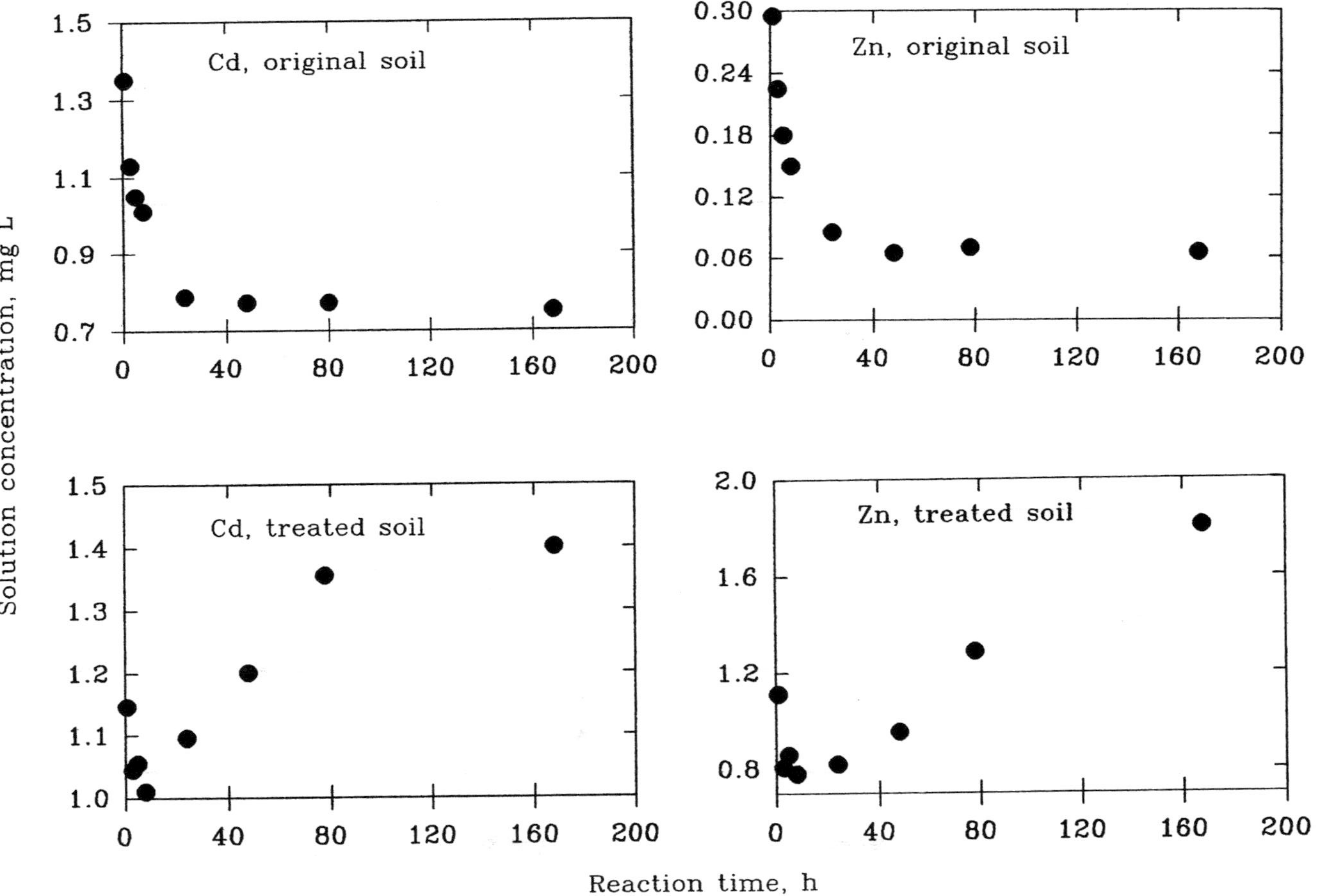


*Figure 1.* Examples of time dependence of the solution concentration in contact with the original and treated soil of initial concentration 20 mg L$^{-1}$ Cd or Zn.

*Table 1.* Estimated values of the parameters for each system.

| System | Initial conc. (mg L$^{-1}$) | K$_d$ (mg kg$^{-1}$) | $b$ | $k_1$ (h$^{-1}$) | $k_2$ (h$^{-1}$) | $n$ | $k_s$ (h$^{-1}$) | $R^2$ |
|---|---|---|---|---|---|---|---|---|
| | | $S_E$ | | $S_1$ | | | $S_{irr}$ | |
| Orig./Cd | 10 | 0.53 | 0.41 | 3.52 | 0.21 | 1.07 | – | 0.99 |
| | 20 | 1.02 | 0.39 | 5.43 | 0.38 | 0.82 | – | 0.97 |
| | 40 | 1.54 | 0.89 | 5.49 | 0.85 | 0.88 | – | 0.99 |
| | 60 | 1.07 | 0.15 | 3.15 | 1.15 | 1.02 | – | 0.98 |
| OMW/Cd | 40 | 14.5 | 0.65 | 0.24 | 0.20 | 2.01 | – | 0.97 |
| | 60 | 4.08 | 0.10 | 2.81 | 1.10 | 1.03 | – | 0.98 |
| Orig./Cu | 40 | 27.9 | 0.45 | 10.0 | 0.07 | 0.92 | 1.20 | 0.98 |
| | 60 | 36.8 | 0.61 | 9.43 | 0.08 | 0.80 | 1.20 | 0.99 |
| OMW/Cu | 40 | 48.2 | 0.36 | 2.94 | 0.20 | 0.98 | – | 0.99 |
| | 60 | 32.5 | 0.42 | 2.83 | 0.15 | 1.02 | 0.08 | 0.99 |
| Orig./Zn | 10 | 6.07 | 0.41 | 5.41 | – | 1.12 | 0.13 | 0.94 |
| | 20 | 14.4 | 0.84 | 3.37 | – | 1.05 | – | 0.97 |
| | 40 | 20.9 | 0.53 | 0.82 | – | 1.17 | – | 0.99 |
| | 60 | 26.7 | 0.71 | 0.27 | – | 1.49 | – | 0.99 |
| OMW/Zn | 60 | 1.00 | 0.73 | 0.42 | 0.04 | 0.59 | – | 1.00 |

we combined this program with a subroutine based on the SIMPLEX method (Barrow, 1979) which has been frequently used for similar purposes.

Table 1 shows a summary of the parameters obtained for each system. In all cases, the percentage of variance accounted for did not increase with the inclusion of fractions $S_2$ and $S_3$, therefore these fractions were considered which did not contribute to the reaction and no values are given for $k_3$, $k_4$, $k_5$ or $k_6$. The values of K$_d$ and of the rate coefficients were affected by coefficients of variation between 20 and 40%, and this must be kept in mind when comparing data. This large error is unavoidable due to the rather large number of parameters.

The values of K$_d$ are clearly higher for Cu than for the other metals, with little difference between the two soils. This means that the residue has no effect on the fast, 'equilibrium' adsorption of Cu. Instead, the K$_d$ for Cd increased noticeably in the presence of OMW, and decreased for Zn. When calculating the values of each adsorbed fraction at each reaction time (which is done with the computer program), we found that $S_E$, 'equilibrium' adsorption, never accounted for more that 2–3% of the total adsorbed. In any case the results suggest a different origin of this fraction in each metal: the residue seems to favour the fast adsorption of Cd and prevents it in the case of Zn, in this latter case perhaps through blocking of the surfaces involved. However, it is convenient to remember the doubtful meaning of a Freundlich constant, and therefore we must be careful in interpreting this 'instantaneous' parameter, which accounts for only a very low proportion of the amount adsorbed.

The forward rate coefficient of the 'kinetic' fraction $S_1$ is generally decreased in the presence of OMW. Moreover, the 'irreversible' fraction, which is significant only in the case of Cu, clearly disappears when OMW is present. We have suggested

(Madrid and Díaz-Barrientos, 1994) the importance of carbonate surfaces in the adsorption of these three metals, and the likely effect of blocking of such surfaces by the organic residue. If we assume that the kinetic adsorption ($S_1$) is related with these surfaces, the behaviour of $k_1$ would agree with such interpretation. The irreversible fraction in Cu experiments, $S_{irr}$, could be attributed to precipitation of sparingly soluble compounds, perhaps a basic carbonate, which would not form in the presence of OMW.

The increase in solution concentration in most experiments with the treated soil is likely to be due to a slow complexing reaction of the polymers present in the OMW. Modelling this reaction is beyond the possibilities of the model used as it stands, although we are now trying to include it. However, the parameters estimated here agree with the idea that the residue causes a decrease in the retention properties of the soil for the three metals as described by the model. As a consequence, the solution concentration values after a few days of reaction become 2–3 times higher for Cd and about 10–20 times higher for Cu and Zn when the soil receives OMW. This increase is not quantitatively strong (a few mg per litre at the most), but can be significant from the environmental standpoint, limiting the behaviour of some soils as 'sink' for heavy metals.

## Acknowledgement

This research was supported by the Dirección General de Investigación Científica y Técnica (Spain) as part of Project No. PB92-0052.

## References

Barrow N J 1979 Computer programmes for calculating charge and adsorption of ions in variable charge surfaces. CSIRO Division of Land Resources Management, Tech. Memorandum 79/3, Perth, Australia. 24 pp.

Bejarano M and Madrid L 1992 Solubilization of heavy metals from a river sediment by a residue from olive oil industry. Environ. Technol. 13, 979–985.

Bowden J W, Nagarajah S, Barrow N J, Posner A M and Quirk J P 1980 Describing the adsorption of phosphate, citrate and selenite on a variable charge mineral surface. Aust. J. Soil Res. 18, 49–60.

Madrid L and Díaz-Barrientos E 1994 Retention of heavy metals by soils in the presence of a residue from the olive-oil industry. Eur. J. Soil Sci. 45, 71–77.

Roldán I, Hermosín M C and Cornejo J 1993 Tricyclazol retention by soils. Sci. Total Environ. 132, 217–228.

Selim H M 1992 Modeling the transport and retention of inorganics in soils. Adv. Agron. 47, 331–384.

Selim H M, Amacher M C and Iskandar I K 1990 Modeling the Transport of Heavy Metals in Soils. CRREL Monograph 90-2, Hanover, NH, USA. 156 pp.

Sposito G 1980 Derivation of the Freundlich equation for ion exchange reactions in soils. Soil Sci. Soc. Am. J. 44, 652–654.

# 37
# Soil microbial biomass as a marker of heavy metal contamination and bioavailability

L. LEITA, G. MUHLBACHOVA, C. MONDINI, L. MARCHIOL AND M. DE NOBILI

## 1. Introduction

The fertility of natural soil ecosystems depends on the rate of turnover of soil organic matter, mediated by the soil microbial biomass. The upset of microbiological equilibria therefore affects nutrient cycling with serious consequences for the soil–plant system. Heavy metals are the most important, as they persist in soil indefinitely once they enter it and their effects upon the microbial population can be essentially permanent. There is evidence that large amounts of heavy metals introduced in soil depress the turnover rate of organic matter, because of their toxicity on the microbial biomass (Brookes and McGrath, 1984; Chander and Brookes, 1991, 1993). It therefore seems reasonable to place particular enphasis upon biological processes in assessing the potential of microbiological properties as indicators of soil pollution. In fact, microbes, having both mass and activity, and being in intimate contact with soil micro-environment are, in many ways, ideal monitors of soil pollution (Brookes, 1992).

The determination of microbial respiration and biomass carbon are widely used by ecologists as an index both of microbial activity and microbial biomass itself (Debosz *et al.*, 1985; Doelmann, 1985). In most cases, heavy metal pollution has little effect on biomass C and soil respiration, expressed as $CO_2$-C evolved per gram of soil, at low levels of contamination, but with larger doses these parameters were found to decrease (Baath, 1989; Chander and Brookes, 1993; Hattori, 1991, 1992; Tyler, 1981), probably because of the overall effect of the decrease in soil microbial biomass. In contrast, Ausmus *et al.* (1978) and Chaney *et al.* (1978) found an increase in $CO_2$ evolution after adding Pb, Zn and Cd to the soil; these observations were confirmed by Chander and Brookes (1991) who found that more $CO_2$ per unit biomass C was evolved from soil which had received contaminated sludge. Anderson and Domsch (1990) suggested that an ecophysiological parameter, such as the metabolic quotient ($qCO_2$) of the soil microbial biomass, could be useful to study environmental effects on microbial pool in terms of energy demand for survival. The metabolic quotient, in fact, expresses the $CO_2$-C produced per unit of biomass and time and is therefore more specifically useful than the soil respiration rate.

*D. Rosen* et al. *(eds): Modern Agriculture and the Environment, 449–457.*

Interactions between metals and the soil environment are complex. Soil is a complex porous matrix usually possessing a high metal-binding capacity, primarily due to its small, highly surface active components such as clays, organic materials and hydrous metal oxides. The fate of toxic metal cations in the soil environment depends largely on the interactions of these metals with inorganic and organic surfaces, and a considerable amount of work has been done to evaluate the adsorption or complexation of heavy metals by soil constituents, such as clays and organic matter. One potentially important organic surface which has received little attention is that of the microbial cells as a possible factor involved in heavy metal bioavailability (Mullen *et al.*, 1989). In particular, organic matter including soluble and particulate fractions from plants, animals, microbes, and their products binds large amounts of metals in natural systems, where they accumulate at the soil surface (Flemming *et al.*, 1990; Walker *et al.*, 1989). Soil microorganisms which are typically associated with the organic fractions of the soil microenvironment would be expected to participate actively in the mobilization-immobilization equilibria of heavy metals. In particular, bacteria have a high surface area-to-volume ratio and, as strictly physical cellular interface, should have a high capacity for sorbing and releasing metals from and into the solution (Beveridge, 1988). The aim of the present study was to test the possibility of a relationship between soil microbial biomass and bioavailability of heavy metals.

## 2. Materials and methods

### *2.1 Soil*

Soil taken at 0–30 cm depth was sieved at 5 mm. The soil was composed of 18.44% clay, 32.72% lime and 48.44% sand. The soil pH was 6.1 with an organic C content of 1.72%, C/N ratio was 10, and CEC was 17.6 mEq 100 $g^{-1}$. Before starting the experiment the soil was adjusted to 50% of its water holding capacity (WHC) and equilibrated for 7 days at 25°C. Three soil aliquots were then sprayed separately with solutions containing appropriate amounts of $ZnSO_4$, $Tl_2SO_4$ or $PbSO_4$ to give a final soil concentration of 600 $\mu$g Zn, 400 $\mu$g Tl or 200 $\mu$g Pb $g^{-1}$ soil. The control and contaminated soil samples were put in polyethylene bags containing 8 kg of soil each, then placed in sealed containers and incubated in the dark for 7 weeks at 25°C. Two jars containing, respectively, 500 ml distilled water and 200 ml NaOH 1M were put into each container to prevent changes in the soil water content and to trap the $CO_2$ evolved.

### *2.2 Soil microbial biomass C determination by fumigation–extraction*

The measurements of the soil microbial biomass C were performed using the fumigation–extraction method (FE) according to Vance *et al.* (1987). Briefly, three replicate aliquots of moist soil corresponding to 50 g of oven dried soil (24 h, 105°C) were placed in 100 ml jars and fumigated with ethanol-free chloroform in the dark for 24 h at 25°C. After fumigant removal the soil was extracted with 200 ml

0.5M $K_2SO_4$ for 30 min. Similarly, three replicates of unfumigated soil were analysed at the time fumigation commenced. Organic C in soil extracts was determined by wet digestion with $K_2Cr_2O_7$ and titration of the excess dichromate with $FeSO_4$. The microbial biomass C was calculated from the relationship: Biomass C = 2.64 $E_c$, where $E_c$ is the difference between organic C extracted from the fumigated and non-fumigated treatments, both expressed as $\mu$g C $g^{-1}$ oven dry soil (Vance *et al.*, 1987).

### 2.3 *Soil respiration*

Soil respiration, expressed as amount of organic C released as $CO_2$ per g of incubated soil, was measured weekly by titration with HCl, after absorption in NaOH and precipitation with $BaCl_2$ (Stotzky, 1965).

### 2.4 *Available fractions of Zn, Tl and Pb*

Available fractions of Zn, Tl and Pb were extracted on 10 g soil with 20 ml extracting solution (0.005M DTPA, 0.01M $CaCl_2$ and 0.1M TEA adjusted to pH 7.3) according to Lindsay and Norwell (1978).

All extracts were centrifuged at 8500 *g*, filtered on Whatman filter 0.2 $\mu$m and clear solutions were then directly analyzed for Zn, Tl and Pb content by atomic absorption spectrophotometry by means of a Perkin–Elmer 2380 equipped with a deuterium lamp as background corrector.

## 3. Results and discussion

### 3.1 *Effect of Zn, Tl and Pb on soil microbial biomass C and soil respiration rate*

The dynamics of microbial biomass C calculated from FE determination is shown in Figure 1. The microbial biomass C, which was 228.6 $\mu$g $g^{-1}$ soil at the beginning of the experiment, declined slightly during the first week of incubation of control and contaminated soils. Thereafter the microbial biomass C remained constant at about 203.4 $\mu$g $g^{-1}$ soil in control soil and decreased progressively in contaminated soil samples throughout the experiment. The addition of Zn, Tl and Pb did not cause significantly different inhibitory effects on microbial biomass C. At the end of the experiment, there was a 45% decrease of the initial value of microbial biomass C in soil contaminated with Zn, and a 41% and 38% decrease in those contaminated with Tl and Pb, respectively

The addition of contaminants to soil generally depresses soil respiration (Baath, 1989; Hattori, 1991, 1992), probably because of the overall effect of the decrease in soil microbial biomass. By contrast, results obtained in our experiment showed an increase in the soil respiration rate, expressed as cumulative $CO_2$-C evolved weekly per gram of soil (Figure 2). Soils contaminated with Zn, Tl and Pb displayed a larger evolution of $CO_2$ during the overall incubation period. One reason for the large

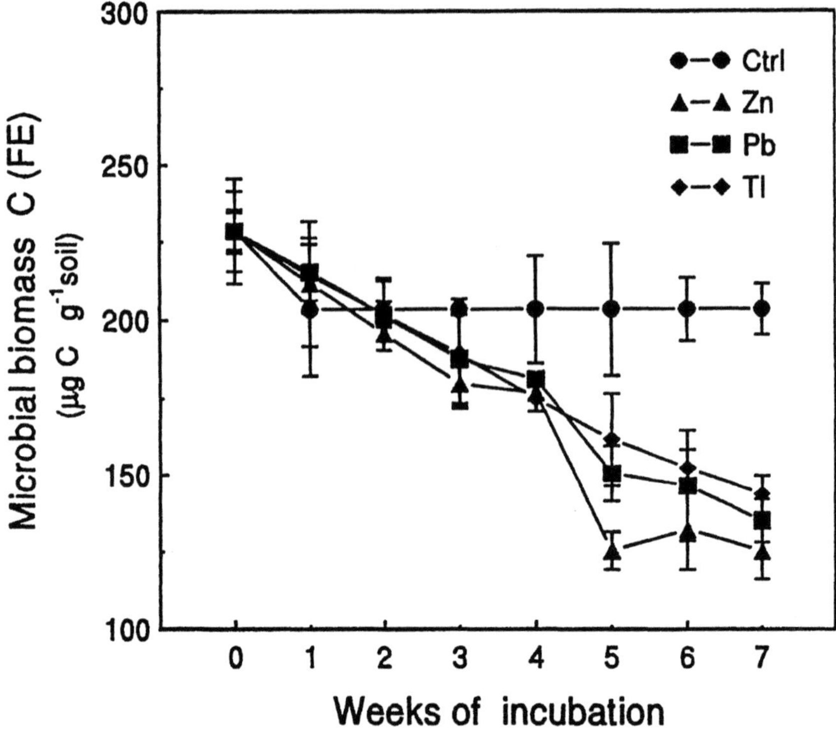


*Figure 1.* Trends in microbial biomass C in control and soil contaminated with Zn, Pb and Tl during 7-week incubation period.

increase in the soil respiration may be the increase in energy expended by the microbial pool to survive under abiotic stress (Ausmus *et al.*, 1978; Chander and Brookes, 1993). The increase in soil respiration rate corresponds to an even larger increase in the specific respiration rate (metabolic quotient) of the surviving microbial biomass. Results of metabolic quotient $qCO_2$ calculation are reported in

*Table 1.* Metabolic quotient ($qCO_2$) of microbial biomass in control and contaminated soil samples (mg $CO_2$-C $mg^{-1}$ mg biomass C $h^{-1}$ $\times$ $10^{-4}$).

| | Week of incubation | | | | | | | Mean |
|---|---|---|---|---|---|---|---|---|
| | 1 | 2 | 3 | 4 | 5 | 6 | 7 | |
| Control | 2.17 | 2.25 | 2.47 | 2.45 | 2.47 | 2.49 | 2.50 | 2.40 |
| Zn | 3.35 | 3.92 | 5.11 | 5.32 | 7.59 | 7.38 | 7.89 | 5.79 |
| Tl | 2.50 | 3.80 | 5.21 | 5.48 | 6.01 | 6.72 | 6.29 | 5.14 |
| Pb | 2.74 | 3.40 | 4.88 | 4.99 | 6.09 | 6.43 | 7.27 | 5.11 |

Table 1 and show a different response of microbial communities in control and contaminated soils. The $qCO_2$ of the control remained constant at about $2.4 \times 10^{-4}$ mg $CO_2$-C $mg^{-1}$ biomass C $h^{-1}$ during the whole incubation period, whereas a net increase occurred in soil contaminated with heavy metals. A significant increase of metabolic quotient of biomass in Zn treated soil was observed after the first week of incubation and was the highest during the overall experiment, confirming the greater demand for energy by microorganisms in order to survive the greater biocide effect displayed by this metal. The microbial biomass in Pb and Tl treated soils showed a net increase of $qCO_2$ only after the second week of incubation, and reached peak levels during the last two weeks. These results suggested that it is reasonable to assume that the surviving biomass responds to soil contamination with Zn, Tl and Pb with an increase of microbial activity, as measured by $CO_2$ evolution and that the addition of contaminants to soil may alter the physiological status of the microbial pool, with consequent changes in microbial death rates.

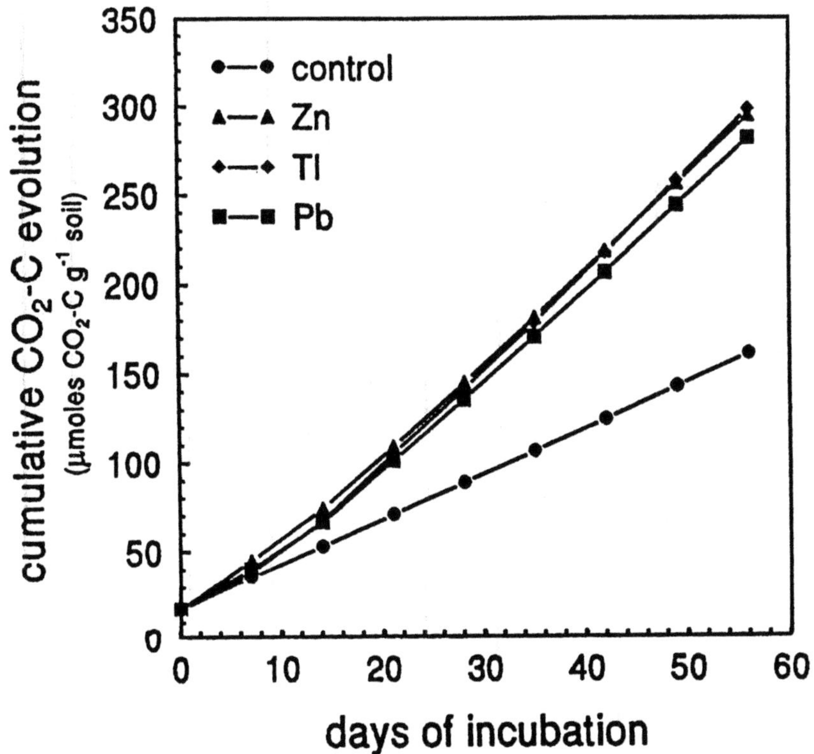


*Figure 2.* Effect of Zn, Tl and Pb on cumulative $CO_2$-C evolution in soil.

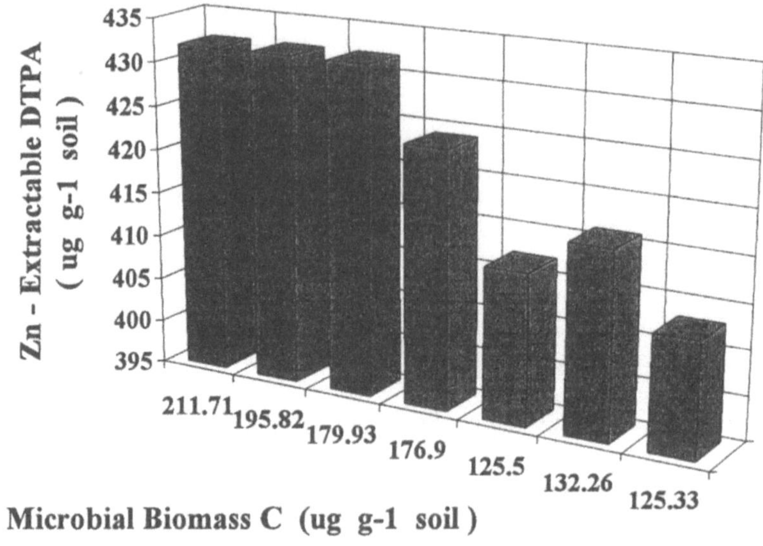


*Figure 3.* Relation between available fraction of Zn and microbial biomass C.

### *3.2 Relations between available fractions of Zn, Tl and Pb and microbial biomass C*

The mobilization/immobilization equilibria of heavy metals in soil are modulated by many chemical–physical and biological factors. The relation between the mobilization of heavy metals and soil microbial biomass C depends on several factors, such as the production of organic chelates, and microbial incorporation in living cells and biopolymers (Mullen *et al.*, 1989). In our experiment, we found significant correlations for available Zn ($r^2 = 0.91$), Tl ($r^2 = 0.88$) fractions and microbial biomass C but not for Pb. The DTPA-extractable fraction of Zn remained fairly constant during the first three weeks of incubation, when the microbial biomass was significantly depressed. Afterwards, the decrease of DTPA-extractable Zn corresponded with the minimum content of the microbial the biomass C (Figure 3), suggesting that Zn could be progressively immobilized in the microbial cell. This behaviour may be explained by the fact that even after death, bacterial cell walls can remain undecomposed in contaminated soils for prolonged periods since autocatalytic enzymes responsible for wall decomposition are inactivated by heavy metals (Ferris *et al.*, 1988).

The opposite behaviour was found for Tl, the availability of which increased from 4.5% of the total content after the first week of incubation to 16% at the end of the experiment. The negatively linear relationship with microbial biomass shown in Figure 4 might be the result of progressive mobilization of Tl by microorganisms, either through the production of chelating substances to reduce its toxicity or by a

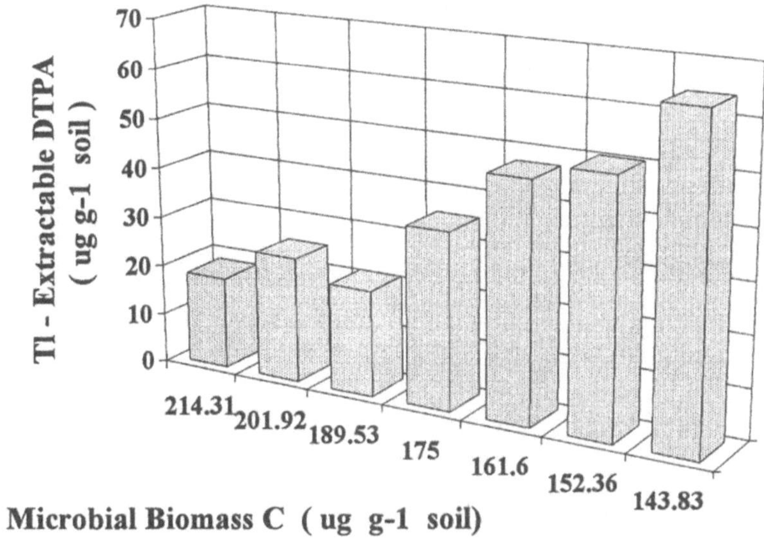


*Figure 4.* Relation between available fraction of Tl and microbial biomass C.

progressive release from decomposed dead cells. One cause of the difference from the Zn pattern might be that Zn is also an essential micro-element for microorganisms.

We did not find any significant relationship between available fraction of Pb and soil microbial biomass C. However, as can be seen in Figure 5, the amount of available Pb decreased during the three weeks of the experiment and increased thereafter, showing a tendency to remain constant at $123 \pm 6\ \mu g\ g^{-1}$ until the end of the experiment. The behaviour of Pb may be in part explained by the greater interference of humic substances on mobilization/immobilization equilibria of this metal with respect to Zn and Tl. Moreover the lack of any relationship between available Pb and microbial biomass C might be caused by the fact that Pb has more dubious effects on soil microorganisms. Results obtained suggest that each metal behaves quite differently and that different mechanisms might be involved at metabolic level. In addition, microbially mediated processes might be involved in determining the bioavailability of heavy metals and to predict the possible toxicological effects of pollution in soils.

**Acknowledgements**

We thank Mr A Bertoni and Mr A Cudini for technical assistance. This research was supported by a grant from the Italian Ministry of Agriculture, Finalised project PANDA, Subproject 3, Series 1.

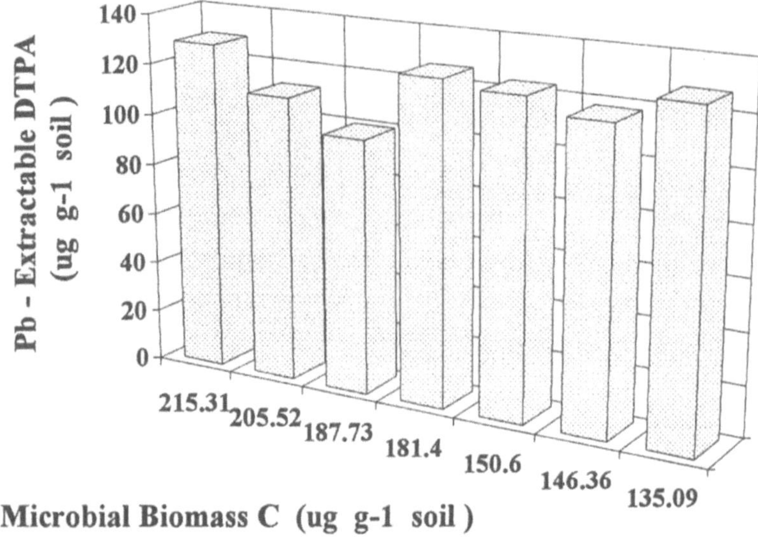


*Figure 5.* Relation between available fraction of Pb and microbial biomass C.

## References


Anderson T H and Domsch K H 1990 Application of eco-physiological quotients ($qCO_2$ and qD) on microbial biomasses from soils of different cropping histories. Soil Biol. Biochem. 22, 251–255.

Ausmus B S, Dodson G J and Jackson D R 1978 Behaviour of heavy metals in forest microcosms. III Effects on litter soil carbon metabolism. Water Air Soil Poll. 10, 19–26.

Baath E 1989 Effects of heavy metals in soil on microbial processes and population (a review). Water Air Soil Poll. 47, 335–379.

Beveridge T J 1988 The bacterial surface: General considerations towards design and function. Can. J. Microbiol. 34, 363–372.

Brookes P C 1992 The potential of microbiological properties as indicators in soil pollution monitoring. In Workshop on Soil Monitoring: Methods for Early Detection and Surveying of Soil Contamination and Degradation, 18–23 October, Monte Verita, Switzerland.

Brookes P C and McGrath S P 1984 Effects of metal toxicity on the size of the soil microbial biomass. J. Soil Sci. 35, 341–346.

Chander K and Brookes P C 1991 Effects of heavy metals from past applications of sewage sludge on microbial biomass and organic matter accumulation in a sandy loam and silty loam UK soil. Soil Biol. Biochem. 23, 927–932.

Chander K and Brookes P C 1993 Residual effects of zinc, copper and nickel in sewage sludge on microbial biomass in a sandy loam. Soil Biol. Biochem. 25, 1231–1239.

Chaney W R, Kelly J M and Strickland R C 1978 Influence of cadmium and zinc on carbon dioxide evolution from litter and soil from a black oak forest. J. Environ. Qual. 7, 115–119.

Debosz K, Babich H and Stotzky G 1985 Toxicity of lead to soil respiration: Mediation by clay, minerals, humic acids and compost. Boll. Environ. Contam. Toxicol. 35, 517–524.

Doelman P 1985 Resistance of soil microbial communities to heavy metals. In Microbial Communities in Soil. Eds V Jensen, A Kjoller, and L H Sorensen. pp 369–384. Elsevier, London.

Ferris F G, Fyfe W S and Beveridge T J 1988 Metallic ion binding by *Bacillus subtilis*: Implications for the fossilization of microorganisms. Geology 16, 149–152.

Flemming C A, Ferris F G, Beveridge T J and Bailey G W 1990 Remobilization of toxic heavy metals adsorbed to bacterial wall-clay composites. Appl. Environ. Microbiol. 56, 3191–3203.

Hattori H 1991 Influence of cadmium on decomposition of glucose and cellulose in soil. Soil Sci. Plant Nutr. 37, 39–45.

Hattori H 1992 Influence of heavy metals on soil microbial activities. Soil Sci. Plant Nutr. 38, 93–100.

Lindsay W L and Norwell W A 1978 Development of a DTPA soil test for zinc, iron, manganese, and copper. Soil Sci. Soc. Am. J. 42, 421–428.

Mullen M D, Wolf D C, Ferris F G, Beveridge T J, Flemming C A and Bailey G W 1989 Bacterial sorption of heavy metals. Appl. Environ. Microbiol. 55, 3143–3149.

Stotzky G 1965 Microbial respiration. In Methods of Soil Analysis. Eds C A Black, D D Evans, J L White, L E Ensminger and F E Clark. pp 1550–1570. American Society of Agronomy, Madison, WI.

Tyler G 1981 Heavy metals in soil biology and biochemistry. Soil Biochem. 5, 371–414.

Vance E D, Brookes P C and Jenkinson D S 1987 An extraction method for measuring soil microbial biomass C. Soil Biol. Biochem. 19, 703–707.

Walker S G, Flemming C A, Ferris F G, Beveridge T J and Bailey G W 1989 Physiological interaction of *Escherichia coli* cell envelopes and *Bacillus subtilis* cell walls with two clays and ability of the composite to immobilize heavy metals from solution. Appl. Environ. Microbiol. 55, 2976–2984.

# 38
# Copper and lead species in soil and their uptake by plants when applied as carbonates

ANNA CHLOPECKA

## 1. Introduction

Root uptake by plants is one of the most important processes by which heavy metals may enter the food chain. Under natural conditions, soils generally contain very low levels of heavy metals, including Cu and Pb. Certain soil parent materials, such as the sedimentary, sulphide-bearing shales may, however, contain indigenously high levels of Cu and Pb. Soils developed on such materials often contain higher amounts of these metals than soils developed on other materials. In addition to high levels of metals occurring naturally, considerable amounts of metals can also be introduced into soils via anthropogenic pathways, such as municipal sewage sludge application (Dudka and Chlopecka, 1990) and the application of commercial fertilizers, especially phosphatic fertilizers (Singh *et al.*, 1988). Trace metals also enrich the soil through atmospheric deposition (Jones, 1990). The Sudbury area in Canada and Lower Silesia in Poland are examples of areas polluted by Cu mining, while Upper Silesia, Poland, is polluted by Pb mining (Gzyl, 1990; Roszyk and Szerszen, 1988; Rutherford and Bray, 1979).

Total soil metal analysis may give information concerning possible enrichment of the soil with metals, but there is virtually a lack of comprehensive criteria for estimating their bioavailability and biological effects. This is because it is the chemical forms of metals in the soil that determine their mobility and behaviour in the environment. An empirical approach commonly used for studying the mobility and bioavailability of metals in soils is to use selective sequential extraction. Several selective sequential extraction schemes have been described for soil and sediments (Gupta and Chen, 1975; McLaren and Crawford, 1973; Shuman, 1985; Tessier *et al.* 1979). Calvet *et al.* (1990) gave an extensive review of the procedures developed up to 1986. The most widely used among these procedures is the one developed by Tessier *et al.* (1979) (Calvet *et al.*, 1990; Chlopecka, 1993; Spevackova and Kucera, 1989). The procedure involves partitioning of metals into exchangeable, bound to carbonates, bound to iron and manganese oxides, bound to organic matter, and residual bound.

Because the chemical forms of metals in soil are affected by pH, organic matter, redox potential (Eh), oxides of Fe and Mn, cation exchange capacity, total concentration of metals in soil, source term and other factors (Adriano, 1986), it is

*D. Rosen* et al. *(eds): Modern Agriculture and the Environment, 459–468.*

also necessary to assay the soils for these parameters and relate them to the chemical forms.

In soil, Cu could be found mostly in the organic form and residual bound fractions (Chlopecka, 1993; Ramos *et al.*, 1994; Shuman, 1991). Very little Cu in the exchangeable form is expected to occur in soils; however, the bulk of available Cu reserves may lie in the organic fraction (Gupta and Chen, 1975; McLaren and Crawford, 1973). Lead in non-contaminated soils is bound mainly to the residual phase, and it may be relatively abundant in the oxide and carbonate fractions (Chlopecka, 1993; Dudka and Chlopecka, 1990).

Anthropogenic activities may not only affect the total metal contents of the soil but also the chemical forms of the metals, which in turn can be influenced by the source term of the metals. Nearly all source terms for Pb (automotive – PbClBr and $PbSO_4$; mining – PbS and $PbCO_3$; paint – $PbCO_3Pb(OH)_2$ and $PbCrO_4$) are transformed in soil by chemical and microbiological processes. It has been generalized that soil has a high capacity to adsorb Pb onto solid phases of hydrous Fe oxide and organic matter (Chaney, 1988).

The main objective of this study was to investigate the influence of Cu and Pb carbonates on the chemical forms of these metals in the soil, and their uptake by cereal crops over time. This experiment was a part of research at the Trace Elements Laboratory, at the Institute of Soil Science and Cultivation of Plants in Pulawy, Poland, on the effect of metal carriers (sulphate, sulphide, carbonate, chloride) on trace element forms simulating contaminated environments.

## 2. Materials and methods

### 2.1 *The pot experiment*

A light sandy soil (pH 6.5; organic matter 1.38%; soil particles having diameter < 0.02 mm – 16%; cation exchange capacity (CEC) – 7.1 mEq 100 $g^{-1}$) was used in the three-year experiment which employed Mitscherlich's pots. The soil had high contents of available K and P, and medium content of available Mg. The soil was treated with copper carbonate (malachite) or lead carbonate (cerussite) at 50, 100, 500 $\mu g\ g^{-1}$. Untreated soil served as control. The treatments were replicated four times in a randomized design.

### 2.2 *Plants*

From 1989 to 1991 three monocotyledonous plant species: barley (*Hordeum vulgar* L.), maize (*Zeal mays* L.), and wheat (*Triticum vulgar* L.) were planted. The potted soils were fertilized with N, P and K, and watered with deionised water during the growth period.

Spring barley was sown at the end of May 1989 (16 plants per pot), grown for 120 days and then harvested at the full maturity stage. Samples of barley straw and grain were collected and analysed separately. Maize was planted at the beginning of May 1990 and harvested after 110 days at the full maturity stage. Samples of the maize plants were separated into young leaves, old leaves, shoots and grain, and analyzed individually. Wheat was grown from 16 April to 9 August 1991 till full maturity. The plants were divided into straw and grain after harvest.

The sampled plants were washed in deionised water, dried at 70°C, weighed and ground in a stainless steel mill. A 5-g subsample was dry ashed at 450°C for 16 hr, dissolved in 10 ml 50% (v/v) HCl and diluted to 50 ml with double-distilled water.

### *2.3 The soil*

Each year after the plant harvest soil samples were taken from each of the pots for chemical analysis. These samples were air-dried, ground in a porcelain mortar and passed through a 2-mm sieve. The air-dried < 2-mm soil samples were stored in paper bags for subsequent analysis. The fraction < 2 mm was used for the determination of pH, organic matter, particle size distribution, and cation exchange capacity. Representative soil samples were air-dried and homogenized in an agate mortar to obtain a fraction with particles < 0.01 mm diameter, which were used for total analysis and sequential extraction. Subsamples of 2 g were digested with $HClO_4$ and HF mixture. Following evaporation to dryness on a hot plate the samples were dissolved in 50% (v/v) HCl and diluted to 100 ml with 1% (v/v) HCl.

The sequential extraction scheme developed by Tessier *et al.* (1979) was adopted to partition Cu and Pb into the following operationally-defined fractions: exchangeable (1.0 M $CaCl_2$), carbonate (1.0 M NaOAc, pH 5), Fe–Mn oxide (0.04 M $NH_4OH{\cdot}HCl$), organic (0.02 M $HNO_3$ + 30% $H_2O_2$; 3.2 M $NH_4{\cdot}OAc$), and residual (conc. HF + conc. $HClO_4$). Extractions were carried out in triplicate using 1-g subsamples in polyethylene centrifuge tubes. Following each extraction the mixtures were centrifuged at 2000 *g* for 30 min at 15°C. After each extraction, the residues were washed with 8 ml of water, centrifuged as before, and then the washwater was discarded. The total metal contents in the plant and soil samples and in the extracts were determined by flame atomic absorption spectroscopy (Perkin Elmer 403). The lower limits of analytical determination were as follows (in $\mu g\ g^{-1}$): soils – Cu, 0.4 and Pb, 0.05; plants – Cu, 0.2 and Pb, 0.04. Analytical precision of the AAS method was validated by reanalyzing several duplicate samples (i.e. 10% of total number of samples).

## 3. Results and discussion

### *3.1 The soil*

The addition of copper and lead carbonates increased the soil pH in the first year (from pH 6.5 to 6.9 and 7.4 with the highest doses of Cu and Pb, respectively). In the ensuing two years of the study, the pH of the treated soil somewhat decreased, most likely due to fertiliser additions. The most drastic drop in pH was observed in the third year (from 6.9 to 4.5) for soil receiving the highest doses of copper carbonate. Over time the pH of the control soil also shifted but did not affect the distribution of the metals among the various forms. The residual and organic fractions were the most abundant forms of Cu in the control soil but the exchangeable form of Cu was

*Table 1.* Copper content ($\mu g\ g^{-1}$) of various soil forms; mean of three replicates with relative standard deviation < 20%.

| Treatment | Form | | | | | Sum of forms | Total |
|---|---|---|---|---|---|---|---|
| | F1 | F2 | F3 | F4 | F5 | | |
| Control | 0.01 | 0.01 | 0.40 | 3.49 | 3.00 | 6.91 | 5.90 |
| | (0.5%) | (0.5%) | (6%) | (50%) | (43%) | (100%) | |
| *First year* | | | | | | | |
| I | 0.73 | 15.41 | 19.27 | 14.56 | 4.30 | 54.27 | 71.20 |
| | (1%) | (28%) | (36%) | (27%) | (8%) | (100%) | |
| II | 1.95 | 46.56 | 41.53 | 25.59 | 5.30 | 120.90 | 119.50 |
| | (2%) | (38.5%) | (34%) | (21%) | (4.5%) | (100%) | |
| III | 4.79 | 250.00 | 186.60 | 79.50 | 10.30 | 534.20 | 548.70 |
| | (1%) | (47%) | (35%) | (15%) | (2%) | (100%) | |
| *Second year* | | | | | | | |
| I | 2.08 | 13.60 | 20.67 | 25.27 | 5.50 | 67.12 | 71.20 |
| | (3%) | (20%) | (31%) | (38%) | (8%) | (100%) | |
| II | 2.96 | 24.53 | 35.60 | 29.91 | 5.80 | 98.80 | 119.50 |
| | (3%) | (25%) | (36%) | (30%) | (6%) | (100%) | |
| III | 24.79 | 206.40 | 194.67 | 85.47 | 11.70 | 522.77 | 548.70 |
| | (5%) | (39%) | (37%) | (16.5%) | (2%) | (100%) | |
| *Third year* | | | | | | | |
| I | 4.56 | 17.20 | 22.70 | 35.50 | 6.70 | 86.70 | 71.20 |
| | (5%) | (20%) | (26%) | (41%) | (8%) | (100%) | |
| II | 11.36 | 41.60 | 34.40 | 59.80 | 9.30 | 156.50 | 119.50 |
| | (7.5%) | (26.5%) | (22%) | (38%) | (6%) | (100%) | |
| III | 61.30 | 173.10 | 174.70 | 113.20 | 13.00 | 535.50 | 548.70 |
| | (11.5%) | (32.5%) | (32.5%) | (21%) | (2.5%) | (100%) | |

F1 – exchangeable; F2 – carbonate; F3 – Fe–Mn oxide; F4 – organic; F5 – residue; values in brackets represent percentage of element in each form; the sums of forms were rounded to 100%; levels of added Cu: I – 50, II – 100, III – 500 $\mu g\ g^{-1}$.

relatively low (Table 1). Shuman (1985), using fractionation technique, reported that Cu was found largely in unavailable forms. As with other metals, absorbed and exchangeable Cu are the smallest fractions in soil (Sandres *et al.*, 1986). In soil treated with three levels of copper carbonate, proportional levels of Cu bound to carbonates increased markedly in the first year of study and virtually remained unchanged over two years (Table 1). Due to the increased carbonate form, corresponding decreases of residual and organic fractions were observed. Other investigators found high amounts of Cu in the carbonate fraction for alkaline soils and in the Fe-oxide fraction for acid soils, probably due to occlusion and strong adsorption (Kuo *et al.*, 1983; Miller *et al.*, 1983; Sims, 1986).

In the third year of the experiment, Cu in the carbonate fraction was lower in the treatments with the highest dose of copper carbonate than in the previous year but the proportion of exchangeable Cu in this treatment increased (from 0.5% up to 11.5%). These changes were most likely effected by the decrease of pH in soil over time. Although pH may have not influenced Cu level in the soil solution, increasing the pH apparently decreased Cu in the exchangeable fraction, making it less

*Table 2.* Lead content ($\mu g\ g^{-1}$) of various soil forms; mean of three replicates with relative standard deviation < 20%.

| Treatment | Form | | | | | Sum of forms | Total |
|---|---|---|---|---|---|---|---|
| | F1 | F2 | F3 | F4 | F5 | | |
| Control | 0.10 (0.5%) | 0.80 (3.5%) | 3.20 (15.2%) | 4.01 (19.2%) | 12.67 (61.6%) | 20.78 (100%) | 19.00 |
| *First year* | | | | | | | |
| I | 0.53 (1%) | 15.95 (19%) | 32.40 (38%) | 22.12 (26%) | 14.00 (16%) | 85.00 (100%) | 82.20 |
| II | 0.54 (0.5%) | 35.50 (28.5%) | 43.70 (35%) | 31.70 (25%) | 14.00 (11%) | 125.40 (100%) | 118.70 |
| III | 1.09 (0.5%) | 205.37 (47.5%) | 150.70 (35%) | 56.15 (13%) | 18.70 (4%) | 431.80 (100%) | 535.00 |
| *Second year* | | | | | | | |
| I | 0.48 (0.5%) | 11.52 (16%) | 18.88 (26%) | 26.60 (37%) | 14.80 (20.5%) | 72.30 (100%) | 82.20 |
| II | 0.85 (0.5%) | 28.27 (20%) | 52.53 (37%) | 41.56 (29.5%) | 18.00 (13%) | 141.20 (100%) | 188.70 |
| III | 2.45 (1.5%) | 189.00 (40.5%) | 192.00 (41.5%) | 59.29 (12.5%) | 18.50 (4%) | 461.20 (100%) | 535.00 |
| *Third year* | | | | | | | |
| I | 0.48 (0.5%) | 6.64 (8%) | 24.0 (29.5%) | 36.86 (45%) | 14.00 (17%) | 82.00 (100%) | 82.20 |
| II | 1.01 (1%) | 15.20 (12.5%) | 44.5 (36.5%) | 49.15 (40%) | 12.00 (10%) | 121.90 (100%) | 118.70 |
| III | 1.44 (0.5%) | 179.73 (32.5%) | 193.30 (35%) | 137.82 (25%) | 38.00 (7%) | 550.30 (100%) | 535.00 |

F1 – exchangeable; F2 – carbonate; F3 – Fe–Mn oxide; F4 – organic; F5 – residue; values in brackets represent percentage of element in each form; the sums of forms were rounded to 100%; levels of added Pb: I – 50, II – 100, III – 500 $\mu g\ g^{-1}$.

bioavailable (El-Kherbawy and Sanders, 1984; Sanders *et al.*, 1986). Increasing soil pH redistributed Cu form in the exchangeable and organic fractions to water-soluble, Mn-oxide and Fe-oxide fractions. This finding was corroborated by Sims (1986), who reported that lime additions reduced exchangeable and sorbed Cu and increased Cu in the more tightly bound fractions.

Sposito *et al.* (1982) found that Cu added in sewage sludge to arid-zone soils increased the carbonate fraction with time. The results of this three year study were consistent with those of Emmerich *et al.* (1982) where little change in Cu forms occurred with time; furthermore the increasing trend of exchangeable Cu form with time in this study is also consistent with the observations by Silivera and Sommers (1977).

Lead occurred mainly in the residual form in the control soil concomitant with relatively high proportions in the organic and oxide forms. Carbonate and exchangeable forms were of minor importance (Table 2). As with copper carbonate, addition of lead carbonate increased the proportion of the carbonate and oxide forms (Table 2) by several times in relation to the control. The highest increases of carbonate and oxide forms of Pb were observed with the maximum level of lead carbonate (up to 205, 189, 179 and 150, 192, 193 $\mu g\ g^{-1}$ for carbonate and oxide Pb in each year, respectively). As expected, the levels of Pb in the carbonate form decreased during the last year as a result of soil acidification from fertilization. Metabolic products of plant roots ($H_2CO_3$, organic acids), which are secreted into soil during plant growth, could lower the pH of the rhizosphere, thereby enhancing dissolution of metals from the carbonates. Hence, the carbonate fraction can be regarded as a pool of potentially phytoavailable metals (Xian, 1989). Higher amounts of Pb were found also in the organic fraction, especially during the third year in treatments with first and second doses of lead carbonate. In another polluted soil having a total amount of Pb at 230 $\mu g\ g^{-1}$, 43% of the total Pb was in the organic form (Xian, 1989). This can be attributed to lead's strong affinity to humic acid (Takenaga and Aso, 1975), responsible for forming stable Pb–organic complexes. The exchangeable form of Pb did not change much with time in treated soils (Table 2). Low concentrations of the exchangeable Pb in soils explain its low mobility and phytoavailability in the soil environment.

### *3.2 Plants*

Concentrations of Cu in straw and grain of barley from treated soils were significantly higher than those from the control soils (Table 3). However, these values are still lower than or comparable with natural background values. During the first year the low uptake of Cu by barley was not surprising in view of the low levels of this element in the exchangeable form. Consequently there was poor correlation between the plant and the soil values (Table 4). Generally, the levels of Cu between the grain and straw were similar, an indication that this element was translocated easily to the grain tissue from the vegetative parts. Barley yield apparently was not adversely affected by Cu addition (Table 5).

In the second year, significantly higher levels of exchangeable Cu in soil were observed, concomitant with the higher uptake of this element by maize (Tables 1 and

*Table 3.* Cu and Pb content ($\mu g\ g^{-1}$) of barley and wheat; mean values of four replicates with relative standard deviation < 20%.

| Treatment | Cu – straw | Cu – grain | Pb – straw | Pb – grain |
|---|---|---|---|---|
| *Barley* | | | | |
| C | $3.7^a$ | $4.7^a$ | $2.9^a$ | $0.74^a$ |
| I | $6.9^b$ | $7.5^b$ | $3.2^a$ | $0.87^a$ |
| II | $7.3^{bc}$ | $7.6^b$ | $3.3^a$ | $0.85^a$ |
| III | $8.4^c$ | $9.6^c$ | $6.7^b$ | $1.20^b$ |
| *Wheat* | | | | |
| C | $2.8^a$ | $2.5^a$ | $2.95^a$ | $0.94^a$ |
| I | $8.1^b$ | $9.1^b$ | $2.45^a$ | $1.02^a$ |
| II | $18.5^c$ | $11.8^b$ | $2.15^a$ | $1.15^a$ |
| III | $19.0^c$ | No grain | $2.80^a$ | $1.25^a$ |

C – control soil; I, II, III – levels of added metals as indicated in Tables 1 and 2; means followed by different letters are significantly different at $P < 0.05$ (Tukey's multiple range test).

*Table 4.* The correlation between the concentration ($\mu g\ g^{-1}$) in barley, maize, and wheat parts and in the soil.

| Plant part | Cu – T | Cu – E | Pb – T | Pb – E |
|---|---|---|---|---|
| *Barley* | | | | |
| Straw | 0.73 | 0.81 | 0.96* | 0.91* |
| Grain | 0.85 | 0.89* | 0.95* | 0.94* |
| *Maize* | | | | |
| Old leaves | 0.92* | 0.88 | 0.93* | 0.84 |
| Young leaves | 0.71 | 0.65 | 0.66 | 0.43 |
| Shoots | 1.00** | 1.00** | 0.92* | 0.96* |
| Grain | 0.96* | 0.95* | 0.28 | 0.02 |
| *Wheat* | | | | |
| Straw | 0.71 | 0.70 | 0.23 | –0.26 |
| Grain | 0.80 | 0.78 | 0.87 | 0.86 |

* and ** represent the correlation coefficients between the metal concentrations in plant parts and soil at $P < 0.05$ and 0.01; T – refer to the soil; E – refer to the exchangeable form.

*Table 5.* Yield of barley, maize, and wheat (g/pot); mean of four replicates with relative standard deviation < 20%.

| Treatment | Barley | Maize | Wheat |
|---|---|---|---|
| Control | $23.5^a$ | $119.3^a$ | $42.0^a$ |
| Cu1 | $22.5^a$ | $126.9^a$ | $35.0^a$ |
| Cu2 | $22.4^a$ | $131.4^a$ | $35.0^a$ |
| Cu3 | $26.4^a$ | $24.0^b$ | $3.5^b$ |
| Pb1 | $23.4^a$ | $138.1^a$ | $38.0^a$ |
| Pb2 | $19.1^a$ | $118.2^a$ | $37.0^a$ |
| Pb3 | $15.9^b$ | $133.0^a$ | $43.0^a$ |

Copper doses ($\mu g\ g^{-1}$): Cu1 – 50, Cu2 – 100, Cu3 – 500; lead doses ($\mu g\ g^{-1}$): Pb1 – 50, Pb2 – 100, Pb3 – 500; means followed by different letters are significantly different at $P < 0.05$.

6). The highest amount of Cu was found for the young leaves and shoots of maize at the highest dose treatment. Copper concentrations in maize shoots and in the soil, for both total and exchangeable Cu, correlated very well (Table 4). From regression equation ($y = 1.30 + 0.56x$; $r^2 = 99.2^{***}$; $P < 0.01$) it can be deduced that for a unit increase of exchangeable Cu in soil, a corresponding 0.56 unit increase of this element in plant tissue can be expected. Significant linear relationships between Cu concentrations in shoots and in the soil indicate that the shoot values could serve as a predictor for the total Cu level in soil ($r^2 = 100^{***}$; $P < 0.01$).

The yield of maize was not affected by the first and second dose of Cu (Table 5). However, the third dose reduced total yield significantly. This dose reduced shoot yield by about 90%, old leaves by about 82%, and young leaves by about 20% based on the control treatment. Due to the poor growth effected by the highest dose, no grain yield was obtained. Severe contamination of soil by Cu, could adversely affect plant yield (Walsh *et al.*, 1972)

*Table 6.* Cu and Pb content ($\mu g\ g^{-1}$) of maize; mean values of four replicates with relative standard deviation < 20%.

| Treatment | Old leaves | Young leaves | Shoots | Grain |
|---|---|---|---|---|
| *Copper* | | | | |
| C | $2.4^a$ | $1.8^a$ | $0.6^a$ | $1.6^a$ |
| I | $5.5^b$ | $8.7^b$ | $2.5^b$ | $3.6^b$ |
| II | $6.6^b$ | $16.9^c$ | $3.7^c$ | $3.5^b$ |
| III | $10.0^c$ | $17.7^c$ | $15.0^d$ | No grain |
| *Lead* | | | | |
| C | $8.06^a$ | $4.25^a$ | $0.90^a$ | $0.37^a$ |
| I | $7.40^a$ | $3.65^a$ | $1.25^a$ | $0.25^a$ |
| II | $8.55^a$ | $4.00^a$ | $1.70^{ab}$ | $0.30^a$ |
| III | $10.25^a$ | $4.50^a$ | $2.30^b$ | $0.35^a$ |

C – control soil; I, II, III – levels of added metals as indicated in Tables 1 and 2; means followed by different letters are significantly different at $P < 0.05$ (Tukey's multiple range test).

Copper concentrations in straw and grain of wheat planted in the third year were higher than for the barley (Table 5). Of the three crops tested, wheat appeared to be the most sensitive to Cu. The highest dose virtually killed the wheat seedlings after two weeks of growth, but with the first and the second doses the yield of wheat was reduced by about 20% compared with the control plant.

Copper accumulated in the straw and grain of cereals, but in maize the highest accumulation was observed in young leaves. These results show that Cu was effectively transported among the various plant parts and further corroborates previous observations (Adriano, 1986; Walsh *et al.*, 1972).

In contrast with Cu, the addition of increasing levels of lead carbonate to the soil hardly changed the concentrations of this metal in plant tissue (Table 3). This was not unexpected since the exchangeable form of this metal in soil was negligible (Table 2). However, the treatment with highest dose of lead carbonate (Pb – 500 mg $kg^{-1}$ soil) caused significant accumulation of Pb in straw and grain of barley (Table 5). It is generally known that Pb is fairly immobile in soil and therefore unavailable for plant root uptake even in heavily Pb polluted soils (Davis and Carlton-Smith, 1980; Thornton and Timothy, 1984). The second and third doses of Pb significantly reduced the yield of barley (Table 5), while high levels of Pb in soil may not cause direct phytotoxicity. Such levels may directly injure root tissues, altering plant nutrition (Pruszynski and Wozny, 1985). With maize, increased concentrations of Pb were produced by the highest lead carbonate dose (Table 6), where concentrations in the shoots were twice as high as in the control plant shoots. However, maize yield was not adversely affected even by the highest dose (Table 5).

Interestingly, the concentrations of Pb in wheat tissues were virtually unchanged by any of the treatments. Hence the yield was not affected by the treatments (Table 5). Because the bioavailable forms of Pb in soils are usually negligible, producing low uptake, correlations between soil levels and plant concentrations were poor (Table 5).

## 4. Summary

Generally, the residual form of metals in soils is the most predominant in non-polluted soils. The distribution of metals among the various forms largely depends on the dominant soil constituent and type of metals. Large fraction of metals might be complexed by organic matter compounds, especially with Cu as indicated in this study. The exchangeable form almost invariably represents the least abundant form of metal in terms of percentage distribution, but represents one of the important forms for plant nutrition. This is the case with Cu in this study, where the least concentration in the exchangeable fraction occurred. However, exchangeable Cu increased with time in this study which is reflected by increased Cu levels in plant tissues. On the other hand, the exchangeable form of Pb in soil did not change with time, causing unchanged levels in plant tissues.

The differing plant response to the excesses of Cu and Pb were apparently related to the mode of metal occurrence in the soil. Soil contamination resulted in an increase in bioavailable forms of Cu and to a much lesser degree of Pb, which was reflected by enhanced metal uptake by the plants. In contrast, both the exchangeable form and plant concentration of Pb did not change much after soil contamination. Additions of the copper and lead carbonates to the soil affected the yields of plants. In the first year, only the highest dose of lead carbonate added significantly reduced the yield of barley. This yield depression may be due to root injury rather than excess Pb uptake. The highest doses of copper carbonate reduced yield of maize and wheat drastically. Lead carbonate did not have any effect on yield of the wheat and maize. The presented data are another strong confirmation of the hypothesis that phytoavailability of metals largely depends on their form in soils.

## Acknowledgment

The author acknowledges financial assistance through award number DE-FC09-96SR18546 from the U.S. Department of Energy to the University of Georgia Research Foundation.

## References

Adriano D C 1986 Trace Elements in the Terrestrial Environment. Springer-Verlag, New York. 517 p.

Calvet R, Bourgeois S and Msaky J J 1990 Some experiments on extraction of heavy metals present in soil. J. Environ. Anal. Chem. 39, 31–45.

Chlopecka A 1993 Forms of trace metals from inorganic sources in soils and amounts found in spring barley. Water Air Soil Pollut. 69, 127–134.

Chaney R L 1988 Metal speciation and interaction among elements affect trace element transfers in agricultural and environmental food-chains. In Metal Speciation: Theory, Analysis and Application. Eds J R Kramer and H E Allen. pp 219–260. Lewis Publishers, Chelsea, MI.

Davis R D and Carlton-Smith C 1980 Crops as Indicators of the Significance of Contamination of Soil by Heavy Metals. Technical Report TR 140, Stevenage Laboratory Water Research Center, Stevenage, 44 p.

Dudka S and Chlopecka A 1990 Effect of solid phase speciation on metal mobility and phytoavailability in sludge-amended soil. Water Air Soil Pollut. 51, 153–160.

Emmerich W E, Lund L J, Page A L and Chang A C 1982 Solid phase forms of heavy metals in sewage sludge-treated soils. J. Environ. Qual. 11, 178–181.

El-Kherbawy M I and Sanders J R 1984 Effect of pH and phosphate status of a silty clay loam on manganese, zinc, and copper concentrations in soil fractions and in clover. J. Sci. Food Agric. 35, 733–739.

Gupta S K and Chen K Y 1975 Partitioning of trace metals in selective chemical fractions of near shore sediments. Environ. Lett. 10, 129–158.

Gzyl J 1990 Lead and cadmium contamination of soils and vegetables in the Upper Silesia region of Poland. Sci. Total Environ. 96, 199–209. Agric. Scand., Sect. B. 42, 88–93.

Jones K C 1990 Atmospheric deposition as a source of heavy metals and organic contaminants to agroecosystemls. In Contaminated Soil '90. Eds F Arendt, M Hinsenveld and W J van den Brink. pp 189–196. Kluwer Academic Publishers, The Netherlands.

Kuo S, Heilman P E and Baker A S 1983 Distribution and forms of copper, zinc, cadmium, iron, and manganese in soils near a copper smelter. Soil Sci. 135, 101–109.

McLaren R G and Crawford D V 1973 Studies on soil copper. I. The fractionation of copper in soils. J. Soil Sci. 24, 172–181.

Miller W P, McFee W W and Kelly J M 1983 Mobility and retention of heavy metals in sandy soils. J. Environ. Qual. 12, 579–584.

Pruszynski R and Wozny A 1985 The reaction of lupin roots on the presence of lead in the medium. Biochem. Physiol. Pflanzen. 180, 309–318.

Ramos L, Hernandez L M and Gonzalez M J 1994 Sequential fractionation of copper, lead, cadmium and zinc in soils from near Donana National Park. J. Environ. Qual. 23, 50–57.

Roszyk E and Szerszen L 1988 Accumulation of heavy metals in the arable layer of soils of the sanitary protection zone in the vicinity of copper metallurgic plants. Part I. 'Legnica'. Rocz. glebozn 4, 135–146 (in Polish).

Rutherford G K and Bray C R 1979 Extent and distribution of soil heavy metal contamination near a nickel smelter at Coniston, Ontario. J. Environ. Qual. 8, 219–222.

Sanders J R, Adams T M and Christensen B T 1986 Extractability and bioavailability of zinc, nickel, cadmium and copper in three Danish soils sampled 5 years after application of sewage sludge. J. Sci. Food Agric. 37, 1155–1164.

Singh J P, Karwasra P S and Singh M 1988 Distribution and forms of copper, iron, manganese, and zinc in calcareous soils of India. Soil Sci. 148, 359–366.

Shuman L M 1985 Fractionation method for soil microelements. Soil Sci. 140, 11–12.

Shuman L M 1991 Chemical Forms of Micronutrients in Soils. In Micronutrients in Agriculture. 2nd ed. Eds J J Mortvedt, F R Cox, L M Shuman and R M Welch. pp 113–144. SSSA Books Series, Madison, WI.

Silviera D J and Sommers L E 1977 Extractability of copper, zinc, cadmium, and lead in soils incubated with sewage sludge. J. Environ. Qual. 6, 47–52.

Sims J T 1986 Soil pH effects on the distribution and plant availability of manganese, copper, and zinc. Soil Sci. Soc. Am. J. 50, 367–373.

Spevackova V and Kucera J 1989 Trace element speciation in contaminated soils studied by atomic absorption spectrometry and neutron activation analysis. Int. J. Environ. Anal. Chem. 35, 241–251.

Sposito G, Lund L J and Chang A C 1982 Trace metal chemistry in arid-zone field soils amended with sewage sludge: I. Fractionation of Ni, Cu, Zn, Cd, and Pb in solid phases. Soil Sci. Soc. Am. J. 46, 260–264.

Takenaga H and Aso S 1975 Studies on the physiological effect of humic acid (part 9). Stability constants of cation-nitrohumic acid chelates. J. Soil Sci. Plant Nutri. 46, 349–354.

Tessier A, Campbell P G C and Bisson M 1979 Sequential extraction procedure for the speciation of particulate trace metals, Anal. Chem. 7, 844–851.

Thornton I and Timothy H J 1984 Sources of lead and associated metals in vegetables grown in British Urban Soils: Uptake from the soil versus air deposition. In Trace Substances in Environmental Health – XVIII. Ed. D D Hemphill. pp 303–310. University of Missouri.

Walsh L M, Erhart W M and Seibel H D 1972 Copper toxicity in snapbeans (*Phaseolus vulgaris* L.). J. Environ. Qual. 1, 197–200.

Xian X 1989 Effect of chemical forms of cadmium, zinc, and lead in polluted soils on their uptake by cabbage plants. Plant Soil 113, 257–264.

39

# Levels of copper, zinc and manganese in the vineyard soils of Dão wine region – Portugal

R.S. DIAS AND J.C. SOVERAL-DIAS

## 1. Introduction

Portugal has a long tradition of wine production, this culture representing an important agricultural sector. Due to long term use of copper fungicides in vineyards, high concentrations of copper are reported in agricultural soils of several Portuguese wine regions, some of them having a significant proportion of old vineyards.

As copper is not easily leached out and is absorbed in small quantities by the culture, it accumulates in the soil surface layer (Delas, 1963). Due to the relative tolerance of the adult vine plant to high copper levels (Anne and Dupuis, 1953), phytotoxicity develops, or is more likely to evolve, in young plants, after vine root-up. This practice has been encouraged in the last years by EU measures, in order to plant new vineyards or change to other crops, hence improving the viticulture competitivity.

In Portugal there are scarce data on contamination of agricultural soils with heavy metals such as copper. Knowledge of the extent of Cu contamination is therefore desirable. It is also of interest to determine manganese and zinc levels in soil, as these micronutrients are included in the composition of some fungicides.

## 2. Objective

The aims of this study were: first, to assess the degree of copper contamination and the vertical distribution of this element in vineyard soils of Dão wine region, in central Portugal; and second to evaluate the manganese and zinc contents of these soils, as well as their distribution throughout the soil profile.

The Dão wine region, delimited in 1908, covers an area of 376 000 ha (Figure 1). Approximately 14 400 ha are exploited with vineyards. The region is quite homogeneous from the lithological point of view, granite being the dominant stone. Ninety-seven per cent of the vineyard soils are derived from granite and vary from humic to dystric Cambisols, according to the FAO soil classification (Loureiro and Cardoso, 1993). They have a light texture, medium acid reaction and a low to medium organic matter content. The type of soil and the age of the vineyards (60% of them older than 60 years) contribute to a higher risk for the development of copper toxicity.

*D. Rosen* et al. *(eds): Modern Agriculture and the Environment, 469–477.*

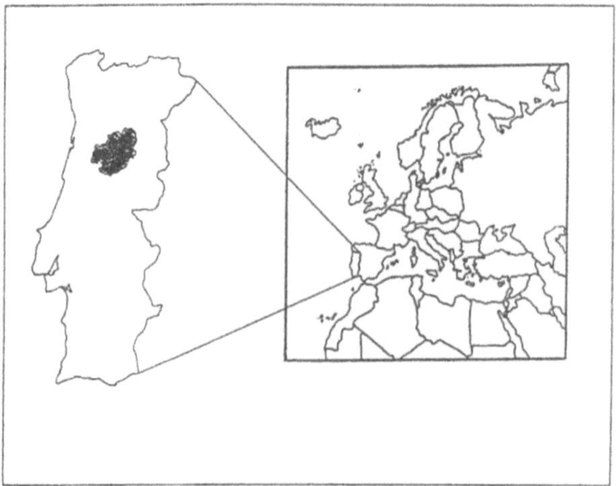

*Figure 1.* Localisation of Dão wine region in Portugal.

## 3. Materials and methods

### *3.1 Soil sampling*

Soil was sampled in 44 vineyards dispersed throughout the Dão wine region, during spring–summer 1992. The following aspects were considered: the vineyard age; the distribution of the vineyard area in the several municipalities of this region; and the soil reaction (we decided to sample only acid soils, which are largely dominant in the region, as in these soils there is a higher risk for the development of copper toxicity).

In each vineyard, soil was sampled at four depths (0–20, 20–40, 40–60 and 60–80 cm). Two soil composite samples were taken per depth, each one of them composed of 20 subsamples. The soil was air dried and sieved through a 2-mm stainless steel mesh.

### *3.2 Analytical methods*

The soil parameters analysed were:

- Total Cu, Mn and Zn, by soil digestion with *aqua regia* (4:1 by vol. solution of 36% by wt HC1: 70% by wt $HNO_3$), followed by atomic absorption spectrophotometry (McGraph and Cunliffe, 1985).
- Extractable Cu, Mn and Zn in a 5 g: 50 $cm^3$ soil: solution of ammonium acetate 0.5M + acetic acid 0.5M + EDTA 0.02M, followed by atomic absorption spectophotometry (Lakanen and Erviö, 1971).
- pH ($H_2O$) and pH ($CaCl_2$), potentiometrically in a 1:2.5 soil:water and soil:$CaCl_2$ 0.01M suspension, respectively, after one hour of contact.

*Table 1.* Copper concentration in the vineyard soils.

| Depth (cm) | Variable | Extractable Cu (mg $kg^{-1}$) | Total Cu (mg $kg^{-1}$) | Extractable Cu/total Cu |
|---|---|---|---|---|
| 0–20 | $x \pm s_e$ | 30.4 ± 4.14 | 93 ± 10.8 | 0.30 ± 0.015 |
| | Median | 22.4 | 85 | 0.31 |
| | Range | 1.7–113.0 | 10–324 | 0.07–0.51 |
| 20–40 | $x \pm s_e$ | 12.1 ± 2.25 | 37 ± 4.7 | 0.25 ± 0.021 |
| | Median | 5.7 | 26 | 0.22 |
| | Range | 0.6–63.8 | 6–165 | 0.05–0.57 |
| 40–60 | $x \pm s_e$ | 3.8 ± 0.67 | 21 ± 1.8 | 0.11 ± 0.014 |
| | Median | 1.8 | 17 | 0.09 |
| | Range | 0.1–20.0 | 4–54 | 0.01–0.45 |
| 60–80 | $x \pm s_e$ | 2.3 ± 0.42 | 17 ± 1.8 | 0.11 ± 0.010 |
| | Median | 1.3 | 13 | 0.11 |
| | Range | 0.1–13.4 | 2–50 | 0.01–0.48 |

$n = 43$; $x \pm s_e$ = mean ± standard error of the mean.

- Total organic carbon (Tinsley, 1950).
- Texture (Silva *et al.*, 1975).

Results presented are the average of the two samples.

## 4. Results and discussion

The sampled soils were dominantly of light texture (loamy sand), medium acid reaction (average pH ($H_2O$) = 5.4 and pH ($CaCl_2$) = 4.5) and with a medium (seldom low to very low) organic matter content. Only the micronutrient content will be discussed.

### *4.1 Copper content in topsoil*

Half of the vineyards investigated had a total Cu content higher than 85 mg $kg^{-1}$, showing the existence of Cu contamination (Table 1). Similarly, the average extractable Cu content indicates a high availability of this element to the plant. The natural total Cu concentration varies from 1–50 mg $kg^{-1}$ (Swaine, 1962) and is usually lower (around 10 mg $kg^{-1}$) in soils derived from granite (Katyal and Randhawa, 1983). In fact, two of the younger vineyards had a total Cu content of 10 and 13 mg $kg^{-1}$.

One of the vineyards was eliminated from the statistical analysis as an outlier, due to a known history of cupric fungicide application, in excess to what is usual in the region, leading to 351 mg $kg^{-1}$ of soil total Cu content.

In Figure 2 the frequency distribution of total Cu for the topsoil is shown. It can be seen that 43% of the vineyards investigated have a total Cu content higher than 100 mg $kg^{-1}$.

### *4.2 Copper distribution with depth*

Results show clearly the accumulation of Cu at the soil surface layer (Table 1 and Figure 3). In fact, even in well drained and slightly acid soils Cu has a reduced mobility, due to its high affinity to soil colloids.

The extractable level of Cu exhibited a similar trend to the total level, but extractability decreased with increasing depth.

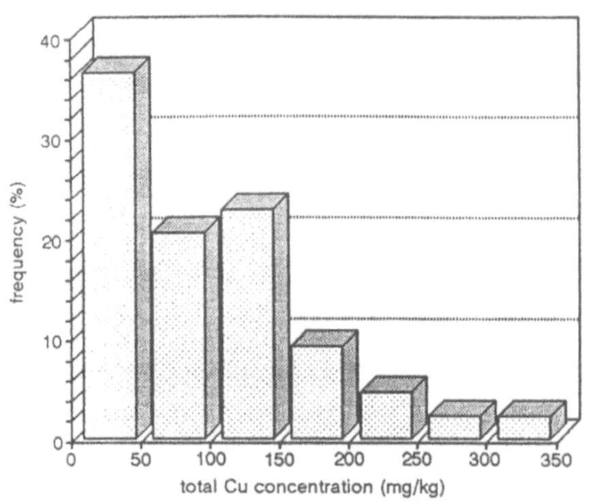


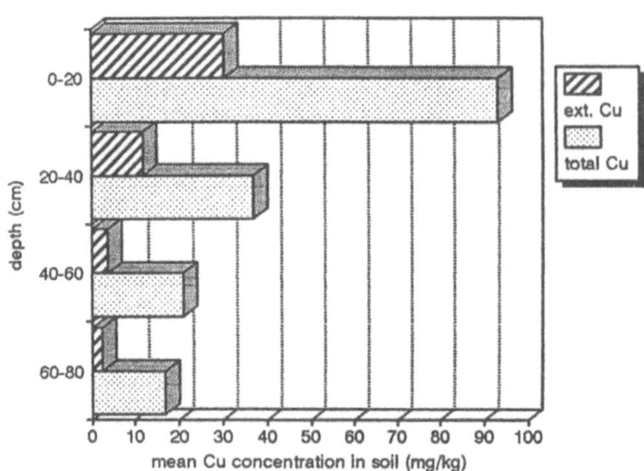


*Figure 3.* Copper distribution with depth.

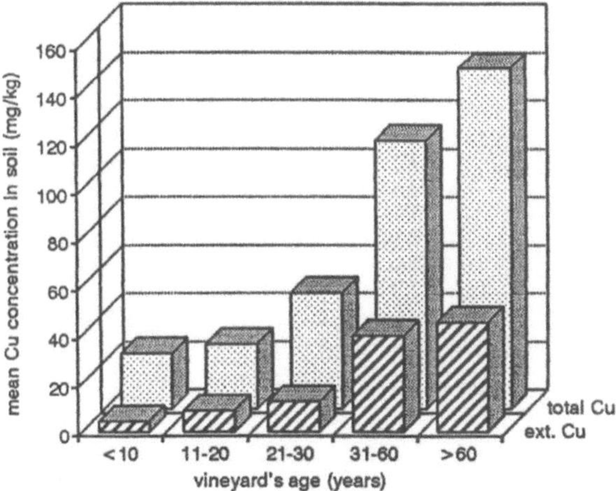


*Figure 4.* Relationship between copper concentration in soil and the age of the vineyard.

## *4.3 Copper variation with the age of the vineyard*

As expected, Cu concentration increases with the age of the vineyard (Figure 4 and Table 2), and in those older than 30 years it was in average higher than 100 mg $kg^{-1}$, although varying with the specific history of fungicide application.

*Table 2.* Relationship between copper concentration in soil and the age of the vineyard.

| Vineyards *n* | (%) | Age (years) | Depth (cm) | Extractable Cu[a] (mg $kg^{-1}$) | Total Cu[a] (mg $kg^{-1}$) |
|---|---|---|---|---|---|
| 5 | 12 | ≤ 10 | | 4.3 ± 1.15 | 23 ± 2.7 |
| 7 | 17 | 11–20 | | 8.9 ± 1.83 | 27 ± 2.3 |
| 5 | 12 | 21–30 | 0–20 | 12.3 ± 4.25 | 48 ± 15.2 |
| 9 | 21 | 31–60 | | 39.9 ± 10.48 | 111 ± 21.6 |
| 16 | 38 | > 60 | | 45.3 ± 5.66 | 141 ± 15.3 |
| | | ≤ 10 | | 2.2 ± 0.81 | 19 ± 3.1 |
| | | 11–20 | | 2.5 ± 0.70 | 14 ± 1.7 |
| | | 21–30 | 20–40 | 3.9 ± 1.85 | 21 ± 6.4 |
| | | 31–60 | | 15.7 ± 4.97 | 45 ± 9.1 |
| | | > 60 | | 18.4 ± 4.34 | 52 ± 9.2 |
| | | ≤ 10 | | 1.9 ± 0.75 | 17 ± 2.8 |
| | | 11–20 | | 1.1 ± 0.20 | 13 ± 1.8 |
| | | 21–30 | 40–60 | 2.4 ± 1.40 | 21 ± 6.9 |
| | | 31–60 | | 4.6 ± 1.66 | 24 ± 4.3 |
| | | > 60 | | 5.4 ± 1.36 | 24 ± 3.2 |
| | | ≤ 10 | | 1.5 ± 0.67 | 16 ± 2.9 |
| | | 11–20 | | 1.0 ± 0.24 | 9 ± 1.8 |
| | | 21–30 | 60–80 | 1.8 ± 1.21 | 17 ± 7.0 |
| | | 31–60 | | 2.2 ± 0.85 | 18 ± 3.9 |
| | | > 60 | | 3.4 ± 0.87 | 20 ± 3.4 |

[a] Results expressed as mean ± standard error of the mean.

*Table 3.* Manganese and zinc concentrations in the vineyards soils.

| Depth (cm) | Variable | Extractable Mn (mg kg$^{-1}$) | Total Mn (mg kg$^{-1}$) | Extractable Zn (mg kg$^{-1}$) | Total Zn (mg kg$^{-1}$) |
|---|---|---|---|---|---|
| 0–20 | $x \pm s_e$ | 16.6 ± 1.35 | 297 ± 17.8 | 3.4 ± 0.39 | 76 ± 2.8 |
| | Range | 5.4–49.2 | 111–630 | 0.6–14.5 | 27–119 |
| 20–40 | $x \pm s_e$ | 9.6 ± 1.11 | 291 ± 18.7 | 1.3 ± 0.13 | 70 ± 2.7 |
| | Range | 1.5–39.7 | 90–583 | 0.2–4.0 | 20–102 |
| 40–60 | $x \pm s_e$ | 8.0 ± 0.96 | 297 ± 20.2 | 0.7 ± 0.07 | 65 ± 2.6 |
| | Range | 0.5–35.5 | 75–645 | 0.1–1.9 | 18–98 |
| 60–80 | $x \pm s_e$ | 6.3 ± 0.57 | 315 ± 19.7 | 0.6 ± 0.04 | 72 ± 2.7 |
| | Range | 1.3–13.9 | 71–579 | 0.2–1.3 | 20–102 |

$n = 44$; $x \pm s_e$ = mean ± standard error of the mean.

*Table 4.* Relationship between manganese and zinc concentrations in the topsoil and the age of the vineyard.

| Vineyards *n* | (%) | Age (years) | Extractable Mn[a] (mg kg$^{-1}$) | Total Mn[a] (mg kg$^{-1}$) | Extractable Zn[a] (mg kg$^{-1}$) | Total Zn[a] (mg kg$^{-1}$) |
|---|---|---|---|---|---|---|
| 5 | 12 | ≤ 10 | 6.9 ± 0.59 | 216 ± 32.1 | 1.4 ± 0.32 | 81 ± 7.0 |
| 7 | 17 | 11–20 | 21.6 ± 5.32 | 306 ± 28.7 | 2.9 ± 0.45 | 68 ± 5.7 |
| 5 | 12 | 21–30 | 19.8 ± 4.68 | 374 ± 70.6 | 3.4 ± 1.52 | 82 ± 9.6 |
| 9 | 21 | 31–60 | 17.1 ± 2.01 | 271 ± 31.0 | 4.2 ± 1.22 | 71 ± 8.7 |
| 16 | 38 | > 60 | 16.2 ± 1.74 | 307 ± 34.6 | 3.6 ± 0.53 | 78 ± 3.3 |

[a] Results expressed as mean ± standard error of the mean.

### *4.4 Manganese and zinc content in topsoil*

Mn and Zn concentrations in the soils investigated are in the range of non-contaminated soils derived from granite (Table 3). According to Teixeira *et al.* (1981), in those soils the average natural concentrations of Mn and Zn are around 330 and 85 mg kg$^{-1}$, respectively.

### *4.5 Manganese and zinc distribution with depth*

There was no significant ($p < 0.05$) vertical variation of total Mn and Zn concentrations with depth (Table 3 and Figures 5 and 6). Concerning the extractable levels, it can be seen that there is a significantly ($p < 0.05$) higher concentration in the soil upper layer. This is related to several factors, namely the higher organic matter content as well as a higher microbial activity at the soil surface (Sillampää, 1972).

### *4.6 Manganese and zinc variation with the age of the vineyard*

There are no significant ($p < 0.05$) increases of total Mn and Zn concentrations with the age of the vineyards (Table 4 and Figures 7 and 8). Yet, the extractable level of

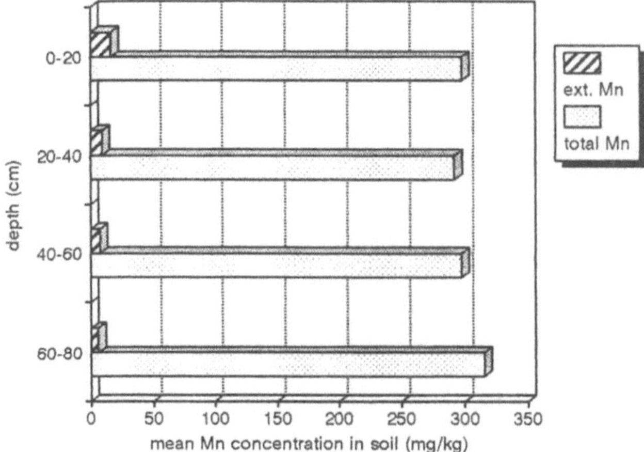


*Figure 5.* Manganese distribution with depth.

both micronutrients is significantly ($p < 0.05$) lower in the younger vineyards, not differing significantly in the other classes.

## 5. Conclusions

The total copper content of 43% of the vineyards investigated was higher than 100 mg $kg^{-1}$ in the soil upper layer (0–20 cm). Also, 57% of the vineyards have an extractable copper level above 20 mg $kg^{-1}$. These numbers show the existence of copper contamination.

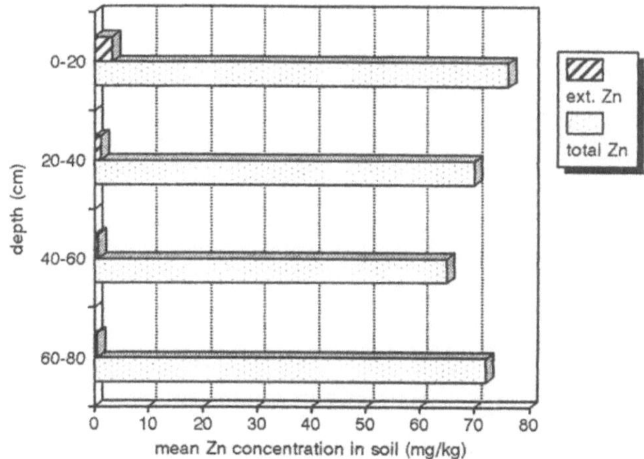


*Figure 6.* Zinc distribution with depth.

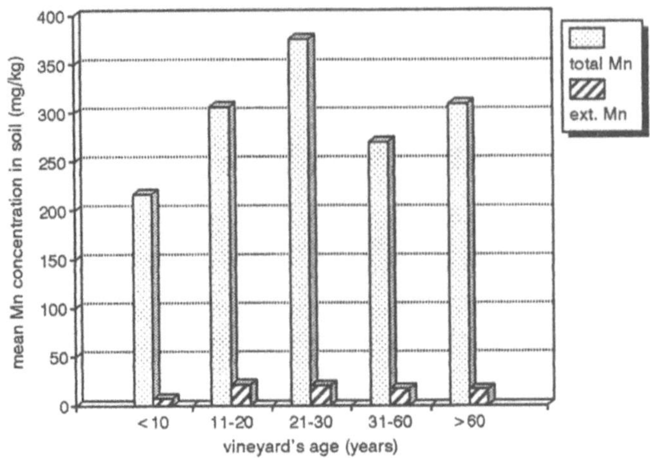


*Figure 7.* Relationship between manganese concentration in soil and the age of the vineyard.

There is a strong accumulation of copper at the soil surface layer.

Copper concentration increases with the age of the vineyard. In those older than 30 years (representing more than 80% of the vineyards in the region), average total and extractable copper are 111 and 40 mg $kg^{-1}$, respectively. These levels are considered high, especially taking in consideration the low buffer capacity of those soils.

The development of copper phytotoxicity is quite probable in the studied region, particularly when non-tolerant species are grown in former old vineyard soils.

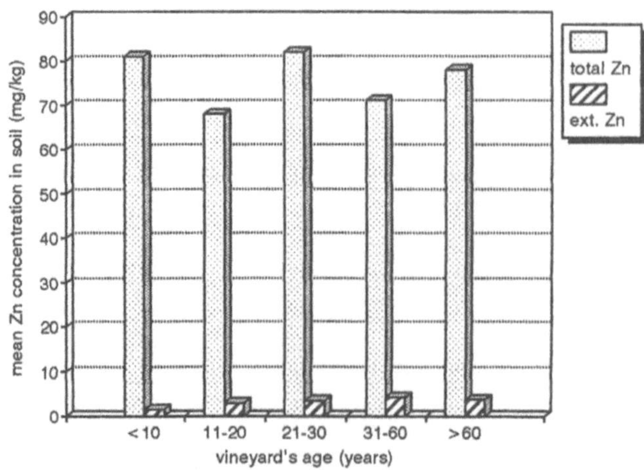


*Figure 8.* Relationship between zinc concentration in soil and the age of the vineyard.

Attention must be paid to this, by the adoption of agricultural techniques (namely liming) that effectively decrease copper availability.

The average total manganese and zinc concentrations in the topsoil are 297 and 76 mg $kg^{-1}$, respectively, which are in the range of non-contaminated soils.

## Acknowledgements

The authors are grateful to Eng. R. Oliveira, Eng. V. Pedroso and Eng. C. Pacheco for their collaboration in the field work, as well as to 'LQARS' staff who carried out the analytical work. They also thank Eng. M.F. Calouro and C. Sempiterno for their valuable advice. This work is partly sponsored by the Portuguese Junta Nacional de Investigação Cientifica e Tecnológica (JNICT).

## References

Anne P and Dupuis M 1953 Toxicité du cuivre a l'egard de quelques plantes cultivées. C.R. Acad. Agric. de France 39, 58–61.

Delas J 1963 La toxicité du cuivre accumulé dans les sols. Agrochimica 7, 258–288.

Katyal J C and Randhawa N S 1983 Micronutrients. FAO Fertilizer and Plant Nutrition Bull. 7, 82 p.

Lakanen E and Erviö R 1971 A comparison of eight extractants for the determination of plant available micronutrients in soils. Acta Agr. Fenn. 23, 223–232.

Loureiro V and Cardoso A H 1993 Os vinhos do Dão. Chaves Ferreira-Pub., SA, Lisboa, Portugal, 160 p.

McGraph S P and Cunliffe C H 1985 A simplified method for the extraction of the metals Fe, Zn, Cu, Ni, Pb, Cr, Co and Mn from soils and sewage sludges. J. Sci. Food Agric. 36, 794–798.

Sillampää M 1972 Trace elements in soils and agriculture. FAO Soil Bull. 17, 67 p.

Silva A A *et al.* 1975 Métodos de análise de solos, plantas e águas. Pedologia 10 (3), 79 p.

Swaine D J 1962 Trace element content of fertilizers. Comm. Bur. Soil Sci., UK, Technical Communication No. 52, 306 p.

Teixeira A J *et al.* 1981 Solos derivados de granitos e xistos da região noroeste de Portugal: Micronutrientes totais e extraíveis. Pedologia 16, 1–99.

Tinsley J 1950 The determination of organic carbon in soils by dichromate mixture. Trans. IV Int. Cong. Soil Sci. I, 161–164.

# SECTION VIII

# Bioremediation

## 40
# Bioremediation in agriculture: dream or reality?

Y. HENIS

## 1. Introduction

The development of modern agriculture has made it possible to increase the supply of food to growing populations all over the world. However, this development has been accompanied by an intensive pollution of the environment, in a way and to an extent similar to the effect of the earlier industrialization process on the environment. Pollutants in agriculture can be divided into four main groups: fertilizers; animal waste; fossil fuel and its combustion products; and pesticides. Whereas fertilizers and waste are subjected to the common recycling processes constantly operating in nature, fuel and pesticide pollution comprise long-term, often persistent problems, due to their xenobiosis, recalcitrance and potential toxicity. Removal and/or detoxification of pollutants can be done by physical, chemical, or biological means, the latter being of special interest because of its low cost, and its low adverse effect on the environment. The biological approach of remediating the environment, termed bioremediation, is defined by Atlas and Pramer (1990) as 'the use of biological agents to reclaim soils and waters polluted by substances hazardous to human health and/or the environment'.

In its wider sense, this definition applies to any pollutant and biological process at any site, either *in situ* or in specific purification plants and sites (Alexander, 1994). However, in its narrower sense, bioremediation mainly refers to xenobiotic and toxic pollutants, either synthetic or alien to the environment, and to remediation processes mainly operating *in situ.* Any ecologically harmful chemical which accumulates in nature due to its introduction by man or to unbalanced activity, can be regarded as a pollutant, and its removal by microorganisms can be considered as bioremediation.

There is nothing new in bioremediation: until the appearance of *Homo sapiens* on Earth, nature has taken care of itself by employing microorganisms in the nitrogen, carbon, phosphorus and sulphur cycles *in situ.* It is the unbalanced activity of the human population, accompanied by the uncontrolled release of recalcitrant, xenobiotic and toxic compounds, which raised the need for development of additional means of remediation. In this chapter, only the basic principles and some representative results will be discussed, in an attempt to assess the future feasibility of the biological approach.

*D. Rosen* et al. *(eds): Modern Agriculture and the Environment, 481–489.*

*Table 1.* Fate of pollutants.

| |
|---|
| *Abiotic processes* |
| Washing |
| Adsorption |
| Photochemical degradation |
| Volatilisation |
| *Biotic processes* |
| Complexation |
| Polymerisation |
| Biomagnification |
| • Adsorption |
| • Uptake |
| • Consumption |
| • Predation |
| Biodegradation |
| • Oxygenation |
| • $\beta$-Oxidation |
| • Hydrolysis |
| • Reduction |

### 2. *Interactions*

The pollutant, the microorganisms capable of degrading it, and the environment, mutually interact with each other. For degradation to occur, organisms capable of degrading the pollutant must be present, along with the appropriate environmental conditions that allow for microbial growth and multiplication. In most cases, the pollutant will be degraded while fulfilling at least one function essential for growth and multiplication, such as carbon, nitrogen or energy source. Adequate environmental conditions include the presence of complementary nutrients containing phosphorus, iron, potassium, sulphur, and microelements as well as adequate moisture, pH and temperature (Dalyan *et al.*, 1990 a,b; Harder and Hopner, 1991; Harder *et al.*, 1991; Wurdemann *et al.*, 1990).

No degradation will ocur without the multiplication of the degrading microflora. However, compounds that do not serve as a nutrient or energy source can also be degraded in the presence of other degradable compounds which support growth by serving as a nutrient and energy source. This process, termed cometabolism, plays an important role in the degradation of chlorinated biphenyls (Dean-Ross, 1987; Focht, 1987).

## 3. Fate of pollutants

Pollutants may be affected in the environment by either biotic or abiotic factors (Table 1). In some cases, abiotic factors such as washing, volatilization or adsorption may play a more prominent role in remediation than the biotic ones (Shannon and Unterman, 1993). On the other hand, the specific adsorption of pollutants by biotic systems has been employed for both the removal and recovery of heavy metals in aquatic environments (Sela *et al.*, 1989; Shabtai and Fleminger, 1994). The most

*Table 2.* Factors involved in bioremediation.

| |
|---|
| Presence of degrader(s) |
| Initial abundance of degrader(s) |
| Accompanying microorganisms |
| Multiplication capacity |
| Pollutant's concentration threshold |
| Pollutant's solubility |
| Pollutant's adsorption to microorganisms |
| Pollutant's recalcitrance |
| Pollutant's enzymatic complexing |
| Presence of other toxic agents |
| Presence of nutrients |
| Contamination period (acclimation) |
| Presence of electron acceptors |
| Physical conditions: moisture, temperature, pH, Eh |

important biotic process in bioremediation of organic pollutants is biodegradation, the mechanism depending on the pollutant's chemical nature. Thus, aromatics are degraded by hydroxylation and oxygenation, aliphatics by $\beta$-oxidation (Alexander, 1981; Dragun, 1988; van der Meer *et al.*, 1992), organophosphates and triazine pesticides by hydrolysis (Cook and Hutter, 1984; Cook *et al.*, 1985), and halogenated compounds by reductive dehalogenation (Mohn and Tiedje, 1992). Other biotic processes include complexation, such as the incorporation of chlorophenols into soil organic matter (Bollag, 1992), polymerization, and biomagnification by means of adsorption, uptake, consumption or predation (Alexander, 1994). Biomagnification may play an adverse role in the environment, due to pollutant accumulation throughout the food chain, causing damage to primary, secondary and tertiary consumers, such as in the case of the effect of biomagnified DDT on egg shells of birds (Peakall, 1970). On the other hand, in some cases biomagnification has been employed for bioremediation for the removal and recovery of heavy metals in aquatic environments (Sela *et al.*, 1989; Shabtai and Fleminger, 1994).

## 4. Factors involved

The conditions which allow for bioremediation of a polluted site are listed in Table 2. First, microorganisms capable of degrading the pollutant must be present. Second, their initial abundance is of prime importance, since biotransformation at a prominent rate will not occur unless the degraders reach a certain level, which depends on the degrader, the pollutant and the environment. This is one of the reasons for the requirement for a preceding acclimation period (Wiggins *et al.*, 1987) prior to the initiation of pollutant degradation at a measurable rate. Growth and multiplication are required for the synthesis of the enzymes involved in the biodegradation, for active uptake of those pollutants which are degraded intracellularly, and for the maintenance and preservation of the microorganisms involved in the biodegradation. Since these activities require a certain minimal amount of energy, degradation will not occur unless the pollutant is present at a certain concentration threshold (Rittman and McCarty, 1980 a,b). However, as shown by Bouwer (1984), many halogenated aromatic and aliphatic pollutants are degraded even at a sub-concentration threshold when acetate is added to the degrading culture as a primary substrate, in addition to the pollutant. Another interesting finding of Bouwer was that whereas the degradation of aromatics (either chlorinated or non-chlorinated ones) at sub-concentration thresholds in the presence of acetate as a primary substrate did occur under aerobic conditions, the degradation of chlorinated aliphatics in the presence of acetate occurred under methanogenic conditions only.

Accompanying microorganisms, even those incapable of degrading the pollutant, often play an important role in the process. Thus, protozoa may slow down biodegradation by feeding on the degraders (Wiggins and Alexander, 1988b). Other microorganisms may play a synergistic role in biodegradation by forming a consortium. Thus, the chlorobenzoate dehalogenating *Desulphomonile tiedjii*, the benzoate fermenting rod BZ-2 and the methanogen *Methanospirillum* sp., form a consortium which fully degrades chlorobenzoate (Mohn and Tiedje, 1992). In this consortium, chlorobenzoate is dehalogenated by *D. tiedjii*; the resulting benzoate is fermented by BZ-2 into acetate, $CO_2$ and $H_2$; the acetate is fermented further by *D. tiedjii*, which also obtains vitamins from its partners; the hydrogen and the carbon dioxide are transformed into methane by the methanogen. Another interesting consortium, involved in bioremediation of oil-polluted sites of the Persian Gulf following the blowing-up of oil wells in Kuwait in 1991, was observed by Sorkoh *et al.* (1992), who reported on the formation of consortia consisting of blue green algae and bacteria on oil patches along the ocean beaches south of Kuwait. The blue green algae produced mats which stuck to the oil patches, along with polysaccharides which allowed the bacteria to remain adhered to the oil patches.

The availability of the pollutant as a substrate is affected by its adsorption onto organic and inorganic fractions as well as by its solubility, complexing potential and recalcitrance. Additional carbon and energy sources may be required to optimize the bioremediation process. This is especially true in the case of cometabolism, in which a compound is being degraded without serving as energy, carbon or any other benefactory source for the degrading microflora (Alexander, 1994). Addition of nutrients to polluted sea-water in the Exxon-Valdez Prince William sound clean-up in Alaska has doubled the rate of bioremediation (Atlas and Bartha, 1992; Prince, 1992; Rosenberg and Ron, 1995). The use of oleophilic fertilizers has proved to be of a special value in bioremediation of oil spills of aquatic environments (Prince, 1992). In other cases, the combined addition of microorganisms and oleophilic fertilizers was very effective in the bioremediation of oil-contaminated beaches (Rosenberg and Ron, 1995).

The presence of an appropriate electron acceptor, or, in the case of anaerobic processes, an electron donor, is also of great importance. Oxygen (Michaelson *et al.*, 1992; Piotrowsky, 1991), or oxygen-yielding compounds such as hydrogen peroxide (Piotrowsky, 1991) have been successfully used for *in situ* bioremediation of hydrocarbon-contaminated soil and groundwater (Piotrowsky, 1991) On the other hand, some bioremediation processes, especially of halogenated hydrocarbons, in which the pollutant serves as an electron acceptor, take place under anaerobic conditions (Bouwer, 1984; Kaake *et al.*, 1992; Mohn and Tiedje, 1992; Wilson and Rees, 1985).

## 5. Technological approach

Any attempt to employ bioremediation *in situ* must be preceded by thorough microbial, chemical and physical analyses. According to the results obtained, the polluted environment should be amended to ensure optimal conditions for the degrading microflora. There are two basic technological approaches to bioremediation: the ecological and the seeding approaches (Figure 1), and they should be chosen according to the resulting analyses. If a degrading microflora is present in adequate amounts and/or can be enriched and activated *in situ* in a relatively short time, no inoculation is necessary, and the ecological approach should be preferred. On the other hand, if a degrading microflora is absent or cannot be enriched, the seeding approach is recommended. According to Piotrowsky (1991), one can either use laboratory-grown local microflora (if present), or 'superbugs' developed and grown in the laboratory.

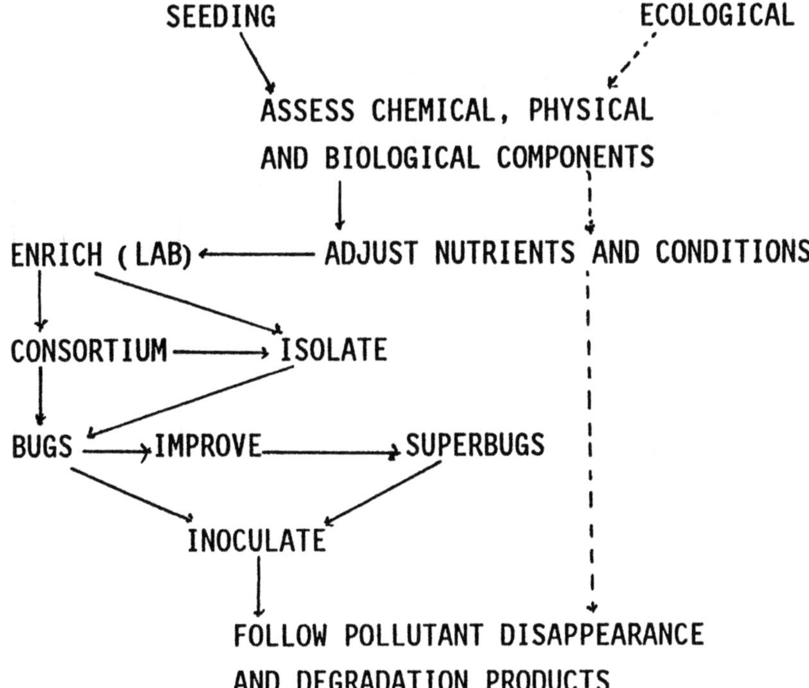


*Figure 1.* Approaches to bioremediation.

*Table 3.* Recorded cases of successful bioremediation.

| Case | Comments | References |
|---|---|---|
| Exxon Valdez spill | Fertilisers added | Alexander (1994); Atlas and Bertha (1992); Raveh *et al.* (1992); Rosenberg and Ron (1995) |
| Alyeska ballast water | Aeration enhanced degradation | Prince (1992) |
| Soil and ground water wood preservatives, Montana | $H_2O_2$ enhanced degradation | Prince (1992) |
| Oil-contaminated beach | Fertilisers and inoculum | Rosenberg and Ron (1995) |

A second question concerns the nature of the degrading microflora: should we use one organism, or a consortium, i.e. a mixed culture obtained by repeated transfers through the enriched contaminated samples? Since preparation of bacterial inoculants is a relatively simple procedure, probably the best solution will be to use a preparation consisting of all the elements mentioned above: nutrients, degraders, and an appropriate electron acceptor/donor.

## 6. Successful and failed treatment cases

There are many reports available on the successful bioremediation of polluted environments. Some representative cases are shown in Table 3. In most cases, the ecological, rather than the seeding approach has been used. The amendments added were either fertilizers, electron acceptors/donors, or both.

The understanding of failed experiments is often as important or even more so, than successful ones, since it enables us to identify mistakes and to avoid them in future operations. Unfortunately, workers are not inclined to report on their own failures. Some rare reports of failures of bioremediation, cited by Alexander (1994), are given in Table 4. In all of these cases, bacterial inoculants capable of degrading pesticides or fuel oil under laboratory conditions, have failed, for unknown reasons, to degrade the pollutant under field conditions.

## 7. Potential inoculants

Some microorganisms having a potential use of inoculants in bioremediation of aquatic and terrestrial environments were isolated and studied in our laboratory. They include a parathion-hydrolysing *Pseudomonas* sp. (Rothschild *et al.*, 1990), three prometryn-degrading isolates of *Bacillus cereus* (Mizrachi *et al.*, 1994), and a formaldehyde-degrading *Halomonas* sp. (Oren *et al.*, 1992). In addition, the degradation of atrazine by the white rot fungus *Pleurotus pulmonarius* (Masaphy *et al.*, 1993), and the reduction of nitroaromatic halophilic anaerobic bacteria (Gurevich *et al.*, 1993), have been demonstrated. *Bacillus cereus* G was shown to degrade prometryn in non-sterile, natural soil (Mizrachi *et al.*, 1994). Parathion-hydrolase, obtained from a parathion-degrading pseudomonad, protected mice from

*Table 4.* *In situ* inoculation: recorded failures[a].

| Compound | Organism and site | Reference[b] |
|---|---|---|
| 4-(2,4-DB)[c] | *Flavobacterium* sp. on alfalfa seeds in soil | MacRae and Alexander (1965) |
| Fuel oil | Aliphatic hydrocarbon degraders added to soil | Lehtomaki and Niemela (1975) |
| 2,4–DCP[d] | *Pseudomonas* sp. added to soil surface or to lake water | Goldstein *et al.* (1985) |
| Metolachlor | *Streptomyces* sp. added to soil | Liu *et al.* (1990) |
| Crude oil | Hydrocarbon-degrading marine consortium | Tagger *et al.* (1983) |

[a] After Alexander (1994).
[b] Cited by Alexander (1994).
[c] 4-(2,4-Dichlorophenoxy)butyric acid.
[d] 2,4-Dichlorophenoxyacetic acid.

the toxic effect of the closely related organophosphate ester, nerve gas tabun (Raveh *et al.*, 1992). The potential use of these organisms in bioremediation has not yet been explored.

## 8. Conclusions

(1) *In situ* bioremediation of sites contaminated with oil spills, pesticides and overused chemicals is possible, and has been successfully employed in aquatic and terrestrial environments.

(2) To ensure a successful bioremediation, treatment should be preceded by the assessment of physical, chemical and biological conditions of the contaminated site, followed by amendment of the conditions whenever possible and necessary.

(3) In most cases, the indigenous microflora alone is capable of carrying out the bioremediation, following an acclimation period.

(4) In the absence of indigenous active microorganisms, the addition of laboratory-grown cultures should be considered.

## References

Alexander M 1981 Biodegradation of chemicals of environmental concern. Science 211, 132–138.
Alexander M 1994 Biodegradation and Bioremediation. Academic Press, San Diego, CA.
Atlas R M and Bartha R 1992 Hydrocarbon biodegradation and oil spill bioremediation. In Advances in Microbial Ecology 12, 287–338.
Atlas R M and Pramer D 1990 Focus on bioremediation. ASM News 56, 352–354.
Bollag J-M 1992 Decontaminating soil with enzymes. Environ. Sci. Technol. 26, 1876–1881.
Bouwer E J 1984 Biotransformation of organic micropollutants in the subsurface. Proc. NWWA/1401

Conf. on Petroleum Hydrocarbons and Organic Chemicals in Ground Water Prevention, Detection and Restoration, Houston, TX, pp 66–81. National Water Well Assoc., Ohio.
Cook A M and Hutter R 1984 Deethylsimazine: Bacterial dechlorination, deamination and complete degradation. J. Agric. Food Chem. 32, 581–585.
Cook A M, Beilstein P, Grossenbacher H and Hutter R 1985 Ring cleavage and degradative pathway of cyanuric acid. Biochem. J. 231, 25–30.
Dalyan U, Harder H and Höpner Th 1990a Hydrocarbon biodegradation in sediments and soils. A systematic examination of physical and chemical conditions – Part II. pH values. Erdol & Kohle, Erdgas, Petrochemie 43, 337–342.
Dalyan U, Harder H and Höpner Th 1990b Hydocarbon biodegradation in sediments and soils. A systematic examination of physical and chemical conditions – Part III. Temperature. Erdol & Kohle, Erdgas, Petrochemie 43, 435–437.
Dean-Ross D 1987 Biodegradation of toxic wastes in soil. ASM News 53, 490–492.
Dragun J 1988 Microbial degradation of petroleum products in soil. In Soil Contaminated by Petroleum. Eds E J Calabrese and P T Kostecki. pp 289–300. John Wiley & Sons, New York.
Focht D D 1987 Ecological and evolutionary considerations of the metabolism of xenobiotic chemicals in soil. In Future Developments in Soil Science Research. Ed. L L Boersma. pp 157–166. Soil Science Society of America Inc., Madison, Wisconsin.
Gurevich P, Oren A, Sarig S and Henis Y 1993 Reduction of aromatic nitro compounds in anaerobic ecosystems. Wat. Sci. Tech. 27, 89–96.
Harder H and Höpner Th 1991 Hydrocarbon biodegradation in sediments and soils. A systematic examination of physical and chemical conditions – Part V. Moisture. Erdol & Kohle, Erdgas, Petrochemie 44, 329–332.
Harder H, Kürzel-Seidel B and Höpner Th 1991 Hydrocarbon biodegradation in sediments and soils. A systematic examination of physical and chemical conditions – Part IV. Special aspects of nutrient demand. Erdol & Kohle, Erdgas, Petrochemie 44, 59–62.
Kaake R H, Roberts D J, Stevens T O, Crawford R L and Crawford D L 1992 Bioremediation of soils contaminated with the herbicide 2-*sec*-butyl-4,6-dinitrophenol (Dinoseb). Appl. Environ. Microbiol. 58, 1683–1689.
Masaphy S, Levanon D, Vaya J and Henis Y 1993 Isolation and characterization of a novel atrazine metabolite produced by the fungus *Pleurotus pulmonarius*, 2-chloro-4-ethylamino-6-(1-hydroxyisopropyl)amino-1,3,5-triazine. Appl. Environ. Microbiol. 59, 4342–4346.
Michaelsen M, Hulsch R, Höpner Th and Berthe-corti L 1992 Hexadecane mineralization in oxygen-controlled sediment–seawater cultivations with autochthonous microorganisms. Appl. Environ. Microbiol. 58, 3072–3077.
Mizrachi R, Gerstl Z and Henis Y 1994 Degradation of the s-triazine herbicides prometryn and simetryn by strains of *Bacillus cereus* and their possible use in soil bioremediation. 7th Int. Congress of Bacteriology and Applied Microbiology Division, Prague, Czech Republic, July 1994. p 528.
Mohn W and Tiedje J M 1992 Microbial reductive dehalogenation. Microbiol. Rev. 56, 482–507.
Oren A, Gurevich P, Azachi M and Henis Y 1992 Microbial degradation of pollutants at high salt concentrations. Biodegradation 3, 387–398.
Peakall D B 1970 Pesticides and reproduction of birds. Sci. Am. 222, 72–78.
Piotrowski M R 1991 Bioremediation of hydrocarbon contaminated surface water, groundwater, and soils: The microbial ecology approach. In Hydrocarbn Contaminated Soils and Groundwater, Vol. 1 Eds P T Kostecki and E J Calabrese. pp 204–238. Lewis publishers, USA.
Prince R C 1992 Bioremediation of oil spills, with particular reference to the spill from the Exxon Valdez. In Microbial Control of Pollution. 48th Symposium of the Society for General Microbiology. Eds E J C Fry, G M Gedd, R Herbert, C W Jones and I A Watson-Craik. pp 19–34. Cambridge University Press, Cambridge, UK.
Raveh L, Segall Y, Leader H *et al.* 1992 Protection against tabun toxicity in mice by prophylaxis with an enzyme hydrolyzing organophosphate esters. Biochem. Pharmacol. 44, 397–400.
Rittman B E and McCarty P L 1980a Model of steady state biofilm kinetics. Biotech. Bioeng. 22, 2343–2357.
Rittman B E and McCarty P L 1980b Evaluation of steady state biofilm kinetics. Biotech. Bioeng. 22, 2359–2373.
Rosenberg E and Ron E 1995 Bioremediation of petroleum. In Bioremediation: Principles and Applications. Eds R L Crawford and D Crawford. Cambridge University Press, Cambridge, UK.
Rothschild N, Levanon D, Gertle A, Sakal E and Henis Y 1990 Purification and characterization of the enzyme parathion hydrolase isolated from *Pseudomonas* sp. Proc. Annu. Meet. Israel Soc. Microbiol. p 23.

Sela M, Garti J and Tel-Or E 1989 The accumulation and the effect of heavy metals on the water fern *Azola filiculoides*. New Phytol. 112, 7–12.

Shabtai Y and Fleminger G 1994 Adsorption of *Rhodococcus* GIN-1 (NCIMB 40340) to titanium dioxide and coal fly ash particles. 7th Int. Congress of Bacteriol. and Appl. Microbiol. Prague, Czech Republic, July 1994. p 282.

Shannon M J R and Unterman R 1993 Evaluating bioeremediation: Distinguishing fact from fiction. Annu. Rev. Microbiol. 47, 715–738.

Sorkhoh N, Al-Hasan R and Radwan S 1992 Self-cleaning of the Gulf. Nature 359, 109.

van der Meer J R, De vos W M, Harayama S and Zehnder A J B 1992 Molecular mechanisms of genetic adaptation to xenobiotic compounds. Microbiol. Rev. 56, 677–694.

Wiggins B A and Alexander M 1988a Role of chemical concentration and second carbon sources in acclimation of microbial communities for biodegradation. Appl. Environ. Microbiol. 54, 2803–2807.

Wiggins B A and Alexander M 1988b Role of protozoa in microbial acclimation for mineralization of organic chemicals in sewage. Can. J. Microbiol. 34, 661–666.

Wiggins B A, Jones S H and Alexander M 1987 Explanations for the acclimation period preceding the mineralization of organic chemicals in aquatic environments. Appl. Environ. Microbiol. 53, 791–796.

Wilson B H and Rees J F 1985 Biotransformation of gasoline hydrocarbons in methanogenic aquifer material. Proc. NWWA/API Conference on Petroleum Hydrocarbons and Organic Chemicals in Ground Water – Prevention, Detection and Restoration, Houston, TX. pp 128–139. National Water Well Assoc., Ohio.

Wurdemann S, Harder H and Höpner Th 1990 Hydrocarbon biodegradation in sediments and soils. A systematic examination of physical and chemical conditions – part I. Grain size, surface area and soiltype. Erdol & Kohle, Erdgas, Petrochemie 43, 217–224.

# 41
# Microbial transformation of pesticides in agricultural soil

D. LEVANON

## Introduction

Non-point source pollution of ground and surface water by pesticides has been recorded in agricultural areas in Europe and the US (Jamet and Delleu, 1993; Ritter, 1990). This phenomenon is causing concern, due to potential risks to public health and environmental quality. This non-point source pollution is characterized by unidentified specific origins and low concentrations of pesticides found in contaminated water. Detailed, laborious and expensive monitoring programmes are required to assess the nature and severity of the pollution. Once water is polluted, its quantity and distribution make it virtually impossible to eradicate the very low concentrations (measured in ppb) of pesticide involved. Efforts must therefore be made to minimize the risk of penetration by the pollutant to levels below the root zone, as a result of leaching or transport by run-off surface water. Such efforts must include the development of 'best management practices' (BMP), to ensure the proper use of pesticides by the grower, with minimal hazard to water quality. Among farm management practices, tillage is one that may have a significant impact on the fate of pesticides in the soil (USDA, 1989).

The practice of reduced tillage in orchards and field crops was developed for a number of reasons. Less intense tillage reduces energy and labour costs and soil compaction. Crop residues left on the surface contribute to conditions that improve water penetration, while minimizing run-off and soil erosion, as a result of higher porosity and water holding capacity (Dick *et al.*, 1989). On the other hand, reduced tillage (also called conservation tillage) requires more intensive use of herbicides, to control weeds, and sometimes greater use of insecticides, fungicides, etc., to control pests and diseases that develop in the crop residues. The combination of improved water penetration and increased use of pesticides can lead to their leaching into ground water (Edwards *et al.*, 1992).

There is also evidence that, under reduced tillage, and especially no-tillage (NT), numerous macropores develop in the soil, resulting in faster movement of water to the sub-surface. This, too, creates the potential for increased leaching of pesticides; but the results of field studies on the topic are ambiguous. Some studies show increased leaching of pesticides under NT (Gish *et al.*, 1991; Isensee *et al.*, 1990; Kanwar, 1991), while others show no significant differences between tilled and untilled soils (Starr, 1990; Starr and Glotfelty, 1990). Laboratory studies of

*D. Rosen* et al. *(eds): Modern Agriculture and the Environment, 491–499.*

undisturbed soil cores from the field have also produced contradictory results (Edwards *et al.*, 1992; Fermanich and Daniel, 1991; Levanon *et al.*, 1993).

The respective roles of biotic and abiotic processes under NT and plough tillage (PT), and their impact on the leaching of pesticides, were studied by Levanon *et al.* (1994). Abiotic processes, such as adsorption/retention of pesticides, resulting from higher organic matter content, and a higher rate of biodegradation, resulting from bigger and more active microbial populations, were among the explanations for a lower degree of leaching in NT soil. Synergy between fungi and bacteria, which enhanced the biotransformation of pesticides in NT soil, was also demonstrated in this study.

The present article is a review, amended with recent results, of the effect of tillage on the biotransformation of pesticides in the soil. The role of tillage is discussed as a case study of how farm management practice can be applied to the manipulation of the desired microbial transformation of organic pollutants.

## 2. Materials and methods

Soil samples were taken, according to Levanon *et al.* (1993) from the upper layer of the soil (0–5 cm) of NT and PT plots, under permanent monoculture of corn since 1974. The properties of the soil are presented in Table 1.

The following pesticides were studied:

(1) *Herbicides*

- Atrazine (6-chloro-N-[ethyl]-N-[isopropyl]-1,3,5 triazine-2,4-diamine);
- Alachlor (2-chloro-N-[2-6-diethylphenyl]-N-[methoxymethyl] acetamide);
- Metolachlor (2-chloro-N-[2-ethyl-6-methylphenyl]-N-[2-methoxy-1-methyl-ethyl] acetamide.

(2) *Insecticides*

- Carbofuran (2,3-dihydro-2,2-dimethyl-7-benzofuranul methyl carbamate);
- Diazinon (0,0-diethyl-0-[6-methyl-2-{1-methylethyl}-4-pyrimidinyl] phosphorothioate);
- Malathion (diethyl mercapto-succinate s-ester 0,0-dimethyl phosphorodithioate).

*Table 1.* Properties of the soil under NT and PT tillage in monoculture of corn since 1974.

| Property | PT | NT |
|---|---|---|
| pH | 6.8 | 5.8 |
| Organic C (%) | 1.15 | 1.65 |
| Biomass C ($\mu$g $g^{-1}$) | 420 | 780 |
| Total N (%) | 0.112 | 0.163 |
| $NO_3^-$-N (mg $kg^{-1}$) | 16.2 | 9.4 |
| $NH_4^+$-N (mg $kg^{-1}$) | 8.1 | 10.9 |
| $P_2O_5$ (mg $kg^{-1}$) | 156.1 | 238.6 |

Silt loam soil: 26% sand; 57% silt; 17% clay.

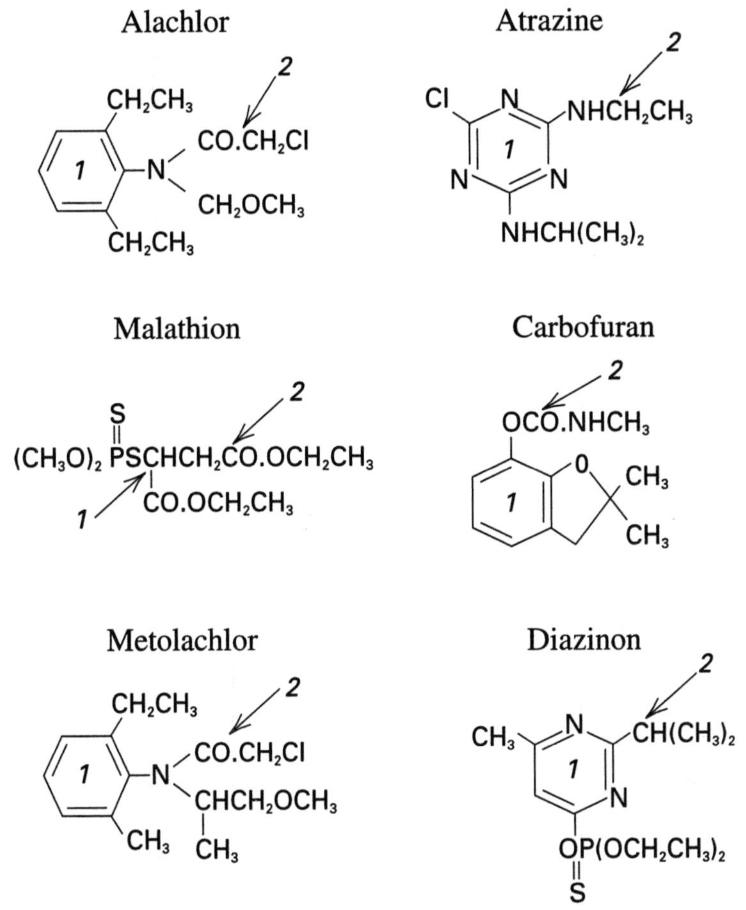

Alachlor

Atrazine

Malathion

Carbofuran

Metolachlor

Diazinon

*Figure 1.* Molecular structure of the six pesticides, with indication of the two positions of radiolabelling with $^{14}C$ (for each molecule) that were used in the present study.

Pure standard and radiolabelled pesticides were obtained from the Environmental Chemistry Lab and the Pesticide Degradation Lab, both at USDA ARS, Beltsville, MD. The molecular structures and the positions of $^{14}C$ labels are shown in Figure 1. Two labelling positions were studied separately for each pesticide: position 1 – uniformly ring-labelled (for malathion-succinate label); and position 2 – side-chain label (for malathion ethyl ester label).

Selective inhibition of soil microorganisms was applied according to Levanon *et al.* (1994) and comprised:

(1) Control – no inhibition.

(2) Inhibition of soil fungi.

(3) Inhibition of soil bacteria.

Microbial transformation of pesticides was studied in biometer flasks, according to Levanon *et al.* (1994). The transformation of each pesticide, added separately to the soil, was measured by trapping evolving $^{14}C\text{-}CO_2$ in a base trap, and measuring recovery by means of a scintillation counter (Packard). Each pesticide was added to 50 g soil at field moisture, to a concentration of 1.0 $\mu g\ g^{-1}$, after amendment with 50 000 DPM of radiolabelled formulation ($^{14}C$) and kept at 25°C for 30 days.

## 3. Results and discussion

The term 'mineralization' is used in this study, since all pesticides studied were uniformly ring labelled (malathion was succinate labelled), as designated by 'position 1' in Figure 1. The evolution of $^{14}C\text{-}CO_2$ from such a formulation is an indication of the complete mineralization of the molecule. Mineralization of all the pesticides studied took place more quickly in NT than in PT soil (Figure 2). This faster mineralization can be explained by the fact that the higher content of organic matter, nitrogen and phosphorus in the upper layer of NT soil led to larger and more active populations of microorganisms, thereby creating conditions favourable to the mineralization of organic compounds, including pesticides (Levanon, 1993; Levanon *et al.*, 1994).

The next stage of this study was to assess the relative importance of fungi and bacteria in this process. This part of the study was based on the assumption that, since populations of both fungi and bacteria are higher in NT soil, both groups probably participate in the enhanced mineralization process. This hypothesis was partly confirmed by experimental results: almost no mineralization of diazinon, alachlor, metolachlor or atrazine took place when either fungi or bacteria were inhibited (Figure 3). On the other hand, carbofuran and malathion were mineralized at the same rate as the control when fungi were inhibited, and hence by the activity of bacteria. In this series of experiments the radiolabel of the pesticide was also in position 1, so that the evolution of $^{14}C\text{-}CO_2$ is an indication of complete mineralization. Hence, four of the six pesticides studied were mineralized in the soil by concomitant activity of fungi and bacteria. The inability of any single strain of microorganism to mineralize xenobiotic chemicals (such as pesticides) completely, under field conditions, has been observed in many cases (Miller and Poindexter, 1994).

Some knowledge of the pathways of biotransformation of pesticides was obtained by the use of formulations radiolabelled in position 2. In this case, the evolution of $^{14}C\text{-}CO_2$ indicated partial mineralization of the molecule. By selective inhibition of fungi and bacteria, their role in the partial degradation (biotransformation) of each pesticide can be elucidated. It was demonstrated that, in the cases of diazinon, alachlor, metolachlor and atrazine, degradation of 'side chains' was mainly due to fungal metabolism (Figure 4). Since these pesticides were not completely mineralized when bacteria were inhibited (Figure 3), it can be deduced that the undegraded central part of the molecule, the ring, is degraded mainly by bacteria.

These findings allow us to formulate the probable pathway of the mineralization of these four pesticides in NT soil as follows: stage 1 – fungal metabolism of the side-chains (position 2 in Figure 1); stage 2 – bacterial degradation of the ring (position 1 in Figure 1). Results from other studies support these findings. Kaufman and Kearny (1976) found that soil fungi are able to cleave the side-chains of triazines

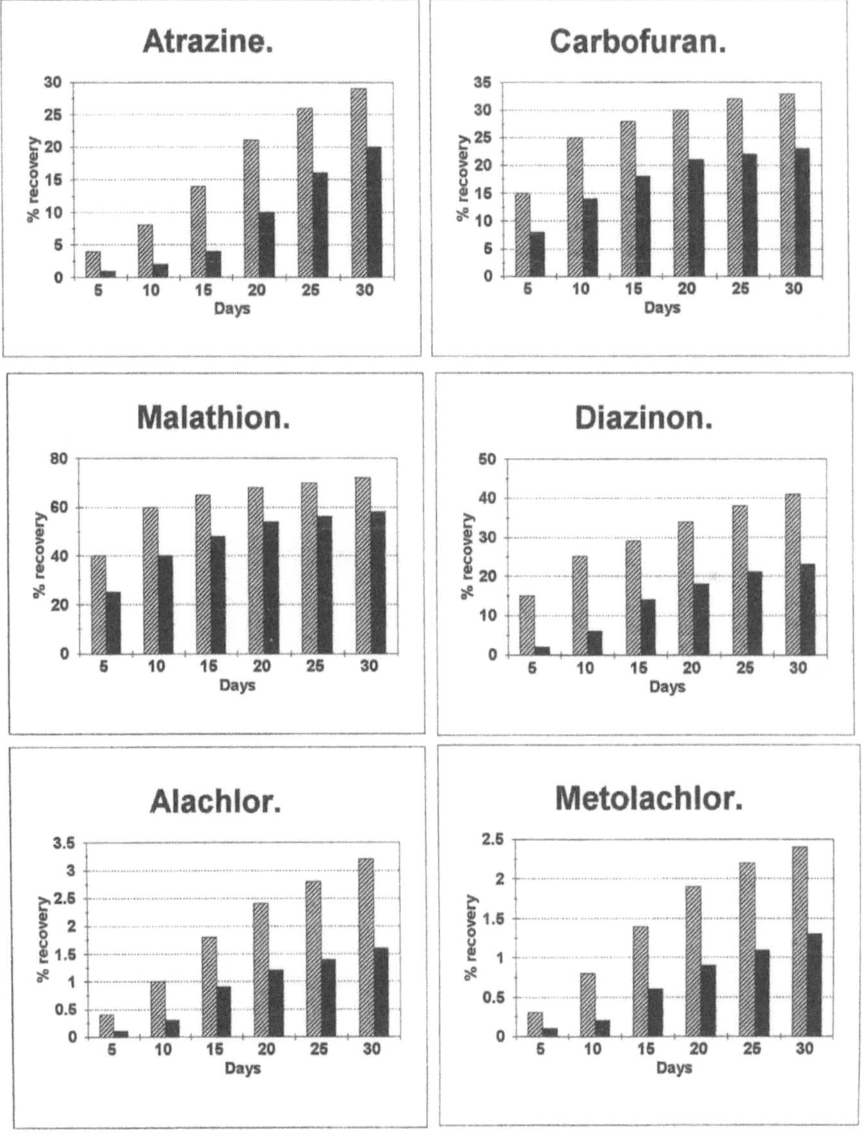


*Figure 2.* Mineralisation of the pesticides in the soil, measured as percentage recovery of $^{14}C$ of the initial amount added to NT (▨) or PT (■) soil.

via N-dealkylation, and more recently Masaphy *et al.* (1993) demonstrated the same activity by white-rot fungi. However, further research is required to characterise the bacterial–fungal consortium, to identify its constituent species and their respective roles in the transformation process. Such data will contribute to the understanding of the biotransformation of pesticides in the soil, allowing prediction and eventually manipulation in different soil environments.

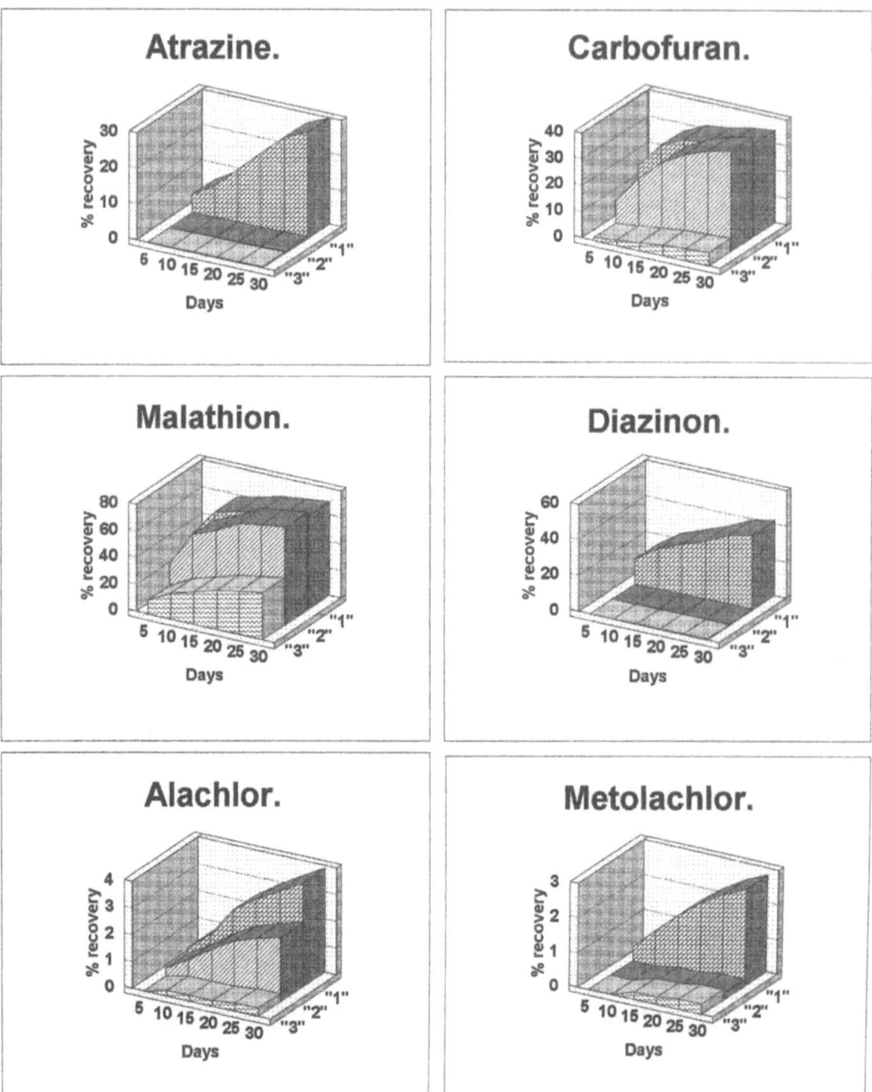


*Figure 3.* Mineralization of the pesticides in NT soil, measured as percentage recovery of the initial amount of $^{14}C$ added to soil supplemented with bactericide ("3"); fungicide ("2"); control ("1").

## 4. Conclusion

The NT system can serve as a case study for the development of farming practices that could enhance the detoxification and mineralization of xenobiotic chemicals in the soil. The NT system in the monoculture of corn causes shifts in physical, chemical and microbiological constituents of the upper layer of the soil. The

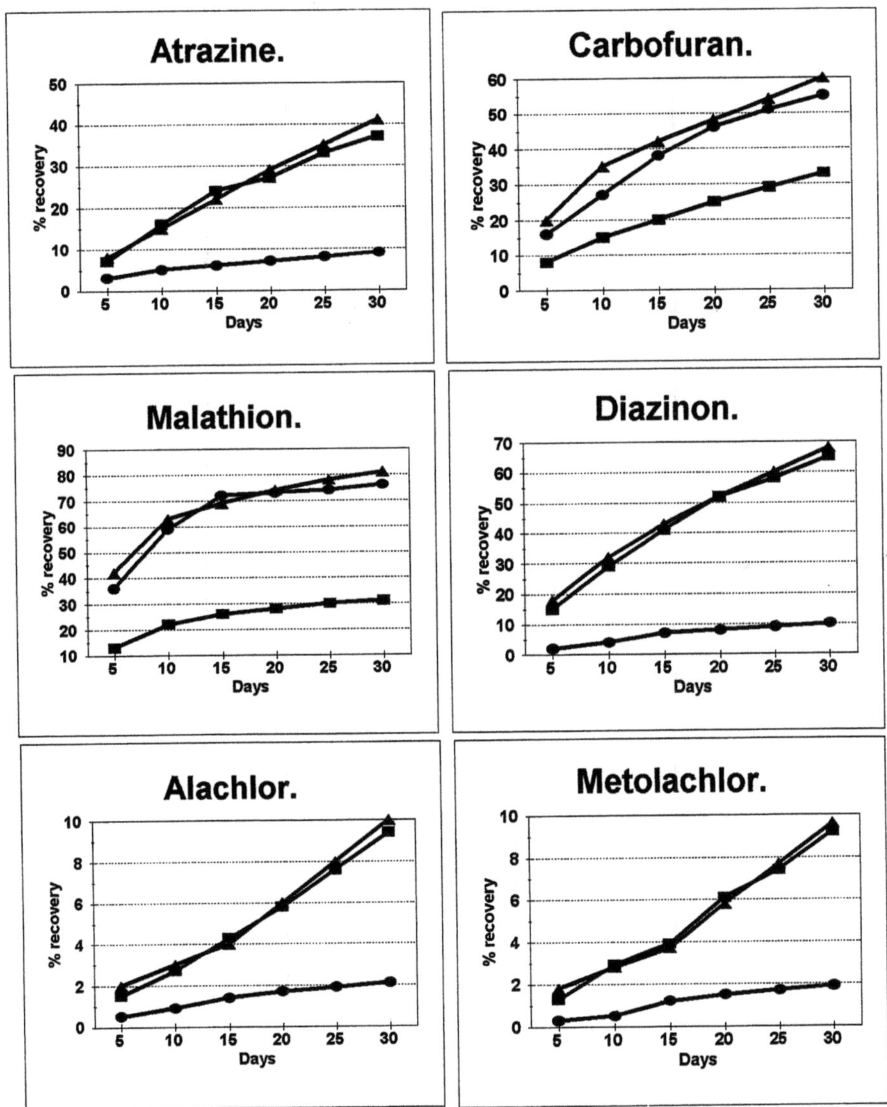


*Figure 4.* Partial mineralization of the pesticide in NT soil, measured as percentage recovery of the initial amount of $^{14}C$ added to soil supplemented with bactericide ( ); fungicide ( ); control (▲).

conditions that develop are favourable to the adsorption and biodegradation of pesticide residues.

Hence, the conditions in NT soil resemble a biofilter for organic chemicals, in a two-stage process: adsorption and biodegradation. Other studies, under field conditions, have also demonstrated that enhanced degradation of xenobiotic organic pollutants can be achieved by the manipulation of environmental conditions, by the addition of C or N sources, minerals, air or water, etc. (Gibson and Syler, 1992). The addition of an N source and water have been used on several occasions to enhance the degradation of such contaminants as crude oil (Atlas, 1994). Hence, recent recommendations for the field evaluation of bioremedial methods have concluded that the enhancement of indigenous microorganisms should be the first option in the bioremediation of a contaminated site, and inoculation by specific degrading microorganisms should only be given second priority (Miller and Poindexter, 1994).

The prevention of pollution caused by agricultural practices should be achieved by the development of techniques that enhance the activity of the soil as a biofilter: NT management is an example by which this goal is attained, in the case of pollution by pesticide residues. Specific methods should be developed for other cases of environmental pollution resulting from farming practices. One of the most important examples, worldwide, is the use of effluents for irrigation. Such effluents contain xenobiotics that include toxic organic chemicals. Some of these, of domestic or industrial origin, are resistant to degradation in sewage treatment plants, and are therefore still found in the treated effluents used for irrigation in arid and semi-arid zones. Such chemicals, introduced by irrigation, can sometimes be found below the root zone of crops, and later reach ground water (Feigin *et al.*, 1990).

Research is currently underway in our laboratories to develop methods to enhance the ability of the soil to act as a biofilter, and thereby to ensure minimal pollution of ground water by toxic organic chemicals that are at present to be found in domestic and industrial effluents.

## Acknowledgment

I would like to thank the administration and staff of the Environmental Chemistry Laboratory, NRI, ARS, USDA in Beltsville, MD, for providing the facilities for most of the research that was conducted towards this study.

## References

Atlas R M 1994 Assessing the efficacy of bioremediation and biodegradation of petroleum hydrocarbons in the marine environment. In Proc. 7th Intl. IUMS Cong., Prague, Czech Republic.

Dick W A, Rosenberg R J, McCoy E L and Edwards W M 1989 Surface hydraulic response to no-tillage. Soil Sci. Soc. Am. J. 53, 1520–1526.

Edwards W M, Shipitalo M J, Dick W A and Owens L B 1992 Rainfall intensity affects transport of water and chemicals through macropores in no-till soil. Soil Sci. Soc. Am. J. 56, 52–58.

Feigin A, Ravina I and Shalhevet J 1990 Irrigation with Treated Sewage Effluents. Springer Verlag, Berlin.

Fermanich K J and Daniel T C 1991 Pesticide mobility and persistence in microlysimeter soil columns from a tilled and no-tilled plot. J. Environ. Qual. 20, 195–202.

Gibson D T and Syler G S 1992 Scientific foundations of bioremediation. Current status and future needs. Am. Acad. Microbiol., Washington, DC.

Gish T J, Isensee A R, Nash R G and Helling C S 1991 Impact of pesticides on shallow groundwater quality. Trans. Am. Soc. Agric. Eng. 34, 1745–1753.

Isensee A R and Sadeghi A M 1993 Impact of tillage practice on runoff and pesticide transport. J Soil Water Cons. 48, 523–527.

Isensee A R, Nash R G and Helling C S 1990 Effect of conventional vs. no-tillage on pesticide leaching to shallow ground water. J. Environ. Qual. 19, 434–440.

Jamet P and Delleu R 1993 Environmental fate of pesticides – behaviour of pesticides in soil. Agro-food Industry Hi-Tech., May–June, 19–21.

Kanwar R S 1991 Preferential movement of nitrate and herbicides to shallow groundwater as affected by tillage and crop rotation. In Proc. Preferential Flow. Eds T J Gish and A Shirmohammadi. pp 328–337. Chicago, IL.

Kaufman D D and Kearney P C 1976 Microbial transformations in soil. In Herbicides: Physiology, Biochemistry, Ecology. 2nd ed., Vol. II. Ed L J Audus. pp 29–64. Academic Press, London.

Levanon D 1993 The role of fungi and bacteria in the mineralization of the pesticides atrazine, alachlor, malathion and carbofuran in the soil. Soil Biol. Biochem. 25, 1097–1105.

Levanon D, Coddling E E, Meisinger J J and Starr J L 1993 Mobility of agrochemicals through soil from two tillage systems. J. Environ. Qual. 22, 155–161.

Levanon D, Coddling E E, Meisinger J J and Starr J C, 1994 Impact of tillage on microbial activity and the fate of pesticides in the soil. Water Air Soil Poll. 72, 179–189.

Masaphy S, Levanon D, Vaya J and Henis Y 1993 Isolation and characterization of a novel atrazine metabolite produced by the fungus *Pleurotus pulmonarius*. Appl. Environ. Microbiol. 59, 4342–4346.

Miller R V and Poindexter J S 1994 Strategies and mechanisms for field research in environmental bioremediation. Am. Acad. Microbiol. Special Report. Washington, DC.

Ritter W F 1990 Pesticide contamination of ground water in the US – a review. J. Environ. Sci. Health 1, 1–29.

Sigua G C, Isensee A R and Sadeghi A M 1993 Influence of rainfall intensity and crop residue on leaching of atrazine through intact no-till soil cores. Soil Sci. 156, 225–232.

Starr J L 1990 Spatial and temporal variation of ponded infiltration. Soil Sci. Soc. Am. J. 54, 629–636.

Starr J L and Glotfelty D E 1990 Atrazine and bromide movement through a silty loam soil. J. Environ. Qual. 19, 552–558.

US Department of Agriculture 1989 Research Plan for Water Quality. Washington, DC.

# 42
# Biological atrazine destruction mediated by bacteria

RAPHI T. MANDELBAUM, NIR SHAFIR AND LARISSA KAUTZKI

## 1. Introduction

Many toxic organic compounds that have entered the environment as a result of man's activities are biodegradable or potentially biodegradable to less toxic organic compounds or to minerals. Pesticides represent a major group of such synthetic organic compounds, with over 45 000 registered pesticides totalling 550 million kg and 6.5 billion dollars annually in the United States alone (Ritter, 1990). Among the pesticides, herbicides are the largest group, comprising some 70% of the total pesticide volume applied in production agriculture. Among the herbicides, atrazine is the most widely utilized member of the large family of the *s*-triazine herbicides. Farmers throughout the world favour atrazine due to its low cost and high efficacy

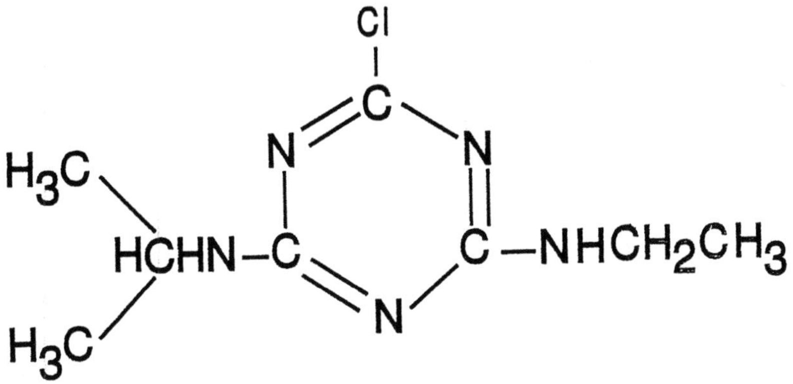

*Figure 1.* Molecular structure of atrazine.

for selective weed control in a variety of agricultural environments. The *s*-triazines are symmetrical, six-membered rings with alternating carbon and nitrogen atoms; each carbon and nitrogen atom nominally contributes three bonds to the ring structure and the ring carries three substitutes, one on each carbon atom. Atrazine carries isopropylamine, ethylamine and a chlorine atom as substitutes (Figure 1).

*D. Rosen* et al. *(eds): Modern Agriculture and the Environment, 501–521.*

Atrazine was first synthesized over 40 years ago, and today is the most widely used herbicide in the US. It is also registered in more than 70 countries worldwide. Approximately 67% of all US corn acreage, 65% of all sorghum acreage and 90% of all sugarcane acreage is treated with atrazine (Parshley, 1995). Atrazine alone accounts for about one-eighth of all pesticide tonnage applied in the United States (Ritter, 1990).

## 2. Problems associated with atrazine

In spite of its popularity, atrazine has a perplexing environmental record. It is only moderately degradable in some soils, with half-lives under cold climate conditions (15°C) of up to 105 and 166 weeks in some clay and sandy loam soils, respectively (Bowmer, 1991). The mobility of atrazine and atrazine metabolites through soils has contributed to the contamination of ground water (Baker and Richards, 1992; Belluck *et al.*, 1991). Moreover, atrazine is resistant to biodegradation in regular activated sludge sewage works, making it is impossible to utilize this standard biological treatment to clean atrazine from wastewater or from manufacturing effluents (Cook, 1987). The physicochemical and biological degradation of *s*-triazines have been reviewed by Cook (1987) and by Erickson and Lee (1989). Biodegradation under limited nitrogen conditions has proved successful for some *s*-triazines such as deisopropylatrazine, deethyldeisopropylatrazine, ammeline, N-isopropyl ammelide, ammelide, melamine and cyanuric acid, but not for the most widely used herbicides such as atrazine and simazine (Cook, 1987). Moreover, treatment of drinking water under conditions of nitrogen, carbon and/or energy stress is difficult to engineer. Indeed, even in drinking water mineral nutrients are still abundant (ppm level) relative to the herbicide levels (Vandepitte *et al.*, 1994). To further complicate the situation, atrazine has been classified as a class C carcinogen by the US Environmental Protection Agency (EPA) with a maximum concentration limit (MCL) set at 3 $\mu g\ L^{-1}$ in drinking water supplies (Federal Register 56: 3552). It has also been suspected that indirect hazards may be involved with atrazine usage. Atrazine has been suspected, for example, of forming *N*-nitrosoatrazine by ingestion of atrazine in conjunction with nitrite and by chemical formation at low pH levels in the presence of nitrite (Weisenburger *et al.*, 1987). *N*-nitrosoatrazine itself has been related to increased incidence of non-Hodgkin's lymphoma (Weisenburger, 1990). The Ciba Geigy Corporation, the manufacturer of atrazine, challenged these findings and experimentally demonstrated that the formation of *N*-nitrosoatrazine occurs only under artificial/non-biological conditions. The presence of food and the rapid metabolism of the *N*-nitrosoatrazine in mammals are the most important factors in preventing the presence of this derivative (Anonymous, 1995).

Due to increased concerns regarding large-scale atrazine usage, considerable worry was voiced about its fate in the environment. Consequently, the EPA notified the Ciba Geigy Corporation in November 1994, of the initiation of a special review of pesticides containing atrazine to determine whether the use of atrazine may pose unreasonable risks to human beings or the environment (Federal Register, 60412). A special review can result in a decision to cancel, restrict, or continue the use of the pesticide in question. Ciba's response to the EPA, based on both existing information and new data, and on analysis by independent laboratories, established unequivocally that atrazine poses no unreasonable risks to humans or the environment, when used according to instructions and in accordance with generally accepted practices (Parshley, 1995). In view of these results, farmers from all around the US sent thousands of letters in support of continuation of atrazine manufacture and use in agriculture. It was assumed that the consequences of using alternative herbicides could include reduced yields, negative impact on conservation tillage, additional trips across the field, increased costs for chemical/non-chemical weed control, increased use of alternative herbicides, inferior weed control, limitation in application flexibility; and in most cases higher risks of crop injury. Therefore, it is most likely that atrazine will remain on the market for years to come, and that the ability to remediate atrazine contamination from soil, water and effluents will be of great value for environmental management.

## 3. Nature of atrazine contaminations as related to biodegradation

The run-off from agricultural fields is the most common source of low-level atrazine contamination of surface water, typically of the order of 1–10 $\mu g\ L^{-1}$. It has been shown that the bioremediation of such trace amounts may be difficult and is probably not cost-effective (Dries *et al.*, 1987). However, European Council (EC) legislation restricts the occurrence of individual pesticides in drinking water to 0.1 $\mu g\ L^{-1}$ and states that the total amount of these compounds may not exceed 0.5 $\mu g\ L^{-1}$ (Council of the European Communities, 1980). Conventional methods of water treatment (e.g. sand filtration, coagulation and adsorption to powdered activated carbon) often fail to reduce *s*-triazine concentrations to the EC maximum admissible concentration (Foster *et al.*, 1991). Nevertheless, the ability to degrade low levels of atrazine bacterially has been demonstrated by inoculation of atrazine degrading bacteria on granulated activated carbon (bio-augmentation) to yield 40% degradation from water containing 0.5 ppb atrazine (Feakin *et al.*, 1994).

Point source contaminations, however, such as those occurring at sites subjected to accidental spills, neglect, rinsate disposal into evaporation pits, back siphoning into wells or from discharge of industrial effluents from herbicide manufacturing sites, are of major concern and are the prime candidates for bioremediation. Point source contamination of groundwater with atrazine can potentially cause pollution far exceeding the regulated drinking water maximum contaminant level (MCL). For example, a spill of 1 kg of atrazine can contaminate one million cubic metres of water to a level of 1 ppb, ten times exceeding the standard MCL level of 0.1 $\mu g\ L^{-1}$ established by EC Directives 80/778/EEC 15.06.80 and 81/858 EEC 19.10.1981. Obviously, when such a spill is discharged to a smaller water body, the concentration of the contaminant may be much higher. Thus, it is of crucial importance to contain and destroy atrazine spills when still present in large concentrations in a relatively limited area, before they contaminate larger bodies of water. Moreover, atrazine recalcitrance and mobility in soil is enhanced at high concentrations (Rao and Davidson, 1979; Winterlin *et al.*, 1989). Therefore, point-source spills on soils may result in groundwater contamination even more quickly than might have been predicted from leaching studies under field concentrations.

Pesticide handling areas such as dealerships, aerial strips, mixing and loading areas, and pesticide manufacturing sites are particularly vulnerable to point source contamination. Typical problems faced by agrochemical facilities were demonstrated at the Galesville Chemical Co. (GCC) in Piatt, Illinois (Felsot *et al.*, 1988). Soil to a depth of 1 m along a railroad-right of way at the GCC facility had been found by the IEPA to be contaminated with unacceptably high levels of herbicides (e.g. 24 000 ppm in the top 10 cm and 100 ppm at a depth of 60 cm). The herbicides were discharged to this area as a waste stream arising from the mixing, loading, and cleaning operations of GCC. Pesticides residues were detected in the ditches lining the streets of Galesville (population less than 500), and in the two water wells. IEPA ordered excavation and clean-up of the site.

Eight case studies have been conducted in Iowa in the vicinity of farm-chemical supply dealerships to follow levels of contamination. Groundwater contamination has occurred in most of these areas. Atrazine concentrations detected in the groundwater and in the pools of water or soil in the loading and handling areas were measured at up to 70 ppm (70 000 ppb) and up to 65 ppb in groundwater (Ritter, 1990). In Madison, MN, the atrazine concentration in soil next to an air strip where atrazine was accidentally discharged exceeded 40 000 ppm (40 g $kg^{-1}$) some ten years after the accident occurred (Mandelbaum, unpublished). In Israel, atrazine concentration was higher than 400 ppb, 26 m below the ground surface in a sandy aquifer next to a pesticide manufacturing plant (Mandelbaum, unpublished). These are only a few examples of heavy contaminations. To further complicate the situation, atrazine point source contaminations are frequently found in association with a variety of man-made pollutants such as N-fertilizers, other pesticides or a variety of anthropogenic organic chemicals which further compound the pollution problem. In addition, pesticides that are exposed for long periods undergo an ageing process that impairs bioavailability and thus may reduce biodegradation (Felsot and Dzantor, 1990). Most soils are to some degree able to decontaminate many pollutants naturally to non-hazardous levels through chemical, physical and biological processes. There are, however, limits to how much a soil can absorb and self-decontaminate without some negative effects. Once it occurs, soil contamination with herbicides may reduce the quality of soils through direct detrimental effects to humans and other animals, via soil transfer or through the food chain. Contaminated soil may become non-culturable or crop yields may be reduced, and herbicides or herbicide degradation products may leach through contaminated soils and pollute ground and surface water.

All of these factors make it clear that point-source contaminations must be remediated as soon as possible after they occur. Strict environmental regulations, high costs of conventional treatments, and potential long-term legal liabilities provide the necessary impetus to develop efficient and relatively inexpensive biotechnological treatments to remediate point-sources.

Indeed, the fate and persistence of *s*-triazines and especially atrazine in soils under a variety of conditions has been extensively studied over the last 20 years (Buser, 1990; Pick *et al.*, 1992), with much research focusing on bacterial and fungal degradation of herbicides to allow for microbial destruction of herbicides at point sources, before they diffuse into the environment.

## 4. Bioremediation as a solution

Bioremediation, the process by which hazardous substances are degraded by microorganisms, is at the forefront of a larger group of innovative remediation technologies being applied to remediate soil and water contaminations worldwide. Although the process of bioremediation has been utilized for decades in the field of wastewater engineering, its application to contaminated soils, groundwater and industrial effluents is fairly new and still undergoing intensive development (Casper and Landsmann, 1992; Sayler *et al.*, 1991)

The principal cause for the heightened interest in bioremediation of pesticide spill sites, is the potential for significantly reducing clean-up costs at contaminated sites. Reported and projected costs have been as low as $75 to $200 per cubic yard, as compared to the more conventional technologies such as incineration or secure landfilling, where costs may be in the range of $200 to $800 per cubic yard (Gabriel, 1991).

In the case of *in-situ* bioremediation, there is the added benefit of minimal site disruption and the reduced potential for public exposure. This is no small consideration on built-up sites, where contaminated soil may extend beneath roadways, railroads or buildings. Another advantage is the relative simplicity of the technology compared to certain other on-site treatment methods. Operational sophistication may be dramatically lower than on-site incineration, solidification, stabilization, vitrification or soil washing systems. Despite the potential benefits and relative simplicity of bioremediation systems, there are a number of pitfalls awaiting would-be practitioners. The most notable problem areas and issues that must be considered are:

- Existence of microorganisms capable of contaminant metabolism.
- Scale-up from bench to pilot to field level.
- The operation requires a multidisciplinary team.
- Valuation of remediation results.
- Unrealistic expectations.
- Regulatory barriers.
- Liability in the event of failure to achieve clean-up goals.

A typical bioremediation team should have expertise in microbiology, engineering, hydrogeology, soil science and chemistry.

So far, the major barrier for bioremediation of atrazine contaminations is the non-availability of appropriate microbial strains on-site. To overcome this problem, bioaugmentation with laboratory strains adopted to atrazine degradation under conditions similar to those prevailing at the bioremediation site should be considered. In cases where the indigenous microflora is capable of limited atrazine metabolism, it might be necessary to add a supplemental food base, electron acceptors, or change the environmental conditions (temperature, pH, moisture ...) to enhance their activity. Legal limitations may be imposed on bioremediation operations. For example the use of bacteria may reduce the major bulk quantity of the herbicide but still fail to meet regulatory standards such as those for drinking water which limit herbicide levels in water to ppb levels. Other legal limitations may be imposed on the use of certain bacterial strains, most notably genetically engineered strains (GEMs).

An area where many operations run into difficulties is the scale-up from bench to pilot to full scale operations. In fact, most studies aimed at the biodegradation of pesticides have failed outdoors even when they have been very successful in the laboratory. The implementation of field scale degradation of atrazine is still not within reach and requires more research inputs from many disciplines, including microbiology, hydrology, geology, civil engineering, chemistry and soil science. Most of the initial key problems are microbiological. A conceptual biodegradation framework based on the principles of Dubey and Holmes (1995) can be partially adapted for atrazine degradation. This framework may include:

(1) Identification of microorganisms able to degrade at least partially atrazine to non-herbicidal, non-toxic metabolites.

(2) Identification and localization of the genes for atrazine degradation, including studies to see whether these genes are chromosomally or plasmid encoded.

(3) Use of gene probe and PCR technology to identify other bacteria possessing atrazine degrading capacities, and adapted to specific environments.

(4) Determination of the optimal conditions for growth and activity of free and immobilized bacteria.

(5) Establishment of the metabolic pathways involved in atrazine degradation, so that the rate-limiting steps can be identified.

(6) Development of microbial strains suitable for practical applications, using genetically modified microorganisms.

(7) Development of cell free enzymatic systems to degrade atrazine in aqueous systems. This is especially valuable for treatment of industrial effluents, where the carrier body may include substances toxic or injurious to microorganisms.

(8) Development of practical engineering processes that are not only suited to the chemical and physiological characteristics of the microorganisms used, but also to the chemical, mechanical and environmental peculiarities of the carrier materials to be processed, with specific reference to industrial effluents.

## 5. Bacterial degradation of atrazine

The first step towards a successful bioremediation operation is the search for a suitable biocatalyst. Bacteria are notorious for their catalytic ability and extensive selection efforts in various laboratories in several countries have yielded bacterial preparations capable of atrazine degradation.

### *5.1 Incentives for atrazine degradation*

Heterotrophic growth and maintenance of bacteria need to be supported by energy supplied by electron donors (substrate). Therefore, energy produced from reduced compounds serves as the incentive for biodegradation. Atrazine reductance degree (calculated by the equivalents of oxygen per mole of carbon that are required to oxidize the molecule to $CO_2$, $H_2O$, $NO_3$, and HCl) is 3.75 and the estimated value of heat of combustion is 106 kJ for an equivalent of available electrons (Erickson and Lee, 1989). Since atrazine ring carbons are in the oxidation state of $CO_2$, only oxidation of the side chains carbons can yield energy. Indeed, many of the isolated bacteria could utilize the side chains but could not cleave the ring. In a number of studies, additional substrates have been used to provide carbon and energy for microbial growth, while atrazine was used as an N-source (Cook, 1987). It is also possible that some bacteria will degrade atrazine as a cosubstrate via non-specific enzymatic reactions.

## 6. Microbial N-dealkylation reactions

Atrazine dealkylating activity was encountered in both fungi and bacteria. Bacterial dealkylation of *s*-triazines was probably first reported by Bryant (1963) who isolated three groups of *Pseudomonas* species from enrichment cultures of soil treated with seven *s*-triazine compounds. However, only one of the species could grow on ethylamine containing *s*-triazines as the sole source of carbon (indicating the use of the side chains as the C and energy source). Dealkylation in soil was initially attributed to fungal activity (Couch *et al.*, 1965; Kaufman and Blake, 1970), but Behki and Khan (1986) have demonstrated that significant atrazine degradation via dealkylation occurs in soil incubated in the presence of cyclohexamide (antifungal antibiotic). Other bacterial species capable of dealkylating atrazine have also been reported. Giardina *et al.* (1980) isolated an actinomyces, *Nocardia* sp., from a soil enrichment culture that could utilise atrazine as the sole source of carbon and nitrogen. Their metabolic studies have shown both N-dealkylation and deamination of atrazine to yield 2-chloro-4,6-diamino-*s*-triazine (Giardina *et al.*, 1982) which was reported to be non-phytotoxic in bioassay with oats.

Several *Pseudomonas* sp. that can dealkylate atrazine have been isolated. Three species of *Pseudomonas* capable of utilising atrazine as a sole source of C were isolated by Behki and Khan (1986) from a soil having a long history of atrazine application. The preferential formation of deisopropylatrazine over deethylatrazine was reported. Interestingly, previous studies by Kaufman and Blake (1970), reported that deethylatrazine, and not deisopropylatrazine, was the major metabolite resulting from fungal degradation of atrazine, indicating the involvement of different dealkylating enzymes.

Since deisopropylatrazine has been previously shown by Cook and Huetter (1984) to be completely mineralised by the combined action of the *Pseudomonas* and *Rhodococcus* isolates obtained in their study, a complete biodegradation of atrazine is imaginable through the combined action of three bacterial species. Recently, another bacterium that could further degrade dealkylated atrazine was reported. Deethylsimazine and deethylatrazine were degraded by a strain of *Klebsiella pneumoniae* which used the dealkylated metabolites as a nitrogen source (Ernst and Rehm, 1995). Deethylatrazine was degraded under both aerobic and anaerobic conditions faster than deisopropylatrazine. This bacterium was also assumed to mineralize completely other *s*-triazines (but not atrazine, simazine or propazine) when present in a mixture simulating industrial effluents.

Dealkylation was widely assumed to be a prerequisite for further dechlorination. For example, *Pseudomonas* strains 192 and 194 which could dechlorinate dealkylated atrazine could not dechlorinate the intact mother compound (Behki and Khan, 1986). Several *Rhodococcus* strains were screened for their ability to degrade atrazine and were found to dealkylate atrazine. The dealkyation was hypothesized to be catalyzed by an oxidative mechanism, as there was no atrazine metabolism in the absence of oxygen (Behki *et al.*, 1993). Some *Rhodococcus* strains seem to have a broad dealkylating spectrum. For example, Strain B-30 was reported to be able to dealkylate atrazine, simazine and propazine, via a supposed oxidative reaction with atrazine being the best substrate (Behki and Kahn, 1994). Recently the gene encoding for atrazine dealkylation in *Rhodococcus* sp. strain TE1, was identified to reside on a 6.2-kb *kpn*I plasmid fragment which could be electroporated and fully expressed in other *Rhodoccocus* strains (Shao and Behki, 1995), but not in *E. coli.*

Dealkylation reactions have also been demonstrated to occur as a result of non-specific cometabolic activity in *Pseudomonas* strains, without apparent benefit to the bacterium. *Pseudomonas flurescens* strain LMG 10140 grown under hydrogenotrophic conditions transformed atrazine to both deisopropyl and desethyl atrazine. The dealkylation products were not further transformed. Also, in the absence of hydrogen, dealkylation did not occur (Vandepitte *et al.*, 1994).

Very slow rates of dealkylation have been reported for an unidentified Gram-negative bacterium and a *Bacillus* isolate RK016. Both species preferentially dealkyated the ethylamine side chain (Korpraditskul *et al.*, 1993).

Recently, two reports on pure bacterial cultures capable of rapid atrazine mineralization when used as the sole C and N source have been published. Radosevitch *et al.* (1995) have isolated an unidentified atrazine degrader from agricultural soil previously impacted with atrazine. The bacterium was a Gram-negative, facultatively anaerobic motile rod that was tentatively related to *Agrobacterium* or *Xanthobacter* sp. and could degrade atrazine in non-growing cultures at a rate of about 113 $\mu$g atrazine mg$^{-1}$ (cell dry weight) h$^{-1}$ (Radosevitch *et al.*, 1995). The metabolic pathway was not elucidated due to rapid metabolism of the transient metabolites. Moreover, it was not clear whether the degradation involves oxidative or hydrolytic enzymes, due to the much reduced, but still existing atrazine degradation under anaerobic conditions.

Yanze-Kontchou and Gschwind (1994), have isolated *Pseudomonas* strain YAYA6 from a mixed culture that was previously reported to degrade atrazine. Complete atrazine degradation by strain YAYA6 proceeded via several pathways. The first steps involved dechlorination with formation of hydroxyatrazine as a major metabolite. However, N-dealkylation with formation of both desethylatrazine and deisopropylatrazine was not ruled out.

## 7. Bacterial atrazine dechlorination

Dechlorination is probably the most important degradation step of atrazine, as dealkylation products still maintain herbicidal activity, while dechlorination abolishes toxicity (Kaufman and Blake, 1970).

Microbial dechlorination of intact atrazine (dechlorination without prior modification of the molecule) has been reported only very recently. Dechlorination has in fact been conservatively reported to occur in soil, mainly through chemical soil catalysis, while only N-dealkylation reactions were reported to be biologically mediated. Mandelbaum *et al.* (1993) demonstrated that bacteria can effectively dechlorinate atrazine and that dechlorinating enzymes occur in soil bacteria. However, their work reported the presence of dechlorination enzymatic activity in mixed bacterial cultures, and it was not clear at that time whether the hydrolytic enzymes were contributed from a single bacterial species or represented the combined effect of proteins from several bacteria. The difficulty in isolating bacterial pure cultures that could dechlorinate atrazine is reflected in the latest available review on atrazine degradation (Erickson and Lee, 1989); the authors conclude that the role of microorganisms in the hydrolysis of atrazine is not clear, and that the presence of both alkyl groups on atrazine appears to inhibit dechlorination. Moreover, the ability to dechlorinate atrazine was suspected to appear only in fungi (Couch *et al.*, 1965). It has been reported only recently that bacteria can mediate hydrolytic dechlorination of atrazine (Mandelbaum *et al.*, 1993, 1995; Yanze-Kontchou and Gschwind, 1994). Behki and Khan (1986) have reported the dechlorination of atrazine by two of their strains, but atrazine was dechlorinated only as a second step after dealkylation took place. A purified bacterial hydrolase (from *Rhodococcus corallinus* NRRL B-15444R) was described by Mulbry (1994), but in this case also, the enzyme could dechlorinate deisopropyl atrazine and desethyl-atrazine, but atrazine and simazine could not be used as substrates. In fact atrazine and simazine inhibited the degradation of dealkylated atrazine.

A breakthrough has recently been made in atrazine dechlorination with the isolation of a new *Pseudomonas* strain, nominated ADP, which has been reported to degrade rapidly very high concentrations of atrazine through dechlorination of intact atrazine as the first degradation step (Mandelbaum *et al.*, 1995). Most recently De Souza *et al.* (1995) cloned and partially characterized the gene region from *Pseudomonas* strain ADP that encodes atrazine dechlorination.

## 8. Atrazine degradation by *Pseudomonas* sp. strain ADP

*Pseudomonas* strains are ubiquitous in soil and their diverse biodegradative capabilities for a variety of natural and xenobiotic compounds have long attracted scientific interest (Galli *et al.*, 1992).

### *8.1 Isolation*

*Pseudomonas* sp. Strain ADP isolated from a herbicide spill site was capable of metabolizing atrazine at very high concentrations (1000 ppm) through dechlorination of the ring as the first degradative step (Mandelbaum *et al.*, 1995). A clearing plate method that has been developed enabled the isolation of the pure culture. The method has been applied to a mix culture LFB6 (Mandelbaum *et al.*, 1993) and yielded an atrazine-metabolizing *Pseudomonas* denoted ADP. Detailed taxonomic studies identified the organism as a fluorescent *Pseudomonas* sp. designated ADP, that did not fit into any currently known species.

### *8.2 Identification*

The isolated bacterium was a Gram-negative motile rod. It contained a single polar flagellum observable by electron microscopy. Bacterial colonies on atrazine solid medium were generally circular with an irregular border and were opaque and light brown with darker coloration in the centre. Growth on King's medium B was very rapid, with the formation of fluorescent colonies in less than 24 h. The bacterium grew well at 42°C. It used glucose as a carbon source but did not utilise fructose, sucrose, galactose, lactose or maltose. It was catalase positive and oxidase positive, reduced nitrate, and it grew non-fermentatively on glucose as indicated on triple sugar iron agar. These observations suggested that the organism might belong to the RNA group 1 of the genus *Pseudomonas*.

Three commonly used identification procedures were further used for the identification of the atrazine degrading culture, but produced confusing results. Biolog identification system (Biolog, Hayward, California) was used according to the manufacturer's protocol. The assay uses a 96 multi-well plate designed to test the ability of the bacterium to use a panel of 95 different carbon sources. When oxidation occurs, reduction of tetrazolium dye produces a purple colour (positive growth). These results are compared to a database of nearly 800 bacteria. The unknown (*Ps.* ADP) was identified as *Pseudomonas citronellolis*, American Type Culture Collection (ATCC) 13674 with a similarity index of 0.554 (positive identification).

Ribosomal RNA (rRNA) appears to be the choice for molecular biological approaches to phylogeny (Lee, 1994). In order to identify the atrazine degrading microorganism further, the 16S rRNA of *Ps.* ADP and that of *Ps. citronellolis* were analyzed. Sequence analysis data have also been compared with existing data from

GenBank on the 16S rRNA sequences of *Ps. aeruginosa*, *Ps. cepacia*, and *Ps. testosteroni* as well as to *Escherichia coli* that served as a quality control standard. Cell lysis and total RNA isolation from the bacteria were made using zirconium beads beating and SDS lysis as previously described by Tsien *et al.* (1990). The resulting digest was extracted with phenol and immediately frozen until analysis. Eight synthetic 16S rRNA molecules of *E. coli* were used as previously described by Tsien *et al.* (1990). The bulk 16S rRNA preparations were used for sequencing analysis without further purification. The primer sequences were elongated by use of the reverse transcriptase method, and the resulting DNA sequences were aligned on a VAX 8600 computer by using published programs (Olsen, 1988) and compared with sequences in GenBank. Homology data analysis is presented in Table 1. The 16S rRNA data showed 96.1% homology to *Ps. citronellolis* and 94.4% to *Ps. aeruginosa*. The other *Pseudomonas* species

*Table 1.* Homology data analysis.

| Sequence pair | Homology | Distance | Total positions | Omitted gaps | Unweighted fractions | | |
|---|---|---|---|---|---|---|---|
| | | | | | Match | Mismatch | Gaps |
| *E. coli*: ADP | 0.828 | 0.19480 | 1486 | 1 | 0.818 | 0.156 | 0.026 |
| *E. coli*: *Ps. citronellolis* | 0.835 | 0.18622 | 1192 | 1 | 0.827 | 0.153 | 0.020 |
| *E. coli*: *Ps. aeruginosa* | 0.842 | 0.17734 | 1502 | 1 | 0.834 | 0.147 | 0.019 |
| *E. coli*: *Ps. testosteroni* | 0.811 | 0.21715 | 1548 | 4 | 0.805 | 0.179 | 0.016 |
| *E. coli*: *Ps. cepacia* | 0.814 | 0.21346 | 1460 | 3 | 0.808 | 0.176 | 0.016 |
| ADP: *Ps. citronellolis* | 0.961 | 0.03964 | 1190 | 1 | 0.952 | 0.029 | 0.019 |
| ADP: *Ps. aeruginosa* | 0.944 | 0.05774 | 1467 | 0 | 0.939 | 0.049 | 0.012 |
| ADP: *Ps. testosteroni* | 0.808 | 0.22158 | 1478 | 5 | 0.798 | 0.177 | 0.024 |
| ADP: *Ps. cepacia* | 0.827 | 0.19674 | 1456 | 6 | 0.816 | 0.157 | 0.027 |
| *Ps. citronellolis*: *Ps. aeruginosa* | 0.934 | 0.06857 | 1185 | 1 | 0.927 | 0.057 | 0.016 |
| *Ps. citronellolis*: *Ps. testosteroni* | 0.806 | 0.22506 | 1191 | 1 | 0.798 | 0.183 | 0.018 |
| *Ps. citronellolis*: *Ps. cepacia* | 0.828 | 0.19570 | 1177 | 3 | 0.819 | 0.160 | 0.021 |
| *Ps. aeruginosa*: *Ps. testosteroni* | 0.822 | 0.20338 | 1493 | 4 | 0.816 | 0.169 | 0.015 |
| *Ps. aeruginosa*: *Ps. cepacia* | 0.833 | 0.18878 | 1450 | 5 | 0.826 | 0.157 | 0.017 |
| *Ps. testosteroni*: *Ps. cepacia* | 0.886 | 0.12397 | 1453 | 4 | 0.880 | 0.107 | 0.012 |

ADP: Atrazine-degrading *Pseudomonas* isolate

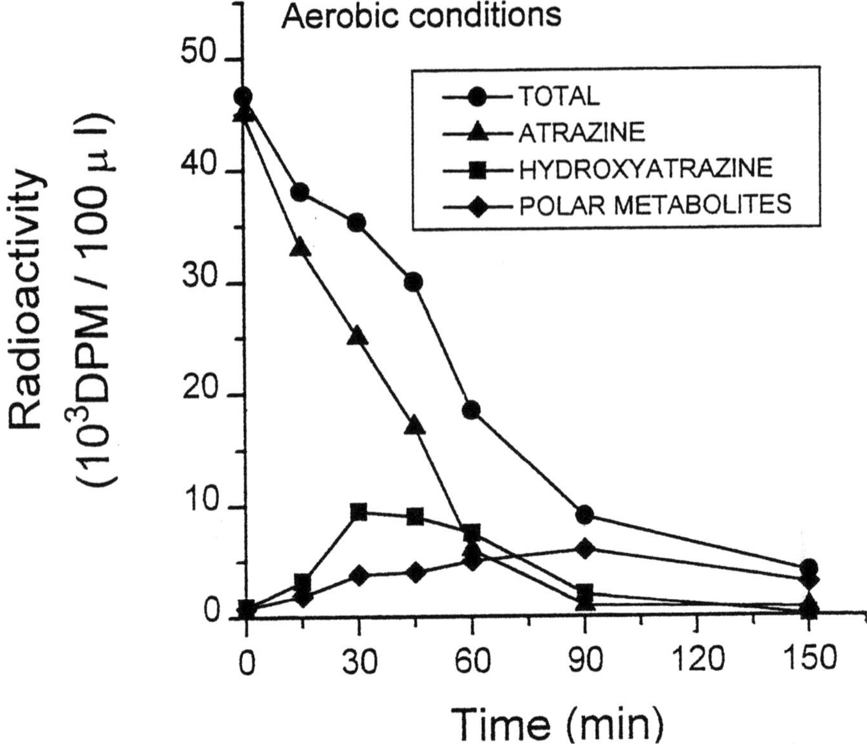


*Figure 2.* Metabolism of 100 ppm of [$^{14}$C] atrazine by resting-cell suspensions of *Pseudomonas* sp. strain ADP under aerobic conditions, showing formation of hydroxyatrazine and other polar water soluble transient metabolites.

from other RNA groups were far less homologous with *Ps.* ADP than *Ps. citronellolis*. From the above results it could be concluded that there is a reasonable match between *Ps.* ADP and *Ps. citronellolis*. However, fatty acid analysis of *Ps.* ADP revealed a close match with *Ps. aeruginosa*. Therefore, the isolate was identified as a new strain of *Pseudomonas* and denoted as strain ADP (atrazine degrading *Pseudomonas*). The isolated bacterium did not precisely resemble any currently defined species, which is not an uncommon finding, given the diversity of soil bacteria and the likelihood that only a small fraction have been characterised to date. However, its rigorous identification here as a *Pseudomonas* sp. in rRNA group 1 is important for developing molecular biological approaches for the study of bacterial atrazine metabolism.

### *8.3 Atrazine metabolism*

A culture of *Ps.* ADP was grown in liquid atrazine medium (Mandelbaum *et al.*, 1993) to yield an active atrazine degrading liquid culture. Uniformly ring labelled [$^{14}$C] atrazine was added to 100 ppm unlabelled atrazine in sterile 0.1 M phosphate buffer to yield a radiolabel concentration of 1.1 pCi ml$^{-1}$. Twenty ml of the buffered medium were aseptically transferred to a 250-ml Erlenmeyer flask. The flasks were inoculated to yield a final concentration of $1.25 \times 10^{10}$ cells $ml^{-1}$, and stoppered after being flushed for a few seconds under a stream of oxygen (99.6% pure) to ensure aerobic conditions and to reduce the amount of $N_2$ in the flask atmosphere. At periodic intervals, a 1-ml sample from the buffer medium was removed with a syringe, and analysed by HPLC as described by Mandelbaum *et al.* (1995). Fractions from the HPLC solvents were collected (200 $\mu$l) and the distribution of radioactivity in the eluting peaks (e.g. polar metabolites, hydroxyatrazine and intact atrazine peaks) was determined. For experiments under oxygen limited conditions, cell suspension containing $1.25 \times 10^{10}$ cells $ml^{-1}$ in an airtight stoppered glass vial (25 ml) were subjected to three consecutive vacuum suction cycles, each followed by flushing with argon. The suspension was then amended with atrazine as in the aerobic cultures. Samples were removed with an airtight syringe and the missing volume replaced with argon.

Under these conditions, 0.46 mM (100 ppm) atrazine was completely metabolized aerobically in less than 150 min (Figure 2) and atrazine or its metabolites could not be recovered from the medium, indicating complete transfer to a volatile form. In other experiments, it was shown that ring-labelled carbons volatilised as carbon dioxide (Figure 3). An analysis of the culture medium by HPLC revealed the

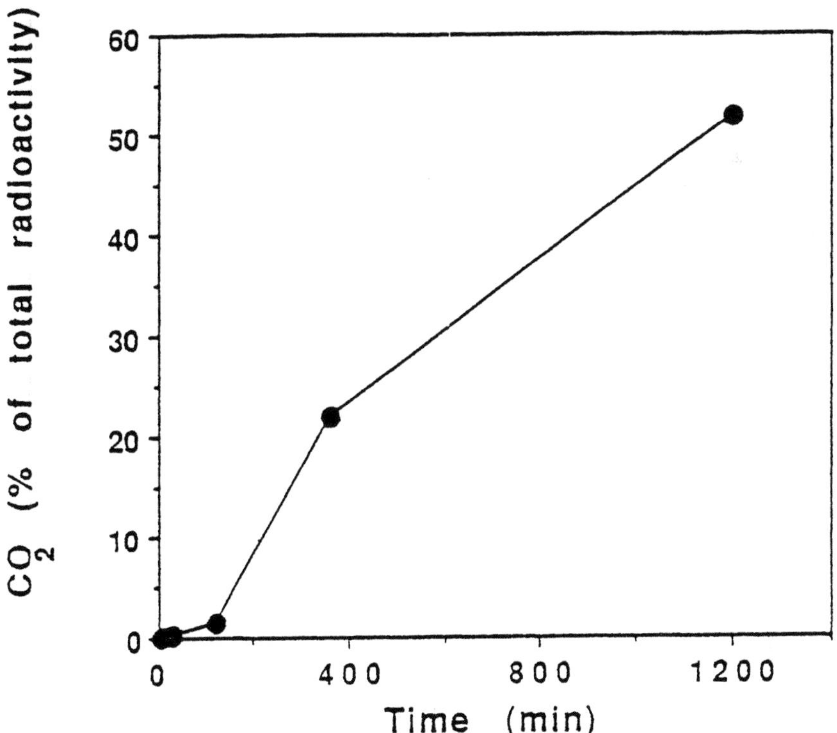


*Figure 3.* [$^{14}$C]-Carbon dioxide evolution from 2000 ppm atrazine in liquid medium inoculated with *Pseudomonas* sp. strain ADP.

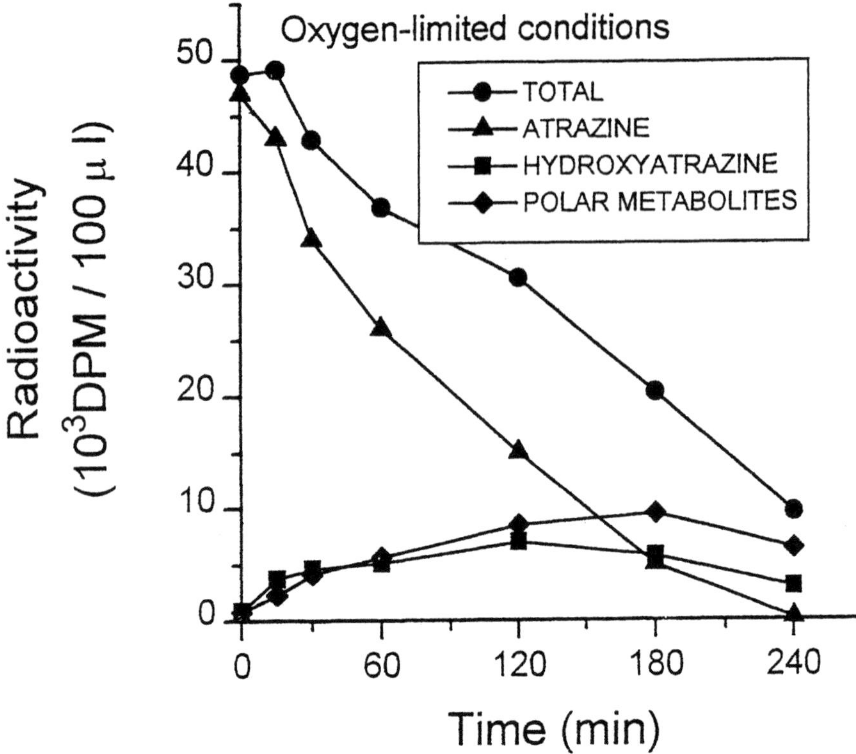


*Figure 4.* Metabolism of 100 ppm of [$^{14}$C] atrazine by resting-cell suspensions of *Pseudomonas* sp. strain ADP under oxygen limited conditions, showing formation of hydroxyatrazine and other polar water soluble transient metabolites.

transient presence of $^{14}$C-compounds more polar than atrazine. The major metabolite at the shorter time points co-migrated with hydroxyatrazine. To confirm the identity of the product, the metabolite was purified and subjected to mass spectrometry. The mass spectrum was identical to that of authentic hydroxyatrazine. Additionally, more polar metabolite(s) were observed to elute approximately 3 min after injection. Given the similar migrations of a number of potential atrazine metabolites (dealkylated hydroxytriazines), the identities of additional metabolites will require further work. The time course for the formation of hydroxyatrazine and other polar compounds is shown in Figure 2 and suggests that hydroxyatrazine is an intermediate in atrazine metabolism by *Pseudomonas* ADP. Separate studies were performed with resting cell suspensions that were not aerated. The metabolism of atrazine under these conditions was similar although the degradation rate was slower (Figure 4).

### *8.4 Atrazine degradation by cell-free extracts*

Cells were harvested by centrifugation at the end of the logarithmic growth, washed twice with sodium phosphate buffer (10 mM, pH 7.2) and resuspended in the same buffer to yield a concentration of $OD_{600} = 5.0$. Crude cell extract was prepared using a French press (operating at 12 000 psi). Cells were passed through the pressure cell three times followed by centrifugation at 30 000 *g* for 30 min at 4°C. The supernatant was filtered through a 0.2 $\mu$m filter and immediately frozen at −70°C. Protein content of the resulting crude extract was assayed with a bicinchoninic acid assay (Pierce Chemical Co., Rockford, IL) and adjusted to yield 1–3 mg protein $ml^{-1}$ crude extract.

[$^{14}$C]-atrazine (100 ppm) was incubated with 1.5 mg $ml^{-1}$ soluble proteins for various lengths of time, the reactions were quenched with organic solvent and the two fractions were analyzed for residual radioactivity. The kinetic course of the metabolism is shown in Figure 5. Within 10 min, 70% of the atrazine extractable

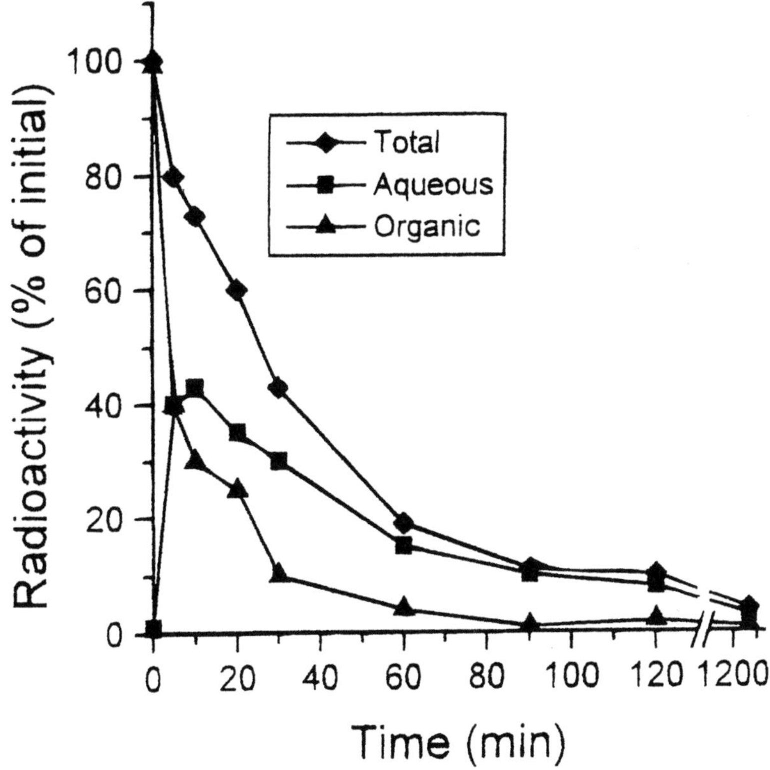


*Figure 5.* Atrazine degradation by *Pseudomonas* ADP crude extract (1.5 mg protein $ml^{-1}$). The cells were harvested at late logarithmic growth phase, French pressed and filtered through a 0.2 $\mu$m filter. The resulting crude extract was amended with 100 ppm [$^{14}$C]atrazine and samples extracted at time intervals with an equal volume of ethyl acetate. Residual radiolabel was measured.

into organic solvent was transformed into water-soluble materials and volatile material. Over the course of two hours, the radioactivity in the soluble fraction was depleted, presumably directed to $CO_2$ that volatilised out of the reaction mixture. After 20 hours most of the radioactivity associated with the ring carbons of atrazine volatilized from the system. The crude extract could be stored at $-70°C$ for several months with no apparent loss of activity.

### 8.5 *Atrazine mineralisation*

Uniformly ring labelled [$^{14}C$]-atrazine (7.8 mCi/mmole, 99.6% radiochemically pure) was added to 2000 ppm (representing possible spill site concentrations) unlabelled atrazine in 20 ml of sterile atrazine medium to yield $1.1 \times 10^{-3}$ mCi/ml ($pH = 7.3$). The medium was aseptically transferred to a 250-ml biometric flask. The assay time course was started with inoculation using a 24-h-old *Ps.* ADP culture (to yield $OD_{600} = 1.0$) and the Erlenmeyer was sealed with an airtight rubber stopper. Incubation was performed at 30°C with agitation at 25 rpm on a platform shaker. At selected times, the flasks were opened and the amount of radiolabelled carbon dioxide trapped in 2 ml of 2N NaOH solution was determined. This experiment demonstrated that over a time course of 20 h, more than 50% of the ring carbons of atrazine were trapped in the form of carbon dioxide in the base traps (Figure 3). It should be noted that the degradation rate was largely influenced by the solubility limit of atrazine (33 ppm) and that lower atrazine concentrations have been mineralized much more quickly (data not shown).

### 8.6 *Atrazine degradation in polluted water*

Polluted water from an aquifer adjacent to an atrazine manufacturing plant in Israel were used to verify the ability of *Ps.* ADP to degrade atrazine in contaminated water containing ppb amounts of atrazine in addition to a variety of other xenobiotics including organochlorinated solvents, other pesticides, and unidentified organic matter. The polluted aquifer was anoxic with dissolved oxygen levels below 5 ppm, electrical conductivity of 700 $\mu$S/cm, $pH = 7.2$, 10 ppm nitrate, 30 ppm sulphate and 260 ppb atrazine. Non-growing cells of *Ps.* ADP, harvested at the early stationary phase, were washed twice in phosphate buffer and added to the polluted water to yield a final bacterial concentration equivalent to 0.5 mg bacterial protein/ml. Tap water spiked with atrazine (200 ppb) was used as a control. Atrazine was degraded more rapidly in the tap water, but after 18 h the atrazine level was less than 30 ppb in both treatments (Figure 6). This laboratory experiment indicates the possible use of *Ps.* ADP under adverse conditions.

### 8.7 *Atrazine degradation in soil slurry*

Inoculation of bacteria into soils has often yielded lower degradation rates than equivalent studies in growth media. Such results have recently been reported, where atrazine dealkylation after bacterial inoculation of soil was very slow in spite of the

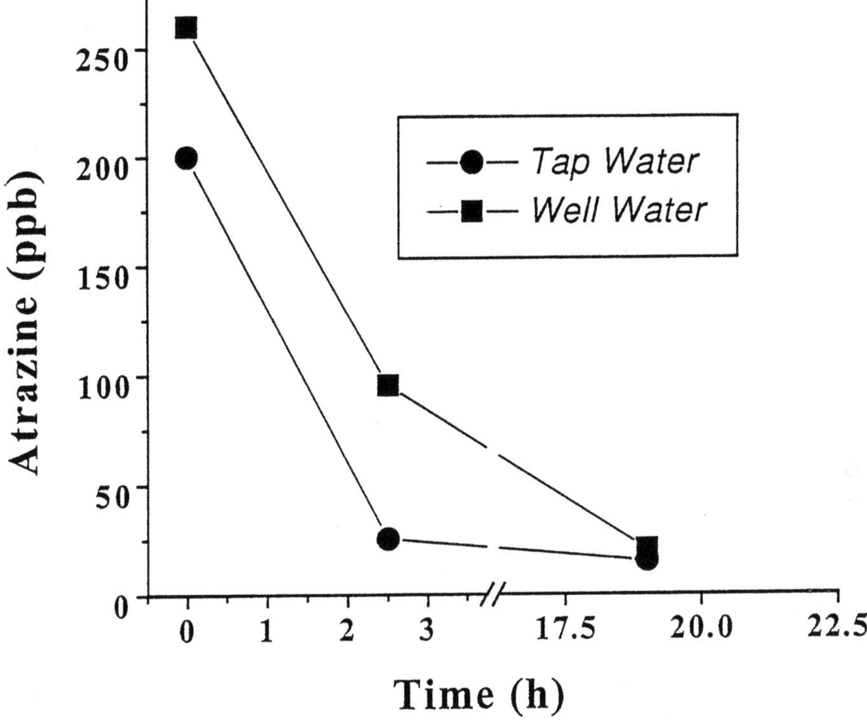


*Figure 6.* Atrazine degradation in polluted water from an aquifer adjacent to an atrazine manufacturing plant. Atrazine spiked tap water was used as control.

presence of atrazine degrading organisms. The authors suggest that insufficient cell density is the cause (Kolpin and Kalkhoff, 1993). In the present study we have demonstrated that atrazine can be quickly degraded in soil under slurry conditions, where *Ps.* ADP was applied to the soil immediately after application of atrazine. Air-dried red Mediterranean clayey soil (50 g) in 1-L Erlenmeyer was mixed with 100 ml of atrazine contaminated water (30 ppm) and *Ps.* ADP cells to yield a final bacterial concentration of 80 $\mu$g bacterial protein (about $2 \times 10^{-9}$ cells) $g^{-1}$ air dried soil. All treatments were triplicated and incubated in a reciprocal incubator at 27°C, operating at 50 rpm. Non-amended control and soil amended with 500 ppm mercuric chloride were used for controls. Slurry samples (5 ml) were removed at various time intervals. Each sample was centrifuged for 3 min at 10 000 *g*. The supernatant was filtered through a fibreglass filter and analysed for atrazine using an HPLC method described in detail elsewhere (Mandelbaum *et al.*, 1993). The pellet was resuspended in 3 ml methanol and extracted overnight in a reciprocal shaker (150 rpm). Atrazine analysis was performed using the same HPLC method and expressed on a dry soil basis. After the first day, atrazine could not be found in the supernatant (Figure 7 A) or in the soil (Figure 7 B) inoculated with *Ps.* ADP. In the poisoned soil, atrazine was not degraded throughout the experiment.

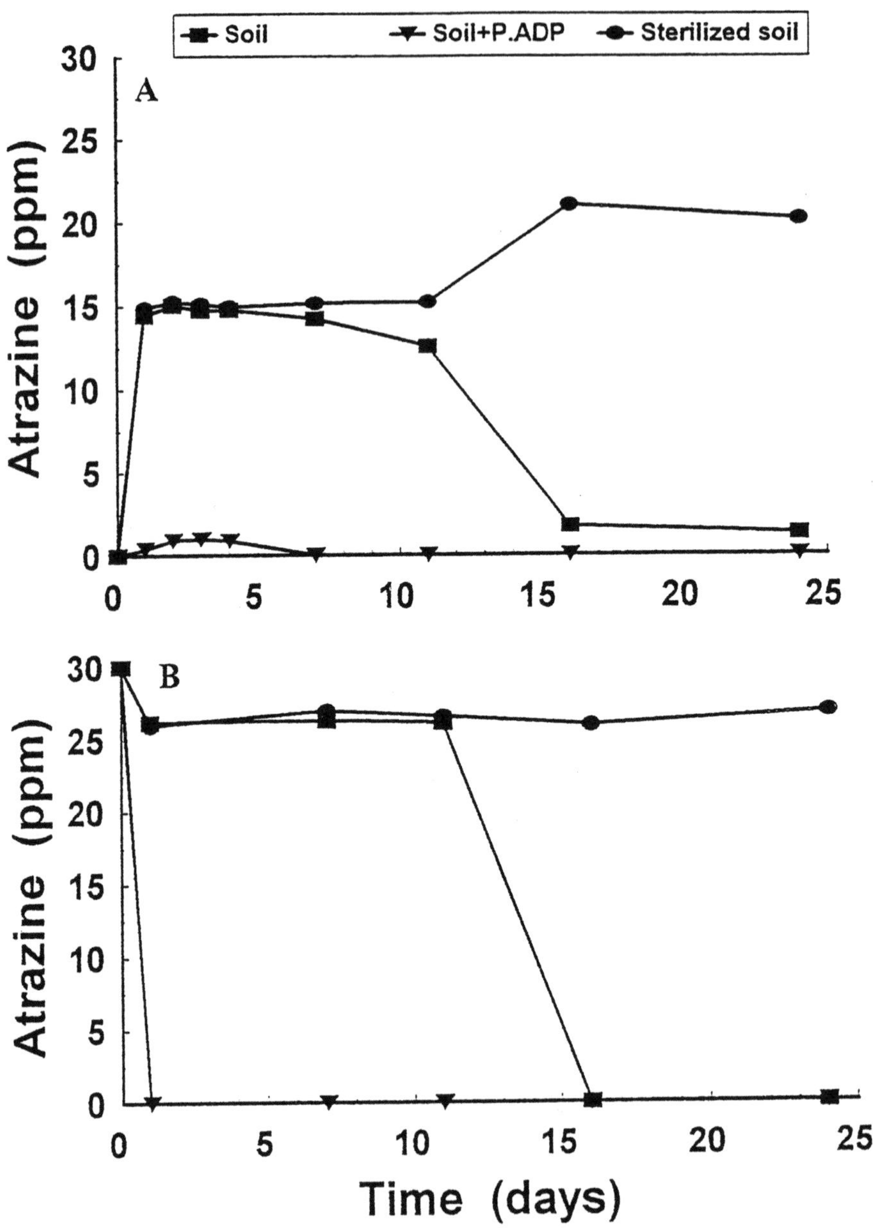


*Figure 7.* Atrazine degradation in slurry of red Mediterranean clayey soil spiked with atrazine. The treatments included native soil, native soil amended with *Pseudomonas* sp. strain ADP and native soil poisoned with 500 ppm (final concentration) mercuric chloride. A = atrazine concentration in the aqueous phase; B = atrazine sorbed to soil.

Interestingly, atrazine was not degraded in the non-poisoned control during a lag time of ten days, followed by vigorous atrazine degradation in the aqueous phase, followed by atrazine decline in the soil fraction.

It was concluded that atrazine could be effectively degraded from soil when *Ps.* ADP was applied shortly after the atrazine. Also, it was evident that atrazine degrading bacteria existed in the soil and could degrade atrazine after a lag time of ten days. The standard deviation in all treatments was less than 10%, and cross-contamination was ruled out, as all samplings were first made in the non-inoculated treatments.

## 9. Summary

Biological treatment of pollution is frequently recognised as being less expensive than physicochemical treatment. Over the years many mixed and pure cultures of atrazine degrading microorganisms isolated from a variety of sources have been described. In many of these studies the isolated microorganisms were suggested for use in biological treatment schemes. However, to the best of our knowledge, until now, none of the described microorganisms was used for actual atrazine bioremediation or biological effluent treatment. Incomplete degradation, slow degradation rate, cost of carbon and oxygen source, hostile environmental conditions (salinity, for example), were among the reasons for the non-commercialisation of previous biodegradation agents.

The isolation of *Ps.* ADP which rapidly mineralizes high concentrations of atrazine in various environments through a novel degradation pathway, may provide a bacterium suitable for actual bioremediation and waste treatment. Further studies will be focused at the genes and proteins which mediate atrazine metabolism.

## Acknowledgments

We thank Janis MacFarland of Ciba-Geigy Corporation for providing the *s*-triazine standards, and the information regarding the atrazine special review by the EPA. We also thank Prof. Larry Wackett and Prof. Mike Sadowsky from the University of Minnesota for their involvement and co-operation throughout the work. This work was supported in part by a grant from Ciba-Geigy Corporation and by grant US-2364-93 from the Israeli-American BARD programme.

## References

Anonymous 1995 Possible formation of nitrosoatrazine in mammalian organisms. In Atrazine. Tier II-section 3- annex II. pp 73–74. Ciba-Geigy Ltd, Greensboro, NC.

Baker B D and Richards P R 1992 Herbicide concentrations in Ohio's drinking water supplies: A quantitative exposure assessment. In Rational Readings on Environmental Concerns, Ed. J H Lehr. pp 84–97. Van-Nostrand Reinhold, New York.

Behki R M and Khan S U 1986 Degradation of atrazine by *Pseudomonas:* N-dealkylation and dehalogenation of atrazine and its metabolites. J. Agric. Food Chem. 34, 746–749.

Behki R M and Khan S U 1994 Degradation of atrazine, propazine and simazine by *Rhodococcus* strain B-30. J Agric. Food. Chem. 42, 1237–1241.

Behki R, Topp E, Dick W and Germon P 1993 Metabolism of the herbicide atrazine by *Rhodococcus* strains. Appl. Environ. Microbiol. 59, 1955–1959.

Belluck D A, Benjamin, S L and Dawson T 1991 Groundwater contamination by atrazine and its metabolites. In Pesticide Transformation Products. Fate and Significance in the Environment. Eds L Somasandarum and J R Coats. pp 254–273. ACS Symposium Series 459, Washington, DC.

Bowmer K H 1991 Atrazine persistence and toxicity in two irrigated soils of Australia. Aust. J. Soil Res. 29, 339–350.

Bryant J B 1963 PhD thesis. pp 31–40. Pennsylvania State University.

Buser H 1990 Atrazine and other *s*-triazine herbicides in lakes and in rain in Switzerland. Environ. Sci. Technol. 24, 1049–1058.

Casper R and Landsmann J 1992 The Biosafety Results of Field Tests of Genetically Modified Plants and Microorganisms. Proceedings of the second International Symposium, Biologiche Bundesanstalt fur Land-und Forstwirtschaft, Coslar, Germany.

Cook A M 1987 Biodegradation of *s*-triazine xenobiotics. FEMS Microbiol. Rev. 46, 93–116.

Cook A M and Huetter R 1984 Desethylsimazine: Bacterial dechlorination, deamination and complete degradation. J. Agric. Food Chem. 32, 581–585.

Couch R W, Gramlich J V, Davis D E and Funderburk H H 1965 The metabolism of atrazine and simazine by soil fungi. Proc. South Weed Conf. 18, 623.

DeSouza M, Wacke H L P, Boundy-Mills K L, Mandelbaum R T and Sadowky M J 1995 Cloning, characterization and expression of a gene region from *Pseudomonas* sp. Strain ADP involved in the dechlorination of atrazine. Appl. Environ. Microbiol. 61, 3373–3378.

Dries D, De Corte B, Liessens J, Steurbaut W, Dejonckheere W and Verstraete W 1987 Recalcitrance of atrazine at low levels to aerobic and hydrogenotrophic microorganisms. Biotechnol. Lett. 9, 811–816.

Dubey S K and Holmes D S 1995 Biological cyanide destruction mediated by microorganisms. World J. Microbiol. Biotechnol. 11, 259–265.

Erickson E L and Lee H K 1989 Degradation of atrazine and related *s*-triazines. Crit. Rev. Environ. Cont. 19, 1–13.

Ernst C and Rehm H J 1995 Utilization of chlorinated *s*-triazines by a new strain of *Klebsiella pneumoniae*. Appl. Microbiol. Biotechnol. 42, 763–768.

Feakin S J, Blackburn E and Burns R G 1994 Biodegradation of *s*-triazine herbicides at low concentrations in surface waters. Water Res. 28, 2289–2296.

Felsot A S and Dzantor K E. 1990 Enhancing biodegradation for detoxification of herbicide waste in soil. In: Enhanced Biodegradation of Pesticides in the Environment. Eds Racke K D and Coats J R pp. 249–268. Am. Chem. Soc. Symp. Ser. No. 42, Am. Chem. Soc., Washington, D C.

Felsot A S, Liebl R and Bicki T 1988 Feasibility of land application of soils contaminated with pesticide waste as a remediation practice. Final Project Report. Illinois Hazardous Waste Research and Information Center, Illinois State Water Survey Division, Savoy, IL.

Foster D M, Rachwal A J and White S L 1991 New treatment processes for pesticides and chlorinated organics control in drinking water. J. Inst. Wat. Envir. Man. 5, 466–477.

Gabriel P F 1991 Innovative technologies for contaminated site remediation: Focus on bioremediation. J. Air Waste Manag. Assoc. 41, 1657–1660.

Galli E, Silver S and Witholt B 1992 *Pseudomonas*, Molecular Biology and Biotechnology. American Soc. Microbiol., Washington, DC.

Giardina M C, Giardi M T and Fliacchioni G 1980a 4-amino-2-chloro-1,3,5,-triazine: A new metabolite of atrazine by a soil bacterium. Agric. Biol. Chem. 44, 2067–2072.

Giardina M C, Giardi M T and Fliacchioni G 1980b Atrazine metabolism by *Nocardia*: Elucidation of initial pathway and synthesis of possible metabolites. Agric. Biol. Chem. 46, 1439.

Kaufman D D and Blake J 1970 Degradation of atrazine by soil fungi. Soil Biol. Biochem. 2, 73–80.

Kolpin D W and Kalkhoff S J 1993 Atrazine degradation in a small stream in Iowa. Environ. Sci. Technol. 27, 134–139.

Korpraditskul R, Katayama A and Kuwatsuka S 1993 Degradation of atrazine by soil bacteria in the stationary phase. J. Pest. Sci. 18, 293–298.

Lee S 1994 Recent advances in the analysis of microbial communities and activity in marine ecosystems. In Beyond the Biomass. Eds K Ritz, J Dighton and K E Giller. pp 95–100. The British Society of Soil Science.

Mandelbaum R M, Wackett L P and Allan D L 1993 Rapid hydrolysis of atrazine to hydroxyatrazine by soil bacteria. Environ. Sci. Technol. 27, 1943–1946.

Mandelbaum R T, Allan D L and Wackett L P 1995 Isolation and characterization of a *Pseudomonas* sp that mineralizes the *s*-triazine herbicide atrazine. Appl. Environ. Microbiol. 61, 1451–1457.

Mulbry W 1994 Purification and characterization of an inducible *s*-triazine hydrolase from *Rhodococcus corallinus* NRRL B-15444R. Appl. Environ. Microbiol. 60, 613–618.

Olsen G H 1988 Phylogenetic analysis using ribosomal RNA. Methods Enzymol. 164, 793–812.

Parshley T J 1995 Atrazine/simazine, response to the United States Environmental Protection Agency's position document 1: Initiation of special review, November 23, 1994. Ciba-Geigy Corporation, Greensboro, NC.

Pick F E, Van Dyke LP and Botha E 1992 Atrazine in ground and surface water in maize production areas on the Transvaal, South Africa. Chemosphere 25, 335–341.

Radosevich M, Traina S J, Hao Y L and Tuovinen O H 1995 Degradation and mineralization of atrazine by a soil bacterial isolate. Appl. Environ. Microbiol. 61, 297–302.

Rao P S C and Davidson J M 1979 Adsorption and movement of selected pesticides at high concentrations in soil. Water Res. 13, 375–380.

Ritter W F 1990 Pesticide contamination of ground water in the United States – review. J. Environ. Sci. Health. B25, 1–29.

Sayler G S, Fox R, Blackburn J W 1991 Environmental Biotechnology for Waste Treatment. Plenum Press, New York.

Schoen S R and Winterlin W L 1987 The effects of various soil factors and amendments on the degradation of pesticide mixtures. J. Environ. Sci. Health. B22, 347–377.

Shao Z Q and Behki R 1995 Cloning of the genes for degradation of the herbicides EPTC (*s*-ethyl dipropylthiocarbamate) and atrazine from *Rhodococcus* sp strain TE1. Appl. Environ. Microbiol. 61, 2061–2065.

Tsien H C, Bratina B J, Tsuji K and Hanson R S 1990 Use of oligonucleotide signature probes for identification of physiological groups of methylotrophic bacteria. Appl. Environ. Microbiol. 56, 2858–2865.

Vandepitte V, Wierinck I, Devos P, Depoorter M P, Houwen F and Verstraete W 1994 N-dealkylation of atrazine by hydrogenotrophic fluorescent pseudomonads. Water Air Soil Pollut. 78, 335–341.

Weisenburger D D 1990 Environmental epidemiology of non-Hodgkin's lymphoma in Eastern Nebraska, Am. J. Ind. Med. 303–305.

Weisenburger D D, Joshi S S, Hickman T I, Babcock K M, Walker B A and Mirvish, S S 1987 N-nitrosoatrazine (NNAT) synthesis, chemical properties, acute toxicity and mutagenicity. Proc. Am. Assoc. Cancer. 28–103.

Winterlin N, Seiber J N, Craigmill A, Baier J, Woodrow J and Walker G 1989 Degradation of pesticide waste taken from a highly contaminated soil evaporation pit in California. Arch. Environ. Contam. Toxicol. 18, 734–747.

Yanze-Kontchou C and Gschwind N 1994 Mineralization of the herbicide atrazine as a carbon source by a *Pseudomonas* strain. Appl. Environ. Microbiol. 60, 4297–4302.

# 43
# Removal of lead from aqueous solutions by a *Brevibacterium* strain

V. ANDREONI, M. COLOMBO, D. DI SIMINE, C. FINOLI, G. ORIGGI, A. VECCHIO AND R. CARZANIGA

## 1. Introduction

Heavy metals are of concern when found in high concentrations in soils, surface waters, and groundwater, due to their toxic nature. The non-biodegradability of these elements is responsible for their accumulation in the biosphere and translocation through the food chain. Some microorganisms have developed different and efficient mechanisms for tolerating high levels of metal content and exhibit a number of enzymatic or non-enzymatic processes for the uptake and accumulation of heavy metals (Gadd, 1990; Silver and Misra, 1988; Silver and Walderhaugh, 1992). Microbial metal uptake is receiving attention for:

- the study of metal nutrition and toxicity at the cellular level;
- the important role of these organisms in biogeochemical cycling and in biological food chains;
- their biotechnological potential in metal removal processes.

Petrol combustion, lead smelters, pesticides, and fertilizers are responsible for a significant input of lead into terrestrial and aquatic ecosystems (Francis and Dodge, 1986). The release of lead into the environment renders necessary the development of low-cost and effective treatments for its removal. The present work was aimed at characterizing a strain of *Brevibacterium* in order to study microbe-lead interactions and to check the removal and the uptake mechanism by resting cells of the metal from dilute solution, in view of the use of the microorganism in biotreatments of aqueous wastes.

## 2. Materials and methods

### 2.1 Bacterial strain and growth conditions

The *Brevibacterium iodinum* strain used in this study was isolated from a soil of a shooting range near Milan, Italy, containing 260 mg/kg $Pb^{2+}$ (Andreoni *et al.*, 1994). The strain was maintained on bacto tryptone glucose (BTG) agar containing bacto-tryptone (Difco) 10.0 g and glucose 2.0 g per litre, and on BTG agar supplemented with lead 500 mg $L^{-1}$, added as lead nitrate.

*D. Rosen* et al. *(eds): Modern Agriculture and the Environment, 523–533.*

Cultures were tested for inducibility to lead resistance. Actively growing cultures of induced and uninduced cells were incubated in BTG broth, pH 7.5, supplemented with suitable amounts of lead (to obtain a final lead concentration of 500, 750 or 1500 mg $L^{-1}$), at 30°C, with shaking at 200 rpm in an incubator shaker. At various time intervals, samples of cultures were aseptically removed and the ATP pool of growing cells was monitored by measuring the adenosine triphosphate (ATP) and protein content. ATP was measured by a luciferin–luciferase assay using a luminometer Biocounter P 1500 Lumac (Jago *et al.*, 1989); protein content was determined by the method of Bradford (Bradford, 1976) after alkaline hydrolysis of the cells.

### 2.2 *Antibiotic and metal resistance*

BTG agar was used as medium to determine the antibiotic resistance of *Brevibacterium.* Sensitivity tests were carried out by the standardized disk method (Bauer *et al.*, 1966). Antibiotics and their concentrations employed were as follows: clindamycin, 0.5 $\mu$g; cefuroxime, 30 $\mu$g; cefotaxime, 30 $\mu$g; cefamandole, 9 $\mu$g; penicillin, 1 IU; nalidixic acid, 30 $\mu$g; cefotetan, 10 $\mu$g; cephalothin, 6 $\mu$g; colistin, 6 $\mu$g; moxalactam, 16 $\mu$g; amoxicillin, 25 $\mu$g; tetracycline, 5$\mu$g; kanamycin, 8 $\mu$g; erythomycin, 3 $\mu$g; and mezlocillin, 30 $\mu$g. Resistance to antibiotics was positive if growth occurred up to the edge of the disk after 48 h.

The resistance of the microorganism to various metals was tested by adding to BTG broth different amounts of lead nitrate, nickel chloride, cobalt chloride, cadmium acetate, zinc sulphate, mercurous chloride, potassium chromate, sodium arsenite and arsenate, and aluminium sulphate. Before addition, the metal solutions were sterilised by filtration (0.22 $\mu$m). The tolerance to metals was determined by monitoring the growth photometrically at 650 nm.

### 2.3 *Lead uptake experiments by resting cells*

Cells of *Brevibacterium* were grown in BTG broth for 17 h, harvested, rinsed and resuspended in assay buffer 0.025 M 1,4-piperazin-bis-(ethansulphonic acid) (PIPES), pH 6.3, to achieve an absorbance of about 1.7 at 540 nm.

#### 2.3.1 Kinetics of lead uptake

One-hundred ml resting cell suspensions was added to 100 ml of 0.025 M PIPES containing 20 mg $L^{-1}$ $Pb^{2+}$. Cells were incubated for 15 min, 30 min, and 1, 2, 4 h respectively to accumulate $Pb^{2+}$. At these time intervals, the cells were harvested by centrifugation (10 000 *g*, 15 min). The supernatants were removed and the cells washed three times with fresh buffer; the washings were combined with the supernatants into a volumetric flask. Lead was determined on the cells and on supernatant aliquots.

2.3.2 Concentration effect on lead uptake

One-hundred ml cell suspensions were added to 100 ml of buffer solution containing lead to give final $Pb^{2+}$ concentrations in the range 5–40 mg $L^{-1}$ and then incubated at 30°C for 1 h. At the end of incubation the cells were harvested by centrifugation and treated as described in the kinetics of lead uptake.

2.3.3 Lead distribution in cells

One-hundred ml resting cell suspensions were added to 100 ml of 0.025 M PIPES containing 40 mg $L^{-1}$ $Pb^{2+}$ and incubated both in the absence and in the presence of glucose 4 g $L^{-1}$. After an hour the cells were harvested by centrifugation and rinsed with a 10 mM EDTA solution, 3 × 10 ml. The phases were separated by centrifugation and the EDTA solutions were combined. After recovery the cells were disrupted in a French pressure cell (American Instruments Co.) using two passes at 20 000 psi. The broken material was centrifuged at 60 000 *g* for 20 min at 4°C to separate the cytoplasm from the cell walls and membranes. Lead was determined in the supernatant, EDTA rinsing, cytoplasm and cell envelopes.

Determination of lead was performed using a Varian SpectrAA 20 absorption spectrometer after $HNO_3$ digestion of samples according to APHA standard methods (APHA, 1992). All glassware was soaked with hot 30% $HNO_3$ and rinsed with metal-free water. All the experiments were performed in duplicate, but only the mean values were plotted. Duplicate data differed from the mean values by less than 5%.

## 2.4 *Bacterial dry weight*

The cells, harvested by centrifugation and rinsed twice with deionised water, were dried overnight at 105°C.

## 2.5 *Transmission electron microscopy*

Electron microscopy was performed on resting cells incubated either in the absence or in the presence of 20 mg $L^{-1}$ $Pb^{2+}$ and harvested after 1 hour by centrifuging. The cells were prefixed with 3% (v/v) glutaraldehyde in 0.1 M phosphate buffer, pH 7.2. Samples were left in the fixative overnight at 4°C, carefully washed with the same buffer, post-fixed for 2 hours in 1% (w/v) $OsO_4$ in phosphate buffer, and then pelleted in 0.9% agar at 45°C. The pellets were cut in 1-$mm^2$ pieces which were then dehydrated in an ethanol series and embedded in London Resin White. Ultrathin sections were stained with 2% (w/v) uranyl acetate aqueous solution and examined in a Jeol 100 SX transmission electron microscope at 80 kV.

### 2.6 *Energy dispersive X-ray analysis*

After 1 h suspension in PIPES buffer containing 20 mg $L^{-1}$ $Pb^{2+}$, the cells were centrifuged, washed twice with distilled water and dried at 105°C. Cells were mounted on a glass slide without gold spattering and analysed by a scanning electron microscope (SEM), operating at 7.5 kV, equipped with an energy dispersive X-ray spectroscope (EDS) using a Cambridge Link under standard conditions with a live time of 200 sec.

### 2.7 *Chemicals*

All chemicals were of analytical grade.

## 3. Results

*Brevibacterium iodinum* was capable of growing in the presence of lead up to 1500 mg $L^{-1}$. The effect of various amounts of $Pb^{2+}$ on the metabolic conditions of growing cells of the microorganism is reported in Figure 1. The ATP pool was affected by $Pb^{2+}$ added at increasing concentrations, which exerted a perturbation on energy-linked functions. A phase of 4 hours was necessary to cells to adapt to the metal before reaching an actively metabolizing phase; the duration of this phase was not dependent on lead concentration. Interestingly, about the same protein yield was obtained in the first 2 hours of incubation when the cells were cultured both in the absence and in the presence of $Pb^{2+}$ and then the protein content of cells cultured in the presence of $Pb^{2+}$was higher than the control (Table 1). Cell cultures, maintained in the presence and in the absence of lead, when inoculated in a Pb-containing

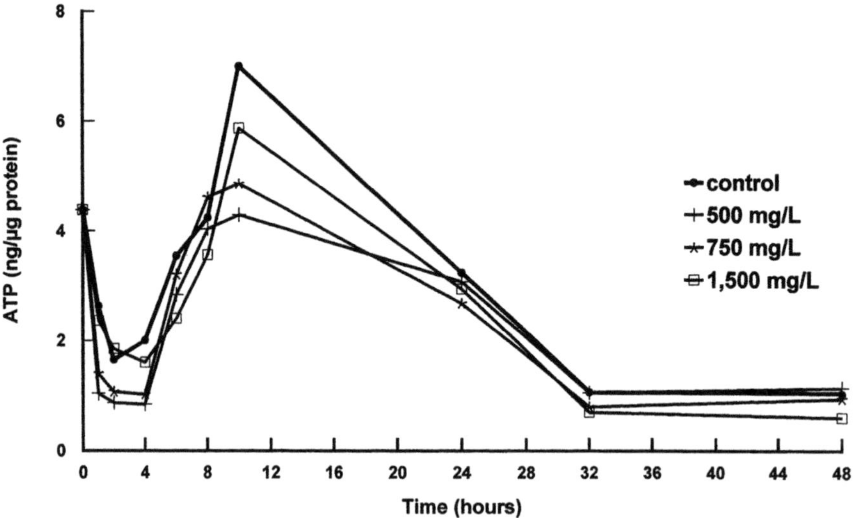


*Figure 1.* Effect of various concentrations of lead on ATP pool of growing cells of *B. iodinum*.

*Table 1.* Effect of lead concentrations on protein content of growing cells.

| | $Pb^{2+}$ added | | | |
|---|---|---|---|---|
| Incubation time (hours) | 0 mg $L^{-1}$ $\mu g$ protein $ml^{-1}$ | 500 mg $L^{-1}$ $\mu g$ protein $ml^{-1}$ | 750 mg $L^{-1}$ $\mu g$ protein $ml^{-1}$ | 1500 mg $L^{-1}$ $\mu g$ protein $ml^{-1}$ |
| 0 | 20 | 20 | 20 | 20 |
| 1 | 21 | 34 | 25 | 21 |
| 2 | 34 | 38 | 30 | 24 |
| 4 | 49 | 82 | 103 | 74 |
| 6 | 145 | 196 | 178 | 154 |
| 8 | 200 | 212 | 248 | 231 |
| 10 | 160 | 200 | 217 | 232 |
| 24 | 316 | 485 | 458 | 463 |
| 32 | 314 | 355 | 430 | 419 |
| 48 | 304 | 333 | 344 | 357 |

For all experiments, approx. $3 \times 10^8$ cells $ml^{-1}$ were inoculated into 150 ml BTG broth either supplemented with $Pb^{2+}$ or unsupplemented.

medium, showed no difference in the lag phase suggesting that the metal resistance was constitutive (data not shown). The strain showed simultaneous resistance to antibiotics and metals (Table 2).

The capacity of *Brevibacterium* strain to sorb lead is reported in Figure 2. Cells demonstrated a rapid accumulation of a considerable amount of lead within 15 minutes, which subsequently was slow and gradual. Resting cells accumulated most of the metal, and removed, in an hour, about 95–97% lead from concentrations ranging from 5 to 40 mg $L^{-1}$. Lead uptake depended on metal concentration and increased almost linearly in the tested concentration range. The maximum uptake, achieved after 1 h of incubation, was 13.10 $\mu g$ of metal per mg of dry weight in the presence of 40 mg $L^{-1}$ lead (Table 3). The relationship between the concentration of residual lead in solution and the amount of lead adsorbed per unit biomass was determined by applying the linearised Freundlich isotherm (Freundlich, 1926) (Figure 3).

In the experiments on cellular lead distribution carried out in the absence of glucose the removed metal was accumulated in the cytoplasm (13.8%) and associated with the cell envelopes (86.2%); about 68% of the latter was released by EDTA rinsing. In the experiments in the presence of glucose the cytoplasmic content increased up to 18.8% (Figure 4).

The electronmicrographs (Figure 5) revealed the existence of electron-dense deposits of lead, which were not present in the control cells incubated in the absence of lead. These deposits were located on the cell wall of the bacteria and within the cells. In the X-ray spectroscopic spectrum, peaks of Mg, P, Pb, and K were present (Figure 6). The P and K peaks were higher than the Pb and Mg peaks. These results showed that the deposits in these bacterial cells consisted mainly of K and P and additionally of Pb.

*Table 2.* Metal and antibiotic resistance of *Brevibacterium iodinum*.

| Metal | mg $L^{-1}$ | | Antibiotic | $\mu g$ | |
|---|---|---|---|---|---|
| Aluminium | 10 | + | Amoxicillin | 8 | R/S |
| | 50 | − | | 25 | S |
| | | | Cefamandole | 9 | R |
| Arsenic(III) | 150 | + | Cefoperazone | 14 | R/S |
| | 500 | − | Cefotaxime | 30 | R |
| | | | Cefotetan | 10 | R |
| Arsenic(V) | 10 000 | + | Cefuroxime | 12 | R |
| | | | Cephalothin | 6 | R |
| Cadmium | 50 | + | Clindamycin | 0.5 | R |
| | 200 | − | | 2 | R/S |
| | | | Colistin | 6 | R |
| Chromium(VI) | 5 000 | + | Erythromycin | 3 | S |
| | | | Kanamycin | 8 | S |
| Cobalt | 75 | + | Mezlocillin | 30 | S |
| | 150 | − | Moxolactam | 16 | R |
| | | | Nalidixic acid | 30 | S |
| Copper | 120 | + | Penicillin | 1 IU | R |
| | | | Tetracycline | 5 | S |
| Mercury(II) | 20 | + | Ticarcillin | 25 | R/S |
| | 50 | − | | | |
| Silver | 5 | + | | | |
| | 15 | − | | | |
| Zinc | 100 | + | | | |
| | 200 | − | | | |

+ growth; − absence of growth.
R, resistant; S, sensitive; R/S, slightly sensitive.

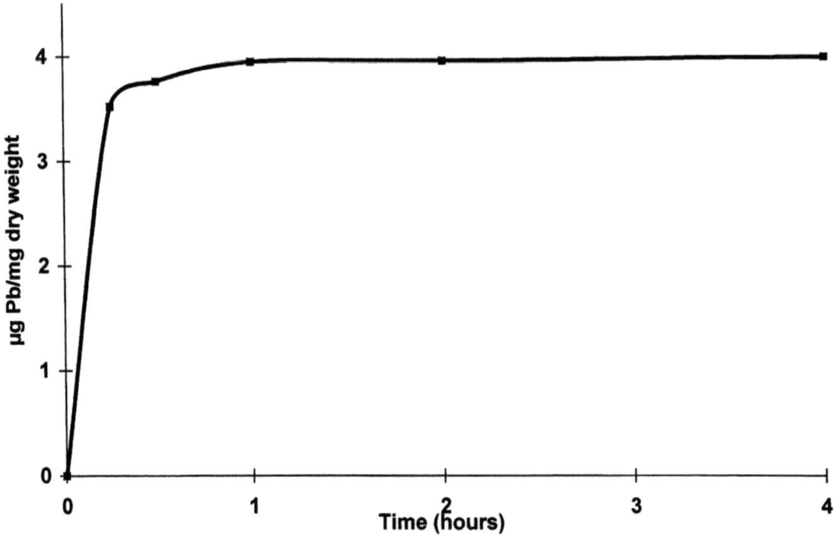


*Figure 2.* Time course of lead uptake by resting cells of *B. iodinum* in the presence of 10 mg $L^{-1}$ lead.

*Table 3.* Effect of lead concentrations on lead uptake by resting cells of *Brevibacterium iodinum.*

| Treatment (mg Pb $L^{-1}$) | Dry weight (DW) (mg $ml^{-1}$) | Metal uptake $\mu g$ Pb mg $DW^{-1}$ | Lead concentration at equilibrium (mg $L^{-1}$) |
|---|---|---|---|
| 5 | 3.24 | 1.50 | 0.14 |
| 10 | 2.57 | 3.72 | 0.33 |
| 20 | 2.90 | 6.72 | 0.52 |
| 30 | 2.57 | 11.06 | 1.52 |
| 40 | 2.90 | 13.10 | 2.00 |

Resting cell suspensions were incubated with lead (5, 10, 20, 30, 40 mg $L^{-1}$) in 0.025 M PIPES buffer, pH 6.3.

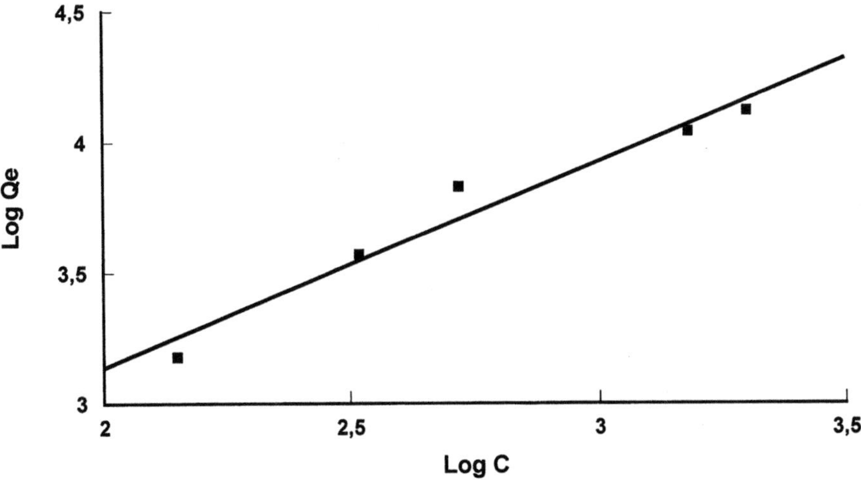


**C: concentration of lead remaining in solution at equilibrium (µg/L)**
**Qe: lead adsorbed per unit biomass (µg/mg dry weight)**

*Figure 3.* Freundlich plot of lead biosorption by *B. iodinum.*

## 4. Discussion

*Brevibacterium iodinum* is able to grow in the presence of lead up to 1500 mg $L^{-1}$ and growing cells seem to resist high levels of lead by synthesizing inducible proteins, whose function has not been investigated. The protein content could increase in consequence of the synthesis of metal binding proteins, or membrane proteins implicated in ion transport and efflux. Production of metallothioneins or proteins showing metallothionein-like properties, involved in metal storage and detoxification, have been reported by Higham *et al.* (1984), and Olafson *et al.* (1988). Many authors report that, in some bacteria, resistance to several divalent cations is based on the action of inducible proteins, ATPase and proton-cation

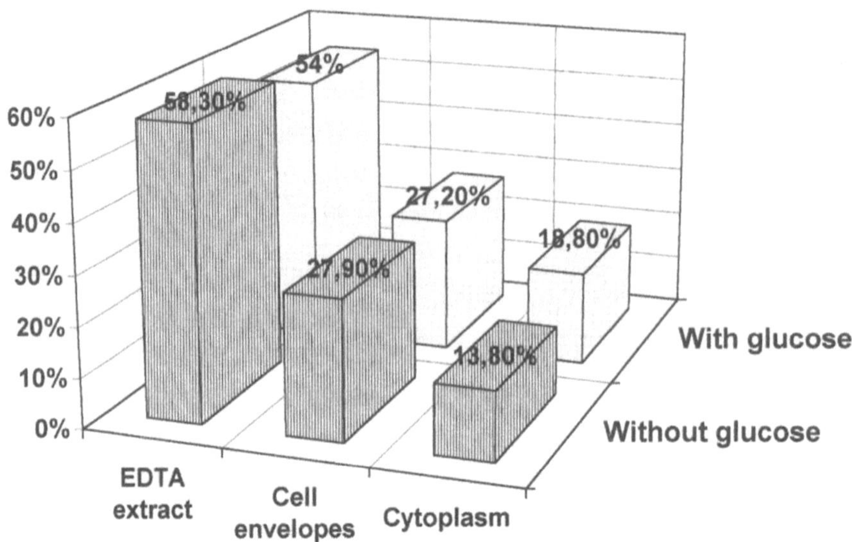


*Figure 4.* Cellular distribution of lead after accumulation by resting cells of *B. iodinum* treated with 20 mg $L^{-1}$ lead.

antiporters, which prevent the internal accumulation of metal (Nies, 1992; Nies *et al.*, 1989; Nucifora *et al.* 1989).

Resting cells of the strain are capable of removing about 95–97% of the metal from solutions ranging from 5–40 mg $L^{-1}$. The biosorption of lead fits the Freundlich isotherm suggesting a linear equilibrium between the metal concentration in solution and that bound to the cell surface, thus indicating a monolayer adsorption.

Lead uptake in this strain seems to occur in two steps: the first is rapid and is energy independent; the second step, the uptake within the cells, might be energy dependent. The sorption of metal ions to resting cells of *Brevibacterium* appears to involve primarily an ion-change type of reaction; data obtained in the kinetics and desorption experiments support this hypothesis. Most of the metal (86%) is associated with the cell envelopes and approximately 68% of the bound metal ions are removed from the *Brevibacterium* cells by the strong chelating agent EDTA. The cytoplasmic content of $Pb^{2+}$ indicates a diffusion of the metal. Lead penetration is likely to be energy-dependent; in fact the presence of glucose enhances the uptake within the cells by stimulating the transport of lead into resting cells. Lead, besides being incorporated, probably into polyphosphate bodies, is precipitated and accumulated on the surface. Lead is accumulated as a compound of lead and phosphorus in an amorphous state whose molar ratio has not been determined. Accumulation of lead as $PbHPO_4$ by a *Pseudomonas fluorescens* and of cadmium as $CdHPO_4$ by a *Citrobacter* sp. have been reported by Aickin *et al.* (1979), and by

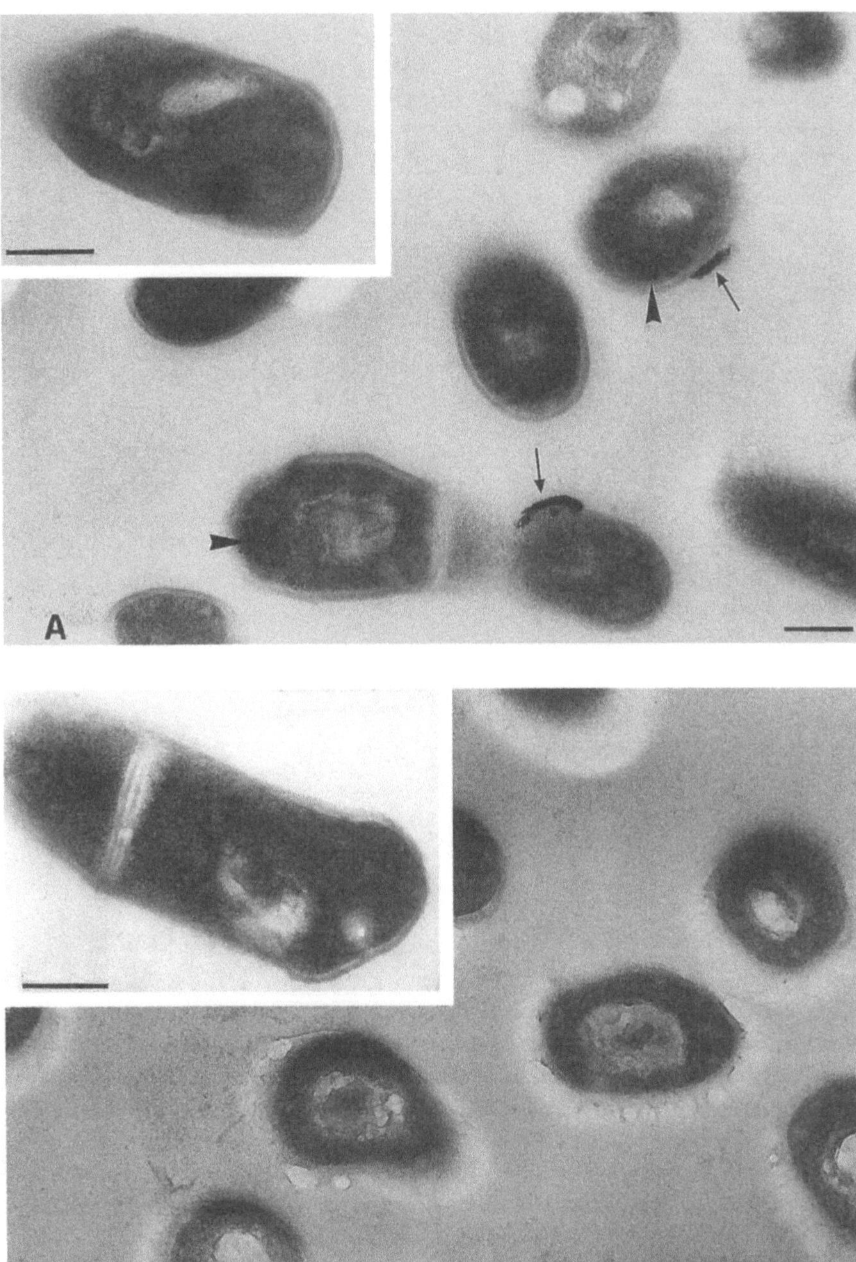


*Figure 5.* Thin sections of *B. iodinum* treated with 20 mg $L^{-1}$ lead (A) and untreated (B). $Pb^{2+}$ deposits are present both in the cytoplasmic regions close to the cell periphery (arrow heads) and on the cell wall (arrows) (all bars = 200 nm).

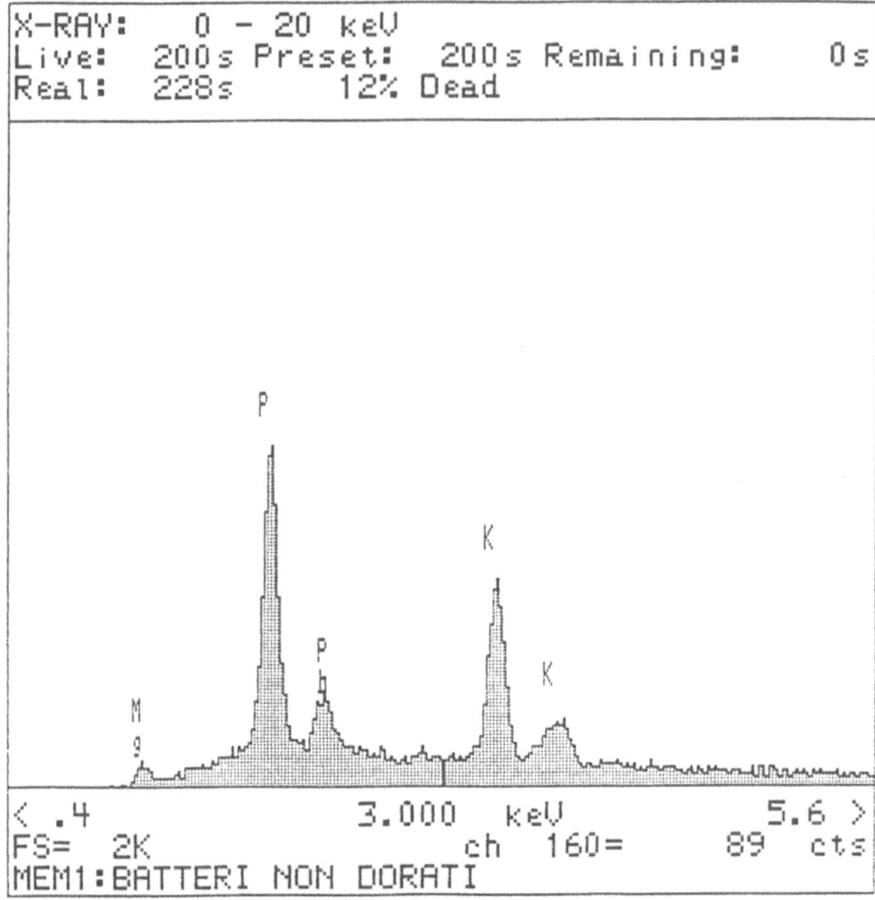


*Figure 6.* EDS spectrum from resting cells of *B. iodinum* treated with 20 mg $L^{-1}$ lead.

Macaskie *et al.* (1987), respectively. Jensen *et al.* (1986) described the polyphosphate bodies as binding sites for many heavy metals in algae.

The application of alternative methods for removing metals from aqueous solutions is a growing field of research and the use of microbial biomass, which exhibits a selective retention of heavy metal ions, represents an inexpensive material to utilize in clean-up processes. The results presented here indicate that the capability of the microorganism isolated is potentially useful for removing lead from solutions. Moreover, the bound lead can be released and separated by bacteria.

## Acknowledgement

This work was supported by a 60% grant from the Ministry of University and Scientific and Technological Research.

## References

Aickin R M, Dean A C R, Cheetham A K and Skarnulis A J 1979 Electron microscope studies on the uptake of lead by a *Citrobacter* sp. Microbios Lett. 9, 7–15.

Andreoni V, Fasoli R, Sorlini C and Finoli C 1994 Investigation of lead resistance in a *Brevibacterium* strain. Ann. Microbiol. Enzimol. 44, 153–161.

APHA, AWWA, WPCF 1992 Standards Methods for the Examination of Water and Wastewater. 18th ed. American Public Health Ass., Washington, DC, 1266 p.

Bauer A W, Kirby W M M, Sherris J C and Turk M 1966 Antibiotic susceptibility testing by a standardized disk method. Am. J. Clin. Pathol. 45, 493–496.

Bradford M M 1976 A rapid and sensitive method for the quantitation of microgram quantities of protein utilizing the principle of protein-dye binding. Anal. Biochem. 72, 248–254.

Francis A J and Dodge C J 1986 Anaerobic bacterial dissolution of lead oxide. Arch. Environ. Contam. Toxicol. 15, 611–616.

Freundlich C H 1926 Colloid and Capillary Chemistry. Methuen, London.

Gadd G M 1990 Heavy metal accumulation by bacteria and other microorganisms. Experientia 46, 834–840.

Higham D, Sadler P J and Scawen M D 1984 Cadmium-resistant *Pseudomonas putida* synthesizes novel cadmium proteins. Science 225, 1043–1046.

Jago P H, Stanfield G, Simson W J and Hammond J R M 1989 The effect of extractant and sample composition on the performance of a commercial firefly luciferase reagent. In ATP Luminescence. Eds P E Stanley *et al.* p 53. Blackwell Scientific Publishers, Oxford, UK.

Jensen T E, Rachlin J W, Jani V and Warkentine B E 1986 Heavy metal uptake in relation to phosphorus nutrition in *Anabaena variabilis* (Cyanophyceae). Environ. Pollut. Ser. A42, 261–271.

Macaskie L E, Dean A C R, Cheetham A K, Jakeman R J B and Skarnulis A J 1987 Cadmium accumulation by a *Citrobacter* sp.: The chemical nature of the accumulated metal precipitate and its location on the bacterial cells. J. Gen. Microbiol. 133, 539–544.

Nies D H 1992 Resistance to cadmium, cobalt, zinc and nickel in microbes. Plasmid 27, 17–28.

Nies D H, Nies A, Chu L and Silver S 1989 Expression and nucleotide sequence of a plasmid-determined divalent cation efflux system from *Alcaligenes eutrophus.* Proc. Natl. Acad. Sci. USA 86, 7351–7355.

Nucifora G, Chu L, Misra T K and Silver S 1989 Cadmium resistance from *Staphylococcus aureus* plasmid pI258 *cadA* gene results from a cadmium-efflux ATPase. Proc. Natl. Acad. Sci. USA 86, 3544–3548.

Olafson R W, McCubbin W D and Kay C M 1988 Primary and secondary-structural analysis of a unique prokaryotic metallothionein from a *Synechoccus* sp. cyanobacterium. Biochem. J. 251, 691–699.

Silver S and Misra T K 1988 Plasmid-mediated heavy metal resistances. Ann. Rev. Microbiol. 42, 717–743.

Silver S and Walderhaugh M 1992 Gene regulation of plasmid- and chromosome-determined inorganic ion transport in bacteria. Microbiol. Rev. 56, 195–228.

# 44
# Polycyclic aromatic hydrocarbon degradation by white rot fungi

M. WOLTER, F. ZADRAZIL, R. MARTENS AND M. BAHADIR

## 1. Introduction

Polycyclic aromatic hydrocarbons (PAH) are ubiquitous environmental pollutants, usually formed by incomplete combustion processes. Many PAH are known to be mutagenic and carcinogenic. In addition, PAH with more than three condensed rings are characterised by microbial recalcitrance.

Extreme contaminations of soils with PAH are found in the vicinity of coal processing plants. These findings require increasing efforts to decontaminate these soils to avoid public health hazards. Besides physicochemical techniques, biological methods can provide an economical and efficient alternative for soil decontamination.

White rot fungi are known to be able to degrade the polymer lignin which is generally resistant to microbial attack. The lignin is oxidised by extracellular enzyme systems (Hammel *et al.*, 1986). Because of the low specificity of the enzymes it is assumed that they are able to oxidise persistent environmental pollutants. It has in fact been confirmed by some scientists that white rot fungi are able to degrade the highly persistent and carcinogenic PAH with four and more condensed rings (Bumpus *et al.*, 1985; Sack and Günther, 1993). The investigations focused mainly on the species *Phanerochaete chrysosporium.* The question arose whether this species is the most suitable white rot fungus to degrade PAH in soil.

About 50 white rot fungi were screened by Martens and Zadrazil (1992) for their ability to degrade PAH. In a sandy loamy agricultural soil artificially contaminated with $^{14}$C-pyrene or $^{14}$C-benzo(a)anthracene (10 mg kg$^{-1}$ each), mineralization rates of up to 70% for $^{14}$C-pyrene and 50% for $^{14}$C-benz(a)anthracene were found. During the experiment two different types of fungi could be observed: one group was growing from the substrate (straw) into the soil; the other group did not show the ability to colonise the soil. Surprisingly the experiments with fungi which were not growing into the soil showed higher mineralization rates than the experiments with fungi growing into the soil. The second observation was a disintegration of straw to a non-structured pulp in the samples with fungi not growing into the soil. It is assumed that from this pulp enzymes or degradation products were transported into the soil which supported the native microflora in the soil in degradation of the PAH.

The aim of this work was to investigate whether the high mineralization rates which were found by use of the agricultural soil could also be achieved with a soil

*D. Rosen* et al. *(eds): Modern Agriculture and the Environment, 535–544.*

from a contaminated site. Furthermore, the mineralization rate in the presence of the native microflora had to be compared with the mineralization rate caused by the white rot fungi alone. The experiments were performed in the same manner as described by Martens and Zadrazil (1992).

Additionally, the degradation of eight selected EPA (Environmental Protection Agency)-PAH in the contaminated site was estimated by HPLC with fluorescence detection.

## 2. Materials and methods

### *2.1 Reagents*

4,5,9,10-$^{14}$C-pyrene (2.07 GBq mol$^{-1}$, purity 98.9%) and 7,10-$^{14}$C-benzo(a)pyrene (2.18 GBq mmol$^{-1}$, purity 99.4%) were purchased from Amersham Buchler (Braunschweig). The following PAH were used as analytical standards: chrysene (Aldrich), benzo(b)fluoranthene (Aldrich), benzo(ghi)perylene (Aldrich), dibenz-(a,h)anthracene (Sigma), indeno(1,2,3-cd)pyrene (Amchro). The purities were at least 97%.

### *2.2 Soil*

The soil sample was taken from the site of a former coal tar factory. It was a sandy loamy soil with the following characteristics: pH 7.5; organic carbon 15.9%; inorganic carbon 4.9%; water holding capacity 42.2%; and N-content 0.209%. The content of EPA-PAH with four and more condensed rings ranged from 3 to 15 mg kg$^{-1}$ per compound.

The sample was air dried, passed through a sieve (2 mm inner diameter) and homogenized.

### *2.3 Fungi*

The species *Ganoderma applanatum, Dichomitus squalens* were used as representatives for the non-soil-colonizing fungi and two *Pleurotus* species were used as representatives for the soil-colonizing fungi. They were obtained from the Institut für Bodenbiologie. The fungi were cultivated in glass tubes (length 230 mm and inner diameter 27 mm) on 22 g (wet weight, 76% water content) of sterilized chopped wheat straw. The utilization of wheat straw as substrate for PAH-degradation by white rot fungi has been described by Morgan *et al.* (1993). Within two to four weeks the fungi grew from the inoculum through the 7-cm long straw column.

### *2.4 Incubation*

After the complete colonization of the straw by the fungi, 30-g samples of the soil were contaminated with $^{14}$C-pyrene or $^{14}$C-benzo(a)pyrene (17 kBq). For this

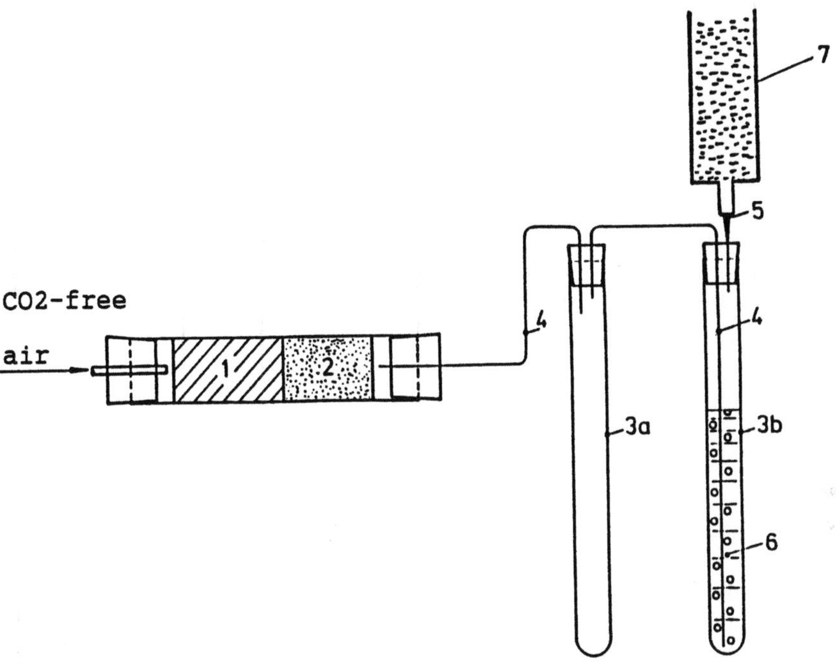


*Figure 1.* Experimental design of degradation experiments. 1, chopped wheat straw; 2, soil; 3a, test tube to prevent return flow; 3b, test tube as $CO_2$-trap; 4, capillary tubings; 5, syringe needle; 6, 25 ml 2N NaOH; 7, syringe filled with glasswool.

purpose the $^{14}C$-PAH were fixed on 0.75 g of the test soil, which was then mixed into the soil sample. These samples were placed on the inner surface of the straw column giving a soil column of 9 cm. Controls with soil only and soil with sterile straw were also run. Additionally, $^{14}C$-pyrene was fixed on 0.75 g soil and added into the inoculated straw column in order to measure the mineralization rate caused by the white rot fungi alone. Figure 1 illustrates the experimental design. The filled columns were flushed continually with $CO_2$-free air at a rate of 8 to 10 ml min$^{-1}$. The experiments were run at 25°C for 20 weeks as quadruplicates.

### 2.5 *Analyses*

The $^{14}CO_2$ released was trapped in 2N NaOH solution and the air was finally passed through glasswool impregnated with paraffin oil in order to trap lipophilic volatile substances. Aliquots of the NaOH solution were measured weekly by liquid scintillation counting (LSC) in order to quantify the $^{14}CO_2$. The paraffin oil was eluted with 30 ml toluene after incubation and the activity was quantified by LSC.

To achieve a reproducible PAH-quantification by HPLC, several extraction and clean-up methods were tested for their efficiency to extract the PAH from the soil of the contaminated site (Table 1). Aliquots of the soil residues after extraction with methods 2, 4, 5 and 6 (Table 1) were combusted in an oxidiser (Biological Oxidizer Ox 300; R.J. Harvey). The $^{14}CO_2$ was trapped and quantified in an alkaline scintillation cocktail.

The incubated soil samples were extracted using method 6. For further clean-up aliquots of the extracts were concentrated and redissolved in cyclohexane. The solution was passed through a silica column (5 g wet silica gel, 15% water content) using 75 ml cyclohexane as mobile phase. For HPLC-analysis the cleaned extracts were redissolved in 50 ml methanol. The recovery of the clean-up was 98% for $^{14}C$-pyrene and $^{14}C$-benzo(a)pyrene.

The PAH were quantified by RP-HPLC with fluorescence detection with programmed wavelength adjustment. Five microlitres of the extracts were injected into a Hewlett-Packard HPLC (main unit 1090 with fluorescence detector 1046A and integration unit QS/20) fitted with a chromsep column (length 200 mm, inner diameter 3 mm, 5 $\mu$m particle diameter, Chrompack). The elution gradient was programmed as follows (given as vol.% of acetonitrile in water): 60% for 6 minutes, followed by a linear gradient to 95% for 15 minutes, followed by a linear gradient to 100% for 1 minute, holding for 16 minutes, followed by linear gradient to 60% for 4 minutes and holding for 5 minutes (Krahl *et al.*, 1994). The temperature was 26.5°C and the flow rate 0.55 ml/min. The wavelengths for excitation and emission were as follows (excitation/emission, nm): pyrene (237/420); benz(a)anthracene, chrysene (277/405); benzo(b)fluoranthene, benzo(k)fluoranthene (297/440); benzo(a)pyrene (380/420); dibenz(a,h)anthracene (290/420); indeno(1,2,3-cd)pyrene (300/500).

The $^{14}C$-activity in the toluene-extract was quantified by LSC and soil residues after extraction were combusted in the oxidiser in order to complete the $^{14}C$-balance.

## 3. Results

### 3.1 *Preliminary investigations*

The extraction efficiencies of the methods investigated for $^{14}C$-pyrene and $^{14}C$-benzo(a)pyrene are shown in Table 1. Soxhlet extraction of freeze dried samples (method 6) with toluene for 5 hours and clean-up with silica gel gave the best results. The extraction efficiency of $^{14}C$-pyrene was 86% and that of $^{14}C$-benzo(a)pyrene was 68%. The efficiencies measured were considerably lower than found by authors who used different soils (Holoubek *et al.*, 1990; Kicinski *et al.*, 1989). Evidently the PAH were strongly adsorbed by the test soil. The mean recoveries of total radioactivity from extracts and soil residues amounted to 94–105% of the applied activity.

### 3.2 *Mineralisation to $^{14}CO_2$*

Within 20 weeks of incubation, up to 16% of the $^{14}C$-pyrene and up to 4% of $^{14}C$-benzo(a)pyrene were mineralized (Table 2). The higher biodegradability of

*Table 1.* Efficiencies of selected methods for the extraction of PAH from a contaminated soil. The means and the standard deviations of triplicates are shown.

| | Method | Extraction efficiency (% activity applied) | |
|---|---|---|---|
| | | $^{14}$C-pyrene | $^{14}$C-benzo(a)pyrene |
| (1) | Saponification (Grimmer *et al.*, 1975): 4 h, 2N KOH, 160 ml water + 40 ml methanol | 28.7 ± 3.98 | 18.3 ± 0.56 |
| (2) | Saponification (Eschenbach *et al.*, 1991): 4 h, 2N KOH, 180 ml water + 20 ml methanol | 45.3 ± 1.55 | 34.6 ± 1.73 |
| (3) | Shaking at ambient temperature (Steinwandter, 1991): 23 h, 120 Upm, 100 ml water + 200 ml acetone | 26.0 ± 2.52 | 13.4 ± 0.30 |
| (4) | Shaking at ambient temperature (Blankenhorn and Hornung, 1992): 0.5 h, 120 Upm, 125 ml cyclohexane + 125 ml acetone | 36.1 ± 1.01 | 19.4 ± 1.19 |
| (5) | Soxhlet (Holoubek *et al.*, 1990): 5 h, 7 cycles/h, 90 ml *n*-hexane + 90 ml acetone | 63.2 ± 2.66 | 46.7 ± 1.40 |
| (6) | Soxhlet (Kicinski *et al.*, 1989): 5 h, 7 cycles/h, 180 ml toluene, freeze drying | 86.2 ± 1.25 | 68.1 ± 2.21 |

pyrene than that of benzo(a)pyrene observed has already been described in the literature (Bossert and Bartha, 1986; Heitkamp and Cerniglia, 1987).

The highest mineralization rates of $^{14}$C-pyrene were measured during the first ten weeks. In Figure 2 the mineralization rates in the experiments with *Pleurotus* species 1 and of controls are shown. With other fungi applied very similar mineralization kinetics were obtained.

The mineralization rates of $^{14}$C-pyrene (11–16%) and $^{14}$C-benzo(a)pyrene (3–4%) in the tubes containing soil and inoculated straw were almost the same as in the controls with soil and sterile straw ($^{14}$C-pyrene 13%, $^{14}$C-benzo(a)pyrene 4%). Obviously the native microflora had a much higher influence on the PAH-degradation in the contaminated soil than the white rot fungi. Soil alone showed a mineralization rate for $^{14}$C-pyrene of 10% but only 1% of the $^{14}$C-benzo(a)pyrene was detected as $^{14}CO_2$. Evidently the mineralization of $^{14}$C-benzo(a)pyrene was supported by the addition of sterile or inoculated straw.

In the tubes containing inoculated straw without soil a much lower mineralization of $^{14}$C-pyrene was observed. Depending on the fungus 0.2 to 4.5% of the applied activity were found as $^{14}CO_2$. When acidifying the sodium hydroxide solutions

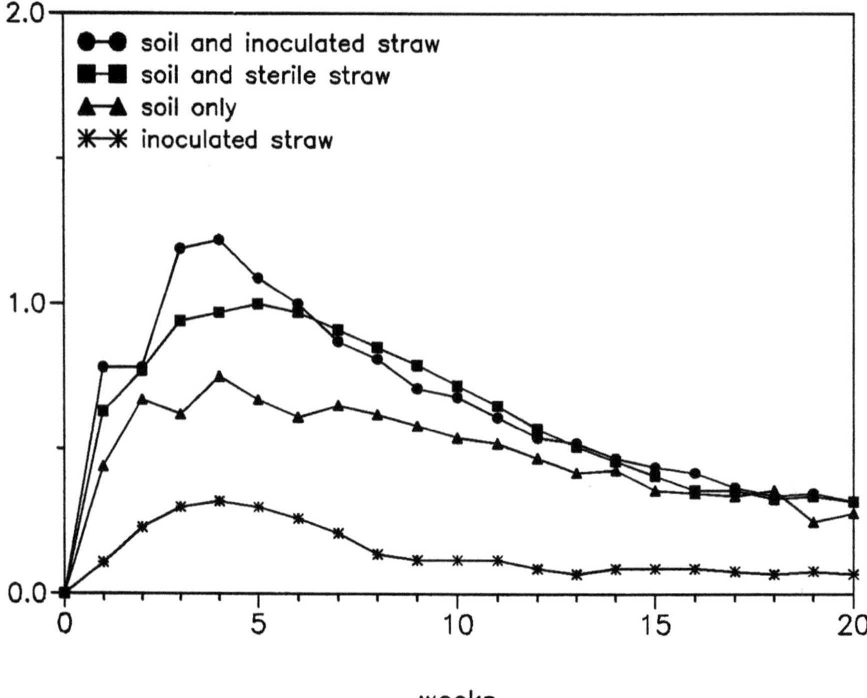


*Figure 2.* Mineralization rate of $^{14}C$-pyrene obtained with *Pleurotus* species 1.

*Table 2.* Mineralization rates of $^{14}C$-pyrene and $^{14}C$-benzo(a)pyrene within 20 weeks incubation with white rot fungi. The means and the standard deviations of quadruplicates are shown.

| Experiment | Mineralization rate (% activity applied) | |
|---|---|---|
| | $^{14}C$-pyrene | $^{14}C$-benzo(a)pyrene |
| Soil and inoculated straw | | |
| *Pleurotus* species 1 | 13.5 ± 0.34 | 3.5 ± 0.84 |
| *Pleurotus* species 2 | 10.9 ± 1.06 | 3.5 ± 0.53 |
| *Ganoderma applanatum* | 13.3 ± 2.62 | 4.0 ± 0.22 |
| *Dichomitus squalens* | 15.7 ± 1.38 | 4.3 ± 0.61 |
| Soil and sterile straw | 12.8 ± 1.43 | 4.1 ± 0.73 |
| Soil only | 9.9 ± 0.51 | 0.9 ± 0.13 |
| Inoculated straw | | |
| *Pleurotus* species 1 | 2.9 ± 0.27 | n.d. |
| *Pleurotus* species 2 | 4.5 ± 1.13 | n.d. |
| *Ganoderma applanatum* | 0.2 ± 0.25 | n.d. |
| *Dichomitus squalens* | 2.0 ± 1.35 | n.d. |

n.d., not determined.

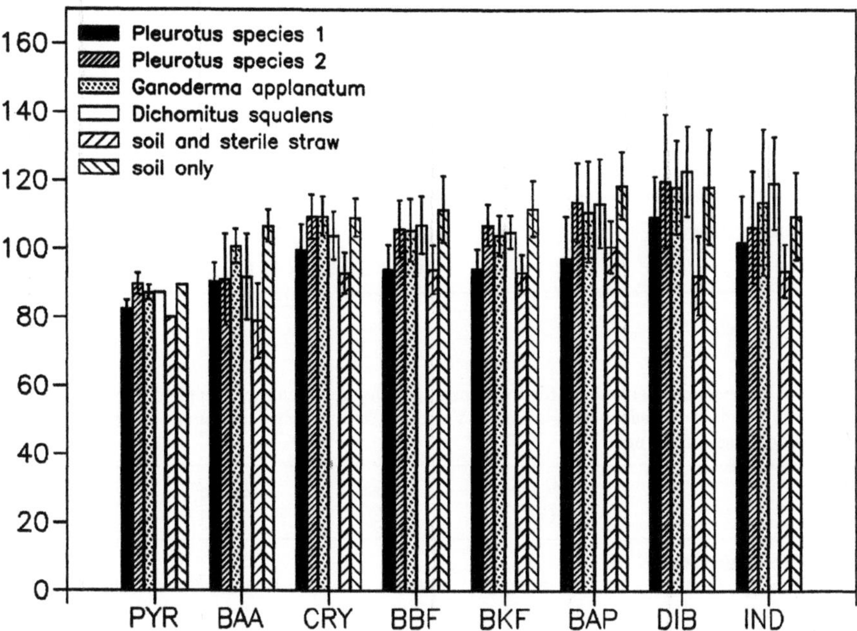


*Figure 3.* Degradation of PAH in soil of a contaminated industrial site. PAH contents after 20 weeks of incubation with white rot fungi as a percentage of PAH-contents at the beginning. PYR, pyrene; BAA, benz(a)anthracene; CRY, chrysene; BBF, benzo(b)fluoranthene; BKF, benzo(k)fluoranthene; BAP, benzo(a)pyrene; DIB, dibenz(a,h)anthracene; IND, indeno(1,2,3-cd)pyrene.

(used as $^{14}CO_2$-trap) no $^{14}$C-activity remained in solution. In the paraffin oil eluted no $^{14}$C-activity was measurable. These observations indicate that no volatile radioactive metabolites left the tubes.

### *3.3 Quantification of PAH by HPLC*

Measurements of the PAH-degradation of the eight PAH selected in the soil samples after 20 weeks incubation by HPLC corresponded to the mineralization of $^{14}$C-pyrene and $^{14}$C-benzo(a)pyrene obtained. Only for pyrene a decrease of approximately 20% was found in samples from each experiment. For the other selected 4–5-ring PAH (inclusively benzo(a)pyrene) no clear degradation was found (Figure 3). Obviously the metabolization of the PAH was very low.

### *3.4 Total balance of activity*

To verify the obtained mineralization rates the soil residues after extraction were combusted in the oxidiser. The mean recoveries of total activity for $^{14}$C-pyrene and

*Table 3.* Recovery of total activity for $^{14}$C-pyrene in percentage of applied activity, shown as means and standard deviations of triplicates.

| | $^{14}CO_2$ (%) | Extract (%) | Residue (%) | Total (%) |
|---|---|---|---|---|
| Soil and inoculated straw | | | | |
| *Pleurotus* species 1 | 13.5 ± 0.41 | 59.4 ± 2.67 | 23.6 ± 1.57 | 96.5 ± 3.12 |
| *Pleurotus* species 2 | 11.2 ± 0.96 | 65.6 ± 2.26 | 18.8 ± 1.01 | 95.6 ± 2.65 |
| *Ganoderma applanatum* | 14.6 ± 0.41 | 62.3 ± 0.44 | 20.1 ± 0.80 | 97.0 ± 1.00 |
| *Dichomitus squalens* | 15.2 ± 1.25 | 64.6 ± 3.41 | 20.1 ± 0.60 | 99.9 ± 3.68 |
| Soil and sterile straw | 12.1 ± 0.37 | 64.3 ± 2.07 | 23.2 ± 0.23 | 99.6 ± 2.12 |
| Soil only | 10.2 ± 0.11 | 68.4 ± 1.39 | 18.6 ± 1.01 | 97.2 ± 1.72 |

*Table 4.* Recovery of total activity for $^{14}$C-benzo(a)pyrene in percentage of applied activity, shown as means and standard deviations of triplicates. Values without standard deviations represent activities obtained with one sample only.

| | $^{14}CO_2$ (%) | Extract (%) | Residue (%) | Total (%) |
|---|---|---|---|---|
| Soil and inoculated straw | | | | |
| *Pleurotus* species 1 | 3.8 ± 0.69 | 46.1 ± 1.89 | 42.4 ± 3.50 | 92.2 ± 4.03 |
| *Pleurotus* species 2 | 3.8 ± 0.25 | 47.5 ± 1.24 | 38.4 ± 2.07 | 89.6 ± 2.43 |
| *Ganoderma applanatum* | 4.1 | 45.3 | 47.0 | 96.2 |
| *Dichomitus squalens* | 4.6 | 37.8 | 52.7 | 95.0 |
| Soil and sterile straw | 3.5 | 50.1 | 42.4 | 96.0 |
| Soil only | 0.7 | 62.7 | 33.9 | 97.3 |

$^{14}$C-benzo(a)pyrene added up to 95 to 100% of the applied activity. The results are shown in Tables 3 and 4. Comparing Table 1 with Tables 3 and 4 it is obvious that an increase of non-extractable residues of the $^{14}$C-labelled PAH took place during incubation. For $^{14}$C-pyrene an increase from 14 to 24% of the applied activity and for $^{14}$C-benzo(a)pyrene an increase from 32 to 53% were measured.

## 4. Discussion

A comparison of the mineralization rates of $^{14}$C-pyrene obtained in this experiment and in that of Martens and Zadrazil (1992) is shown in Figure 4. The mineralization rates in the soil from the contaminated site were considerably lower than in the agricultural soil used by Martens and Zadrazil (1992). This observation can be explained by the extremely high content of organic carbon in the contaminated soil that obviously reduced the bioavailability of the PAH. The carbon content in the agricultural soil was only 1%. Furthermore, the reduced bioavailability is supported by the finding of low extraction efficiencies. The dependence of reduced

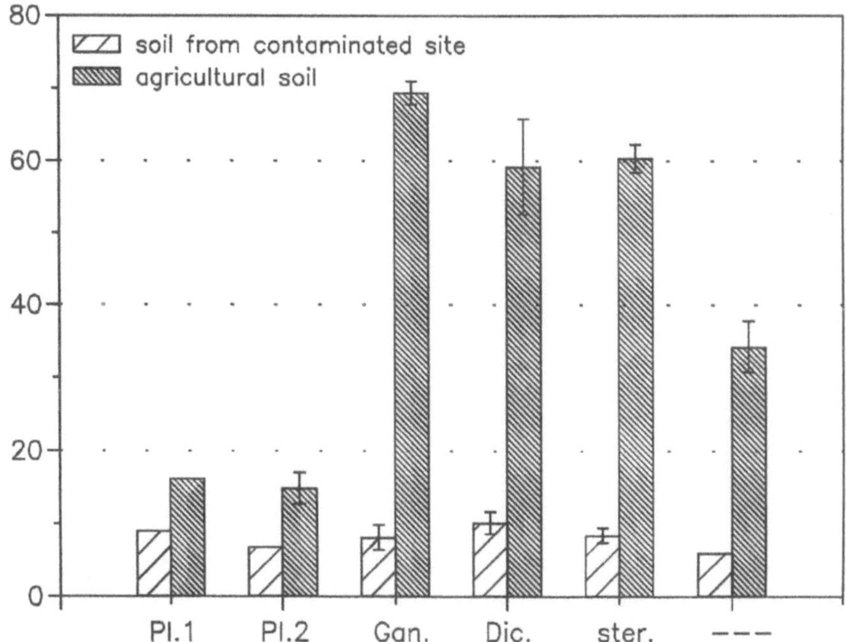


*Figure 4.* Mineralisation of $^{14}$C-pyrene to $^{14}CO_2$ in two soils with different organic carbon content within 10 weeks in percentage of applied activity. P1.1, *Pleurotus* species 1; Pl.2, *Pleurotus* species 2; Gan, *Ganoderma applanatum*; Dic, *Dichomitus squalens*; str., sterile straw; - - -, soil only.

bioavailability of PAH and organic soil carbon has been previously shown by Means *et al.* (1980) and Weißenfels *et al.* (1993).

In addition an inhibition of the activity of the indigenous soil microbial population as well as of inoculated fungi by toxic substances must be taken into account. This explanation is supported by the presence of high contents of phenols and chromium at the contaminated site. For this reason the *Pleurotus* species had not been able to colonise the contaminated soil permanently, as found in the experiments of Martens and Zadrazil (1992).

From all our experiments we draw the conclusion that under suitable environmental conditions, PAH with four condensed rings can be mineralised by microbial communities to a high degree. However, bioavailability reduced by adsorption phenomena is the most important factor regulating the degradation of PAH.

## Acknowledgements

We greatly appreciate the assistance of J. Krahl for the HPLC measurements and A. Kremling for technical assistance.

## References

Blankenhorn I and Hornung C 1992 Richtlinie zur bestimmung von polycyclischen aromatischen kohlenwasserstoffen (PAK) in Boden-, Abfall-, und Altlastenproben -Lfu -PAK 7/92 -. Ed.: Landesanstalt für Umweltschutz Baden-Württemberg.

Bossert I D and Bartha R 1986 Structure-biodegradability relationships of polycyclic aromatic hydrocarbons in soil. Bull. Environ. Contam. Tox. 37, 490–496.

Bumpus J A, Tien, Wright and Aust 1985 Oxidation of persistent environmental pollutants by a white rot fungus. Science 228, 1434–1436.

Eschenbach A, Gehlen P and Bierl R 1991 Untersuchungen zum Einfluβ von Fluoranthen und Benzo(a)pyren auf Bodenmikroorganismen und zum mikrobiellen Abbau dieser Substanzen. Mitteilungen d. Dt. Bodenkundlichen Gesellschaft 63, 91–94.

Grimmer G, Hildebrandt A and Böhnke H 1975 Gaschromatographische profilanalyse von polycyclischen aromatischen kohlenwasserstoffen. Dtsch. Lebensm.-Rundsch. 71, 93–100.

Hammel K E, Kalyanaraman B and Kirk T K 1986 Oxidation of polycyclic aromatic hydrocarbons and dibenzo(p)dioxins by *Phanerochaete chrysosporium* ligninase. J. Biol. Chem. 261, 16948–16952.

Heitkamp M A and Cerniglia C E 1987 Effects of chemical structure and exposure on the microbial degradation of polycyclic aromatic hydrocarbons in freshwater and estuarine ecosystems. Environ. Toxicol. Chem. 6, 513–546.

Holoubek I, Paasivirta J, Maatela P *et al.* 1990 Comparison of extraction methods for polycyclic aromatic hyrdocarbons, determination in sediments. Toxicol. Environ. Chem. 25, 137–154.

Kicinski H G, Adamek S and Kettrup A 1989 Trace enrichment and HPLC analysis of polycyclic aromatic hydrocarbons in environmental samples, using solid phase extraction in connection with UV/VIS diode-array and fluorescence detection. Chromatographia 28, 203–206.

Krahl J, Vellguth G and Bahadir M 1994 Rapid determination of polycyclic aromatic hydrocarbons in the exhaust gases of direct-injecting diesel engines by HPLC. Fresenius J. Anal. Chem. 350, 372–374.

Martens R and Zadrazil F 1992 Screening of white rot fungi for their ability to mineralize polycyclic aromatic hydrocarbons in soil. In Soil Decontamination Using Biological Processes. Ed. DECHEMA, 505–510.

Means J C, Wood S G, Hassett J J and Banwart W L 1980 Sorption of polynuclear aromatic hydrocarbons by sediments and soils. Environ. Sci. Technol. 14, 1524–1528.

Morgan P, Lee S A, Lewis S T, Sheppard A N and Watkinson R J 1993 Growth and biodegradation by white-rot fungi inoculated into soil. Soil Biol. Biochem. 25, 279–287.

Sack U and Günther T 1993 Metabolism of PAH by fungi and correlation with extracellular enzymatic activities. J. Basic Microbiol. 33, 269–277.

Steinwandter H 1991 Contributions to residue analysis in soils. Fresenius J. Anal. Chem. 339, 30–33.

Weißenfels W D, Klewer H J and Berger F 1993 Mikrobielle abbaubarkeit und biotoxizität von polyzyklischen aromatischen kohlenwasserstoffen (PAK) in böden. BioEngineering 9, 29–34.

# 45
# Potential utilization of phenoloxidases immobilized in organic gels for decontamination of polluted sites

C. CRECCHIO, P. RUGGIERO, M.D.R. PIZZIGALLO AND M. CURCI

## 1. Introduction

The overproduction, uncontrolled use, accidental release or incorrect disposal of toxic compounds have often caused significant environmental health hazards.

Among the toxic organic and inorganic compounds, phenols and their derivatives and aromatic amines, originating from industrial activities such as petroleum refining, organic chemicals and dyes, wood treatments, paper bleaching, and from agricultural management, such as pesticide derivatives, are considered priority pollutants and in some cases human carcinogens (US EPA, 1980). Therefore, the removal of such chemicals from the environment and in particular from natural and waste waters is of great practical significance. As the environmental and health consequences of these compounds become better known, more attention must be directed towards the development of innovative technologies for cleaning them up. The efficiency, feasibility and costs of the experimental system employed are particularly important.

At present, some physicochemical technologies are available to treat contaminated sites. These include chemical precipitation or extraction of pollutants, or their adsorption on activated polymers. Many of these techniques, however, do not completely destroy or inactivate the hazardous compounds, and they can give rise to secondary contamination effects; moreover, they are often cumbersome and cost-inefficient.

Among the remediation techniques, biological methods are generally considered the most promising, being based on natural processes which are generally cost-competitive and potentially able to bring about the complete destruction of many hazardous compounds with low environmental impact (Skladany and Blaine Metting, 1992). Although bioremediation, that is, the utilization of microbial cells for the transformation of toxic molecules into non-hazardous forms, has several important advantages, it is not without substantial disadvantages, such as the difficulties involved in isolation and purification of selective microorganisms, the competition phenomena among different strains and metabolic inhibition (Goldstein *et al.*, 1985; Kearney *et al.*, 1987). To overcome these drawbacks, cell-free enzymes directly derived from these microorganisms have been usefully applied in agricultural and environmental management (Dick and Tabatabai, 1993).

*D. Rosen* et al. *(eds): Modern Agriculture and the Environment, 545–552.*

In the case of aromatic compounds, phenoloxidase enzymes such as laccase, tyrosinase and peroxidase, catalyse the oxidation of phenols giving rise to oligomeric and polymeric products usually less toxic than the starting monomers (Alberti and Klibanov, 1981; Dec and Bollag, 1990); sometimes the same enzymes catalyze the co-oligomerization of particularly recalcitrant molecules with naturally occurring phenols (syringic and vanillic acids) in a sort of humification reaction (Bollag and Liu, 1985). A significant improvement in practical employment of cell-free enzymes is obtained through their immobilization upon solid supports or entrapment inside insoluble matrices. Immobilized enzymes are usually more effective than free enzymes, since they can be reused in the same reaction and they are generally more resistant to microbial proteolysis and thermal inactivation (Kennedy and Cabral, 1987; Nannipieri and Bollag, 1991). Natural or synthetic supports such as soils, clays, humic acids, porous glass beads or alginate beads have been employed for the immobilization of phenoloxidases and the polymerization of toxic phenols (Davis and Burns, 1990; Ruggiero and Radogna, 1988; Ruggiero *et al.*, 1989; Sarkar *et al.*, 1989).

The immobilization of laccase from *Trametes versicolor* and tyrosinase from mushroom in a new type of insoluble gel originated from water-in-oil microemulsions has been studied for the purpose of selecting immobilized systems which are highly efficient and easy to prepare (Luisi *et al.*, 1990). The initial ternary system consisted of an organic solvent, a surfactant and water. At the appropriate molar ratio of water and surfactant solubilized in organic solvent, it is possible to obtain reverse micelles and, by adding gelatine, organic gels (Haering and Luisi, 1986) in which different molecules such as proteins, vitamins and dyes may be added and entrapped. By evaporating the organic phase, gel films are obtained. These films, which contain the entrapped biomolecules, are completely insoluble in organic solvents which swell up in aqueous solutions.

This study was initiated to determine the best conditions for a significant and stable immobilization of laccase and tyrosinase. Such a system provided the opportunity to determine the extent of enhanced biotransformation of natural and xenobiotic aromatic compounds from aqueous solutions.

## 2. Materials and methods

Tyrosinase (2430 units $mg^{-1}$) was purchased from Sigma Chem. Co. (St Luis, MO). Laccase was extracted and purified from the basidomicete *Trametes v.* The fungus was grown on potato dextrose medium (4 g $L^{-1}$ potato extract, 20 g $L^{-1}$ dextrose, pH 5.5) for 10 days at 25°C in static conditions, after which *p*-xylidine was added as a laccase inducer (1 mM final concentration) and the fungus was grown for an additional 7 days at 15°C. Laccase was isolated from the culture filtrates, dialysed and concentrated by ultrafiltration, and purified using a DEAE-Sephacel anion exchange resin eluted with 0.1 M citrate phosphate–0.01 M NaCl buffer at pH 5.0 and a gel filtration column (Sephadex G-200) eluted with 0.1 M citrate phosphate buffer pH 5.0.

Reverse micelles were obtained by adding 7.2 ml $L^{-1}$ water to 0.1 M dioctyl sulphosuccinate in isoctane. Gelatine was added to 5 ml of micellar suspension under vigorous stirring and solubilized by bringing the temperature up to 50°C. During the cooling phase (below 30°C), 40 units of laccase or tyrosinase were added under continuous stirring. Before complete gelation, the suspension was poured on plates to allow isoctane evaporation, producing gel films completely insoluble in organic solvents and swelling up in aqueous solutions. These gels were resuspended in buffer and submitted to an assay of enzyme activity.

Free and immobilized laccase and tyrosinase activity was measured polarographically. One unit was defined as the amount of catalyst which caused the uptake of 1 $\mu$mol of oxygen per minute under the following conditions: reaction temperature 25°C; final volume 3 ml; 0.1 M citrate phosphate buffer pH 3.8 and 1 mM 2,6-DMP for laccase; 0.1 M phosphate buffer pH 7.0 and 6 mM catechol for tyrosinase.

To study the effect of gelatine concentration, gel films at different concentrations (1.5–4.0%) and containing the same units of enzyme were prepared and resuspended in 3 ml of buffered solutions. These suspensions were incubated for 16 h at 4°C and centrifuged at 8000 *g* for 20 minutes at 4°C. Aqueous and gel phase were separately analysed for enzyme activity.

To test the reusability of the enzyme, a 10 ml column was packed with gel containing entrapped tyrosinase (50 units). After equilibration with 0.1 M phosphate buffer at pH 7.0, 10 ml of 0.3 mM catechol in the same buffer was loaded on the column. The catechol concentration remaining in the solution was calculated by spectrophotometric analysis of the eluted fractions. After several washings, which eliminated coloured products of enzyme reaction from the gel, a fresh catechol solution was loaded on the column. These cycles were repeated ten times. A column packed with gel without tyrosinase and another containing the enzyme subjected to thermal inactivation after the first cycle, were used as controls.

The activity of gel-immobilized laccase and tyrosinase towards various substrates was also tested. Phenols and aromatic amines (vanillic acid, syringic acid, 2,6-DMP, 2,4-DCP, 1-naphthol, caffeic acid, catechol, 4-choloroaniline) were solubilised in phosphate buffer pH 7.0 and citrate phosphate pH 3.8 at a final concentration of 100 ppm. Five ml of each of these solutions was incubated with different amounts of immobilized tyrosinase or laccase (60 units and 0.95 units, respectively). The suspensions were incubated under continuous stirring at 25°C for varying lengths of time (from 15 min to 16 h), depending on the enzyme and the substrate under investigation. After centrifugation, aqueous phases were analyzed by HPLC in order to determine the amount of each substrate removed by laccase or tyrosinase with respect to control without enzyme.

All the experiments were conducted in triplicate.

## 3. Results

Table 1 shows the enzyme activities released from the gel in the aqueous phase and the activities maintained inside the gel as a function of the gelatine concentration.

*Table 1.* Effect of gelatine concentration on tyrosinase and laccase activity. Results are expressed as percentages of the initial activities.

| | Tyrosinase activity (%) | | Laccase activity (%) | |
|---|---|---|---|---|
| Gelatine (%) | Aqueous phase | Gel phase | Aqueous phase | Gel phase |
| 1.5 | 82.0 | 16.9 | 25.0 | 44.9 |
| 2.0 | 6.1 | 65.8 | 11.2 | 57.0 |
| 2.5 | 4.5 | 73.3 | 7.2 | 67.2 |
| 3.0 | 8.2 | 64.3 | 16.0 | 55.1 |
| 3.5 | 5.3 | 56.4 | 14.8 | 56.6 |
| 4.0 | 3.0 | 11.3 | 14.2 | 50.8 |

The results obtained indicate that the amount of gelatine influences the immobilization and the enzyme activity, especially in the case of tyrosinase. In fact, at a gelatine concentration of less than 2%, about 80% of the initial activity was released in the aqueous phase, whilst at values higher than 2.5% there was a constant decrease in the activity of the gel. Immobilized laccase activity appeared less dependent on the gelatine concentration in the range of gelatine concentration from 2.0 to 4.0%, and soluble laccase activity was higher than that of soluble tyrosinase. For both enzymes the best immobilization conditions were observed at a gelatine concentration value of 2.5%. Table 1 also shows that the immobilization process brings about a loss of activity. In fact, considering the total activity recovered in aqueous and in gel phases, the loss of tyrosinase activity ranged from 1.1 to 85.7% and laccase activity was reduced by 25.0 to 31.8% in the range of gelatine concentrations employed.

To study the reusability of the immobilized tyrosinase in the same reaction, a column was packed with gel-immobilized tyrosinase. A catechol solution was then loaded on the column and its recovery in the eluted fractions was monitored. After the removal from the gel of the reaction products by washing, a fresh catechol solution was loaded, and the procedure repeated for ten cycles. As shown in Figure 1, about 50% of catechol disappeared from the eluted solution in each cycle. A column control with gel-immobilized tyrosinase inactivated for 1 h at 70°C and another control with gel without enzyme were used to demonstrate that the partial disappearance of the substrate was due only to the tyrosinase activity and not to non-specific adsorption on the gel.

Substrate specificity of immobilized laccase and tyrosinase was tested by using some naturally occurring phenols and amines. With the exception of vanillic acid, which proved almost recalcitrant to the tyrosinase reaction, each substrate was significantly removed by gel-immobilised laccase and tyrosinase (5–95%) under the experimental conditions used here (Figure 2). Similar experiments were carried out using 4-chloroaniline, 1-naphthol and 2,4-dichlorophenol as examples of xenobiotic compounds arising from the biological and non-biological degradation of various pesticides and considered hazardous pollutants of soil and water. Significant results

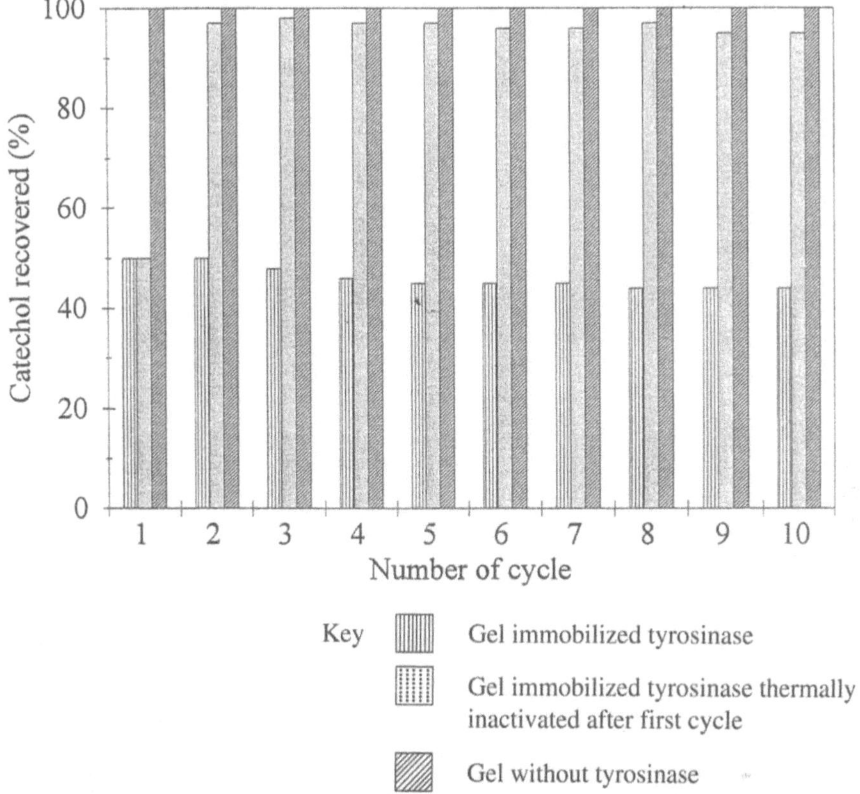


*Figure 1.* Repeated elutions of catechol through a column packed with gel immobilised tyrosinase, gel immobilized tyrosinase thermally inactivated after the first cycle, and gel without tyrosinase.

were obtained for both enzymes (Figure 3). Worthy of note is the almost complete oxidation of 1-naphthol and 4-chloroaniline obtained by the gel-immobilized laccase.

## 4. Discussion

The addition of water to an organic solution of a surfactant, such as dioctylsulphosuccinate in isoctane, gives rise to reversed micellar suspensions or microemulsions consisting of spherical aggregates in which the hydrophilic heads are directed inwards and the hydrophobic tails of the surfactant are oriented towards the bulk solvent (Haering, 1989). These reversed micelles are gelified, solubilizing gelatine by warming up the microemulsions and evaporating the organic solvent. The gelation process is strictly dependent on gelatine concentration and water content (Haering and Luisi, 1986). Organic gels so formed have been employed as a matrix for immobilization of two different phenoloxidases. Since this type of immobilization does not involve covalent bonds but the trapping of the enzyme inside the matrix, the role of the gelatine in this process is obviously important. On one hand,

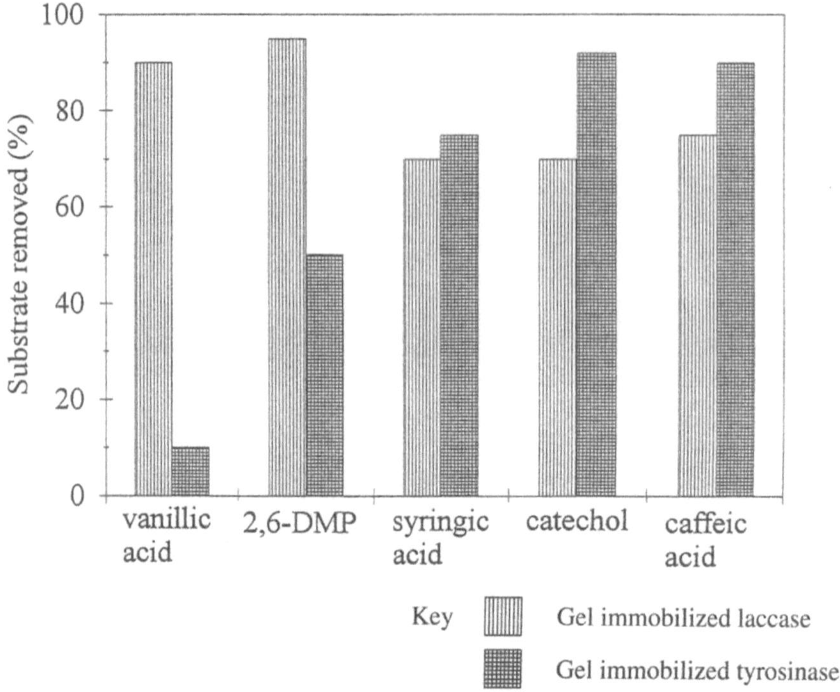


*Figure 2.* Activity of gel immobilized laccase and tyrosinase towards naturally occurring substrates.

the gels have to retain the molecules in a stable manner, while on the other hand possible inhibition phenomena of enzyme activity have to be avoided. The best balance between enzyme activity and gel retention ability was observed at gelatine concentration of 2.5%. Differences in the enzyme immobilization as a function of gelatine concentration might be due to the different structures of the two enzymes, whilst the partial loss of catalytic power, usually shown in many immobilization techniques, is probably due to decreased diffusion of the substrate or to inhibition phenomena and to a likely partial protein denaturation due to the presence of an organic phase.

The loss of catalytic efficiency of the gel-immobilized enzymes in comparison to that of free enzymes is balanced by some other significant advantages. Data not reported here indicate that both laccase and tyrosinase immobilized in organic gels are significantly more resistant to proteolytic enzymes and to thermal inactivation, both of which typically limit the use of enzymes as powerful catalysts in bioremediation reactions. Furthermore, one of the remarkable properties of the gel-immobilized tyrosinase is that it can be separated from the reaction mixture and used repeatedly to transform the substrate with about the same efficiency. Catechol is not retained through the column in the absence of the enzyme, and its retention through the column in the presence of the inactivated enzyme is quite negligible, thus confirming

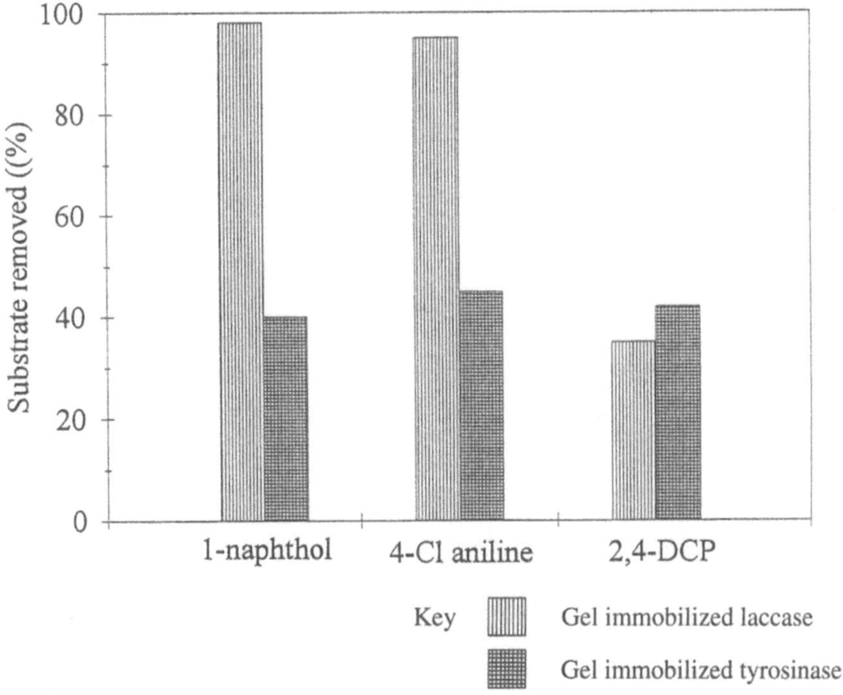


*Figure 3.* Activity of gel immobilized laccase and tyrosinase towards xenobiotic compounds.

that the disappearance of catechol from eluted solution is due to its oxidation catalysed by the enzyme and not to its partial entrapment inside the gel. By varying the experimental conditions (different enzyme/substrate ratios, length of the column, flow rate through the column), a more complete removal of the substrate could be obtained.

Both immobilized enzymes were capable of oxidising various organic compounds such as syringic, vanillic and caffeic acids, 2,6-dimethoxyphenol and catechol, all of which are widespread in the environment where they are involved in the turnover of lignin and humic substances. The differences in the catalytic power of each enzyme towards the substrates depend not only on the substrate specificity, but also on the unknown degree of saturation of each substrate and on the enzyme/substrate ratio that was about 60 times higher for tyrosinase than for laccase.

The same conclusion can be drawn when xenobiotic substrates, such as 1-naphthol, 4-chloroaniline and 2,4-DCP, are considered. Therefore, the immobilised phenoloxidases can also serve as a means to reduce toxicity of toxic compounds in terrestrial and aquatic environments.

In conclusion, the use of gelatine gels is a new and very simple method to immobilize phenoloxidases, important enzymes of the carbon cycle. The relatively high enzyme activity immobilised, the efficiency of the oxidation power towards various substrates shown by both enzymes, the inertness of the support, the possibility of reusing the same catalyst several times in the same reaction, as well as other important characteristics of the immobilized enzymes, suggest that this technique is a simple and inexpensive alternative approach for bioremediation of polluted sites, especially for waste and deep waters.

Nonetheless, further improvements and large scale experiments are needed before these enzyme systems can be implemented as a water treatment method.

## Acknowledgments

This research was supported by Ministero dell'Universitá e della Ricerca Scientifica e Tecnologica (MURST)-Progetti Nazionali 40% and by Consiglio Nazionale delle Ricerche (CNR).

## References

Alberti B A and Klibanov A M 1981 Enzymatic removal of dissolved aromatics from industrial aqueous effluents. Biotechnol. Bioeng. 11, 373–379.

Bollag J-M and Liu S-J 1985 Copolymerization of halogenated phenols and syringic acid. Pestic. Biochem. Physiol. 23, 261–272.

Davis S and Burns R G 1990 Decolorization of phenolic effluents by soluble and immobilized phenol oxidases. Appl. Microbiol. Biotechnol. 32, 721–726.

Dec J and Bollag J-M 1990 Detoxification of substituted phenols by oxidoreductive enzymes through polymerization reactions. Arch. Environ. Contam. Toxicol. 19, 534–541.

Dick W A and Tabatabai M 1993 Significance and potential utilization of soil enzymes. In Soil Microbial Ecology. Applications in Agricultural and Environmental Management. Ed. F Baine Metting. pp 95–127. Marcel Dekker, New York.

Goldstein R M, Malory L M and Alexender M 1985 Reasons for possible failure of inoculant to enhance biodegradation. Appl. Environ. Microbiol. 50, 977–983.

Haering G 1989 PhD thesis, ETH, Zurich, Switzerland.

Haering G and Luisi P L 1986 Hydrocarbon gels from water in oil microemulsions. J. Phys. Chem. 90, 5892–5895.

Kearney P C, Karns J S and Mulbry W W 1987 Engineering soil microorganisms for pesticide degradation. In Pesticide Science and Biotechnology. Ed. R Greenhalgh. pp 591–596. Blackwell Scientific Publishers, Boston, MA.

Kennedy J F and Cabral J M S 1987 Enzyme immobilization. In Enzyme Technology, Biotechnology, Vol.7a. Ed. J F Kennedy. pp 347–404. Verlagsgesellschaft GmbH, Weinheim, Germany.

Luisi P L, Scartezzini R, Haering G and Schurtenberger P 1990 Organogels from water in oil microemulsions. Colloids Polym. Sci. 168, 356–374.

Nannipieri P and Bollag J-M 1991 Use of enzymes to detoxify pesticide-contaminated soils and waters. J. Environ. Qual. 20, 510–517.

Ruggiero P and Radogna V M 1988 Humic acid–tyrosinase interactions as a model of soil humic–enzyme complexes. Soil Biol. Biochem. 20, 353–359.

Ruggiero P, Sarkar J-M and Bollag J-M 1989 Detoxification of 2,4-dichlorophenol by a laccase immobilized on soil or clay. Soil Sci. 147, 361–370.

Sarkar J-M, Leonowicz A and Bollag J-M 1989 Immobilization of enzymes on clays and soils. Soil Biol. Biochem. 21, 223–230.

Skladany M J R and Blaine Metting F 1992 Bioremediation of contaminated soils. In Soil Microbial Ecology. Applications in Agricultural and Environmental Management. Ed. F Blaine Metting. pp 483–513. Marcel Dekker, New York.

US Environmental Protection Agency 1980 Industrial Wastewater. Report No. 600/8-80-026 Washington, DC.

# SECTION IX

# Economic and Policy Aspects of Natural Resources

# 46
# Economics and sustainable development

DAVID ZILBERMAN

The last two decades have witnessed two major problems in agriculture. First, excess supply of food. One manifestation of this problem is that many farmers are being paid to refrain from production. Another manifestation is that countries accumulate large inventories of food products and they cannot get rid of them. For example, the US government has an inventory of cheese that may last for several years. The second problem is severe environmental side effects caused by agriculture. This is manifested by the risks to human health and ecosystems posed by the accumulation of chemical residues in water and soil (Pimentel and Lehman, 1993); soil erosion; elimination of fisheries; and depletion of forests. These environmental concerns have led to the realization that the existing paradigm of agricultural management requires some sort of change.

Demand for change is often characterised by the term 'sustainability'. All members of the United Nations have agreed that sustainable development is an important policy objective (Batie, 1989). However, the interpretation of this concept is subject to debate. The term means all things to all people.

There are three avenues by which people, especially those from the natural sciences, approach the interpretation and implementation of the goals of sustainability. The first can be described as 'back to the good old days'. Here there is an underlying assumption that the 'bad' modern monocultural paradigm has eliminated many useful practices, and if we can only recapture them, we will be able to produce all the output we need without negative side effects. There is a nostalgic and idealistic appreciation for traditional crop rotation and indigenous cultural practices, and a suspicion of anything modern, especially chemical fertilizers and pesticides. Traditional practices are seen as both safer and less dehumanising.

The second approach is that of 'command and control'. Here, the assumption is that science knows best, and experts should tell farmers what to do to manage their production better. Accordingly, educated, all-knowing decision makers are to identify solutions and direct the ignorant greedy farmers.

The third approach, to which I subscribe, is based on incentives. Here, unlike the first approach of trying to reduce the role of science in addressing agricultural problems, the aim is to increase it. But, unlike the second approach, instead of forcing people, policy makers try to align the incentives of individuals with the incentives of society at large. This last approach is based on the use of economics in

*D. Rosen* et al. *(eds): Modern Agriculture and the Environment, 555–566.*

solving environmental problems in agriculture and attaining sustainable and desirable outcomes.

This chapter argues that economics should play a major role in shaping the future of agriculture. In particular, it is crucial for both measuring the environmental impacts of modern agriculture and identifying solutions to environmental problems. However, the contribution of economics depends on collaboration with other disciplines, and the design of new agricultural policies should be an interdisciplinary effort. The chapter explains the economic perspective on problems of agriculture and the environment and what types of input economists expect from biologists and ecologists.

## 1. Why 'back to the past' has no future

To understand the role and perspective of economics in addressing environmental problems in agriculture, let us examine the three approaches to sustainability outlined above and, in particular, what they mean to agricultural economists.

First, let us examine the 'back to the past' approach which attempts to disregard modern input and tries to return to traditional agriculture. In spite of the emotional appeal of this approach, the bottom line is that modernisation is irreversible. The increases in yields associated with modernisation cannot be ignored. We take abundance of food for granted, but the biggest achievement of modern agriculture is that we are capable of feeding a much larger population with the same amount of land. Furthermore, the output–labour relationship has declined drastically and there has been an incredible movement of people from the country to the cities. Traditional agriculture is typified by labour intensity. So, if the goal is to produce the same levels of output with traditional methods, we will require a far greater rural workforce. Attempts to relocate people from the cities to the country may be quite painful, as the recent experiment in Cambodia demonstrates.

I do not believe that we are serious about returning to the past, but people are disillusioned with the performance of modern science. The argument is that scientifically based agriculture has resulted in the use of 'bad' inputs, monoculture, and a lot of other ills. But, let us not forget that modern science is relatively young and the use of science in agriculture is a very recent phenomenon, dating back only 100 years at most. The unhappiness with science is the result of impatience because scientific agriculture is only in its infancy.

Modern agriculture relies on monoculture, not because it is the only scientific way to do things, but because it is much easier to deal systematically and analytically with the production of one crop than with the production of many crops. It is misleading to assume that reliance on monoculture and many inputs is integral to scientific agriculture. Application of science to agriculture was motivated by the desire to increase yields and reduce costs, and at the early stages of this enterprise, the most obvious opportunities were in monocultural systems. As our knowledge increases, it can be expected that modern science will be harnessed to steer agriculture towards multicrop systems. Naturally, sustainable ecosystems are multispecies, and it is reasonable that such systems hold better potential for improved productivity and environmental quality than monocultural systems. Ecologists and biologists are challenged to develop an understanding of ecosystems and to learn how different species coexist and interact. Economists, agronomists, engineers, and resource managers are challenged to use such knowledge to develop policies, technologies, and management practices that are economically efficient and environmentally sound. These are no small tasks since we are not yet very good at developing accurate models for single crops. Nonetheless, it is reasonable to believe that within the next several hundred years, a systematic and scientific agricultural production system will develop that relies on the co-production of several crops.

## 2. The wrong presumptions behind 'command and control'

Now let us move on to discuss some of the problems with the 'command and control' approach to sustainability. The implicit presumption behind this approach is that farmers are ignorant and greedy, basing their decisions solely on profitability considerations with complete disregard for the environment.

These implicit presumptions are wrong for several reasons. First, formal education is not akin to wisdom, and many farmers who are not formally educated have the capacity to make rational choices. Furthermore, they possess practical and specific knowledge that may not be available to policy makers or scientists, and that knowledge may lead them to take actions that seem unsound to outsiders. Second, many modern farmers are highly educated. Furthermore, at least in California, farmers hire or consult experts before making major farming decisions. Third, farmers like anyone else pursue profits, but their choices are affected by their considerations. They are likely to be the first victims of toxic agricultural chemicals (since they apply such chemicals and, in general, are more exposed to them) and, therefore, are interested in increased safety. Recent studies explaining choices of chemicals by farmers (Wiebers, 1992) found that increased toxicity of chemicals has a negative effect on the use of agricultural chemicals. In addition, rural lifestyles entail much interaction with nature and enjoyment of environmental amenities, so farmers are not inherently hostile to environmental preservation and actually may pursue it.

Farmers balance profitability, safety, and environmental quality considerations in their daily life and, while their choices may be suboptimal from a social perspective, they do the best they can given their incentives and constraints. Farmers resent regulations imposed from afar by regulators who the farmers perceive to be hostile and uninformed on the situation in the field. This resentment may lead to behaviour that may negate the original intent of many environmental regulations. For example, regulations aimed to protect endangered species in California impose drastic constraints on farming activities once an endangered species is being observed on a farm. I found that this mode of regulation has led farmers to engage in practices that will reduce the likelihood that endangered species will be found on their property.

Some of the 'politically correct' individuals who are behind the command and control approach differentiate between agribusiness (evil, greedy, and environmentally destructive) and small-scale family farms (unpolluting, unwasteful, and environmentally conscientious). In fact, these generalizations are often unfounded. In my work I have found that wealthy farmers have the resources to try alternative production methods (for example, they may hire an entomologist for IPM), while smaller farmers, because they are under such financial pressure, tend to pollute much more. This is not surprising. Generally, the houses of rich people are likely to be cleaner than those of the poor, not because the rich are inherently cleaner, but because they have more resources to hire help or buy cleaning products. Thus, command and control policies that are biased in favour of small farmers do not always tend to improve environmental quality, and in some cases they may even harm it.

Another presumption that some proponents of command and control hold is that environmental problems can be addressed by imposing uniform regulations. Such regulations may impose the use of a 'best management practice' (BMP). Heterogeneity and variability hamper the implementation of such command and control policies. Because of heterogeneity, optimal regulations are not likely to be uniform. They should vary to reflect differences in location, personality, ability, etc. Therefore, BMP regulations are not likely to work very well. Zilberman *et al.* (1991) show the inefficiency associated with uniform bans of pesticides. The productivity and risk associated with the use of pesticides vary significantly across locations. Policies that lead to reduction in pesticide use, so that most of their risks are eliminated while most of their benefits are preserved, are much more sound than across-the-board bans.

Similarly, the coexistence of several technologies at the same time does not imply suboptimality and require intervention imposing the best technology. Variations in conditions may lead to optimal outcomes where several technologies coexist. For example, Dinar and Zilberman (1991) show that drip, sprinkler, and furrow irrigation may coexist as part of an efficient strategy for water and drainage management in California. Each technology is optimal under different conditions.

Imposition of a unique best management practice is also suboptimal because of dynamic considerations. New technologies and innovations are introduced over time. As technologies change, some new practices may be better than the required best practice. Furthermore, the diffusion of a new practice is gradual. Early in its life, a new technology may be suiting best only a small set of situations. Over time, the process of 'learning by doing' (Arrow, 1962) will reduce the cost of the new technology, and experience with its use will improve its productivity so it will become most appropriate for a larger (but not necessarily all) set of circumstances. Strict imposition of 'best practices' will hamper the beneficial processes of technological change.

I sense anti-economics bias among the proponents of 'command and control'. Some non-economists may have the perception that economics is inherently biased against the environment. I believe that economics offers tools to analyse human choices and allocation of resources and can accommodate many perspectives. The outcome of economic analysis depends on the input in terms of assumptions and values of key parameters such as the rate of discount. One can use economics to develop policies that will attain environmental quality objectives at least cost or with least impacts on income distribution and under many other constraints. I find the strength of economics to be its realistic perspective about human behaviour. The key assumption of economic analysis is that people pursue their self-interest. To me this is self-evident. Even the most ardent environmentalist in academia is likely to worry more about his tenure than about the environment. Economics does not judge the pursuit of self-interest to be morally good or bad but takes it as given. Policy designs by economists aim to provide incentives that align the private self-interest with the societal good rather than condemn self-interest and tell people what to do.

Command and control was the mode of operation of the centrally planned economies of the Eastern bloc. These policies failed miserably in the farm sector and have led to environmental disasters. They failed because decision makers do not have enough information to make the 'correct' decisions, and it is more difficult to enforce decisions, especially in the farm sector. Furthermore, when a small group of people make centralized decisions, they lose a lot of the wisdom and initiative available among the population at large. Social scientists have accumulated enough evidence to argue that when you want people to do something well, you have to learn their incentive scheme, understand how they operate, and design your incentive accordingly.

## 3. Economic perspective on sustainability

There is not an established economic perspective on sustainability, but the literature on environmental and resource economics (Carlson *et al.*, 1993) provides a base on which I rely when presenting an economic perspective on sustainability. My view of the notion of sustainability is that we have four objectives: efficiency, namely, more is better than less; equity (more even income distribution is preferable to less even income distribution); environmental quality; and safety. The economy and the environment are subject to randomness and uncertainty, and a sustainable policy should provide some degree of protection, especially against very dangerous events. Sustainable policies may vary in the weight given to these objectives when they conflict.

Jean Paul Chavas (1993) views survival and safety as key features of sustainability, and he presented an excellent framework for assessing policies mostly according to their impact on survival of species. Howarth and Norgaard's (1990) approach to sustainability emphasises environmental quality and inter-generational equity. They suggest that societies will commit themselves to the preservation of essential environmental amenities, and the economic system will operate to attain efficiency subject to these preservation constraints. Solow (1992) does not perceive us as having an environmental crisis. He argues that the reliance on the free enterprise system and modern science over the last 200 years have brought unparallelled rates of growth and improvement in human conditions. He recommends continuing on the same path and believes that research and economic incentives will lead to development of solutions if and when resource scarcity problems will arise in the future.

I believe that the economic models for the management of exhaustible resources under uncertainty (Carlson *et al.*, 1993; Dasgupta and Heal, 1974; Dixit and Pindyck, 1994; Fisher, 1981) provide a very useful framework to attain policies for sustainable development. The capacity of the environment to absorb pollution is exhaustible in many cases.[1] One conclusion of the literature of exhaustible resources is that private actions that cause environmental degradation have to be affected by government policies. These policies may include taxation of activities that damage the environment, subsidisation of activities that improve environmental quality, and support of research and extension activities leading to development and use of environmentally sound practices. In some circumstances the prescribed policies may be the establishment of markets in pollution rights (where the total volume of pollution is controlled by the government and individuals are given tradable pollution rights), the privatisation of resources that are overused because of open access, and even direct controls. Economic models of exhaustible resources provide a framework to identify data needed for quantitative analysis and values of key policy parameters. The implementation of policies prescribed by these models may require legislative effort, legal codification, and establishment of procedures for monitoring and enforcement. The emerging bodies of literature on law and economics and incentive compatibility provide guidelines that will allow the establishment and implementation of incentives which will lead people to manage the environment in a more reasonable manner. This will assure its survival in the long run and prevent irreversible damages.

## 4. Sustainability and growth

Daly (1968) suggests that economic development is the primary cause of environmental problems, and his analysis implies that under sustainable policies economic growth will be halted and economic activities will be stabilized at levels that are sustainable. As I argued before, I do not see growth *per se* as the problem but rather a lack of incentives to discourage pollution and environmental contamination. I do not foresee that the pursuit of environmental protection policies will result in steady state outcomes with stable and constant production and consumption levels. In my mind the only certain thing is change. The evolutionary process will continue, presenting us with new surprises and challenges. Research

[1] Actually it is more accurate to distinguish between models of renewable resources and non-renewable resources. Both types of resources are exhaustible. Models of renewable resources under uncertainty are appropriate to address problems of forestry, water sources, and fishery management. The capacity of the environment to absorb certain types of toxins may be non-renewable.

and development activities will lead to the discovery of new technologies and solutions. Finally, I believe that change is desirable. It adds interest and excitement to life. Sustainable policies can protect the environment and natural resources and can reduce environmental risks, but economic growth and change will continue.

There is growing evidence that the pursuit of environmental objectives does not necessarily lead to reduction in production levels. There are many situations in which increased production and environmental quality improvements can go hand in hand. Some of these situations occur when production technologies have relatively low technical efficiency. In many production systems, not all the applied variable inputs are actually consumed in the production process. For example, in the case of irrigation with traditional gravitational technologies, often only 60% of the applied water is actually consumed by the plants. The remaining 40% ends up as run-off or percolates underground. Technical efficiency is measured by the percentage of applied input that is consumed in the production process. With modern drip irrigation, the irrigation efficiency (the ratio of efficient to applied water) is up to 90%. Similarly, in the case of aerial spraying of pesticides, only 25% of the sprayed materials actually reach the target crop, and the remaining 75% is spillover and represents a potential environmental problem. Ground spraying increases the accuracy of applications significantly, and the spillover is much smaller (Pimentel and Lehman, 1993).

These examples suggest that application efficiencies can be increased by adopting technologies with greater precision. In the case of water, this means modern irrigation technologies such as drip and sprinkler. For pesticides, ground spraying and pest scouting increase application efficiency and reduce drift and misuse. In most cases increased application efficiency results in reduction of negative spillovers. Greater application precision is likely to be associated with higher investment or other fixed application costs. In many cases the transition from low-efficiency to high-efficiency technologies results in reductions in input use and increases in yields. Caswell and Zilberman (1986) show that, in the case of irrigation, transition from furrow to drip or sprinkler irrigation is likely to save water and increase yield (per unit of land). Thus, farmers face trade-offs between more precise technologies that save inputs and increase yields and less precise technologies that cost much less. The reasons that producers use technology with low application efficiency is that input prices are low and there are minimal (or no) penalties for spillovers. Caswell and Zilberman (1986) show that in the case of irrigation and drainage, changes in incentives (higher water prices, drainage penalties) will lead to adoption of modern irrigation technologies, and this will increase output levels and reduce water use and drainage. This result can be generalized: incentives that increase the prices of resources and penalize pollution may lead to adoption of more precise technologies that are environmentally safer and produce more output.

Economic incentives, such as pollution taxes, reduce pollution and increase conservation, not only because of their impact on farmers, but also because of their impact on researchers and developers of technology. Hayami and Ruttan (1985) introduce the notion of induced innovation. They argue that research and development activities are affected by economic incentives just like other activities in the economy. Research and development activities are not conducted in a vacuum. Researchers are aware of societal problems and want to help solve them. Societal needs and economic realities play an important role in the allocation of research funds. The role of the private sector in research and development is increasing, and private investment in research is motivated by profitability considerations. It is no wonder that Israel developed drip irrigation because of the relative high price of water in Israel and the national need to settle arid areas, such as the Negev and the Arava (Hochman *et al.*, 1985). These incentives led many talented individuals to devote much time and resources to the development of water conservation technologies, in particular, drip irrigation.

Thus, strong anti-pollution and environmental conservation policies induce researchers to look for greener technologies. Such technologies are more likely to be developed if research grants and appointments are dedicated to the study of environmental science and the environmental side effects of agricultural activities. Society can signal appreciation for environmental quality and preservation through economic incentives. Economic incentives can be used to spur the development of environmentally-friendly technologies and to encourage farmers to adopt these technologies while, at the same time, maintaining or even increasing production levels.

The development of drip irrigation demonstrates that scarcity provides a powerful incentive to innovate. The success stories of the Arava (Hochman *et al.*, 1985) demonstrate that it is essential to have a sufficiently educated workforce to carry out research and experiment with new technologies. Thus, a strong educational infrastructure at all levels, including adult education and extension, is crucial for developing the capacity to innovate and adopt new technologies that improve environmental quality and economic well-being.

## 5. Population considerations

Problems of sustainability and growth cannot be discussed without mentioning the crucial issue of population growth. If there is one element that has to be controlled in order to preserve the environment, it is population growth. Yet, there is relatively very little research in this area. While population policies are of major importance in many countries, they are hardly discussed and the lack of attention to this matter makes much of our other research less relevant. Economics can and has contributed to understanding population growth. Decisions about family size and responsibility towards raising and educating children are affected by economic incentives. Therefore, any environmental policies should have an explicit component that addresses population growth as well as education. A better scientific understanding of population dynamics is a major challenge that requires collaboration and inputs from physical, biological, and social disciplines.

## 6. On the implementation of economic solutions to environmental problems

An important criterion for the choice of policy tools is simplicity. One reason policy makers favour the use of uniform technology regulations to control environmental problems is their simplicity. But as we argued earlier, uniform technology standards are inefficient because of heterogeneity considerations. This heterogeneity is in terms of land quality, weather conditions, and, in particular, human capital.

Zilberman *et al.* (1991) find that 20% of the water used in California generates about 70% of the economic benefits, while the least efficiently used 10% of the water generates less than 2% of the economic benefits. Thus, if the government aims to reduce agricultural water use by say, 10%, it will be efficient to use a policy tool that will eliminate water use at the least-efficient activities, and then the water supply reduction will result in less than a 2% reduction of the economic benefits of the agricultural water use. A uniform (and thus simple) incentive that will result in the efficient outcome is taxation of water use. Farmers will not use water if the water tax is higher than the economic benefit the water generates. Thus, taxation eliminates water use at the least-efficient activities, and the policy maker will have to estimate the tax level that reduces water use by 10% and impose it.

Zilberman and Marra (1993) provide several examples from empirical studies that have compared the performance of pollution or resource taxes with direct controls. In some cases, the tax will achieve the same levels of improvement in environmental quality at less than 50% of the costs. In spite of the efficiency of pollution taxes, they have not been used frequently. Buchanan and Tullock (1975) use political economic argumentation to explain the infrequent use of pollution taxes. While taxes may attain environmental policy targets at least cost to society, they are not a least-cost policy from the polluters' perspective. Polluters suffer losses both because they reduce their pollution levels and because they pay taxes on the pollution they continue to generate. Because they entail less costs to polluters, polluters may even prefer policies such as direct controls that are the worst from society's perspective. Since polluters are significantly affected by pollution regulations, they may engage in political lobbying to affect the choice of policy instruments.

Tradable pollution rights are as efficient as taxes but have different distribution effects. One example is pollution licences. For example, the United States allows the use of 2000 tons of methyl bromide. The government licenses the use of methyl bromide and people trade the licences. For example, if users cut methyl bromide use by 70%, they will receive initial allotments that are 30% of their traditional use, and then they can trade. Trading the right to pollute is similar to taxation. The people who benefit most from the polluting agents will buy, and then an efficient resource allocation can be reached. By recognizing heterogeneity and introducing taxation or mechanisms like tradable permits, pollution can be reduced at least cost.

Pesticide taxation has been tried in Europe, but generally taxation of pollution and scarce resources has been mild in the United States. Last year, the Clinton Administration lost the opportunity to introduce an energy tax because they presented it as a revenue-generating device and underemphasized its capacity to preserve the environment. As economists who care about the environment, we face a challenge to educate the public about the power of incentives and to move towards a taxation system that provides incentives to preserve the environment. In order for taxation to have a significant and immediate effect, it has to be drastic. If we have a 10% pesticide tax, its impact will be slow because the costs of adjustment and transition are quite substantial and adjustment to the tax will be gradual. However, a 100% tax on certain chemicals may make a difference and cause people to switch to alternative chemicals or at least to reduce the use of the undesirable chemicals significantly. Obviously, we have to educate the politicians and the public about the value of such taxes, but sometimes it is not unreasonable and may be justified on economic grounds.

How can one justify a 100% pesticide tax? A recent study by David Pimentel from Cornell showed that, on average, the productivity of pesticides is four times their cost.[2] This study also shows that the social cost of pesticides is twice the private cost of pesticides. What do I mean? The cost of pesticides in terms of health, endangered species, and ground water contamination is twice the private cost – the cost of buying the material. I do not believe that a 100% tax on pesticides will reduce productivity very much. It will make farmers think twice before they apply chemicals, and it will provide incentives to develop application technologies that are more efficient. We will see continued pesticide use without much side effects. Introducing a drastic pesticide tax will allow keeping the baby and keeping the environment. Currently, there is pressure to ban pesticides, and that is much more costly and inefficient.

One encouraging development is the tendency to convert existing agricultural policies so they can be effective in meeting environmental objectives. For example, the Conservation Reserve Program is, in essence, a set-aside programme, namely, a programme whereby farmers are paid for setting aside their land and not producing a product. This programme helps reduce supply so farmers can get higher prices. Under the Conservation Reserve Program, environmental quality considerations determine which lands are set aside. The exact procedures of this programme and its dimensions are subject to policy debate, but from an economic perspective, it is desirable that the lands that will be set aside are the lands that provide the greatest environmental benefits per dollar spent.

## 7. Economists and the environment

Thus far, I have presented an economic perspective to address sustainability problems and environmental policies in general. I believe that economists can play a major and critical role in the environmental policy arena, but their perception by other disciplines, as well as the practice of economics, has to change. The major contribution of economics is that it is a discipline that provides rules for

[2] While I have some problems with the methodology used, I think that the example is useful for expository purposes.

co-ordination of activities and choices. In order to be effective addressing environmental problems, as well as any other set of microeconomic problems, economists have to know the details of the specific enterprises. They have to integrate physical and biological models into economics.

There is a tendency among economists to rely on secondary data. In many microeconomic applications, economists obtain such data and run statistical analyses without studying the basic physical relationships, and if the statistical model has high $R^2$, it assumes that modellers capture basic phenomena. But, obviously in many situations, that is not the case, and economists lose credibility because they may suffer from lack of knowledge of the subject matter. Therefore, applied environmental economists have to be knowledgeable about the subject matter. Thus, students of the field should be encouraged to learn biology, to understand biological processes, and to integrate biology into economics. I do not believe that an economist, even with the best education, can become an instant expert on any problem. It took me about five years to try to understand pesticides, and I am still only a beginner. I admire Dan Yaron because he has devoted his career to studying water systems and he really knows water. Economists are challenged to develop career paths where they become knowledgeable in subject matter, understand the perspectives of other disciplines, and communicate with people from other disciplines; and then they can make meaningful contributions and be entitled to be called 'experts'.

The input of economists in the management of natural resources is underutilized. Obviously, economists have to be more committed to these areas, learn them better, and improve their ability to communicate with other disciplines. On the other hand, biologists sometimes do not recognise the data information needs of economists. Policy-relevant data that economists need seem sometimes trivial to biologists. Some biological studies may discard the performance data that are most relevant from an economic perspective. Economists need data on input–output relationships. They need better information about production and environmental management practices. They need quantitative measures of quality characteristics and how they are affected by production activities. To have effective environmental and health policies, it is important to have better quantitative measurements of damage to crops, the environment, and humans from pests and pesticides. This may not be very interesting from a biological perspective, but it is essential for policy. Therefore, policy-relevant research is needed in the biological sciences. If such research is carried out and a dialogue among economists, decision scientists, and natural scientists is established, we will be able to develop better policies and that will allow society to achieve efficiency with environmental quality.

## References

Arrow K J 1962 The economic implications of learning by doing. Rev. Econ. Stud. 81, 155–173.

Batie S 1989 Sustainable development: Challenges to the profession of agricultural economics. Am. J. Agric. Econ. 71, 1083–1101.

Buchanan J M and Tullock G 1975 Polluter's profits and political response: Direct control vs. taxes. Am. Econ. Rev. 65, 139–147.
Carlson G, Zilberman D and Miranowski J Eds 1993 Agricultural and Environmental Resource Economics. Oxford University Press, New York.
Caswell M and Zilberman D 1986 The effects of well depth and land quality on the choice of irrigation technology. Am. J. Agric. Econ. 68, 798–811.
Chavas J P 1993 On sustainability and the economics of survival. Am. J. Agric. Econ. 75, 72–83.
Daly H E 1968 On economics as a life science. J. Polit. Econ. 76, 392–406.
Dasgupta P and Heal G 1974 The optimal depletion of exhaustible resources. Rev. Econ. Stud. 41, 3–28.
Dinar A and Zilberman D 1991 Effects of input quality and environmental conditions on selection of irrigation technologies. In The Economics and Management of Water and Drainage in Agriculture. Eds A Dinar and D Zilberman. Kluwer Academic Publishers, Norwell, MA.
Dixit A and Pindyck R 1994 Investment Under Uncertainty. Princeton University Press, Princeton, NJ.
Fisher A C 1981 Resource and Environmental Economics. Cambridge University Press, New York.
Hayami Y and Ruttan V M 1985 Agricultural Development: An International Perspective. Johns Hopkins University Press, Baltimore.
Hochman E, Vitkon G, Just R and Zilberman D 1985 The dynamics of agricultural development in sparsely populated areas: The case of the Arava. In Desert Development: Man and Technology in Sparselands. Ed. Y Gradus. D. Reidel Publishing Company, Dordrecht, Holland.
Howarth R B and Norgaard R 1990 Intergenerational resource rights, efficiency, and social optimality. Land Econ. 66, 1–11.
Pimentel D and Lehman H Eds 1993 The Pesticide Question. Chapman and Hall, New York.
Solow R 1992 An almost practical step towards sustainability. Invited Lecture, Resources for the Future, Washington, DC.
Wiebers C 1992 Economic and environmental effects of pest management information and pesticides: The case of processing tomatoes in California. Unpublished PhD Dissertation, Technischen Universitat, Berlin, Germany.
Zilberman D and Marra M 1993 Agricultural externalities. In Agricultural and Environmental Resource Economics. Eds G Carlson, D Zilberman and J Miranowski. pp 221–267. Oxford University Press, New York.
Zilberman D, Schmitz A, Casterline G, Lichtenberg E and Siebert J 1991 The economics of pesticide use and regulation. Science 253, 518–522.

# 47
# Homogeneity and heterogeneity of bio-resources in economic models

EITHAN HOCHMAN AND EYAL BRILL

## 1. Introduction

Israeli agriculture in the last decade has been characterized by structural changes. It has changed from administrative systems based on centralized allocation, into decentralized systems, operating through markets. This transformation has been interrelated with technological and institutional changes. This chapter will present the hypothesis that these changes are part and parcel of the evolution of resource allocation systems.

According to the framework that will be presented, homogeneity and heterogeneity play an alternating role over time both as a catalyst and as a constraint in the process.

Several stepping stones can be identified in the development of a bio-economic system. The first stage of hunting and fishing is characterised by random harvest of yield with high variety. The second stage of controlled production systems is characterised by homogeneity and uniformity. These characteristics serve as a leverage for the development process. This stage is also characterised by increased collective action. Availability of resources (land and water) and friendly environment enable equal allocation of resources among individuals even though this allocation is inefficient. Accordingly, the institutions that deal with the allocation of the resources follow egalitarian rules for allotment of water, land and other resources.

At the second stage the agricultural production units specialize according to their initial endowment of physical and human capital and the environmental conditions. The old institutional framework is still the basis for resource allocation. However, the specialization of the economic units generates heterogeneity in the distribution of physical and human capital. At this stage the efficiency of the egalitarian allocation is decreasing.

Scarcity of resources and environmental entropy are the trigger of the third stage. At this stage the search for efficient utilization of bio-resources generates allocation mechanisms based on the comparative advantages of the economic units. The solution may be technological or institutional.

Three case studies are examined. The first deals with renewable resources using the example of aquaculture. The second case study deals with exhaustible resources

*D. Rosen* et al. *(eds): Modern Agriculture and the Environment, 567–582.*

using the case of water resources. The third deals with the impact of the modern agriculture industry on the environment.

## 2. The technological paradigm: appropriate technologies in aquaculture.[1]

Aquaculture is an example of an industry that went through the three stages during the last decades. The scarcity of bio-resources in the open seas, and the need for control over the quality and size of the harvested yield, motivated the transformation into both semi-intensive and super-intensive systems of controlled production. Thus, in the case of aquaculture, the intensive systems are closely connected with the second and the third stages that are described above.

The case of intensive aquaculture or mariculture is an example of a stochastic growing inventory problem (Rausser and Hochman, 1979; Karp *et al.*, 1986). The features characterizing the problem are as follows:

1. Time enters at two levels: (a) the age of the growing crop and (b) the calendar date of harvesting.
2. Decisions take place in future time, which implies uncertainty with respect to the surrounding physical environment as well as to the market environment.

The *round pond* technology developed in the Oceanic Institute in Hawaii (OI) is an example of a super-intensive production system of the Equadorian white shrimp, *Penaeus vannaei*, based on the 'all in all out' principle and characterized by homogeneity and uniformity of the product (see Hochman *et al.*, 1990). The shrimps are stocked at densities ranging from 45 to 150 animals per square metre. The pond incorporates features of intensive systems from around the world. It is round in shape, 1 metre deep, slightly sloped towards the centre, 337 square metres in area, and has a large centre sump/drain. The sides of the pond are cement block and the bottom is compacted dirt. A paddle wheel aerates and mixes the pond water in a circular pattern that causes organic wastes to accumulate in the centre drain for disposal.

The bio-economic model consists of the following:

The biological growth function of the shrimp which is depicted in Figure 1.

An estimation of such a function based on the OI data yields

$$ln(W) = 4.59 - \underset{(32.20)}{238.25}(1/x) + \underset{(4.42)}{18.53}DT(1/x) + \underset{(1.16)}{5.72}DD1(1/x) + \underset{(2.34)}{14.53}DD2(1/x) \tag{1}$$

$R^2 = 0.96$

where $W$ is the average weight of the animal in grams, and $x$ is the animal age in days. $DT$ is a dummy variable that takes the value of 1 for temperature greater or equal to 26°C and 0 otherwise. DDI and DD2 are dummy variables for density

[1] This case study is based on two research projects. One was conducted in the Oceanic Institute in Hawaii, and reported in Hochman et al. (1990). The second one is an on going research conducted in the Arava region in Israel in cooperation with Avi Koren and his R&D team. The second research is partially reported in Hochman and Brill (1993). We are grateful to Mr. Avi Koren and his team for their helpful cooperation.

*Table 1.* NPV for continuous production as a result of weight of marketing.

| Weight | 17.3 | 18.6 | 19.9 | 21.2 | 22.4 | 23.6 | 24.8 | 26.0 | 27.1 | 28.2 |
|---|---|---|---|---|---|---|---|---|---|---|
| NPV | 7.46 | 7.59 | 7.69 | 7.76 | 7.81 | 7.84 | 7.85 | 7.84 | 7.82 | 7.79 |

effects representing stocking densities of 100 and 75 animals per square metre respectively. The numbers in parentheses are $t$-statistics. When DD1 and DD2 are both set to zero, it represents stocking density of 150 animals per square metre.

The second element is the price–size relationship, i.e. the average price per pound, depending on the size of the animal. The price–weight relationship that was estimated from the OI date is given in Equation 2:

$$P = -0.008 + \underset{(37.05)}{0.161\,W} - \underset{(11.30)}{0.0008\,W^2} + \underset{(0.02)}{0.0006\,\mathrm{DS}} + \underset{(6.02)}{0.200\,\mathrm{DF}} + \underset{(2.85)}{0.0928\,\mathrm{DW}} \quad (2)$$

$$R^2 = 0.96$$

where $P$ is the price per pound and $W$ is mean weight in grams per animal. DS is a dummy variable for summer, DF is a dummy variable for fall, and DW is a dummy variable for winter. The numbers in parentheses are $t$-statistics.

The estimated cost figures were based on the OI *round pond* farm calculations. They were based on a hypothetical farm of 24 ponds, 0.2 hectare (0.5 acre) each with stocking density of 100 animals per square metre. The feed intake assumed feeding rates based on wet weight of shrimp. Feed costs were calculated at $0.00125 per gram of feed. Labour costs were calculated at $203 per week per pond for regular sampling and feeding. Energy costs were calculated at $56 per week per

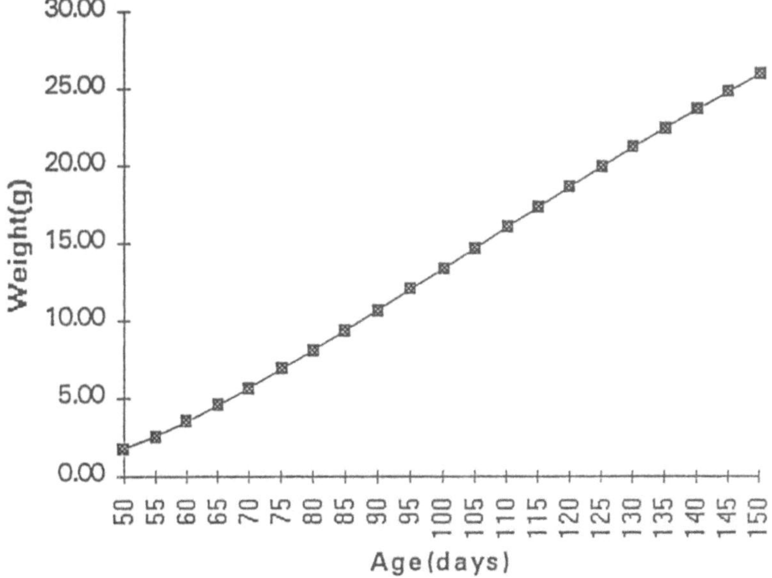


*Figure 1.* Growth function of the shrimp.

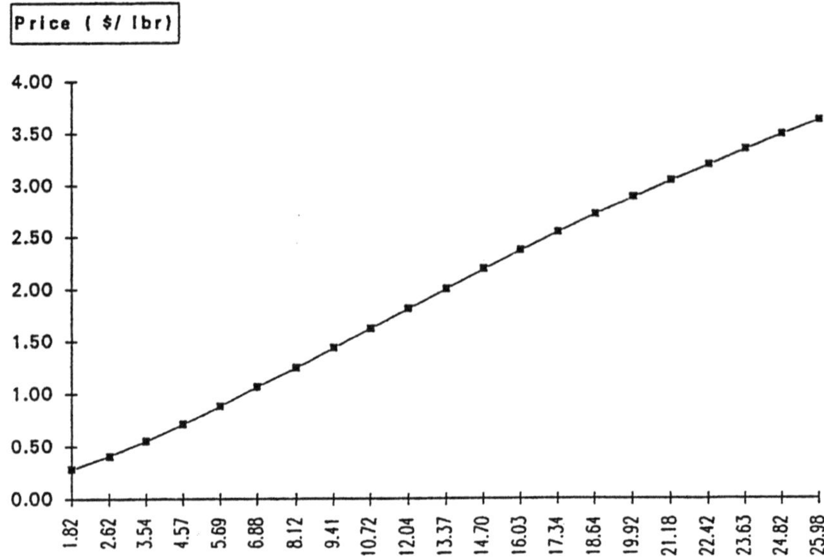


*Figure 2.* Price–size relationship of shrimp.

pond. Stocking costs were calculated at $3000 per pond, for buying post larva and feeding it in a nursery for 65 days.

The problem of the producer, assuming a continuous production process, is to determine the optimal length of the time period between sequential harvests. Note that the decision to postpone the marketing of the crop to the next period involves a trade-off between the benefits resulting from the positive price–size relationship and the loss resulting from a delay in the introduction of the new batch of shrimps.

Table 1 depicts the net present value (NPV) of several policies of harvesting. The best results are obtained from a marketing weight of 24 grams per animal, which corresponds to an optimal length of 145 days between harvests.

Aqua production systems that are based on 'all in all out' growing inventory management (used in shrimps and broilers), suffer from the increased density characterizing the last stage of growth. Such high densities may bring a collapse of the biological equilibrium of the system and hence big losses. The adoption of the appropriate technology should recognise the biological rules and constraints, while pursuing economic profits.

Recently, a desert climate technology named the *bubble pond*, has been developed in the southern Israeli desert – the Arava region (by the R&D team headed by Avi Koren). The bubble pond consists of a completely controlled environment, achieved by use of a PVC sleeve. The lower half of the sleeve is sunk in the ground and filled with 300 cubic metres of water. The upper half forms a bubble by the introduction of pressurised air. The PVC roof allows a selective penetration of the solar radiation. The aeration system is controlled by a circular ventilator using entrance and exit departures on both sides of the dome.

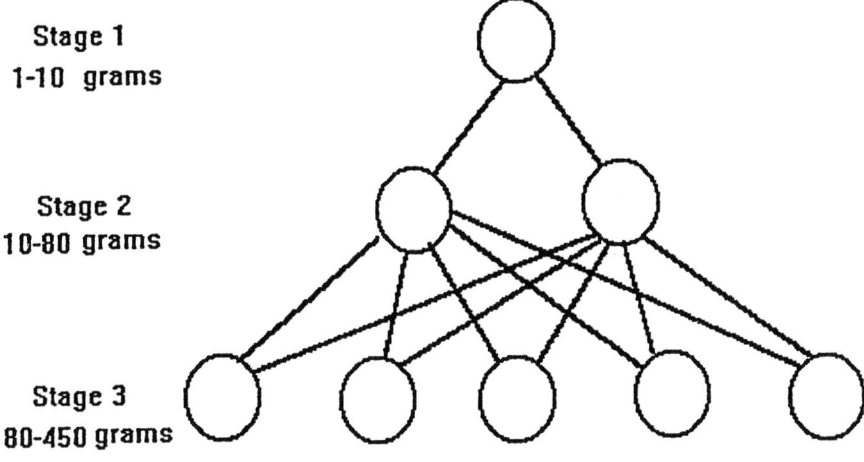


*Figure 3.* Growth stages in the bubble pond technology.

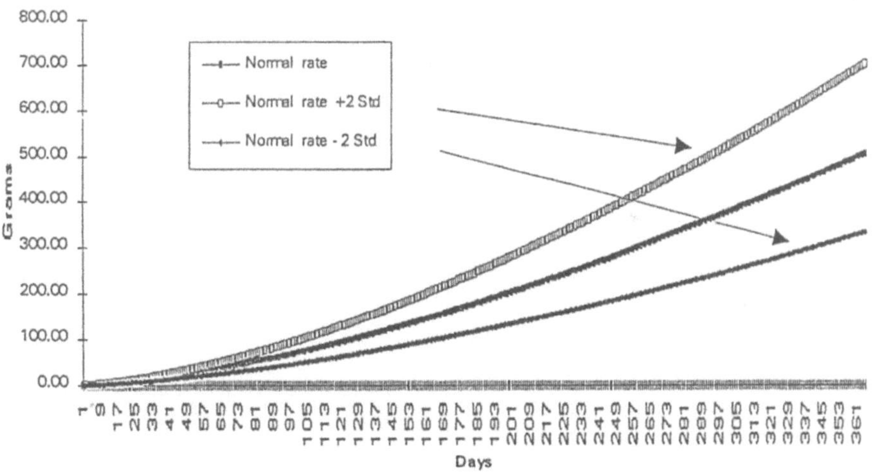


*Figure 4.* Growth function of the Red Tilapia (Based on data collected by Arava R&D).

This is a super-intensive production technology used in the Red Tilapia aqua farms. The production process is based on a 'skimming' mechanism (see Figure 3). The fish life cycle production involves three stages, each in a different pond location. In the first stage the rearing of fry takes place, until the fish reach a weight of about ten grams. The juvenile fish (over ten grams) are transferred into the second stage where they are reared until they reach 80–100 grams. Then the fingerlings are transferred into the maturation ponds. They remain in the growing out ponds for a year to a year and a half. The average market weight of the Tilapia is between 400

*Table 2.* Mean weight and coefficient of variation for the two policies.

| Pond no. | Policy 1 | | | Policy 2 | | |
|---|---|---|---|---|---|---|
| | Average weight (g) | Coefficient of variation | Batch size | Average weight (g) | Coefficient of variation | Batch size |
| 1 | 402.73 | 2.87% | 10 710 | 278.71 | 31.48% | 9810 |
| 2 | 294.98 | 5.08% | 8730 | 241.90 | 41.43% | 9980 |
| 3 | 195.47 | 6.16% | 15 580 | 199.20 | 47.76% | 7640 |
| 4 | 287.36 | 24.85% | 7360 | 223.11 | 45.22% | 8100 |
| 5 | 107.08 | 8.12% | 15 240 | 212.96 | 44.62% | 9610 |

and 500 grams. Marketing is based on 'skimming' of the larger fish from a certain weight and above. Each stage may involve several ponds according to the size of the farm. Naturally the number of the ponds is increased at each stage.

Using a computer simulation model two different policies were examined. The first one is an 'all in all out' policy adjusted to the skimming mechanism. The second one is a policy which takes advantage of the heterogeneity of the growth rate. The length of the Tilapia growth time from hatchery to marketing size may vary from 300 days to 450 days, and accordingly varies with the distribution of weights at a given age.

The difference between the two policies is that following policy 1 each cropped batch at stage 2 is matched by a pond at stage 3, while in policy 2 each batch cropped from a pond of stage 2 is allocated among several ponds of stage 3.

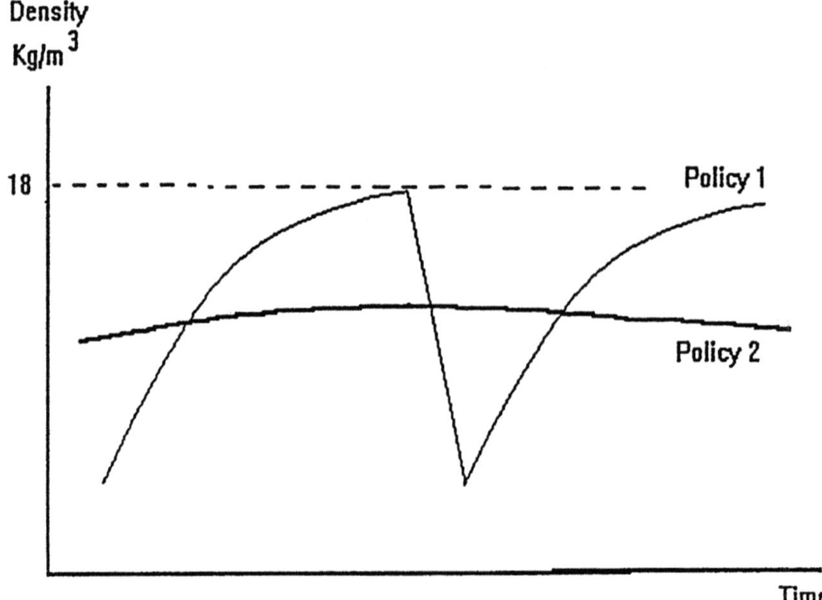


*Figure 5.* Density distribution under the two policies.

*Table 3.* Yields and densities for the different policies.

| Policy | Batch size (fingerlings) | Annual biomass (Tons) | Densities | | |
|---|---|---|---|---|---|
| | | | Average (kg m$^{-3}$) | Minimum (kg m$^{-3}$) | Maximum (kg m$^{-3}$) |
| 1 | 20 000 | 42 080 | 10.22 | 2.5 | 20.5 |
| 2 | 20 000 | 46 239 | 8.53 | 6.8 | 11.5 |
| 2a | 35 000 | 64 661 | 12.24 | 8.0 | 18.3 |
| 2b | 45 000 | 65 817 | 14.46 | 9.6 | 18.8 |

The steady state results of computer simulation of the two policies are depicted in Table 2 for five grow out ponds (stage 3).

The different skimming policies result in a higher variability of the average weight and number of fish in policy 1 as compared to policy 2, and a lower coefficient of variation within the ponds in policy 1 with respect to policy 2, (i.e. the coefficient of variation in policy 1 is around 5% while in policy 2 it is around 40%). In other words, policy 1 results in variability between the ponds of the average weight and the number of fish, while policy 2 results in variability within the ponds and homogeneity between the ponds.

Figure 5 depicts the distribution of the density per cubic metre over the growth period in a typical pond at stage 3 under the two policies. If the regime is of policy 1 the density may shift from 3 kg m$^{-3}$ during early stages of the growth, to 16–17 kg m$^{-3}$ at a later stage, right before the cropping date. On the other hand, policy 2 results in a constant flow of fish in and out of the pond, thus the density is stable at a constant density per cubic metre.

The above production profiles allow for a more efficient use of the ponds at stage 3. Table 3 demonstrates that following policy 2 instead of policy 1 it is possible to increase the yield by about 40% by eliminating the density bottleneck. For a detailed discussion see Hochman and Brill (1994).

## 3. The institutional paradigm: the case of 'common property' in water[2]

In the case of a natural resource such as water, the combination of 'common property rights' (i.e. collective ownership) and free access (i.e. each individual can utilize the resource without constraint), will result in what is named 'the tragedy of commons', i.e. the depletion and the waste of the commonly owned resource (Hardin 1968).

Thus, in the case of water, the first stage of the process is identified with free mining of wells or free diversion of water from the main stream. The transformation from the first stage of the process to the second stage can be characterized by forming various types of organisation to deal with the phenomenon described above.

The institutional set-up of the water industry in Israel (as in many other states) is based on quotas allocated to water users by a central agency which is administered

[2] This case study is part of Eyal Brill's PhD dissertation.

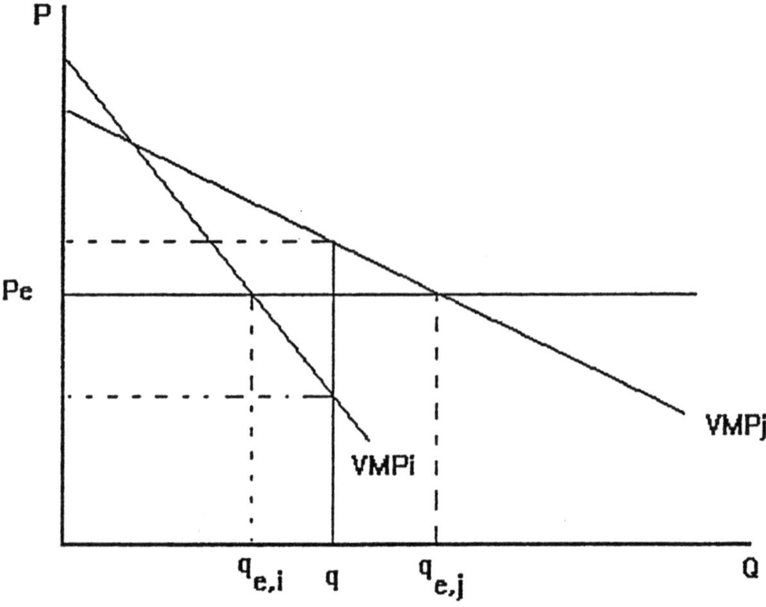


*Figure 6.* The outcome of a fixed percentage allocation.

by the government. The main obstacle in achieving an efficient allocation of water in such an organization is the asymmetry between the information about aggregate available water which is known to the central decision maker and the information at the farm level (e.g. the efficient quantities of water needed for each crop) that the individual farmers do not tend to reveal (Zusman, 1991).

Lack of information in many cases yields an equal allocation among users. Figure 6 describes the result of such an allocation. Let $q$ be the equal water use right of each user. For some, the value of marginal productivity of water, at the allocated water use rights, will be higher than the price of water, and thus they will be potential buyers of water (see Figure 6, curve VMPj). However, for some of the water users, at their given water use rights, the value of marginal productivity of water will be lower than the price of water, and they will be potential sellers of water (see Figure 6, curve VMPi).

The existence of potential buyers and sellers indicates that the allocation is not an efficient one.

Severe droughts that took place in the second half of the last decade, as well as an increased demand for water by the urban sector, led to the recognition of the need for a policy change in the allocation of water in the agricultural sector.

A possible change in the allocation mechanism of water is the transformation into a market mechanism. The market mechanism is characterized by private tradeable property rights for water.

The appearance of water market mechanisms as an institutional set-up for water allocation is the transformation into the third stage in the case of the water. The water market as an allocation mechanism is discussed by Burness and Quirk (1979), Howe *et al.* (1986) and Shah and Zilberman (1993). In the following, an example of a mechanism design, which is based on the theory of co-operatives (Zusman, 1988), is suggested to imitate a trading mechanism among water co-operative members at the regional level. The suggested mechanism is used in the case of an Israeli water co-operative (see Brill and Hochman, 1994).

This mechanism makes it possible to bridge between the heterogeneity of the individual demands for water, which reflects different agricultural skills of the farmers, and the egalitarian water quota allocation system.

### 3.1 *The market mechanism*

The market mechanism is achieved by constructing four rules.

*Rule 1*

Each member of the co-operative has an annual water use right. This right is determined according to his historical right and according to the aggregate quantity used by the co-operative. Thus, the annual use right ($q_n^a$), for member $n$ is given by:

$$q_n^a = \left(\frac{q_n^h}{Q^h}\right) Qe \tag{3}$$

where $q_n^h$ is the historical right of member $n$, $Q^h$ is a summation of the historical rights of all co-operative members and $Qe$ is the efficient amount of water, i.e. the reduced amount of water that should be used by the co-operative members when adopting a saving policy of water allocation.

*Rule 2*

Water consumption, by each individual user within the framework of his annual water use rights, will be charged the price of $P'$, which equals the regional average costs of water production:

$$P' = \left\{ \int_0^{Qe} Mc(q)dq - S + K \right\} \times 1/Qe = Ac(Qe) \tag{4}$$

where $Mc(q)$ is the marginal costs of generating water to the cooperative, $S$ is a lump sum transfer from the central government to the co-operative, and $K$ is the fixed cost of generating the water.

*Rule 3*

Water consumption, by each individual user, in excess of the annual water use right, will be charged the efficiency price $Pe = Mc(Qe)$.

*Rule 4*

In the case where the water consumption by the individual user is short of his water rights, the consumer will be compensated for the difference by:

$$R = r \times \left(q_n^a - q_n\right) \tag{5}$$

where $r$ is calculated by:

$$r = Mc(Qe) - Ac(Qe). \tag{6}$$

The above four basic rules of allocation and pricing of the regional water district, differentiate between the allocation of equal (homogeneous) shares of water use rights among the co-operative members and the efficient (heterogeneous) application of water among the crops, supported by efficient pricing. In the following application, we compare the welfare and the water use under two policies. The first policy consists of administrative allocation of water quotas, while the second one is based on the application of the above four rules.

### 3.2 An application

Empirical data collected from a region in Israel (Hasharon region) are used in a simulation of the suggested mechanism. The table in the Appendix contains data for 50 major crops in the region for the year 1991. Each row in the table depicts the data for one crop. The crops are listed in descending order according to the average rent per cubic metre of water. The average rent per cubic metre of water is based on data made available by the Israeli Ministry of Agriculture. Note that out of 79 million $m^3$ allocated water, about 6 million $m^3$ water are applied to marginal crops with negative average rent.

The farmers in the Hasharon are organized as a water co-operative. Out of a total consumption of 79 million $m^3$ per year, about 30 million $m^3$ are generated by wells within the region. The average cost of pumping a cubic metre of water from a local well is 0.43 NIS. Surface water is imported from outside the region via the national aqueduct at the price of 0.65 NIS per cubic metre of water.

The quasi-rent which results from the application of 1 $m^3$ water to a given crop is the profits derived by deducting the average costs from the average revenue of a given crop. The obtained quasi-rent is multiplied by the total amount of water applied to the crop, and is registered in column seven. The welfare generated by the current policy depicted in Table 4, 20 203 000 NIS, is obtained by summing over the crops in column 7 in the Appendix.

In the following analysis we assume a fixed water–land ratio for each individual crop. We examine the simulations of two policies aimed at achieving the use of 31 000 000 $m^3$ water which is the reduced amount of water dictated by the binding environmental policy.

Table 4 depicts the current allocation and the simulated allocation under the two policies.

*Table 4.* Different policies of water allocation.

| | | Current situation | Policy 1 | Policy 2 |
|---|---|---|---|---|
| Total water used | (1000 $m^3$) | 79 051 | 31 557 | 31 562 |
| Irrigated area | (dunam) | 117 133 | 46 759 | 48 119 |
| Welfare | (1000 NIS) | 20 203 | 12 030 | 52 116 |
| Profits | (1000 NIS) | 20 203 | 12 030 | 52 116 |
| Profit/$m^3$ | (NIS) | 0.25 | 0.38 | 1.65 |
| Profit/dunam | (NIS) | 0.17 | 0.25 | 1.08 |
| Efficient price | (NIS) | – | – | 0.65 |
| Average costs | (NIS) | 0.57 | 0.44 | 0.44 |
| $Q \times r$ | (NIS) | 6324 | 6626 | 6628 |
| Compensation | (NIS/$m^3$) | – | – | 0.21 |

### 3.2.1 Policy 1: administrative use of equal reduction of the water quotas

The allocation of water in the region (31 557 000 $m^3$) is obtained through the administrative distribution of the water quotas. In order to achieve this goal, the water for each of the crops is reduced by a fixed proportion. This yields a total irrigated area of 46 759 dunams and a total welfare of 12 030 000 NIS (see policy 1 in Table 4). Note that this policy involves a reduction in the welfare of the region from the previous value of 20 203 000 NIS.

### 3.2.2 Policy 2: the implementation of the suggested mechanism

An efficient pricing mechanism, which is based on the different individual demands (according to the different crops), results in a quantity demanded of 31 562 000 $m^3$ water for a price of 0.65 NIS. Also, total irrigated area and total welfare will be 48 119 dunams and 52 116 000 NIS respectively.

Each set aside cubic metre of water right is compensated by 0.21 NIS, while the price of a cubic metre used within the water use right is 0.44 NIS. Thus, the reservation price is equal to 0.65 NIS, which is also the efficient price.

Note that policy 2 is superior to policy 1, yielding higher total profits and higher water rents and land rents. Therefore, policy 2 generates the least resistance.

## 4. Agricultural production externalities paradigm: the case of heterogeneous industry[3]

In the case of environmental resources the first stage is characterized by an unpolluted, friendly environment. The development of production industries, including the agricultural industry, which pollutes the environment, required government intervention. The heterogeneous structure of the agricultural production units results in different goals achieved by different policy tools, e.g., taxes vs. standards. In the following it is shown that in the case of an industry which consists

[3] This case study is based on Hochman and Zilberman (1978) and on Carlson *et al.* (1993) Ch. 6.

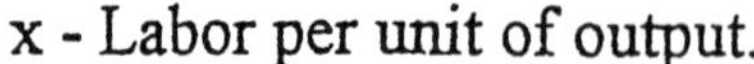


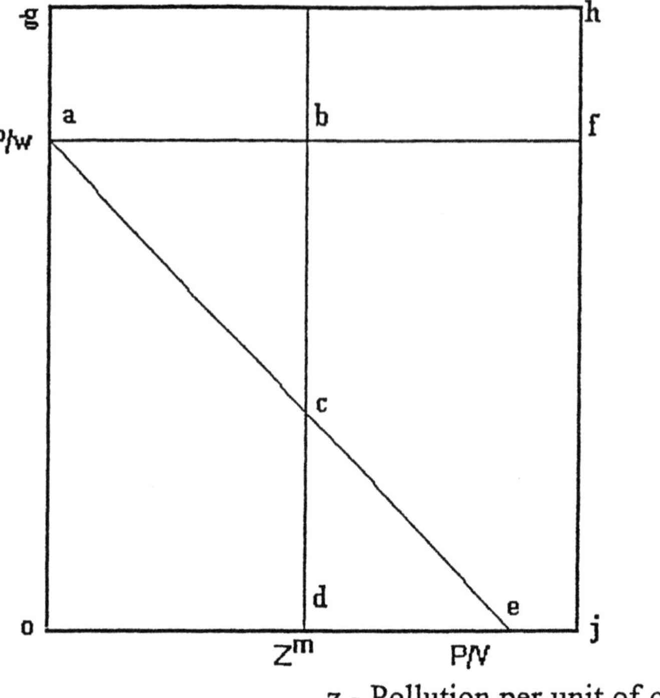


*Figure 7.* The effect of firm heterogeneity on the relative impact of taxes and standards.

of heterogeneous farms, a given level of environmental quality is achieved more efficiently by levying taxes. However, imposing standards to achieve the same level of environmental quality results in higher output.

Hochman and Zilberman (1978) introduced a framework which incorporates the choice of policy tools with technological heterogeneity. They consider an industry with many production units where each unit produces output and generates a pollution with a fixed coefficient technology. Technology coefficients are allowed to vary across firms, for example, the dairy industry, which produces milk and pollutes the environment with its solid waste.

Two alternative policy tools – anti-pollution taxes versus direct control through pollution standards – are used to achieve two conflicting goals, maximum output vs. minimum pollution. Their approach is presented here for the case where output is produced with one input, i.e. labour, and generates pollution.

Figure 7 may be used to describe the impact of regulating the industry by using alternatively each of the two policy tools. The vertical axis measures the labour/output ratio denoted by $x$, while the horizontal axis measures the pollution/output ratio denoted by $z$. The area *afjo* measures the geographical location of all the production units with positive profits. Thus, if $p$ denotes the price of the product and $w$ denotes wages, then the area *afjo* includes all the production units with $x < p/w$. If the industry is regulated by an anti-pollution tax, then the area *oae* measures all the production units that remain profitable, i.e. the production units within the constraint line $wx + vz < p$ ($v$ is the anti-pollution tax per unit of pollution). A pollution standard that restricts the pollution/output ratio not to exceed $z^m$ reduces the region of operating units to *oabd.*

An anti-pollution tax may force relatively inefficient but low polluting units (the ones in *abc*) to stop operating while allowing very efficient but intensive polluting units to continue to operate (units in *cde*). On the other hand, direct regulation setting an upper limit on the pollution/output ratio may cause efficient but pollution intensive facilities to stop operation while allowing relatively low pollution units to stay in business.

The motivation in this case is the prevention of the degradation of the environmental resources. Therefore the overall effect of the two policies should be compared while attaining a target level of Z.

Baumol and Oates (1971) proved that tax is an efficient policy and thus attains the target with least cost. Using Figure 7 it can be demonstrated, however, that when standards are used, aggregate output is greater than under taxes. The proof is simply by observing that if both policies attain the same pollution level, then the production units in *abc* that operate only under a standard generate the same pollution as in *cde* (that operate only under tax). But since the pollution/output ratio of the production units in *abc* is smaller than that of the production units in *cde*, the production units in *abc* produce more output than the production units in *cde*. Hence, aggregate output under the standard regulation (produced by production units in *oacd* and *abc*) is greater than the output under the anti-pollution tax policy (produced by production units in *oacd* and *cde*).

This case demonstrates again (as did the previous one), that in the case of heterogeneity, when choosing between prices and quantities, resource and environmental policies vary in their impact on the industry's performance.

The assumption of fixed proportion technology associated with 'putty clay' models may be quite restrictive even for short run analysis. Just and Antle (1990) expanded the framework for analysis of impacts of environmental policies on heterogeneous industries to allow two types of responses to pollution taxation. Some production units may find operation under pollution taxes impossible and stop operating, while others may modify their operation, reducing the generation of pollution by substituting pollution with other inputs or using pollution abatement strategies.

## 5. Concluding remarks

This chapter examined the evolution of a bio-economic utilisation of natural resources as a three-stage process in which homogeneity and heterogeneity play an important role.

Modern agriculture moves from production with abundant resources into facing the challenge of scarcity. Thus, the efficient use of the bio-resource depends upon detailed study of the bio-technological properties and their integration into the decision and policy making models.

In the first case study the different growth rates of the Tilapia were used in a computerized cropping management model in order to increase the yield of a given aqua farm from 42 tons to 60 tons.

In the second case study, a payment mechanism that recognises the different crop production structures, was designed to save more than half the quantity of the water while at the same time the welfare of the region was increased from 20 million NIS to 52 million NIS.

In the third case study where two conflicting goals were defined – yield vs. environmental quality – it was shown that when the structure of the industry is heterogeneous, the selection of policy tool is not unambiguous. Even though operating through the price mechanism (via taxation) is still more efficient it has opportunity costs in terms of output, when compared to direct control (standards).

*Table A1.* Individual crops budget data – Hasharon region 1992.

| Crop[a] | (1) | (2) | (3) | (4) | (5) | (6) | (7) |
|---|---|---|---|---|---|---|---|
| 1 | 4.90 | 920 | 600 | 552 | 552 | 2392 | 2705 |
| 2 | 4.41 | 920 | 1000 | 920 | 1472 | 3536 | 4057 |
| 3 | 3.31 | 89 | 400 | 36 | 1508 | 98 | 118 |
| 4 | 2.99 | 17 130 | 630 | 10 792 | 12 300 | 26 154 | 32 268 |
| 5 | 2.98 | 121 | 750 | 91 | 12 390 | 219 | 270 |
| 6 | 2.82 | 228 | 600 | 137 | 12 527 | 308 | 386 |
| 7 | 2.60 | 228 | 600 | 137 | 12 664 | 278 | 356 |
| 8 | 2.56 | 920 | 900 | 828 | 13 492 | 1651 | 2120 |
| 9 | 2.14 | 89 | 500 | 45 | 13 536 | 70 | 95 |
| 10 | 2.13 | 121 | 750 | 91 | 13 627 | 142 | 193 |
| 11 | 1.83 | 920 | 1600 | 1472 | 15 099 | 1860 | 2694 |
| 12 | 1.80 | 1001 | 400 | 400 | 15 500 | 494 | 721 |
| 13 | 1.77 | 920 | 965 | 888 | 16 387 | 1068 | 1571 |
| 14 | 1.67 | 406 | 150 | 61 | 16 448 | 67 | 102 |
| 15 | 1.63 | 1476 | 700 | 1033 | 17 481 | 1099 | 1684 |
| 16 | 1.52 | 121 | 750 | 91 | 17 572 | 87 | 138 |
| 17 | 1.44 | 1000 | 500 | 500 | 18 072 | 437 | 720 |
| 18 | 1.35 | 9253 | 750 | 6940 | 25 012 | 5437 | 9369 |
| 19 | 1.31 | 89 | 500 | 45 | 25 056 | 33 | 58 |
| 20 | 1.30 | 89 | 550 | 49 | 25 105 | 36 | 64 |
| 21 | 1.29 | 1781 | 650 | 1158 | 26 263 | 838 | 1493 |
| 22 | 1.23 | 405 | 100 | 41 | 26 304 | 27 | 50 |
| 23 | 1.23 | 861 | 350 | 301 | 26 605 | 200 | 371 |
| 24 | 1.09 | 1476 | 700 | 1033 | 27 638 | 541 | 1126 |
| 25 | 1.07 | 672 | 120 | 81 | 27 719 | 41 | 86 |
| 26 | 0.96 | 1476 | 700 | 1033 | 28 752 | 407 | 992 |
| 27 | 0.85 | 1476 | 700 | 1033 | 29 785 | 293 | 878 |
| 28 | 0.81 | 672 | 120 | 81 | 29 866 | 20 | 65 |
| 29 | 0.80 | 1111 | 500 | 556 | 30 421 | 130 | 444 |
| 30 | 0.74 | 1476 | 700 | 1033 | 31 454 | 179 | 765 |

*Table A1.* cont.

| Crop[a] | (1) | (2) | (3) | (4) | (5) | (6) | (7) |
|---|---|---|---|---|---|---|---|
| 31 | 0.68 | 672 | 160 | 108 | 31 562 | 12 | 73 |
| 32 | 0.62 | 71 | 350 | 25 | 31 587 | 1 | 15 |
| 33 | 0.60 | 2124 | 750 | 1593 | 33 180 | 53 | 956 |
| 34 | 0.51 | 672 | 100 | 67 | 33 247 | −4 | 34 |
| 35 | 0.40 | 1042 | 520 | 542 | 33 789 | −90 | 217 |
| 36 | 0.39 | 1042 | 400 | 417 | 34 206 | −74 | 163 |
| 37 | 0.38 | 121 | 750 | 91 | 34 296 | −17 | 34 |
| 38 | 0.35 | 1042 | 480 | 500 | 34 797 | −108 | 175 |
| 39 | 0.33 | 1476 | 700 | 1033 | 35 830 | −244 | 341 |
| 40 | 0.26 | 672 | 200 | 134 | 35 964 | −41 | 35 |
| 41 | 0.23 | 504 | 220 | 111 | 36 075 | −37 | 26 |
| 42 | 0.05 | 41 155 | 700 | 28 809 | 64 884 | −14 880 | 1440 |
| 43 | −0.14 | 12 095 | 700 | 8467 | 73 350 | −5982 | −1185 |
| 44 | −0.20 | 2281 | 700 | 1597 | 74 947 | −1224 | −319 |
| 45 | −0.22 | 121 | 750 | 91 | 75 037 | −71 | −20 |
| 46 | −0.34 | 121 | 750 | 91 | 75 128 | −82 | −31 |
| 47 | −0.35 | 2687 | 700 | 1881 | 77 009 | −1724 | −658 |
| 48 | −0.54 | 121 | 500 | 61 | 77 070 | −67 | −33 |
| 49 | −0.71 | 89 | 550 | 49 | 77 119 | −62 | −35 |
| 50 | −0.75 | 658 | 700 | 461 | 77 579 | −606 | −345 |
| 51 | −1.26 | 920 | 1600 | 1472 | 79 051 | −2689 | −1855 |

[a] Detailed list of the crop names is available upon request.

Column definition:
(1) Average rent per $m^3$ (NIS).
(2) Total cultivated land for a crop (dunams).
(3) Applied water for 1 dunam of crop ($m^3$).
(4) Total applied water for a crop ($m^3 \times 1000$), [(2) × (3)].
(5) Accumulated water for the region ($m^3 \times 1000$).
(6) Income per crop (NIS × 1000), [(1) × (4) – water costs].
(7) Contribution of the crop to the region overall welfare (NIS × 1000) [(1) × (4)].

## Appendix

Budget data for individual crops in the Hasharon region in 1992 are given in Table A1.

## References


Baumol W J and Oates W E 1971 The use of standards and prices for protection of the environment. Swed. J. Econ. 73, 42–54.

Brill E and Hochman E 1994 Allocation and pricing of water with common property at the regional level. The Center For Agricultural Economic Research. Hebrew University. Working paper No. 9404 (Hebrew).

Burness H, Stuart and Quirk J P 1979 Appropriative water rights and the efficient allocation of resources. Am. Econ. Rev. March.

Carlson A, Zilberman G D and Miranowski J A 1993 Agricultural and Environmental Resource Economics. Oxford University Press, Oxford.

Hardin G 1968 The tragedy of the commons. Science 162, 1243–1248.

Hochman E and Brill E 1994 Introduction of aquaculture in a desert climate: The case of the 'Arava' region in Israel. Presented at International Symposium on Socio-Economic Aspects of Aquaculture. Taiwan 14–17 December 1993.

Hochman E and Zilberman D 1978 Examination of environmental policies using production and pollution micro-parameter distributions. Econometrica, July, 739–760.

Hochman E, Leung P, Rowland L W, and Wyban J A 1990 Optimal scheduling in shrimp mariculture: A stochastic growing inventory problem. Am. J. Agric. Econ. May, 382–393.

Howe W C, Schurmeier D R and Douglas W 1986 Innovative approaches to water allocation: The potential for water markets. Water Resour. Res. 22, 439–445.

Just R and Antle J 1990 Interactions between agricultural and environmental policies. Am. Econ. Rev. 80, 197–202.

Karp L, Sadeh A and Griffin W L 1968 Cycles in agricultural production: The case of aquaculture. Am. J. Agric. Econ. 68, 553–561.

Rausser G C and Hochman E 1979 Dynamic Agricultural Systems: Economic Prediction and Control. North-Holland Publishing Co., Amsterdam.

Shah F and Zilberman D 1993 Queuing vs markets in water resource allocation. Department of Agricultural and Resource Economics, University of California at Berkeley, (unpublished).

Zusman P 1988 Individual Behavior and Social Choice in a Cooperative Settlement. pp 58–64. The Magnes Press, The Hebrew University.

Zusman P 1991 A conceptual framework for a regulatory policy of the Israeli water resources. The Center For Agricultural Economic Research. Hebrew University. Working paper no. 9108.

# 48
# Problems and prospects in the political economy of trans-boundary water issues

RICHARD E. JUST, JOHN K. HOROWITZ AND SINAIA NETANYAHU

## 1. Introduction

The problem of water scarcity has become more acute in many regions of the world because of economic and population growth and because of degradation of historic water resources. New sources and more efficient use of old ones are needed but large investments are required and effective institutions are needed for allocating water, monitoring its use, and ensuring its quality. Because most water projects involve water drawn from and used by multiple jurisdictions, regional or international co-operation is required. Improvement in demand and supply management holds potential for increasing water availability domestically but political realities do not always permit economic optimization.

In many countries, socio-political issues such as food security, tradition, settlement policy, development of arid regions, and equity dominate the cost of water supply among national priorities. Economic improvements must be considered in the context of socio-political goals. Identifying economic measures that are feasible in the context of national objectives can reduce the cost of maintaining ideologies but can also reduce the need to pursue some non-economic goals. For example, international co-operation made possible by peace can make water supply enhancing projects feasible while the feedback effects of improved international trust can reduce the need for security which may motivate some water pricing inefficiency.

This chapter considers the potential for trans-boundary co-operation in water resource sharing with an application to Israel and its neighbours. Past conflicts and co-operation over water resources in the Jordan Basin and the Mountain Aquifer are discussed along with the potential for new international projects. Domestic water issues in Israel, Jordan, and Judea and Samaria are also considered as a standard of comparison. The economic potential for new international projects is discussed in the context of existing domestic water pricing inefficiencies imposed by socio-political factors. The benefits of these projects are compared to the benefits of domestic measures that improve economic efficiency. The ongoing peace process is considered as a catalyst that can enable a combination of these measures given the associated feedback effects.

*D. Rosen* et al. *(eds): Modern Agriculture and the Environment, 583–612.*

The discussion of international projects considers the strategic possibilities that complicate international agreements. This necessitates viewing international co-operation as a two-stage process where in the first stage governments negotiate a sharing arrangement and, possibly with World Bank help, design and pay for large scale projects such as reservoirs and dams. In the second stage, the question of compliance with the sharing arrangement must be addressed given strategic behaviour that becomes possible once investments are in place. Because these decisions are separated over time, water allocation rules must be robust to changes in governments, economic variables, political forces, and water availability.

## 2. Overview of trans-boundary water issues in Israel and the Jordan River Basin

### *2.1 Increasing scarcity and alternative options*

Water scarcity is a major problem in the Jordan River Basin. Per capita water availability is far below the 'water stress zone' in Israel, Judea, Samaria, the Gaza Strip, and Jordan (Falkenmark *et al.*, 1990). This scarcity has led to major disputes over water use. In Israel, Jewish and Arab populations face severe conflicts over the exploitation of the Mountain Aquifer water in Judea and Samaria. High population growth and lack of water resource planning have created severe water scarcity in Gaza as well. At the international level, Israel, Jordan, Syria, and Lebanon have severe and long-standing disputes over water allocation in the Jordan Basin.

For the most part to date, countries have attempted to solve scarcity problems by domestic means. Options on the supply side include reuse of sewage effluent, use of marginal floodwater and saline sources, seawater desalination, cloud seeding, internal transfer of water surplus, and separation of drinking water from irrigation and industrial water permitting use of recycled waste water in the latter uses (Rhoades and Dinar, 1991). All of these options are either expensive or have an uncertain pay-off. Many of these options are currently being pursued on an experimental basis but potential is limited unless new technology can be developed that lowers cost and/or produces more certain benefits. On the demand side, domestic water conserving measures include reducing agriculture production, changing agricultural cropping choices, and employing a price mechanism to reduce marginal price inequities among sectors and regions after accounting for public water cost differences. These options have more certain economic benefits but are possibly contrary to other national objectives.

With little doubt, water is not allocated efficiently within any of the Middle East countries. Simple economic efficiency analysis suggests that domestic reallocation of water could well postpone water scarcity problems. For example, the agricultural sector is the main water user and pays a water price below other sectors. This pricing structure induces farmers to use water to the point where its marginal value is below marginal values of water in other sectors. A reallocation of water away from agriculture holds potential for raising the aggregate benefits of water use and making available the water needed for further industrial and urban growth. Certainly, the increase in water scarcity and decline in water quality that has resulted from increased industrial and urban demands raises questions regarding social efficiency of the agricultural sector under its present structure (Dinar and Zilberman, 1991; Kanazawa, 1991). The lack of efficient management has been underscored recently by consecutive years of drought.

Reduced water allocation to agriculture will induce costly structural changes including:

(1) adoption of more efficient, capital-intensive irrigation practices (Tuijl, 1993);

(2) a shift to crops that require less irrigation and inferior water quality; and

(3) elimination of low-value crops (e.g. field crops) and/or water-intensive crops (e.g. cotton) that likely do not pay the full cost of water.

Further urban and industrial growth and/or reductions in agricultural water use will likely result in declines in overall cultivation and a relative shift in cropping towards fresh flowers, fruits, and vegetables that earn a high return in European markets sufficient to cover the full cost of water. Such restructuring, however, would necessitate abandoning food self-sufficiency policies such as pursued by the Israeli Ministry of Agriculture.

While obvious possibilities exist for cutting water use by restructuring agricultural cropping patterns or reducing total agricultural activity, these options may not be consistent with other national objectives. Economic aspects of water are not necessarily a top priority for many nations. For example, government policies may give higher priority to ideological and national security considerations. Subsidised agricultural water prices may simply reflect a comparative advantage in pursuing legitimate socio-political goals such as settlement of remote areas for security purposes. Thus, the potential gains revealed by simple economic efficiency analysis may not be practical in the larger scheme of things.

A brief assessment of domestic options suggests that, unless new domestic supply-enhancing technologies surface or national objectives change substantially, water shortages are likely to become acute and force major restructuring of domestic water use. Even if feasible domestic conservation possibilities are pursued, increases in the demand for water due to population growth and rising living standards are likely to cause water self-sufficiency to be lost in the foreseeable future. Alternatively, effective water supply can be increased by international co-operation in a variety of projects – although some domestic water reallocation with accompanying structural changes of the agricultural sectors is a likely part of any efficient solution (Frederick, 1993). Given the growing dimensions of scarcity, international co-operation in water resource development appears to be imperative. International water co-operation, however, will involve both high initial investments in infrastructure, and an increase in international goodwill and technological interdependency.

### 2.2 *Water allocation by markets versus bargaining*

The creation of water markets is widely propounded as a means of achieving economic efficiency in water resource allocation within jurisdictions. Similarly, establishment of international water markets is often proposed as a means of facilitating efficient trans-boundary water reallocation and establishing incentives for development of international water projects. However, as experience in many forms of international trade has shown, international markets are not likely to operate free of regulations and restrictions that reduce the efficiency of competition.

In addition, water markets require institutions, laws, regulations, financial arrangements, specific capital investments, monitoring capability, and quality control. Currently, most of these institutional features do not fully exist in trans-boundary Middle East water problems. The presence of transaction costs and the specificity of capital related to water projects in the absence of stable institutional infrastructure are likely to decrease the cost advantages of markets and increase governance costs.

Finally, international water projects that enhance water supply typically carry high investment costs in specific capital items that have few alternative uses. When assets become semi-specific, bilateral contracting becomes preferable to market arrangements, and when assets become highly specific, an internal organization dominates free markets (Williamson, 1981). In the context of Middle East water problems, these results suggest that governments should negotiate contracts for reallocation of water resources and development of new international water projects because of the highly specific nature of required capital investments. In other words, bargaining appears to offer a more suitable mechanism than water markets for facilitating international water co-operation. Moreover, bargaining may be the only initial means of facilitating water transfer given the current lack of effective trading relationships.

Increasing opportunities for voluntary Middle East water co-operation among Israel, Jordan, and the Palestinians have surfaced as a result of recent peace talks and agreements. These opportunities are broadened by World Bank involvement (in research, data collection, and funding) and have attracted academic interest including research groups at the University of Pennsylvania, the Kennedy School at Harvard, the Hammer Fund, the Foerder Institute, the Sapir Center for Development at Tel Aviv University, and Ben Gurion University. Recent research results are reported by Naff and Matson (1984), Lowi (1993), Clarke (1993), Beschorner (1992/93), Kally (1993), Soffer (1992), Kliot (1994), Eckstein *et al.* (1994) and Zeitouni *et al.* (1994), among others. Many conferences are also contributing to the debate including those in Kfar Blum (1992), Zurich (1992), Waterloo (May 1992), and Cyprus (July 1994), to name a few. However, no general agreement among Israel, Jordan, Lebanon, Syria, and the Palestinian Autonomy (all of which claim rights over the Jordan Basin) appears to be close. While some partial agreements have been reached for parts of the Basin, the potential of limiting future bargaining positions deters piecemeal bilateral agreements.

## *2.3 Disputes and disagreements to date*

Before discussing the potential for international co-operation in water resource development, a short history of experience to date is useful. This discussion is facilitated by an overview of the water resources in the region. The upper Jordan and its tributaries originate in Israel and Lebanon. Jordan, Syria, and the Palestinians also have interests in the River's flows. The upper Jordan flows into Lake Kinneret (also known as the Sea of Galilee and Lake Tiberias) which captures about 500 million cubic metres per year (Kally, 1993). Water from the Kinneret is diverted into the Israeli National Water Carrier. The upper Jordan River is fed by three main sources: Dan Spring in Israel (250 million $m^3$ $yr^{-1}$); the Hasbani River which is fed by springs in Lebanon (85–105 million $m^3$ $yr^{-1}$) and Israel (45 million $m^3$ $yr^{-1}$); and Hermon (Banias) Spring in the Israeli-annexed Golan Heights, which prior to 1967 was part of Syria (110–120 million $m^3$ $yr^{-1}$).

The main tributary (40 kilometres in length) in the Basin is the Yarmouk River which forms part of the Syrian–Jordanian border upstream and the Jordanian–Golan Heights border downstream. The Yarmouk River joins the Jordan River 10 kilometres below the Kinneret. Syria and Jordan use the Yarmouk River to irrigate about 15 000 hectares and 13 800 hectares, respectively. Syrian and Jordanian flows into the Yarmouk account for 375 million $m^3$ $yr^{-1}$ and 100 million $m^3$ $yr^{-1}$, respectively. Two other water sources in the Jordan Basin are the Eastern Rim (South Yarmouk) that originates in Jordan (200–220 million $m^3$ $yr^{-1}$) and the Western Rim (South Sea of Galilee) that originates in Israel (45–55 million $m^3$ $yr^{-1}$).

In the lower part of the Jordan River, irrigation returns from both sides of the bank feed into the Jordan River below the Kinneret before it discharges into the Dead Sea. These flows are heavily polluted. In addition, Israel diverts saline springs to the lower Jordan River which would otherwise pollute the Kinneret and more than double its chloride level. For example, about 16.25 million $m^3$ were diverted in 1992 (Hydrological Service, 1994). The costs associated with improving this low quality water prevent the riparian countries from exploiting lower Jordan River waters.

Historic-use rights in the Jordan Basin are extremely complex, as they are throughout the world. Johnston, who was President Eisenhower's personal representative to the Jordan River Valley Development Project, authored a water plan for the Middle East which, although accepted by Basin experts, was never carried out (Johnston, 1954). However, a partial agreement based on Johnston's plan exists between Israel and Jordan, where the United States serves as mediator. The Johnston Plan for Yarmouk River water allocated 275 million $m^3$ $yr^{-1}$ to Jordan, 90 million $m^3$ $yr^{-1}$ to Syria, and 25 million $m^3$ $yr^{-1}$ to Israel. Jordan now uses 120 million $m^3$ $yr^{-1}$, Syria 170 million $m^3$ $yr^{-1}$, and Israel 100 million $m^3$ $yr^{-1}$. Israel uses virtually all of the 500 million $m^3$ $yr^{-1}$ captured by Lake Kinneret from the upper Jordan River.

The problem in implementing the plan is that 'rights' bear little weight in the absence of the means and infrastructure to use them. For example, Syria as an upstream user has been able use more than its share. Jordan, on the other hand, has not had sufficient infrastructure to use all of its allocated water. Finally, Israel uses the Yarmouk at its lowermost end and has been able to use more water than allocated under the Johnston plan because the sum of Syrian and Jordanian use has been less than planned. With respect to the Jordan River, Israel has been able to exploit virtually all of the capacity because salinity problems in the lower Jordan and lack of alternative transportation facilities from the upper Jordan/Kinneret prevent access by Jordan.

Not surprisingly, countries dispute the current share allocations. Jordan does not accept the current level of water use by Israel. Israel, on the other hand, demands an additional 140–150 million $m^3$ $yr^{-1}$ from Yarmouk water to be used by the Palestinians residing in Judea and Samaria which were under Jordanian responsibility when Johnston's plan was written. Jordan opposes Israel's demand and the United States supports Jordan in the matter (see Fisseha (1981) and Soffer (1992) for further discussion).

In addition to the Israeli–Jordanian conflict, Syria and Lebanon reject Johnston's plan and some of Israel's rights to Jordan–Yarmouk water. Syria claims rights to the Banias as part of its claim on the Golan Heights, and Lebanon demands rights to the Hasbani in the Israeli Security Zone. Additionally, Palestinians dispute Israel's pumping of water emerging in the coastal plain from the Yarkon–Tanninim aquifer that originates in Judea and Samaria (Lowi, 1991). This dispute is indirectly related to the Israeli–Jordanian conflict over water in the Jordan Basin and adds to the tension among countries in the region. (For a summary of the history of water conflict and co-operation in the area, see Kliot (1994) and Wolf (1993)).

During the long period of these disputes, several changes have opened possibilities for renegotiating Jordan–Yarmouk water allocation. Lebanon's economy does not depend on Jordan water. Thus, possibilities likely exist for transferring its rights in return for adequate compensation. Southeastern Syrian agriculture is heavily dependent on water and would likely be reluctant to give up rights although Syria has a major alternative water source in the Euphrates River. Jordan depends heavily on water in the Jordan Basin but its position is relatively inferior because it is located in the lower part of the Basin and presently has little control over quantity or quality of the water. Finally, Israel is unlikely to give up rights because the Jordan supplies up to 25–35% of its water (Kliot, 1994; Naff and Matson, 1984; Shamir *et al.*, 1985; Soffer, 1992).

Although negotiation among countries seems to be unavoidable, political barriers complicate the process substantially (LeMarquand, 1977). Many potential projects hold promise for improving water availability and permitting reallocation. However, costs in terms of financial investment, time to establish infrastructure, and dependency on other countries constitute crucial obstacles.

## 2.4 *Potential international water projects*

With negotiation and co-operation among riparian countries, several trans-boundary water projects have potential to ease current water constraints. A number of such projects are listed in Table 1. Two of the most important possibilities which are considered further in the chapter for illustrative purposes are as follows.

### 2.4.1 Diversion of Yarmouk River water to Lake Kinneret

Currently Jordan exploits the Yarmouk's summer flow to irrigate the Ghor Valley leaving a substantial quantity of winter floodwater unutilized. Several research projects (Ben-Shahar *et al.*, 1989; Kally, 1993) suggest that the most economical regional water project is to construct a canal to divert winter floodwater from the Yarmouk River to Lake Kinneret. The water would be diverted back to the riverbed through a second canal during the summer to facilitate flow by gravitation to the Ghor Valley. Because the Yarmouk River lies in Jordan and the Kinneret lies in Israel, however, international co-operation between the two countries is necessary.

Under this plan, 80–100 million $m^3\ yr^{-1}$ out of the 180 million $m^3\ yr^{-1}$ of the Yarmouk water captured in the Kinneret would be allocated to Jordan. However, the remaining share of the 180 million $m^3\ yr^{-1}$ currently unutilized by Jordan due to capacity limitations of the Ghor Canal, could be transferred by Israel to storage in the coastal aquifer. This water could make up for the water deficit currently occurring in the coastal aquifer, which is becoming salinised due to overuse. Alternatively, Israel's share of the water could be diverted to the National Water Carrier in peak-demand periods in lieu of coastal aquifer water diversion. But this may require widening the National Water Carrier.

This project apparently offers lower cost additional water than other proposed projects (Table 1). To demonstrate the potential benefits of the project, note that the cost of the proposed project is approximately \$28 million.[1] Assume that water at its point of use is valued at, say, an average of \$0.20 $m^{-3}$, which corresponds roughly to a weighted average of marginal prices in Israel over uses in agriculture, industry, and households. Transportation costs to Jordan from Israel are approximately \$0.08 (Fisher, 1994). Transportation costs to Israel from Jordan are at most \$0.05 (Eckstein *et al.*, 1994). If Jordan receives 80 million $m^3\ yr^{-1}$ and Israel receives 100 million $m^3\ yr^{-1}$, then total net benefits would be \$24.6 million $yr^{-1}$. Thus, net benefits from the project are highly positive. Investment costs are almost recovered in a single year of operation.

The project also yields quality and other reduced-variability benefits. Chloride levels in the Kinneret which captures roughly 500 million $m^3\ yr^{-1}$ had reached 248 parts per million by 1991 (Hydrological Service, 1994). By running an additional 180 million $m^3\ yr^{-1}$ of high quality Yarmouk water through the Kinneret, its salinity would be reduced by 20% (Ben-Shahar *et al.*, 1989). Since the value of water is a downward-sloping, concave function of salinity (at low levels of salinity), such an averaging of qualities is preferred. Reduced-variability benefits are possible because storage in the Kinneret can not only increase summer flows, but make them more uniform throughout the growing season and, to a lesser extent, across years.

[1] The reported cost is \$21 million in 1983 dollars (Kally, 1993). Note that values and prices are stated in 1990 dollars throughout the paper. We use the GDP price deflator (IMF, 1993) to convert prices to 1990 dollars. All water prices are per cubic metre.

*Table 1.* Potential international water projects affecting the Jordan–Yarmouk River Basin[a].

| Project | Participants | Investment ($1 000 000s) | Quantity of water (million $m^3$ $yr^{-1}$) | Price of water ($ $cm^{-1}$) | Description |
|---|---|---|---|---|---|
| An international desalination project | Israel<br>Jordan<br>Palestinians | 3000–5000 | 128 | 0.88–1.20 | Various plans with various desalination technologies are proposed. Independent desalination projects are also possible |
| Building a 650 million $m^3$ pipeline to transfer water from the Euphrates | Jordan<br>Iraq | 2000 | 120–160 | 0.50 East Jordan<br>1.00 West Jordan | Transferring water is costly and complex |
| Diverting Nile River water to Gaza and the Negev | Egypt<br>Israel<br>Gaza | 136–347 | 91–149 | 0.22–0.247 | Water to be supplied to Judea and Samaria instead of Gaza and the Negev by the National Water Carrier. Conveyance cost is reduced but Nile River water quality and ideological issues present problem |
| Mukheiba Dam: a Jordanian dam on the Yarmouk (1960s) | Israel<br>Jordan | 552.4 | 180 | 0.074–0.131 | Water would be supplied to Ghor Valley farmers, who currently pay less than one cent per cubic metre |
| Maqarin Dam: a Jordanian–Syrian dam on the Yarmouk (1978) | Jordan<br>Syria | 1031 | 275 | 0.262–0.455 | Total capacity of 320 million $m^3$. Syria gains 15 megawatts of electricity. The US has offered financial support providing Israel's interests are considered |
| Unity Dam: a Jordanian–Syrian dam on the Yarmouk (1987) | Jordan<br>Syria | 460.2 | 225 | 0.34[b] | Half the size of the proposed Maqarin Dam. Syria would get all the electric power generated. Water (50 million $m^3$) to be transported to Amman/Zarqa |
| Storing Yarmouk River water in Lake Kinneret | Jordan<br>Israel | 27.6 | 100–240 | 0.014–0.022 | Winter floodwater would be diverted by canal to Lake Kinneret and pumped to the Jordanian Ghor Valley in the summer months and to the Israeli National Water Carrier in the winter |
| Litani–Hasbani Project | Lebanon<br>Israel | 10 | 100 | 0.088 | Would also produce 70 million kilowatts per year |

[a] All monetary values are given in 1990 US dollars with prices deflated by the US Consumer Price Index.
[b] Year of price is not known for certain but applies to years 1990 to 1993.
Sources: Kally (1993); Lowi (1993); *Statistical Abstract of the United States* (1993); and World Bank (1994).

### 2.4.2 Construction of the Jordanian/Syrian Unity (Wahda) Dam

Construction of a dam on the upper Yarmouk River has also been proposed as a means of stabilizing water flows. To stabilise flows, currently unused water flows in the high-flow winter season would be captured by the dam and released during summer. The water would be released directly into the River and then channelled into the Ghor Canal downstream. The Kinneret alternative for storing floodwater is significantly cheaper than building a dam but the political infrastructure for co-operation between Syria and Jordan was much stronger than between Israel and Jordan when the Unity Dam was first considered in 1987.

The currently proposed Unity Dam is a scaled back version of the much larger but considerably more expensive Maqarin Dam that had been proposed earlier. This dam would require Jordanian–Syrian co-operation because it would be located on the Yarmouk River where it serves as an international border between Jordan and Syria. The dam would generate not only water storage and regulation facilities but electricity generation capacity. Because of Syrian and Jordanian infrastructure and terrain, the additional water would be most profitably used by Jordan whereas Syria would benefit from electricity generation.

The World Bank has a loan pending to finance building the Unity Dam, but the loan can be granted only if Israel removes its objection. The Yarmouk River currently contributes 3%–7% of Israel's national water supply and the dam project could threaten Israel's ability to meet growing water demands (Starr, 1992). One possibility for negotiating a removal of Israel's objection and thereby freeing pending loans and investments is to offer a share of the associated additional water resources to Israel. According to Shuval (1992), an agreement between Israel, Jordan, and Syria was negotiated unsuccessfully prior to the 1991 Gulf War under US guidance.

One problem with the Unity Dam proposal is its expense ($500 million). The capitalised cost of additional water provided by the dam is above the current agricultural use value. Estimates for the capitalized cost of Unity Dam water are \$0.26–0.45 $cm^{-1}$. If users had to bear this full cost, the water would not be used for agricultural production in the Ghor Valley, Jordan's main agricultural area. However, water at this price would be considerably below the estimated market clearing price in other sectors and regions which exceeds \$1.00 $cm^{-1}$ (World Bank, 1994). Furthermore, with sufficient construction subsidies from international agencies, use for Ghor Valley agriculture may also become attractive privately.

## *2.5 Summary*

In view of mutual incentives for international projects, most notably between Israel and Jordan on the Yarmouk–Kinneret canals, and the increasingly desperate condition of water particularly in Jordan, the potential for co-operation appears to be enormous. However, any useful approach to co-operation must take account of non-economic and non-efficiency objectives of all participating parties sufficiently to generate incentives to co-operate and sustain co-operation (Biswas, 1993). Obviously, these benefits of co-operation can only be gained if agreements provide positive benefits to all co-operating countries and a fair distribution of the benefits between countries.

## 3. Potential for co-operation in the Jordan Basin[2]

### *3.1 Mutual incentives for Israeli–Jordanian co-operation*

Consider first potential co-operation on the proposed canals connecting the Yarmouk River, the Kinneret, and the Ghor Valley (Table 1). This project contains two important components each of which provides benefits for one of the co-operating parties. One canal carries currently unutilized water from the Yarmouk River increasing Israeli water supply. The other canal carries part of the water back to Jordan at a time when Jordan can make use of it, thus increasing effective Jordanian water supply. Both parties have incentives to co-operate. Apparently, current infrastructure limitations of the Ghor Canal would allow Jordan to use only 80 million $m^3$ $yr^{-1}$ (in addition to its current use) leaving Israel free to use the other 100 million $m^3$ $yr^{-1}$.

Pursuing 'half' of the proposed project by building the canal from the Kinneret to the Ghor Canal (without building the canal to divert water from the Yarmouk River) would allow Jordan to have access to the Kinneret. But Israel's historic-use rights to that water apparently preclude Jordan's use of any of the existing water. In addition, valuable winter flood water would continue to be wasted. Similarly, pursuing the other half of the proposed project by building a canal diverting Yarmouk River water to Israel without providing for a return of water to Jordan only provides one-sided benefits for Israel. Jordan would have no incentive to reduce its claim to Yarmouk River water even though some of it is unused. A potential bargain can be negotiated by combining and balancing these one-sided benefits so that individual benefits to each country are agreeable.

### *3.2 Implications of bargaining theory for co-operation on the Kinneret–Yarmouk project*

Prospects for determining a fair and agreeable bargain can be investigated using game theory. Results from game theory show that if certain standard axioms are satisfied and a bargaining solution can be reached, then it must be the Nash bargaining solution.[3] This sections considers implications of the Nash bargaining solution for sharing the benefits of a Yarmouk River diversion project. To use this

[2] Details on the applications of bargaining theory to co-operation discussed in this chapter can be found in Just *et al.* (1994a).

[3] These axioms require the parties to be rational and treated symmetrically, and that the result is independent of irrelevant alternatives and not affected by linear transformations of each party's preferences.

framework with limited empirical information, we consider some initial plausible assumptions and then examine sensitivity to assumptions. The implications of the solution are robust with respect to assumptions so this approach is informative.

To illustrate, suppose elasticities of water demand are 0.25 in both Israel and Jordan under current marginal pricing schemes. Consider weighted average current prices of \$0.25 $cm^{-1}$ in Israel and \$0.40 $cm^{-1}$ in Jordan with aggregate current water quantities of 1700 million $m^3$ $yr^{-1}$ in Israel and 880 million $m^3$ $yr^{-1}$ in Jordan. Now consider Israeli–Jordanian bargaining over a new 180 million $m^3$ $yr^{-1}$ of water. Total water supply is then 1700 + 880 + 180 million $m^3$ $yr^{-1}$. The Nash bargaining solution of this problem is an equal split of the additional Yarmouk water, with 90 million $m^3$ $yr^{-1}$ for each party.[4] The equal-split solution is surprisingly robust to changes in demand elasticities and to changes in assumptions about the current prices of water. An exactly equal split is predicted for a large range of parameter values.

If a feasibility constraint is added reflecting that the Ghor Canal can handle only 80 million $m^3$ $yr^{-1}$ of additional water, then the constraint is binding. As a result, Jordan gets 80 million $m^3$ $yr^{-1}$ and Israel gets 100 million $m^3$ $yr^{-1}$. Again, this solution is insensitive to assumptions on parameter values. This constrained solution increases water supply by 6% in Israel and 9% in Jordan. Simple analysis of demand implies that this change will decrease the marginal value of water by about 2% in both countries. Fisher (1994) draws a similar conclusion.

### 3.3 *Stability of the co-operative solution*

As this example illustrates, physical constraints caused by available infrastructure can play a significant role. If the Ghor Canal can handle no more than an additional 80 million $m^3$ $yr^{-1}$ and the marginal value of water is higher in Jordan than Israel, then this constraint is binding for a large range of parameter values. The robustness of the unconstrained solution, on the other hand, is related to the assumption of equal elasticities of demand in the two countries. This condition may not hold.

If the Ghor Canal capacity is limited and the marginal value of water in Jordan is higher than in Israel, then the game theory solution is the same as the water markets

[4] *Mathematically, water demands under these assumptions follow:*

$$p_i = \beta_i q_i^{-\alpha_i}, \qquad p_j = \beta_j q_j^{-\alpha_j}$$

where $i$ denotes Israel and $j$ denotes Jordan. Using these demand functions, the benefits derived from water for the respective countries follow:

$$U_i = \frac{\beta_i q_i^{1-\alpha_i}}{1-\alpha_i}, \qquad U_j = \frac{\beta_j q_j^{1-\alpha_j}}{1-\alpha_j}.$$

Reservation points for purposes of bargaining $\overline{U}_i$ and $\overline{U}_j$, are calculated by substituting current use levels in these expressions using existing water use levels. Under the assumptions above, $\alpha_i = \alpha_j = 0.25$, $p_i =$ \$0.25, $p_j =$ \$0.40, $q_i = 1700$ million $m^3$ $yr^{-1}$, $q_j = 880$ million $m^3$ $yr^{-1}$, $\beta_i = 1.61$, and $\beta_j = 2.18$. The Nash bargaining solution is obtained by maximising:

$$\left[\frac{\beta_i q_i^{1-\alpha_i}}{1-\alpha_i} - \overline{U}_i\right]\left[\frac{\beta_j (K-q_i)^{1-\alpha_j}}{1-\alpha_j} - \overline{U}_j\right].$$

where $K = 1700 + 880 + 180$ million $m^3$ $yr^{-1}$.

solution. In both solutions, the capacity constraint is binding. If the capacity constraint is not binding, however, then the solutions can be quite different. The water market solution would allocate all of the additional water to the country with highest marginal value of water compared to the Nash bargaining solution where allocations are equal. This suggests that future investment in water transportation infrastructure in the Jordan Valley could cause a future conflict in water allocation once the Yarmouk canals are in place. This is one reason why Jordan may have been reluctant to negotiate an international agreement with Israel.

This case also illustrates why a market solution may be unacceptable in a trans-boundary water problem. A market solution typically allocates all of an increase in water availability to the country with the highest marginal value of water. Understandably, countries are reluctant to give up all of their claim to a constraining resource given uncertainties about future contingencies and demand growth.

How well any agreement will perform depends on continuing incentives to follow its provisions. Further consideration of strategic behaviour that may become advantageous once investments are in place reveals further potential problems that may be forestalling an agreement. That is, Israel may have a greater opportunity and higher incentives than Jordan to renege on any co-operative agreement once investments are in place. Jordan has little incentive to divert less than the 180 million $m^3$ $yr^{-1}$ because the water is currently unutilized winter flood water. Jordan can get increased summer water only by carrying through with the agreement.

The incentive not to adhere to the agreement may be larger for Israel. Once the water is captured, Israel could refuse to divert the entire agreed-upon amount back to Jordan. Israel could justify its action by arguing that Jordan was wasting the water and that the higher value use was in Israel. Such an argument would be difficult to verify and difficult to refute. Retaliation by Jordan would have to take place the following season and would not be credible because Jordan gains nothing from not diverting the water as long as Israel is sending back some substantive share of the agreed amount. These considerations suggest another reason why Jordan may have been reluctant to negotiate an international agreement with Israel. Protection against this type of conflict is best considered before investment takes place. In fact, without such protection, an agreement may never be reached.

### *3.4 Prospects and problems associated with multilateral co-operation*

Another obstacle to co-operation on the Yarmouk–Kinneret canals is that benefits and incentives for co-operation may be significantly affected by the Unity Dam on the upper Yarmouk.

The Unity Dam promises to deliver an additional 225 million $m^3$ $yr^{-1}$ of water to Jordan (in addition to electricity to Syria) at a capitalized cost of \$0.34 $cm^{-1}$ (see Table 1). This cost is higher than the cost of water made available by the Yarmouk–Kinneret canals, but the cost appears to be lower than the current marginal value of water in Jordan and Jordan's vulnerability to strategic behaviour of another country is less. Because Syria and Jordan could each benefit from the dam and it is located on a common border, vulnerability to *ex post* strategic behaviour does not appear to be a problem. Apparently, only Israel's objection is a major obstacle.

From Jordan's viewpoint, the dam provides a substitute for the Yarmouk–Kinneret canals. If the Unity Dam were already in place, the value to Jordan of the canals is close to zero. On the other hand, if the Yarmouk canals were already in place, the value to Jordan of the Unity Dam is much less and possibly close to zero because the dam's capacity exceeds the capacity of the Ghor Canal.

From Israel's viewpoint, the Unity Dam is important only if it affects the construction of the Yarmouk–Kinneret canals. If the dam is constructed before the canals, then the value of the canals derives solely from the value of the water to Israel. This value is smaller than the value when both Jordan and Israel use the project, but the project still would be likely to pass a benefit–cost test. In this case, Jordan's incentive to co-operate with Israel must derive from other means such as monetary compensation. More importantly, once both projects are constructed, Israel becomes subject to strategic behaviour by Jordan.

For Israel to receive water, Jordan must divert water that it will not use. If an international agreement is likely to be carried out as agreed, then Israel would presumably be willing to support the Unity Dam in lieu of the canals if it were guaranteed 100 million $m^3\ yr^{-1}$ at costs comparable to the canal cost. Presumably, Jordan would be willing to agree to this amount given current Jordanian capacity in the Ghor Canal. However, consider strategic behaviour that becomes possible once either both projects are in place or the Unity Dam is in place. In this case, the water could be diverted either during the summer, after storage behind the Unity Dam, or during the winter, in which case the water would have passed through potential upstream storage at the Unity Dam. Because Jordan would already have storage capacity in this case, there would be no point to shipping water into the Kinneret for later return to Jordan. Furthermore, Jordan would have no incentive to continue shipping water to Israel unless other forms of compensation were developed. Jordan could simply renege on the agreement without consequence. Furthermore, if Jordan were to increase its capacity for utilization, say, by expanding the Ghor Canal, then Jordan would face an incentive to break the agreement to provide water to Israel.

These considerations demonstrate how future contingencies can change incentives to participate in agreements once fixed investments are made. Anticipation of these consequences explains why parties have been unable to reach agreement on projects that offer substantial net economic gains. Israel refuses to remove its objection to the Unity Dam because that would give Jordan a strategic incentive to break any agreement acceptable to Israel. Jordan refuses to participate in the Yarmouk–Kinneret canals because that would give Israel a strategic incentive to break any agreement acceptable to Jordan. Project designs are needed that not only offer economic gains but provide sufficient incentives to follow through on agreements once investments are in place, i.e., build in sufficient remedies for each side to make strategic behaviour unattractive to the other.

## 4. Potential co-operation between Israel and the Palestinians on the Mountain Aquifer

### *4.1 The Mountain Aquifer*

Use of the Mountain Aquifer in Israel represents another important trans-boundary water problem. The Mountain Aquifer lies under the mountains of Judea and Samaria. It consists of two hydrologically disconnected basins: the Yarkon–Taninnim (YT) Basin and the Eastern Mountain (EM) Basin. A large portion of water caught in this area emerges in a long line of springs within Israel extending from the Tanninim to the Yarkon in the YT Basin. The YT Basin includes six sub-aquifers collectively called the Western Aquifer. Maximum sustainable yield is 350 million $m^3$ $yr^{-1}$. Two aquifers of the EM Basin underlie Judea and Samaria; these are the Eastern (sustainable yield of 125 million $m^3$ $yr^{-1}$) and Northern (sustainable yield of 140 million $m^3$ $yr^{-1}$) Aquifers (Lowi, 1993).

During recent years, an average of 331 million $m^3$ $yr^{-1}$ has been pumped from the YT Basin, and the parties have used an additional 51 million $m^3$ $yr^{-1}$ of YT spring flows (Hydrological Service, 1994). Use of the YT has historically been dominated by Israel; on average, Israel has used 362 million $m^3$ $yr^{-1}$ and Palestinians have used 20 million $m^3$ $yr^{-1}$ (Isaac *et al.*, 1994; Lowi, 1993). The quantity consumed by Palestinians has not been changed since 1967 (Gvirtzman, 1994). Israel has also been the primary user of the Northern Aquifer of the EM basin. Palestinians have relied mainly on the Eastern Aquifers. Their total pumping from all three sources has averaged 100–125 million $m^3$ $yr^{-1}$. Pumping quotas allotted to the Palestinians by the Israeli Civil Administration in 1990 were limited to 125 million $m^3$ $yr^{-1}$. A small margin for growth in personal use is permitted. Jewish settlers have pumped 40–60 million $m^3$ $yr^{-1}$ mainly from the Eastern Aquifer.

Israel's exploitation appears to be slightly higher than the maximum sustainable yield.[5] However, current information regarding maximum sustainable yields is subject to uncertainties so it is difficult to determine how much each of the aquifers are currently being overdrawn. More importantly, it is not clear how the quantity withdrawn from one sub-aquifer affects the sustainable yield of the other sub-aquifers within each of the two Basins. For example, if less water were withdrawn from the Northern Aquifer, perhaps more water could be sustainably pumped from the Eastern Aquifer.

Water rights are disputed. The question is how much water each of the parties should extract and whether Israel should compensate the Palestinians for Israel's pumping from the YT because this water originates in Judea and Samaria. Agreement between Israel and the Palestinians may be more difficult to achieve than in the Israeli–Jordanian cases above because the contested groundwater is harder to monitor and unforeseen (and unverifiable) problems may arise in its extraction. Also, the hydrology of the region does not establish a clear 'owner' or even a hierarchy of use rights.

[5] There appears to be some salinity in the south due to overpumping in southwestern margins of the Basin, and in the north due to transport of saline water from the centre. This salinity is, apparently, the result of pumping over the maximum sustainable yield.

In the higher elevations and on the eastern slope, which is where Palestinian water supplies are drawn, extraction can be quite costly and is an increasing function of the amount pumped. Pumping costs also increase as the amount of water remaining in the aquifer decreases. Depending on the location, pumping costs are between \$0.16 $cm^{-1}$ to \$0.34 $cm^{-1}$ (Fisher, 1994). By comparison, the wells in the lower-elevation YT basin where Israel draws water are gravity fed. They are located at the storage area, where the pumping rate is stable and pumping is cheapest (Gvirtzman, 1994). Pumping costs range from \$0.15 $cm^{-1}$ to \$0.20 $cm^{-1}$ (Fisher, 1994). As a result of this relationship, Israel's pumping costs are largely unaffected by Palestinians' water use, but Palestinian pumping costs appear to be raised, even in the short run, by Israel's withdrawals. Therefore, Israel appears to impose an externality on the Palestinians.

### *4.2 Implications of bargaining theory for co-operation on aquifer withdrawal*

Again, bargaining theory offers insights into efficient and equitable solutions. An efficient solution would likely include a tax on water pumped by Israel that would compensate Palestinians for increased pumping costs; nevertheless, Israel would extract and use most of the water because of its lower extraction costs and higher value. However, the parties still would have to come to an agreement over how much water each of them had an ownership right to. This section considers the application of bargaining theory to the Israeli–Palestinian problem in the YT Basin of the Mountain Aquifer.

To simplify the problem, suppose that the parties agree to extract no more than 350 million $m^3$ $yr^{-1}$, the maximum sustainable yield, and that Israel can extract up to this amount at a cost of \$0.15 $cm^{-1}$. Suppose Palestinians extract water at a cost that increases quadratically in quantity pumped by Palestinians and multiplicatively in quantities pumped by both Israelis and Palestinians.

We consider bargaining over a per unit water charge paid to the Palestinians. Bargaining is undertaken to determine the amount of the unit charge and the quantities used by each party.

The Nash bargaining solution is for the Palestinians to pump 122 million $m^3$ $yr^{-1}$ and for Israel to pump 228 million $m^3$ $yr^{-1}$ and pay a charge \$0.07 $cm^{-1}$.[6] The marginal cost for the water extracted by Palestinians would then be \$0.39. This

[6] Suppose the cost of pumping water quantity $q_p$ is:

$$c(q_p, q_i) = \alpha q_i q_p + \gamma q_p^2$$

and suppose Israeli and Palestinian benefits from water are:

$$U_i = \frac{\beta_i (Q_i + q_i)^{1-\alpha_i}}{1-\alpha_i} - \tau q_i - 0.15 q_i, \quad U_p = \frac{\beta_P (Q_p + q_P)^{1-\alpha_P}}{1-\alpha_P} + \tau q_i - c(q_p, q_i).$$

The first terms are surplus for constant elasticity demands where $\alpha_i = 0.407$, $\alpha_p = 0.44$, $\beta_i = 7.81$, $\beta_p = 4.29$, $\gamma = 0.00128$, and $\alpha = 0.0003347$. The parameter values are based on an average of Fisher's (1994) sectoral demand elasticities weighted by sectoral uses. The second terms represent the charge paid by Israelis to Palestinians, and the third terms represent the respective party's cost of pumping. For purposes of discussion, the reservation utilities are 117 for the Palestinians and 976 for Israel. See Netanyahu *et al.* (1994) for further discussion.

solution is sensitive to both Israeli and Palestinian pumping costs and the reservation utilities (the level of benefits below which a party is better off without an agreement). However, reservation utilities are harder to model in this case because current (non-co-operative) extraction levels are not sustainable. The solution is only mildly sensitive to the overall pumping constraint. If the pumping constraint were relaxed, Israel would pump more, at least in the short run, but the externality imposed on Palestinians would rise too, as would the charge. The charge also implicitly includes compensation for Palestinian 'ownership' but this amount is extremely sensitive to the reservation utilities, which should reflect Palestinian ability to exercise its ownership in the absence of any trans-boundary agreement.

### *4.3 Stability of the co-operative solution*

As with the Yarmouk–Kinneret canal project, watershed geography drives most of the solution. If Israel can extract 350 million $m^3$ $yr^{-1}$ from the YT springs at minimal cost and if Palestinian use is more costly, almost any solution will allow Israel to extract most of this amount. Furthermore, the low elevation user of an aquifer, like the upstream user of a river, has first claim on the water when no agreement or control is in force. Thus, more precise modelling of the reservation utilities would likely further favour allocating most of the water to Israel.

In this case, the bargaining solution does not differ radically from a market solution. With a market solution, Israel should compensate the Palestinians for the externality that Israel's extraction imposes on Palestinian pumping costs. This charge is likely small as in the bargaining solution. Additionally, an efficiency solution would address the common-property overuse of the aquifer by means of a tax or quota on pumping paralleling the quota imposed in the bargaining solution.

Again, however, the bargaining analysis reveals that the solution may not be stable given the incentives for co-operation by the individual parties. It is not clear that either party would have sufficient incentive to abide by the agreement even if an appropriate pumping limit could be determined. First, both parties have an incentive to renege because monitoring is difficult. Since Israel's pumping is at low cost, it can deviate with little consequence. Palestinian deviation is not as advantageous because its future pumping costs are raised thereby. Further understanding of these considerations requires a non-co-operative model. Clearly also, this example as well as the others points out that a market allocation of water benefits may be so one-sided that international conflicts cannot be resolved thereby. This difficulty of negotiation is illuminated by a game theoretic analysis. For further analysis of this problem, see Netanyahu *et al.* (1994).

## 5. Current water situation in Israel

### *5.1 Overview of water pricing and allocation*

The remainder of this chapter turns to water allocation problems within individual countries. Clearly, the value of water forthcoming from international projects depends on how it is allocated within individual countries because different users pay different prices. Thus, the value of additional water cannot be divorced from how water is allocated within individual countries. This analysis also reveals a comparison of the value of improved internal allocation with the value of improved water availability through international projects. This section looks at the current water situation in Israel. These considerations are then taken into account in assessing the marginal value of water for Israel.

The Israeli economy can be divided into three sectors for the purposes of discussing water use: domestic (households), agriculture, and industry. Prices paid by each of these sectors for water in 1990 are shown in Table 2. Each sector faces its own block tariff schedule for water. A similar price schedule has been in effect every year since 1990. Farms and some industrial firms have allotted quotas. They can, in general, use more than their quota but must pay a higher price for over-quota use for which availability is not guaranteed.

The price schedule and observed agricultural demands together demonstrate that the marginal value of water in agriculture may have declined between 1990 and 1992 (Table 3). During this same period, the real price of water (at each quantity) rose by 33%, reducing the indirect water subsidy. However, many farmers apparently attained a lower water price by reducing water use relative to their quota. If water use in 1994 is similar to 1992, then the current (national average) marginal private value of water in agriculture could be as low as $0.12 cm$^{-1}$ in 1990 dollars.[7]

It is not hard to understand why the private value of water may have declined in agriculture; it is due to more efficient irrigation. Whether agricultural demand for water at any given price will continue to decline depends both on opportunities to expand the use of currently available technologies and on what new technologies are discovered. The degree of accessibility to the European market may play a role as well. The demand curve could shift out if product prices rise, but product prices are likely to be less important than changes in irrigation technologies.

*Table 2.* Water prices in Israel, 1990[a].

| | |
|---|---|
| Domestic (households)[b] | $0.32 cm$^{-1}$ for 0–8 cm |
| | $0.75 cm$^{-1}$ for 8–16 cm |
| | $1.23 cm$^{-1}$ for more than 16 cm |
| | $0.58 cm$^{-1}$ for outside water use (e.g. gardening) |
| Agriculture | $0.125 cm$^{-1}$ for the first 80% of quota |
| | $0.20 cm$^{-1}$ for the next 20% of quota |
| | $0.26 cm$^{-1}$ for water used above quota |
| Industry | $0.15 cm$^{-1}$ (average price paid) plus sewage charges |

[a] Prices are given in 1990 dollars.
[b] Households must also pay a sewage fee of $0.32 cm$^{-1}$.
Source: Tahal (1990).

[7] This is the proposed price for 1994 in 1990 dollars (Mekorot, 1993).

*Table 3.* Recent increases in agricultural water prices in Israel[a].

| Farm use as a share of quota | 1990 ($) | 1991 ($) | 1992 ($) |
|---|---|---|---|
| Substantially below quota | 0.125 | 0.142 | 0.167 |
| Up to quota | 0.200 | 0.227 | 0.266 |
| Over quota | 0.260 | 0.296 | 0.433 |
| Substantially over quota | | | 0.699 |

[a] Prices are given in 1990 dollars per cubic metre. 'Substantially below quota' represents use less than 80% of quota in 1990 and 1991 and less than 50% of 1989 quota for 1992. 'Up to quota' represents use between 80% and 100% of quota in 1990 and 1991 and use between 50% of 1989 quota and 100% of quota for 1992. 'Over quota' represents use between 100% and 130% of quota for 1990 and 1991 and use between 100% and 110% of quota for 1992. 'Substantially over quota' represents use above 110% of quota.

Sources: Tahal (1990); Government of Israel, *Register*, various issues; IMF (1993).

Differences in water prices among sectors suggest that sector incomes are weighted differently in policy formulation. If each of the three sectors is characterised by constant elasticity demand (Fisher, 1994), then the implicit relative weights are 2, 1.7, and 1 for agriculture, industry, and households, respectively.[8] Typically, the differences in weights among sectors are attributed to differences in political power of individual sectors. Alternatively, the higher weight on agriculture may reflect a long-standing priority to support food self-sufficiency based on war-time conditions. Similarly, the higher weight on industry relative to households may be motivated by an infant industry argument. Considering the benefits to future generations of economic development, subsidies for industry may be economically rational. Furthermore, household water charges are a major income for the municipalities. Part of the higher price to households may be a form of taxation to provide residential public goods.

### 5.2 *Domestic trans-boundary water allocation issues*

The percentage of quota used by farmers varies substantially by region (Figure 1). Given the increasing block rate pricing structure, this imposes a regional variation in water prices following Table 3 as well. What explains these regional differences in pricing and what are the prospects for more economical pricing and allocation? With efficient allocation, prices among regions should differ by no more than the costs of transportation. However, the price differences suggested by Figure 1 are hard to explain on the basis of costs of water transportation. The highest prices are being paid along the National Carrier closest to the source waters of Lake Kinneret.

[8] With stated elasticities, the multiplicative constant terms corresponding to 1990 use levels are 1200, 100, and 450 million $m^3$ $yr^{-1}$ at prices $0.17, $0.20, and $0.34 for the three sectors, respectively (Water Commission, 1994). These prices correspond to current prices where household water is valued at municipality gate prices. The implicit relative weights are determined by choosing weights such that observed quantities (or prices) maximise a weighted sum of sectoral water benefits.

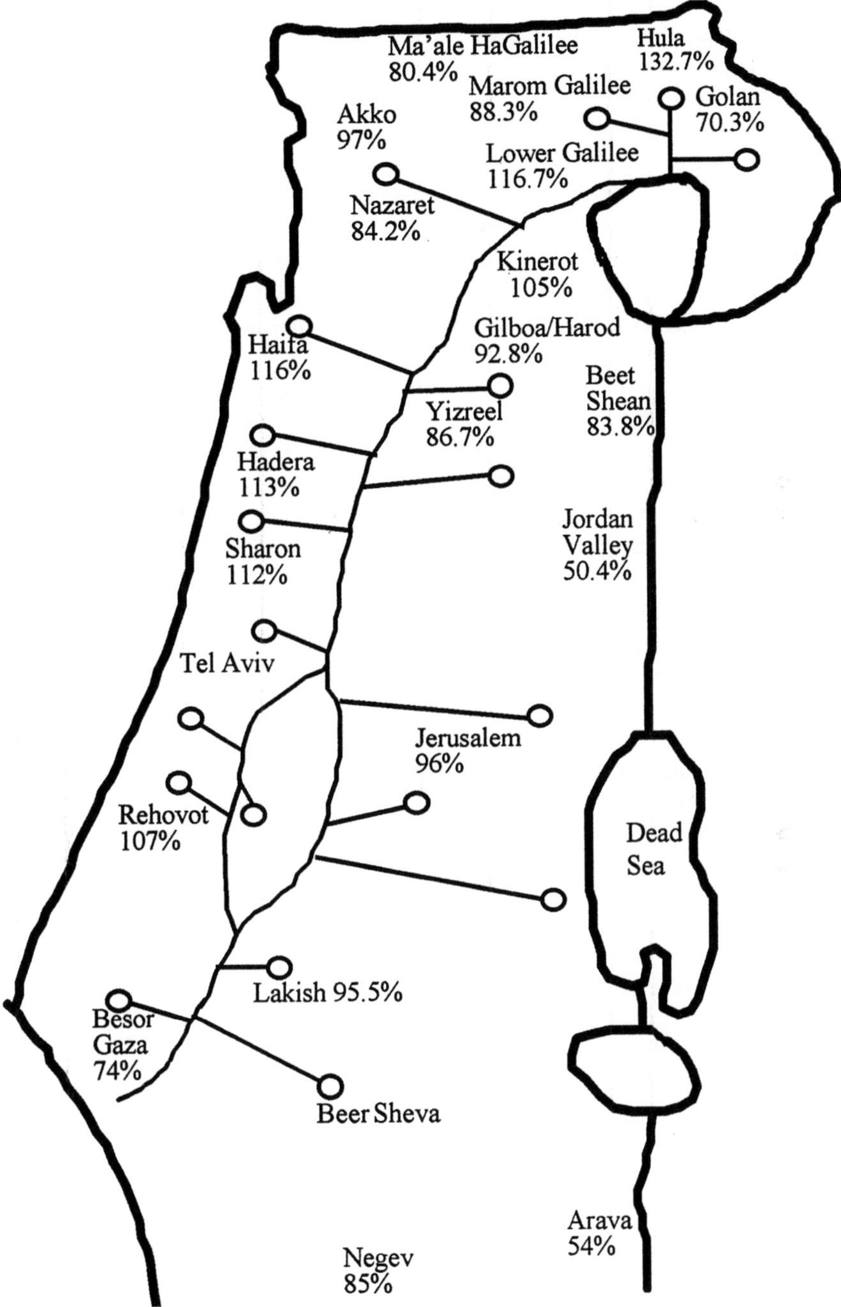


*Figure 1.* Regional agricultural water use as percentage of quota in Israel, 1990.

Typically, inefficient water price variations (usually among sectors) have been explained by political power of advantaged groups. However, this variation in prices does not appear to be consistent with a political power explanation because it does not favour the concentration of farmers in the Northern Coastal regions. Alternatively, this variation in prices is consistent with setting water quotas so as to provide lower water prices in regions where settlement is a national goal. Interestingly, the regions which use less than 80% of quota (and thus pay the most favourable agricultural water price) have been the remote regions in which the government has encouraged settlement over the past few decades for national security and development reasons. The regions which use over 100% of quota (and thus pay the highest agricultural water price) are the most secure regions of the Northern Coast.

A further analysis of quota variations among regions also tends to support this conclusion. If quotas represent true water availability, then the regional use pattern in Figure 1 suggests the possibility that water is over-allocated and, in fact, questions whether water is in shortage at all. A simple comparison of use to quota shows that water is available in surplus in the Arava and the Negev regions of the South and only in shortage in the Northern Coast. However, the layout and direction of flow in the National Carrier and associated pipelines suggests that reallocation towards the North is possible and less costly so that Northern shortages need not exist.[9]

Alternatively, if quotas are assigned, at least partially, to provide subsidies to settlement of remote areas, then comparison of water use to quotas is meaningless as an assessment of water shortage. As evidence in favour of this interpretation, note that water use in the Arava is filled primarily from groundwater and groundwater in the Arava has been overdrafted, i.e., used beyond sustainable levels (Ministry of Agriculture, 1994; The Hydrological Service, 1994), at the same time that use has been below quota (Figure 1). Thus, at least in the case of the Arava, the quota apparently does not reflect true supply. Alternatively, the quotas are high relative to use levels, offering a substantial incentive encouraging settlement of a remote area.

## 6. Current water situation in Jordan

### *6.1 Water shortages and limited potential for market allocation*

Some aspects of water allocation in Jordan are similar to Israel. Industry and households connected to the water system in the Jordan Valley pay for water and

[9] The possibility exists that local use and availability patterns make reallocation unfeasible. For example, some farms are connected by local pipelines to the National Carrier, but some are not. Excess supply may be associated with unconnected farms. However, supply of fresh water in the remote regions of the South comes primarily from groundwater for which availability conditions are broadly applicable. At least at the margin, reallocation should be possible by simply shipping less water down the National Carrier. Differences in water quality might also explain why seemingly feasible reallocation cannot be carried out profitably. However, approximately 70% of farm water use is relatively high quality surface water so adjustment at the margin appears to be feasible in this case as well. Industry uses a similar mix of water qualities (Tahal, 1990).

*Table 4.* Household and industrial water prices, Jordan 1993[a]

| Location | Water quantities | Water and sewage prices |
|---|---|---|
| Amman | 0–20 cm | $0.20 $cm^{-1}$ |
| | 21–40 cm | $0.34 $cm^{-1}$ |
| | 41–70 cm | $0.75 $cm^{-1}$ |
| | 71–100 cm | $1.05 $cm^{-1}$ |
| | More than 100 cm | $1.27 $cm^{-1}$ |
| Jordan Valley | 0–40 cm | $0.14 $cm^{-1}$ |
| | 41–70 cm | $0.23 $cm^{-1}$ |
| | 71–100 cm | $0.52 $cm^{-1}$ |
| | 101–150 cm | $0.90 $cm^{-1}$ |
| | More than 150 cm | $1.27 $cm^{-1}$ |
| Rest of the country | 0–20 cm | $0.14 $cm^{-1}$ |
| | 21–40 cm | $0.19 $cm^{-1}$ |
| | 41–70 cm | $0.60 $cm^{-1}$ |
| | 71–100 cm | $1.05 $cm^{-1}$ |
| | More than 100 cm | $1.27 $cm^{-1}$ |

[a] Prices are given in 1990 dollars and are for water plus sewage.
[b] Charges are based on a three-month billing cycle.
Source: World Bank (1993).

sewage according to a block tariff structure much like in Israel. However, households and industry face identical price schedules as shown in Table 4. Moreover, water supplies are erratic. Firms and households are not necessarily able to purchase the water they want at stated prices. On some days, no water is available at all. Therefore, determining the marginal value of water is difficult. A recent World Bank publication (1994) suggests that the market clearing price in the Highland is higher than $1.00, but less than the $4.50 charged by private water tanker owners. This very high marginal value of water suggests why an international water market linking Jordan and Israel would tend to allocate new sources of water entirely to Jordan.

As in Israel, the marginal value of water is higher in non-agricultural uses than in agriculture.[10] Reallocation away from agriculture would likely require compensating farmers as well as determining the allocation (or rationing) of remaining water. The establishment of a water market for farmers could improve efficiency of allocation. However, developing an effective water market in Jordan would involve substantial costs for infrastructure to facilitate monitoring and repair of leaky canals.

Moreover, reallocation of water away from the agricultural sector may contradict other national goals. As in the case of Israel, the subsidy for agriculture is possibly rooted in a national interest in food self-sufficiency that takes precedence over water issues. Objectives have been to turn the fertile land of the Jordan Valley into the breadbasket of Jordan (Lowi, 1993). Moreover, reallocation of water away from the agricultural sector apparently contradicts other national goals related to the

[10] For example, farmers have water on days when the cities do not, and when water is available farmers pay a price below the non-agricultural price.

development of the Jordan Valley for the settlement of refugees. Therefore, inter-sectoral and inter-regional price equalization may be an inappropriate water policy. Alternatively, means are needed for improving efficiency of water allocation in the context of these national goals.

### *6.2 Domestic trans-boundary water allocation issues*

Unlike in Israel, little infrastructure exists to permit inter-regional water transfer. To the extent that sectoral activity is regionally based, possibilities for inter-sectoral reallocation are also limited. There are two hydrological areas: the Highland and the Jordan Valley. In the Highland, farmers are not connected to a water system. Instead, they pump water from privately owned wells and bear only energy costs of \$0.11–0.22 $cm^{-1}$. From 1987 to 1993, irrigated acreage expanded greatly due to an increase in groundwater licences issued. Currently, more than 85% of Highland irrigation water comes from groundwater.

Highland industrial and domestic uses are also served primarily by groundwater. However, predictions are that demands from these sources will shortly exceed sustainable groundwater supplies (again assuming users pay only for extraction), even if agricultural demands were to cease completely. After recognising this problem in 1993, water authorities stopped issuing new drilling licences.

In the Jordan Valley, farmers receive water through a series of channels fed primarily by the Yarmouk River. Water is rationed. Delivery is regulated by 'ditch riders' who open and close channels according to farmers' weekly water orders made to the Jordan Valley Authority. Limits on each farm's allocation are based on the type of crop. Unfortunately, 50–60% of the water flow is unaccounted due to broken meters and leaky channel gates. In other areas of the Jordan Valley, water is delivered through pressurised pipes where water is more tightly monitored. Again water is rationed.

In both systems in the Jordan Valley the price is negligible, i.e., less than \$0.01, which is perhaps one-fifth of the operating and maintenance cost. An increase to \$0.04, which would at least cover a large proportion of the delivery costs, has been recommended. Agricultural water prices have begun to be increased recently, but the adjustments have been minor. In October 1994, Jordan Valley farmers were notified of an increase in water prices to \$0.011–\$0.02 $cm^{-1}$ depending on the type of crop each farmer grows. Even this meagre increase provoked a substantial adverse reaction. Farmers responded by going on strike and by not sending their children to school.

Because the Valley and the Highland are so poorly connected hydrologically, water projects involving the Jordan Valley with the nearby and accessible water resources of the Yarmouk and Israel offer more promising prospects for trans-boundary water co-operation than do domestic possibilities. Furthermore, because of the growing Highland shortage, prospects for improving Highland water availability may depend critically on enhancing water availability in the Valley through international co-operation and then augmenting the infrastructure to transport it to the Highland (e.g. via the Dayr Alla pumping station).

## 7. Current water situation in Judea and Samaria

Judea and Samaria obtain almost all of their water from wells and springs that tap into the three underground sub-aquifers of the overall Mountain Aquifer. Only about 7% of current water comes from surface run-off or purchases from Mekorot. Therefore, almost all possibilities for trans-boundary water sharing are tied to recharge, pumping costs, and use of the associated aquifers. A rudimentary market appears to operate for well water. The average price was $0.17 in 1990. Average consumption is 110–133 million $m^3$ $yr^{-1}$. Pumping rates, quantities, and period are determined and monitored by the Israeli Civil Administration.

Judea and Samaria are relatively undeveloped and, unlike Jordan and Israel, future water may be much more valuable than current prices suggest. Investment in drilling and irrigation technology could be quite valuable. However, this investment would lower, not raise, the value of water. Increases in demand are likely to be caused by increasing population and increasing income which may accompany investment.

## 8. Domestic water allocation and potential for reallocation[11]

### *8.1 Internal water reallocation*

Given the wide variation in prices among users, the value of water forthcoming from new international projects and the potential for bargaining depends critically on its allocation. Because current water allocation, at least in Israel and Jordan, appears to be based at least in part on rational non-economic and non-efficiency objectives that influence price ratios among users, the value of marginal increments in water availability must be evaluated accordingly. Additionally, the potential for changing price ratios among users in response to new water availability or changes in other external factors must also be considered. One such external factor that could result in a change in price ratios is new prospects for peace. The importance of goals such as food self-sufficiency and settlement of remote areas may be reduced by successful peace accords. If so, then partial equalisation of water prices (across regions and sectors) may result. To set the stage for considering increments in water availability under such alternative scenarios, this section first considers the implications of partial price equalisation at current water availability levels. Israel is used as an example.

Water pricing policies in Israel currently favour agriculture and industry by giving them lower prices than household consumers. An increase in agricultural and/or industry prices will reallocate water to the household sector, which has a higher marginal value, and can raise total (unweighted) economic benefits from water. The cost to agriculture or industry depends on the price rise, total water availability, and demand elasticities in all three sectors. To demonstrate, assume that agriculture (A), industry (I), and households (H) have respective elasticities of demand of 0.5, 0.3, and 0.2 following Fisher (1994), and suppose that total water availability is

[11] Details on the economic analysis of domestic water allocation and pricing presented in this section can be found in Just *et al.* (1994b).

*Table 5.* Water allocation with sectoral price equalisation, 1750 million m³ supply.

| | Sectoral water use | Sectoral surplus from water[b] | Average price |
|---|---|---|---|
| $\gamma_A = 0.50$, $\gamma_I = 0.56$[a] | $q_A = 1160$ | \$401 | \$0.17 |
| (Current scenario) | $q_I = 111$ | \$ 31 | \$0.20 |
| | $q_H = 478$ | \$207 | \$0.35 |
| Total | | \$639 | MV[c] = \$0.25 |
| $\gamma_A = 0.56$, $\gamma_I = 0.56$ | $q_A = 1016$ | \$375 | \$0.18 |
| | $q_I = 130$ | \$ 34 | \$0.18 |
| | $q_H = 604$ | \$249 | \$0.33 |
| Total | | \$659 | MV = \$0.26 |
| $\gamma_A = 0.50$, $\gamma_I = 0.60$ | $q_A = 1171$ | \$403 | \$0.17 |
| | $q_I = 90$ | \$ 27 | \$0.21 |
| | $q_H = 489$ | \$210 | \$0.34 |
| Total | | \$640 | MV = \$0.26 |
| $\gamma_A = 1.00$, $\gamma_I = 0.56$ | $q_A = 411$ | \$239 | \$0.29 |
| | $q_I = 199$ | \$ 46 | \$0.16 |
| | $q_H = 1141$ | \$414 | \$0.29 |
| Total | | \$699 | MV = \$0.28 |
| $\gamma_A = 1.00$, $\gamma_I = 1.00$ | $q_A = 431$ | \$245 | \$0.28 |
| | $q_I = 31$ | \$ 13 | \$0.28 |
| | $q_H = 1288$ | \$457 | \$0.28 |
| Total | | \$714 | MV = \$0.28 |

[a] Household price receives weight 1. Household price = $p_H$, agricultural price = $\gamma_A p_H$, and industry price = $\gamma_I p_H$.
[b] Surplus is $S_i = \beta_j q_j^{1-\alpha_j}/(1-\alpha_j)$ and $q_j$ is market-clearing for price $\gamma_j p_H$.
[c] Marginal Value of water = $\Delta$ Total surplus/$\Delta$ Total water use where Total surplus = $\Sigma_i S_i$.

1750 million m³ yr$^{-1}$ which corresponds roughly to the pre-drought year 1990. Let sectoral price relationships be represented by $p_A = \gamma_A p_H$ and $p_I = \gamma_I p_H$, where $p_j$ is the price in sector $j$ and $\gamma_j$ is the price ratio, a choice variable; $j$ = A, I, H. The $\gamma_j$s may be regarded as indirect choice variables reflecting premiums placed on certain types of economic activity due to non-economic and non-efficiency objectives such as food security concerns, development regards for future generations, etc. If $\gamma_A = \gamma_I = 1$, then prices are fully equalized and sectors face identical prices. Current price ratios are approximately $\gamma_A = 0.50$ and $\gamma_I = 0.56$, which corresponds to prices $p_A = \$0.17$, $P_I = \$0.20$, and $p_H = \$0.34$ (see footnote 3).

Table 5 shows water use, prices, and sectoral benefits of water for different values of $\gamma_A$ and $\gamma_I$ where water supply is 1750 million m³ yr$^{-1}$. Table 6 shows similar outcomes when water supply is increased to 1850 million m³ yr$^{-1}$. Total benefits are unambiguously increasing as weights approach unity and, correspondingly, prices approach equality. Industry is helped by an increase in agriculture's price relative to households' price even when industry's price relative to households' price is unchanged.

Now consider an internal reallocation away from agriculture resulting from a 12% increase in agricultural water prices (which leaves agricultural prices 38% below average household prices). This adjustment could occur, for example, in response to

*Table 6.* Water allocation with sectoral price equalisation, 1850 million m³ supply.

| | Sectoral water use | Sectoral surplus from water[b] | Average price |
|---|---|---|---|
| $\gamma_A = 0.50, \gamma_I = 0.56$[a] | $q_A = 1205$ | \$409 | \$0.17 |
| | $q_I = 119$ | \$ 32 | \$0.19 |
| | $q_H = 526$ | \$223 | \$0.34 |
| Total | | \$664 | |
| $\gamma_A = 0.56, \gamma_I = 0.56$ | $q_A = 1052$ | \$382 | \$0.18 |
| | $q_I = 138$ | \$ 36 | \$0.18 |
| | $q_H = 660$ | \$267 | \$0.32 |
| Total | | \$685 | |
| $\gamma_A = 0.50, \gamma_I = 0.60$ | $q_A = 1216$ | \$411 | \$0.17 |
| | $q_I = 96$ | \$ 28 | \$0.20 |
| | $q_H = 538$ | \$227 | \$0.34 |
| Total | | \$666 | |
| $\gamma_A = 1.00, \gamma_I = 0.56$ | $q_A = 422$ | \$242 | \$0.29 |
| | $q_I = 208$ | \$ 48 | \$0.16 |
| | $q_H = 1220$ | \$437 | \$0.29 |
| Total | | \$727 | |
| $\gamma_A = 1.00, \gamma_I = 1.00$ | $q_A = 443$ | \$248 | \$0.28 |
| | $q_I = 33$ | \$ 13 | \$0.28 |
| | $q_H = 1375$ | \$481 | \$0.28 |
| Total | | \$742 | |

[a] Household price receives weight 1. If household price $= p_H$, then agricultural price $= \gamma_A p_H$, and industry price $= \gamma_I p_H$

[b] Surplus is $S_i = \beta_j q_j^{1-\alpha_j}/(1-\alpha_j)$ and $q_j$ is market-clearing for price $\gamma_j p_H$. Surplus does not include costs of the project.

successful experience with new peace accords if agricultural subsidies are motivated by interests in food and border security. The associated change in pricing parameters to $\gamma_A = 0.56$ and $\gamma_I = 0.56$ raises total benefits by 3.1% although agricultural benefits decline by 6.5%. By comparison, a minor reallocation from current prices to $\gamma_A = 0.5$ and $\gamma_I = 0.6$, which reduces some of the preference towards industry, raises aggregate benefits by a much smaller 0.16% over current benefits.

Complete equalisation of agricultural and domestic prices keeping the industry break in place raises aggregate benefits by 9.4% while reducing agricultural benefits by 40.4%. Complete reallocation implied by equal prices across all sectors raises aggregate benefits by 11.7%. Thus, the subsidization of agricultural water is apparently responsible for most of the water use inefficiency. These results and similar exercises with parameters potentially reflecting Jordanian demands suggest that water conflicts in the Middle East are exacerbated by internal divisions of water that favour agriculture through low water prices.

### *8.2 Can new water change old coalitions?*

Now consider the possibility that new water availability may be able to weaken the political coalitions that lead to unequal pricing by reducing the sectoral costs of price equalisation. Compare, for example, the results with initial price weights $\gamma_A = 0.50$ and $\gamma_I = 0.56$, and 1750 million $m^3$ $yr^{-1}$ aggregate supply (Table 5), to the results with partial price equalization represented by weights $\gamma_A = 0.56$ and $\gamma_I = 0.56$, and the higher 1850 million $m^3$ $yr^{-1}$ aggregate supply (Table 6) available after completion of the Yarmouk–Kinneret canals. In this comparison, the additional supply is allocated relatively more to domestic and industrial uses. Agriculture is less affected than when weights are adjusted without an increase in supply. Agricultural benefits decline by 4.7% ($401 million to $382 million) in this case compared to the 6.5% ($401 million to $375 million) decline when aggregate supply remains constant. Thus, the cost for agriculture to adjust weights is considerably reduced. This also permits a much larger aggregate benefits gain from the project, 7.2% ($639 million to $685 million) compared to 3.9% ($639 million to $664 million) when the weights are not adjusted along with the supply increase.

Table 5 also shows the change in total benefits from a change in aggregate supply computed by comparing the aggregate benefits in Tables 5 and 6. These figures reflect the marginal value of water for the country. The marginal value of water is higher the more equal are prices. This occurs because increments to water supply are assumed to be allocated among the sectors to maintain the given price weights, and more equal prices imply that higher value sectors implicitly receive more weight in water allocation.

An alternative to these allocation rules is to allocate increases in supply only to industry and households, e.g. the high value users. Consider the objective of raising benefits for industry and households subject to agricultural benefits not decreasing. Because industry and households face higher prices than agriculture, allocating 100 million $m^3$ $yr^{-1}$ of new water to them (leaving agriculture's water unchanged) raises aggregate benefits more than allocating 100 million $m^3$ $yr^{-1}$ among all three sectors according to a given weighting scheme. These results demonstrate how critically the value of a marginal increase in water depends on who gets the water.

### 8.3 *Internal reallocation versus external co-operation*

The results in Tables 5 and 6 demonstrate that internal reallocation of water is more valuable to the Israeli economy than external co-operation over the Yarmouk–Kinneret canals, provided that the political parties can effect sufficient internal reallocation. Under current price weights ($\gamma_A = 0.50$, $\gamma_I = 0.56$), benefits are increased by 3.9% by the 100 million $m^3$ $yr^{-1}$ additional water supply available to Israel from co-operation and completion of the canal project.[12] In contrast, even minor internal reallocation (changing the weights to $\gamma_A = 0.56$ and $\gamma_I = 0.56$) raises aggregate benefits by 3.1% holding aggregate supply constant at 1750 million $m^3$ $yr^{-1}$. A much larger 9.4% gain is possible by equating agricultural and domestic

[12] Compare the first entries of Tables 5 and 6. Benefit calculations at 1850 million $m^3$ $yr^{-1}$ do not include costs of the Yarmouk project, but these one-time costs are small relative to the effects shown in Table 5 and 6, which accrue yearly.

weights ($\gamma_A = 1.0$ and $\gamma_I = 0.56$) and an 11.7% gain is possible by equating all prices. These reallocations provide more to the Israeli economy than the additional water of the canal project. Again, however, one must bear in mind that these internal reallocations may not be appropriate in the context of other national objectives. If not, then international co-operation may offer the only rational alternative for easing water scarcity problems.

### 8.4 *Interdependence with current peace prospects*

If, however, the agricultural water subsidy has a food security basis that will be eliminated with new peace agreements, then price equalising adjustments may become feasible even if they have not been previously. Conversely, international co-operation on enhancing water supply may be important both for reducing internal political objections (the adverse welfare effect on agriculture is reduced) and for increasing confidence in peace and co-operation.

If lasting agreements on water cannot be reached, the ultimate success of peace is threatened. On the other hand, if lasting international water agreements can be found, then the long-term success of peace is more likely which, in turn, facilitates reduced security needs and more internal water efficiency. These considerations suggest that internal water price equalization, international water projects, and peace have synergistic effects that complement one another. Progress may not be possible on any of these three objectives individually but large gains may be possible when the three are taken together.

These considerations imply that political power of the agricultural sector should not be over-emphasized as an explanation for current water pricing inefficiency. Strengthening the agricultural sector is not a declared government policy *per se* but is embedded in socio-political policies that have emerged in part from Zionist spirit and in part from national security considerations. Food security, settlement, and development of the Negev clearly are playing a major role in dictating the direction of planning and development in Israel.

Similarly, the norm of economic efficiency should not be overemphasized as a standard of comparison for water allocation schemes and mechanisms. Dominance of economic considerations by ideological, political, and nationalistic objectives clearly characterized the period from establishment of the state of Israel until at least the completion of the National Water Carrier in 1964. The years since then are characterized by an increasing conflict between economic and socio-political agendas. When pursuit of non-economic and non-efficiency objectives represents legitimate and rational public concerns, then economic efficiency may become an inappropriate goal. To the extent that rational public concerns impose inequitable price ratios among users, international projects and the new water they produce must be valued accordingly. Similarly, the potential implications of new water and of other changes in external factors for price ratios imposed by rational public concerns must be evaluated as well.

## 9. Concluding remarks

All of the countries in the Middle East face constraints to growth because of limited water supply, degradation and overexploitation of historic water resources, and population growth. International conflicts have hindered the adoption of potential solutions. External conflicts may also have exacerbated internal conflicts over water. Fragile diplomatic relations may have increased the desire for agricultural self-sufficiency, leading to subsidization of water to agriculture. Political rent seeking may have been more effective in these tense circumstances. However, it is not clear that political power models are necessary to explain current water pricing structures. Governments' water policies are often secondary to other socio-political goals. Considerations such as industrial growth, intra-sectoral equity, rationing, food security, settlement, and development of arid regions all have high priorities in national agendas.

Co-operation in water planning may allow reallocation of existing water resources or development of new resources that will benefit all parties. New resources can often be used more efficiently than existing resources because inefficient prior claims need not take precedence. More importantly, the presence of new resources may facilitate overcoming existing barriers to more efficient use of existing water resources because certain national objectives may be appeased thereby. Better use of existing resources can occur because the political and economic costs of changing existing water use patterns are reduced when supply is higher.

Several major international projects have been proposed that apparently offer significant possibilities for efficiency gains. However, studies have typically failed to address the problem of how to facilitate needed co-operation. Demonstration of potential economic efficiency gains only implies that welfare could be improved if competitive solutions could be approximated. When such solutions involve lopsided sharing of the water forthcoming from particular international projects, international co-operation is not likely. Yet, because of terrain and transportation possibilities, such problems appear to result from most leading but yet undeveloped project proposals. This chapter has demonstrated that standard bargaining considerations may cause obstacles to co-operation in spite of large potential efficiency gains. Further work in the context of bargaining is needed to identify provisions that will assure continued co-operation once costly initial investments are undertaken. Once all parties can be assured that promised future benefits are protected, co-operation efforts can be expected to be more fruitful.

Much of the debate about Middle East water is focused on projections of future demand. It is difficult in these debates to understand the potential for internal reallocation and to understand how reallocation opportunities affect the value of international projects. This chapter has concentrated on analysing these relationships. Results show that internal reallocation is potentially very effective in postponing scarcity problems. However, such solutions may not be politically feasible or may become feasible only with peace agreements and accumulated experience contributing to confidence in those agreements.

Once peace is achieved and international water projects become part of the regional development agenda, an increase in benefits is expected. With additional water supplies available to each country, the loss to specific sectors (especially to the agricultural sector) from price equalisation will be lower than if water were simply reallocated with unchanged aggregate water availability. However, complete price equalisation may continue to be inappropriate given other national concerns such as national security, land settlement, arid region development, tradition, and religion.

The empirical work in this chapter is clearly highly tentative because: estimation of water supply and demand is at a primitive stage; and because some of the physical hydrological relationships are not clear. Nevertheless, the results are sufficient to demonstrate that simple economic modelling of efficiency potential may be of little value in international negotiations. Water markets tend to allocate all incremental water to the country with highest marginal value and, hence, do not reflect the potential of international projects given political realities of international negotiations.

## Acknowledgements

This research was made possible in part by Grant 58-319R-4-046 from the Research and Scientific Exchange Division, Foreign Agricultural Service, United States Department of Agriculture, and in part by support provided by the Office of Education and Institutional Development, Bureau for Research and Development, U.S. Agency for International Development, under the terms of Grant No. PDC#0095-A-00-1126-00 through the Institute for Policy Reform. The opinions expressed herein are those of the authors and do not necessarily reflect the views of the U.S. Department of Agriculture, the U.S. Agency for International Development or the Institute for Policy Reform.

## References

Ben-Shahar H, Fishelson G and Hirsch S 1989 Economic Cooperation and Middle East Peace. Weidenfeld and Nicolson, London.

Beschorner N 1992/93 Water and Instability in the Middle East. Adelphi Paper 273. International Institute of Strategic Studies, Winter 1992/93.

Biswas A K 1993 Management of international waters: Problems and perspective. Int. J. Wat. Resour. Dev. 9, 167–188.

Clarke R 1993 Water: The International Crisis. The MIT Press, Cambridge.

Dinar A and Zilberman D 1991 Effects of input quality and environmental conditions on selection of irrigation technologies. In The Economics and Management of Water and Drainage in Agriculture. Eds A Diner and D Zilberman. Kluwer Academic Publishers Norwell, Ma.

Eckstein Z and Fishelson G 1994 Israeli Country Paper. Prepared for the Harvard Water Project, June 1994.

Eckstein Z, Zakai D, Nachtom Y and Fishelson G 1994 The Allocation of Water Sources Between Israel, the West Bank and Gaza: The Economic Viewpoint. The Armand Hammer Fund and the Sapir Center for Development, Tel Aviv University.

Falkenmark M, Lunkqvist J and Widstrand A 1990 Water Scarcity – An Ultimate Constraint in Third World Development. Tema V, Report 14, University of Linkoping, Department of Water and Environmental Studies, Linkoping, Sweden.

Fisher F M 1994 An Economic Framework for Water Negotiation and Management. ISEPME, MIT and Harvard, June 1994.

Fisseha Y 1981 State succession and the legal status of international rivers. In The Legal Regime of International Rivers and Lakes. Eds R Zecklin and L Caflisch. Martinus Nijhoff Publishers, The Hague.

Frederick K D 1993 Balancing Water Demands with Supplies; The Role of Management in a world of Increasing Scarcity. The World Bank, Technical Paper No. 189, The World Bank, Washington, DC.

Government of Israel. Register, various issues.

Gvirtzman H 1994 Groundwater allocation in Judea and Samaria. In Water and Peace in the Middle East. Eds J Isace and H Shuvel. Elsevier, Amsterdam.

International Monetary Fund 1993 Government Finance Statistics Yearbook. Washington, DC.

Isaac J, Youssef O, Ja'fari M, Shawwa I, DeShazo J R and Newell R 1994 Water Supply and Demand in Palestine: 1990 Baseline Estimates and Projections for 2000, 2010, and 2020. Prepared for the Harvard Middle East Water Project, Harvard University, 19 January 1994.

Johnston E 1954 Jordan River Valley Development. USA Department of State Publication 5361.

Just R E, Horowitz J K and Netanyahu S 1994a Peace and Prospects for International Water Projects in the Jordan–Yarmouk River Basin. University of Maryland, December 1994.

Just R E, Netanyahu S and Horowitz J K 1994b The Political Economy of Domestic Water Allocation: The Cases of Israel and Jordan. University of Maryland, December 1994.

Kally E 1993 Water and Peace. Praeger, Westport.

Kanazawa M 1991 Water quality and the economic efficiency of appropriative water rights. In The Economics and Management of Water and Drainage in Agriculture. Eds A Diner and D Zilberman, Kluwer Academic Publishers, Norwell, MA.

Kliot N 1994 Water Resources and Conflict in the Middle East. Routledge, London.

Kuffner U 1993 Water transfer and distribution schemes. Wat. Int. 18.

LeMarquand D G 1977 International Rivers; The Politics of Cooperation. Westwater Research Center, University of British Columbia.

Lowi M R 1991 West Bank Water Resources and the Resolution of Conflict in the Middle East. Occasional Paper Series of the Project on Environmental Change and Acute Conflict. pp 1–17. Toronto.

Lowi M 1993 Water and Power; The Politics of Scarce Resources in the Jordan River Basin. Cambridge University Press, Cambridge.

Mekorot Water Company 1993 The Company's 1994 Budget. Israel, December 1993 (in Hebrew).

Ministry of Agriculture and The Jewish Agency 1994 Agricultural Sectors in 1991–1992. Tel Aviv, January 1994.

Naff T, and Matson R Eds 1984 Water in the Middle East – Conflict or Cooperation. Westview, Boulder, Co.

Netanyahu S, Just R E and Horowitz J K 1994 Possibilities and Limitations in Sharing the Mountain Aquifer between Israel and the Palestinians. Working paper, University of Maryland, College Park, MD.

Rhoades J D and Dinar A 1991 Reuse of agricultural drainage water to maximize the beneficial use of multiple water supplies for irrigation. In The Economics and Management of Water and Drainage in Agriculture. Eds A Diner and D Zilberman. Kluwer Academic Publishers, Norwell, MA.

Shamir U, Bear J and Arad N 1985 Water Policy for Israel. The Technion, Haifa (in Hebrew).

Shuval H 1992 Approaches to resolving the water conflicts between Israel and her neighbors – a regional water-for-peace plan. Wat. Int. 17, 133–143.

Soffer A 1992 Rivers of Fire: The Conflict of Water in the Middle East. Am Oved, Israel.

Starr J R 1992 Water wars. Foreign Policy 86, 17–35.

Tahal Consulting Engineers Ltd 1990 Israel Water Sector Review: Past Achievement, Current Problems, Future Options. Tel Aviv.

Tahal Consulting Engineers Ltd 1993 Israel Water Study for the World Bank. Working paper, draft, Israel.

The Hydrological Service 1994 The Development of the Exploitation and the Status of Groundwater Resources in Israel Until Fall 1993. Ministry of Agriculture, Water Commission.

Tuijl W V 1993 Improving Water Use in Agriculture; Experiences in the Middle East and North America. The World Bank Technical Paper No. 201. The World Bank, Washington, DC.

Williamson O E 1981 The modern corporation: Origins, evolution, attributes. J. Econ. Lit. 19, 1537–1568.

Wolf A 1993 The Jordan watershed: Past attempts of cooperation and lessons for the future. Wat. Int. 18, 5–17.

World Bank 1993 Jordan, Agricultural Sector Adjustment Loan; Water Component. Working paper.

World Bank 1994 Jordan, Agricultural Sector Adjustment Loan; Working Paper: Issues in Water Pricing. Draft. 16 February 1994.

Zeitouni N, Becker N and Shechter M 1994 Two Models of Water Market Mechanisms with an Illustrative Application to the Middle East. Natural Resources and Environmental Research Center and the Department of Economics, University of Haifa.

49
# Australian decentralized water allocation

NORMAN J. DUDLEY

## 1. Introduction

Many environmentally sensitive sites in Australia are dependent on water for their continued existence. These include streams containing flora and fauna as well as on-stream and off-stream wetlands. A large number of these sites have been severely degraded by diversion of water for alternative uses (Boyd, 1994; Lloyd *et al.*, 1994; McCosker and Duggin, 1993; McGregor *et al.*, 1994; Wettin *et al.*, 1994). The overwhelmingly dominant cost of restoring most of these degraded environmental sites is the opportunity cost of the water required. That is, the value of the required water in its best non-environmental use.

Many environmentalists now accept that the environmental first-best solution of restoring original, before-white-settlement water supplies to environmental uses is infeasible in practice. However, effects of less than full restoration of original flow regimes to the environment are still poorly understood, in spite of growing research into the topic. Because the effects of reallocating less than the original quantities of water to environmental sites are so poorly understood, many environmental managers and researchers emphasise the need for flexibility and adaptability of environmental water allocations as a better understanding of the water requirements of such systems emerges slowly from ongoing research.

Efficient resource use by non-environmental water users requires that they have secure title to property rights to water supplies with known probabilities of supply (Dudley, 1990). Their requirement for secure title to water with constant probabilities of supply puts non-environmental water users on an apparent collision course with environmental uses requiring flexible and adaptable supplies while information on environmental requirements improves. That is, on the one hand is the requirement for non-changing rules for sharing the scarce and variable-supply resource; on the other is the requirement for sharing arrangements which are flexible and adaptable through time.

Reconciling these opposing requirements by centralized command and control, or even centralized demand management via centralized pricing, is extremely complex in highly variable environments (Dudley, 1991a). However, a form of decentralized management exists which is an appealing alternative. Its introduction requires a few steps. First, establish sound property rights for water resources. Second, establish markets for water resources. Third, make an initial allocation to each of the parties.

*D. Rosen* et al. *(eds): Modern Agriculture and the Environment, 613–624.*

These steps allow water resource allocation through time to adapt to suit evolving environmental requirements while guaranteeing that other users will not forgo their secure titles to fixed-probability water supplies unless they receive compensation which they perceive to be adequate. This perception of fixed sharing of water resources, or full compensation for any reduced sharing, by other users will ensure that they have incentives to manage the water resources under their control in an economically and technically efficient way.

But efficient water resource use requires more than the security of long-term supply probabilities – it also requires the short-term and long-term integration of supply-side and demand-side management.

## 2. Integrating supply-side and demand-side management

The very high climatic uncertainty under which water resources planning and management decisions are made in the northern two-thirds of Australia drew attention, in the late 1960s, to the need to integrate the supply-side and demand-side management of water resource systems (Dudley, 1990; Dudley and Musgrave, 1993). The resulting early Australian work integrated the management of headworks water storages (supply) and irrigation farm area-to-irrigate and water application-rate decisions (demand), under uncertainty. This integration involved optimization of short-, intermediate-, and long-run decision making for irrigation systems based on surface storage by using computer models combining the operations research tools of computer simulation and stochastic dynamic programming (Dudley and Burt, 1973). Failure to take an integrated approach can result in greatly reduced regional income from irrigation (Dudley, 1988).

Short-run (SR) decisions concern the amount of water to save at points within irrigation seasons for possible later use, given the condition of the irrigated crop, its area currently irrigated, and the reservoir water available. Water is saved by delaying irrigation and/or relegating some of the irrigated area to rain-fed mode for the rest of that season. Intermediate-run (IR) decisions concern the area to plant to irrigated crops at the season's start, depending on available reservoir contents, the size of the area available and its soil water status. Long-run (LR) or investment decisions concern the best size of land area to develop and equip to be suitable for irrigation.

Obtaining an optimal balance between supply-side and demand-side management is the purpose of Integrated Resource Planning, a concept now supported by the US EPA and the American Water Works Association (Heaney, 1993).

Integration of supply-side and demand-side management of water resources is slow to occur in practice. Supply-side management is still largely proactive with demand-side management being reactive. For example, the allocation of reservoir contents between consumption in the current season and carry-over storage for use in future seasons is still largely made by supply-side managers in most irrigation areas in Australia. Whereas some of this may be due to the reluctance of supply-side managers to relinquish control, demand-side managers and researchers have been slow to become involved in integrated supply-side and demand-side management.

Both irrigation researchers and farmers have tended to concentrate on 'making two blades of grass grow where one grew before' instead of, for example, examining the effects of changing land/water ratios through time in response to current outlooks for water demand and supply. Perhaps failure to realise that reservoir management can change not only the 'when' of water deliveries, but also the long-term 'how much', is a contributing factor. For whatever reason, both irrigation farmers and researchers have tended to concentrate on what happens within the farm boundary, at least with respect to regulated supplies. New emphasis on community-based water management decision making will be unlikely to redress this imbalance in the short term because water resource management models, data and institutional arrangements will still be needed – and most of what is currently available stems from a supply-side perspective.

Of course, there are exceptions. Notable on the research side is work of Hearn and co-workers (see Hearn, 1994). Notable on the irrigation management side is the funding, by the Cotton Research and Development Corporation and the Cooperative Research Centre for Sustainable Cotton Production, of joint work with the NSW Department of Water Resources to examine the process of replacing the current release-sharing system with capacity sharing in northern NSW.

In the predominantly winter rainfall areas of southern Australia, integration of supply-side and demand-side management is an easier task than in the northern, summer-rainfall areas. In the south, relatively accurate forecasts of both supplies of, and demands for, irrigation water are available at the start of each season. This eliminates the need for short-run decisions and means that only long-run decisions are made under climate-induced uncertainty. The more common systems analysis computer methods, with their much lighter computational burdens, are much more appropriate in these higher latitudes (Dudley, 1990; Dudley and Musgrave, 1993).

However, the success of earlier stochastic methods under the uncertainty of lower latitudes depended on the assumption of one water user controlling both reservoir and farm operational decisions. This is at variance with the existing situation with irrigation areas comprising many independent farm decision makers, as well as other users of reservoir water. To preserve the integration of supply-side and demand-side management for multiple water users, the concept of 'capacity sharing' was introduced.

## 3. Difficulties with centralizing the integration of supply-side and demand-side management problems with centralized management

A trade-off exists within integrated system management between aims of low versus high reliability of supply meeting demand. That is, for systems management integrating supply-side and demand-side decision making, there is a trade-off between: curtailing average demand, and hence average net benefits, so that uncertain, highly-variable supply meets the limited but still uncertain, highly-variable demand with a high probability; and expanding average demand so that average net benefits are increased at the cost of reduced reliability of those net benefits. Maintenance of high reservoir carry-over levels to achieve stable, reliable water supplies inevitably implies large reservoir spills on average, which means lower average reservoir releases. This is especially true in the newer, northern irrigation areas in Australia with their relatively high ratio of peak or high flows to reservoir capacities. The increased evaporation and other losses associated with high carry-over are less important than increased spills.

Trade-offs between high mean–high variability releases and low mean–low variability releases can be far from small. Computer modelling has shown that mean annual net revenue to irrigation in a region is about twice as large for high mean–high variance systems as for low mean–low variance systems.

Achieving an acceptable balance or trade-off between irrigation returns with high mean–high variability, on one hand, versus low mean–low variability on the other, becomes a very difficult task for centralized management as users' preferences become more diverse. While most users in systems we have modelled are cotton growers with similar aversions to risk, centralized integration of supply-side and demand-side management is relatively straightforward. However, as irrigators become less homogeneous and non-irrigation uses become more important, such as environmental requirements, trying to derive common reservoir carry-over rules that simultaneously suit every user becomes increasingly complex.

Grappling with the problem of moving from the situation of one decision maker controlling both reservoir and farm management decisions, to a system with multiple water users, led to the concept of capacity sharing.

## 4. Capacity sharing–decentralized integration of supply-side and demand-side management

Capacity sharing (CS) is an institutional arrangement providing each user, or group of users, of reservoir water with perpetual or long-term rights to a percentage of reservoir inflows and a percentage of empty reservoir capacity or space in which to store those inflows, and from which to control releases (Dudley, 1994). CS is like each user (individual or group) owning their own small reservoir on their own small stream. Their reservoir releases through time can be managed according to their particular supply reliability preferences. The concept can also be extended to include user rights to percentage shares of tributary flows downstream of the reservoir. Under CS, each user's supply-side and demand-side decisions are made by each individual decision-making entity, and the probabilities of (non-transferred) water supply can be calculated directly from historical or synthesized streamflow data alone (Dudley, 1990, 1992a; Dudley and Musgrave, 1993).

CS is best described with the aid of diagrams, as in Dudley (1992a), but they require too much space to reproduce here. Instead, consider the analogy of the reservoir to a money savings bank in which each account has a maximum to which the account can accrue (which is a percentage of the maximum sum of deposits which the bank can contain); negative balances are ruled out; withdrawals are under the control of the account holder; but deposits are made by a stochastic process over which the account holder has no control, although the probabilities of deposits are known. Each depositor seeks a preferred balance between running out of funds periodically due to keeping their average account funds low, and forgoing potential deposits periodically due to high average account contents. Each makes their own choice and each reaps the consequences. Funds can be transferred from one account to another by a joint request by the affected account holders. As well, two account holders may agree to transfer, from one to the other, all or a portion of one's rights to deposits and/or account capacity. The contents of each account are known through time by monitoring withdrawals, transfers, interest and deposits with the aid of computers. Similarly, users' entitlements to reservoir inflows would be added to users' reservoir 'accounts' (unless full), withdrawals and share of evaporation and seepage losses subtracted, and periodic adjustments made to ensure that the sum of 'deposits' so computed equals the actual reservoir contents (Dudley and Musgrave, 1988). The concept of CS has been developed in Australia to overcome limitations of the current water allocation methods, especially in the predominantly summer and uniform precipitation areas characterized by highly unreliable water supplies. In these areas an almost continuous task of water management is the reassessment of the desirable balance between allocating currently stored water to the immediate season versus saving some for future use in the form of reservoir 'carryover'. The more water saved for future use, the greater the reliability of water supply from the system, but at the cost of reduced average supplies because of increased reservoir spills and evaporation losses. This trade-off between mean quantity and reliability of supply is very important in most of Australia. The main problem with the traditional methods of sharing water supplies among users is that all but a few special users are forced to share equally in whatever trade-off point is chosen, regardless of their preferences. It is extremely difficult to tailor supply reliability to individual user needs on a large scale, under traditional allocation methods, while preserving flexibility to take emerging demands into account. However, CS easily allows the individual users or groups to choose their own mean/reliability trade-off at a point in time, and to re-choose through time in light of changing conditions and preferences (Dudley, 1990).

It can be shown algebraically that CS is at least as good as the commonly-used method (release sharing) for achieving an efficiency objective in the predominantly winter rainfall area of southern Australia (Alaouze, 1991; Musgrave *et al.*, 1989). Recent legislation in a southern state, Victoria, incorporates provision for its use (Dudley and Musgrave, 1993).

One of the features of capacity sharing is that it can operate at many levels. This is explained in a further quote from Dudley (1994).

> The foregoing discussion has presented CS as a single-level division of reservoir resources, as if each individual or group of final water users would have 'their own small reservoir on their own small stream'. However, there can be multiple layers of such divisions, with groups at subsequently lower levels having 'their own smaller reservoirs on their own smaller streams' (Dudley, 1992b; Dudley and Musgrave, 1993). For example, political jurisdictions within a country, such as

states, provinces or counties, may each hold first-level shares. Those entities may then use CS to allocate water among, say, various multi-purpose water storage uses, such as irrigation, flood management, instream flow maintenance, and urban and industrial use. For some of those uses, such as irrigation and urban/industrial, CS may be used again to allocate water among individual end-users or small groups of users in similar locations, with similar use patterns, or who have traditionally managed their allocation as a common property resource. The managers of shares at each level would have security of water supply probabilities with independence from the supply-reliability effects of the reservoir management decisions of others. Each level would use similar techniques for debiting, crediting and monitoring the water 'account' of each shareholder within its level of management.

Two types of short-run opportunity costs need to be taken into account when integrating supply-side and demand-side management if the most beneficial use is to be obtained from water resources. The first type is the opportunity cost of each single user using their own water in a current time period, expressed as the value of that water to them if saved for their own use in the future. The second type is the value of water to other users, both now and in the future. That is, before using a unit of water in the current period, both the value of that water to the same user in later periods, and the value of that water to other users in both the same and later periods, should be taken into account (Dudley and Scott, 1994).

The first type of opportunity cost – the value of water to the same user if used later rather than now – would be taken into account by users under capacity sharing as described so far. That is, its inclusion would be part of the integrated supply-side and demand-side decision making of each holder of a capacity share, whether that share is held by an individual user or group of users. However, the independent integrated decision making fostered by the capacity sharing concept described so far in this chapter does not encourage incorporation of the second type of opportunity cost. That is, when specific shareholders, whether individuals or groups, are deciding whether or not to use an extra unit of water in the current period, they have no incentive to take into account its value to other users.

However, a feature of capacity sharing is its specification of water property rights which are explicit, exclusive, and enforceable. This provides a very sound basis for a market for permanent transfer of water already in the system, whether reservoir contents or stream flows; permanent transfer of rights to future stream flows; or temporary transfers of rights to future stream flows. The existence of such markets gives an incentive for water users to abandon their individualistic supply-side and demand-side management and, via their market transactions, integrate such management across the range of water uses and users in the valley. That is, water will move to its highest valued use both through time and across uses.

It may be inferred from the above discussion that capacity sharing promotes efficient supply-side and demand-side management because the individual decision makers stand to reap the consequences of their actions in a very direct way. That is, they will be better managers and stewards of resources to which they have individual rather than collective rights. It was noted above that capacity sharing can operate at many levels. It may be used to allocate water between nations or states, with subsequent lower-level allocations also being made by capacity sharing or some other means. In many Western-type societies, more efficient resource use is likely to result when capacity sharing is carried through to lower levels. This would be because individual users are more directly exposed to the results of their actions than where capacity sharing ceases at 'bulk' or 'wholesale' levels with 'retail' allocations being made by some other means. However, this is not necessarily the case in all cultures. In more traditional societies, common property rules and mores are often very well established and appear to foster socially efficient water use (Dudley, 1992b). Such common property rules and mores may lead to ecologically sustainable management as well.

## 5. Some advantages of opportunity-cost pricing

Centralized water pricing is often expected to perform a number of functions – to be multi-purpose. One is to achieve an economically efficient allocation of water both through time and across a range of users. A second it to repay at least part of the capital and operating costs of the water supply system. The first becomes extremely complex, if not impossible, in practice in northern-Australian situations where both supply of, and demand for, water are highly variable and unpredictable (Dudley, 1991a). In such situations, the second cost-recovery role also has problems. Whereas it is still possible to use pricing for cost recovery, the resulting revenues stream is quite unstable, fluctuating with the supplies received by the water users. Also, the average level of prices through time required to recover costs varies according to the reliability of the system selected by the centralized reservoir management. High-reliability systems management would require higher mean prices per unit of water supplied than low-reliability systems management. This is due to the higher mean quantities supplied by the latter than the former.

These two roles of centralized water pricing can be separated under capacity sharing. As indicated above, the efficient allocation role can be achieved by markets. Then, the cost-recovery role can be filled by charges levied on the capacity shareholders in accordance with their share of reservoir capacity held, and their use of distribution and other attributes of the water supply system. That is, according to their long-term rights held, and regardless of their participation in short-term water markets. These costs could be fixed per year to the advantage of the supply authority, or could vary with supply conditions, while maintaining the same long-term average, to the advantage of water users.

Decisions about timing of system capacity expansion could also be decentralized under capacity sharing. Potential holders of reservoir capacity shares could make conditional bids for them prior to construction. Only if the total of the bids matched or exceeded construction costs would the bids become binding and construction commence.

Capacity sharing can be introduced gradually. To begin, a small percentage of reservoir contents and reservoir inflows, and perhaps downstream unregulated flows as well, could be operated under capacity sharing, with the remainder of the water resources of the area managed as before. Nevertheless, some jurisdictions favour even more gradual approaches as indicated in the following section, possibly because of their lack of political power.

## 6. Another means of decentralizing water management

In most large Australian irrigation areas, irrigators' water rights specify a notional quantity of water to be received each year. This quantity reflects the area irrigated by the holder at a point in the past. However, this notional quantity will be supplied in an irrigation season only if the total water supply position for that season is sufficiently favourable. In some seasons it may be exceeded; in some it will not be met. It depends on the quantity of water in the system at the start of the irrigation season, and the portion of that which is committed for carry-over to the next season. As noted above, centralised determination of the quantity to carry over to the next season becomes very complex as water users become more diverse in their water supply reliability requirements. (One of the main advantages of capacity sharing is that each shareholder makes their own carry-over decision in light of their preferences for reliability of supply.)

One alternative to capacity sharing is to allow each water user to make an individual decision to carry over a portion of the water allocated to them in the current season, for their use in future seasons. That is, there would be private or individual carry-over water in the reservoir beyond the public or common carry-over. However, to date individuals have only been allowed to carry water over from one season to early in the next without penalty. Carry-over water not used by the middle of next season is lost to individuals at a progressive rate until all is lost by season's end. This is a recipe for promotion of inefficient water use, and seems likely to be modified in the future.

With this correction, there may be few problems with such private carry-over while few users take advantage of the mechanism. However, as more users do so, there will need to be complex rules governing the fate of individual and in-common carry-overs should the reservoir fill and spill. This is in contrast to the simple, transparent rules of capacity sharing. It appears to this author that the joint/individual carry-over system will become more like capacity sharing through time, with individual carry-overs replacing the in-common carry-over and simple, transparent rules governing reservoir spill losses replacing complex ones.

## 7. Capacity shares for both irrigation and environmental uses

A preliminary study into allocation of the scarce water resources in a river valley between extractive irrigation and in-stream environmental flows by capacity sharing was conducted in 1990 (Dudley, 1991b). This was followed by an exploratory study into combining capacity sharing with transferable discharge permits for which the allowable discharge per permit held by dischargers is a function of stream conditions. If low stream flows restricted discharges of water containing wastes, a discharger holding both capacity shares of headworks reservoir capacity and transferable discharge permits can increase permitted discharge by buying extra permits and/or releasing water from their reservoir capacity share. It shows considerable potential for highly variable stream flows, as in Australia (Dudley *et al.*, 1993).

A further project modelling the holding of capacity shares by both irrigators and wetlands managers has been completed. The division of the water resources of the valley by capacity shareholding allows both irrigators and wetlands managers to manage water as if they had their own small reservoir on their own small stream. Different allocations of reservoir inflows, reservoir capacity and downstream unregulated tributary flows allow the quantification of trade-offs between allocations favouring wetlands and those favouring irrigation (Dudley and Scott, 1994; Dudley *et al.*, 1994)

A new project will develop models to quantify such trade-offs between irrigation and in-stream environmental uses. This is a joint project between the Centre for Water Policy Research and ecosystems researchers of the Centre for Catchment and In-stream Management at Griffith University, Queensland, Australia. A main feature of this new work will be better quantification of the effects of different levels of in-stream flows than in Dudley (1991b). Farmer net revenues will remain the measure of benefits of water allocated to irrigation.

In each of these studies, capacity sharing allows the individual shareholders to integrate the supply-side and demand-side management of the water resources under their control with minimal interference from other users. That is, they are free to take full account of the opportunity costs of using their own water now rather than later. But capacity sharing also provides a very sound basis for trading water as well, as noted above. This promotes the integration of supply-side and demand-side operational management across users. Trading of long-term rights to water resources should promote the efficient use of water resources in the long term.

## 8. Resource allocation decisions reduced to arms-length overseeing

Traditional command and control short-term and long-term management of Australia's water resources is characterized by centralized control of water allocation decisions. For example, the determination of the portion of reservoir contents to be allocated to the current season and the quantity to carry-over to future seasons, at the start of each season and as valley water supply conditions change during each season, is highly centralized in two ways. First, the decisions are made by the central water agency. Second, they are made by weighing up the current- and future-season water requirements of all of the diverse water users grouped into only two classes – high-security users (livestock and domestic users; irrigators of perennial crops) and other users (mainly irrigation and environmental uses).

Capacity sharing first decentralizes both of these classes of water use decisions by having individual decision makers themselves control the integration of their own supply-side and demand-side decisions. That is, decisions are made: locally by the shareholders; and for each shareholder individually. Then, via markets for water resources, supply-side and demand-side decisions are integrated across the individual uses. But such across-user integration decisions are still made by the individual shareholders looking after their own interests.

However, there may be occasions when, in the public interest, water needs to be reallocated from users who can afford to pay the market price to those who cannot. It is essential that current users know that they will be compensated sufficiently to provide adequate incentive for them to manage the water resources under their control in an efficient way. If there is a 'thick' market for water resources with many participants, so that the market prices do not reflect the monopoly power of a few water users, such reallocation may best be performed by the appropriate government authority entering the market and buying water for the new or extra use. If, because of market distortions such as monopoly power, market prices exceed those which would be expected from a 'thick' market with many buyers and sellers, other forms of acquisition with correct compensation may need to be in place.

In any case, the management role of the government is reduced to such long-term periodic reallocation to correct for failures of the marketing system rather than the present continual, detailed short-term management and allocation. In effect, the agency management is changed from 'hands-on' to 'arms-length'.

There are some movements towards decentralized water management apart from capacity sharing, as indicated in the following section.

## 9. Geographical decentralization

In the southern mainland state of Victoria, the Rural Water Corporation (RWC) has been the state agency managing the states water resources for many years. However, the geographic area of the state and the RWC are now being sub-divided into five or six main water resources divisions. The central RWC is being disbanded. Each of the new divisions is autonomous, with the opportunity to trade water across the divisional boundaries to their mutual advantage. Also, bulk or wholesale level capacity sharing is being introduced. Hence, water management is being decentralized geographically and, at the same time, integration of supply-side and demand-side decision making is being made by smaller units closer to individual, final water users. Other states are examining or planning various forms of decentralization.

## 10. Conclusion

Water resources policy research in Australia, especially in areas where summer and uniform rainfall is dominant, has been pointing to potential advantages of a particular form of decentralized water resources management. It is advocated that this be achieved by integrating supply-side and demand-side decisions by management devolved as closely as feasible to the final water users. Such management requires the introduction of new property rights or institutional arrangements which define users' control of water closer to its source, and include the marketing of short-term and long-term rights to water. How far this integrated management can be devolved towards the final water users in an area depends largely on the characteristics of its water distribution system. This form of decentralization promotes efficient use of water and related resources by confronting water users with the individual, social, short-term and long-term opportunity costs of the water they use. Affected water uses include agricultural, urban, industrial, energy, in-stream and off-stream environmental uses. Different sections of the water industry in Australia are moving in this general direction at different speeds.

## Acknowledgements

The comments of Professor Warren Musgrave on earlier drafts is gratefully acknowledged, with the usual caveat. Funding from the (Australian) Land and Water Resources Research and Development Corporation is gratefully acknowledged.

## References

Alaouze C 1991 The optimality of capacity sharing in stochastic dynamic programming problems of shared reservoir operation. Water Resour. Bull. 27, 381–386.

Boyd K 1994 Getting environmental flows. In Proceedings of Environmental Flows Seminar. pp 36–43. Australian Water and Wastewater Association, Artarmon, NSW, Australia.

Dudley N 1988 Volume sharing of reservoir water. Water Resour. Res. 24, 641–648.

Dudley N 1990 Alternative institutional arrangements for water supply probabilities and transfers. In Transferability of Water Entitlements – An International Seminar and Workshop. Eds J Pigram and B Hooper. pp 79–90. University of New England Centre for Water Policy Research, Armidale, NSW, Australia.

Dudley N 1991a Toward Optimal allocation of Surface Water by Pricing. Discussion Paper No 1. University of New England Centre for Water Policy Research, Armidale, NSW, Australia. 20 p

Dudley N 1991b Management models for integrating competing and conflicting demands for water. In Water Allocation for the Environment – An International Seminar and Workshop. Eds J Pigram and B Hooper. pp 169–182. University of New England Centre for Water Policy Research, Armidale, NSW, Australia.

Dudley N 1992a An innovative institutional arrangement which incorporates the risk preferences of water users. In Risk-based Decision Making in Water Resources V. Eds Y Haimes, D Moser and E Stakhiv. pp 174–199. American Society of Civil Engineers, New York.

Dudley N 1992b Water allocation by markets, common property and capacity sharing: companions or competitors? Natural Resour. J. 32, 757–778.

Dudley N 1994 An innovative institutional arrangement with potential for improving the management of international water resources. In Water and Peace in the Middle East. Eds J Issac and H Shuval. pp 469–480. Elsevier, Amsterdam, The Netherlands.

Dudley N and Burt O 1973 Stochastic reservoir management and system design for irrigation. Water Resour. Res. 9, 507–522.

Dudley N and Musgrave W 1988 Capacity sharing of water reservoirs. Water Resour. Res. 24, 649–658.

Dudley N and Musgrave W 1993 The economics of irrigation water allocation under uncertain conditions. In Water for Sustainable Development in the 21st Century. Eds A K Biswas, M Jellali and G E Stout. pp 126–137. Oxford University Press, New Delhi and Oxford.

Dudley N and Scott B 1994 Quantifying Trade-offs Between In-stream and Environmental Uses Under Weather Uncertainty. Paper to International Conference on Coordination and Decentralization in Water Resource Management, Rehovot, Israel, October.

Dudley N, Coelli M and Pigram J 1993 An Integrated Approach to Tradeable Discharge Permits and Capacity Sharing under Australian Conditions. Discussion paper prepared for The Environmental Research Trust and The Sydney Water Board, NSW, University of New England Centre for Water Policy Research, Armidale. 116 p

Dudley N, Scott B and Coelli M 1994 Tradeoffs Between Effectiveness and Opportunity Costs of Water Supplies to Wetlands; an Explanatory Manual. University of New England Centre for Water Policy Research. 68 p

Heaney J 1993 New directions in watershed planning and management. Water Resour. Update 93, 3–8.

Hearn A 1994 OZCOT: A simulation model for cotton crop management. Ag. Syst. 44, 257–299.

Lloyd L, Atkins B, Boon P, Jacobs T and Roberts J 1994 Natural processes within Murray–Darling floodplain wetlands. In Murray–Darling Floodplain Wetlands, Proceedings of the Floodplains Wetlands Management Workshop, Albury. Eds T Sharley and C Huggan. pp 6–23. Murray Darling Basin Commission, Canberra, Australia.

McCosker R and Duggin J 1993 Gingham Watercourse Management Plan Final Report. Department of Ecosystem Management, University of New England. 45 p

McGregor M, Harrison S and Tisdell C 1994 Assessing the impact of research projects related to natural resources. Land and Water Resources Research and Development Corporation Occasional Paper Series No 08/94, Canberra, Australia. 111 p

Musgrave W, Alaouze C and Dudley N 1989 Capacity sharing and its implications for system reliability. In Proceedings of the National Workshop on Planning and Management of Water Resource Systems: Risk and Reliability. Eds G Dandy and A Simpson. pp 176–185. Australian Water Resources Council, Conference Series No. 17, Aust. Gov. Publishing Service, Canberra.

Wettin P, Jacobs T, Johnson P and Ryan N 1994 Water sharing and floodplain wetlands in the Murray–Darling Basin. In Murray–Darling Floodplain Wetlands, Proceedings of the Floodplains Wetlands Management Workshop, Albury. Eds T Sharley and C Huggan. pp 24–41. Murray Darling Basin Commission, Canberra, Australia.

# 50
# The practical impact of ecological chemical principles

F. KORTE

In the middle of the last century, the rapid development of the sciences, and especially of the natural sciences, added a new dimension to life. This was the beginning of a period rich in fantasy and optimism. During that period, important statements were often published in the form of letters such as this one written by J. von Liebig: 'The study of chemistry does not only serve the promotion of the material interests of mankind but gives us in addition an insight into the miracles of creation surrounding us, upon which depend in a high degree the existence, the survival, and the further development of mankind. No other scientific discipline offers us such fertile ground for thinking and reflection and incessantly new discoveries. The industry has profited enormously from the knowledge of chemistry' (Liebig, 1865).

With this philosophy in mind, the development of new inventions accelerated throughout the industrialized world, resulting in the registration of an enormous number of new scientific facts and laws. Some of them had practical applications which were used by developing industry. The First and Second World Wars gave further impetus to this process of practical application of scientific findings for the development of, for example, the telephone, television, automobile, train and aeroplane industry, use of nuclear energy, discovery and application of antibiotics, insecticides such as DDT, and many more. These achievements were made possible by the availability of inexpensive energy in the form of coal, oil and natural gas. Production trends in the chemical industry are well documented worldwide (Iliff, 1973; Korte, 1976).

Because of this rapid, unforeseen trend of application of new scientific knowledge, people began to worry whether technical progress could be kept under proper control and damage to the environment kept to a minimum (Fritsch, 1993; Gershinowitz, 1972; Schmidt-Bleek and von Weizsäcker, 1994).

In order to minimise environmental damage by chemicals, concepts such as allowable concentrations and ADIs were established. In the process, an experimental method for evaluating environmental damage was developed (Freitag *et al.*, 1985). This method shows that on the basis of our present knowledge, approximately 30% of the chemicals currently on the market should be replaced by safer ones because of their undesirable ecostability.

Polychlorinated hydrocarbons have been used for many years with great success to combat pests. As the use of these compounds became more widespread, evidence showed that they were extremely stable and accumulated in the tissues of birds, fish

*D. Rosen* et al. *(eds): Modern Agriculture and the Environment, 625–631.*

and mammals, including, presumably, man. Efforts have been made to replace these compounds by the synthesis of materials with insecticidal properties comparable to those of cyclodienes or toxaphene, but with a high rate of biodegradation (Coulston Patent, 1993).

Although present criteria refer primarily to aquatic ecosystems, it is recognised that certain substances may affect other ecosystems populated by anything from soil microflora and microfauna to primates. Current tests for the evaluation of ecotoxicity of chemicals released into the environment are inadequate because of the huge numbers of these chemicals, and their often unknown physicochemical properties. The information derived from these tests may be insufficient for classification of the compounds. In an attempt to solve this problem, a number of recommendations have been made by the Forum für Wissenschaft, Wirtschaft und Politik (International Academy of Environmental Safety) e.V., Bonn (1972):

(1) Reliable priorities for investigating certain products can be established only when the amount, nature and place of use are known.

(2) After the establishment of well adopted evaluation parameters, the ecotoxicological evaluation of environmental chemicals is most important.

(3) As regards the financial effects of the deterioration of environmental quality and the possible maintenance costs, standards for financial estimates have to be set.

These recommendations suggest that not all possible damages should be regulated. First the environmental impact should be defined and characterized, then appropriate regulations should be established.

Recommendations by the EU Scientific Advisory Committee to Examine the Toxicity and Ecotoxicity of Chemical Compounds, Luxembourg, established standards for concentrations of chemicals in water in $\mu g\ L^{-1}$ for water quality objectives, which have since been modified for political and economic reasons (Rasmussen, 1994). It is nevertheless unlikely that the impact of chemicals introduced into the environment can be controlled by scientific approaches alone. There can hardly be any doubt remaining that the so-called maximum allowable concentrations and other regulatory decisions are often based on politics rather than science. These values are accepted by the political community. No matter how reliable the scientific basis for judging the safety of products and practices, the final decision is primarily the result of sociopolitical circumstances. Public acceptance is obviously capable of overriding scientifically established data.

There is, however, growing recognition of the potentially substantial impact of even modest changes in atmospheric chemistry on global environmental parameters, such as might be expected to accompany preindustrial human activity. Recent examination of new Greenland ice cores (Chemosphere Special Volume, 1994; Trends '93, 1994) indicates variation in $CO_2$ concentrations over an 8–20 year period during the Pleistocene that correlate with changes in atmospheric dust concentrations. This suggests that this measurement is capable of resolving seasonal events, a degree of accuracy which would be necessary to test the preindustrial anthropogenic hypothesis which deals with changes caused by agricultural practices. An additional impetus comes from recent reconstructions of temperatures and flood regimes during the period of 1250–1450 AD from the sediment record along the Mississippi River. In this record, large variations in flooding are apparently related to 1–2°C fluctuations in temperature and, perhaps more ominously for the anthropogenic model, to 10–20% changes in annual precipitation. Changes in rainfall on tropical islands are a documented result of agricultural expansion, deforestation, and other human environmental impacts that clearly do not require advanced technology. Other examples are also given in the literature cited above.

Perhaps we should concentrate our efforts not only on establishing tolerance data for all the millions of chemicals introduced into the environment, but should utilize the knowledge of different scientific disciplines to deal with environmental issues properly and efficiently, as suggested by the Forum für Wissenschaft, Wirtschaft und Politik in 1972. If these recommendations were adopted by responsible persons in science, politics and society, our situation could be comparable to that of J. von Liebig in 1865 – full of optimism – and we could once again hope to take all necessary steps.

During recent decades, concern has been expressed about the impact of both agricultural and industrial chemicals on the environment. Some steps have been taken by industrialists and politicians to improve the situation by defining tolerance levels for some chemicals. However, the environmental impact of industrial activities and its assessment in more sparsely inhabited areas cannot be compared to its effects in urban areas. The recovery of gold using the cyanide leaching procedure is just one example.

Gold was once isolated from a gravel stream bed by panning, which relied solely on gravity. Later, mercury amalgamation was used to extract gold (60%) from rocky ores after crushing them in mills. The residues from mercury amalgamation remain in evidence at many stream sites in the US. Although this method is still in use in the Amazon region, industrialists sought more efficient methods, and the cyanide process was developed and described by G. Bödlönder in 1896 in the Zeitschrift für Angewandte Chemie. In the cyanide heap leach process, the ore is crushed and placed on pads lined with compacted clay, or on polyethylene sheets. These heaps can stand from 10 to 200 feet high, span several hundred acres and involve from a thousand to a million tons of ore. A cyanide solution (0.25%, 1–4 pounds of cyanide per ton of ore) which dissolves gold is allowed to percolate through the heap. The leaching can take from several days to a few months, depending on the procedure and the type of ore (Sternkamp, 1992). The cyanide–gold complex formed is stored in ponds. In order to remove the gold, the liquid undergoes several treatments, including active coal treatment, heat, chemical processes and electrolysis.

Before turning our attention to the impact of cyanide on the environment, the routes leading to contamination should be noted. Cyanide from leach processing can leak through tears and/or punctures in pad liners. In addition to reports about these leaks beneath heaps by Fulkerson (1990), ABC News broadcast a documentary film in 1989 about the 'new gold rush in the USA' which showed quite vividly the leaks and punctures in the lines carrying the cyanide solution. The other sources are open piles and solution ponds which can overflow and spill cyanide solution into different environmental compartments. Heap leach operations threaten ecosystems with additional pollutants such as nitrogen compounds released during cyanide degradation. In addition, toxic heavy metals such as lead, cadmium, and mercury commonly present in ore can be mobilized during crushing or leaching. These metals accumulate to varying degrees in plants, mammals and fish, thus entering the food chain. Cyanide contamination of rivers, lakes and seas can permanently damage the swimming ability of some species. In 1990, a leak in Colorado was reported to have destroyed aquatic life along a 17-mile stretch of one river, and in the same year, 10 million gallons of cyanide solution spilled into the South Carolina River, killing thousands of fish (Young, 1992). The EPA (1986) permits free cyanide levels of 3.5 mg $L^{-1}$ in this area for any 24-hour average, with an upper limit for freshwater life of 52 mg $L^{-1}$ at any given time (Heming and Thurston).

Animals can ingest cyanide, inhale it, or absorb it through the skin. Birds and mammals are drawn to cyanide solution storage pools as a source of water. In Nevada, 10 000 animals died between 1986–1991 as a result of ingesting the solution (Albersworth *et al.*, 1992).

The effect of mining on the Ok Tedi River is another example of environmental damage. Since mining was initiated in 1984, the river has become a receptacle for mining waste, causing the death of several types of fish. It has been polluted by mine sediment, and the river bed is filling up, causing the level of the river to rise. The social and economic effects of this situation for local inhabitants have been documented by Maun (1992).

Since cyanide is highly toxic, the mining industry takes heavy precautions and no accidental human deaths due to cyanide poisoning have been reported. Toxicological studies (ATSDR, 1993) have shown that short-term exposure to high levels of cyanide can harm nervous, respiratory and cardiovascular systems. The lethal dose for humans is 1–3 mg $kg^{-1}$ body weight. Long-term exposure to sublethal doses causes irritation of the eyes, loss of appetite, headaches and dizziness, and damage to the nervous system and thyroid gland.

The cyanide heap leach process produces vast quantities of waste during the process of recovering gold at a rate of approximately 3 g per ton of ore. Fear of environmental damage is not unfounded, as the following example points out. One of the methods of extracting gold present in ore is as a complex in an aqueous solution of NaCN, and regenerating it involves passing the solution through coal:

$$2\,\mathrm{Au} + \mathrm{NaCN} + \frac{1}{2}\,\mathrm{O_2} \rightarrow 2\,\mathrm{Na[Au(CN)]} + 2\,\mathrm{NaOH}$$

$$\mathrm{Na[Au(CN)]} \xrightarrow{\mathrm{C}} \mathrm{Au} + \mathrm{NaCN}.$$

The amount of NaCn generated annually, the land area needed to dump the waste, and the quantity of water used for processing $250 \times 10^3$ tons of ore (1–10 ppm gold) per year (Korte, 1993) are:

- *NaCN* : 122.5 tons per year
- *Land* : 1.56 hectares per year
- *Water*: 70371 $m^3$ per year.

This huge quantity of cyanide is distributed into the air, water and soil. The amount of HCN that escapes into the atmosphere is estimated at 22 tons per year, or 60 kg per day, depending on the cyanide used. Gold production was 2200 tons in 1992. This means a worldwide emission of 20 000 tons of HCN into the atmosphere, where it is quite stable – HCN half-life = 267 days (Atkinson, 1985). A 0.05% solution of NaCN leads to residual free cyanide concentration of 177 mg $L^{-1}$, and it is known that the cyanide concentration at tailings impoundment is more than 80 mg $L^{-1}$. The toxicity of both HCN and NaCN is well documented (Marquardt and Schäfer, 1994). There are so far no recommended safety levels of $CN^-$ in water, soils or air, although the European Chemical Industry Council is taking action in this direction. In a recent meeting (CEFIC, 1994) where environmental regulations for HCN were discussed, no recommendations were made. Due to the complexity of the subject, the Council is now collecting data to be analyzed and discussed in the spring of 1996 at the 7th European HCN Forum. Very little is known about the stability of NaCN in soil, but the half-life of total cyanide calculated on the basis of anaerobic degradation is estimated at 1–2 years (Altlastensanierung, 1988). A recently recommended process (INCO $SO_2$/air process) for detoxification of HCN may bring about an improvement in the situation, but risk-benefit evaluations are not yet available.

In addition to standard terrestrial mining operations, research aimed at exploiting the mineral resources of the seabed is underway. Before such operations can be used for commercial purposes, a thorough investigation of their effect on marine flora and fauna is essential (Herzig, 1994).

These few examples, however they are interpreted, show that our current technology does not include a general concept of the degree to which environmental considerations should be incorporated into our technological activities. Debate about the use of nuclear energy further illustrates this point. Perhaps the concept of consensus among society, science, politics and all involved groups requires further development.

In order to maintain and improve our present civilization, we must bring regulators in all countries to understand that they can be of immense service to mankind if they take into consideration the statement of the Heidelberg Appeal to the Heads of State and Governments presented in 1992 at the Meeting on the Environment in Rio de Janeiro, Brazil. It was signed by 425 members of the scientific and intellectual communities. The last paragraph reads: 'The greatest evils which stalk our Earth are ignorance and oppression, and not Science, Technology and Industry, whose instruments, when adequately managed, are indispensable tools of a future shaped by Humanity, by itself and for itself, overcoming major problems like overpopulation, starvation and worldwide diseases' (Korte and Coulston, 1994).

## Acknowledgments

The author gratefully thanks Mrs Gabriela Falbo for her co-operation, criticism and the linguistic revision of this manuscript, and Dr Pran Nath Moza for his valuable technical assistance.

## References

Albersworth D *et al.* 1992 Poisoned Profits: Cyanide Heap Leach Mining and its Impact of the Environment. National Wildlife Federation, Washington, DC.

Altlastensanierung 1988 Zweiter Internationaler TNO/BMFT Kongress über Altlastensanierung, 11–15.4.88, Bundesrepublik Deutschland. Eds K Wolf, W J Van den Brink and F J Colon. Kluwer Academic Publishers, Dodrecht/Boston/London.

Atkinson R 1985 Kinetics and mechanisms of the gas-phase reactions of the hydroxyl radical with organic compounds under atmospheric conditions. Chem. Rev. 85, 159–161.

ATSDR (Agency for Toxic Substances and Disease Registry) 1993 Toxicological Profile for Cyanide. US Department of Health and Human Services.

Chemosphere Special Volume 1994 Preindustrial Human Environmental Impacts: Are there lessons for global change? Sci. Policy 29, 827–1143.

CEFIC (European Chemical Industry Council, Brussels) 1994 European HCN Forum, Marseille, 15–16.9.94.

Coulston F 1993 Non-bioaccumulable pesticides. United States Patent No. 5270345. 14.12.93.

Documentary film 1986 The New Gold Rush. Mineral Policy Center, Washington, DC. ABC News 23.6.89.

EPA 1986 Heap Leach Technology and Potential Effects in the Black Hills. 30 September.

Forum für Wissenschaft, Wirtschaft und Politik e.V. Bonn 1972 Recommendations of the 2nd International Symposium, 'Chemical and toxicological aspects of environmental quality.' In Global Aspects of Chemistry, Toxicology and Technology as Applied to the Environment. Environmental Quality and Safety 1, 252–253. Georg Thieme Publishers, Stuttgart.

Freitag D, Ballhorn L, Geyer H and Korte F 1985 Environmental hazard profile of organic chemicals. An experimental method for the assessment of the behavior of organic chemicals in the ecosphere by means of simple laboratory test with $^{14}C$ labelled chemicals. Chemosphere 14, 1589–1616.

Fritsch B 1993 Mensch-Umwelt-Wissen: Evolutionsgeschichtliche Aspekte des Umweltproblems. pp 103–181. Verlag der Fachvereine an den schweizerischen Hochschulen und Techniken AG, Zürich, and B.G. Teubner Verlag, Stuttgart.

Fulkerson B 1990 Cyanide Spills Common in Nevada Mining. Citizen Alert.

Gershinowitz H 1972 The complexity of environmental problems. In Global Aspects of Chemistry, Toxicology and Technology as Applied to the Environment. Environmental Quality and Safety 1, 1–9. Georg Thieme Publishers, Stuttgart.

Heidelberg Appeal to Heads of States and Governments 1992 United Nations Conference on Environment and Development (UNCED). Rio de Janeiro, Brazil, 3–14.6.92. pp 72–73.

Herzig P M 1994 Erzfabriken in der Tiefsee – Die Erforschung mariner Rohstoffvorkommen. Alexander von Humboldt Stiftung – Magazin Nr. 64, 27–36.

Iliff N A 1973 Organic chemicals (except agricultural animal health chemicals and food additives). In Global Aspects of Chemistry, Toxicology and Technology as Applied to the Environment. Environmental Quality and Safety. 2, 64–71. Georg Thieme Publishers, Stuttgart.

Korte F 1976 Global inputs and trends of chemical residues in the biosphere. In Global Aspects of Chemistry, Toxicology and Technology as Applied to the Environment. Environmental Quality and Safety. 5, 183–196. Georg Thieme Publishers, Stuttgart.

Korte F 1993 Ökologisch-Chemische Betrachtung zum vorgeschlagenen Goldgewinnungsprojekt unter Anwendung der Cyanidlaugung in der Türkei durch die Tüprag Metal Mining Ltd.

Korte F and Coulston F 1994 Some considerations of the impact of energy and chemicals on the environment. Reg. Tox. Pharm. 19, 219–227.

Liebig J V 1865 Chemische Briefe. CF Wintersche Verlagshandlung, Leipzig und Heidelberg.

Maun A 1992 Spokelsman of Ok Tedi/ Fly River Eneco Association, Dokumentation epd-Entwicklungspolitik, 13.6.92.

Marquardt H and Schäfer S G 1994 Lehrbuch der Toxikolgie. pp 555–558. B.E. Wissenschafts-Verlag, Mannheim, Leipzig, Wien, Zürich.

Rasmussen F B 1994 EEC water quality objectives for chemicals dangerous to aquatic environments. Reviews of Environmental Contamination and Toxicology 137, 83–110.

Schmidt-Bleek F and von Weizsäcker E U 1994 Wieviel Umwelt braucht der Mensch? MIPS–das Maß für ökologisches Wirtschaften. pp 139–144, 215–266. Birkhäuser Verlag, Berlin.

Sternkamp T J 1992 Gold: A watershed year? In Annual Survey and Outlook. Eng. Mining J. April.

Trends '93 1994 A Compendium of Data on Global Change. Eds. T A Boden, D P Kaiser, R J Sepanski and F W Stoss. Production Ed. G M Logsdon. Carbon Dioxide Information Analysis Center Publication No. ORNL/CDIAC-65, ESD Publication No. 4195.

Young J 1992 Mining the Earth. Worldwatch Paper 109, July. p 24.

# 51
# Reuse of effluent: Israel as a model

R. LASTER

> "And when they came to Marah, they could not drink of the waters of Marah, for they were bitter. Therefore the name of it was Marah. And the people murmured against Moses saying 'What shall we drink?' And he cried unto the Lord, and the Lord showed him a tree, and cast it into the waters, and the waters were made sweet" (Exodus XV. 3).

Moses had the uncanny ability to transform sweet water into brackish, and brackish into sweet. Generations of people have tried to repeat this trick, with little success. Yet with all his ability, Moses never reached the land of milk and honey. It's too bad; the world could certainly have gained from his expertise.

## 1. Introduction

Israel may be a land flowing with milk and honey but its water flow is not abundant. As a result, since the founding of the state, Israel's leaders have looked for ways to develop its water potential to the fullest extent. This development required: the capturing of head waters of the perennial flowing streams; pumping water from Lake Kinneret, Israel's only sweet water lake, to arid regions in the southern part of the country; overpumping of ground water; the desalinization of salt water in Eilat; and reuse of sewage water throughout the country. This chapter will concentrate on the reuse of sewage water using Israel as a model.

Reuse of sewage has been practised throughout the centuries by various communities, but Israel has been one of the few to declare reclamation and reuse of sewage as a national goal. Since the late 1950s Israel has determined that it must include in its water cycle reclamation of waste water and its reuse in order to meet the potential needs of the country (Grinwald, 1980).

Reuse of sewage is not without its problems (Feigin and Shalhevet, 1991), but it also has its advantages. First, the inclusion of waste water as part of the water cycle is a recognition of the ecological concept that all water is to be treated as a national resource. Second, Israel's Water Law declares that all water sources, including sewage, are to be under the control of the state. Third, Israel's expanding population over the years and limited sources of water demand that efforts be made to protect existing water sources and enable the reuse of sewage for agricultural use in order to free fresh water for its growing population. Finally, reclamation of waste water and

*D. Rosen* et al. *(eds): Modern Agriculture and the Environment, 633–646.*

its reuse is an excellent method to protect Israel's surface and ground water sources, as well as eliminating environmental nuisances.

Yet there are problems in reuse of sewage that need to be resolved in order for Israel to achieve its national goal. First, it must be recognized that there are psychological elements in the reuse of sewage and that these obstacles must be overcome. Second, in order to use reclaimed waste water for irrigation, treatment methods that purify it to a high degree must be developed. Third, there is a need to separate pipes for reclaimed sewage from potable water. Israel, for instance, originally designed its water supply system on the basis of a 'single pipe system' of drinking water quality to all parts of the country, including agricultural uses. People have become accustomed to drinking water in the fields from any pipe found near their working station. This must be changed if sewage is to be reused. Next, there are problems of potential bacteria infecting agricultural crops, as well as bacteria reaching households located near agricultural crops irrigated with purified sewage. Finally, one does not want to increase the problem of ground water pollution from seepage of poor quality sewage. This chapter will describe Israel's experience with the reuse of purified effluent from the founding of the country in 1948 until today.

## 2. Facts and figures

Israel's engineers and hydrologists tell us that the total amount of water replenished in Israel annually is 3 billion cubic metres. Of this annual replenishment volume, 80% of the water is used for agricultural use, some 12–15% for domestic, and the remainder for industrial use. The latest figures show that sewage water has reached approximately 250 million cubic metres annually, with a potential for the year 2005 of 400–450 million cubic metres annually (Ministry of Agriculture, 1992). The amount of sewage water purified in Israel to secondary or higher treatment levels reaches approximately 130 million cubic metres annually, with the remainder of the sewage water being treated at a lower quality. The highest level of treatment of sewage is presently done at the Gush Dan Treatment Center, called the Shafdan, which treats approximately 90 million cubic metres of sewage annually to a level which could be considered almost of potable quality (Ministry of Agriculture, 1992).

## 3. Government policy: fits and starts

The policy for reuse of sewage has gone through a number of stages during the development of the country (Shuval, 1980). For the most part the government has adopted policy in a series of fits and starts beginning in 1948. As soon as the country was founded, the Sanitation Department of the Ministry of Health, together with several kibbutzim and moshavim in Israel, began to pressure the government for a policy for land treatment of wastewater coupled with rational agricultural irrigation. The theory here was that there would be a solution to a serious sanitary problem on the one hand with a built-in incentive for the reuse of sewage for agricultural purposes on the other. It must be remembered that in the early stages of the founding of Israel, government policy was to encourage settlers to farm the land, as there was a feeling that those Jews coming from the diaspora would better serve their country if they were engaged in agriculture. This ideology combined with the economic and professional needs of agricultural and sanitation engineers led to Israel's first large-scale attempt at the reuse of purified effluent.

As early as 1956, the Government embarked on its first overall plan for reuse of sewage for agriculture. Yet, since Israel's Water Law came into effect in 1959, there was then in existence no legal vehicle to carry forward government policy. There was in effect, however, a British mandate ordinance requiring that business establishments meet certain criteria in order to be licensed. The Government decided to use this vehicle to publish guidelines for reuse of purified effluent. (These special conditions have never been published as regulations. The guidelines were used, as their name implies, to guide the consumer in the reuse of purified effluent.)

The guidelines were published by the Ministry of Health and they included the following. First, it was decided that there could be no reuse of raw effluent; a purification process, at that time primary treatment, was deemed sufficient. Second, it was determined that purified effluent could be used only on industrial or fodder crops, not crops eaten raw, or fruit trees, or other crops, to avoid direct sewage contamination of the crops. Finally, there were also permitted types of irrigation: ridge and farrow could be used for fruit trees with spray to be used on industrial or fodder crops. In addition, it was permitted to use purified effluent in fish ponds for raising of fish.

Over time scientific research made discoveries that led to a need to upgrade these guidelines. In 1965, they were upgraded to require that primary treatment include the minimum use of an oxidation pond and no spray irrigation within 200 metres of a residential district or 50 metres from a road (Shuval, 1980). This latter condition came about as studies began to show the development of enteric diseases in those agricultural communities where spray irrigation of purified effluent was employed.

In 1970 a cholera epidemic set off a demand for a whole new concept of reuse of purified effluent. The cholera epidemic occurred as a result of primitive farming methods: non-purified effluent from the city of Jerusalem was used to grow raw fruits and vegetables. As a result, the Government stepped in and began the development of what came to be known as the National Sewerage Project. The National Sewerage Project was a result not only of the cholera epidemic but also it was clear that Israel had to discover new sources of water for its residents. At that time the population of the country had grown and Israel was using almost all of its natural sources of water. In addition, and probably more important, the Government had just completed the National Water Carrier, a huge engineering project bringing water from the north of the country to the semi-arid south. With the completion of the national water project, hundreds of engineers and scientists were without work for lack of a hydrological project. Therefore it was decided that the Government would embark on a national sewerage project and transfer this personnel problem into a national resource (Shuval, 1980).

## 4. The National Water and Sewerage Administration

### *4.1 The Shafdan*

The project decided upon was the Dan Region Wastewater Treatment Project (the Shafdan) to reclaim waste water from the Gush Dan region, which includes Tel Aviv and its satellite cities. The project aimed at purifying 60 million $m^3$ of water by the year 2000. Originally it was decided that the project would raise the level of purified effluent to that of potable water. This was to be done by a series of treatment processes including polishing ponds and recharge of the ground water so that the water could be used for any purpose. It was discovered, however, that in recharging the ground water over time it was still susceptible to pollution from pathogens, bacteria, viruses and traces of toxic metals. Also at the same time there began the development of standards in the United States and other parts of the Western world, limiting reuse of sewage water.

The debate was finally resolved with a decision in 1976 that purified effluent from the Dan reclamation project would be used only for agricultural and industrial use. Therefore, after a series of the most careful treatment processes, a special pipe was designed and built to carry the water from the Dan reclamation project to the northern Negev for irrigation on permitted crops. The Government further enhanced its decision by converting the guidelines for purified effluent into regulations formally issued in 1981.

The Shafdan, therefore, stands out as an enlightened approach to reuse of purified effluent, and it can serve as a model to other countries. It involves taking wastewater from Israel's largest centre of population, purifying it under the most rigorous controls, monitoring its quality, constructing special pipes and pumping stations, providing irrigation for permitted crops to an area outside the Tel-Aviv megalopolis, and publishing regulations for reuse of purified effluent. All these measures ensure that there is a continual flow of purified effluent to a semi-arid region of the country. In short, a national project run locally but in accordance with published national standards.

### *4.2 The 'Dark Ages'*

While the Shafdan stands out as an enlightened approach to reuse of purified effluents, a less enlightened approach overtook the National Sewerage Administration during the period 1980 to 1990. During this period, the power to grant government loans to local authorities for the design and construction of sewage purification plants was in the hands of a single bureaucrat. This bureaucrat decided he would fund local authority sewage projects only if the projects met his criteria. The criteria he set were based on low cost design and operation of treatment plants with the purified effluent used for non-edible agricultural crops in the nearby vicinity. At that time cotton was a favourite of Israel's farmers and the system designed by this bureaucrat was based on primitive settling ponds for sewage with the purified effluent piped to cotton fields nearby for irrigation. The result of this policy was to set Israel's national policy for wastewater reuse back in quality and quantity.

It was 'discovered' that local authorities expanded at the expense of agricultural land nearby. This made agricultural use of purified effluent more erratic and far more remote. Second, with several winters of heavy rain, there were no takers for low quality effluent; farmers built their own reservoirs for rain and flood waters. Third, with no consumers for the low level effluent, it trickled its way into streams and wadis, causing environmental nuisances. And finally, cotton was king up until 1990, when the world market for cotton was glutted.

### *4.3 The 'Enlightenment'*

At this point, the Government acted to revamp the entire National Sewerage Administration. Beginning in 1992, a new staff of professionals replaced the single bureaucrat approach. This staff has encouraged the design and construction of modern sewage purification plants. Encouragement comes in the form of government loans and grants to communities which can prove that they have the capability of creating the economic and professional basis for modern wastewater purification systems. The National Sewerage Administration requires that the sewage unit of the local authority be operated as a closed economic system, funded by sewage charges as published in municipal by-laws. In addition, no grants are issued for the construction of a sewage purification plant unless the plant is designed to meet standards for purified effluent set by the Ministry of Health. It is these standards which also resolved the 'price wars' over the sale of purified effluent, discussed below.

## 5. Pricing

Modern eco-economics posits that the best way to create conditions for sustainable development is through a pricing mechanism for externalities. If a market could be developed for the sale and purchase of rights limiting pollution, market forces will lead to a cleaner environment at an optimal price. Can Israel's performance in purifying effluent for agricultural use serve as an example of eco-economics?

This question, posed to anyone familiar with Israel's economy, seems facetious. Israel can hardly be characterized as a *laissez-faire* economy. The Government intervenes at all levels of economic decision making. Yet the following discussion of pricing of purified effluent shows just how strong market forces are, even in a government-dominated economy. It may also serve as a model to other countries.

### *5.1 The problem: the water system*

In Israel, all water is metered and all water is charged to consumers based on consumption. The Minister of Agriculture sets water prices, and he has determined that agricultural users pay less than domestic users who pay less than industrial users. Yet, even with regulated prices, farmers will purchase purified effluent only if the price of the effluent is lower than that being paid for potable water. The problem has been exacerbated by the fact that, except for the Shafdan, Israel has a national water grid and a one-pipe system for all users, agricultural, industrial and domestic. Therefore if one wanted to use purified effluent for agricultural purposes, it would be necessary to design and build special pipes for this use. In addition, these pipes would have to be so designed to prevent their accidental use by persons unaware of their final destination.

A number of papers were written on the subject in an effort to set policy for the price of purified effluent (Selbst, 1980). Regulations were published in 1965 but even these regulations did not resolve the issue (Water Regulations, 1965). The general practice has been that agricultural users of purified sewage pay 75% of the value of potable water. In most cases even this is not done because the agricultural users balk at using the purified effluent as a replacement for potable water. In addition, it is often encumbent on them to pay for the laying of the pipes for the separate system for the purified effluent. Can law help to resolve this impasse?

### 5.2 *The second problem: the legal system*

Israel's Water Law of 1959 gave control over policy planning and implementation for water use and reuse to the Ministry of Agriculture – the Water Commission. The same law made supply of water to consumers the responsibility of the National Water Company – Mekorot. The law however did not change the status quo whereby local authorities remain responsible for collection and disposal of sewage. This responsibility, which had its origins in the British Mandate period, was later recognised by the Knesset in legislation. The Local Authorities (Sewage) Law of 1962 granted local authorities the right to collect sewage and even went so far as to grant local authorities ownership rights over purified effluents. One would think that with the publication of the Local Authorities (Sewage) Law, the question of sale of purified effluent had been resolved. As it turns out, this property right is a limited one because of the bifurcated nature of Water Law in Israel.

First, the local authorities cannot determine the quality of effluent to be reused. That quality is set by the Ministry of Health, with help from the Ministry of Environment. Second, the Water Commission licenses water suppliers, and under the Water Law, it is forbidden to supply water, including waste water, without a licence. Finally, planning and building of sewage treatment plants, laying water pipes and building pumping stations requires planning approval and authorisation from the Ministry of Interior, the Ministry of Agriculture, the Ministry of Health and the Ministry of Environment. Finally, water prices are set by the Ministry of Agriculture.

To make matters even more difficult for the local authority and market mechanisms, Israel created the National Sewerage Administration to fund local authority sewage projects. Local authorities naturally flock to the National Sewerage Administration to get cheap loans to design and build sewage purification plants, but these loans come with government strings attached. Among other things, the National Sewage Administration dictates the quality of purified effluents to be produced from a sewage purification plant that it bankrolls.

One would imagine that under the thick blanket of central government intervention, economic considerations would be smothered. But the resilience of market forces is something to be reckoned with. Once the municipality receives its grant to build a sewage purification plant; once the Government approves the design and construction of the plant; once the effluent is produced; a curious phenomenon occurs. The local authority ignores government legislation for pricing of water, government warnings and sometimes even contractual obligations with the National Sewage Administration and begins to bargain with agricultural users over the price of purified effluent.

In order to enhance its bargaining position, the local authorities used a mechanism often associated with piracy – taking hostages and demanding ransom. The hostage in this case is not a person but a water source: a stream or wadi. The local authority responds to government intervention by declaring that they have collected the sewage, thereby removing a local nuisance, and they will just as freely pour that nuisance into the nearest stream if no one is interested in buying the effluent at the price they suggest. Yet the farmers balk at buying the effluent unless it is cheaper than the price they are paying for potable water, which is always cheap due to government subsidies. By bickering, the two parties freeze government policy for reuse.

Inadvertently, the Ministry of Health may have broken the deadlock between the local authorities and the farmers by the publication of standards for purified effluents in 1991. The health standard led to the Water Commissioners' introduction of a policy known as the zero alternative method discussed below.

### *5.3 The 'zero alternative method'*

When the farmers and the local authorities square off, the Water Commission moves in to 'arbitrate' the issue (Tamir, 1993). The Water Commission points out to the local authorities that since 1991 all local authorities purify their sewage to a minimum standard promulgated by the Ministry of Health. The local authorities must, therefore, design their sewage purification systems to reach this standard. Each local authority must set its sewage charges at a rate sufficient to compensate for the design capability of the sewage purifications system. Once the local authority has brought its sewage to the standard set for purified effluent it may discharge its effluent anywhere: into a local stream or wadi or sell it to agricultural users. This point is called by the Water Commission the zero, or beginning, point because at the point where the quality of purified effluent has reached the government standard, the cost of doing so is covered by sewage charges. From that point and beyond, if agricultural users require special piping or additional treatment, the cost should be borne by them and not by the local authority (Shelef, 1993). Armed with this argument, the Water Commission has successfully arbitrated arguments between local authorities and farmers.

The above discussion has been presented for several reasons. First, one would expect that with government control over all water sources, including sewage, the Government would dictate to local authorities as to disposal and reuse of waste water. In fact, this has not been the case because local authorities acted in their own economic interest and not in the interest of serving government policy. As long as there were no regulations setting the quality for discharge of municipal sewage, local authorities could 'blackmail' the Government. The passage of the Ministry of Health regulations setting quality standards for effluent discharge, in effect acted as an economic control to incorporate 'the externality' in the local authority decision-making process, making the bargaining position of the local authority equal to the farmers, thereby leading to a resolution of a stalemate and enabling reuse of sewage in accordance with government policy. Although the zero alternative method has not made its way into legislation, its common sense logic could be used by other countries with economies and government controls over water similar to Israel's.

## 6. The legislative framework

It was originally contemplated in 1959, with the publication of the Water Law, that all water, potable, run-off and sewage, would be controlled by the Ministry of Agriculture and particularly by the Water Commission, which is a part of the Ministry of Agriculture. Yet this legislative act was not followed by elimination of authority granted to various groups already engaged in water distribution and purification. For example, sewage was being treated by local authorities, and the Knesset should have taken the power to purify domestic waste water out of the hands of local authorities and placed it in the hands of regional water authorities (Laster, 1980). This was not done, however, and over time this led to serious problems, discussed previously, regarding local authority control over sewage purification.

In addition, the Water Commission, by concentrating its energies on water supply, ignored the gradual accumulation of environmental nuisances resulting from unpurified sewage. The climax of this one-sided approach was the cholera epidemic of 1970, when the Ministry of Health demanded that it be given control over the regulation of waste water quality. The Knesset acceded to this request and in 1970 the Minister of Health received the power to determine the quality of waste water for irrigation purposes and other commercial uses (Law to Amend the Public Health Ordinance, (1970).

The Minister has exercised his power twice since that time: in 1981, and, under pressure from the Minister of Environment, again in 1990 (Public Health Rules, 1981). The regulations are set out in the Appendix. They basically follow the guidelines for reuse of purified effluent used in the 1960s and 1970s in Israel. First the regulations define purified effluent and require a minimal purification process to reach the defined quality of purified effluent. Second they determine the type of irrigation to be used and the type of crop to be used for purified effluent. If one looks at the regulations, it can be seen that the major permitted uses are for crops for industrial use, not to be eaten raw and not to be used in the food industry.

## 7. The Ministry of Environment

With the creation of the Ministry of Environment, the Government could have redesigned government control over water uses. Instead, the hydrological cycle has been further dissected. The Minister of the Environment demanded that he be given power over the protection of Israel's water sources. The Government and the Knesset acceded to his request and the Minister of the Environment was given power to determine the quality of Israel's water sources and intervene in the decision-making process for the approved uses of wastewater for irrigation. Yet the Minister of Agriculture maintains his responsibility over water supply and the Water Commissioner has maintained his power to prevent water pollution.

With his new found power, the Minister of Environment has forced the Minister of Health to amend the 1981 regulations for reuse of purified effluent (Public Health Rules, 1981). Under the new amendment, before a representative of the Ministry of Health can issue a permit for a permitted use of effluent or for an approved laboratory, he must first consult with a representative of the Ministry of Environment. The problem with this approach, however, is that the use of purified effluent demands attention by one Ministry at all times. Introducing an additional bureaucratic cog in the wheels will not produce an efficient regulatory system. Criticism of this unwieldy approval process has already been voiced both by the Ministry of Health and the Ministry of Environment (in interviews with Yishyahu Bar-or, Head of Department of Water Quality, Ministry of Environment, and Uri Aloni, Ministry of Health, on 2 September 1993, the two ministries were seen to have different interpretations of the regulations.)

## 8. The Ministry of Infrastructure

As a result of the 1996 elections, a new Ministry of Infrastructure was created. On his appointment, the Minister of Infrastructure requested that all authority for aspects of water, sewage and drainage be transferred to the new ministry. Yet the government balked at the transfer, and therefore has further convoluted the management of water sources in Israel.

To date, the Ministry of Infrastructure has responsibility, together with the Water Commission for abstraction, billing, and recharging licences. The Minister of Infrastructure determines water fees, except for water used for agricultural purposes. In addition, the new Ministry is responsible for the law for measuring water and its related sections. Furthermore, the Water Commission has been transferred to the Ministry of Infrastucture and the National Sewage Administration, as well.

In short, there are now five ministries involved in creating, planning and executing water policy in Israel, instead of the original four. One could sum up the situation by using the typical Middle East expression, "Why make things simple, if you can make them complicated?"

## 9. Policy for the future

In the late 1980s, following several years of drought and a reduction in the ground water table, the Government decided to reassess the need for reuse of sewage water as a replacement for potable water in agriculture. The Water Commissioner enlisted the help of Israel's water planners to prepare a national plan for the reuse of sewage. Two plans were prepared, one by Tahal, Israel's largest group of engineers for water planning, and the other by a consortium of water planners. Both groups came to the following conclusion (Ministry of Agriculture, 1992).

The population of Israel is growing due to the increase of immigrants from the Soviet Union and natural childbirth. The need for water of drinking water quality in the cities and towns will require a reduction in that water for agricultural use. There is an increasing amount of sewage water being produced by the growing populations in towns and communities which can be purified and used for agriculture. Israel should aim towards the highest quality of purification of this water to enable unlimited use for agricultural crops. Achieving unlimited use for agricultural crops will require a huge investment in water purification, which could be offset by the tremendous gains in the flexibility of the use of water in all regions rather than transferring water from one region to another at great cost.

The potential for reuse of sewage by the year 2010 is 450 000 000 cubic metres of sewage annually; the amount of potable water supplied to agriculture should be reduced from 80% today to 55% by the year 2010; and there is a need for heavy investment in the purification of wastewater (up to $550 000 000) over the next ten years. With this investment in purified wastewater and reduction in agricultural use of drinking water, the average cost of a cubic metre of purified water should be between 20 and 50 cents. Therefore both groups recommended that the Government develop a national policy for reuse of sewage water based on the following principles.

(1) Sewage and wastewater are part of the water sources of Israel.

(2) The treatment and purification of sewage water to its highest level will promote irrigation with no limitations.

(3) It is important that all pollution sources be removed from water sources.

(4) Agricultural users are to receive purified sewage water to replace drinking water now in use.

(5) There will be a need for co-ordination between those purifying sewage and those using purified sewage.

(6) There should be a plan for reuse of sewage water to the largest extent possible in the areas in which it is produced in order to prevent transfer of sewage over large areas.

(7) There is a need for pretreatment of industrial sewage in order to reduce the necessity for reducing pollutants in local sewage water to be reused for agriculture.

## 10. Analysis and conclusion

At first sight, Israel's approach to reuse of purified effluent is not exemplary. Too many authorities chasing too little effluent. On the other hand, Israel can serve as an example to other countries in its failures, as in its successes. Let us first look at the successes.

First, Israel has made it national policy that reuse of purified effluent is a recognised marketable water resource. Second, Israel has promulgated regulations with a health based standard for agriculture reuse. Third, Israel has designed one of the world's most modern plants for purification recharge and distribution of effluent from its largest population centre. Fourth, Israel has created an economic incentive for agricultural reuse of effluent by reducing the amount of potable water available to agricultural users. This bodes well for the past and the present. The problem is with the future.

Israel has not exploited its early legislative advantage over other countries to control the use and misuse of its water sources. As early as 1959 Israel had passed four water laws culminating with the Water Law published that year. The country did not capitalise on this legislative achievement. The pressures of development spurred on by water engineers blinded policy makers into thinking that Israel could expand at will, with water potential increasing as needs developed. One could cynically call this period 'obtainable development' instead of 'sustainable development'. Wells were dug, pipes were laid, water was pumped, pulled and pushed. Gradually, however, communities grew and sewage changed from a nuisance to a blight and from a blight to a hazard. Policy did not grow with it. But, just as standing water grows stagnant, so does unmoving government policy.

To meet the growing needs of a developing country, Israel should have looked outside its borders to see how other, more developed countries cope with sewage purification. For one, national policy must determine a designated use for each type of purified effluent. The Government could determine that purified effluent be used for irrigation, recreation, or simply poured into the sea. In any of these eventualities a certain quality for the water must be set and maintained by the Government.

The success of any water policy is based on a government's ability to connect the sections of the hydrological cycle, not separate its parts. Modern policy requires an overall review of water supply, water use and reuse. It is recommended that this policy be set as part of a watershed approach and not, as it is done today, on the individual municipality level. It should also be noted that Israel does not live in a vacuum. Its agricultural crops are sold all over the world and they must meet not only national standards but international ones as well. This means that reuse of purified effluent for agricultural use must be carefully controlled so as to meet standards set by other countries, including standards set for residual toxics found in fruits and vegetables.

Israel has been a world leader in the reuse of purified effluent for irrigation. In order to maintain its leadership, however, Israel cannot rest on its past achievements. Reuse is a challenge that can be met, provided that proper precautions are taken within a workable legal framework. Precautions are within the expertise of scientists and planners; creating the proper framework, the work of lawyers and legislators. Both of these disciplines must work together to ensure Israel's advancement in the field of recycling of liquid waste, to protect the present generation without burdening the future.

## Appendix

*Public Health Rules (Purification of Wastewater for Irrigation) 5741–1981*

By the power invested in me by section 65A of the Public Health Ordinance 1940, and after consultation with the Ministry of Agriculture, I promulgate the following rules:

Definitions

In these rules:
'Examination': laboratory examination or field examination;
'Permit': permit given under section 65A of the ordinance;
'Engineer': the district sanitary engineer of the Ministry of Health;
'Director': the Director General of the Ministry of Health or a person he has authorized for purposes of these rules, or parts of them;
'the Book': the latest edition of 'Standard Methods for Examination of Water and Wastewater' published by the American Association for Public Health, Washington, DC, a copy of which will be available for public inspection...;
'Wastewater': water after being used for household, industrial, trade or agricultural use, including wastewater which contains particles floating or dissolved solids and whose source is from humans or animals whether mixed with water or not, whether flowing in a pipe or in a stream or located in any other place;
'Environmental Representative': a person who the Minister of the Environment has authorized for purpose of these rules;
'an Authorized Laboratory': a laboratory recognized by the Director for purposes of these rules;
'Purification System': a series of devices aimed at purifying sewage to transform it to purified wastewater;
'Purified Wastewater': (1) wastewater or sewage which has been treated in a secondary aerobic system, approved by the Engineer or the Director in consultation with the environmental representative, which operates in a satisfactory manner;
(2) sewage which has been purified naturally to the satisfaction of the Engineer or the Director, and he has approved them in writing after consultation with the Environmental Representative as available for irrigation;
(3) water mixed with purified wastewater;
(4) runoff from swimming pools to which was not added in any fashion sewage water.

2. No one will irrigate agricultural crops with wastewater unless it is purified wastewater in accordance with these rules.

3.a. A request to grant approval of a device for purification of wastewater for irrigation purposes (hereinafter 'the request') will be filed with the Engineer with the office in his district or where the purification device is located.

b. If the Minister decides after consultation with the Environmental Representative to grant the request or reject it he will so notify the person requesting of the approval or rejection.

c. The person making the request may appeal the decision before the Director within 30 days of the day he recieved notice.

d. The Director's decision will be final.

4.a. A license will remain in effect for one year only.

b. A request for a new license will be presented to the Engineer in his office in the district where the purification device exists three months before the termination date of the license in effect.

c. Once a request for a new license has been filed the person making a request may continue to use wastewater until the Engineer makes his decision.

5. Without derogating from section 6, purified wastewater will only be used for irrigation of those agricultural crops so categorized in the supplement, and this under the following conditions:

1. A person using purified waste holds a license in effect and carries out its conditions;

2. The concentration of dissolved oxygen in the purified wastewater is at least one-half of a milligram per liter;

3. The purified wastewater does not contain toxic substances which might, in the opinion of the Director, endanger the health of those who come in contact with the purified wastewater or with the irrigated crop whether in the field or after cultivated.

6.a. A crop not categorized in the supplement will not be irrigated by purified wastewater unless the person using wastewater has received permission in writing and in advance from the Director, and the conditions of section 5 are carried out.

b. The Director will decide in his aforesaid agreement in this section on the conditions or without them or he may refuse to give the license only after he has consulted with the Water Commissioner and the Environmental Representative.

c. The Director will not give his aforesaid agreement in this section unless he is convinced that irrigation with purified sewage will not harm public health; in his decision the Director will set out the type of approved crops for irrigation with purified sewage, the quality of the approved purified sewage, and the type of irrigation to be used, etc.

7.a. Once a permit has required field checks, a field check will be carried out according to a system approved by the Director in the license, by a person (hereinafter 'the sampler') who is approved for this purpose by the Director General of the Ministry of Environment.

a.a. A system of field sampling will be approved only after consultation with the Environmental Representative.

b. The sampler will write down immediately after finishing his field check, in a special book (hereinafter 'the fieldbook') which is designated for this purpose only the following details:

1. The date of the sample;
2. The hour of the sample;
3. The site of the sample;
4. The results of the field sample;
5. Comments.

c. The fieldbook will be open for inspection by the Engineer, the Director, the Environmental Representative or one who has been appointed for such purposes, at regular work hours, the book will be kept for at least two years from the date of the last recording in it.

8. If the license granted requires a laboratory test, the test will be carried out on the dates so set out in the license, in an approved laboratory and according to a system set out in the book or any other system which was so approved by the Director.

*Supplement – Types of Crops Approved for Irrigation With Purified Sewage*

1. Agricultural crops that will not be used for food for humans, such as cotton, seed crops and dry alfalfa.
2. Agricultural crops to be used for industry which will go through an industrial process to prevent the transfer of micro-organisms such as wheat crops and sugarbeets, but not those crops to be used in the preserves industry.
3. Woods and forests and those places not to be used for recreation or camping.

## References

Feigin I and Shalhevet R J 1991 Irrigation with Treated Sewage Effluent: Management for Environmental Protection.

Grinwald Z 1980 Water and Israel pp 159–170 (published in Hebrew).

Laster L 1980 Legal aspects of water quality management in Israel. In Water Quality Management under Conditions of Scarcity: Israel as a Case Study. Ed. H. Shuval.

Law to Amend the Public Health Ordinance 1970 (No. 4), 5730–1970.

Ministry of Agriculture – Water Commission, A National Framework Plan for Reuse of Purified Effluent in Israel, April 1992, Tahal (published in Hebrew).

Public Health Rules (Purification of Waste Water for Irrigation) 1981 5741–1981, as amended 5750–1990.

Selbst N 1980 Economic, social, and administrative considerations in reuse of waste water. In Water Quality Management under Conditions of Scarcity: Israel as a Case Study. Ed. H Shuval.

Shelef E 1993 Interview with Mr Eleazer Shelef, Head of the Department of Water Quality and Reuse of Sewage, Water Commission, 30 August 1993.

Shuval H 1980 Quality management aspects of waste water reuse in israel. In Water Quality Management under Conditions of Scarcity: Israel as a Case Study. Ed. H Shuval. Academic Press.

Tamir O 1993 Interview with Mrs Ora Tamir, Legal Adviser to the Water Commission, 30 August 1993.

Water Regulations (Computation of Water Cost) 1965 5752–1991, K. T. 224.

MIX
Papier aus verantwortungsvollen Quellen
Paper from responsible sources
FSC® C105338

If you have any concerns about our products,
you can contact us on
**ProductSafety@springernature.com**

In case Publisher is established outside the EU,
the EU authorized representative is:
**Springer Nature Customer Service Center GmbH**
**Europaplatz 3, 69115 Heidelberg, Germany**

Printed by Libri Plureos GmbH
in Hamburg, Germany